306361828.

D0315385

Principles and Techniques of
Biochemistry and Molecular Biology
Seventh edition

EDITED BY KEITH WILSON AND JOHN WALKER

This new edition of the bestselling textbook integrates the theoretical principles and experimental techniques common to all undergraduate courses in the bio- and medical sciences. Three of the 16 chapters have new authors and have been totally rewritten. The others have been updated and extended to reflect developments in their field exemplified by a new section on stem cells. Two new chapters have been added. One on clinical biochemistry discusses the principles underlying the diagnosis and management of common biochemical disorders. The second one on drug discovery and development illustrates how the principles and techniques covered in the book are fundamental to the design and development of new drugs. In-text worked examples are again used to enhance student understanding of each topic and case studies are selectively used to illustrate important examples. Experimental design, quality assurance and the statistical analysis of quantitative data are emphasised throughout the book.

- Motivates students by including cutting-edge topics and techniques, such as drug discovery, as well as the methods they will encounter in their own lab classes
- Promotes problem solving by setting students a challenge and then guiding them through the solution
- Integrates theory and practise to ensure students understand why and how each technique is used.

KEITH WILSON is Professor Emeritus of Pharmacological Biochemistry and former Head of the Department of Biosciences, Dean of the Faculty of Natural Sciences, and Director of Research at the University of Hertfordshire.

JOHN WALKER is Professor Emeritus and former Head of the School of Life Sciences at the University of Hertfordshire.

Cover illustration

Main image Electrophoresis gel showing recombinant protein. Photographer: J.C. Revy. Courtesy of Science Photo Library.

Top inset Transcription factor and DNA molecule. Courtesy of: Laguna Design/Science Photo Library.

Second inset Microtubes, pipettor (pipette) tip & DNA sequence. Courtesy of Tek Image/Science Photo Library.

Third inset Stem cell culture, light micrograph. Photographer: Philippe Plailly. Courtesy of Science Photo Library.

Fourth inset Embryonic stem cells. Courtesy of Science Photo Library.

Bottom inset Herceptin breast cancer drug, molecular model. Photographer: Tim Evans. Courtesy of Science Photo Library.

Principles and Techniques of
Biochemistry and
Molecular Biology

Seventh edition

Edited by

KEITH WILSON AND JOHN WALKER

CAMBRIDGE
UNIVERSITY PRESS

CAMBRIDGE UNIVERSITY PRESS
Cambridge, New York, Melbourne, Madrid, Cape Town, Singapore,
São Paulo, Delhi, Tokyo, Mexico City

Cambridge University Press
The Edinburgh Building, Cambridge CB2 8RU, UK

Published in the United States of America by
Cambridge University Press, New York

www.cambridge.org
Information on this title: www.cambridge.org/9780521516358

First published by Edward Arnold 1975 as *A Biologist's Guide to Principles and Techniques of Practical Biochemistry*
Second edition 1981; Third edition 1986
Third edition first published by Cambridge University Press 1992; Reprinted 1993
Fourth edition published by Cambridge University Press 1994 as *Principles and Techniques of Practical Biochemistry*; Reprinted 1995, 1997; Fifth edition 2000
Sixth edition first published by Cambridge University Press 2005 as *Principles and Techniques of Biochemistry and Molecular Biology*; Reprinted 2006, 2007
Seventh edition first published by Cambridge University Press 2010

Printed in the United Kingdom at the University Press, Cambridge

A catalogue record for this publication is available from the British Library

Library of Congress Cataloging-in-Publication Data

Principles and techniques of biochemistry and molecular biology / edited by Keith Wilson,
John Walker. – 7th ed.
    p. cm.
  ISBN 978-0-521-51635-8 (hardback) – ISBN 978-0-521-73167-6 (pbk.)
    1. Biochemistry–Textbooks.   2. Molecular biology–Textbooks.   I. Wilson, Keith, 1936–   II. Walker,
John M., 1948–   III. Title.
  QP519.7.P75 2009
  612′.015–dc22   2009043277

ISBN 978-0-521-51635-8 Hardback
ISBN 978-0-521-73167-6 Paperback

CONTENTS

The colour figure section is between pages 128 and 129

PREFACE TO THE SEVENTH EDITION

In designing the content of this latest edition we continued our previous policy of placing emphasis on the recommendations we have received from colleagues and academics outside our university. Above all, we have attempted to respond to the invaluable feedback from student users of our book both in the UK and abroad. In this seventh edition we have retained all 16 chapters from the previous edition. All have been appropriately updated to reflect recent developments in their fields, as exemplified by the inclusion of a section on stem cells in the cell culture chapter. Three of these chapters have new authors and have been completely rewritten. Robert Burns, Scottish Agricultural Science Agency, Edinburgh has written the chapter on immunochemical techniques, and Andreas Hofmann, Eskitis Institute of Molecular Therapies, Griffith University, Brisbane, Australia has written the two chapters on spectroscopic techniques. We are delighted to welcome both authors to our team of contributors.

In addition to these changes of authors, two new chapters have been added to the book. Our decision taken for the sixth edition to include a section on the biochemical principles underlying clinical biochemistry has been well received and so we have extended our coverage of the subject and have devoted a whole chapter (16) to this subject. Written in collaboration with Dr John Fyffe, Consultant Biochemist, Royal Hospital for Sick Children, Yorkhill, Glasgow, new topics that are discussed in the chapter include the diagnosis and management of kidney disease, diabetes, endocrine disorders including thyroid dysfunction, conditions of the hypothalamus–pituitary–adrenal axis such as pregnancy, and pathologies of plasma proteins such as myeloma. Case studies are included to illustrate how the principles discussed apply to the diagnosis and treatment of individual patients with the conditions.

Our second major innovation for this new edition is the introduction of a new chapter on drug discovery and development. The strategic approaches to the discovery of new drugs has been revolutionised by developments in molecular biology. Pharmaceutical companies now rely on many of the principles and experimental techniques discussed in the chapters throughout the book to identify potential drug targets, screen chemical libraries and to evaluate the safety and efficacy of selected candidate drugs. The new chapter illustrates the principles of target selection by reference to current drugs used in the treatment of atherosclerosis and HIV/AIDS, emphasises the strategic decisions to be taken during the various stages of drug discovery and development and discusses the issues involved in clinical trials and the registration of new drugs.

We continue to welcome constructive comments from all students who use our book as part of their studies and academics who adopt the book to complement their teaching. Finally, we wish to express our gratitude to the authors and publishers who have granted us permission to reproduce their copyright figures and our thanks to Katrina Halliday and her colleagues at Cambridge University Press who have been so supportive in the production of this new edition.

KEITH WILSON AND JOHN WALKER

CONTRIBUTORS

PROFESSOR A. AITKEN
Division of Biomedical & Clinical Laboratory Sciences
University of Edinburgh
George Square
Edinburgh EH8 9XD
Scotland, UK

DR A.R. BAYDOUN
School of Life Sciences
University of Hertfordshire
College Lane
Hatfield
Herts AL10 9AB, UK

DR R. BURNS
Scottish Agricultural Science Agency
1 Roddinglaw Road
Edinburgh EH12 9FJ
Scotland, UK

DR J. FYFFE
Consultant Clinical Biochemist
Department of Clinical Biochemistry
Royal Hospital for Sick Children
Yorkhill
Glasgow G3 8SF
Scotland, UK

PROFESSOR ANDREAS HOFMANN
Structural Chemistry
Eskitis Institute for Cell & Molecular Therapeutics
Griffith University
Nathan
Brisbane, Qld 4111
Australia

PROFESSOR K. OHLENDIECK
Department of Biology
National University of Ireland
Maynooth
Co. Kildare
Ireland

DR S.W. PADDOCK
Howard Hughes Medical Institute
Department of Molecular Biology
University of Wisconsin
1525 Linden Drive
Madison, WI 53706
USA

DR R. RAPLEY
School of Life Sciences
University of Hertfordshire
College Lane
Hatfield
Herts AL10 9AB, UK

PROFESSOR R.J. SLATER
School of Life Sciences
University of Hertfordshire
College Lane
Hatfield
Herts AL10 9AB, UK

PROFESSOR J.M. WALKER
School of Life Sciences
University of Hertfordshire
College Lane
Hatfield
Herts AL10 9AB, UK

PROFESSOR K. WILSON
Emeritus Professor of Pharmacological Biochemistry
School of Life Sciences
University of Hertfordshire
College Lane
Hatfield
Herts AL10 9AB, UK

The following abbreviations have been used throughout this book.

AMP	adenosine 5′-monophosphate
ADP	adenosine 5′-diphosphate
ATP	adenosine 5′-triphosphate
bp	base-pairs
cAMP	cyclic AMP
CHAPS	3-[(3-chloroamidopropyl)dimethylamino]-1-propanesulphonic acid
c.p.m.	counts per minute
CTP	cytidine triphosphate
DDT	2,2-bis-(p-chlorophenyl)-1,1,1-trichloroethane
DMSO	dimethylsulphoxide
DNA	deoxyribonucleic acid
e^-	electron
EDTA	ethylenediaminetetra-acetate
ELISA	enzyme-linked immunosorbent assay
FAD	flavin adenine dinucleotide (oxidised)
$FADH_2$	flavin adenine dinucleotide (reduced)
FMN	flavin mononucleotide (oxidised)
$FMNH_2$	flavin mononucleotide (reduced)
GC	gas chromatography
GTP	guanosine triphosphate
HAT	hypoxanthine, aminopterin, thymidine medium
Hepes	4(2-hydroxyethyl)-1-piperazine-ethanesulphonic acid
HPLC	high-performance liquid chromatography
kb	kilobase-pairs
M_r	relative molecular mass
min	minute
NAD^+	nicotinamide adenine dinucleotide (oxidised)
NADH	nicotinamide adenine dinucleotide (reduced)
$NADP^+$	nicotinamide adenine dinucleotide phosphate (oxidised)
NADPH	nicotinamide adenine dinucleotide phosphate (reduced)
Pipes	1,4-piperazinebis(ethanesulphonic acid)

P_i	inorganic phosphate
p.p.m.	parts per million
p.p.b.	parts per billion
PP_i	inorganic pyrophosphate
RNA	ribonucleic acid
r.p.m.	revolutions per minute
SDS	sodium dodecyl sulphate
Tris	2-amino-2-hydroxymethylpropane-1,3-diol

1 Basic principles

K. WILSON

1.1 BIOCHEMICAL AND MOLECULAR BIOLOGY STUDIES

1.1.1 Aims of laboratory investigations

Biochemistry involves the study of the chemical processes that occur in living organisms with the ultimate aim of understanding the nature of life in molecular terms. Biochemical studies rely on the availability of appropriate analytical techniques and on the application of these techniques to the advancement of knowledge of the nature of, and relationships between, biological molecules, especially proteins and nucleic acids, and cellular function. In recent years huge advances have been made in our understanding of gene structure and expression and in the application of techniques such as mass spectrometry to the study of protein structure and function. The Human Genome Project in particular has been the stimulus for major developments in our understanding of many human diseases especially cancer and for the identification of strategies that might be used to combat these diseases. The discipline of molecular biology overlaps with that of biochemistry and in many respects the aims of the two disciplines complement each other. Molecular biology is focussed on the molecular understanding of the processes of replication, transcription and translation of genetic material whereas biochemistry exploits the techniques and findings of molecular biology to advance our understanding of such cellular processes as cell signalling and apoptosis. The result is that the two disciplines now have the opportunity to address issues such as:

- the structure and function of the total protein component of the cell (*proteomics*) and of all the small molecules in the cell (*metabolomics*);
- the mechanisms involved in the control of gene expression;

- the identification of genes associated with a wide range of human diseases;
- the development of gene therapy strategies for the treatment of human diseases;
- the characterisation of the large number of 'orphan' receptors, whose physiological role and natural agonist are currently unknown, present in the human genome and their exploitation for the development of new therapeutic agents;
- the identification of novel disease-specific markers for the improvement of clinical diagnosis;
- the engineering of cells, especially stem cells, to treat human diseases;
- the understanding of the functioning of the immune system in order to develop strategies for the protection against invading pathogens;
- the development of our knowledge of the molecular biology of plants in order to engineer crop improvements, pathogen resistance and stress tolerance;
- the application of molecular biology techniques to the nature and treatment of bacterial, fungal and viral diseases.

The remaining chapters in this book address the major experimental strategies and analytical techniques that are routinely used to address issues such as these.

1.1.2 Experimental design

Advances in biochemistry and molecular biology, as in all the sciences, are based on the careful design, execution and data analysis of experiments designed to address specific questions or hypotheses. Such experimental design involves a discrete number of compulsory stages:

- the identification of the subject for experimental investigation;
- the critical evaluation of the current state of knowledge (the 'literature') of the chosen subject area noting the strengths and weaknesses of the methodologies previously applied and the new hypotheses which emerged from the studies;
- the formulation of the question or hypothesis to be addressed by the planned experiment;
- the careful selection of the biological system (species, *in vivo* or *in vitro*) to be used for the study;
- the identification of the variable that is to be studied; the consideration of the other variables that will need to be controlled so that the selected variable is the only factor that will determine the experimental outcome;
- the design of the experiment including the statistical analysis of the results, careful evaluation of the materials and apparatus to be used and the consequential potential safety aspects of the study;
- the execution of the experiment including appropriate calibrations and controls, with a carefully written record of the outcomes;
- the replication of the experiment as necessary for the unambiguous analysis of the outcomes;

- the evaluation of the outcomes including the application of appropriate statistical tests to quantitative data where applicable;
- the formulation of the main conclusions that can be drawn from the results;
- the formulation of new hypotheses and of future experiments that emerge from the study.

The results of well-designed and analysed studies are finally published in the scientific literature after being subject to independent peer review, and one of the major challenges facing professional biochemists and molecular biologists is to keep abreast of current advances in the literature. Fortunately, the advent of the web has made access to the literature easier than it once was.

1.2 UNITS OF MEASUREMENT

1.2.1 SI units

The French Système International d'Unités (the SI system) is the accepted convention for all units of measurement. Table 1.1 lists basic and derived SI units. Table 1.2 lists numerical values for some physical constants in SI units. Table 1.3 lists the commonly used prefixes associated with quantitative terms. Table 1.4 gives the interconversion of non-SI units of volume.

1.2.2 Molarity – the expression of concentration

In practical terms one mole of a substance is equal to its molecular mass expressed in grams, where the molecular mass is the sum of the atomic masses of the constituent atoms. Note that the term molecular mass is preferred to the older term molecular weight. The SI unit of concentration is expressed in terms of moles per cubic metre ($mol\,m^{-3}$) (see Table 1.1). In practice this is far too large for normal laboratory purposes and a unit based on a cubic decimetre (dm^3, $10^{-3}\,m$) is preferred. However, some textbooks and journals, especially those of North American origin, tend to use the older unit of volume, namely the litre and its subunits (see Table 1.4) rather than cubic decimetres. In this book, volumes will be expressed in cubic decimetres or its smaller counterparts (Table 1.4). The molarity of a solution of a substance expresses the number of moles of the substance in one cubic decimetre of solution. It is expressed by the symbol M.

It should be noted that atomic and molecular masses are both expressed in daltons (Da) or kilodaltons (kDa), where one dalton is an atomic mass unit equal to one-twelfth of the mass of one atom of the ^{12}C isotope. However, biochemists prefer to use the term relative molecular mass (M_r). This is defined as the molecular mass of a substance relative to one-twelfth of the atomic mass of the ^{12}C isotope. M_r therefore has no units. Thus the relative molecular mass of sodium chloride is 23 (Na) plus

Table 1.1 **SI units – basic and derived units**

Quantity	SI unit	Symbol (basic SI units)	Definition of SI unit	Equivalent in SI units
Basic units				
Length	metre	m		
Mass	kilogram	kg		
Time	second	s		
Electric current	ampere	A		
Temperature	kelvin	K		
Luminous intensity	candela	cd		
Amount of substance	mole	mol		
Derived units				
Force	newton	N	$kg\,m\,s^{-2}$	$J\,m^{-1}$
Energy, work, heat	joule	J	$kg\,m^2\,s^{-2}$	$N\,m$
Power, radiant flux	watt	W	$kg\,m^2\,s^{-3}$	$J\,s^{-1}$
Electric charge, quantity	coulomb	C	$A\,s$	$J\,V^{-1}$
Electric potential difference	volt	V	$kg\,m^2\,s^{-3}A^{-1}$	$J\,C^{-1}$
Electric resistance	ohm	Ω	$kg\,m^2\,s^{-3}A^{-2}$	$V\,A^{-1}$
Pressure	pascal	Pa	$kg\,m^{-1}\,s^{-2}$	$N\,m^{-2}$
Frequency	hertz	Hz	s^{-1}	
Magnetic flux density	tesla	T	$kg\,s^{-2}\,A^{-1}$	$V\,s\,m^{-2}$
Other units based on SI				
Area	square metre	m^2		
Volume	cubic metre	m^3		
Density	kilogram per cubic metre	$kg\,m^{-3}$		
Concentration	mole per cubic metre	$mol\,m^{-3}$		

Table 1.2 **SI units – conversion factors for non-SI units**

Unit	Symbol	SI equivalent
Avogadro constant	L or N_A	$6.022 \times 10^{23}\,\text{mol}^{-1}$
Faraday constant	F	$9.648 \times 10^{4}\,\text{C}\,\text{mol}^{-1}$
Planck constant	h	$6.626 \times 10^{-34}\,\text{J}\,\text{s}$
Universal or molar gas constant	R	$8.314\,\text{J}\,\text{K}^{-1}\,\text{mol}^{-1}$
Molar volume of an ideal gas at s.t.p.		$22.41\,\text{dm}^{3}\,\text{mol}^{-1}$
Velocity of light in a vacuum	c	$2.997 \times 10^{8}\,\text{m}\,\text{s}^{-1}$
Energy		
calorie	cal	$4.184\,\text{J}$
erg	erg	$10^{-7}\,\text{J}$
electron volt	eV	$1.602 \times 10^{-19}\,\text{J}$
Pressure		
atmosphere	atm	$101\,325\,\text{Pa}$
bar	bar	$10^{5}\,\text{Pa}$
millimetres of Hg	mm Hg	$133.322\,\text{Pa}$
Temperature		
centigrade	°C	$(t\,°\text{C} + 273.15)\,\text{K}$
Fahrenheit	°F	$(t\,°\text{F} - 32)5/9 + 273.15\,\text{K}$
Length		
Ångström	Å	$10^{-10}\,\text{m}$
inch	in	$0.0254\,\text{m}$
Mass		
pound	lb	$0.4536\,\text{kg}$

Note: s.t.p., standard temperature and pressure.

35.5 (Cl) i.e. 58.5, so that one mole is 58.5 grams. If this was dissolved in water and adjusted to a total volume of $1\,\text{dm}^{3}$ the solution would be one molar (1 M).

Biological substances are most frequently found at relatively low concentrations and in *in vitro* model systems the volumes of stock solutions regularly used for experimental purposes are also small. The consequence is that experimental solutions are usually in the mM, μM and nM range rather than molar. Table 1.5 shows the interconversion of these units.

Table 1.3 **Common unit prefixes associated with quantitative terms**

Multiple	Prefix	Symbol	Multiple	Prefix	Symbol
10^{24}	yotta	Y	10^{-1}	deci	d
10^{21}	zetta	Z	10^{-2}	centi	c
10^{18}	exa	E	10^{-3}	milli	m
10^{15}	peta	P	10^{-6}	micro	μ
10^{12}	tera	T	10^{-9}	nano	n
10^{9}	giga	G	10^{-12}	pico	p
10^{6}	mega	M	10^{-15}	femto	f
10^{3}	kilo	k	10^{-18}	atto	a
10^{2}	hecto	h	10^{-21}	zepto	z
10^{1}	deca	da	10^{-24}	yocto	y

Table 1.4 **Interconversion of non-SI and SI units of volume**

Non-SI unit	Non-SI subunit	SI subunit	SI unit
1 litre (l)	$10^3\,ml$	$= 1\,dm^3$	$= 10^{-3}\,m^3$
1 millilitre (ml)	1 ml	$= 1\,cm^3$	$= 10^{-6}\,m^3$
1 microlitre (μl)	$10^{-3}\,ml$	$= 1\,mm^3$	$= 10^{-9}\,m^3$
1 nanolitre (nl)	$10^{-6}\,ml$	$= 1\,nm^3$	$= 10^{-12}\,m^3$

Table 1.5 **Interconversion of mol, mmol and μmol in different volumes to give different concentrations**

Molar (M)	Millimolar (mM)	Micromolar (μM)
$1\,mol\,dm^{-3}$	$1\,mmol\,dm^{-3}$	$1\,\mu mol\,dm^{-3}$
$1\,mmol\,cm^{-3}$	$1\,\mu mol\,cm^{-3}$	$1\,nmol\,cm^{-3}$
$1\,\mu mol\,mm^{-3}$	$1\,nmol\,mm^{-3}$	$1\,pmol\,mm^{-3}$

1.3 WEAK ELECTROLYTES

1.3.1 The biochemical importance of weak electrolytes

Many molecules of biochemical importance are weak electrolytes in that they are acids or bases that are only partially ionised in aqueous solution. Examples include the amino acids, peptides, proteins, nucleosides, nucleotides and nucleic acids. It also includes the reagents used in the preparation of buffers such as ethanoic (acetic) acid and phosphoric acid. The biochemical function of many of these molecules is dependent upon their precise state of ionisation at the prevailing cellular or extracellular pH. The catalytic sites of enzymes, for example, contain functional carboxyl and amino groups, from the side chains of constituent amino acids in the protein chain, which need to be in a specific ionised state to enable the catalytic function of the enzyme to be realised. Before the ionisation of these compounds is discussed in detail, it is necessary to appreciate the importance of the ionisation of water.

1.3.2   Ionisation of weak acids and bases

One of the most important weak electrolytes is water since it ionises to a small extent to give hydrogen ions and hydroxyl ions. In fact there is no such species as a free hydrogen ion in aqueous solution as it reacts with water to give a hydronium ion (H_3O^+):

$$H_2O \rightleftharpoons H^+ + HO^-$$
$$H^+ + H_2O \rightleftharpoons H_3O^+$$

Even though free hydrogen ions do not exist it is conventional to refer to them rather than hydronium ions. The equilibrium constant (K_{eq}) for the ionisation of water has a value of 1.8×10^{16} at $24\,°C$:

$$K_{eq} = \frac{[H^+][OH^-]}{[H_2O]} = 1.8 \times 10^{16} \tag{1.1}$$

The molarity of pure water is 55.6 M. This can be incorporated into a new constant, K_w:

$$1.8 \times 10^{-16} \times 55.6 = [H^+][HO^-] = 1.0 \times 10^{-14} = K_w \tag{1.2}$$

K_w is known as the autoprotolysis constant of water and does not include an expression for the concentration of water. Its numerical value of exactly 10^{-14} relates specifically to $24\,°C$. At $0\,°C$ K_w has a value of 1.14×10^{-15} and at $100\,°C$ a value of 5.45×10^{-13}. The stoichiometry in equation 1.2 shows that hydrogen ions and hydroxyl ions are produced in a $1:1$ ratio, hence both of them must be present at a concentration of 1.0×10^{-7} M. Since the Sörensen definition of pH is that it is equal to the negative logarithm of the hydrogen ion concentration, it follows that the pH of pure water is 7.0. This is the definition of neutrality.

Ionisation of carboxylic acids and amines

As previously stressed, many biochemically important compounds contain a carboxyl group (-COOH) or a primary (RNH_2), secondary (R_2NH) or tertiary (R_3N) amine which can donate or accept a hydrogen ion on ionisation. The tendency of a weak acid, generically represented as HA, to ionise is expressed by the equilibrium reaction:

$$\begin{array}{ccc} HA & \rightleftharpoons & H^+ + \qquad A^- \\ \text{weak acid} & & \text{conjugate base (anion)} \end{array}$$

This reversible reaction can be represented by an equilibrium constant, K_a, known as the acid dissociation constant (equation 1.3). Numerically, it is very small.

$$K_a = \frac{[H^+][A^-]}{[HA]} \qquad (1.3)$$

Note that the ionisation of a weak acid results in the release of a hydrogen ion and the conjugate base of the acid, both of which are ionic in nature.

Similarly, amino groups (primary, secondary and tertiary) as weak bases can exist in ionised and unionised forms and the concomitant ionisation process is represented by an equilibrium constant, K_b (equation 1.4):

$$RNH_2 + H_2O \rightleftharpoons RNH_3^+ + HO^-$$
$$\text{weak base} \qquad \text{conjugate acid}$$
$$\text{(primary amine)} \quad \text{(substituted ammonium ion)}$$

$$K_b = \frac{[RNH_3^+][HO^-]}{[RNH_2][H_2O]} \qquad (1.4)$$

In this case, the non-ionised form of the base abstracts a hydrogen ion from water to produce the conjugate acid that is ionised. If this equation is viewed from the reverse direction it is of a similar format to that of equation 1.3. Equally, equation 1.3 viewed in reverse is similar in format to equation 1.4.

A specific and simple example of the ionisation of a weak acid is that of acetic (ethanoic) acid, CH_3COOH:

$$CH_3COOH \rightleftharpoons CH_3COO^- + H^+$$
$$\text{acetic acid} \qquad \text{acetate anion}$$

Acetic acid and its conjugate base, the acetate anion, are known as a conjugate acid–base pair. The acid dissociation constant can be written in the following way:

$$K_a = \frac{[CH_3COO^-][H^+]}{[CH_3COOH]} = \frac{[\text{conjugate base}][H^+]}{[\text{weak acid}]} \qquad (1.5a)$$

K_a has a value of 1.75×10^{-5} M. In practice it is far more common to express the K_a value in terms of its negative logarithm (i.e. $-\log K_a$) referred to as pK_a. Thus in this case pK_a is equal to 4.75. It can be seen from equation 1.3 that pK_a is numerically equal to the pH at which 50% of the acid is protonated (unionised) and 50% is deprotonated (ionised).

It is possible to write an expression for the K_b of the acetate anion as a conjugate base:

$$CH_3COO_3^- + H_2O \rightleftharpoons CH_3COOH + HO^-$$

$$K_b = \frac{[CH_3COOH][HO^-]}{[CH_3COO^-]} = \frac{[\text{weak acid}][OH^-]}{[\text{conjugate base}]} \qquad (1.5b)$$

K_b has a value of 1.77×10^{-10} M, hence its pK_b (i.e. $-\log K_b$) = 9.25.

Multiplying these two expressions together results in the important relationship:

$$K_a \times K_b = [H^+][OH^-] = K_w = 1.0 \times 10^{-14} \text{ at } 24\,^\circ C$$

Table 1.6 pK_a values of some acids and bases that are commonly used as buffer solutions

Acid or base	pK_a
Acetic acid	4.75
Barbituric acid	3.98
Carbonic acid	6.10, 10.22
Citric acid	3.10, 4.76, 5.40
Glycylglycine	3.06, 8.13
Hepes[a]	7.50
Phosphoric acid	1.96, 6.70, 12.30
Phthalic acid	2.90, 5.51
Pipes[a]	6.80
Succinic acid	4.18, 5.56
Tartaric acid	2.96, 4.16
Tris[a]	8.14

Note: [a]See list of abbreviations at the front of the book.

hence

$$pK_a + pK_b - pK_w - 14 \tag{1.6}$$

This relationship holds for all acid–base pairs and enables one pK_a value to be calculated from knowledge of the other. Biologically important examples of conjugate acid–base pairs are lactic acid/lactate, pyruvic acid/pyruvate, carbonic acid/bicarbonate and ammonium/ammonia.

In the case of the ionisation of weak bases the most common convention is to quote the K_a or the pK_a of the conjugate acid rather than the K_b or pK_b of the weak base itself. Examples of the pK_a values of some weak acids and bases are given in Table 1.6. Remember that the smaller the numerical value of pK_a the stronger the acid (more ionised) and the weaker its conjugate base. Weak acids will be predominantly unionised at low pH values and ionised at high values. In contrast, weak bases will be predominantly ionised at low pH values and unionised at high values. This sensitivity to pH of the state of ionisation of weak electrolytes is important both physiologically and in *in vitro* biochemical studies employing such analytical techniques as electrophoresis and ion-exchange chromatography.

Ionisation of polyprotic weak acids and bases

Polyprotic weak acids and bases are capable of donating or accepting more than one hydrogen ion. Each ionisation stage can be represented by a K_a value using the convention that K_a^1 refers to the acid with the most ionisable hydrogen atoms and K_a^n the acid with the least number of ionisable hydrogen atoms. One of the most important biochemical examples is phosphoric acid, H_3PO_4, as it is widely used as the basis of a buffer in the pH region of 6.70 (see below):

$$H_3PO_4 \rightleftharpoons H^+ + H_2PO_4^- \quad pK_a^1 \; 1.96$$
$$H_2PO_4^- \rightleftharpoons H^+ + HPO_4^{2-} \quad pK_a^2 \; 6.70$$
$$HPO_4^{2-} \rightleftharpoons H^+ + PO_4^{3-} \quad pK_a^3 \; 12.30$$

Example 1 **CALCULATION OF pH AND THE EXTENT OF IONISATION OF A WEAK ELECTROLYTE**

Question Calculate the pH of a 0.01 M solution of acetic acid and its fractional ionisation given that its K_a is 1.75×10^{-5}.

Answer To calculate the pH we can write:

$$K_a = \frac{[\text{acetate}^-][H^+]}{[\text{acetic acid}]} = 1.75 \times 10^{-5}$$

Since acetate and hydrogen ions are produced in equal quantities, if $x =$ the concentration of each then the concentration of unionised acetic acid remaining will be $0.01 - x$. Hence:

$$1.75 \times 10^{-5} = \frac{(x)(x)}{0.01 - x}$$
$$1.75 \times 10^{-7} - 1.75 \times 10^{-5}x = x^2$$

This can now be solved either by use of the quadratic formula or, more easily, by neglecting the x term since it is so small. Adopting the latter alternative gives:

$$x^2 = 1.75 \times 10^{-7}$$

hence

$$x = 4.18 \times 10^{-4} \text{ M}$$

hence

$$\text{pH} = 3.38$$

The fractional ionisation (α) of the acetic acid is defined as the fraction of the acetic acid that is in the form of acetate and is therefore given by the equation:

$$\alpha = \frac{[\text{acetate}]}{[\text{acetate}] + [\text{acetic acid}]}$$
$$= \frac{4.18 \times 10^{-4}}{4.18 \times 10^{-4} + 0.01 - 4.18 \times 10^{-4}}$$
$$= \frac{4.18 \times 10^{-4}}{0.01}$$
$$= 4.18 \times 10^{-2} \text{ or } 4.18\%$$

Thus the majority of the acetic acid is present as the unionised form. If the pH is increased above 3.38 the proportion of acetate present will increase in accordance with the Henderson–Hasselbalch equation.

1.3.3 Buffer solutions

A buffer solution is one that resists a change in pH on the addition of either acid or base. They are of enormous importance in practical biochemical work as so many biochemical molecules are weak electrolytes so that their ionic status varies with pH so there is a need to stabilise this ionic status during the course of a practical experiment. In practice, a buffer solution consists of an aqueous mixture of a weak acid and its conjugate base. The conjugate base component would neutralise any hydrogen ions generated during an experiment whilst the unionised acid would neutralise any base generated. The Henderson–Hasselbalch equation is of central importance in the preparation of buffer solutions. It can be expressed in a variety of forms. For a buffer based on a weak acid:

$$pH = pK_a + \log \frac{[\text{conjugate base}]}{[\text{weak acid}]} \tag{1.7}$$

or

$$pH = pK_a + \log \frac{[\text{ionised form}]}{[\text{unionised form}]}$$

For a buffer based on the conjugate acid of a weak base:

$$pH = pK_a + \log \frac{[\text{weak base}]}{[\text{conjugate acid}]} \tag{1.8}$$

or

$$pH = pK_a + \log \frac{[\text{unionised form}]}{[\text{ionised form}]}$$

Table 1.6 lists some weak acids and bases commonly used in the preparation of buffer solutions. Phosphate, Hepes and Pipes are commonly used because of their optimum pH being close to 7.4. The buffer action and pH of blood is illustrated in Example 2 and the preparation of a phosphate buffer is given in Example 3.

Buffer capacity

It can be seen from the Henderson–Hasselbalch equations that when the concentration (or more strictly the activity) of the weak acid and base is equal, their ratio is one and their logarithm zero so that $pH = pK_a$. The ability of a buffer solution to resist a change in pH on the addition of strong acid or alkali is expressed by its buffer capacity (β). This is defined as the amount (moles) of acid or base required to change the pH by one unit i.e.

$$\beta = \frac{db}{dpH} = \frac{-da}{dpH} \tag{1.9}$$

where db and da are the amount of base and acid respectively and dpH is the resulting change in pH. In practice, β is largest within the pH range $pK_a \pm 1$.

Example 2 **BUFFER ACTION AND pH OF BLOOD**

The normal pH of blood is 7.4 and is maintained at this value by buffer action in particular by the action of HCO_3^- and CO_2 resulting from gaseous CO_2 dissolved in blood and the resulting ionisation of carbonic acid:

$$CO_2 + H_2O \rightleftharpoons H_2CO_3$$
$$H_2CO_3 \rightleftharpoons H^+ + HCO_3^-$$

It is possible to calculate an overall equilibrium constant (K_{eq}) for these two consecutive reactions and to incorporate the concentration of water (55.6 M) into the value:

$$K_{eq} = \frac{[H^+][HCO_3^-]}{[CO_2]} = 7.95 \times 10^{-7} \quad \text{hence } pK_{eq} = 6.1$$

Rearranging:

$$pH = pK_{eq} + \log \frac{[HCO_3^-]}{[CO_2]}$$

When the pH of blood falls due to the metabolic production of H^+, these equilibria shift in favour of increased production of H_2CO_3 that in turn ionises to give increased CO_2 that is then expired. When the pH of blood rises, more HCO_3^- is produced and breathing is adjusted to retain more CO_2 in the blood thus maintaining blood pH. Some disease states may change this pH causing either acidosis or alkalosis and this may cause serious problems and in extreme cases, death. For example, obstructive lung disease may cause acidosis and hyperventilation alkalosis. Clinical biochemists routinely monitor patient's acid–base balance in blood, in particular the ratio of HCO_3^- and CO_2. Reference ranges for these at pH 7.4 are $[HCO_3^-] = 18.0 - 26.0 \, mM$ and $pCO_2 = 4.6$–6.9 kPa, which gives $[CO_2]$ in the range of 1.20 mM.

Question A patient suffering from acidosis had a blood pH of 7.15 and $[CO_2]$ of 1.15 mM. What was the patient's $[HCO_3^-]$ and what are the implications of its value to the buffer capacity of the blood?

Answer Applying the above equation we get:

$$pH = pK_{eq} + \log \frac{[HCO_3^-]}{[CO_2]}$$
$$7.15 = 6.10 + \log \frac{[HCO_3^-]}{1.15}$$
$$1.05 = \log \frac{[HCO_3^-]}{1.15}$$

Taking the antilog of this equation we get $11.22 = [HCO_3^-]/1.15$

Therefore $[HCO_3^-] = 12.90 \, mM$ indicating that the bicarbonate concentration in the patient's blood had decreased by 11.1 mM i.e. 47% thereby severely reducing the buffer capacity of the patient's blood so that any further significant production of acid would have serious implications for the patient.

Example 3 **PREPARATION OF A PHOSPHATE BUFFER**

Question How would you prepare $1 \, dm^3$ of 0.1 M phosphate buffer, pH 7.1, given that pK_a^2 for phosphoric acid is 6.8 and that the atomic masses for Na, P and O are 23, 31 and 16 daltons respectively?

Answer The buffer will be based on the ionisation:

$$H_2PO_4^- \rightleftharpoons HPO_4^{2-} + H^+ \quad pK_a^2 = 6.8$$

and will therefore involve the use of solid sodium dihydrogen phosphate (NaH_2PO_4) and disodium hydrogen phosphate (Na_2HPO_4).

Applying the appropriate Henderson–Hasselbalch equation (equation 1.7) gives:

$$7.1 = 6.8 + \log \frac{[HPO_4^{2-}]}{[H_2PO_4^-]}$$

$$0.3 = \log \frac{[HPO_4^{2-}]}{[H_2PO_4^-]}$$

$$2.0 = \frac{[HPO_4^{2-}]}{[H_2PO_4^-]}$$

Since the total concentration of the two species needs to be 0.1 M it follows that $[HPO_4^{2-}]$ must be 0.067 M and $[H_2PO_4^-]$ 0.033 M. Their molecular masses are 142 and 120 daltons respectively; hence the weight of each required is $0.067 \times 143 = 9.46 \, g$ (Na_2HPO_4) and $0.033 \times 120 = 4.00 \, g$ (NaH_2PO_4). These weights would be dissolved in approximately $800 \, cm^3$ pure water, the pH measured and adjusted as necessary, and the volume finally made up to $1 \, dm^3$.

Selection of a buffer

When selecting a buffer for a particular experimental study, several factors should be taken into account:

- select the one with a pK_a as near as possible to the required experimental pH and within the range $pK_a \pm 1$, as outside this range there will be too little weak acid or weak base present to maintain an effective buffer capacity;
- select an appropriate concentration of buffer to have adequate buffer capacity for the particular experiment. Buffers are most commonly used in the range 0.05–0.5 M;
- ensure that the selected buffer does not form insoluble complexes with any anions or cations essential to the reaction being studied (phosphate buffers tend to precipitate polyvalent cations, for example, and may be a metabolite or inhibitor of the reaction);

- ensure that the proposed buffer has other desirable properties such as being non-toxic, able to penetrate membranes, and does not absorb in the visible or ultraviolet region.

1.3.4 Measurement of pH – the pH electrode

The pH electrode is an example of an ion-selective electrode (ISE) that responds to one specific ion in solution, in this case the hydrogen ion. The electrode consists of a thin glass porous membrane sealed at the end of a hard glass tube containing 0.1 M hydrochloric acid into which is immersed a silver wire coated with silver chloride. This silver/silver chloride electrode acts as an internal reference that generates a constant potential. The porous membrane is typically 0.1 mm thick, the outer and inner 10 nm consisting of a hydrated gel layer containing exchange-binding sites for hydrogen or sodium ions. On the inside of the membrane the exchange sites are predominantly occupied by hydrogen ions from the hydrochloric acid whilst on the outside the exchange sites are occupied by sodium and hydrogen ions. The bulk of the membrane is a dry silicate layer in which all exchange sites are occupied by sodium ions. Most of the coordinated ions in both hydrated layers are free to diffuse into the surrounding solution whilst hydrogen ions in the test solution can diffuse in the opposite direction replacing bound sodium ions in a process called ion-exchange equilibrium. Any other types of cations present in the test solution are unable to bind to the exchange sites thus ensuring the high specificity of the electrode. Note that hydrogen ions do not diffuse across the dry glass layer but sodium ions can. Thus effectively the membrane consists of two hydrated layers containing different hydrogen ion activities separated by a sodium ion transport system.

The principle of operation of the pH electrode is based upon the fact that if there is a gradient of hydrogen ion activity across the membrane this will generate a potential the size of which is determined by the hydrogen ion gradient across the membrane. Moreover, since the hydrogen ion concentration on the inside is constant (due to the use of 0.1 M hydrochloric acid) the observed potential is directly dependent upon the hydrogen ion concentration of the test solution. In practice a small junction or asymmetry potential (E^*) is also created in part as a result of linking the glass electrode to a reference electrode. The observed potential across the membrane is therefore given by the equation:

$$E = E^* + 0.059 \text{ pH}$$

Since the precise composition of the porous membrane varies with time so too does the asymmetry potential. This contributes to the need for the frequent recalibration of the electrode commonly using two standard buffers of known pH. For each 10-fold change in the hydrogen ion concentration across the membrane (equivalent to a pH change of 1 in the test solution) there will be a potential difference change of 59.2 mV across the membrane. The sensitivity of pH measurements is influenced by the prevailing absolute temperature.

The most common forms of pH electrode are the glass electrode (Fig. 1.1a) and the combination electrode (Fig. 1.1b) which contains an in-built calomel reference electrode.

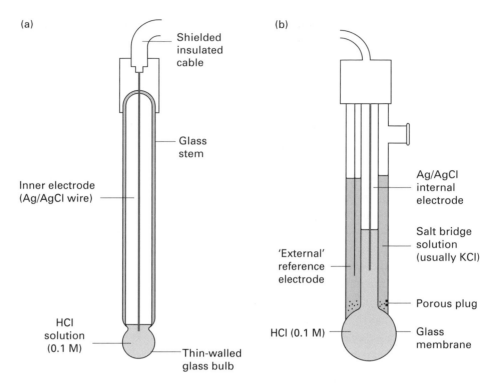

(a)

Shielded
insulated
cable

Glass
stem

Inner electrode
(Ag/AgCl wire)

HCl
solution
(0.1 M)

Thin-walled
glass bulb

(b)

Ag/AgCl
internal
electrode

Salt bridge
solution
(usually KCl)

'External'
reference
electrode

Porous plug

HCl (0.1 M)

Glass
membrane

Fig. 1.1 Common pH electrodes: (a) glass electrode; (b) combination electrode.

1.3.5 Other electrodes

Electrodes exist for the measurement of many other ions such as Li^+, K^+, Na^+, Ca^{2+}, Cl^- and NO_3^- in addition to H^+. The principle of operation of these ion-selective electrodes (ISEs) is very similar to that of the pH electrode in that permeable membranes specific for the ion to be measured are used. They lack absolute specificity and their selectivity is expressed by a selectivity coefficient that expresses the ratio of the response to the competing ions relative to that for the desired ion. Most ISEs have a good linear response to the desired ion and a fast response time. Biosensors are derived from ISEs by incorporating an immobilised enzyme onto the surface of the electrode. An important example is the glucose electrode that utilises glucose oxidase to oxidise glucose (Section 15.3.5) in the test sample to generate hydrogen peroxide that is reduced at the anode causing a current to flow that is then measured amperometrically. Micro sensor versions of these electrodes are of great importance in clinical biochemistry laboratories (Section 16.2.2). The oxygen electrode measures molecular oxygen in solution rather than an ion. It works by reducing the oxygen at the platinum cathode that is separated from the test solution by an oxygen-permeable membrane. The electrons consumed in the process are compensated by the generation of electrons at the silver anode hence the oxygen tension in the test sample is directly proportional to the current flow between the two electrodes. Optical sensors use the enzyme luciferase (Section 15.3.2) to measure ATP by generating light and detecting it with a photomultiplier.

1.4 QUANTITATIVE BIOCHEMICAL MEASUREMENTS

1.4.1 Analytical considerations and experimental error

Many biochemical investigations involve the quantitative determination of the concentration and/or amount of a particular component (the analyte) present in a test sample. For example, in studies of the mode of action of enzymes, trans-membrane transport and cell signalling, the measurement of a particular reactant or product is investigated as a function of a range of experimental conditions and the data used to calculate kinetic or thermodynamic constants. These in turn are used to deduce details of the mechanism of the biological process taking place. Irrespective of the experimental rationale for undertaking such quantitative studies, all quantitative experimental data must first be questioned and validated in order to give credibility to the derived data and the conclusions that can be drawn from them. This is particularly important in the field of clinical biochemistry in which quantitative measurements on a patient's blood and urine samples are used to aid a clinical diagnosis and monitor the patient's recovery from a particular disease. This requires that the experimental data be assessed and confirmed as an acceptable estimate of the 'true' values by the application of one or more standard statistical tests. Evidence of the validation of quantitative data by the application of such tests is required by the editors of refereed journals for the acceptance for publication of draft research papers. The following sections will address the theoretical and practical considerations behind these statistical tests.

Selecting an analytical method

The nature of the quantitative analysis to be carried out will require a decision to be taken on the analytical technique to be employed. A variety of methods may be capable of achieving the desired analysis and the decision to select one may depend on a variety of issues. These include:

- the availability of specific pieces of apparatus;
- the precision, accuracy and detection limits of the competing methods;
- the precision, accuracy and detection limit acceptable for the particular analysis;
- the number of other compounds present in the sample that may interfere with the analysis;
- the potential cost of the method (particularly important for repetitive analysis);
- the possible hazards inherent in the method and the appropriate precautions needed to minimise risk;
- the published literature method of choice;
- personal preference.

The most common biochemical quantitative analytical methods are visible, ultraviolet and fluorimetric spectrophotometry, chromatographic techniques such as HPLC and GC coupled to spectrophotometry or mass spectrometry, ion-selective electrodes and immunological methods such as ELISA. Once a method has been selected it must be developed and/or validated using the approaches discussed in the following sections. If it is to be used over a prolonged period of time, measures will need to be put in place to ensure that there is no drift in response. This normally entails an internal quality control approach using reference test samples covering the analytical range that are measured each time the method is applied to test samples. Any deviation from the known values for these reference samples will require the whole batch of test samples to be re-assayed.

The nature of experimental errors

Every quantitative measurement has some uncertainty associated with it. This uncertainty is referred to as the experimental error which is a measure of the difference between the 'true' value and the experimental value. The 'true' value normally remains unknown except in cases where a standard sample (i.e. one of known composition) is being analysed. In other cases it has to be estimated from the analytical data by the methods that will be discussed later. The consequence of the existence of experimental errors is that the measurements recorded can be accepted with a high, medium or low degree of confidence depending upon the sophistication of the technique employed, but seldom, if ever, with absolute certainty.

Experimental error may be of two kinds: systematic error and random error.

Systematic error (also called determinate error)

Systematic errors are consistent errors that can be identified and either eliminated or reduced. They are most commonly caused by a fault or inherent limitation in the apparatus being used but may also be influenced by poor experimental design. Common causes include the misuse of manual or automatic pipettes, the incorrect preparation of stock solutions, and the incorrect calibration and use of pH meters. They may be constant (i.e. have a fixed value irrespective of the amount of test analyte present in the test sample under investigation) or proportional (i.e. the size of the error is dependent upon the amount of test analyte present). Thus the overall effect of the two types in a given experimental result will differ. Both of these types of systematic error have three common causes:

- *Analyst error*: This is best minimised by good training and/or by the automation of the method.
- *Instrument error*: This may not be eliminable and hence alternative methods should be considered. Instrument error may be electronic in origin or may be linked to the matrix of the sample.
- *Method error*: This can be identified by comparison of the experimental data with that obtained by the use of alternative methods.

Identification of systematic errors

Systematic errors are always reproducible and may be positive or negative i.e. they increase or decrease the experimental value relative to the 'true' value. The crucial characteristic, however, is that their cause can be identified and corrected. There are four common means of identifying this type of error:

- *Use of a 'blank' sample*: This is a sample that you know contains none of the analyte under test so that if the method gives a non-zero answer then it must be responding in some unintended way. The use of blank samples is difficult in cases where the matrix of the test sample is complex, for example, serum.
- *Use of a standard reference sample*: This is a sample of the test analyte of known composition so the method under evaluation must reproduce the known answer.
- *Use of an alternative method*: If the test and alternative methods give different results for a given test sample then at least one of the methods must have an inbuilt flaw.
- *Use of an external quality assessment sample*: This is a standard reference sample that is analysed by other investigators based in different laboratories employing the same or different methods. Their results are compared and any differences in excess of random errors (see below) identify the systematic error for each analyst. The use of external quality assessment schemes is standard practice in clinical biochemistry laboratories (see Section 16.2.3).

Random error (also called indeterminate error)

Random errors are caused by unpredictable and often uncontrollable inaccuracies in the various manipulations involved in the method. Such errors may be variably positive or negative and are caused by such factors as difficulty in the process of sampling, random electrical 'noise' in an instrument or by the analyst being inconsistent in the operation of the instrument or in recording readings from it.

Standard operating procedures

The minimisation of both systematic and random errors is essential in cases where the analytical data are used as the basis for a crucial diagnostic or prognostic decision as is common, for example, in routine clinical biochemical investigations and in the development of new drugs. In such cases it is normal for the analyses to be conducted in accordance with standard operating procedures (SOPs) that define in full detail the quality of the reagents, the preparation of standard solutions, the calibration of instruments and the methodology of the actual analytical procedure which must be followed.

1.4.2 Assessment of the performance of an analytical method

All analytical methods can be characterised by a number of performance indicators that define how the selected method performs under specified conditions. Knowledge of these performance indicators allows the analyst to decide whether or not the method is acceptable for the particular application. The major performance indicators are:

- *Precision* (also called *imprecision* and *variability*): This is a measure of the reproducibility of a particular set of analytical measurements on the same sample of test analyte. If the replicated values agree closely with each other, the measurements are said to be of *high precision* (or *low imprecision*). In contrast, if the values diverge, the measurements are said to be of *poor* or *low precision* (or *high imprecision*). In analytical biochemical work the normal aim is to develop a method that has as high a precision as possible within the general objectives of the investigation. However, precision commonly varies over the analytical range (see below) and over periods of time. As a consequence, precision may be expressed as either *within-batch* or *between-batch*. Within-batch precision is the variability when the same test sample is analysed repeatedly during the same batch of analyses on the same day. Between-batch precision is the variability when the same test sample is analysed repeatedly during different batches of analyses over a period of time. Since there is more opportunity for the analytical conditions to change for the assessment of between-batch precision, it is the higher of the two types of assessment. Results that are of high precision may nevertheless be a poor estimate of the 'true' value (i.e. of *low accuracy* or *high bias*) because of the presence of unidentified errors. Methods for the assessment of precision of a data set are discussed below. The term imprecision is preferred in particular by clinical biochemists since they believe that it best describes the variability that occurs in replicated analyses.

- *Accuracy* (also called *trueness*, *bias* and *inaccuracy*): This is the difference between the mean of a set of analytical measurements on the same sample of test analyte and the 'true' value for the test sample. As previously pointed out, the 'true' value is normally unknown except in the case of standard measurements. In other cases accuracy has to be assessed indirectly by use of an internationally agreed reference method and/or by the use of external quality assessment schemes (see above) and/or by the use of population statistics that are discussed below.

- *Detection limit* (also called *sensitivity*): This is the smallest concentration of the test analyte that can be distinguished from zero with a defined degree of confidence. Concentrations below this limit should simply be reported as 'less than the detection limit'. All methods have their individual detection limits for a given analyte and this may be one of the factors that influence the choice of a specific analytical method for a given study. Thus the Bradford, Lowry and bicinchoninic acid methods for the measurements of proteins have detection limits of 20, 10 and 0.5 μg protein cm^{-3} respectively. In clinical biochemical measurements, sensitivity is often defined as the ability of the method to detect the analyte without giving false negatives (see Section 16.1.2).

- *Analytical range:* This is the range of concentrations of the test analyte that can be measured reproducibly, the lower end of the range being the detection limit. In most cases the analytical range is defined by an appropriate calibration curve (see Section 1.4.6). As previously pointed out, the precision of the method may vary across the range.

- *Analytical specificity* (also called *selectivity*): This is a measure of the extent to which other substances that may be present in the sample of test analyte may interfere with the analysis and therefore lead to a falsely high or low value. A simple example is the ability of a method to measure glucose in the presence of other hexoses such as mannose and galactose. In clinical biochemical measurements, selectivity is an index of the ability of the method to give a consistent negative result for known negatives (see Section 16.1.2)

- *Analytical sensitivity*: This is a measure of the change in response of the method to a defined change in the quantity of analyte present. In many cases analytical sensitivity is expressed as the slope of a linear calibration curve.
- *Robustness*: This is a measure of the ability of the method to give a consistent result in spite of small changes in experimental parameters such as pH, temperature and amount of reagents added. For routine analysis, the robustness of a method is an important practical consideration.

These performance indicators are established by the use of well-characterised test and reference analyte samples. The order in which they are evaluated will depend on the immediate analytical priorities, but initially the three most important may be specificity, detection limit and analytical range. Once a method is in routine use, the question of assuring the quality of analytical data by the implementation of quality assessment procedures comes into play.

1.4.3 Assessment of precision

After a quantitative study has been completed and an experimental value for the amount and/or concentration of the test analyte in the test sample obtained, the experimenter must ask the question 'How confident can I be that my result is an acceptable estimate of the 'true' value?' (i.e. is it accurate?). An additional question may be 'Is the quality of my analytical data comparable with that in the published scientific literature for the particular analytical method?' (i.e. is it precise?). Once the answers to such questions are known, a result that has a high probability of being correct can be accepted and used as a basis for the design of further studies whilst a result that is subject to unacceptable error can be rejected. Unfortunately it is not possible to assess the precision of a single quantitative determination. Rather, it is necessary to carry out analyses in replicate (i.e. the experiment is repeated several times on the same sample of test analyte) and to subject the resulting data set to some basic statistical tests.

If a particular experimental determination is repeated numerous times and a graph constructed of the number of times a particular result occurs against its value, it is normally bell-shaped with the results clustering symmetrically about a mean value. This type of distribution is called a Gaussian or normal distribution. In such cases the precision of the data set is a reflection of random error. However, if the plot is skewed to one side of the mean value, then systematic errors have not been eliminated. Assuming that the data set is of the normal distribution type, there are three statistical parameters that can be used to quantify precision.

Standard deviation, coefficient of variation and variance – measures of precision

These three statistical terms are alternative ways of expressing the scatter of the values within a data set about the mean, $\bar{x}$, calculated by summing their total value and dividing by the number of individual values. Each term has its individual merit. In all three cases the term is actually measuring the width of the normal distribution curve such that the narrower the curve the smaller the value of the term and the higher the precision of the analytical data set.

The standard deviation (s) of a data set is a measure of the variability of the population from which the data set was drawn. It is calculated by use of equation 1.10 or 1.11:

$$s = \sqrt{\frac{\Sigma(x_i - \bar{x})^2}{n - 1}} \tag{1.10}$$

$$s = \sqrt{\frac{\Sigma x_i^2 - (\Sigma x_i)^2/n}{n - 1}} \tag{1.11}$$

$(x_i - \bar{x})$ is the difference between an individual experimental value (x_i) and the calculated mean $\bar{x}$ of the individual values. Since these differences may be positive or negative, and since the distribution of experimental values about the mean is symmetrical, if they were simply added together they would cancel out each other. The differences are therefore squared to give consistent positive values. To compensate for this, the square root of the resulting calculation has to be taken to obtain the standard deviation.

Standard deviation has the same units as the actual measurements and this is one of its attractions. The mathematical nature of a normal distribution curve is such that 68.2% of the area under the curve (and hence 68.2% of the individual values within the data set) is within one standard deviation either side of the mean, 95.5% of the area under the curve is within two standard deviations and 99.7% within three standard deviations. Exactly 95% of the area under the curve falls between the mean and 1.96 standard deviations. The precision (or imprecision) of a data set is commonly expressed as ±1 SD of the mean.

The term $(n - 1)$ is called the degrees of freedom of the data set and is an important variable. The initial number of degrees of freedom possessed by a data set is equal to the number of results (n) in the set. However, when another quantity characterising the data set, such as the mean or standard deviation, is calculated, the number of degrees of freedom of the set is reduced by 1 and by 1 again for each new derivation made. Many modern calculators and computers include programs for the calculation of standard deviation. However, some use variants of equation 1.10 in that they use n as the denominator rather than $n - 1$ as the basis for the calculation. If n is large, greater than 30 for example, then the difference between the two calculations is small, but if n is small, and certainly if it is less than 10, the use of n rather than $n - 1$ will significantly underestimate the standard deviation. This may lead to false conclusions being drawn about the precision of the data set. Thus for most analytical biochemical studies it is imperative that the calculation of standard deviation is based on the use of $n - 1$.

The coefficient of variation (CV) (also known as relative standard deviation) of a data set is the standard deviation expressed as a percentage of the mean as shown in equation 1.12.

$$CV = \frac{s100\%}{\bar{x}} \qquad\qquad (1.12)$$

Since the mean and standard deviation have the same units, coefficient of variation is simply a percentage. This independence of the unit of measurement allows methods based on different units to be compared.

The variance of a data set is the mean of the squares of the differences between each value and the mean of the values. It is also the square of the standard deviation, hence the symbol s^2. It has units that are the square of the original units and this makes it appear rather cumbersome which explains why standard deviation and coefficient of variation are the preferred ways of expressing the variability of data sets. The importance of variance will be evident in later discussions of the ways of making a statistical comparison of two data sets.

To appreciate the relative merits of standard deviation and coefficient of variation as measures of precision, consider the following scenario. Suppose that two serum samples, A and B, were each analysed 20 times for serum glucose by the glucose oxidase method (see Section 15.3.5) such that sample A gave a mean value of 2.00 mM with a standard deviation of $\pm$0.10 mM and sample B a mean of 8.00 mM and a standard deviation of $\pm$0.41 mM. On the basis of the standard deviation values it might be concluded that the method had given a better precision for sample A than for B. However, this ignores the absolute values of the two samples. If this is taken into account by calculating the coefficient of variation, the two values are 5.0% and 5.1% respectively showing that the method had shown the same precision for both samples. This illustrates the fact that standard deviation is an acceptable assessment of precision for a given data set but if it is necessary to compare the precision of two or more data sets, particularly ones with different mean values, then coefficient of variation should be used. The majority of well-developed analytical methods have a coefficient of variation within the analytical range of less than 5% and many, especially automated methods, of less than 2%.

1.4.4 Assessment of accuracy

Population statistics

Whilst standard deviation and coefficient of variation give a measure of the variability of the data set they do not quantify how well the mean of the data set approaches the 'true' value. To address this issue it is necessary to introduce the concepts of population statistics and confidence limit and confidence interval. If a data set is made up of a very large number of individual values so that n is a large number, then the mean of the set would be equal to the population mean mu (μ) and the standard deviation would equal the population standard deviation sigma (σ). Note that Greek letters represent the population parameters and the common alphabet the sample parameters. These two population parameters are the best estimates of the 'true' values since they are based on the largest number of individual measurements so that the influence of random errors is minimised. In practice the population parameters are seldom measured for obvious practicality reasons and the sample parameters have

Example 4   **ASSESSMENT OF THE PRECISION OF AN ANALYTICAL DATA SET**

Question   Five measurements of the fasting serum glucose concentration were made on the same sample taken from a diabetic patient. The values obtained were 2.3, 2.5, 2.2, 2.6 and 2.5 mM. Calculate the precision of the data set.

Answer   Precision is normally expressed either as one standard deviation of the mean or as the coefficient of variation of the mean. These statistical parameters therefore need to be calculated.

Mean

$$\bar{x} = \frac{2.2 + 2.3 + 2.5 + 2.5 + 2.6}{5} = 2.42 \, \text{mM}$$

Standard deviation
Using both equations (1.10) and (1.11) to calculate the value of s:

x_i	$x_i - \bar{x}$	$(x_i - \bar{x})^2$	x_i^2
2.2	−0.22	0.0484	4.84
2.3	−0.12	0.0144	5.29
2.5	+0.08	0.0064	6.25
2.5	+0.08	0.0064	6.25
2.6	+0.18	0.0324	6.75
Σx_i 12.1	$\Sigma 0.00$	$\Sigma 0.1080$	$\Sigma 29.39$

Using equation 1.10

$$s = \sqrt{0.108/4} = 0.164 \, \text{mM}$$

Using equation 1.11

$$s = \sqrt{\frac{29.39 - (12.1)^2/5}{4}} = \sqrt{\frac{29.39 - 29.28}{4}} = 0.166 \, \text{mM}$$

Coefficient of variation
Using equation 1.12

$$CV = \frac{0.165 \times 100\%}{2.42}$$
$$= 6.82\%$$

Discussion   In this case it is easier to appreciate the precision of the data set by considering the coefficient of variation. The value 6.82% is moderately high for this type of analysis. Automation of the method would certainly reduce it by at least half. Note that it is legitimate to quote the answers to these calculations to one more digit than was present in the original data set. In practice, it is advisable to carry out the statistical analysis on a far larger data set than that presented in this example.

a larger uncertainty associated with them. The uncertainty of the sample mean deviating from the population mean decreases in the proportion of the reciprocal of the square root of the number of values in the data set i.e. $1/\sqrt{n}$.

Thus to decrease the uncertainty by a factor of two the number of experimental values would have to be increased four-fold and for a factor of 10 the number of measurements would need to be increased 100-fold. The nature of this relationship again emphasises the importance of evaluating the acceptable degree of uncertainty of the experimental result before the design of the experiment is completed and the practical analysis begun. Modern automated analytical instruments recognise the importance of multiple results by facilitating repeat analyses at maximum speed. It is good practice to report the number of measurements on which the mean and standard deviation are based as this gives a clear indication of the quality of the calculated data.

Confidence intervals, confidence limits and the Student's t factor

Accepting that the population mean is the best estimate of the 'true' value, the question arises 'How can I relate my experimental sample mean to the population mean?' The answer is by using the concept of confidence. Confidence level expresses the level of confidence, expressed as a percentage, that can be attached to the data. Its value has to be set by the experimenter to achieve the objectives of the study. Confidence interval is a mathematical statement relating the sample mean to the population mean. A confidence interval gives a range of values about the sample mean within which there is a given probability (determined by the confidence level) that the population mean lies. The relationship between the two means is expressed in terms of the standard deviation of the data set, the square root of the number of values in the data set and a factor known as *Student's* t (equation 1.13):

$$\mu = \bar{x} \pm \frac{ts}{\sqrt{n}} \tag{1.13}$$

where $\bar{x}$ is the measured mean, μ is the population mean, s is the measured standard deviation, n is the number of measurements and t is the Student's t factor. The term $s/\sqrt{n}$ is known as the standard error of the mean and is a measure of the precision of the sample mean. Unlike standard deviation, standard error depends on the sample size and will fall as the sample size increases. The two measurements are sometimes confused, but in essence, standard deviation should be used if we want to know how widely scattered are the measurements and standard error should be used if we want to indicate the uncertainty around a mean measurement.

Confidence level can be set at any value up to 100%. For example, it may be that a confidence level of only 50% would be acceptable for a particular experiment. However, a 50% level means that that there is a one in two chance that the sample mean is not an acceptable estimate of the population mean. In contrast, the choice of a 95% or 99% confidence level would mean that there was only a one in 20 or a one in 100 chance respectively that the best estimate had not been achieved. In practice, most analytical biochemists choose a confidence level in the range 90–99% and most commonly 95%.

Student's t is a way of linking probability with the size of the data set and is used in a number of statistical tests. Student's t values for varying numbers in a data set

Table 1.7 **Values of Student's *t***

Degrees of freedom	Confidence level (%)					
	50	90	95	98	99	99.9
2	0.816	2.920	4.303	6.965	9.925	31.598
3	0.765	2.353	3.182	4.541	5.841	12.924
4	0.741	2.132	2.776	3.747	4.604	8.610
5	0.727	2.015	2.571	3.365	4.032	6.869
6	0.718	1.943	2.447	3.143	3.707	5.959
7	0.711	1.895	2.365	2.998	3.500	5.408
8	0.706	1.860	2.306	2.896	3.355	5.041
9	0.703	1.833	2.262	2.821	3.250	4.798
10	0.700	1.812	2.228	2.764	3.169	4.587
15	0.691	1.753	2.131	2.602	2.947	4.073
20	0.687	1.725	2.086	2.528	2.845	3.850
30	0.683	1.697	2.042	2.457	2.750	3.646

(and hence with the varying degrees of freedom) at selected confidence levels are available in statistical tables. Some values are shown in Table 1.7. The numerical value of t is equal to the number of standard errors of the mean that must be added and subtracted from the mean to give the confidence interval at a given confidence level. Note that as the sample size (and hence the degrees of freedom) increases, the confidence levels converge. When n is large and if we wish to calculate the 95% confidence interval, the value of t approximates to 1.96 and some texts quote equation 1.13 in this form. The term Student's t factor may give the impression that it was devised specifically with students' needs in mind. In fact 'Student' was the pseudonym of a statistician, by the name of W. S. Gossett, who in 1908 first devised the term and who was not permitted by his employer to publish his work under his own name.

Criteria for the rejection of outlier experimental data – *Q*-test

A very common problem in quantitative biochemical analysis is the need to decide whether or not a particular result is an *outlier* and should therefore be rejected before the remainder of the data set are subjected to statistical analysis. It is important to identify such data as they have a disproportionate effect on the calculation of the mean and standard deviation of the data set. When faced with this problem, the first action should be to check that the suspected outlier is not due to a simple experimental or mathematical error. Once the suspect figure has been confirmed its validity is checked by application of Dixon's *Q*-test. Like other tests to be described later, the

Example 5   **ASSESSMENT OF THE ACCURACY OF AN ANALYTICAL DATA SET**

Question   Calculate the confidence intervals at the 50%, 95% and 99% confidence levels of the fasting serum glucose concentrations given in the previous worked example.

Answer   Accuracy in this type of situation is expressed in terms of confidence intervals that express a range of values over which there is a given probability that the 'true' value lies.

As previously calculated, x = 2.42 mM and s = 0.16 mM. Inspection of Table 1.8 reveals that for four degrees of freedom (the number of experimental values minus one) and a confidence level of 50%, t = 0.741 so that the confidence interval for the population mean is given by:

$$\text{confidence interval} = 2.42 \pm \frac{(0.741)(0.16)}{\sqrt{5}}$$

$$= 2.42 \pm 0.05 \text{ mM}$$

For the 95% confidence level and the same number of degrees of freedom, t = 2.776, hence the confidence interval for the population mean is given by:

$$\text{confidence interval} = 2.42 \pm \frac{(2.776)(0.16)}{\sqrt{5}}$$

$$= 2.42 \pm 0.20 \text{ mM}$$

For the 99% confidence level and the same number of degrees of freedom, t = 4.604, hence the confidence interval for the population mean is given by:

$$\text{confidence interval} = 2.42 \pm \frac{(4.604)(0.16)}{\sqrt{5}}$$

$$= 2.42 \pm 0.33 \text{ mM}$$

Discussion   These calculations show that there is a 50% chance that the population mean lies in the range 2.37 to 2.47 mM, a 95% chance that the population mean lies within the range 2.22 to 2.62 mM and a 99% chance that it lies in the range 2.09 to 2.75 mM. Note that as the confidence level increases the range of potential values for the population mean also increases. You can calculate for yourself that if the mean and standard deviation had been based on 20 measurements (i.e. a four-fold increase in the number of measurements) then the 50% and 95% confidence intervals would have been reduced to 2.42 ± 0.02 mM and 2.42 ± 0.07 mM respectively. This re-emphasises the beneficial impact of multiple experimental determinations but at the same time highlights the need to balance the value of multiple determinations against the accuracy with which the experimental mean is required within the objectives of the individual study.

Table 1.8 **Values of Q for the rejection of outliers**

Number of observations	Q (95% confidence)
4	0.83
5	0.72
6	0.62
7	0.57
8	0.52

test is based on a null hypothesis, namely that there is no difference in the values being compared. If the hypothesis is proved to be correct then the suspect value cannot be rejected. The suspect value is used to calculate an experimental rejection quotient, Q_{exp}. Q_{exp} is then compared with tabulated critical rejection quotients, Q_{table}, for a given confidence level and the number of experimental results (Table 1.8). If Q_{exp} is less than Q_{table} the null hypothesis is confirmed and the suspect value should not be rejected, but if it is greater then the value can be rejected. The basis of the test is the fact that in a normal distribution 95.5% of the values are within the range of two standard deviations of the mean. In setting limits for the acceptability or rejection of data, a compromise has to be made on the confidence level chosen. If a high confidence level is chosen the limits of acceptability are set wide and therefore there is a risk of accepting values that are subject to error. If the confidence level is set too low, the acceptability limits will be too narrow and therefore there will be a risk of rejecting legitimate data. In practice a confidence level of 90% or 95% is most commonly applied. The Q_{table} values in Table 1.8 are based on a 95% confidence level.

The calculation of Q_{exp} is based upon equation 1.14 that requires the calculation of the separation of the questionable value from the nearest acceptable value (*gap*) coupled with knowledge of the *range* covered by the data set:

$$Q_{exp} = \frac{x_n - x_{n-1}}{x_n - x_1} = \frac{\text{gap}}{\text{range}} \tag{1.14}$$

where x is the value under investigation in the series $x_1, x_2, x_3, \ldots x_{n-1}, x_n$.

1.4.5 **Validation of an analytical method – the use of t-tests**

A t-test in general is used to address the question as to whether or not two data sets have the same mean. Both data sets need to have a normal distribution and equal variances. There are three types:

- *Unpaired* t-*test*: Used to test whether two data sets have the same mean.
- *Paired* t-*test*: Used to test whether two data sets have the same mean where each value in one set is paired with a value in the other set.
- *One-sample* t-*test*: Used to test whether the mean of a data set is equal to a particular value.

Example 6   **IDENTIFICATION OF AN OUTLIER EXPERIMENTAL RESULT**

Question If the data set in Example 5 contained an additional value of 3.0 mM, could this value be regarded as an outlier point at the 95% confidence level?

Answer From equation 1.15

$$Q_{exp} = \frac{3.0 - 2.6}{3.0 - 2.2} = \frac{0.4}{0.8} = 0.5$$

Using Table 1.11 for six data points Q_{table} is equal to 0.62.

Since Q_{exp} is smaller than Q_{table} the point should not be rejected as there is more than a 95% chance that it is part of the same data set as the other five values. It is easy to show that an additional data point of 3.3 rather than 3.0 mM would give a Q_{exp} of 0.64 and could be rejected.

Each test is based on a null hypothesis, which is that there is no difference between the means of the two data sets. The tests measure how likely the hypothesis is to be true. The attraction of such tests is that they are easy to carry out and interpret.

Analysis of a standard solution – one-sample t-test

Once the choice of the analytical method to be used for a particular biochemical assay has been made, the normal first step is to carry out an evaluation of the method in the laboratory. This evaluation entails the replicated analysis of a known standard solution of the test analyte and the calculation of the mean and standard deviation of the resulting data set. The question is then asked 'Does the mean of the analytical results agree with the known value of the standard solution within experimental error?' To answer this question a t-test is applied.

In the case of the analysis of a standard solution the calculated mean and standard deviation of the analytical results are used to calculate a value of the Student's t (t_{calc}) using equation 1.15. It is then compared with table values of t (t_{table}) for the particular degrees of freedom of the data set and at the required confidence level (Table 1.7).

$$t_{calc} = \frac{(\text{known value} - \bar{x})\sqrt{n}}{s} \tag{1.15}$$

These table values of t represent critical values that separate the border between different probability levels. If t_{calc} is greater than t_{table} the analytical results are deemed not to be from the same data set as the known standard solution at the selected confidence level. In such cases the conclusion is therefore drawn that the analytical results do not agree with the standard solution and hence that there are unidentified errors in them. There would be no point in applying the analytical method to unknown test analyte samples until the problem has been resolved.

Example 7 **VALIDATING AN ANALYTICAL METHOD**

Question A standard solution of glucose is known to be 5.05 mM. Samples of it were analysed by the glucose oxidase method (see Section 15.3.2 for details) that was being used in the laboratory for the first time. A calibration curve obtained using least mean square linear regression was used to calculate the concentration of glucose in the test sample. The following experimental values were obtained: 5.12, 4.96, 5.21, 5.18 and 5.26 mM. Does the experimental data set for the glucose solution agree with the known value within experimental error?

Answer It is first necessary to calculate the mean and standard deviation for the set and then to use it to calculate a value for Student's t.

Applying equations 1.10 and 1.11 to the data set gives $\bar{x} = 5.15$ mM and $s = \pm 0.1$ mM

Now applying equation 1.16 to give t_{calc}:

$$t_{calc} = \frac{(5.05 - 5.15)}{0.1} \sqrt{5} = 2.236$$

Note that the negative difference between the two mean values in this calculation is ignored. From Table 1.10 at the 95% confidence level with four degrees of freedom, $t_{table} = 2.776$. t_{calc} is therefore less than t_{table} and the conclusion can be drawn that the measured mean value does agree with the known value. Using equation 1.13, the coefficient of variation for the measured values can be calculated to be 1.96%.

Comparing two competitive analytical methods – unpaired t-test

In quantitative biochemical analysis it is frequently helpful to compare the perform-ance of two alternative methods of analysis in order to establish whether or not they give the same quantitative result within experimental error. To address this need, each method is used to analyse the same test sample using replicated analysis. The mean and standard deviation for each set of analytical data is then calculated and a Student's t-test applied. In this case the t-test measures the overlap between the data sets such that the smaller the value of t_{calc} the greater the overlap between the two data sets. This is an example of an unpaired t-test.

In using the tables of critical t values, the relevant degrees of freedom is the sum of the number of values in the two data sets (i.e. $n_1 + n_2$) minus 2. The larger the number of degrees of freedom the smaller the value of t_{calc} needs to be to exceed the critical value at a given confidence level. The formulae for calculating t_{calc} depend on whether or not the standard deviations of the two data sets are the same. This is often obvious by inspection, the two standard deviations being similar. However, if in doubt, an F-test, named after Fisher who introduced it, can be applied. An F-test is based on the null hypothesis that there is no difference between the two variances. The test calculates a value for F (F_{calc}), which is the ratio of the larger of the two variances to the smaller variance. It is then compared with critical F values (F_{table}) available in statistical tables

Table 1.9 Critical values of F at the 95% confidence level

Degrees of freedom for S_2	Degrees of freedom for S_1							
	2	3	4	6	10	15	30	∞
2	19.0	19.2	19.2	19.3	19.4	19.4	19.5	19.5
3	9.55	9.28	9.12	8.94	8.79	8.70	8.62	8.53
4	6.94	6.59	6.39	6.16	5.96	5.86	5.75	5.63
5	5.79	5.41	5.19	4.95	4.74	4.62	4.50	4.36
6	5.14	4.76	4.53	4.28	4.06	3.94	3.81	3.67
7	4.74	4.35	4.12	3.87	3.64	3.51	3.38	3.23
8	4.46	4.07	3.84	3.58	3.35	3.22	3.08	2.93
9	4.26	3.86	3.63	3.37	3.14	3.01	2.86	2.71
10	4.10	3.71	3.48	3.22	2.98	2.84	2.70	2.54
15	3.68	3.29	3.06	2.79	2.54	2.40	2.25	2.07
20	3.49	2.10	2.87	2.60	2.35	2.20	2.04	1.84
30	3.32	2.92	2.69	2.42	2.16	2.01	1.84	1.62
∞	3.00	2.60	2.37	2.10	1.83	1.67	1.46	1.00

or computer packages (Table 1.9). If the calculated value of F is less than the table value, the null hypothesis is proved and the two standard deviations are considered to be similar. If the two variances are of the same order, then equations 1.16 and 1.17 are used to calculate t_{calc} for the two data sets. If not, equations 1.18 and 1.19 are used.

$$t_{calc} = \frac{\bar{x}_1 - \bar{x}_2}{s_{pooled}} \sqrt{\frac{n_1 n_2}{n_1 + n_2}} \tag{1.16}$$

$$s_{pooled} = \sqrt{\frac{s_1^2 (n_1 - 1) + s_2^2 (n_2 - 1)}{n_1 + n_2 - 2}} \tag{1.17}$$

$$t_{calc} = \frac{\bar{x}_1 - \bar{x}_2}{\sqrt{(s_1^2 / n_1 + s_2^2 / n_2)}} \tag{1.18}$$

$$\text{degrees of freedom} = \left\{ \frac{(s_1^2 / n_1 + s_2^2 / n_2)^2}{(s_1^2 / n_1)^2 / (n_1 + 1) + (s_2^2 / n_2)^2 / (n_2 + 1)} \right\} - 2 \tag{1.19}$$

where $\bar{x}_1$ and $\bar{x}_2$ are the calculated means of the two methods, s_1^2 and s_2^2 are the calculated standard deviations of the two methods and n_1 and n_2 are the number of measurements in the two methods.

At first sight these four equations may appear daunting, but closer inspection reveals that they are simply based on variance (s^2), mean ($\bar{x}$) and number of analytical measurements (n) and that the mathematical manipulation of the data is relatively easy.

Example 8 **COMPARISON OF TWO ANALYTICAL METHODS USING REPLICATED ANALYSIS OF A SINGLE TEST SAMPLE**

Question A sample of fasting serum was used to evaluate the performance of the glucose oxidase and hexokinase methods for the quantification of serum glucose concentrations (for details see Section 15.3.5). The following replicated values were obtained: for the glucose oxidase method 2.3, 2.5, 2.2, 2.6 and 2.5 mM and for the hexokinase method 2.1, 2.7, 2.4, 2.4 and 2.2 mM.

Establish whether or not the two methods gave the same results at the 95% confidence level.

Answer Using the standard formulae we can calculate the mean, standard deviation and variance for each data set.

Glucose oxidase method

$$\bar{x} = 2.42\,\text{mM}, \ s = 0.16\,\text{mM}, \ s^2 = 0.026\,(\text{mM})^2$$

Hexokinase method

$$\bar{x} = 2.36\,\text{mM}, \ s = 0.23\,\text{mM}, \ s^2 = 0.053\,(\text{mM})^2$$

We can then apply the F-test to the two variances to establish whether or not they are the same:

$$F_{\text{calc}} = \frac{0.053}{0.026} = 2.04$$

F_{table} for the two sets of data each with four degrees of freedom and for the 95% confidence level is 6.39 (Table 1.11).

Since F_{calc} is less than F_{table} we can conclude that the two variances are not significantly different. Therefore using equations 1.17 and 1.18 we can calculate that:

$$s_{\text{pooled}} = \sqrt{\frac{(0.16)^2(4) + (0.23)^2 4}{8}} = \sqrt{\frac{0.102 + 0.212}{8}} = \sqrt{0.039} = 0.198$$

$$t_{\text{calc}} = \frac{2.42 - 2.36}{0.198}\sqrt{\frac{(5)(5)}{10}} = (0.303)(1.58) = 0.48$$

Using Table 1.10 at the 95% confidence level and for eight degrees of freedom t_{table} is 2.306. Thus t_{calc} is far less than t_{table} and so the two sets of data are not significantly different, i.e. the two methods have given the same result at the 95% confidence level.

Comparison of two competitive analytical methods – paired *t*-test

A variant of the previous type of comparison of two analytical methods based upon the analysis of a common standard sample, is the case in which a series of test samples is analysed once by the two different analytical methods. In this case there is no replication of analysis of any test sample by either method. The *t*-test is applied to the differences between the results of each method for each test sample. This is an example of a paired t-test. The formula for calculating t_{calc} in this case is given by equation 1.20:

$$t_{calc} = \frac{\bar{d}}{s_d}\sqrt{n} \tag{1.20}$$

$$s_d = \sqrt{\frac{\Sigma(d_i - \bar{d})^2}{n - 1}} \tag{1.21}$$

where d_i is the difference between the paired results, $\bar{d}$ is the mean difference between the paired results, n is the number of paired results and s_d is the standard deviation of the differences between the pairs.

1.4.6 Calibration methods

Quantitative biochemical analyses often involve the use of a calibration curve produced by the use of known amounts of the analyte using the selected analytical procedure. A calibration curve is a record of the measurement (absorbance, peak area, etc.) produced by the analytical procedure in response to a range of known quantities of the standard analyte. It involves the preparation of a standard solution of the analyte and the use of a range of aliquots in the test analytical procedure. It is good practice to replicate each calibration point and to use the mean ± one standard deviation for the construction of the calibration plot. Inspection of the compiled data usually reveals a scatter of the points about a linear relationship but such that there are several options for the 'best' fit. The technique of fitting the best fit 'by eye' is not recommended, as it is highly subjective and irreproducible. The method of least mean squares linear regression (LMSLR) is the most common mathematical way of fitting a straight line to data but in applying the method, it is important to realise that the accuracy of the values for slope and intercept that it gives are determined by experimental error built into the x and y values.

The mathematical basis of LMSLR is complex and will not be considered here, but the principles upon which it is based are simple. If the relationship between the two variables, such as the concentration or amount of analyte and response, is linear, then the 'best' straight line will have the general form $y = mx + c$ where x and y are the two variables, m is the slope of the line and c is the intercept on the y-axis. It is assumed, correctly in most cases, that the errors in the measurement of y are much greater than those for x (it does not assume that there are no errors in the x values) and secondly that uncertainties (standard deviations) in the y values are all of the same magnitude. The method uses two criteria. The first is that the line will pass through the point $(\bar{x}, \bar{y})$ where $\bar{x}$ and $\bar{y}$ are the mean of the x and y values respectively. The second is that the slope (m) is based on the calculation of the optimum values of m and c that give minimum variation between individual experimental y values and their corresponding values as predicted by the 'best' straight line. Since these variations can be positive or negative (i.e. the experimental values can be greater or smaller than those predicted by the 'best' straight line), in the process of arriving at the best slope the method measures the deviations between the experimental and candidate straight line values, squares

Example 9 COMPARISON OF TWO ANALYTICAL METHODS USING DIFFERENT TEST SAMPLES

Question Ten fasting serum samples were each analysed by the glucose oxidase and the hexokinase methods. The following results, in mM, were obtained:

Glucose oxidase (mM)	Hexokinase (mM)	Difference d_i	Difference minus mean of difference	(Difference minus mean of difference)^2
1.1	0.9	0.2	0.08	0.0064
2.0	2.1	−0.1	−0.22	0.0484
3.2	2.9	0.3	0.18	0.0324
3.7	3.5	0.2	0.08	0.0064
5.1	4.8	0.3	0.18	0.0324
8.6	8.7	−0.1	−0.22	0.0484
10.4	10.6	−0.2	−0.32	0.1024
15.2	14.9	0.3	0.18	0.0324
18.7	18.7	0.0	−0.12	0.0144
25.3	25.0	0.3	0.18	0.0324
		Mean ($\bar{d}$) 0.12		$\sum$ 0.3560

Do the two methods give the same results at the 95% confidence level?

Answer Before addressing the main question, note that the ten samples analysed by the two methods were chosen to cover the whole analytical range for the methods. To assess whether or not the two methods have given the same result at the chosen confidence level, it is necessary to calculate a value for t_{calc} and to compare it with t_{table} for the nine degrees of freedom in the study. To calculate t_{calc}, it is first necessary to calculate the value of s_d in equation 1.21. The appropriate calculations are shown in the table above.

$$s_d = \sqrt{(\Sigma(d_i - \bar{d})^2)/(n-1)}$$
$$= \sqrt{(0.356/9)}$$
$$= 0.199$$

From equation 1.20

$$t_{calc} = \frac{\bar{d}\sqrt{n}}{s_d}$$
$$= (0.12\sqrt{10})/0.199$$
$$= 1.907$$

Using Table 1.10, t_{table} at the 95% confidence level and for nine degrees of freedom is 2.262. Since t_{calc} is smaller than t_{table} the two methods do give the same results at the 95% confidence level. Inspection of the two data sets shows that the glucose oxidase method gave a slightly high value for seven of the ten samples analysed.

Example 9 (*cont.*)

An alternative approach to the comparison of the two methods is to plot the two data sets as an *x/y* plot and to carry out a regression analysis of the data. If this is done using the glucose oxidase data as the *y* variable, the following results are obtained:

Slope: 1.0016, intercept: 0.1057, correlation coefficient *r*: 0.9997.

The slope of very nearly one confirms the similarity of the two data sets, whilst the small positive intercept on the *y*-axis confirms that the glucose oxidase method gives a slightly higher, but insignificantly different, value to that of the hexokinase method.

them (so they are all positive), sums them and then selects the values of m and c that give the minimum deviations. The end result of the regression analysis is the equation for the best-fit straight line for the experimental data set. This is then used to construct the calibration curve and subsequently to analyse the test analyte(s). Most modern calculators will carry out this type of analysis and will simultaneously report the 95% confidence limits for the m and c values and/or the standard deviation associated with the two values together with the 'goodness-of-fit' of the data as expressed by a correlation coefficient, r or a coefficient of determination, r^2. The stronger the correlation between the two variables, the closer the value of r approaches $+1$ or -1. Values of r are quoted to four decimal places and for good correlations commonly exceed 0.99. Values of 0.98 and less should be considered with care since even slight curvature can give r-values of this order.

In the routine construction of a calibration curve, a number of points have to be borne in mind:

- *Selection of standard values*: A range of standard analyte amounts/concentrations should be selected to cover the expected values for the test analyte(s) in such a way that the values are equally distributed along the calibration curve. Test samples should not be estimated outside this selected range, as there is no evidence that the regression analysis relationship holds outside the range. It is good practice to establish the analytical range and the limit of detection for the method. It is also advisable to determine the precision (standard deviation) of the method at different points across the analytical range and to present the values on the calibration curve. Such a plot is referred to as a *precision profile*. It is common for the precision to decrease (standard deviation to increase) at the two ends of the curve and this may have implications for the routine use of the curve. For example, the determination of testosterone in male and female serum requires the use of different methods since the two values (reference range 10–30 nM for males, <3 nM for females) cannot be accommodated with acceptable precision on one calibration curve.
- *Use of a 'blank' sample*: This is one in which no standard analyte is present. One should be included in the experimental design when possible (it will not be possible, for example, with analyses based on serum or plasma). Any experimental value, e.g. absorbance, obtained for it must be deducted from all other measurements.

This may be achieved automatically in spectrophotometric measurements by the use of a double-beam spectrophotometer in which the blank sample is placed in the reference cell.

- *Shape of curve*: It should not be assumed that all calibration curves are linear. They may be curved and best represented by a quadratic equation of the type $y = ax^2 + bx + c$ where a, b and c are constants or they may be logarithmic.
- *Recalibration*: A new calibration curve should be constructed on a regular basis. It is not acceptable to rely on a calibration curve produced on a much earlier occasion.

1.4.7 Internal standards

An additional approach to the control of time-related minor changes in a calibration curve and the quantification of an analyte in a test sample is the use of an internal standard. An ideal internal standard is a compound that has a molecular structure and physical properties as similar as possible to the test analyte and which gives a similar response to the analytical method as the test analyte. This response, expressed on a unit quantity basis, may be different from that for the test analyte but provided that the relative response of the two compounds is constant, the advantages of the use of the internal standard are not compromised. Quite commonly the internal standard is a structural or geometrical isomer of the test analyte.

A known fixed quantity of the standard is added to each test sample and analysed alongside the test analyte by the standard analytical procedure. The resulting response for the standard and the range of amounts or concentrations of the test analyte is used to calculate a relative response for the test analyte and used in the construction of the calibration curve. The curve therefore consists of a plot of the relative response to the test analyte against the range of quantities of the analyte.

Internal standards are commonly used in liquid and gas chromatography since they help to compensate for small temporal variations in the flow of liquid or gas through the chromatographic column. In such applications it is, of course, essential that the internal standard chromatographs are near to, but distinct from, the test analyte.

If the analytical procedure involves preliminary sampling procedures, such as solid-phase extraction, it is important that a known amount of the internal standard is introduced into the test sample at as early a stage as possible and is therefore taken through the preliminary procedures. This ensures that any loss of the test analyte during these preliminary stages will be compensated by identical losses to the internal standard so that the final relative response of the method to the two compounds is a true reflection of the quantity of the test analyte.

1.5 SAFETY IN THE LABORATORY

Virtually all experiments conducted in a biochemistry laboratory present a potential risk to the well-being of the investigator. In planning any experiment it is essential that careful thought be given to all aspects of safety before the experimental design is finalised. Health hazards come from a variety of sources:

- *Chemical hazards*: All chemicals are, to varying extents, capable of causing damage to the body. They may be irritants and cause a short-term effect on exposure. Alternatively they may be corrosive and cause severe and often irreversible damage to the skin. Examples include strong acids and alkalis. Thirdly they may be toxic once they have gained access to the body by ingestion, inhalation or absorption across the skin. Once in the body their effect may range from slight to the extremes of being a poison (e.g. cyanide), a carcinogen (e.g. benzene and vinyl chloride) or a teratogen (e.g. thalidomide). Finally there is the special case of the use of radioactive compounds that are discussed in detail in Chapter 14.
- *Biological hazards*: Examples include human body fluids that may carry infections such as HIV, laboratory animals that may cause allergic reactions or transmit certain diseases, pathogenic animal and cell tissue cultures, and all microorganisms including genetically engineered forms. In the UK, animal experiments must be conducted in accordance with Home Office regulations and guidelines. All experiments with tissue and cell cultures should be conducted in microbiological cabinets that are provided with a sterile airflow away from the operator (Section 2.2).
- *Electrical and mechanical hazards*: All electrical apparatus should be used and maintained in accordance with the manufacturers' instructions. Electrophoresis equipment presents a particular potential for safety problems. Centrifuges, especially high-speed varieties, also need careful use especially in the correct use and balance of the rotors.
- *General laboratory hazards*: Common examples include syringe needles, broken glassware and liquid nitrogen flasks.

   Routine precautions that should be taken to minimise personal exposure to these hazards include the wearing of laboratory coats, which should be of the high-necked buttoned variety for work with microorganisms, safety spectacles and lightweight disposable gloves. It is also good practice not to work alone in a laboratory so that help is to hand if needed. In the UK, laboratory work is subject to legislation including the *Health and Safety at Work Act 1974*, the *Control of Substances Hazardous to Health (COSHH) Regulations 1994* and the *Management of Health and Safety at Work Regulations 1999*. This legislation requires a *risk assessment* to be carried out prior to undertaking laboratory work. As the name implies, a risk assessment requires potential hazards to be identified and an assessment made of their potential severity and probability of occurrence. Action must be taken in cases where the potential severity and probability are medium to high. Such assessments require knowledge of the toxicity of all the chemicals used in the study. Toxicity data are widely available via computer packages and published handbooks and should be on reference in all laboratories. Once the toxicity data are known, consideration may be given to the use of alternative and less toxic compounds or, if it is decided to proceed with the use of toxic compounds, precautions taken to minimise their risk and plans laid for dealing with an accident should one occur. These include arranging access to first-aiders and other emergency services. It is normal for all laboratories to have a nominated Safety Officer whose responsibility it is to give advice on safety issues. To facilitate good practice, procedures for the disposal of organic solvents, radioactive residues, body fluids, tissue and cell cultures and microbiological cultures are posted in all laboratories.

1.6 SUGGESTIONS FOR FURTHER READING

Analytical methodology and quality assurance

Burns, M. (2004). Current practice in the assessment and control of measurement uncertainty in bio-analytical chemistry. *Trends in Analytical Chemistry*, **23**, 393–397.

Carson, P. A. and Dent, N. (eds.) (2007). *Good Clinical, Laboratory and Manufacturing Practices: Techniques for the QA Professional.* London: RSC. (A comprehensive but easy-to-read book aimed at both newcomers and professionals involved in laboratory quality assurance issues.)

Fesling, M. F. W. (2003). Principles: the need for better experimental design. *Trends in Pharmacological Sciences*, **24**, 341–345.

Safety Control of Substances Hazardous to Health Regulations 2002: Approved Code of Practice and Guidance. Kingston-upon-Thames: HSE Books. (A step-by-step approach to understanding the practical implications of COSHH.)

2 Cell culture techniques

A. R. BAYDOUN

2.1 INTRODUCTION

Cell culture is a technique that involves the isolation and maintenance *in vitro* of cells isolated from tissues or whole organs derived from animals, microbes or plants. In general, animal cells have more complex nutritional requirements and usually need more stringent conditions for growth and maintenance. By comparison, microbes and plants require less rigorous conditions and grow effectively with the minimum of needs. Regardless of the source of material used, practical cell culture is governed by the same general principles, requiring a sterile pure culture of cells, the need to adopt appropriate aseptic techniques and the utilisation of suitable conditions for optimal viable growth of cells.

Once established, cells in culture can be exploited in many different ways. For instance, they are ideal for studying intracellular processes including protein synthesis, signal transduction mechanisms and drug metabolism. They have also been widely used to understand the mechanisms of drug actions, cell–cell interaction and genetics. Additionally, cell culture technology has been adopted in medicine, where genetic abnormalities can be determined by chromosomal analysis of cells derived, for example, from expectant mothers. Similarly, viral infections can be assayed both qualitatively and quantitatively on isolated cells in culture. In industry, cultured cells are used routinely to test both the pharmacological and toxicological effects of pharmaceutical compounds. This technology thus provides a valuable tool to scientists, offering a user-friendly system that is relatively cheap to run and the exploitation of which avoids the legal, moral and ethical questions generally associated with animal experimentation. More importantly, cell culture also presents a tremendous potential for future exploitation in disease treatment, where, for instance, defective or malfunctioning genes could be corrected in the host's own cells and transplanted back into the host to treat a disease. Furthermore, successful development of culture techniques for stem cells will provide a much needed cell-based strategy for treating diseases where organ transplant is currently the only available option.

In this chapter, fundamental information required for standard cell culture, together with a series of principles and outline protocols used routinely in growing animal and bacterial cells are discussed. Additionally, a section has been dedicated to human embryonic stem cell culture, an emerging field where protocols to be used routinely are still being developed. The discussion in this chapter is thus limited to techniques that are now becoming routine for stem cell culture and should therefore provide the basic knowledge for those new to the field of cell culture and act as a revision aid for those with limited experience in the field. Throughout the chapter, particular attention is paid to the importance of the work environment, outlining safety considerations together with adequate description and hints on the essential techniques required for tissue culture work.

2.2 THE CELL CULTURE LABORATORY AND EQUIPMENT

2.2.1 The cell culture laboratory

The design and maintenance of the cell culture laboratory is perhaps the most important aspect of cell culture, since a sterile surrounding is critical for handling of cells and culture media, which should be free from contaminating microorganisms. Such organisms, if left unchecked, would outgrow the cells being cultured, eventually resulting in culture-cell demise owing to the release of toxins and/or depletion of nutrient from the culture medium.

Where possible, a cell culture laboratory should be designed in such a way that it facilitates preparation of media and allows for the isolation, examination, evaluation and maintenance of cultures under controlled sterile conditions. In an ideal situation, there should be a room dedicated to each of the above tasks. However, many cell culture facilities, especially in academia, form part of an open-plan laboratory and as such are limited in space. It is not unusual therefore to find an open-plan area where places are designated for each of the above functions. This is not a serious problem as long as a few basic guidelines are adopted. For instance, good aseptic techniques (discussed below) should be used at all times. There should also be adequate facilities for media preparation and sterilisation, and all cell culture materials should be maintained under sterile conditions until used. In addition, all surfaces within the culture area should be non-porous to prevent adsorption of media and other materials that may provide a good breeding ground for microorganisms, resulting in the infection of the cultures. Surfaces should also be easy to clean and all waste generated should be disposed of immediately. The disposal procedure may require prior autoclaving of the waste, which can be carried out using pressurised steam at 121 °C under 105 kPa for a defined period of time. These conditions are required to destroy microorganisms.

For smooth running of the facilities, daily checks should be made of the temperature in incubators, and of the gas supply to the incubators by checking the CO_2 cylinder pressure. Water baths should be kept clean at all times and areas under the work surfaces of the flow cabinets cleaned of any spills.

2.2.2 Equipment for cell culture

Several pieces of equipment are essential. These include a tissue culture hood, incubator(s), autoclave and microscope. A brief description will be given of these and other essential equipments.

Cell culture hoods

The cell culture hood is the central piece of equipment where all the cell handling is carried out and is designed not only to protect the cultures from the operator but in some cases to protect the operator from the cultures. These hoods are generally referred to as laminar flow hoods as they generate a smooth uninterrupted streamlined flow (laminar flow) of sterile air which has been filtered through a high-efficiency particulate air (HEPA) filter. There are two types of laminar flow hood classified as either vertical or horizontal. The horizontal hoods allow air to flow directly at the operator and as a result are generally used for media preparation or when one is working with non-infectious materials, including those derived from plants. The vertical hoods (also known as biology safety cabinets) are best for working with hazardous organisms, since air within the hood is filtered before it passes into the surrounding environment.

Currently, there are at least three different classes of hood used which all offer various levels of protection to the cultures, the operator or both and these are described below.

Class I hoods

These hoods, as with the class II type, have a screen at the front that provides a barrier between the operator and the cells but yet allows access into the hood through an opening at the bottom of the screen (Fig. 2.1). This barrier prevents too much turbulence to air flow from the outside and, more importantly, provides good protection for the operator. Cultures are also protected but to a lesser extent when compared to the class II hoods as the air drawn in from the outside is sucked through the inner cabinet to the top of the hood. These hoods are suitable for use with low-risk organisms and when operator protection only is required.

Class II hoods

Class II hoods are the most common units found in tissue culture laboratories. These hoods offer good protection to both the operator and the cell culture. Unlike class I hoods, air drawn from the outside is passed through the grill in the front of the work area and filtered through the HEPA filter at the top of the hood before streaming down

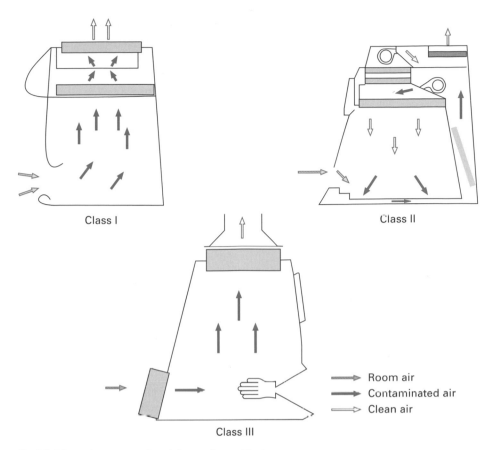

Fig. 2.1 Schematic representation of tissue culture cabinets.

over the tissue culture (Fig. 2.1). This mechanism protects the operator and ensures that the air over the cultures is largely sterile. These hoods are adequate for animal cell culture, which involves low to moderate toxic or infectious agents, but are not suitable for use with high-risk pathogens, which may require a higher level of containment.

Class III hoods

Class III safety cabinets are required when the highest levels of operator and product protection are required. These hoods are completely sealed, providing two glove pockets through which the operator can work with material inside the cabinet (Fig. 2.1). Thus the operator is completely shielded, making class III hoods suitable for work with highly pathogenic organisms including tissue samples carrying known human pathogens.

Practical hints and safety aspects of using cell culture hoods

All hoods must be maintained in a clutter-free and clean state at all times as too much clutter may affect air flow and contamination will introduce infections. Thus, as a rule of thumb, put only items that are required inside the cabinet and clean all work surfaces before and after use with industrial methylated spirit (IMS). The latter is used at an effective concentration of 70% (prepared by adding 70% v/v IMS to 30% Milli-Q water), which acts against bacteria and fungal spores by dehydrating and fixing cells, thus preventing contamination of cultures.

Some cabinets may be equipped with a short-wave ultraviolet light that can be used to irradiate the interior of the hood to kill microorganisms. When present, switch on the ultraviolet light for at least 15 min to sterilise the inside of the cabinet, including the work area. Note, however, that ultraviolet radiation can cause adverse damage to the skin and eyes and precaution should be taken at all times to ensure that the operator is not in direct contact with the ultraviolet light when using this option to sterilise the hood. Once finished, ensure that the front panel door (class I and II hoods) is replaced securely after use. In addition always turn the hood on for at least 10 min before starting work to allow the flow of air to stabilise. During this period, monitor the air flow and check all dials in the control panel at the front of the hood to ensure that they are within the safe margin.

CO$_2$ incubators

Water-jacketed incubators are required to facilitate optimal cell growth under strictly maintained and regulated conditions, normally requiring a constant temperature of 37 °C and an atmosphere of 5–10% CO$_2$ plus air. The purpose of the CO$_2$ is to ensure that the culture medium is maintained at the required physiological pH (usually pH 7.2–7.4). This is achieved by the supply of CO$_2$ from a gas cylinder into the incubator through a valve that is triggered to draw in CO$_2$ whenever the level falls below the set value of 5% or 10%. The CO$_2$ that enters the inner chamber of the incubator dissolves into the culture medium containing bicarbonate. The latter reacts with H$^+$ (generated from cellular metabolism), forming carbonic acid, which is in equilibrium with water and CO$_2$, thereby maintaining the pH in the medium at approximately pH 7.2.

$$HCO_3^- + H^+ \rightleftharpoons H_2CO_3 \rightleftharpoons CO_2 + H_2O$$

These incubators are generally humidified by the inclusion of a tray of sterile water on the bottom deck. The evaporation of water creates a highly humidified atmosphere, which helps to prevent evaporation of medium from the cultures.

An alternative to humidified incubators is the dry non-gassed unit which is not humidified and relies on the use of alternative buffering systems such as 4(2-hydroxy-ethyl)-1-piperazine-ethanesulphonic acid (Hepes) or morpholinopropane sulphonic acid (Mops) for maintaining a balanced pH within the culture medium. The advantage of this system is that it eliminates the risk from infections that can be posed by the tray of water in the humidified unit. The disadvantage, however, is that the culture medium will evaporate rapidly, thereby stressing the cells. One way round this problem is to place the cell culture plate in a sandwich box containing little pots of sterile water. With the sandwich box lid partially closed, evaporation of water from the pots will create a humidified atmosphere within the sandwich box, thus reducing the risk of evaporation of medium from the culture plate.

Practical hints and safety aspects of using cell culture incubators

The incubator should be maintained at 37 °C and supplied with 5% CO$_2$ at all times. A constant temperature can be maintained by keeping a thermometer in the incubator, preferably on the inside of the inner glass door. This can then be checked on a regular basis and adjustments made as required. CO_2 levels inside the unit can be monitored and adjusted by using a gas analyser such as the Fyrite Reader. Regular checks should also be made on the levels of CO_2 in the gas cylinders that supply CO_2 to the incubators and these should be replaced when levels are very low. Most incubators are designed with an inbuilt alarm that sounds when the CO_2 level inside the chamber drops. At this point the gas cylinder must be replaced immediately to avoid stressing or killing the cultures. It is now possible to connect two gas cylinders to a cylinder changeover unit that switches automatically to the second source of gas supply when the first is empty. It is advisable therefore to use this device where possible.

When one is using a humidified incubator, it is essential that the water tray is maintained and kept free from microorganisms. This can be achieved by adding various agents to the water such as the antimicrobial agent Roccal at a concentration of 1% (w/v). Other products such as Thimerosal or SigmaClean from Sigma-Aldrich can also be used. Proper care and maintenance of the incubator should, however, include regular cleaning of the interior of the unit using any of the above reagents then swabbing with 70% IMS. More recently, copper-coated incubators have been introduced which, due to the antimicrobial properties of copper, are reported to reduce microbial contamination.

Microscopes

Inverted phase contrast microscopes (see Chapter 4) are routinely used for visualising cells in culture. These are expensive but easy to operate, with a light source located above and the objective lenses below the stage on which the cells are placed. Visualisation of cells by microscopy can provide useful information about the morphology and state of the cells. Early signs of cell stress may be easily identified and appropriate action taken to prevent loss of cultures.

Other general equipment

Several other pieces of equipment are required in cell culture. These include a centrifuge to spin down cells, a water bath for thawing frozen samples of cells and warming media to 37 °C before use, and a fridge and freezer for storage of media and other materials required for cell culture. Some cells need to attach onto a surface in order to grow and are therefore referred to as adherent. These cells are cultured in non-toxic polystyrene plastics that contain a biologically inert surface on which the cells attach and grow. Various types of plastics are available for this purpose and include Petri dishes, multi-well plates (with either 96, 24, 12 or 6 wells per plate) and screwcap flasks classified according to their surface areas: T-25, T-75, T-225 (cm^2 of surface area). A selection of these plastics is shown in Fig. 2.2.

2.3   SAFETY CONSIDERATIONS IN CELL CULTURE

Because of the nature of the work, safety in the cell culture laboratory must be of a major concern to the operator. This is particularly the case when one is working with pathogenic

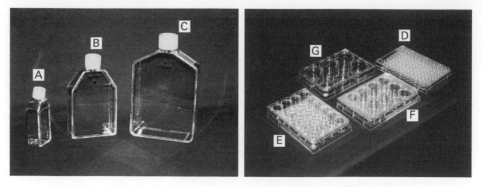

Fig. 2.2 Tissue culture plastics used generally for cell culture. (A–C) T-flasks; (D–G) representative of multi-well plates. (A) T–25 (25 cm^2), (B) T-75 (75 cm^2), (C) T-225 (225 cm^2), (D) 96-well plate, (E) 24-well plate, (F) 12-well plate and (G) 6-well plate.

microbes or with fresh primate or human tissues or cells which may contain agents that use humans as hosts. One very good example of this would be working with fresh human lymphocytes, which may contain infectious agents such as the human immunodeficiency virus (HIV) and/or hepatitis B virus. Thus, when one is working with fresh human tissue, it is essential that the infection status of the donor is determined in advance of use and all necessary precautions taken to eliminate or limit the risks to which the operator is exposed. A recirculation class II cabinet would be a minimum requirement for this type of cell culture work and the operator should be provided with protective clothing including latex gloves and a face mask if required. Such work should also be carried out under the guidelines laid down by the UK Advisory Committee on Dangerous Pathogens (ACDP).

Apart from the risks posed by the biological material being used, the operator should also be aware of his or her work environment and be fairly conversant with the equipment being used, as these may also pose a serious hazard. The culture cabinet should be serviced routinely and checked (approximately every 6 months) to ensure its safety to the operator. Additionally the operator could ensure his or her own safety by adopting some common precautionary measures such as refraining from eating or drinking whilst working in the cabinet and using a pipette aid as opposed to mouth pipetting to prevent ingestion of unwanted substances. Gloves and adequate protective clothing such as a clean laboratory coat should be worn at all times and gloves must be discarded after handling of non-sterile or contaminated material.

2.4 ASEPTIC TECHNIQUES AND GOOD CELL CULTURE PRACTICE

2.4.1 Good practice

In order to maintain a clean and safe culture environment, adequate aseptic or sterile technique should be adopted at all times. This simply involves working under conditions that prevent contaminating microorganisms from the environment from entering the cultures. Part of the precaution taken involves washing hands with antiseptic soap and ensuring that all work surfaces are kept clean and sterile by swabbing with 70% IMS before starting work. Moreover, all procedures, including media preparation and cell handling, should be carried out in a cell culture cabinet that is maintained in a clean and sterile condition.

Other essential precautions should include avoiding talking, sneezing or coughing into the cabinet or over the cultures. A clean pipette should be used for each different procedure and under no circumstance should the same pipette be used between different bottles of media, as this will significantly increase the risk of cross-contamination. All spillages must be cleaned quickly to avoid contamination from microorganisms that may be present in the air. Failing to do so may result in infections to the cultures, which may be reduced by using antibiotics. However, this is not always guaranteed and good aseptic techniques should eliminate the need for antibiotics. In the event of cultures becoming contaminated, these should be removed immediately from the laboratory, disinfected and autoclaved to prevent the contamination spreading. Under no circumstance can an infected culture be opened inside the cell culture cabinet or incubator. Moreover, all waste generated must be decontaminated and disposed of immediately after completing the work. This should be carried out in accordance with the national legislative requirements, which state that cell culture waste including media be inactivated using a disinfectant before disposal and that all contaminated materials and waste be autoclaved before being discarded or incinerated.

The risk from infections is the most common cause for concern in cell culture. Various factors can contribute to this, including poor work environment, poor aseptic techniques and indeed poor hygiene of the operator. The last of these is important, since most of the common sources of infections such as bacteria, yeast and fungus originate from the worker. Maintaining a clean environment and adopting good laboratory practice and aseptic techniques should, therefore, help to reduce the risks of infection. However, should infections occur, it is advisable to address this immediately and eradicate the problem. To do this, it helps to know the types of infection to expect and what to look for.

In animal cell cultures, bacterial and fungal infections are relatively easy to identify and isolate. The other most common contamination originates from mycoplasma. These are the smallest (approximately 0.3 mm in diameter) self-replicating prokaryotes in existence. They lack a rigid cell wall and generally infect the cytoplasm of mammalian cells. There are at least five species known to contaminate cells in culture: *Mycoplasma hyorhinis, Mycoplasma arginini, Mycoplasma orale, Mycoplasma fermentans* and *Acholeplasma laidlawii*. Infections caused by these organisms are more problematic and not easily identified or eliminated. Moreover, if left unchecked, mycoplasma contamination will cause subtle but adverse effects on cultures, including changes in metabolism, DNA, RNA and protein synthesis, morphology and growth. This can lead to non-reproducible, unreliable experimental results and unsafe biological products.

2.4.2 Identification and eradication of bacterial and fungal infections

Both bacterial and fungal contaminations are easily identified as the infective agents are readily visible to the naked eye even in the early stages. This is usually made noticeable by the increase in turbidity and the change in colour of the culture medium owing to the change in pH caused by the infection. In addition, bacteria can be easily identified under microscopic examination as motile round bodies. Fungi on the other hand are distinctive by their long hyphal growth and by the fuzzy colonies they form in the medium. In most cases the simplest solution to these infections is to remove and dispose of the contaminated cultures. In the early stages of an infection, attempts can be made to eliminate the infecting microorganism using repeated washes and incubations with antibiotics or antifungal agents. This is however not advisable as handling infected cultures in the sterile work environment increases the chances of the infection spreading.

As part of the good laboratory practice, sterile testing of cultures should be carried out regularly to ensure that cultures are free from microbial organisms. This is particularly important when preparing cell culture products or generating cells for storage. Generally, the presence of these organisms can be detected much earlier and necessary precautions taken to avoid a full-blown contamination crisis in the laboratory. The testing procedure usually involves culturing a suspension of cells or products in an appropriate medium such as tryptone soya broth (TSB) for bacterial or thioglycollate medium (TGM) for fungal detection. The mixture is incubated for up to 14 days but examined daily for turbidity, which is used as an indication of microbial growth. It is essential that both positive and negative controls are set up in parallel with the sample to be tested. For this purpose a suspension of bacteria such as *Bacillus subtilis* or fungus such as *Clostridium sporogenes* is used instead of the cells or product to be tested. Uninoculated flasks containing only the growth medium are used as negative controls. Any contamination in the cell cultures will result in the broth appearing turbid, as would the positive controls. The negative controls should remain clear. Infected cultures should be discarded, whilst clear cultures would be safe to use or keep.

2.4.3 Identification of mycoplasma infections

Mycoplasma contaminations are more prevalent in cell culture than many workers realise. The reason for this is that mycoplasma contaminations are not evident under light microscopy nor do they result in a turbid growth in culture. Instead the changes induced are more subtle and manifest themselves mainly as a slowdown in growth and in changes in cellular metabolism and functions. However, cells generally return to their native morphology and normal proliferation rates relatively rapidly after eradication of mycoplasma.

The presence of mycoplasma contamination in cultures has, until recently, been difficult to determine and samples had to be analysed by specialist laboratories. There are, however, improved techniques now available for detection of mycoplasma in cell culture laboratories. These include microbiological cultures of infected cells, an

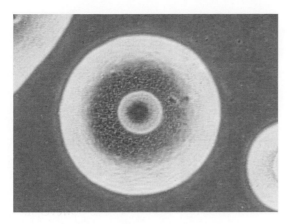

Fig. 2.3 Photograph of mycoplasma, showing the characteristic opaque granular central zone surrounded by a translucent border, giving a 'fried egg' appearance.

indirect DNA staining technique using the fluorochrome dye Hoechst 33258, enzyme-linked immunosorbent assay (ELISA) or polymerase chain reaction (PCR).

With the microbiological culture technique, cells in suspension are inoculated into liquid broth and then incubated under aerobic conditions at 37 °C for 14 days. A non-inoculated flask of broth is used as a negative control. Aliquots of broth are taken every 3 days and inoculated onto an agar plate, which is incubated anaerobically as above. All plates are then examined under an inverted microscope at a magnification of 300× after 14 days of incubation. Positive cultures will show the typical myco-plasma colony formation, which has an opaque granular central zone surrounded by a translucent border, giving a 'fried egg' appearance (Fig. 2.3). It may be necessary to set up positive controls in parallel, in which case plates and broth should be inoculated with a known strain of mycoplasma such as *Mycoplasma orale* or *Mycoplasma pneumoniae*.

The DNA binding method offers a rapid alternative for detecting mycoplasma and works on the principle that Hoechst 33258 fluoresces under ultraviolet light once bound to DNA. Thus, in contaminated cells, the fluorescence will be fairly dispersed in the cytoplasm of the cells owing to the presence of mycoplasma. In contrast, uncontaminated cells will show localised fluorescence in their nucleus only.

The Hoechst 33258 assay, although rapid, is relatively less sensitive when compared with the culture technique described above. For this assay, an aliquot of the culture to be tested is placed on a sterile coverslip in a 35-mm culture dish and incubated at 37 °C in a cell culture incubator to allow cells to adhere. The coverslip is then fixed by adding a fixative consisting of 1 part glacial acetic acid and 3 parts methanol, prepared fresh on the day. A freshly prepared solution of Hoechst 33258 stain is added to the fixed coverslip, incubated in the dark at room temperature to allow the dye to bind to the DNA and then viewed under ultraviolet fluorescence at 1000×. All positive cultures will show fluorescence of mycoplasma DNA, which will appear as small cocci or filaments in the cytoplasm of the contaminated cells (Fig. 2.4b, see also colour section). Negative cultures will show only fluorescing nuclei of

(a)   (b)

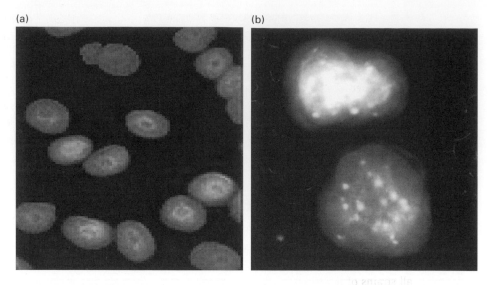

Fig. 2.4 Hoechst 33258 staining of mycoplasma in cells. (a) A Hoechst-negative stain, with the dye staining cellular DNA in the nucleus and thus showing nuclear fluorescence. (b) A Hoechst-positive stain, showing staining of mycoplasma DNA in the cytoplasm of the cells. (See also colour plate.)

uncontaminated cells against a dark cytoplasmic background (Fig. 2.4a, see also colour section). However, this technique is prone to errors, including false-negative results. To avoid the latter, cells should be cultured in antibiotic-free medium for two to three passages before being used. A positive control using a strain of mycoplasma seeded onto a coverslip is essential. Such controls should be handled away from the cell culture laboratory to avoid contaminating clean cultures of cells. It is also important to ensure that the fluorescence detected is not due to the presence of bacterial contamination or debris embedded into the plastics during manufacture. The former normally appear larger than the fluorescing cocci or filaments of mycoplasma. Debris, on the other hand, would show a non-uniform fluorescence owing to the variation in size of the particles usually found in plastics.

ELISA detection of mycoplasma is now becoming more commonly used and can be carried out using specifically designed kits following the manufacturer's protocol and reagents supplied. In this assay, 96-well plates are coated with the antibodies against different mycoplasma species. Each plate is then incubated at 37 °C for 2 h with the required antibody or antibodies before blocking with the appropriate blocking solution and incubating with the test sample(s). A negative control, which is simply media with sample buffer, and a positive control normally provided with the kits, should also be included in each assay. A detection antibody is subsequently added to the samples, incubated for a further 2 h at 37 °C before washing and incubating with a streptavidin solution for 1 h at 37 °C. Each plate is then detected for mycoplasma by adding the substrate solution and read on a plate reader at 405 nm after a further 30 min incubation at room temperature. This method is apparently suitable for detecting high levels of mycoplasma and could also be used to identify several species in one assay.

As with the ELISAs, commercial kits are also available for PCR detection of mycoplasma which contain the required primers, internal control template, positive control template and all the relevant buffers. Samples are generated and set up in a reaction mix as instructed in the manufacturer's protocol. The PCR is performed, again using the defined conditions outlined in the manufacturer's protocol, and the products generated analysed by electrophoresis on a high-grade 2% agarose gel. Although sensitive, PCR detection of mycoplasma is not always the protocol of choice because it has been shown to be prone to false-negative results, presumably due to the presence of ingredients in the kit which may inhibit PCR amplification of the target gene. In addition, this method is time-consuming and expensive.

2.4.4 Eradication of mycoplasma

Until recently, the most common approach for eradicating mycoplasma has been the use of antibiotics such as gentamycin. This approach is, however, not always effective, as not all strains of mycoplasma are susceptible to this antibiotic. Moreover antibiotic therapy does not always result in long-lasting successful elimination and most drugs can be cytotoxic to the cell culture. More recently, a new generation of bactericidal antibiotic preparation referred to as Plasmocin™ was introduced and has been shown to be effective against mycoplasma even at relatively low, non-cytotoxic concentrations. The antibiotics contained in this product are actively transported into cells, thus facilitating killing of intracellular mycoplasma but without any adverse effects on actual cellular metabolism.

Apart from antibiotics, various products have also been introduced into the cell culture market that the manufacturers claim eradicate mycoplasma efficiently and quickly without causing any adverse effects to the cells. One such product is Mynox®, a biological agent that integrates into the membrane of mycoplasma, compromising its integrity and eventually initiating its disintegration. This process apparently occurs within an hour of applying Mynox® and may have the added advantage that it is not an antibiotic and as a result will not lead to the development of resistant strains. It is safe to cultures and eliminated once the medium has been replaced. Moreover, this reagent is highly sensitive, detecting as little as 1–5 fg of mycoplasma DNA, which corresponds to two to five mycoplasma per sample and is effective against many of the common mycoplasma contaminations encountered in cell culture.

2.5 TYPES OF ANIMAL CELL, CHARACTERISTICS AND MAINTENANCE IN CULTURE

The cell types used in cell culture fall into two categories generally referred to as either a primary culture or a cell line.

2.5.1 Primary cell cultures

Primary cultures are cells derived directly from tissues following enzymatic dissociation or from tissue fragments referred to as explants. These are usually the cells of preference, since it is argued that primary cultures retain their characteristics and reflect the true activity of the cell type *in vivo*. The disadvantage in using primary cultures, however, is that their isolation can be labour-intensive and may produce a heterogeneous population of cells. Moreover, primary cultures have a relatively limited lifespan and can be used over only a limited period of time in culture.

Primary cultures can be obtained from many different tissues and the source of tissue used generally defines the cell type isolated. For instance, cells isolated from the endothelium of blood vessels are referred to as endothelial cells whilst those isolated from the medial layer of the blood vessels and other similar tissues are smooth muscle cells. Although both can be obtained from the same vessels, endothelial cells are different in morphology and function, generally growing as a single monolayer characterised by a cobble-stoned morphology. Smooth muscle cells on the other hand are elongated, with spindle-like projections at either end, and grow in layers even when maintained in culture. In addition to these cell types there are several other widely used primary cultures derived from a diverse range of tissues, including fibroblasts from connective tissue, lymphocytes from blood, neurons from nervous tissues and hepatocytes from liver tissue.

2.5.2 Continuous cell lines

Cell lines consist of a single cell type that has gained the ability for infinite growth. This usually occurs after transformation of cells by one of several means that include treatment with carcinogens or exposure to viruses such as the monkey simian virus 40 (SV40), Epstein–Barr virus (EBV) or Abelson murine leukaemia virus (A-MuLV) amongst others. These treatments cause the cells to lose their ability to regulate growth. As a result, transformed cells grow continuously and, unlike primary culture, have an infinite lifespan (become 'immortalised'). The drawback to this is that trans-formed cells generally lose some of their original *in vivo* characteristics. For instance, certain established cell lines do not express particular tissue-specific genes. One good example of this is the inability of liver cell lines to produce clotting factors. Continuous cell lines, however, have several advantages over primary cultures, not least because they are immortalised. In addition, they require less serum for growth, have a shorter doubling time and can grow without necessarily needing to attach or adhere to the surface of the flask.

Many different cell lines are currently available from various cell banks, which makes it easier to obtain these cells without having to generate them. One of the largest organisations that supplies cell lines is the **European Collection of Animal Cell Cultures (ECACC)** based in Salisbury, UK. A selection of the different cell lines supplied by this organisation is listed in Table 2.1.

2.5.3 Cell culture media and growth requirements for animal cells

The cell culture medium used for animal cell growth is a complex mixture of nutrients (amino acids, a carbohydrate such as glucose, and vitamins), inorganic salts (e.g. containing magnesium, sodium, potassium, calcium, phosphate, chloride, sulphate,

Table 2.1 **Examples of cell lines supplied by commercial sources**

Cell line	Morphology	Species	Tissue origin
BAE-1	Endothelial	Bovine	Aorta
BHK-21	Fibroblast	Syrian hamster	Kidney
CHO	Fibroblast	Chinese hamster	Ovary
COS-1/7	Fibroblast	African green monkey	Kidney
HeLa	Epithelial	Human	Cervix
HEK-293	Epithelial	Human	Kidney
HT-29	Epithelial	Human	Colon
MRC-5	Fibroblast	Human	Lung
NCI-H660	Epithelial	Human	Lung
NIH/3T3	Fibroblast	Mouse	Embryo
THP-1	Monocytic	Human	Blood
V-79	Fibroblast	Chinese hamster	Lung
HEP1	Hepatocytes	Human	Liver

and bicarbonate ions) and broad-spectrum antibiotics. In certain situations it may be essential to include a fungicide such as amphotericin B, although this may not always be necessary. For convenience and ease of monitoring the status of the medium, the pH indicator phenol red may also be included. This will change from red at pH 7.2–7.4 to yellow or fuchsia as the pH becomes either acidic or alkaline, respectively.

The other key basic ingredient in the cell culture medium is serum, usually bovine or fetal calf. This is used to provide a buffer for the culture medium, but, more importantly, enhances cell attachment and provides additional nutrients and hormone-like growth factors that promote healthy growth of cells. An attempt to culture cells in the absence of serum does not usually result in successful or healthy cultures, even though cells can produce growth factors of their own. However, despite these benefits, the use of serum is increasingly being questioned not least because of many of the other unknowns that can be introduced, including infectious agents such as viruses and mycoplasma. The recent resurgence of 'mad cow disease' (bovine spongiform encephalitis) has introduced an additional drawback, posing a particular risk for the cell culturist, and has increased the need for alternative products. In this regard, several cell culture reagent manufacturers have now developed serum-free medium supplemented with various components including albumin, transferrin, insulin, growth factors and other essential elements required for optimal cell growth. This is proving very useful, particularly for the pharmaceutical and biotechnology companies involved in the manufacture of drugs or biological products for human and animal consumption.

2.5.4 Preparation of animal cell culture medium

Preparation of the culture medium is perhaps taken for granted as a simple straight-forward procedure that is often not given due care and attention. As a result, most infections in cell culture laboratories originate from infected media. Following the simple yet effective procedures outlined in Section 2.4.1 should prevent or minimise the risk of infecting the media when they are being prepared.

Preparation of the medium itself should also be carried out inside the culture cabinet and usually involves adding a required amount of serum together with anti-biotics to a fixed volume of medium. The amount of serum used will depend on the cell type but usually varies between 10% and 20%. The most common antibiotics used are penicillin and streptomycin, which inhibit a wide spectrum of Gram-positive and Gram-negative bacteria. Penicillin acts by inhibiting the last step in bacterial cell wall synthesis whilst streptomycin blocks protein synthesis.

Once prepared, the mixture, which is referred to as complete growth medium, should be kept at 4 °C until used. To minimise wastage and risk of contamination it is advisable to make just the required volume of medium and use this within a short period of time. As an added precaution it is also advisable always to check the clarity of the medium before use. Any infected medium, which will appear cloudy or turbid, should be discarded immediately. In addition to checking the clarity, a close eye should also be kept on the colour of the medium, which should be red at physiological pH owing to the presence of phenol red. Media that looks acidic (yellow) or alkaline (fuchsia) should be discarded, as these extremes will affect the viability and thus growth of the cells.

2.5.5 Subculture of cells

Subculturing is the process by which cells are harvested, diluted in fresh growth medium and replaced in a new culture flask to promote further growth. This process, also known as passaging, is essential if the cells are to be maintained in a healthy and viable state, otherwise they may die after a certain period in continuous culture. The reason for this is that adherent cells grow in a continuous layer that eventually occupies the whole surface of the culture dish and at this point they are said to be confluent. Once confluent, the cells stop dividing and go into a resting state where they stop growing (senesce) and eventually die. Thus, to keep cells viable and facilitate efficient trans-formation, they must be subcultured before they reach full contact inhibition. Ideally, cells should be harvested just before they reach a confluent state.

Cells can be harvested and subcultured using one of several techniques. The precise method used is dependent to a large extent on whether the cells are adherent or in suspension.

Subculture of adherent cells

Adherent cells can be harvested either mechanically, using a rubber spatula (also referred to as a 'rubber policeman') or enzymatically using proteolytic enzymes. Cells in suspension are simply diluted in fresh medium by taking a given volume of cell suspension and adding an equal volume of medium.

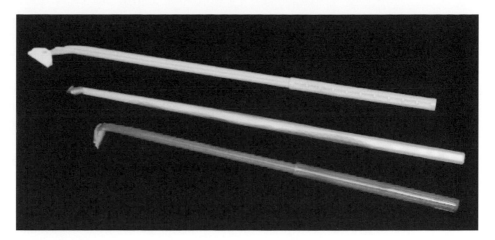

Fig. 2.5 Cell scrapers.

Harvesting of cells mechanically

This method is simple and easy. It involves gently scraping cells from the growth surface into the culture medium using a rubber spatula that has a rigid polystyrene handle with a soft polyethylene scraping blade (Fig. 2.5). This method is not suitable for all cell types as the scraping may result in membrane damage and significant cell death. Before adopting this approach it is important to carry out some test runs where cell viability and growth are monitored in a small sample of cells following harvesting.

Harvesting of cells using proteolytic enzymes

Several different proteolytic enzymes can be exploited including trypsin, a proteolytic enzyme that destroys proteinaceous connections between cells and between cells and the surface of the flask in which they grow. As a result, harvesting of cells using this enzyme results in the release of single cells, which is ideal for subculturing as each cell will then divide and grow, thus enhancing the propagation of the cultures.

Trypsin is commonly used in combination with EDTA, which enhances the action of the enzyme. EDTA alone can also be effective in detaching adherent cells as it chelates the Ca^{2+} required by some adhesion molecules that facilitate cell–cell or cell–matrix interactions. Although EDTA alone is much gentler on the cells than trypsin, some cell types may adhere strongly to the plastic, requiring trypsin to detach.

The standard procedure for detaching adherent cells using trypsin and EDTA involves making a working solution of 0.1% trypsin plus 0.02% EDTA in Ca^{2+}/Mg^{2+}-free phosphate-buffered saline. The growth medium is aspirated from confluent cultures and washed at least twice with a serum-free medium such as Ca^{2+} or Mg^{2+}-free PBS to remove traces of serum that may inactivate the trypsin. The trypsin–EDTA solution (approximately $1\,cm^3$ per $25\,cm^2$ of surface area) is then added to the cell monolayer and swirled around for a few seconds. Excess trypsin–EDTA is aspirated, leaving just enough to form a thin film over the monolayer. The flask is then incubated at $37\,°C$ in a cell culture incubator for 2–5 min but monitored under an inverted light microscope at intervals to detect when the cells are beginning to round up and detach. This is to ensure that the cells are not overexposed to trypsin, as this may result in extensive damage to the cell surface, eventually resulting in cell death. It is important therefore that the proteolysis reaction is quickly terminated by the addition of complete medium containing serum that will inactivate the trypsin. The suspension of cells is collected into a sterile centrifuge tube and spun at 1000 r.p.m. for 10 min to pellet the cells, which are then resuspended in a known volume of fresh complete culture medium to give a required density of cells per cubic centimetre volume.

As with all tissue culture procedures, aseptic techniques should be adopted at all times. This means that all the above procedures should be carried out in a tissue culture cabinet under sterile conditions. Other precautions worth noting include the handling of the trypsin stock. This should be stored frozen at $-20\,^{\circ}$C and, when needed, placed in a water bath just to the point where it thaws. Any additional time in the $37\,^{\circ}$C water bath will inactivate the enzymatic activity of the trypsin. The working solution should be kept at $4\,^{\circ}$C once made and can be stored for up to 3 months.

Subculture of cells in suspension

For cells in suspension it is important initially to examine an aliquot of cells under a microscope to establish whether cultures are growing as single cells or clumps. If cultures are growing as single cells, an aliquot is counted as described in Section 2.5.6 below and then reseeded at the desired seeding density in a new flask by simply diluting the cell suspension with fresh medium, provided the original medium in which the cells were growing is not spent. However, if the medium is spent and appears acidic, then the cells must be centrifuged at 1000 r.p.m. for 10 min, resuspended in fresh medium and transferred into a new flask. Cells that grow in clumps should first be centrifuged and resuspended in fresh medium as single cells using a glass Pasteur or fine-bore pipette.

2.5.6 Cell quantification

It is essential that when cells are subcultured they are seeded at the appropriate seeding density that will facilitate optimum growth. If cells are seeded at a lower seeding density they may take longer to reach confluency and some may expire before getting to this point. On the other hand, if seeded at a high density, cells will reach confluency too quickly, resulting in irreproducible experimental results. This is because trypsin can digest surface proteins, including receptors for drugs, and these will need time (sometimes several days) to renew. Failure to allow these proteins to be regenerated on the cell surface may therefore result in variable responses to drugs specific for such receptors.

Several techniques are now available for quantification of cells and of these the most common method involves the use of a haemocytometer. This has the added advantage of being simple and cheap to use. The haemocytometer itself is a thickened glass slide that has a small chamber of grids cut into the glass. The chamber has a fixed volume and is etched into nine large squares, of which the large corner squares

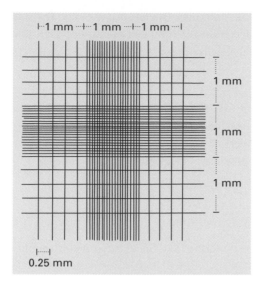

Fig. 2.6 Haemocytometer.

contain 16 small squares each; each large square measures $1\,mm \times 1\,mm$ and is 0.1 mm deep (see Fig. 2.6).

Thus, with a coverslip in place, each square represents a volume of $0.1\,mm^3$ $(1.0\,mm^2$ area $\times 0.1\,mm$ depth) or $10^{-4}\,cm^3$. Knowing this, the cell concentration (and the total number of cells) can therefore be determined and expressed per cubic centimetre. The general procedure involves loading approximately $10\,\mu l$ of a cell suspension into a clean haemocytometer chamber and counting the cells within the four corner squares with the aid of a microscope set at $20\times$ magnification. The count is mathematically converted to the number of cells per cm^3 of suspension.

To ensure accuracy, the coverslip must be firmly in place and this can be achieved by moistening a coverslip with exhaled breath and gently sliding it over the haemocytometer chamber, pressing firmly until Newton's refraction rings (usually rainbow-like) appear under the coverslip. The total number of cells in each of the four $1\text{-}mm^3$ corner squares should be counted, with the proviso that only cells touching the top or left borders but not those touching the bottom and right borders are counted. Moreover, cells outside the large squares, even if they are within the field of view, should not be counted. When present, clumps should be counted as one cell. Ideally $\sim$100 cells should be counted to ensure a high degree of accuracy in counting. If the total cell count is less than 100 or if more than 10% of the cells counted appear to be clustered, then the original cell suspension should be thoroughly mixed and the counting procedure repeated. Similarly, if the total cell count is greater than 400, the suspension should be diluted further to get counts of between 100 and 400 cells.

Since some cells may not survive the trypsinisation procedure it is usually advisable to add an equal volume of the dye trypan blue to a small aliquot of the cell suspension before counting. This dye is excluded by viable cells but taken up by dead cells. Thus, when viewed under the microscope, viable cells will appear as bright translucent structures while dead cells will stain blue (see Section 2.5.12). The number of dead cells can therefore be excluded from the total cell count, ensuring that the seeding density accurately reflects viable cells.

Calculating cell number

Cell number is usually expressed per cm^3 and is determined by multiplying the average of the number of cells counted by a conversion factor which is constant for the haemocytometer. The conversion factor is estimated at 1000, based on the fact that each large square counted represents a total volume of $10^{-4}\,cm^3$.

Thus:

$$\text{cells cm}^{-3} = \frac{\text{number of cells counted}}{\text{number of squares counted}} \times \text{conversion factor}$$

If the cells were diluted before counting then the dilution factor should also be taken into account.

Therefore:

$$\text{cells cm}^{-3} = \frac{\text{number of cells counted}}{\text{number of squares counted}} \times \text{conversion factor} \times \text{dilution factor}$$

To get the total number of cells harvested the number of cells determined per cm^3 should be multiplied by the original volume of fluid from which the cell sample was removed,

i.e.:

$$\text{total cells} = \text{cells cm}^{-3} \times \text{total volume of cell suspension}$$

Example 1 CALCULATION OF CELL NUMBER

Question Calculate the total number of cells suspended in a final volume of 5 ml, taking into account that the cells were diluted 1 : 2 before counting and the number of cells counted with the haemocytometer was 400.

Answer
$$\text{Cells cm}^{-3} = \frac{\text{number of cells counted}}{\text{large squares counted}} \times \text{conversion factor}$$

$$= \frac{400}{4} \times 1000$$

$$= 100\,000 \text{ cells cm}^{-3}$$

Because there is a dilution factor of 2, the correct number of cells cm^{-3} is given as:

$$100\,000 \times 2 = 200\,000 \text{ cells cm}^{-3}$$

Thus in a final volume of 5 cm^3 the total number of cells present is:

$$200\,000 \times 5 = 1\,000\,000 \text{ cells}$$

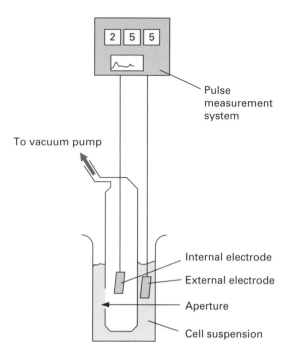

Fig. 2.7 Coulter counter. Cells entering the aperture create a pulse of resistance between the internal and external electrodes that is recorded on the oscilloscope.

Alternative methods for determination of cell number

Several other methods are available for quantifying cells in culture, including direct measurement using an electronic Coulter counter. This is an automated method of counting and measuring the size of microscopic particles. The instrument itself consists of a glass probe with an electrode that is connected to an oscilloscope (Fig. 2.7). The probe has a small aperture of fixed diameter near its bottom end. When immersed in a solution of cell suspension, cells are flushed through the aperture causing a brief increase in resistance owing to a partial interruption of current flow. This will result in spikes being recorded on the oscilloscope and each spike is counted as a cell. One disadvantage of this method, however, is that it does not distinguish between viable and dead cells.

Indirectly, cells can be counted by determining total cell protein and using a protein versus cell number standard curve to determine cell number in test samples. However, protein content per cell can vary during culture and may not give a true reflection of cell number. Alternatively, the DNA content of cells may be used as an indicator of cell number, since the DNA content of diploid cells is usually constant. However, the DNA content of cells may change during the cell cycle and therefore not give an accurate estimate of cell number.

2.5.7 **Seeding cells onto culture plates**

Once counted, cells should then be seeded at a density that promotes optimal cell growth. It is essential therefore that when cells are subcultured they are seeded at the appropriate seeding density. If cells are seeded at a lower density they may take longer to reach confluency and some may die before getting to this point. On the other hand, if seeded at too high a density cells will reach confluency too quickly, resulting in irreproducible experimental results as already discussed above (see Section 2.5.6). The seeding density will vary depending on the cell type and on the surface area of the culture flask into which the cells will be placed. These factors should therefore be taken into account when deciding on the seeding density of any given cell type and the purpose of the experiments carried out.

2.5.8 Maintenance of cells in culture

It is important that after seeding, flasks are clearly labelled with the date, cell type and the number of times the cells have been subcultured or passaged. Moreover, a strict regime of feeding and subculturing should be established that permits cells to be fed at regular intervals without allowing the medium to be depleted of nutrients or the cells to overgrow or become super confluent. This can be achieved by following a standard but routine procedure for maintaining cells in a viable state under optimum growth conditions. In addition, cultures should be examined daily under an inverted microscope, looking particularly for changes in morphology and cell density. Cell shape can be an important guide when determining the status of growing cultures. Round or floating cells in subconfluent cultures are not usually a good sign and may indicate distressed or dying cells. The presence of abnormally large cells can also be useful in determining the well-being of the cells, since the number of such cells increases as a culture ages or becomes less viable. Extremes in pH should be avoided by regularly replacing spent medium with fresh medium. This may be carried out on alternate days until the cultures are approximately 90% confluent, at which point the cells are either used for experimentation or trypsinised and subcultured following the procedures outlined in Section 2.5.5.

The volume of medium added to the cultures will depend on the confluency of the cells and the surface area of the flasks in which the cells are grown. As a guide, cells which are under 25% confluent may be cultured in approximately $1\,cm^3$ of medium per $5\,cm^2$ and those between 25% and 40% or $\geqq$ 45% confluency should be supplemented with $1.5\,cm^3$ or $2\,cm^3$ culture medium per $5\,cm^2$, respectively. When changing the medium it is advisable to pipette the latter on to the sides or the opposite surface of the flask from where the cells are attached. This is to avoid making direct contact with the monolayers as this will damage or dislodge the cells.

2.5.9 Growth kinetics of animal cells in culture

When maintained under optimum culture conditions, cells follow a characteristic growth pattern (Fig. 2.8), exhibiting an initial lag phase in which there is enhanced cellular activity but no apparent increase in cell growth. The duration of this phase is dependent on several factors including the viability of the cells, the density at which the cells are plated and the media component.

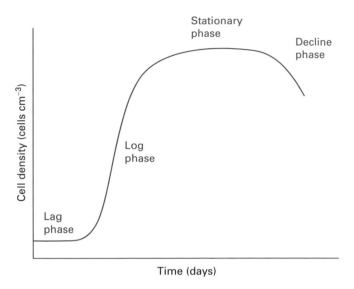

Fig. 2.8 Growth curve showing the phases of cell growth in culture.

The lag phase is followed by a log phase in which there is an exponential increase in cell number with high metabolic activity. These cells eventually reach a stationary phase where there is no further increase in growth due to depletion of nutrients in the medium, accumulation of toxic metabolic waste or a limitation in available growth space. If left unattended, cells in the stationary phase will eventually begin to die, resulting in the decline phase on the growth curve.

2.5.10 Cryopreservation of cells

Cells can be preserved for later use by freezing stocks in liquid nitrogen. This process is referred to as cryopreservation and is an efficient way of sustaining stocks. Indeed, it is advisable that, when good cultures are available, aliquots of cells should be stored in the frozen state. This provides a renewable source of cells that could be used in future without necessarily having to culture new batches from tissues. Freezing can, however, result in several lethal changes within the cells, including formation of ice crystals and changes in the concentration of electrolytes and in pH. To minimise these risks a cryoprotective agent such as DMSO is usually added to the cells prior to freezing in order to lower the freezing point and prevent ice crystals from forming inside the cells. In addition, the freezing process is carried out in stages, allowing the cells initially to cool down slowly from room temperature to $-80\,^\circ$C at a rate of 1–$3\,^\circ$C min^{-1}. This initial stage can be carried out using a freezing chamber or alternatively a cryo freezing container ('Mr Frosty') filled with isopropanol, which provides the critical, repeatable $-1\,^\circ$C min^{-1} cooling rate required for successful cell cryopreservation. When this process is complete, the cryogenic vials, which are polypropylene tubes that can withstand temperatures as low as $-190\,^\circ$C, are removed and immediately placed in a liquid nitrogen storage tank where they can remain for an indefinite period or until required.

The actual cryogenic procedure is itself relatively straightforward. It involves harvesting cells as described in Section 2.5.5 and resuspending them in $1\,cm^3$ of freezing medium, which is basically culture medium containing 40% serum. The cell suspension is counted and appropriately diluted to give a final cell count of between 10^6 and $10^7\,cells\,cm^{-3}$. A 0.9-cm^3 aliquot is transferred into a cryogenic vial labelled with the cell type, passage number and date harvested. This is then made up to $1\,cm^3$ by adding $100\,mm^3$ of DMSO to give a final concentration of 10%. The cells should then be mixed gently by rotating or inverting the vial and placed in a 'Mr Frosty' cryo freezing container. The container and cells are placed in a $-80\,°C$ freezer and allowed to freeze overnight. The frozen vials may then be transferred into a liquid nitrogen storage container. At this stage cells can be stored frozen until required for use.

All procedures should be carried out under sterile conditions to avoid contaminating cultures as this will appear once the frozen stocks are recultured. As an added precaution it is advisable to replace the growth medium in the 24-h period prior to harvesting cells for freezing. Moreover, cells used for freezing should be in the log phase of growth and not too confluent in case they may already be in growth arrest.

2.5.11 Resuscitation of frozen cells

When required, frozen stocks of cells may be revived by removing the cryogenic vial from storage in liquid nitrogen and placing in a water bath at $37\,°C$ for 1–2 min or until the ice crystals melt. It is important that the vials are not allowed to warm up to $37\,°C$ as this may cause the cells to rapidly die. The thawed cell suspension may then be transferred into a centrifuge tube, to which fresh medium is added and centrifuged at 1000 r.p.m. for 10 min. The supernatant should be discarded to remove the DMSO used in the freezing process and the cell pellet resuspended in $1\,cm^3$ of fresh medium, ensuring that clumps are dispersed into single cells or much smaller clusters using a glass Pasteur pipette. The required amount of fresh pre-warmed growth medium is placed in a culture flask and the cells pipetted into the flask, which is then placed in a cell culture incubator and the cells allowed to adhere and grow.

Practical hints and tips in resuscitation of frozen cells

It is important to handle resuscitated cells delicately after thawing as these may be fairly fragile and could degenerate quite readily if not treated correctly. In addition, it is important to dilute the freezing medium immediately after thawing to reduce the concentration of DMSO or freezing agent to which the cells are exposed.

2.5.12 Determination of cell viability

Determination of cell viability is extremely important, since the survival and growth of the cells may depend on the density at which they are seeded. The degree of viability is most commonly determined by differentiating living from dead cells using the dye exclusion method. Basically, living cells exclude certain dyes that are readily taken up by dead cells. As a result, dead cells stain the colour of the dye used whilst living cells remain refractile owing to the inability of the dye to penetrate into the cytoplasm. One of the most commonly used dyes in such assays is trypan blue. This is incubated at a concentration of 0.4% with cells in suspension and applied to a haemocytometer. The haemocytometer is then viewed under an inverted microscope set at $100 \times$ magnification and the cells counted as described in Section 2.5.6, keeping separate counts for viable and non-viable cells.

The total number of cells is calculated using the following equation as described previously:

$$\text{cells cm}^{-3} = \frac{\text{number of cells counted}}{\text{number of squares counted}} \times \text{conversion factor} \times \text{dilution factor}$$

and the percentage of viable cells determined using the following formula:

$$\% \text{ viability} = \frac{\text{number of unstained cells counted}}{\text{total number of cells counted}} \times 100$$

To avoid underestimating cell viability it is important that the cells are not exposed to the dye for more than 5 min before counting. This is because uptake of trypan blue is time sensitive and the dye may be taken up by viable cells during prolonged incubation periods. Additionally, trypan blue has a high affinity for serum proteins and as such may produce a high background staining. The cells should therefore be free from serum, which can be achieved by washing the cells with PBS before counting.

2.6 STEM CELL CULTURE

Stem cells are unspecialised cells which have the ability to undergo self-renewal, replicating many times over prolonged periods, thereby generating new unspecialised cells. More importantly, stem cells have the potential to give rise to specialised cells with specific functions by the process of differentiation. Because of this property, stem cells are now being developed and exploited for cell-based therapies in various disease states. It has therefore become essential to be able to isolate, maintain and grow these cells in culture. This is however an emerging field where protocols to be used routinely are still being developed. This section of the chapter will focus on techniques that are now becoming routine for stem cell culture, focussing essentially on human embryonic stem cells (hESCs). The latter are cells derived from the inner cell mass of the blastocyst which is a hollow microscopic ball made up of an outer layer of cells (the trophoblast), a fluid-filled cavity (the blastocoel) and the cluster of inner cell mass.

Culturing of hESCs can be carried out in a standard cell culture laboratory using equipment already described earlier in the chapter. As with normal cell culture, the important criteria are that good aseptic techniques are adopted together with good laboratory practice. Unlike normal specialised cells, however, culture of hESCs requires certain conditions specifically aimed at maintaining these cells in a viable undifferentiated state. Historically, hESCs, and indeed other stem cells, have been cultured on what are referred to as feeders which act to sustain growth and maintain cells in the undifferentiated state without allowing them to lose their pluripotency (i.e. ability to differentiate, when needed, into specialised cell types of the three germ layers). The most common feeder cultures used are fibroblasts derived from embryos. The methodology for this together with other techniques for successful maintenance and propagation of hESCs are described below. Other protocols such as freezing and resuscitation of frozen cells are similar to those already described and the reader is therefore referred to the relevant sections above.

2.6.1 Preparation of embryonic fibroblasts

Typically, fibroblasts are isolated under sterile conditions in a tissue culture cabinet from embryos obtained from mice at 13.5 days of gestation. Each embryo is minced into very fine pieces using sterile scissors and incubated in a cell culture incubator at 37 °C with trypsin/EDTA (0.25% (w/v)/5 mM) for 20 minutes. The mixture is then pipetted vigorously using a fine-bore pipette until it develops a sludgy consistency. This process is repeated, returning the digest into the incubator if necessary, until the embryos have been virtually digested. The trypsin is subsequently neutralised with culture medium containing 10% serum ensuring that the volume of medium is at least twice that of the trypsin used. The minced tissue is plated onto a tissue culture flask and incubated overnight at 37 °C in a tissue culture incubator. The medium is subsequently removed after 24 h and the cell monolayer washed to remove any tissue debris and non-adherent cells. Adherent cells are cultured to 80–90% confluency before being passaged using trypsin as described in Section 2.5.5. If needed, the trypsinised cells could be propagated, otherwise they should be frozen as described in Section 2.5.10 and used as stock. If the latter is preferred, ensure that cells are frozen at no higher than passage three.

Practical hints and tips in using fibroblast feeders

Mouse fibroblasts should be used as feeders for stem cell culture between passages three and five. This is to ensure that fibroblasts support the growth of undifferentiated cells. After passage five the cells may begin to senesce and could also potentially fail to maintain stem cells in the undifferentiated state. Each batch of feeders prepared should be tested for their ability to support cells in an undifferentiated state.

2.6.2 Inactivation of fibroblast cells for use as feeders

Fibroblasts isolated should be inactivated before they can be used as feeders in order to prevent their proliferation and expansion during culture. This can be achieved using one of two protocols which include either irradiation or treatment with the antibiotic DNA cross-linker mitomycin C. With the former, cells in suspension are exposed to 80 Gy of irradiation using a caesium-source gamma irradiator. This is the dose of irradiation normally used for mouse fibroblasts; however, the radiation dose and exposure time may vary between batches of fibroblasts. As a result, a dose curve should be performed to determine the effective irradiation that is sufficient to stop cell division without cellular toxicity. Once irradiated, cells are spun at 1000 r.p.m. before resuspending the pellet using the appropriate medium and at the appropriate density for freezing or plating on gelatin-coated plates.

With the mitomycin procedure, cells are normally incubated with the compound at a concentration of $10\,\mu g\,cm^{-3}$ for 2–3 h at 37 °C in a cell culture incubator. After this, the mitomycin solution is aspirated and the cells washed several times with phosphate buffered saline or serum-free culture medium to ensure that there are no trace amounts of mitomycin that could affect the stem cells. The cells are then trypsinised, neutralised with serum containing medium, centrifuged and re-plated onto gelatin-coated dishes at the appropriate cell density.

Practical hints and tips with feeders

Of the two methods, exposure of cells to a gamma irradiation is the much preferred methodology because this gives a more consistent and reliable inactivation of cells. More importantly, mitomycin can be harmful and toxic, with embryonic cells showing particular sensitivity to this compound. Use of mitomycin-inactivated fibroblasts should therefore generally be avoided if irradiated feeders can be obtained. If frozen stocks are required of inactivated feeders, these can be prepared as described in Section 2.5.10. It is, however, important to ensure that stocks are not kept frozen for periods exceeding 4 months to avoid degeneration of cells. In addition, once plated, feeders should be used for stem cell culture within 24 h or no longer than 5 days after plating.

2.6.3 Plating of feeder cells

As with standard cell culture, fibroblast feeders are plated on tissue culture grade plastics but usually in the presence of a substrate such as gelatin, to provide the extracellular matrix component needed for cell attachment of the inactivated fibroblasts. In brief, the plates or flasks are incubated for 1 h at room temperature or overnight at 4 °C with the appropriate volume of 0.1% sterile gelatin. Excess gelatin is subsequently removed and the feeder cells plated at the approriate density for each cell line, e.g. 3.5×10^5 cells per 25-cm^2 flask. Feeders should be ready for use after 5–6 h but are best left to establish overnight for better results.

Practical hints and tips in plating feeders

It is important to ensure that the seeding density is optimal for each cell line otherwise feeders may fail to maintain the hESCs in the undifferentiated state. If frozen stocks of feeders are used for plating, these should be resuscitated, resuspended in fresh growth medium and plated on gelatin-coated plates as described in Section 2.5.11. Again the density of post-thaw feeders required to support the cells in an undifferentiated state should be established for each batch of frozen feeders since there is cell loss during the freeze–thaw process.

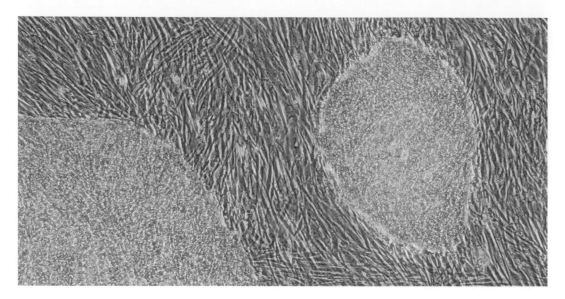

Fig. 2.9 Undifferentiated hESCs on mouse feeder cells.

2.6.4 Culture of human embryonic stem cells

Once the feeders are ready, hESCs can be plated directly by depositing the suspension of hESC onto the feeder layer. The dishes are placed in a cell culture incubator and the cells allowed to attach and establish over a 24-h period. Any non-adherent cells are removed during the first culture medium change. The cells are monitored and fed on a daily basis until the colonies are ready to be passaged. Depending on the conditions of growth, this can usually take up to 6 days.

As with the feeders, frozen stocks of hESCs should be resuscitated and diluted in fresh growth medium as described in Section 2.5.11.

Practical hints and tips in hESC culture

It is important to ensure that the colonies do not grow too large and to the point where adjacent colonies touch each other as this will initiate their differentiation. Similarly, the seeding density should be high enough to sustain growth otherwise sparsely plated colonies will grow very slowly and may never establish fully.

Colonies should be plated on healthy feeders that are not more than 4 days old. More importantly, only tightly packed colonies containing cells with the typical hESC morphology should be passaged (see Fig. 2.9). Any colony that has a less defined border (see Fig. 2.10) at the periphery, with loose cells spreading out or cells with atypical morphology, should not be passaged because these characteristics are evidence of cell differentiation. Should cells differentiate, these should be excised or aspirated before passaging the undifferentiated cells. Alternatively, if the majority of the colonies appear differentiated and no colonies display the characteristic morphology of undifferentiated cells, then it is advisable to discard the cultures and start with a new batch of undifferentiated hESCs.

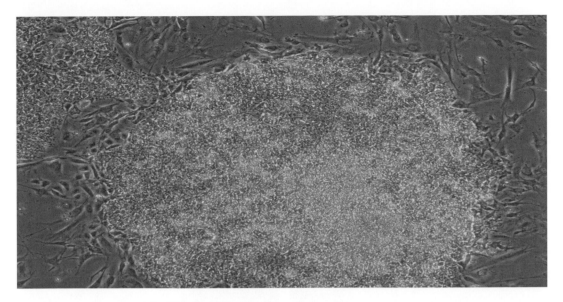

Fig. 2.10 Partially differentiated hESCs on mouse feeder cells.

2.6.5 Enzymatic subculture of hESCs

As with standard cell culture, hESCs can be passaged using enzymes but in this case an enzyme that does not disperse clusters of cells into single cells is preferred. This is because hESCs need to grow in colonies since single cells may not adhere to the feeders and may differentiate easily.

One of the most commonly used enzymes for subculturing hESCs is collagenase. When employed, hESC colonies are washed with phosphate-buffered saline and then incubated for 8–10 min with collagenase IV made up in serum-free medium at a concentration of $1\,mg\,cm^{-3}$. Curled up colonies can then be dislodged with gentle pipetting using a 5-ml pipette to break large clumps. Alternatively, colonies can be fragmented using glass beads. These are then washed with culture medium to remove the enzyme which may otherwise impair the attachment and growth of the cells, thus reducing the plating efficiency. hESCs can be washed by allowing the colonies to sediment slowly over 5–10 mins, leaving any residual feeder cells in the supernatant which are removed by aspiration. The colonies are subsequently resuspended in growth medium and are usually plated at a ratio of between 1 : 3 and 1 : 6. Alternatively, fragmented colonies could be frozen as described in Section 2.5.10 and stored for later use.

2.6.6 Mechanical subculture of hESCs

An alternative to the enzymatic method of subculturing hESCs is to manually cut colonies into appropriate size fragments using a fine-bore needle or a specially designed cutter such as the STEMPRO® EZPassage™ disposable stem cell passaging tool from Invitrogen. To do this, the dish of hESCs is placed under a dissecting

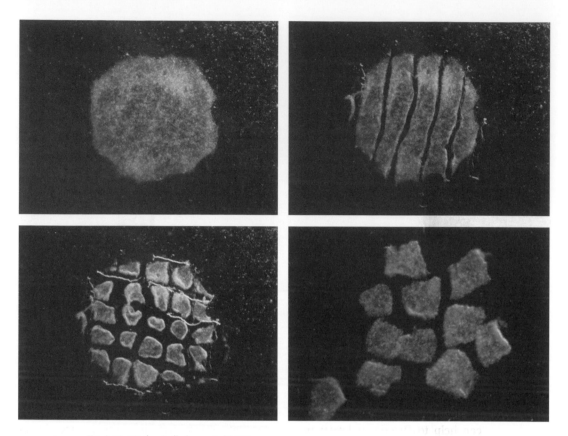

Fig. 2.11 Mechanically harvested hESCs.

microscope in a tissue culture hood. Undifferentiated colonies are identified by their morphology and then cut into grids (see Fig. 2.11) by scoring across and perpendicular to the first cut. Using a 1-ml pipette or pastette, the cut segments are transferred to dishes containing fresh feeders and culture medium. The colony fragments are placed evenly across the feeders (see Fig. 2.12) to avoid the colonies clumping together and attaching to the dish as one mass of cells. The dishes are then carefully transferred to a tissue culture incubator and left undisturbed for 1 day before replacing the spent medium with fresh. Established colonies are then fed every day until subcultured.

2.6.7 Feeder-free culture of hESCs

Although culture of hESCs on feeders has been extensively used, there have been concerns over this procedure when stem cells are being considered for clinical use in humans. One of the main drawbacks of using feeders is the concern over potential transmission of animal pathogens to humans and the possibility of expression of immunogenic antigens. Feeders are also inconvenient, expensive, and time-consuming to generate and inactivate. As a result of these limitations, there has been a drive towards developing a feeder-free culture system using feeder-conditioned media or media supplemented with different growth factors and other signalling molecules essential

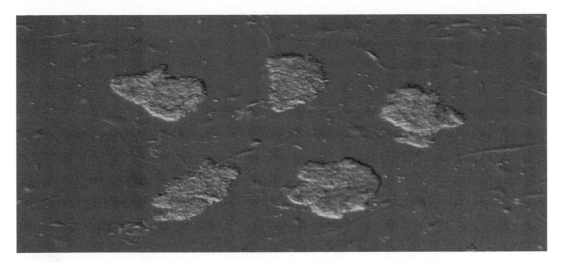

Fig. 2.12 Plating of hESCs onto feeder layer.

for sustaining growth. The conditioned medium can be generated by incubating normal growth medium with feeder cells for 24 h before use.

Feeder-free culture of hESCs is often carried out on tissue culture plastics coated with **Matrigel**, a substrate derived from mouse tumour and rich in **extracellular matrix** proteins such as **laminin, collagen and hepran sulphate proteoglycan**. It is also rich in growth factors such as basic fibroblast growth factor (bFGF) which can help to sustain and promote stem cell growth whilst maintaining them in an undifferentiated state.

Practically, dishes are coated with 5% Matrigel made up in culture medium. Just prior to use, the Matrigel is removed and replaced with culture medium before plating cells. The hESCs, subcultured from feeders or obtained from frozen stocks, are resuspended in conditioned medium supplemented often with bFGF at a concentration of $4 \, ng \, ml^{-1}$ before seeding. Alternatively, normal growth medium could be used but this will require a much higher concentration usually around $100 \, ng \, ml^{-1}$ bFGF. Once established, hESCs are fed every day with fresh growth medium. Colonies on Matrigel tend to show a different morphology to those on feeders; they tend to be larger and less packed initially than when cultured on feeders.

Practical hints and tips in using Matrigel

All work with Matrigel, other than plating of the hESCs, should be carried out at $4 \, °C$. Thus, when coating tissue culture plastics with Matrigel, all the plates and pipette tips should be kept on ice and used cold to prevent the Matrigel solidifying. Stock Matrigel is usually in the solid form and should be placed on ice or in the fridge at $4 \, °C$ overnight until it liquefies. Once liquefied, the Matrigel should be diluted in ice-cold culture medium at a final concentration of 5%. Each plate should have a smooth even layer of Matrigel and if this is not the case, the plates should be incubated at $4 \, °C$ until the Matrigel liquefies and settles as a uniform layer. Once coated, Matrigel plates should be used within 7 days of preparation.

2.7 BACTERIAL CELL CULTURE

As with animal cells, pure bacterial cultures (cultures that contain only one species of organism) are cultivated routinely and maintained indefinitely using standard sterile techniques that are now well defined. However, since bacterial cells exhibit a much wider degree of diversity in terms of both their nutritional and environmental requirements, conditions for their cultivation are diverse and the precise requirements highly dependent on the species being cultivated. Outlined below are general procedures and precautions adopted in bacterial cell culture.

2.7.1 Safety considerations for bacterial cell culture

Culture of microbial cells, like that involving cells of animal origin, requires care and sterile techniques, not least of all to prevent accidental contamination of pure cultures with other organisms. More importantly, utmost care should be given towards protecting the operator, especially from potentially harmful organisms. Aseptic techniques and safety conditions described for animal cell culture should be adopted at all times. Additionally, instruments used during the culturing procedures should be sterilised before and after use by heating in a Bunsen burner flame. Moreover, to avoid spread of bacteria, areas of work must be decontaminated after use using germicidal sprays and/or ultraviolet radiation. This is to prevent airborne bacteria from spreading rapidly. In line with these precautions, all materials used in microbial cell culture work must be disposed of appropriately; for instance, autoclaving of all plastics and tissue culture waste before disposal is usually essential.

2.7.2 Nutritional requirements of bacteria

The growth of bacteria requires much simpler conditions than those described for animal cells. However, due to their diversity, the composition of the medium used may be variable and largely determined by the nutritional classification of the organisms to be cultured. These generally fall into two main categories classified as either autotrophs (self-feeding organisms that synthesise food in the form of sugars using light energy from the sun) or heterotrophs (non-self-feeding organisms that derive chemical energy by breaking down organic molecules consumed). These in turn are subgrouped into chemo- or photoautotrophs or heterotrophs. Both chemo- and photoautotrophs rely on carbon dioxide as a source of carbon but derive energy from completely different sources, with the chemoautotrophs utilising inorganic substances whilst the photoautotrophs use light. Chemoheterotrophs and photoheterotrophs both use organic compounds as the main source of carbon with the photoheterotrophs using light for energy and the chemo subgroup getting their energy from the metabolism of organic substances.

2.7.3 Culture media for bacterial cell culture

Several different types of medium are used to culture bacteria and these can be categorised as either complex or defined. The former usually consist of natural substances, including meat and yeast extract, and as a result are less well defined, since their precise composition is largely unknown. Such media are, however, rich in nutrients and therefore generally suitable for culturing fastidious organisms that require a mixture of nutrients for growth. Defined media, by contrast, are relatively simple. These are usually designed to the specific needs of the bacterial species to be cultivated and as a result are made up of known components put together in the required amounts. This flexibility is usually exploited to select or eliminate certain species by taking advantage of their distinguishing nutritional requirements. For instance, bile salts may be included in media when selective cultivation of enteric bacteria (rod-shaped Gram-negative bacteria such as *Salmonella* or *Shigella*) is required, since growth of most other Gram-positive and Gram-negative bacteria will be inhibited.

2.7.4 Culture procedures for bacterial cells

Bacteria can be cultured in the laboratory using either liquid or solid media. Liquid media are normally dispensed into flasks and inoculated with an aliquot of the organism to be grown. This is then agitated continuously on a shaker that rotates in an orbital manner, mixing and ensuring that cultures are kept in suspension. For such cultures, sufficient space should be allowed above the medium to facilitate adequate diffusion of oxygen into the solution. Thus, as a rule of thumb, the volume of medium added to the flasks should not exceed more than 20% of the total volume of the flask. This is particularly important for aerobic bacteria and less so for anaerobic microorganisms.

In large-scale culture, fermenters or bioreactors equipped with stirring devices for improved mixing and gas exchange may be used. The device (Fig. 2.13) is usually fitted with probes that monitor changes in pH, oxygen concentration and temperature. In addition most systems are surrounded by a water jacket with fast-flowing cold water to reduce the heat generated during fermentation. Outlets are also included to release CO_2 and other gases produced by cell metabolism.

When fermenters are used, precautions should be taken to reduce potential contamination with airborne microorganisms when air is bubbled through the cultures. Sterilisation of the air may therefore be necessary and can be achieved by introducing a filter (pore size of approximately 0.2 mm) at the point of entry of the air flow into the chamber.

Solid medium is usually prepared by solidifying the selected medium with 1–2% of the seaweed extract agar, which, although organic, is not degraded by most microbes thereby providing an inert gelling medium on which bacteria can grow. Solid agar media are widely used to separate mixed cultures and form the basis for isolation of pure cultures of bacteria. This is achieved by streaking diluted cultures of bacteria onto the surface of an agar plate by using a sterile inoculating loop. Cells streaked across the plate will eventually grow into a colony, each colony being the product of a single cell and thus of a single species.

Once isolated, cells can be cultivated either in batch or continuous cultures. Of these, batch cultures are the most commonly used for routine liquid growth and entail

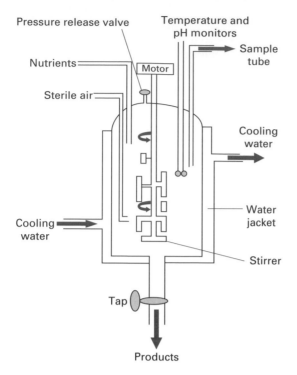

Fig. 2.13 Schematic representation of a fermenter.

inoculating an aliquot of cells into a sterile flask containing a finite amount of medium. Such systems are referred to as closed, since nutrient supply is limited to that provided at the start of culture. Under these conditions, growth will continue until the medium is depleted of nutrients or there is an excessive build-up of toxic waste products generated by the microbes. Thus, in this system, the cellular composition and physiological status of the cells will vary throughout the growth cycle.

In continuous cultures (also referred to as **open systems**) the medium is refreshed regularly to replace that spent by the cells. The objective of this system is to maintain the cells in the exponential growth phase by enabling nutrients, biomass and waste products to be controlled through varying the dilution rate of the cultures. Continuous cultures, although more complex to set up, offer certain advantages over batch cultures in that they facilitate growth under steady-state conditions in which there is tight coupling between cell division and biosynthesis. As a result, the physiological status of the cultures is more clearly defined, with very little variation in the cellular composition of the cells during the growth cycle. The main concern with the open system is the high risk of contamination associated with the dilution of the cultures. However, applying strict aseptic techniques during feeding or harvesting cells may help to reduce the risk of such contaminations. In addition, the whole system can be automated by connecting the culture vessels to their reservoirs through solenoid valves that can be triggered to open when required. This minimises direct contact with the operator or outside environment and thus reduces the risk of contamination.

2.7.5 **Determination of growth of bacterial cultures**

Several methods are available for determining the growth of bacterial cells in culture, including directly counting cells using a haemocytometer as described (Section 2.5.6). This is, however, suitable only for cells in suspension. When cells are grown on solid agar plates, colony counting can be used instead to estimate growth. This method assumes that each colony is derived from a single cell, which may not always be the case, since errors in dilution and/or streaking may result in clumps rather than single cells producing colonies. In addition, suboptimal culture conditions may cause poor growth, thus leading to an underestimation of the true cell count. When cells are grown in suspension, changes in the turbidity of the growth medium could be determined using a spectrophotometer and the absorbance value converted to cell number using a standard curve of absorbance versus cell number. This should be constructed for each cell type by taking the readings of a series of known numbers of cells in suspension (see also Section 12.4.1).

2.8 POTENTIAL USE OF CELL CULTURES

Cell cultures of various sorts from animal and microbes are becoming increasingly exploited not only by scientists for studying the activity of cells in isolation, but also by various biotechnology and pharmaceutical companies for the production of valuable biological products including viral vaccines (e.g. polio vaccine), antibodies (e.g. OKT3 used in suppressing immunological organ rejection in transplant surgery) and various recombinant proteins. The application of recombinant DNA techniques has led to an ever-expanding list of improved products, both from mammalian and bacterial cells, for therapeutic use in humans. These products include the commercial production of factor VIII for haemophilia, insulin for diabetes, interferon-α and β for anticancer chemotherapy and erythropoietin for anaemia. Bacterial cultures have also been widely used for other industrial purposes including the large-scale production of cell proteins, growth regulators, organic acids, alcohols, solvents, sterols, surfactants, vitamins, amino acids and many more products. In addition, degradation of waste products particularly those from the agricultural and food industries is another important industrial application of microbial cells. They are also exploited in the bioconversion of waste to useful end products, and in toxicological studies where some of these organisms are rapidly replacing animals in preliminary toxicological testing of xenobiotics. The advent of stem cell culture now provides the possibility of treating diseases using cell-based therapy. This would be particularly important in regenerating diseased or damaged tissues by transplanting stem cells programmed to differentiate into a specific cell type specialised in carrying out a specific function.

ACKNOWLEDGEMENTS

Images courtesy of Lesley Young and Paula M Timmons, UK Stem Cell Bank, NIBSC, United Kingdom. Thanks also to Lyn Healy, UK Stem Cell Bank, NIBSC, United Kingdom for valuable comments and advice on stem cell culture.

2.9 SUGGESTIONS FOR FURTHER READING

Ball, A. S. (1997). *Bacterial Cell Culture: Essential Data*. John Wiley & Sons, Inc., New York (Gives an adequate background into bacterial cell culture and techniques.)

Davis, J. M. (2002). *Basic Cell Culture: A Practical Approach*, 2nd edn. Oxford University Press, Oxford. (A comprehensive coverage of basic cell culture techniques.)

Freshney, R. I. (2005). *Culture of Animal Cells: A Manual of Basic Technique*, (5th edition). John Wiley & Sons, Inc., New York. (A comprehensive coverage of animal cell culture techniques and applications.)

Furr, A. K. (ed.) (2001). *CRC Handbook of Laboratory Safety*, 5th edn. CRC Press, Boca Raton, FL. (A complete guide to laboratory safety.)

HSC advisory committee on dangerous pathogens (2001). *The Management Design and Operation of Microbiological Containment Laboratories*. HSE books, Sudbury. (Provides guidance, legal requirements and detailed technical information on the design, management and operation of containment laboratories.)

Parekh, S. R. and Vinci, V. A. (2003). *Handbook of Industrial Cell Culture: Mammalian, Microbial, and Plant Cells*. Humana Press, Totowa, NJ. (Provides a good coverage of state-of-the-art techniques for industrial screening, cultivation and scale-up of mammalian, microbial, and plant cells.).

3 Centrifugation

K. OHLENDIECK

3.1 INTRODUCTION

Biological centrifugation is a process that uses centrifugal force to separate and purify mixtures of biological particles in a liquid medium. It is a key technique for isolating and analysing cells, subcellular fractions, supramolecular complexes and isolated macromolecules such as proteins or nucleic acids. The development of the first analytical ultracentrifuge by Svedberg in the late 1920s and the technical refinement of the preparative centrifugation technique by Claude and colleagues in the 1940s positioned centrifugation technology at the centre of biological and bio medical research for many decades. Today, centrifugation techniques represent a critical tool for modern biochemistry and are employed in almost all invasive sub-cellular studies. While analytical centrifugation is mainly concerned with the study of purified macromolecules or isolated supramolecular assemblies, preparative centri-fugation methodology is devoted to the actual separation of tissues, cells, subcellular structures, membrane vesicles and other particles of biochemical interest.

Most undergraduate students will be exposed to preparative centrifugation protocols during practical classes and might also experience a demonstration of analytical centrifugation techniques. This chapter is accordingly divided into a short introduction into the theoretical background of sedimentation, an overview of practical aspects of using centrifuges in the biochemical laboratory, an outline of preparative centrifugation and a description of the usefulness of ultracentrifugation techniques in the biochemical characterisation of macromolecules. To aid in the understanding of the basic principles of centrifugation, the general design of various rotors and separation processes is diagrammatically represented. Often the learning process of undergraduate students is hampered by the lack of a proper linkage between theoretical knowledge and practical applications. To overcome this problem, the description of preparative centrifugation techniques is accompanied by an explanatory flow chart and the detailed discussion of the subcellular fractionation protocol of a specific tissue preparation. Taking the isolation of fractions from skeletal muscle homogenates as an example, the rationale behind individual preparative steps is explained. Since affinity isolation methods not only represent an extremely powerful tool in purifying biomolecules (see Chapter 11), but can also be utilised to separate intact organelles and membrane vesicles by centrifugation, lectin affinity agglutination of highly purified plasmalemma vesicles from skeletal muscle is described. Traditionally, marker enzyme activities are used to determine the overall yield and enrichment of particular structures within subcellular fractions following centrifugation. As an example, the distribution of key enzyme activities in mitochondrial subfractions from liver is given. However, most modern fractionation procedures are evaluated by more convenient methods, such as protein gel analysis in conjunction with immunoblot analysis. Miniature gel and blotting equipment can produce highly reliable results within a few hours making it an ideal analytical tool for high-throughput testing. Since electrophoretic techniques are introduced in Chapter 10 and are used routinely in biochemical laboratories, the protein gel analysis of the distribution of typical marker proteins in affinity isolated plasmalemma fractions is graphically represented and discussed.

Although monomeric peptides and proteins are capable of performing complex biochemical reactions, many physiologically important elements do not exist in isolation under native conditions. Therefore, if one considers individual proteins as the basic units of the proteome (see Chapter 8), protein complexes actually form the functional units of cell biology. This gives investigations into the supramolecular structure of protein complexes a central place in biochemical research. To illustrate this point, the sedimentation analysis of a high-molecular-mass membrane assembly, the dystrophin–glycoprotein complex of skeletal muscle, is shown and the use of sucrose gradient centrifugation explained.

3.2 BASIC PRINCIPLES OF SEDIMENTATION

From everyday experience, the effect of sedimentation due to the influence of the Earth's gravitational field ($g = 981 \, \mathrm{cm \, s^{-2}}$) versus the increased rate of sedimentation in a centrifugal field ($g > 981 \, \mathrm{cm \, s^{-2}}$) is apparent. To give a simple but illustrative example, crude sand particles added to a bucket of water travel slowly to the bottom of the bucket by gravitation, but sediment much faster when the bucket is swung around in a circle. Similarly, biological structures exhibit a drastic increase in sedimentation when they undergo acceleration in a centrifugal field. The relative centrifugal field is usually expressed as a multiple of the acceleration due to gravity. Below is a short description of equations used in practical centrifugation classes.

When designing a centrifugation protocol, it is important to keep in mind that:

- the more dense a biological structure is, the faster it sediments in a centrifugal field;
- the more massive a biological particle is, the faster it moves in a centrifugal field;

- the denser the biological buffer system is, the slower the particle will move in a centrifugal field;
- the greater the frictional coefficient is, the slower a particle will move;
- the greater the centrifugal force is, the faster the particle sediments;
- the sedimentation rate of a given particle will be zero when the density of the particle and the surrounding medium are equal.

Biological particles moving through a viscous medium experience a frictional drag, whereby the frictional force acts in the opposite direction to sedimentation and equals the velocity of the particle multiplied by the frictional coefficient. The frictional coefficient depends on the size and shape of the biological particle. As the sample moves towards the bottom of a centrifuge tube in swing-out or fixed-angle rotors, its velocity will increase due to the increase in radial distance. At the same time the particles also encounter a frictional drag that is proportional to their velocity. The frictional force of a particle moving through a viscous fluid is the product of its velocity and its frictional coefficient, and acts in the opposite direction to sedimentation.

From the equation (3.1) for the calculation of the relative centrifugal field it becomes apparent that when the conditions for the centrifugal separation of a biological particle are described, a detailed listing of rotor speed, radial dimensions and duration of centrifugation has to be provided. Essentially, the rate of sedimentation is dependent upon the applied centrifugal field (cm s^{-2}), G, that is determined by the radial distance, r, of the particle from the axis of rotation (in cm) and the square of the angular velocity, ω, of the rotor (in radians per second):

$$G = \omega^2 r \qquad (3.1)$$

The average angular velocity of a rigid body that rotates about a fixed axis is defined as the ratio of the angular displacement in a given time interval. One radian, usually abbreviated as 1 rad, represents the angle subtended at the centre of a circle by an arc with a length equal to the radius of the circle. Since 360^0 equals 2π radians, one revolution of the rotor can be expressed as 2π rad. Accordingly, the angular velocity in rads per second of the rotor can be expressed in terms of rotor speed s as:

$$\omega = \frac{2\pi s}{60} \qquad (3.2)$$

Example 1 **CALCULATION OF CENTRIFUGAL FIELD**

Question What is the applied centrifugal field at a point equivalent to 5 cm from the centre of rotation and an angular velocity of 3000 rad s^{-1}?

Answer The centrifugal field, G, at a point 5 cm from the centre of rotation may be calculated using the equation
$G = \omega^2 r = (3000)^2 \times 5 \text{ cm s}^{-2} = 4.5 \times 10^7 \text{ cm s}^{-2}$

and therefore the centrifugal field can be expressed as:

$$G = \frac{4\pi^2 (\text{rev min}^{-1})^2 r}{3600} = \frac{4\pi^2 s^2 r}{3600} \tag{3.3}$$

Example 2 **CALCULATION OF ANGULAR VELOCITY**

Question For the pelleting of the microsomal fraction from a liver homogenate, an ultracentrifuge is operated at a speed of 40 000 r.p.m. Calculate the angular velocity, ω, in radians per second.

Answer The angular velocity, ω, may be calculated using the equation:

$$\omega = \frac{2\pi \, \text{rev min}^{-1}}{60}$$

$$\omega = 2 \times 3.1416 \times 40{,}000/60 \text{ rad s}^{-1} = 4188.8 \text{ rad s}^{-1}$$

The centrifugal field is generally expressed in multiples of the gravitational field, g (981 cm s^{-2}). The relative centrifugal field (g), RCF, which is the ratio of the centrifugal acceleration at a specified radius and the speed to the standard acceleration of gravity, can be calculated from the following equation:

$$\text{RCF} = \frac{4\pi^2 (\text{rev min}^{-1})^2 r}{3600 \times 981} = \frac{G}{g} \tag{3.4}$$

RCF units are therefore dimensionless (denoting multiples of g) and revolutions per minute are usually abbreviated as r.p.m.: $\text{RCF} = 1.12 \times 10^{-5} \text{ r.p.m.}^2 r$.

Although the relative centrifugal force can easily be calculated, centrifugation manuals usually contain a nomograph for the convenient conversion between relative centrifugal force and speed of the centrifuge at different radii of the centrifugation spindle to a point along the centrifuge tube. A nomograph consists of three columns representing the radial distance (in mm), the relative centrifugal field and the rotor speed (in r.p.m.). For the conversion between relative centrifugal force and speed of the centrifuge spindle in r.p.m. at different radii, a straight-edge is aligned through known values in two columns, then the desired figure is read where the straight-edge intersects the third column. See Figure 3.1 for an illustration of the usage of a nomograph.

In a suspension of biological particles, the rate of sedimentation is dependent not only upon the applied centrifugal field, but also on the nature of the particle, i.e. its density and radius, and also the viscosity of the surrounding medium. *Stokes' Law* describes these relationships for the sedimentation of a rigid spherical particle:

$$\nu = \frac{2}{9} \frac{r^2 (\rho_p - \rho_m)}{\eta} \times g \tag{3.5}$$

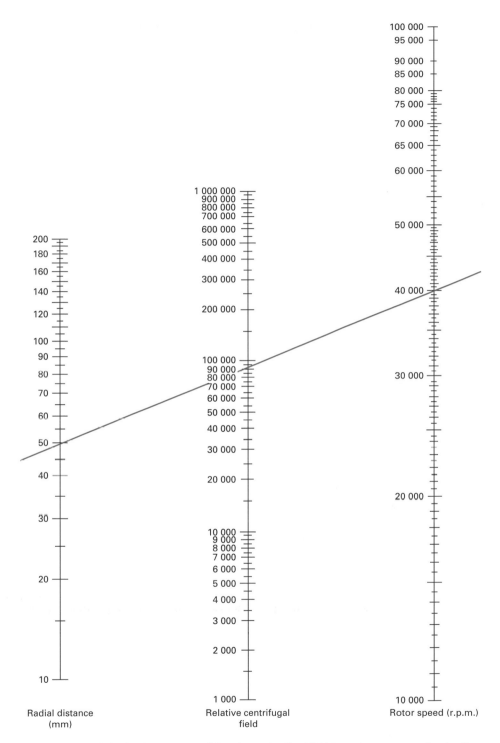

Fig. 3.1 Nomograph for the determination of the relative centrifugal field for a given rotor speed and radius. The three columns represent the radial distance (in mm), the relative centrifugal field and the rotor speed (in r.p.m.). For the conversion between relative centrifugal force and speed of the centrifuge spindle in revolutions per minute at different radii, draw a straight-edge through known values in two columns. The desired figure can then be read where the straight-edge intersects the third column. (Courtesy of Beckman-Coulter.)

where ν is the sedimentation rate of the sphere, 2/9 is the shape factor constant for a sphere, r is the radius of particle, ρ_p is the density of particle, ρ_m is the density of medium, g is the gravitational acceleration and η is the viscosity of the medium.

Example 3 CALCULATION OF RELATIVE CENTRIFUGAL FIELD

Question A fixed-angle rotor exhibits a minimum radius, r_{min}, at the top of the centrifuge tube of 3.5 cm, and a maximum radius, r_{max}, at the bottom of the tube of 7.0 cm. See Fig. 3.2a for a cross-sectional diagram of a fixed-angle rotor illustrating the position of the minimum and maximum radius. If the rotor is operated at a speed of 20 000 r.p.m., what is the relative centrifugal field, RCF, at the top and bottom of the centrifuge tube?

Answer The relative centrifugal field may be calculated using the equation:

$$RCF = 1{,}12 \times 10^{-5} \text{ r.p.m.}^2 r$$

Top of centrifuge tube:

$$RCF = 1{,}12 \times 10^{-5} \times (20\,000)^2 \times 3.5 = 15\,680$$

Bottom of centrifuge tube:

$$RCF = 1{,}12 \times 10^{-5} \times (20\,000)^2 \times 7.0 = 31\,360$$

This calculation illustrates that with fixed-angle rotors the centrifugal field at the top and bottom of the centrifuge tube might differ considerably, in this case exactly two-fold.

Accordingly a mixture of biological particles exhibiting an approximately spherical shape can be separated in a centrifugal field based on their density and/or their size. The time of sedimentation (in seconds) for a spherical particle is:

$$t = \frac{9}{2} \frac{\eta}{\omega^2 r_p^2 (\rho_p - \rho_m)} \times \ln \frac{r_b}{r_t} \tag{3.6}$$

where t is the sedimentation time, η is the viscosity of medium, r_p is the radius of particle, r_b is the radial distance from the centre of rotation to bottom of tube, r_t is the radial distance from the centre of rotation to liquid meniscus, ρ_p is the density of the particle, ρ_m is the density of the medium and ω is the angular velocity of rotor.

The sedimentation rate or velocity of a biological particle can also be expressed as its sedimentation coefficient *(s)*, whereby:

$$s = \frac{\nu}{\omega^2 r} \tag{3.7}$$

Since the sedimentation rate per unit centrifugal field can be determined at different temperatures and with various media, experimental values of the sedimentation coefficient are corrected to a sedimentation constant theoretically obtainable in water at 20 °C, yielding the $S_{20,W}$ value. The sedimentation coefficients of biological macromolecules are relatively small, and are usually expressed (see Section 3.5), as Svedberg units, S. One Svedberg unit equals 10^{-13} s.

3.3 TYPES, CARE AND SAFETY ASPECTS OF CENTRIFUGES

3.3.1 Types of centrifuges

Centrifugation techniques take a central position in modern biochemical, cellular and molecular biological studies. Depending on the particular application, centrifuges differ in their overall design and size. However, a common feature in all centrifuges is the central motor that spins a rotor containing the samples to be separated. Particles of biochemical interest are usually suspended in a liquid buffer system contained in specific tubes or separation chambers that are located in specialised rotors. The biological medium is chosen for the specific centrifugal application and may differ considerably between preparative and analytical approaches. As outlined below, the optimum pH value, salt concentration, stabilising cofactors and protective ingredients such as protease inhibitors have to be carefully evaluated in order to preserve biological function. The most obvious differences between centrifuges are:

- the maximum speed at which biological specimens are subjected to increased sedimentation;
- the presence or absence of a vacuum;
- the potential for refrigeration or general manipulation of the temperature during a centrifugation run; and
- the maximum volume of samples and capacity for individual centrifugation tubes.

Many different types of centrifuges are commercially available including:

- large-capacity low-speed **preparative centrifuges**;
- refrigerated high-speed preparative centrifuges;
- analytical ultracentrifuges;
- preparative ultracentrifuges;
- large-scale clinical centrifuges; and
- small-scale laboratory **microfuges**.

Some large-volume centrifuge models are quite demanding on space and also generate considerable amounts of heat and noise, and are therefore often centrally positioned in special instrument rooms in biochemistry departments. However, the development of small-capacity **bench-top centrifuges** for biochemical applications, even in the case of ultracentrifuges, has led to the introduction of these models in many individual research laboratories.

The main types of centrifuge encountered by undergraduate students during introductory practicals may be divided into microfuges (so called because they centrifuge small volume samples in Eppendorf tubes), large-capacity preparative centrifuges, high-speed refrigerated centrifuges and ultracentrifuges. Simple bench-top centrifuges vary in design and are mainly used to collect small amounts of biological material, such as blood cells. To prevent denaturation of sensitive protein samples, refrigerated centrifuges should be employed. Modern refrigerated microfuges are equipped with adapters to accommodate standardised plastic tubes for the sedimentation of 0.5 to 1.5 cm^3 volumes. They can provide centrifugal fields of approximately 10 000 g and sediment biological samples in minutes, making microfuges an indispensable separation tool for many biochemical methods. Microfuges can also be used to concentrate protein samples. For example, the dilution of protein samples, eluted by column chromatography, can often represent a challenge for subsequent analyses. Accelerated ultrafiltration with the help of plastic tube-associated filter units, spun at low g-forces in a microfuge, can overcome this problem. Depending on the proteins of interest, the biological buffers used and the molecular mass cut-off point of the particular filters, a 10- to 20-fold concentration of samples can be achieved within minutes. Larger preparative bench-top centrifuges develop maximum centrifugal fields of 3000 to 7000 g and can be used for the spinning of various types of containers. Depending on the range of available adapters, considerable quantities of 5 to 250 cm^3 plastic tubes or 96-well ELISA plates can be accommodated. This gives simple and relatively inexpensive bench centrifuges a central place in many high-throughput biochemical assays where the quick and efficient separation of coarse precipitates or whole cells is of importance.

High-speed refrigerated centrifuges are absolutely essential for the sedimentation of protein precipitates, large intact organelles, cellular debris derived from tissue homogenisation and microorganisms. As outlined in Section 3.4, the initial bulk separation of cellular elements prior to preparative ultracentrifugation is performed by these kinds of centrifuges. They operate at maximum centrifugal fields of approximately 100 000 g. Such centrifugal force is not sufficient to sediment smaller microsomal vesicles or ribosomes, but can be employed to differentially separate nuclei, mitochondria or chloroplasts. In addition, bulky protein aggregates can be sedimented using high-speed refrigerated centrifuges. An example is the contractile apparatus released from muscle fibres by homogenisation, mostly consisting of myosin and actin macromolecules aggregated in filaments. In order to harvest yeast cells or bacteria from large volumes of culture media, high-speed centrifugation may also be used in a continuous flow mode with zonal rotors. This approach does not therefore use centrifuge tubes but a continuous flow of medium. As the medium enters the moving rotor, biological particles are sedimented against the rotor periphery and excess liquid removed through a special outlet port.

Ultracentrifugation has decisively advanced the detailed biochemical analysis of subcellular structures and isolated biomolecules. Preparative ultracentrifugation can be operated at relative centrifugal fields of up to 900 000 g. In order to minimise excessive rotor temperatures generated by frictional resistance between the spinning rotor and air, the rotor chamber is sealed, evacuated and refrigerated. Depending on the type, age and condition of a particular ultracentrifuge, cooling to the required running temperature and the generation of a stable vacuum might take a considerable amount of time. To avoid delays during biochemical procedures involving ultracentrifugation, the cooling and evacuation system of older centrifuge models should be switched on at least an hour prior to the centrifugation run. On the other hand, modern ultracentrifuges can be started even without a fully established vacuum and will proceed in the evacuation of the rotor chamber during the initial acceleration process. For safety reasons, heavy armour plating encapsulates the ultracentrifuge to prevent injury to the user in case of uncontrolled rotor movements or dangerous vibrations. A centrifugation run cannot be initiated without proper closing of the chamber system. To prevent unfavourable fluctuations in chamber temperature, excessive vibrations or operation of rotors above their maximum rated speed, newer models of ultracentrifuges contain sophisticated temperature regulation systems, flexible drive shafts and an over-speed control device. Although slight rotor imbalances can be absorbed by modern ultracentrifuges, a more severe misbalance of tubes will cause the centrifuge to switch off automatically. This is especially true for swinging-bucket rotors. The many safety features incorporated into modern ultracentrifuges make them a robust piece of equipment that tolerates a certain degree of misuse by an inexperienced operator (see Sections 3.3.2 and 3.3.4 for a more detailed discussion of safety and centrifugation). In contrast to preparative ultracentrifuges, analytical ultracentrifuges contain a solid rotor which in its simplest form incorporates one analytical cell and one counterbalancing cell. An optical system enables the sedimenting material to be observed throughout the duration of centrifugation. Using a light absorption system, a Schlieren system or a Raleigh interferometric system, concentration distributions in the biological sample are determined at any time during ultracentrifugation. The Raleigh and Schlieren optical systems detect changes in the refractive index of the solution caused by concentration changes and can thus be used for sedimentation equilibrium analysis. This makes analytical ultracentrifugation a relatively accurate tool for the determination of the molecular mass of an isolated macromolecule. It can also provide crucial information about the thermodynamic properties of a protein or other large biomolecules.

3.3.2 Types of rotors

To illustrate the difference in design of fixed-angle rotors, vertical tube rotors and swinging-bucket rotors, Fig. 3.2 outlines cross-sectional diagrams of these three main types of rotors. Companies usually name rotors according to their type of design, the maximum allowable speed and sometimes the material composition. Depending on the use in a simple low-speed centrifuge, a high-speed centrifuge or an ultracentrifuge, different centrifugal forces are encountered by a spinning rotor. Accordingly different types of rotors are made from different materials. Low-speed rotors are usually made of steel or brass, while high-speed rotors consist of aluminium, titanium or fibre-reinforced composites. The exterior of specific rotors might be finished with protective paints. For example, rotors for ultracentrifugation made out of titanium alloy are covered with a polyurethane layer. Aluminium rotors are protected from corrosion by an electrochemically formed tough layer of aluminium oxide. In order to avoid damaging these protective layers, care should be taken during rotor handling.

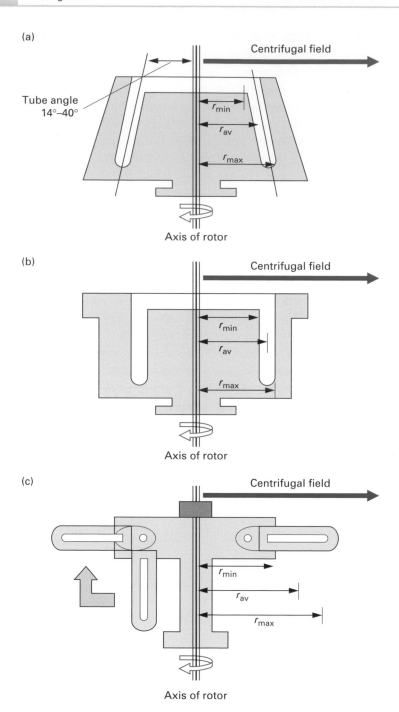

Fig. 3.2 Design of the three main types of rotors used in routine biochemical centrifugation techniques. Shown is a cross-sectional diagram of a fixed-angle rotor (a), a vertical tube rotor (b), and a swinging-bucket rotor (c). A fourth type of rotor is represented by the class of near-vertical rotors (not shown).

Fixed-angle rotors are an ideal tool for pelleting during the differential separation of biological particles where sedimentation rates differ significantly, for example when separating nuclei, mitochondria and microsomes. In addition, isopycnic banding may also be routinely performed with fixed-angle rotors. For isopycnic separation, centrifugation is continued until the biological particles of interest have reached their isopycnic position in a gradient. This means that the particle has reached a position where the sedimentation rate is zero because the density of the biological particle and the surrounding medium are equal. Centrifugation tubes are held at a fixed angle of between 14° and 40° to the vertical in this class of rotors (Fig. 3.2a). Particles move radially outwards and since the centrifugal field is exerted at an angle, they only have to travel a short distance until they reach their isopycnic position in a gradient using an isodensity technique or before colliding with the outer wall of the centrifuge tube using a differential centrifugation method. Vertical rotors (Fig. 3.2b) may be divided into true vertical rotors and near-vertical rotors. Sealed centrifuge tubes are held parallel to the axis of rotation in vertical rotors and are restrained in the rotor cavities by screws, special washers and plugs. Since samples are not separated down the length of the centrifuge tube, but across the diameter of the tube, isopycnic separation time is significantly shorter as compared to swinging-bucket rotors. In contrast to fixed-angle rotors, near-vertical rotors exhibit a reduced tube angle of 7° to 10° and also employ quick-seal tubes. The reduced angle results in much shorter run times as compared to fixed-angle rotors. Near-vertical rotors are useful for gradient centrifugation of biological elements that do not properly participate in conventional gradients. Hinge pins or a crossbar is used to attach rotor buckets in swinging-bucket rotors (Fig. 3.2c). They are loaded in a vertical position and during the initial acceleration phase, rotor buckets swing out horizontally and then position themselves at the rotor body for support.

To illustrate the separation of particles in the three main types of rotors, Fig. 3.3 outlines the path of biological samples during the initial acceleration stage, the main centrifugal separation phase, de-acceleration and the final harvesting of separated particles in the rotor at rest. In the case of isopycnic centrifugation in a fixed angle rotor, the centrifuge tubes are gradually filled with a suitable gradient, the sample carefully loaded on top of this solution and then the tubes placed at a specific fixed-angle into the rotor cavities. During rotor acceleration, the sample solution and the gradient undergo reorientation in the centrifugal field, followed by the separation of particles with different sedimentation properties (Fig. 3.3a). The gradient returns to its original position during the de-acceleration phase and separated particle bands can be taken from the tubes once the rotor is at rest. In analogy, similar reorientation of gradients and banding of particles occurs in a vertical rotor system (Fig. 3.3b). Although run times are reduced and this kind of rotor can usually hold a large number of tubes, resolution of separated bands during isopycnic centrifugation is less when compared with swinging-bucket applications. Since a greater variety of gradients exhibiting different steepness can be used with swinging-bucket rotors, they are the method of choice when maximum resolution of banding zones is required (Fig. 3.3c), such as in rate zonal studies based on the separation of biological particles as a function of sedimentation coefficient.

(a)

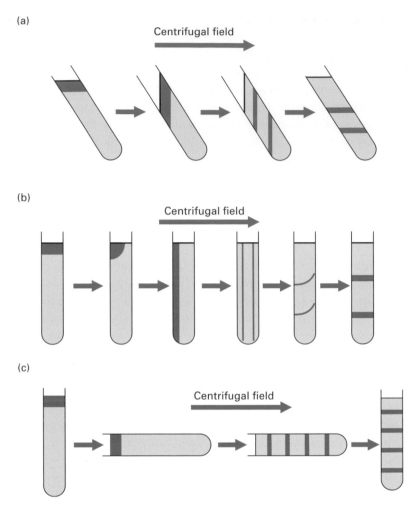

Fig. 3.3 Operation of the three main types of rotors used in routine biochemical centrifugation techniques. Shown is a cross-sectional diagram of a centrifuge tube positioned in a fixed-angle rotor (a), a vertical tube rotor (b), and a swinging-bucket rotor (c). The diagrams illustrate the movement of biological samples during the initial acceleration stage, the main centrifugal separation phase, de-acceleration and the final harvesting of separated particles in the rotor at rest. Using a fixed-angle rotor, the tubes are filled with a gradient, the sample loaded on top of this solution and then the tubes placed at a specific fixed-angle into the rotor cavities. The sample and the gradient undergo reorientation in the centrifugal field during rotor acceleration, resulting in the separation of particles with different sedimentation properties. Similar reorientation of gradients and banding of particles occurs in a vertical rotor system. A great variety of gradients can be used with swinging-bucket rotors, making them the method of choice when maximum resolution of banding zones is required.

3.3.3 Care and maintenance of centrifuges

Corrosion and degradation due to biological buffer systems used within rotors or contamination of the interior or exterior of the centrifuge via spillage may seriously affect the lifetime of this equipment. Another important point is the proper balancing of centrifuge tubes. This is not only important with respect to safety, as outlined below, but might also cause vibration-induced damage to the rotor itself and the drive shaft of the centrifuge. Thus, proper handling and care, as well as regular maintenance of both centrifuges and rotors is an important part of keeping this biochemical method available in the laboratory. In order to avoid damaging the protective layers of rotors, such as polyurethane paint or aluminium oxide, care should be taken in the cleaning of the rotor exterior. Coarse brushes that may scratch the finish should not be used and only non-corrosive detergents employed. Corrosion may be triggered by long-term exposure of rotors to alkaline solutions, acidic buffers, aggressive detergents or salt. Thus, rotors should be thoroughly washed with distilled or deionised water after every run. For overnight storage, rotors should be first left upside down to drain excess liquid and then positioned in a safe and dry place. To avoid damage to the hinge pins of swinging-bucket rotors, they should be dried with tissue paper following removal of biological buffers and washing with water. Centrifuge rotors are often not properly stored in a clean environment; this can quickly lead to the destruction of the protective rotor coating and should thus be avoided. It is advisable to keep rotors in a special clean room, physically separated from the actual centrifugation facility, with dedicated places for individual types of rotors. Some researchers might prefer to pre-cool their rotors prior to centrifugation by transferring them to a cold room. Although this is an acceptable practice and might keep proteolytic degradation to a minimum, rotors should not undergo long-term storage in a wet and cold environment. Regular maintenance of rotors and centrifuges by engineers is important for ensuring the safe operation of a centralised centrifugation facility. In order to judge properly the need for replacement of a rotor or parts of a centrifuge, it is essential that all users of core centrifuge equipment participate in proper book-keeping. Accurate record-keeping of run times and centrifugal speeds is important, since cyclic acceleration and deacceleration of rotors may lead to metal fatigue.

3.3.4 Safety and centrifugation

Modern centrifuges are not only highly sophisticated but also relatively sturdy pieces of biochemical equipment that incorporate many safety features. Rotor chambers of high-speed and ultracentrifuges are always enclosed in heavy armour plating. Most centrifuges are designed to buffer a certain degree of imbalance and are usually equipped with an automatic switch-off mode. However, even in a well-balanced rotor, tube cracking during a centrifugation run might cause severe imbalance resulting in dangerous vibrations. When the rotor can only be partially loaded, the order of tubes must be organised according to the manufacturer's instructions, so that the load is correctly distributed. This is important not only for ultracentrifugation with enormous centrifugal fields, but also for both small- and large-capacity bench centrifuges where the rotors are usually mounted on a more rigid suspension. When using swinging-bucket rotors, it is important always to load all buckets with their caps properly screwed on. Even if only two tubes are loaded with solutions, the empty swinging buckets also have to be assembled since they form an integral part of the overall balance of the rotor system. In some swinging-bucket rotors, individual rotor buckets are numbered and should not be interchanged between their designated positions on similarly numbered hinge pins. Centrifugation runs using swinging-bucket rotors are usually set up with low acceleration and deceleration rates, as to avoid any disturbance of delicate gradients, and reduce the risk of disturbing bucket attachment. This practice also avoids the occurrence of sudden imbalances due to tube deformation or cracking and thus eliminates potentially dangerous vibrations.

Generally, safety and good laboratory practice are important aspects of all research projects and the awareness of the exposure to potentially harmful substances should be a concern for every biochemist. If you use dangerous chemicals, potentially infectious material or radioactive substances during centrifugation protocols, refer to up-to-date safety manuals and the safety statement of your individual department. Perform mock runs of important experiments in order to avoid the loss of precious specimens or expensive chemicals. As with all other biochemical procedures, experiments should never be rushed, and protective clothing should be worn at all times. Centrifuge tubes should be handled slowly and carefully so as not to disturb pellets, bands of separated particles or unstable gradients. To help you choose the right kind of centrifuge tube for a particular application, the manufacturers of rotors usually give detailed recommendation of suitable materials. For safety reasons and to guarantee experimental success, it is important to make sure that individual centrifuge tubes are chemically resistant to solvents used, have the right capacity for sample loading, can be used in the designated type of rotor and are able to withstand the maximum centrifugal forces and temperature range of a particular centrifuge. In fixed-angle rotors, large centrifugal forces tend to cause a collapse of centrifuge tubes, making thick-walled tubes the choice for these rotors. The volume of liquid and the sealing mechanisms of these tubes are very important for the integrity of the run and should be done according to manufacturer's instructions. In contrast, swinging-bucket rotor tubes are better protected from deformation and usually thin-walled polyallomer tubes are used. An important safety aspect is the proper handling of separated biological particles following centrifugation. In order to perform post-centrifugation analysis of individual fractions, centrifugation tubes often have to be punctured or sliced. For example, separated vesicle bands can be harvested from the pierced bottom of the centrifuge tube or can be collected by slicing of the tube following quick-freezing. If samples have been pre-incubated with radioactive markers or toxic ligands, contamination of the centrifugation chamber and rotor cavities or buckets should be avoided. If centrifugal separation processes have to be performed routinely with a potentially harmful substance, it makes sense to dedicate a particular centrifuge and accompanying rotors for this work and thereby eliminate the potential of cross-contamination.

3.4　PREPARATIVE CENTRIFUGATION

3.4.1　Differential centrifugation

Cellular and subcellular fractionation techniques are indispensable methods used in biochemical research. Although the proper separation of many subcellular structures is absolutely dependent on preparative ultracentrifugation, the isolation of large

(a)

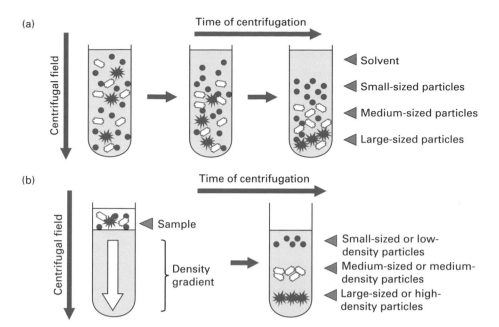

Fig. 3.4 Diagram of particle behaviour during differential and isopycnic separation. During differential sedimentation (a) of a particulate suspension in a centrifugal field, the movement of particles is dependent upon their density, shape and size. For separation of biological particles using a density gradient (b), samples are carefully layered on top of a preformed density gradient prior to centrifugation. For isopycnic separation, centrifugation is continued until the desired particles have reached their isopycnic position in the liquid density gradient. In contrast, during rate separation, the required fraction does not reach its isopycnic position during the centrifugation run.

cellular structures, the nuclear fraction, mitochondria, chloroplasts or large protein precipitates can be achieved by conventional high-speed refrigerated centrifugation. Differential centrifugation is based upon the differences in the sedimentation rate of biological particles of different size and density. Crude tissue homogenates containing organelles, membrane vesicles and other structural fragments are divided into different fractions by the stepwise increase of the applied centrifugal field. Following the initial sedimentation of the largest particles of a homogenate (such as cellular debris) by centrifugation, various biological structures or aggregates are separated into pellet and supernatant fractions, depending upon the speed and time of individual centrifugation steps and the density and relative size of the particles. To increase the yield of membrane structures and protein aggregates released, cellular debris pellets are often rehomogenised several times and then recentrifuged. This is especially important in the case of rigid biological structures such as muscular or connective tissues, or in the case of small tissue samples as is the case with human biopsy material or primary cell cultures.

The differential sedimentation of a particulate suspension in a centrifugal field is diagrammatically shown in Fig. 3.4a. Initially all particles of a homogenate are evenly distributed throughout the centrifuge tube and then move down the tube at their respective sedimentation rate during centrifugation. The largest class of particles forms a pellet on the bottom of the centrifuge tube, leaving smaller-sized structures within the supernatant. However, during the initial centrifugation step smaller particles also become entrapped in the pellet causing a certain degree of contamination. At the end of each differential centrifugation step, the pellet and supernatant fraction are carefully separated from each other. To minimise cross-contamination, pellets are usually washed several times by resuspension in buffer and recentrifugation under the same conditions. However, repeated washing steps may considerably reduce the yield of the final pellet fraction, and are therefore omitted in preparations with limiting starting material. Resulting supernatant fractions are centrifuged at a higher speed and for a longer time to separate medium-sized and small-sized particles. With respect to the separation of organelles and membrane vesicles, crude differential centrifugation techniques can be conveniently employed to isolate intact mitochondria and microsomes.

3.4.2 Density-gradient centrifugation

To further separate biological particles of similar size but differing density, ultracentrifugation with preformed or self-establishing density gradients is the method of choice. Both rate separation or equilibrium methods can be used. In Fig. 3.4b, the preparative ultracentrifugation of low- to high-density particles is shown. A mixture of particles, such as is present in a heterogeneous microsomal membrane preparation, is layered on top of a preformed liquid density gradient. Depending on the particular biological application, a great variety of gradient materials are available. Caesium chloride is widely used for the banding of DNA and the isolation of plasmids, nucleoproteins and viruses. Sodium bromide and sodium iodide are employed for the fractionation of lipoproteins and the banding of DNA or RNA molecules, respectively. Various companies offer a range of gradient material for the separation of whole cells and subcellular particles, e.g. Percoll, Ficoll, Dextran, Metrizamide and Nycodenz. For the separation of membrane vesicles derived from tissue homogenates, ultra-pure DNase-, RNase and protease-free sucrose represents a suitable and widely employed medium for the preparation of stable gradients. If one wants to separate all membrane species spanning the whole range of particle densities, the maximum density of the gradient must exceed the density of the most dense vesicle species. Both step gradient and continuous gradient systems are employed to achieve this. If automated gradient makers are not available, which is probably the case in most undergraduate practical classes, the manual pouring of a stepwise gradient with the help of a pipette is not so time-consuming or difficult. In contrast, the formation of a stable continuous gradient is much more challenging and requires a commercially available gradient maker. Following pouring, gradients are usually kept in a cold room for temperature equilibration and are moved extremely slowly in special holders so as to avoid mixing of different gradient layers. For rate separation of subcellular particles, the required fraction does not reach its isopycnic position within the gradient. For *isopycnic separation*, density centrifugation is continued until the buoyant density of the particle of interest and the density of the gradient are equal.

3.4.3 **Practical applications of preparative centrifugation**

To illustrate practical applications of differential centrifugation, density gradient ultracentrifugation and affinity methodology, the isolation of the microsomal fraction from muscle homogenates and subsequent separation of membrane vesicles with a differing density is described (Fig. 3.5), the isolation of highly purified sarcolemma vesicles outlined (Fig. 3.6), and the subfractionation of liver mitochondrial membrane systems shown (Fig. 3.7). Skeletal muscle fibres are highly specialised structures involved in contraction and the membrane systems that maintain the regulation of excitation–contraction coupling, energy metabolism and the stabilisation of the cell periphery are diagrammatically shown in Fig. 3.5a. The surface membrane consists of the sarcolemma and its invaginations, the transverse tubular membrane system. The transverse tubules may be subdivided into the non-junctional region and the triad part that forms contact zones with the terminal cisternae of the sarcoplasmic reticulum. Motor neuron-induced depolarisation of the sarcolemma travels into the transverse tubules and activates a voltage-sensing receptor complex that directly initiates the transient opening of a junctional calcium release channel. The membrane system that provides the luminal ion reservoir for the regulatory calcium cycling process is represented by the specialised endoplasmic reticulum. It forms membranous sheaths around the contractile apparatus whereby the longitudinal tubules are mainly involved in the uptake of calcium ions during muscle relaxation and the terminal cisternae provide the rapid calcium release mechanism that initiates muscle contraction. Mitochondria are the site of oxidative phosphorylation and exhibit a complex system of inner and outer membranes involved in energy metabolism.

For the optimum homogenisation of tissue specimens, mincing of tissue has to be performed in the presence of a biological buffer system that exhibits the right pH value, salt concentration, stabilising co-factors and chelating agents. The optimum ratio between the wet weight of tissue and buffer volume as well as the temperature (usually 4 °C) and presence of a protease inhibitor cocktail is also essential to minimise proteolytic degradation. Prior to the 1970s, researchers did not widely use protease inhibitors or chelating agents in their homogenisation buffers. This resulted in the degradation of many high-molecular-mass proteins. Since protective measures against endogenous enzymes have been routinely introduced into subcellular fractionation protocols, extremely large proteins have been isolated in their intact form, such as 427 kDa dystrophin, the 565 kDa ryanodine receptor, 800 kDa nebulin and the longest known polypeptide, of 2200 kDa, named titin. Commercially available protease inhibitor cocktails usually exhibit a broad specificity for the inhibition of cysteine-proteases, serine-proteases, aspartic-proteases, metallo-proteases and amino-peptidases. They are used in the micromolar concentration range and are best added to buffer systems just prior to the tissue homogenisation process. Depending on the half-life of specific protease inhibitors, the length of a subcellular fractionation protocol and the amount of endogenous enzymes present in individual fractions, tissue suspensions might have to be replenished with a fresh aliquot of a protease inhibitor cocktail. Protease inhibitor kits for the creation of individualised cocktails are also available

(a) Subcellular membrane systems that can be isolated by differential centrifugation

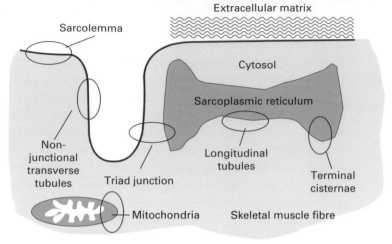

(b) Scheme of subcellular fractionation of membranes from muscle homogenates

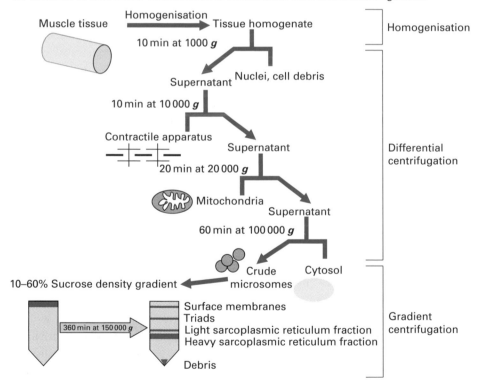

Fig. 3.5 Scheme of the fractionation of skeletal muscle homogenate into various subcellular fractions. Shown is a diagrammatic presentation of the subcellular membrane system from skeletal muscle fibres (a) and a flow chart of the fractionation protocol of these membranes from tissue homogenates using differential centrifugation and density gradient methodology (b).

and consist of substances such as trypsin inhibitor, E-64, aminoethyl-benzenesulfonyl-fluoride, antipain, aprotinin, benzamidine, bestatin, chymostatin, ϵ-aminocaproic acid, N-ethylmaleimide, leupeptin, phosphoramidon and pepstatin. The most commonly used chelators of divalent cations for the inhibition of degrading enzymes such as metallo-proteases are EDTA and EGTA.

3.4.4 Subcellular fractionation

A typical flow chart outlining a subcellular fractionation protocol is shown in Fig. 3.5b. Depending on the amount of starting material, which would usually range between 1 and 500 g in the case of skeletal muscle preparations, a particular type of rotor and size of centrifuge tubes is chosen for individual stages of the isolation procedure. The repeated centrifugation at progressively higher speeds and longer centrifugation periods will divide the muscle homogenate into distinct fractions. Typical values for centrifugation steps are 10 min for 1000 g to pellet nuclei and cellular debris, 10 min for 10 000 g to pellet the contractile apparatus, 20 min at 20 000 g to pellet a fraction enriched in mitochondria, and 1 h at 100 000 g to separate the microsomal and cytosolic fractions. Mild salt washes can be carried out to remove myosin contamination of membrane preparations. Sucrose gradient centrifugation is then used to further separate microsomal subfractions derived from different muscle membranes. Using a vertical rotor or swinging-bucket rotor system at a sufficiently high g-force, the crude surface membrane fraction, triad junctions, longitudinal tubules and terminal cisternae membrane vesicles can be separated. To collect bands of fractions, the careful removal of fractions from the top can be achieved manually with a pipette. Alternatively, in the case of relatively unstable gradients or tight banding patterns, membrane vesicles can be harvested from the bottom by an automated fraction collector. In this case, the centrifuge tube is pierced and fractions collected by gravity or slowly forced out of the tube by a replacing liquid of higher density. Another method for collecting fractions from unstable gradients is the slicing of the centrifuge tube after freezing. Both latter methods destroy the centrifuge tubes and are routinely used in research laboratories.

Cross-contamination of vesicular membrane populations is an inevitable problem during subcellular fractionation procedures. The technical reason for this is the lack of adequate control in the formation of various types of membrane species during tissue homogenisation. Membrane domains originally derived from a similar subcellular location might form a variety of structures including inside-out vesicles, right-side-out vesicles, sealed structures, leaky vesicles and/or membrane sheets. In addition, smaller vesicles might become entrapped in larger vesicles. Different membrane systems might aggregate non-specifically or bind to or entrap abundant solubilised proteins. Hence, if highly purified membrane preparations are needed for sophisticated cell biological or biochemical studies, affinity separation methodology has to be employed. The flow chart and immunoblotting diagram in Fig. 3.6 illustrates both the preparative and analytical principles underlying such a biochemical approach. Modern preparative affinity techniques using centrifugation steps can be performed

(a) Scheme of subcellular fractionation of muscle sarcolemma

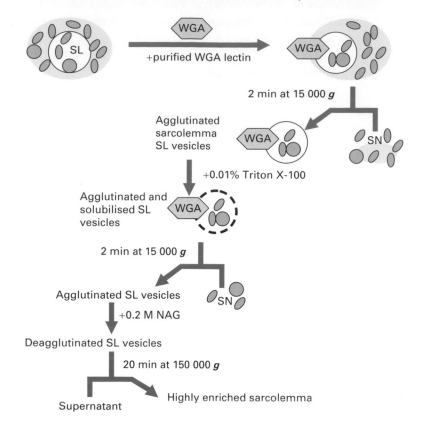

(b) Diagram of immunoblot analysis of subcellular fractionation procedures

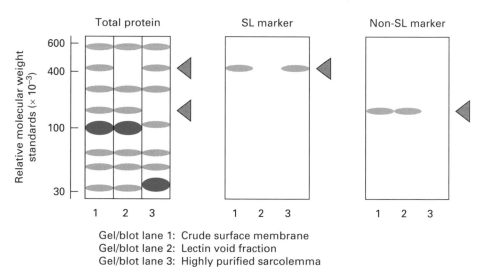

Gel/blot lane 1: Crude surface membrane
Gel/blot lane 2: Lectin void fraction
Gel/blot lane 3: Highly purified sarcolemma

Fig. 3.6 Affinity separation method using centrifugation of lectin-agglutinated surface membrane vesicles from skeletal muscle. Shown is a flow chart of the various preparative steps in the isolation of highly purified sarcolemma vesicles (a) and a diagram of the immunoblot analysis of this subcellular fractionation procedure (b). The sarcolemma (SL) and non-SL markers are surface-associated dystrophin of 427 kDa and the transverse-tubular α_{1S}-subunit of the dihydropyridine receptor of 170 kDa, respectively.

with various biological or chemical ligands. In the case of immuno affinity purification, antibodies are used to specifically bind to their respective antigen.

3.4.5   Affinity purification of membrane vesicles

In Fig. 3.6a is shown a widely employed lectin agglutination method. Lectins are plant proteins that bind tightly to specific carbohydrate structures. The rationale behind using purified wheat germ agglutinin (WGA) lectin for the affinity purification of sarcolemma vesicles is the fact that the muscle plasmalemma forms mostly right-side-out vesicles following homogenisation. By contrast, vesicles derived from the transverse tubules are mostly inside out and thus do not expose their carbohydrates. Glycoproteins from the abundant sarcoplasmic reticulum do not exhibit carbohydrate moieties that are recognised by this particular lectin species. Therefore only sarcolemma vesicles are agglutinated by the wheat germ lectin and the aggregate can be separated from the transverse tubular fraction by centrifugation for 2 min at $15\,000\,g$. The electron microscopical characterisation of agglutinated surface membranes revealed large smooth sarcolemma vesicles that had electron-dense entrapments. To remove these vesicular contaminants, originally derived from the sarcoplasmic reticulum, immobilised surface vesicles are treated with low concentrations of the non-ionic detergent Triton X-100. This procedure does not solubilise integral membrane proteins, but introduces openings in the sarcolemma vesicles for the release of the much smaller sarcoplasmic reticulum vesicles. Low g-force centrifugation is then used to separate the agglutinated sarcolemma vesicles and the contaminants. To remove the lectin from the purified vesicles, the fraction is incubated with the competitive sugar N-acetylglucosamine that eliminates the bonds between the surface glycoproteins and the lectin. A final centrifugation step for 20 min at $150\,000\,g$ results in a pellet of highly purified sarcolemma vesicles. A quick and convenient analytical method of confirming whether this subcellular fractionation procedure has resulted in the isolation of the muscle plasmalemma is immunoblotting with a mini electrophoresis unit. Figure 3.6b shows a diagram of the protein and antigen banding pattern of crude surface membranes, the lectin void fraction and the highly purified sarcolemma fraction. Using antibodies to markers of the transverse tubules and the sarcolemma, such as the α_{1S}-subunit of the dihydropyridine receptor of 170 kDa and dystrophin of 427 kDa, respectively, the separation of both membrane species can be monitored. This analytical method is especially useful for the characterisation of membrane vesicles, when no simple and fast assay systems for testing marker enzyme activities are available.

In the case of the separation of mitochondrial membranes, the distribution of enzyme activities rather than immunoblotting is routinely used for determining the distribution of the inner membrane, contact zones and the outer membrane in density gradients. Binding assays or enzyme testing represents the more traditional way of characterising subcellular fractions following centrifugation. Figure 3.7a outlines diagrammatically the micro compartments of liver mitochondria and the associated marker enzymes. While the monoamino oxidase (MAO) is enriched in the outer membrane, the enzyme succinate dehydrogenase (SDH) is associated with the inner membrane system and a representative marker of contact sites between both

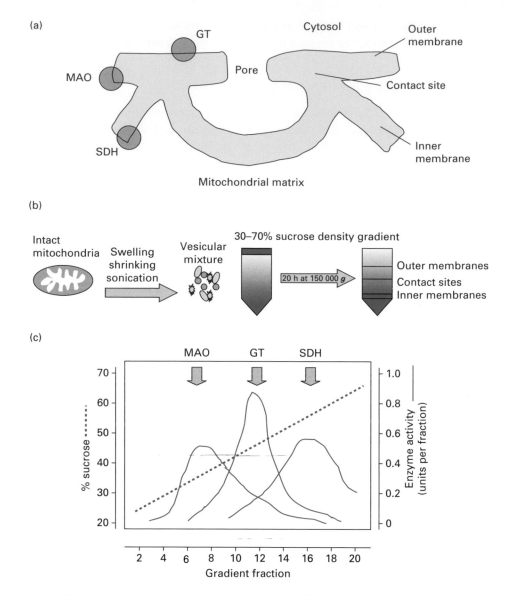

Fig. 3.7 Scheme of the fractionation of membranes derived from liver mitochondria. Shown is the distribution of marker enzymes in the micro compartments of liver mitochondria (MAO, monoamino oxidase; SDH, succinate dehydrogenase; GT, glutathione transferase) (a), the separation method to isolate fractions highly enriched in the inner cristae membrane, contact zones and the outer mitochondrial membrane (b), as well as the distribution of mitochondrial membranes after density gradient centrifugation (c).

membranes is glutathione transferase (GT). Membrane vesicles from intact mitochondria can be generated by consecutive swelling, shrinking and sonication of the suspended organelles. The vesicular mixture is then separated by sucrose density centrifugation into the three main types of mitochondrial membranes (Fig. 3.7b). The distribution of marker enzyme activities in the various fractions demonstrates that the outer membrane has a lower density compared to the inner membrane. The glutathione transferase-containing contact zones are positioned in a band between the inner and outer mitochondrial membrane and contain enzyme activities characteristic for both systems (Fig. 3.7c). Routinely used enzymes as subcellular markers would be the Na^+/K^+-ATPase for the plasmalemma, glucose-6-phosphatase for the endoplasmic reticulum, galactosyl transferase for the Golgi apparatus, succinate dehydrogenase for mitochondria, acid phosphatase for lysosomes, catalase for peroxisomes and lactate dehydrogenase for the cytosol.

3.5 ANALYTICAL CENTRIFUGATION

3.5.1 Applications of analytical ultracentrifugation

As biological macromolecules exhibit random thermal motion, their relative uniform distribution in an aqueous environment is not significantly affected by the Earth's gravitational field. Isolated biomolecules in solution only exhibit distinguishable sedimentation when they undergo immense accelerations, e.g. in an ultracentrifugal field. A typical analytical ultracentrifuge can generate a centrifugal field of $250\,000\,g$ in its analytical cell. Within these extremely high gravitational fields, the ultracentrifuge cell has to allow light passage through the biological particles for proper measurement of the concentration distribution. The schematic diagram of Fig. 3.8 outlines the optical system of a modern analytical ultracentrifuge. The availability of high-intensity xenon flash lamps and the advance in instrumental sensitivity and wavelength range has made the accurate measurement of highly dilute protein samples below 230 nm possible. Analytical ultracentrifuges such as the Beckman Optima XL-A allow the use of wavelengths between 190 nm and 800 nm. Sedimentation of isolated proteins or nucleic acids can be useful in the determination of the relative molecular mass, purity and shape of these biomolecules. Analytical ultracentrifugation for the determination of the relative molecular mass of a macromolecule can be performed by a sedimentation velocity approach or sedimentation equilibrium methodology. The hydrodynamic properties of macromolecules are described by their sedimentation coefficients and can be determined from the rate that a concentration boundary of the particular biomolecules moves in the gravitational field. Such studies on the solution behaviour of macromolecules can give detailed insight into the properties of large aggregates and thereby confirm results from biochemical analyses on complex formation. The sedimentation coefficient can be used to characterise changes in the size and shape of macromolecules with changing experimental conditions. This allows for the detailed biophysical analysis of the effect of variations in the pH value, temperature or co-factors on molecular shape.

Analytical ultracentrifugation is most often employed in

- the determination of the purity of macromolecules;
- the determination of the relative molecular mass of solutes in their native state;
- the examination of changes in the molecular mass of supramolecular complexes;
- the detection of conformational changes; and in
- ligand-binding studies (Section 17.3.2).

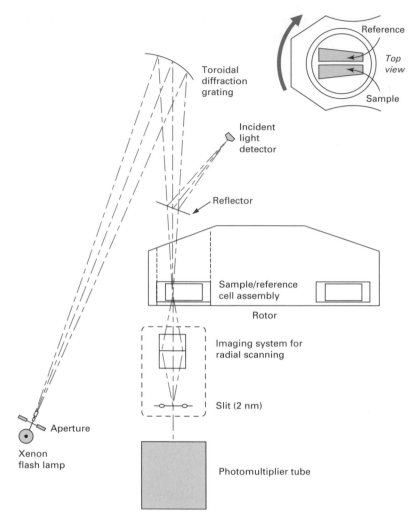

Fig. 3.8 Schematic diagram of the optical system of an analytical ultracentrifuge. The high-intensity xenon flash lamp of the Beckman Optima XL-A analytical ultracentrifuge shown here allows the use of wavelengths between 190 nm and 800 nm. The high sensitivity of the absorbance optics allows the measurement of highly dilute protein samples below 230 nm. (Courtesy of Beckman-Coulter.)

The sedimentation velocity method can be employed to estimate sample purity. Sedimentation patterns can be obtained using the Schlieren optical system. This method measures the refractive index gradient at each point in the ultracentrifugation cell at varying time intervals. During the entire duration of the sedimentation velocity analysis, a homogeneous preparation forms a single sharp symmetrical sedimenting boundary. Such a result demonstrates that the biological macromolecules analysed exhibit the same molecular mass, shape and size. However, one can not assume that the analysed particles exhibit an identical electrical charge or biological activity. Only additional biochemical studies using electrophoretic techniques and enzyme/bioassays can differentiate between these minor subtypes of macromolecules with similar molecular mass. The great advantage of the sedimentation velocity method is that smaller or larger contaminants can be clearly recognised as shoulders on the main peak, asymmetry of the main peak and/or additional peaks. For a list of references outlining the applicability of ultracentrifugation to the characterisation of macromolecular behaviour in complex solution, please consult the review articles listed in Section 3.6. In addition, manufacturers of analytical ultracentrifuges make a large range of excellent brochures on the theoretical background of this method and its specific applications available. These introductory texts are usually written by research biochemists and are well worth reading to become familiar with this field.

3.5.2 Relative molecular mass determination

For the accurate determination of the molecular mass of solutes in their native state, analytical ultracentrifugation represents an unrivalled technique. The method requires only small sample sizes (20–120 mm^3) and low particle concentrations (0.01–1 g dm^{-3}) and biological molecules with a wide range of molecular masses can be characterised. In conjunction with electrophoretic, chromatographic, crystallographic and sequencing data, the biochemical properties of a biological particle of interest can be determined in great detail. As long as the absorbance of the biomolecules to be investigated (such as proteins, carbohydrates or nucleic acids) is different from that of the surrounding solvent, analytical ultracentrifugation can be applied. At the start of an experiment using the boundary sedimentation method, the biological particles are uniformly distributed throughout the solution in the analytical cell. The application of a centrifugal field then causes a migration of the randomly distributed biomolecules through the solvent radially outwards from the centre of rotation. The solvent that has been cleared of particles and the solvent still containing the sedimenting material form a sharp boundary. The movement of the boundary with time is a measure of the rate of sedimentation of the biomolecules. The sedimentation coefficient depends directly on the mass of the biological particle. The concentration distribution is dependent on the buoyant molecular mass. The movement of biomolecules in a centrifugal field can be determined and a plot of the natural logarithm of the solute concentration versus the squared radial distance from the centre of rotation (ln c vs. r^2) yields a straight line with a slope proportional to the monomer molecular mass. Alternatively, the relative molecular mass of a biological macromolecule can be determined by the band sedimentation technique. In this case, the sample is layered on top of a denser solvent. During centrifugation, the solvent forms its own density gradient and the migration of the particle band is followed in the analytical cell. Molecular mass determination by analytical ultracentrifugation is applicable to values from a few hundred to several millions. It is therefore used for the analysis of small carbohydrates, proteins, nucleic acid macromolecules, viruses and subcellular particles such as mitochondria.

3.5.3 Sedimentation coefficient

Biochemical studies over the last few decades have clearly demonstrated that biological macromolecules do not perform their biochemical and physiological functions in isolation. Many proteins have been shown to be multifunctional and their activity is regulated by complex interactions within homogeneous and heterogeneous complexes. Co-operative kinetics and the influence of micro-domains have been recognised to play a major role in the regulation of biochemical processes. Since conformational changes in biological macromolecules may cause differences in their sedimentation rates, analytical ultracentrifugation represents an ideal experimental tool for the determination of such structural modifications. For example, a macro-molecule that changes its conformation into a more compact structure decreases its frictional resistance in the solvent. In contrast, the frictional resistance increases when a molecular assembly becomes more disorganised. The binding of ligands (such as inhibitors, activators or substrates) or a change in temperature or buffering conditions may induce conformational changes in subunits of biomolecules that in turn can result in major changes in the supramolecular structure of complexes. Such modifi-cations can be determined by distinct differences in the sedimentation velocity of the molecular species. Sedimentation equilibrium experiments can be used to determine the relative size of individual subunits participating in complex formation, the stoi-chiometry and size of a complex assembly under different physiological conditions and the strength of interactions between subunits.

When a new protein species is identified that appears to exist under native condi-tions in a large complex, several biochemical techniques are available to evaluate the oligomeric status of such a macromolecule. Gel filtration analysis, blot overlay assays, affinity chromatography, differential immuno precipitation and chemical cross-linking are typical examples of such techniques. With respect to centrifugation, sedimentation analysis using a density gradient is an ideal method to support such biochemical data. For the initial determination of the size of a complex, the sedimen-tation of known marker proteins is compared to the novel protein complex. Biological particles with a different molecular mass, shape or size migrate with different veloci-ties in a centrifugal field (Section 3.1). As can be seen in equation 3.7, the sedimenta-tion coefficient has dimensions of seconds. The value of Svedberg units ($S = 10^{-13}$ s) lies for many macromolecules of biochemical interest typically between 1 and 20, and for larger biological particles such as ribosomes, microsomes and mitochondria between 80 and several thousand. The prototype of a soluble protein, serum albumin of apparent 66 kDa, has a sedimentation coefficient of 4.5 S. Figure 3.9 illustrates the sedimentation analysis of the dystrophin–glycoprotein complex (DGC) from skeletal muscle fibres. The size of this complex was estimated to be approximately 18 S by comparing its migration to that of the standards β-galactosidase (16S) and thyro-globulin (19 S). When the membrane cytoskeletal element dystrophin was first identi-fied, it was shown to bind to a lectin column, although it does not exhibit any carbohydrate chains. This suggested that dystrophin might exist in a complex with surface glycoproteins. Sedimentation analysis confirmed the existence of such a dystrophin–glycoprotein complex and centrifugation following various biochemical modifications of the protein assembly led to a detailed understanding of its compos-ition. Alkaline extraction, acid treatment or incubation with different types of deter-gent causes the differential disintegration of the dystrophin–glycoprotein complex. It is now known that dystrophin is tightly associated with at least 10 different surface proteins that are involved in membrane stabilisation, receptor anchoring and signal

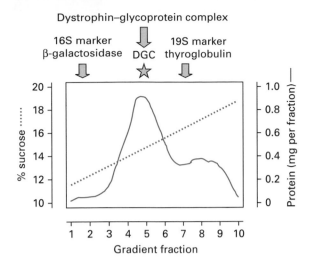

Fig. 3.9 Sedimentation analysis of a supramolecular protein complex. Shown is the sedimentation of the dystrophin–glycoprotein complex (DGC). Its size was estimated to be approximately 18 S by comparing its migration to that of the standards β-galactosidase (16 S) and thyroglobulin (19 S). Since the sedimentation coefficients of biological macromolecules are relatively small, they are expressed as Svedberg units, S, whereby 1 Svedberg unit equals 10^{-13} s.

transduction processes. The successful characterisation of the dystrophin–glycoprotein complex by sedimentation analysis is an excellent example of how centrifugation methodology can be exploited to gain biochemical knowledge of a newly discovered protein quickly.

3.6 SUGGESTIONS FOR FURTHER READING

Burgess, N. K., Stanley, A. M. and Fleming, K. G. (2008). Determination of membrane protein molecular weights and association equilibrium constants using sedimentation equilibrium and sedimentation velocity. *Methods in Cell Biology*, **84**, 181–211. (Focuses on the centrifugal analysis of interactions between integral membrane proteins.)

Cole, J. L., Lary, J. W., Moody, T. P. and Laue, T. M. (2008). Analytical ultracentrifugation: sedimentation velocity and sedimentation equilibrium. *Methods in Cell Biology*, **84**, 143–179. (Provides an excellent synopsis of the applicability of ultracentrifugation to the characterisation of macromolecular behaviour in complex solution.)

Cox, B. and Emili, A. (2006). Tissue subcellular fractionation and protein extraction for use in mass-spectrometry-based proteomics. *Nature Protocols*, **1**, 1872–1878. (Outlines differential centrifugation protocols for the isolation of the nuclear, cytosolic, mitochondrial and microsomal fraction.)

Girard, M., Allaire, P. D., Blondeau, F. and McPherson, P. S (2005). Isolation of clathrin-coated vesicles by differential and density gradient centrifugation. *Current Protocols in Cell Biology*, Chapter 3, Unit 3.13. (Describes a typical subcellular fractionation protocol used in modern biochemical applications.)

Klassen, R., Fricke, J., Pfeiffer, A. and Meinhardt, F. (2008). A modified DNA isolation protocol for obtaining pure RT-PCR grade RNA. *Biotechnology Letters*, **30**, 1041–1044. (Describes typical centrifugation protocol used for the isolation of DNA and RNA molecules.)

4 Microscopy

S. W. PADDOCK

4.1 INTRODUCTION

Biochemical analysis is frequently accompanied by microscopic examination of tissue, cell or organelle preparations. Such examinations are used in many different applications, for example: to evaluate the integrity of samples during an experiment; to map the fine details of the spatial distribution of macromolecules within cells; to directly measure biochemical events within living tissues.

There are two fundamentally different types of microscope: the light microscope and the electron microscope (Fig. 4.1). Light microscopes use a series of glass lenses to focus light in order to form an image whereas electron microscopes use electromagnetic lenses to focus a beam of electrons. Light microscopes are able to magnify to a maximum of approximately 1500 times whereas electron microscopes are capable of magnifying to a maximum of approximately 200 000 times.

Magnification is not the best measure of a microscope, however. Rather, resolution, the ability to distinguish between two closely spaced points in a specimen, is a much more reliable estimate of a microscope's utility. Standard light microscopes have a lateral resolution limit of about 0.5 micrometers (μm) for routine analysis. In contrast, electron microscopes have a lateral resolution of up to 1 nanometer (nm). Both living and dead specimens are viewed with a light microscope, and often in real colour, whereas only dead ones are viewed with an electron microscope, and never in real colour. Computer enhancement methods have improved upon the 0.5 μm resolution limit of the light microscope down to 20 nm resolution in some

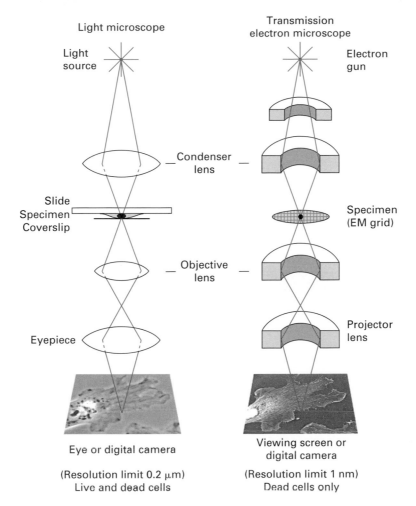

Light microscope

Transmission
electron microscope

Light
source

Electron
gun

Condenser
lens

Slide
Specimen
Coverslip

Specimen
(EM grid)

Objective
lens

Projector
lens

Eyepiece

Eye or digital camera

(Resolution limit 0.2 μm)
Live and dead cells

Viewing screen or
digital camera

(Resolution limit 1 nm)
Dead cells only

Fig. 4.1 Light and electron microscopy. Schematic that compares the path of light through a compound light microscope (LM) with the path of electrons through a transmission electron microscope (TEM). Light from a lamp (LM) or a beam of electrons from an electron gun (TEM) is focussed at the specimen by a glass condenser lens (LM) or electromagnetic lenses (TEM). For the LM the specimen is mounted on a glass slide with a coverslip placed on top, and for the TEM the specimen is placed on a copper or gold electron microscope grid. The image is magnified with an objective lens, glass in the LM and electromagnetic lens in the TEM, and projected onto a detector with the eyepiece lens in the LM or the projector lens in the TEM. The detector can be the eye or a digital camera in the LM or a phosphorescent viewing screen or digital camera in the TEM. (Light and EM images courtesy of Tatyana Svitkina, University of Pennsylvania, USA.)

specialised applications, for example using total internal reflection microscopy (TIRF) (Section 4.3.5).

Applications of the microscope in biomedical research may be relatively simple and routine; for example, a quick check of the status of a preparation or of the health of cells growing in a plastic dish in tissue culture. Here, a simple bench-top light microscope is perfectly adequate. On the other hand, the application may be more involved, for example, measuring the concentration of calcium in a living embryo

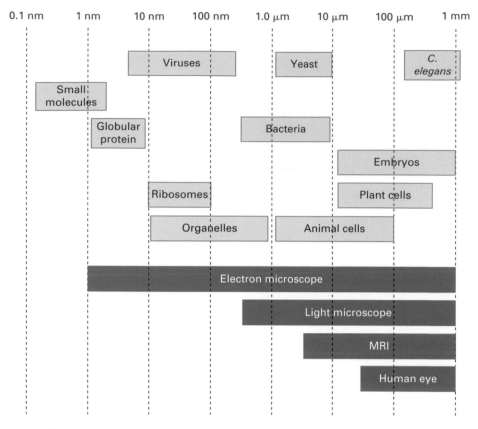

Fig. 4.2 The relative sizes of a selection of biological specimens and some of the devices used to image them. The range of resolution for each instrument is included in the dark bars at the base of the figure. MRI, magnetic resonance imaging.

over a millisecond timescale. Here a more advanced light microscope (often called an imaging system) is required.

Some microscopes are more suited to specific applications than others. There may be constraints imposed by the specimen. Images may be required from specimens of vastly different sizes and magnifications (Fig. 4.2). For example, for imaging whole animals (metres), through tissues and embryos (micrometres), and down to cells, proteins and DNA (nm). The study of living cells may require time resolution from days, for example, when imaging neuronal development or disease processes to milliseconds, for example, when imaging cell signalling events.

The field of microscopy has undergone a renaissance over the past 20 years with many technological improvements to the instruments. Most images produced by microscopes are now recorded electronically using digital imaging techniques – digital cameras, digital image acquisition software, digital printing and digital display methods. In addition, vast improvements have been made in the biological aspects of specimen preparation. These advancements on both fronts have fostered many more applications of the microscope in biomedical research.

4.2 THE LIGHT MICROSCOPE

4.2.1 Basic components of the light microscope

The simplest form of light microscope consists of a single glass lens mounted in a metal frame – a magnifying glass. Here the specimen requires very little preparation, and is usually held close to the eye in the hand. Focussing of the region of interest is achieved by moving the lens and the specimen relative to one another. The source of light is usually the Sun or ambient indoor light. The detector is the human eye. The recording device is a hand drawing or an anecdote.

Compound microscopes

All modern light microscopes are made up of more than one glass lens in combination. The major components are the condenser lens, the objective lens and the eyepiece lens, and, such instruments are therefore called compound microscopes (Fig. 4.1). Each of these components is in turn made up of combinations of lenses, which are necessary to produce magnified images with reduced artifacts and aberrations. For example, chromatic aberration occurs when different wavelengths of light are separated and pass through a lens at different angles. This results in rainbow colours around the edges of objects in the image. This problem was encountered in the early microscopes of van Leeuwenhoek and Hooke, for example. All modern lenses are now *corrected* to some degree in order to avoid this problem.

The main components of the compound light microscope include a light source that is focussed at the specimen by a condenser lens. Light that either passes through the specimen (transmitted light) or is reflected back from the specimen (reflected light) is focussed by the objective lens into the eyepiece lens. The image is either viewed directly by eye in the eyepiece or it is most often projected onto a detector, for example photographic film or, more likely, a digital camera. The images are displayed on the screen of a computer imaging system, stored in a digital format and reproduced using digital methods.

The part of the microscope that holds all of the components firmly in position is called the stand. There are two basic types of compound light microscope stand – an upright or an inverted microscope (Fig. 4.3). The light source is below the condenser lens in the upright microscope and the objectives are above the specimen stage. This is the most commonly used format for viewing specimens. The inverted microscope is engineered so that the light source and the condenser lens are above the specimen stage, and the objective lenses are beneath it. Moreover, the condenser and light source can often be swung out of the light path. This allows additional room for manipulating the specimen directly on the stage, for example, for the microinjection of macromolecules into tissue culture cells, for *in vitro* fertilisation of eggs or for viewing developing embryos over time.

The correct illumination of the specimen is critical for achieving high-quality images and photomicrographs. This is achieved using a light source. Typically light sources are mercury lamps, xenon lamps, lasers or light-emitting diodes (LEDs).

(a)   (b)

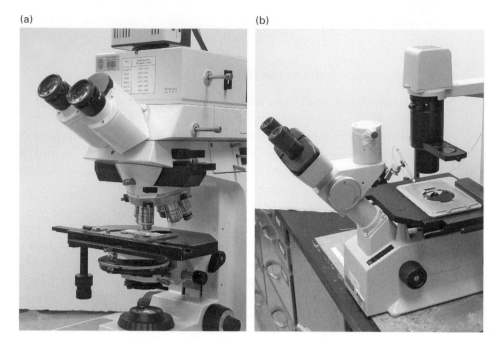

Fig. 4.3 Two basic types of compound light microscope. An upright light microscope (a) and an inverted light microscope (b). Note how there is more room available on the stage of the inverted microscope (b). This instrument is set up for microinjection with a needle holder to the left of the stage.

Light from the light source passes into the condenser lens, which is mounted beneath the microscope stage in an upright microscope (and above the stage in an inverted microscope) in a bracket that can be raised and lowered for focussing (Fig. 4.3). The condenser focusses light from the light source and illuminates the specimen with parallel beams of light. A correctly positioned condenser lens produces illumination that is uniformly bright and free from glare across the viewing area of the specimen (Koehler illumination). Condenser misalignment and an improperly adjusted condenser aperture diaphragm are major sources of poor images in the light microscope.

The specimen stage is a mechanical device that is finely engineered to hold the specimen firmly in place (Fig. 4.4). Any movement or vibration will be detrimental to the final image. The stage enables the specimen to be moved and positioned in fine and smooth increments, both horizontally and transversely, in the X and the Y directions, for locating a region of interest. The stage is moved vertically in the Z direction for focussing the specimen or for inverted microscopes, the objectives themselves are moved and the stage remains fixed. There are usually coarse and fine focussing controls for low magnification and high magnification viewing respectively. The fine focus control can be moved in increments of 1 μm or better in the best research microscopes. The specimen stage can either be moved by hand or by a stepper motor attached to the fine focus control of the microscope, and controlled by a computer.

The objective lens is responsible for producing the magnified image, and can be the most expensive component of the light microscope (Fig. 4.4). Objectives are available

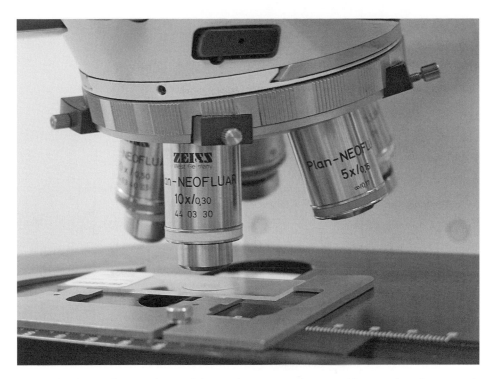

Fig. 4.4 The objective lens. A selection of objective lenses mounted on an upright research grade compound light microscope. From the inscription on the two lenses in focus they are relatively low magnification 10× and 5× of numerical aperture (NA) 0.3 and 0.16 respectively. Both lenses are Plan Neofluar, which means they are relatively well corrected. The 10× lens is directly above a specimen mounted on a slide and coverslip, and held in place on the specimen stage.

in many different varieties, and there is a wealth of information inscribed on each one. This may include the manufacturer, magnification (4×, 10×, 20×, 40×, 60×, 100×), immersion requirements (air, oil or water), coverslip thickness (usually 0.17 mm) and often more-specialised optical properties of the lens (Section 4.2.3). In addition, lens corrections for optical artifacts such as chromatic aberration and flatness of field may also be included in the lens description. For example, words such as fluorite, the least corrected (often shortened to 'fluo'), or plan apochromat, the most highly corrected (often shortened to 'plan' or 'plan apo'), may appear somewhere on the lens.

Objective lenses can either be dry (glass/air/coverslip) or immersion lenses (glass/oil or water/coverslip). As a rule of thumb, most objectives below 40× are air (dry) objectives, and those of 40× and above are immersion (oil, glycerol or water). Should the objective be designed to operate in oil it will be labelled 'OIL' or 'OEL'. Other immersion media include glycerol and water, and the lens will be marked to indicate this. Many lenses are colour-coded to a manufacturer's specifications. Dipping lenses are specially designed to work without a coverslip, and are dipped directly into water or tissue culture medium. These are used for physiological experiments.

The numerical aperture (NA) is always marked on the lens. This is a number usually between 0.04 and 1.4. The NA is a measure of the ability of a lens to collect light from the specimen. Lenses with a low NA collect less light than those with a high NA.

Table 4.1 **Resolution in optical imaging**

	xy	*z*
Standard microscope	0.5 μm	1.6 μm
Confocal/multiple photon	0.25 μm	0.7 μm
TIRF – evanescent wave	0.5 μm	0.3 μm

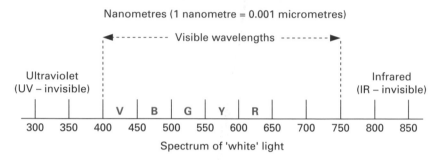

Fig. 4.5 The visible spectrum – the spectrum of white light visible to the human eye. Our eyes are able to detect colour in the visible wavelengths of the spectrum, usually in the region between 400 nm (violet) and 750 nm (red). Most modern electronic detectors are sensitive beyond the visible spectrum of the human eye.

Resolution varies inversely with NA, which implies that higher NA objectives yield the best resolution. Generally speaking the higher-power objectives have a higher NA and better resolution than the lower-power lenses with lower NAs. For example, 0.2 μm resolution can only be achieved using a 100× plan-apochromat oil immersion lens with a NA of 1.4. Should there be a choice between two lenses of the same magnification, then it is usually best to choose the one of higher NA.

The objective lens is also the part of the microscope that can most easily be damaged by mishandling. Many lenses are coated with a protective coating but even so, one scratch on the front of the lens can result in serious image degradation. Therefore, great care should be taken when handling objective lenses. Objective lenses must be cleaned using a protocol recommended by the manufacturer, and only by a qualified person. A dirty objective lens is a major source of poor images.

The resolution achieved by a lens is a measure of its ability to distinguish between two objects in the specimen. The shorter the wavelengths of illuminating light the higher the resolving power of the microscope (Fig. 4.5). The limit of resolution for a microscope that uses visible light is about 300 nm with a dry lens (in air) and 200 nm with an oil immersion lens. By using ultraviolet light (UV) as a light source the resolution can be improved to 100 nm because of the shorter wavelength of the light (200–300 nm). These limits of resolution are often difficult to achieve practically because of aberrations in the lenses and the poor optical properties of many biological specimens. The lateral resolution is usually higher than the axial resolution for any given objective lens (Table 4.1).

The eyepiece (sometimes referred to as the ocular) works in combination with the objective lens to further magnify the image, and allows it to be detected by eye or more

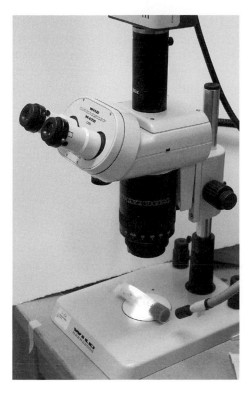

Fig. 4.6 A research-grade stereomicroscope. Note the light source is from the side, which can give a shadow effect to the specimen; in this example a vial of fruit flies. The large objective lens above the specimen can be rotated to zoom the image.

usually to project the image into a digital camera for recording purposes. Eyepieces usually magnify by 10× since an eyepiece of higher magnification merely enlarges the image with no improvement in resolution. There is an upper boundary to the useful magnification of the collection of lenses in a microscope. For each objective lens the magnification can be increased above a point where it is impossible to resolve any more detail in the specimen. Any magnification above this point is often called empty magnification. The best way to improve magnification is to use a higher magnification and higher NA objective lens. Should sufficient resolution not be achieved using the light microscope, then it will be necessary to use the electron microscope (Section 4.6).

In addition to the human eye and photographic film there are two types of electronic detectors employed on modern light microscopes. These are area detectors that actual form an image directly, for example video cameras and charge-coupled devices (CCDs). Alternatively, point detectors can be used to measure intensities in the image; for example photomultiplier tubes (PMTs) and photodiodes. Point detectors are capable of producing images in scanning microscopy (Section 4.3).

Stereomicroscopes

A second type of light microscope, the stereomicroscope, is used for the observation of the surfaces of large specimens (Fig. 4.6). The microscope is used when 3D information is required, for example for the routine observation of whole organisms, for example for screening through vials of fruit flies. Stereomicroscopes are useful for micromanipulation and dissection where the wide field of view and the ability to zoom in and out in magnification is invaluable. A wide range of objectives and eyepieces are available for different applications. The light sources can be from above, from below the specimen, encircling the specimen using a ring light or from the side giving a darkfield effect (Section 4.2.3). These different light angles serve to add contrast or shadow relief to the images.

4.2.2 The specimen

The specimen (sometimes called the sample) can be the entire organism or a dissected organ (whole mount); an aliquot collected during a biochemical protocol for a quick check of the preparation; or a small part of an organism (biopsy) or smear of blood or spermatozoa. In order to collect images from it, the specimen must be in a form that is compatible with the microscope. This is achieved using a published protocol. The end product of a protocol is a relatively thin and somewhat transparent piece of tissue mounted on a piece of glass (slide) in a mounting medium (water, tissue culture medium or glycerol) with a thin square of glass mounted on top (coverslip).

Coverslips are graded by their thickness. The thinnest ones are labelled #1, which corresponds to a thickness of approximately 0.17 mm. The coverslip side of the specimen is always placed closest to the objective lens. It is essential to use a coverslip that is optically matched to the objective lens in order to achieve optimal resolution. This is critical for high-magnification imaging because if the coverslip is too thick it will be impossible to achieve an image.

The goal of a specimen preparation protocol is to render the tissue of interest into a form for optimal study in the microscope. This usually involves placing the specimen in a suitable medium on a glass slide with a coverslip over it. Such protocols can be relatively simple or they may involve a lengthy series of many steps that take several days to complete (Table 4.2). An example of a simple protocol would be taking an aliquot of a biological preparation, for example, isolating living spermatozoa into a balanced salt solution, placing an aliquot of it onto a slide and gently placing a clean coverslip onto the top. The entire protocol would take less than a minute. The coverslip is sealed to the glass slide in some way, for example, using nail polish for dead cells or perhaps a mixture of beeswax and Vaseline for living cells. Shear forces from the movement of the coverslip over the glass slide can cause damage to the specimen or the objective lens. In order to keep cells alive on the stage of the microscope, they are usually mounted in some form of chamber, and if necessary heated.

Many specimens are too thick to be mounted directly onto a slide, and these are cut into thin sections using a device called a microtome. The tissue is usually mounted in a block of wax and cut with the knife of the microtome into thin sections (between 100 µm and 500 µm in thickness). The sections are then placed onto a glass slide, stained and sealed with mounting medium with a coverslip. Some samples are frozen, and cut on a cryostat, which is basically a microtome that can

Table 4.2 **Generalised indirect immunofluorescence protocol**

1. Fix in 1% formaldehyde for 30 min

2. Rinse in cold buffer

3. Block buffer

4. Incubate in primary antibody e.g. mouse anti-tubulin

5. Wash 4× in buffer

6. Incubate in secondary antibody e.g. fluorescein-labelled rabbit anti-mouse

7. Wash 4× in buffer

8. Incubate in anti-fade reagent e.g. Vectashield

9. Mount on slide with a coverslip

10. View using epifluorescence microscopy

keep a specimen in the frozen state, and produce frozen sections more suitable for immunolabelling (Section 4.2.3).

Prior to sectioning, the tissue is usually treated with a chemical agent called a fixative to preserve it. Popular fixatives include formaldehyde and glutaraldehyde, which act by cross-linking proteins, or alcohols, which act by precipitation. All of these fixatives are designed to maintain the structural integrity of the cell. After fixation the specimen is usually permeabilised in order to allow a stain to infiltrate the entire tissue. The amount of permeabilisation (time and severity) depends upon several factors; for example, the size of the stain or the density of the tissue. These parameters are found by trial and error for a new specimen, but are usually available in published protocols. The goal is to infiltrate the entire tissue with a uniform staining.

4.2.3 **Contrast in the light microscope**

Most cells and tissues are colourless and almost transparent, and lack contrast when viewed in a light microscope. Therefore to visualise any details of cellular components it is necessary to introduce contrast into the specimen. This is achieved either by optical means using a specific configuration of microscope components, or by staining the specimen with a dye or, more usually, using a combination of optical and staining methods. Different regions of the cell can be stained selectively with different stains.

Optical contrast

Contrast is achieved optically by introducing various elements into the light path of the microscope and using lenses and filters that change the pattern of light passing through the specimen and the optical system. This can be as simple as adding a piece of coloured glass or a neutral density filter into the illuminating light path; by changing the light intensity; or by adjusting the diameter of a condenser aperture. Usually all of these operations are adjusted until an acceptable level of contrast is achieved for imaging.

The most basic mode of the light microscope is called brightfield (bright background), which can be achieved with the minimum of optical elements. Contrast in brightfield images is usually produced by the colour of the specimen itself. Brightfield is therefore used most often to collect images from pigmented tissues or histological sections or tissue culture cells that have been stained with colourful dyes (Figs. 4.7a, 4.8b).

Several configurations of the light microscope have been introduced over the years specifically to add contrast to the final image. Darkfield illumination produces images of brightly illuminated objects on a black background (Figs. 4.7b, 4.8a). This technique has traditionally been used for viewing the outlines of objects in liquid media such as living spermatozoa, microorganisms or cells growing in tissue culture, or for a quick check of the status of a biochemical preparation. For lower magnifications, a simple darkfield setting on the condenser will be sufficient. For more critical darkfield imaging at a higher magnification, a darkfield condenser with a darkfield objective lens will be required.

Phase contrast is used for viewing unstained cells growing in tissue culture and for testing cell and organelle preparations for lysis (Fig. 4.7c,d). The method images differences in the refractive index of cellular structures. Light that passes through thicker parts of the cell is held up relative to the light that passes through thinner parts of the cytoplasm. It requires a specialised phase condenser and phase objective lenses (both labelled 'ph'). Each phase setting of the condenser lens is matched with the phase setting of the objective lens. These are usually numbered as Phase 1, Phase 2 and Phase 3, and are found on both the condenser and the objective lens.

Differential interference contrast (DIC) is a form of interference microscopy that produces images with a shadow relief (Fig. 4.7e, f). It is used for viewing unstained cells in tissue culture, eggs and embryos, and in combination with some stains. Here the overall shape and relief of the structure is viewed using DIC and a subset of the structure is stained with a coloured dye (Fig. 4.8c).

Fluorescence microscopy is currently the most widely used contrast technique since it gives superior signal-to-noise ratios (typically white on a black background) for many applications (Fig. 4.9). The most commonly used fluorescence technique is called epifluorescence light microscopy, where 'epi' simply means 'from above'. Here the light source comes from above the sample, and the objective lens acts as both condenser and objective lens (Fig. 4.10). Fluorescence is popular because of the ability to achieve highly specific labelling of cellular compartments. The images usually consist of distinct regions of fluorescence (white) over large regions of no fluorescence (black), which gives excellent signal-to-noise ratios.

The light source is usually a high-pressure mercury or xenon vapour lamp, and more recently lasers and LED sources, which emit from the UV into the red wavelengths (Fig. 4.5). A specific wavelength of light is used to excite a fluorescent

(a)

(b)

(c)

(d)

(e)

(f)

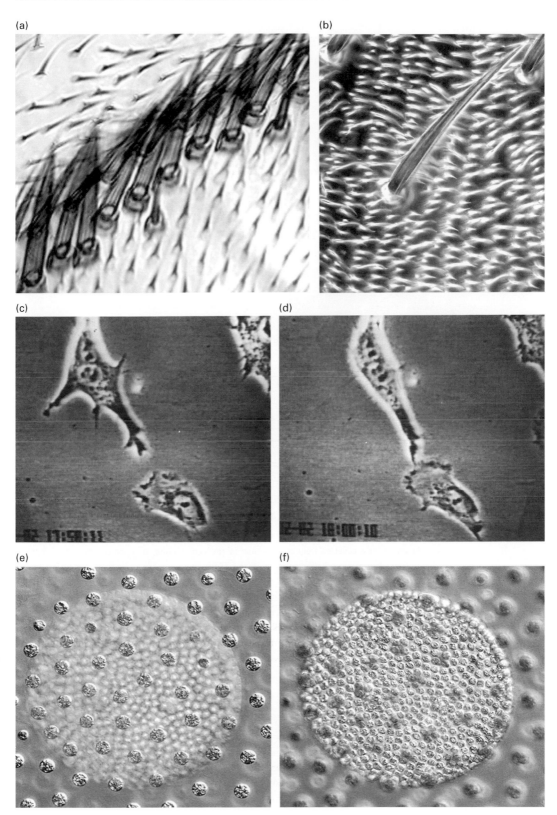

Fig. 4.7 Contrast methods in the light microscope. (a) and (b) A comparison of brightfield (a) and darkfield images (b). Here the sensory bristles on the surface of the fly appear dark on a white background in the bright

(a)

(b)

(c)

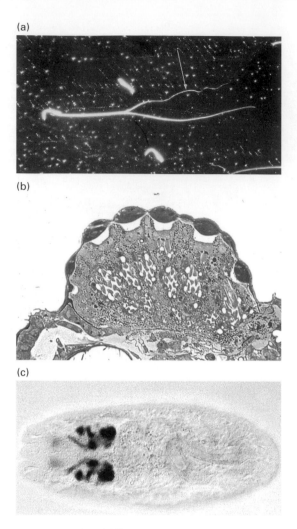

Fig. 4.8 Examples of different preparations in the light microscope. (a) Darkfield image of rat sperm preparation. An aliquot was collected from an experimental protocol in order to assess the amount of damage incurred during sonication of a population of spermatozoa. Many sperm heads can be seen in the preparation, and the fibres of the tail are starting to fray (arrowed). (b) A brightfield image of total protein staining on a section of a fly eye cut on a microtome, and stained with Coomassie blue. (c) DIC image of a stained *Drosophila* embryo – the DIC image shows the outline of the embryo with darker regions of neuronal staining. The DIC image of the whole embryo provides structural landmarks for placing the specific neuronal staining in context of the anatomy.

Caption for Fig. 4.7 (*cont.*)
field image (a) and white on a black background in a dark field image (b). The dark colour in the larger bristles in (a) is produced by pigment. (c) and (d) Phase contrast view of cells growing in tissue culture. Two images extracted from a time-lapse video sequence (time between each frame is 5 min). The sequence shows the movement of a mouse 3T3 fibrosarcoma cell and a chick heart fibroblast. Note the bright 'phase halo' around the cells. (e) and (f) Differential interference contrast (DIC) image of two focal planes of the multicellular alga *Volvox*. (Images (e) and (f) courtesy of Michael Davidson, Florida State University, USA.)

(a)   (b)   (c)

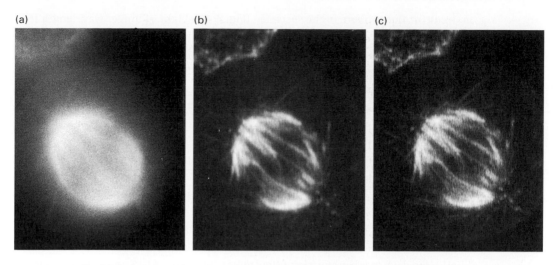

Fig. 4.9 Fluorescence microscopy. Comparison of epifluorescence and confocal fluorescence imaging of a mitotic spindle labelled using indirect immunofluorescence labelling with anti-tubulin (primary antibody) and a fluorescently labelled secondary antibody. The specimen was imaged using (a) conventional epifluorescence light microscopy or (b) and (c) using laser scanning confocal microscopy. Note the improved resolution of microtubules in the two confocal images (b) and (c) as compared with the conventional image (a). (b) and (c) represent two different resolution settings of the confocal microscope. Image (b) was collected with the pinhole set to a wider aperture than (c). (Images kindly provided by Brad Amos, University of Cambridge, UK.)

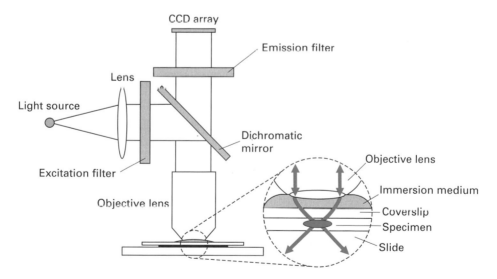

Fig. 4.10 Epifluorescence microscopy. Light from a xenon or mercury arc lamp (Light source) passes through a lens and the excitation filter and reflects off the dichromatic mirror into the objective lens. The objective lens focusses the light at the specimen via the immersion medium (usually immersion oil) and the glass coverslip (see insert). Any light resulting from the fluorescence excitation in the specimen passes back through the objective lens, and since it is of longer wavelength than the excitation light, it passes through the dichromatic mirror. The emission filter only allows light of the specific emission wavelength of the fluorochrome of interest to pass through to the CCD array, where an image is formed.

molecule or fluorophore in the specimen (Fig. 4.10). Light of longer wavelength from the excitation of the fluorophore is then imaged. This is achieved in the fluorescence microscope using combinations of filters that are specific for the excitation and emission characteristics of the fluorophore of interest. There are usually three main filters: an excitation, a dichromatic mirror (often called a dichroic) and a barrier filter, mounted in a single housing above the objective lens. For example, the commonly used fluorophore fluorescein is optimally excited at a wavelength of 488 nm, and emits maximally at 518 nm (Table 4.3).

A set of glass filters for viewing fluorescein requires that all wavelengths of light from the lamp be blocked except for the 488 nm light. A filter is available that allows a maximum amount of 488 nm light to pass through it (the exciter filter). The 488 nm light is then directed to the specimen via the dichromatic mirror. Any fluorescein label in the specimen is excited by the 488 nm light, and the resulting 518 nm light that returns from the specimen passes through both the dichromatic mirror and the barrier filter to the detector. The emission filters only allow light of 518 nm to pass through to the detector, and ensure that only the signal emitted from the fluorochrome of interest reaches it.

Chromatic mirrors and filters can be designed to filter two or three specific wavelengths for imaging specimens labelled with two or more fluorochromes (multiple labelling). The fluorescence emitted from the specimen is often too low to be detected by the human eye or it may be out of the wavelength range of detection of the eye, for example, in the far-red wavelengths (Fig. 4.6). A sensitive digital camera easily detects such signals; for example a CCD or a PMT.

Specimen stains

Contrast can be introduced into the specimen using one or more coloured dyes or stains. These can be non-specific stains, for example, a general protein stain such as Coomassie blue (Fig. 4.8) or a stain that specifically labels an organelle for example, the nucleus, mitochondria etc. Combinations of such dyes may be used to stain different organelles in contrasting colours. Many of these histological stains are usually observed using brightfield imaging. Other light microscopy techniques may also be employed in order to view the entire tissue along with the stained tissue. For example, one can use DIC to view the entire morphology of an embryo and a coloured stain to image the spatial distribution of the protein of interest within the embryo (Fig. 4.8).

More specific dyes are usually used in conjunction with fluorescence microscopy. Immunofluorescence microscopy is used to map the spatial distribution of macro-molecules in cells and tissues. The method takes advantage of the highly specific binding of antibodies to proteins. Antibodies are raised to the protein of interest and labelled with a fluorescent probe. This probe is then used to label the protein of interest in the cell and can be imaged using fluorescence microscopy. In practice, cells are usually labelled using indirect immunofluorescence. Here the antibody to the protein of interest (primary antibody) is further labelled with a second antibody carrying the fluorescent tag (secondary antibody). Such a protocol gives a higher fluorescent signal than using a single fluorescently labelled antibody (Table 4.2).

Table 4.3 **Table of fluorophores**

Dye	Excitation max. (nm)	Emission max. (nm)
Commonly used fluorophores		
Fluorescein (FITC)	496	518
Bodipy	503	511
CY3	554	568
Tetramethylrhodamine	554	576
Lissamine rhodamine	572	590
Texas red	592	610
CY5	652	672
Nuclear dyes		
Hoechst 33342	346	460
DAPI	359	461
Acridine orange	502	526
Propidium iodide	536	617
TOTO3	642	661
Ethidium bromide	510	595
Feulgen	570	625
Calcium indicators		
Fluo-3	506	526
Calcium green	506	533
Reporter molecules		
CFP (cyan fluorescent protein)	443/445	475/503
GFP (green fluorescent protein)	395/489	509
YFP (yellow fluorescent protein)	514	527
DsRed	558	583
Mitochondria		
JC-1	514	529
Rhodamine 123	507	529

Additional methods are available for amplifying the fluorescence signal in the specimen, for example using the tyramide amplification method or at the microscope, for example by using a more sensitive detector.

A related technique, fluorescence *in situ* hybridisation (FISH), employs the specificity of fluorescently labelled DNA or RNA sequences. The nucleic acid probes are hybridised to chromosomes, nuclei or cellular preparations. Regions that bind the probe are imaged using fluorescence microscopy. Many different probes can be labelled with different fluorochromes in the same preparation. Multiple-colour FISH is used extensively for clinical diagnoses of inherited genetic diseases. This technique has been applied to rapid screening of chromosomal and nuclear abnormalities in inherited diseases, for example, Down's syndrome.

There are many different types of fluorescent molecules that can be attached to antibodies, DNA or RNA probes for fluorescence analysis (Table 4.3). All of these reagents including primary antibodies are available commercially or often from the laboratories that produced them. An active area of development is the production of the brightest fluorescent probes that are excited by the narrowest wavelength band and that are not damaged by light excitation (photobleaching). Traditional examples of such fluorescent probes include fluorescein, rhodamine, the Alexa range of dyes and the cyanine dyes. A recent addition to the extensive list of probes for imaging is the quantum dot. Quantum dots do not fluoresce per se but they rather are nanocrystals of different sizes that glow in different colours in laser light. The colours depend on the size of the dots, and they have the advantage that they are not photobleached.

4.3 OPTICAL SECTIONING

Many images collected from relatively thick specimens produced using epifluorescence microscopy are not very clear. This is because the image is made up of the optical plane of interest together with contributions from fluorescence above and below the focal plane of interest. Since the conventional epifluorescence microscope collects all of the information from the specimen, it is often referred to as a wide field microscope. The 'out-of-focus fluorescence' can be removed using a variety of optical and electronic techniques to produce optical sections (Fig. 4.9).

The term *optical section* refers to a microscope's ability to produce sharper images of specimens than those produced using a standard wide field epifluorescence microscope by removing the contribution from out-of-focus light to the image, and in most cases, without resorting to physically sectioning the tissue. Such methods have revolutionised the ability to collect images from thick and fluorescently labelled specimens such as eggs, embryos and tissues. Optical sections can also be produced using high-resolution DIC optics (Fig. 4.7e, f), micro computerised tomography (CT) scanning or optical projection tomography. However, currently by far the most prevalent method is using some form of confocal or associated microscopical approach.

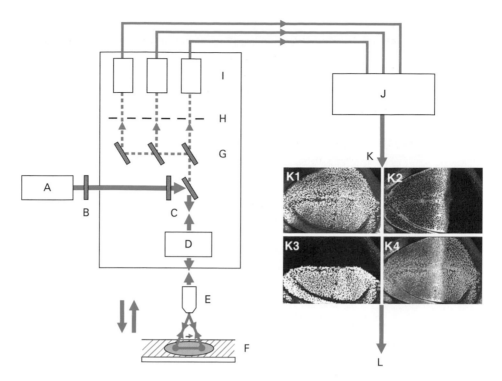

Fig. 4.11 Information flow in a generic LSCM. Light from the laser (A) passes through a neutral density filter (B) and an exciter filter (C) on its way to the scanning unit (D). The scanning unit produces a scanned beam at the back focal plane of the objective lens (E) which focusses the light at the specimen (F). The specimen is scanned in the X and the Y directions in a raster pattern and in the Z direction by fine focussing (arrows). Any fluorescence from the specimen passes back through the objective lens and the scanning unit and is directed via dichromatic mirrors (G) to three pinholes (H). The pinholes act as spatial filters to block any light from above or below the plane of focus in the specimen. The point of light in the specimen is confocal with the pinhole aperture. This means that only distinct regions of the specimen are sampled. Light that passes through the pinholes strikes the PMT detectors (I) and the signal from the PMT is built into an image in the computer (J). The image is displayed on the computer screen (K) often as three greyscale images (K1, K2 and K3) together with a merged colour image of the three greyscale images (K4 and Fig. 4.13a, see colour section). The computer synchronises the scanning mirrors with the build-up of the image in the computer framestore. The computer also controls a variety of peripheral devices. For example, the computer controls and correlates movement of a stepper motor connected to the fine focus of the microscope with image acquisition in order to produce a Z-series. Furthermore the computer controls the area of the specimen to be scanned by the scanning unit so that zooming is easily achieved by scanning a smaller region of the specimen. In this way, a range of magnifications is imparted to a single objective lens so that the specimen does not have to be moved when changing magnification. Images are written to the hard disk of the computer or exported to various devices for viewing, hardcopy production or archiving (L).

4.3.1 Laser scanning confocal microscopes (LSCM)

Optical sections are produced in the laser scanning confocal microscope by scanning the specimen point by point with a laser beam focussed in the specimen, and using a spatial filter, usually a pinhole (or a slit), to remove unwanted fluorescence from above and below the focal plane of interest (Fig. 4.11). The power of the confocal

(a)   (b)

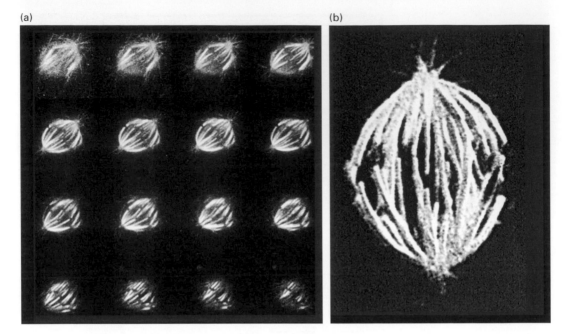

Fig. 4.12 Computer 3D reconstruction of confocal images. (a) Sixteen serial optical sections collected at 0.3 μm intervals through a mitotic spindle of a PtK1 cell stained with anti-tubulin and a second rhodamine-labelled antibody. Using the Z-series macro program a preset number of frames can be summed, and the images transferred into a file on the hard disk. The stepper motor moves the fine focus control of the microscope by a preset increment. (b) Three-dimensional reconstruction of the data set produced using computer 3D reconstruction software. Such software can be used to view the data set from any specified angle or to produce movies of the structure rotating in 3D.

approach lies in the ability to image structures at discrete levels within an intact biological specimen.

There are two major advantages of using the LSCM in preference to conventional epifluorescence light microscopy. Glare from out-of-focus structures in the specimen is reduced and resolution is increased both laterally in the X and the Y directions (0.14 μm) and axially in the Z direction (0.23 μm). Image quality of some relatively thin specimens, for example, chromosome spreads and the leading lamellipodium of cells growing in tissue culture (<0.2 μm thick) is not dramatically improved by the LSCM whereas thicker specimens such as fluorescently labelled multicellular embryos can only be imaged using the LSCM. For successful confocal imaging, a minimum number of photons should be used to efficiently excite each fluorescent probe labelling the specimen, and as many of the emitted photons from the fluorochromes as possible should make it through the light path of the instrument to the detector.

The LSCM has found many different applications in biomedical imaging. Some of these applications have been made possible by the ability of the instrument to produce a series of optical sections at discrete steps through the specimen (Fig. 4.12). This Z series of optical sections collected with a confocal microscope are all in register with each other, and can be merged together to form a single projection of the image (Z **projection**) or a 3D representation of the image (3D reconstruction).

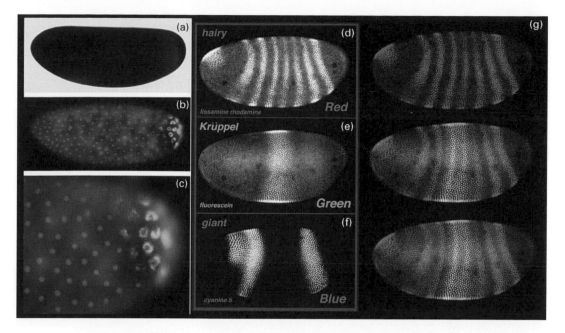

Fig. 4.13 Optical sectioning. Optical sections produced using laser scanning confocal microscopy. Comparison of alkaline phosphatase (a) and tyramide-amplified detection of mRNAs (b,c). Staining patterns obtained using DIG-labelled antisense probes directed against the CG14217 mRNAs, through conventional AP-based detection (a) or tyramide signal amplification (b), using tyramide–Alexa Fluor 488 (green fluorescence). Close-up images of tyramide-amplified samples are also shown (c). In (b) and (c), nuclei were labelled in red with propidium iodide. (d, e, f, g) Triple-labelled *Drosophila* embryo at the cellular blastoderm stage. The images were produced using an air-cooled 25 mW krypton argon laser which has three major lines at 488 nm (blue), 568 nm (yellow) and 647 nm (red). The three fluorochromes used were fluorescein (exc. 496 nm; em. 518 nm), lissamine rhodamine (exc. 572 nm; em. 590 nm) and cyanine 5 (exc. 649 nm; em. 666 nm). The images were collected simultaneously as single optical sections into the red, the green and the blue channels respectively, and merged as a three-colour (red/green/blue) image (Fig. 4.11). The image shows the expression of three genes: *hairy* (in red), *Krüppel* (in green) and *giant* (in blue). Regions of overlap of gene expression appear as an additive colour in the image, for example, the two yellow stripes of *hairy* expression in the *Krüppel* domain (g). (Images (a), (b) and (c) were kindly provided by Henry Krause, University of Toronto, Canada.) (See also colour plate.)

Multiple-label images can be collected from a specimen labelled with more than one fluorescent probe using multiple laser light sources for excitation (Fig. 4.13, see also colour section). Since all of the images collected at different excitation wavelengths are in register it is relatively easy to combine them into a single multicoloured image. Here any overlap of staining is viewed as an additive colour change. Most confocal microscopes are able to routinely image three or four different wavelengths simultaneously.

The scanning speed of most laser scanning systems is around one full frame per second. This is designed for collecting images from fixed and brightly labelled fluorescent specimens. Such scan speeds are not optimal for living specimens, and laser scanning instruments are available that scan at faster rates for more optimal live cell imaging. In addition to point scanning, swept field scanning rapidly moves a μm-thin beam of light horizontally and vertically through the specimen.

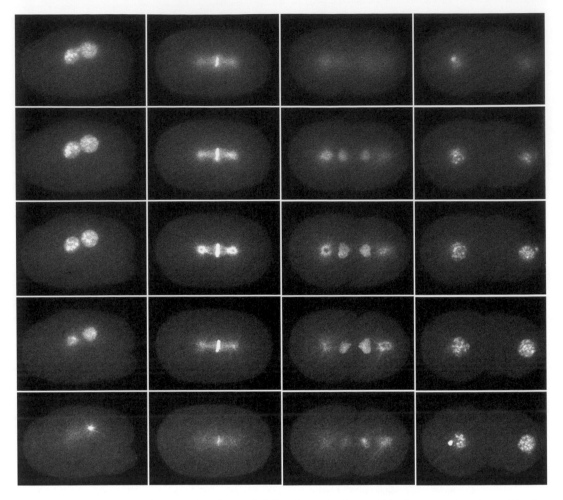

Fig. 4.14 Time-lapse imaging of *Caenorhabditis elegans* development. *Z*-series were collected every 90 s of a developing *C. elegans* embryo genetically labelled with GFP-histone (nuclear material) and GFP-alpha tubulin (microtubules – cytoskeleton) and imaged with a spinning disk confocal microscope. Each column consists of six optical sections collected 2 μm apart, and the columns are separated by 90 s increments of time. (Image kindly provided by Dr Kevin O'Connell, National Institutes of Health, USA.)

4.3.2 Spinning disk confocal microscopes

The spinning disk confocal microscope employs a different scanning system from the LSCM. Rather than scanning the specimen with a single beam, multiple beams scan the specimen simultaneously, and optical sections are viewed in real time. Modern spinning disk microscopes have been improved significantly by the addition of laser light sources and high-quality CCD detectors to the instrument. Spinning disk systems are generally used in experiments where high-resolution images are collected at a fast rate (high spatial and temporal resolution), and are used to follow the dynamics of fluorescently labelled proteins in living cells (Fig. 4.14).

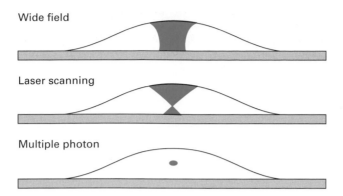

Wide field

Laser scanning

Multiple photon

Fig. 4.15 Illumination in a wide field, a confocal and a multiple photon microscope. The diagram shows a schematic of a side view of a fluorescently labelled cell on a coverslip. The shaded green areas in each cell represent the volume of fluorescent excitation produced by each of the different microscopes in the cell. Conventional epifluorescence microscopy illuminates throughout the cell. In the LSCM fluorescence illumination is throughout the cell but the pinhole in front of the detector excludes the out-of-focus light from the image. In the multiple photon microscope, excitation only occurs at the point of focus where the light flux is high enough.

4.3.3   **Multiple photon microscopes**

The multiple photon microscope has evolved from the confocal microscope. In fact, many of the instruments use the same scanning system as the LSCM. The difference is that the light source is a high-energy pulsed laser with tunable wavelengths, and the fluorochromes are excited by multiple rather than single photons. Optical sections are produced simply by focussing the laser beam in the specimen since multiple photon excitation of a fluorophore only occurs where energy levels are high enough – statistically confined to the point of focus of the objective lens (Fig. 4.15).

Since red light is used in multiple photon microscopes, optical sections can be collected from deeper within the specimen than those collected with the LSCM. Multiple photon imaging is generally chosen for imaging fluorescently labelled living cells because red light is less damaging to living cells than the shorter wavelengths usually employed by confocal microscopes. In addition, since the excitation of the fluorophore is restricted to the point of focus in the specimen, there is less chance of over exciting (photobleaching) the fluorescent probe and causing photodamage to the specimen itself (Fig. 4.15).

4.3.4   **Deconvolution**

Optical sections can be produced using an image processing method called deconvolution to remove the out-of-focus information from the digital image. Such images are computed from conventional wide field microscope images. There are two basic types of deconvolution algorithm: deblurring and restoration. The approach relies upon knowledge of the point spread function of the imaging system. This is usually

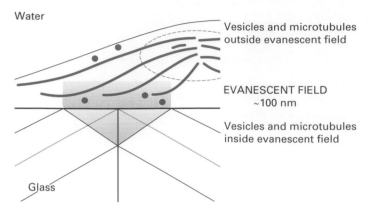

Fig. 4.16 Total internal reflection microscopy (TIRF). A 100-nm thick region of excitation is produced at the glass–water interface when illumination conditions are right for internal reflection. In this example only those vesicles and microtubules within the evanescent field will contribute to the fluorescence image at 100 nm Z-resolution.

measured by imaging a point source, for example, a small sub-resolution fluorescent bead (0.1μm), and imaging how the point is spread out in the microscope. Since it is assumed that the real image of the bead should be a point, it is possible to calculate the amount of distortion in the image of the bead imposed by the imaging system. The actual image of the point can then be restored using a mathematical function, which can be applied to any subsequent images collected under identical settings of the microscope.

Early versions of the deconvolution method were relatively slow; for example, it could take some algorithms in the order of hours to compute a single optical section. Deconvolution is now much faster using today's fast computers and improved software, and the method compares favourably with the confocal approach for producing optical sections. Deconvolution is practical for multiple-label imaging of both fixed and living cells, and excels over the scanning methods for imaging relatively dim and thin specimens, for example yeast cells. The method can also be used to remove additional background from images that were collected with the LSCM, the spinning disk microscope or a multiple photon microscope.

4.3.5 **Total internal reflection microscopy**

Another area of active research is in the development of single molecule detection techniques. For example total internal reflection microscopy (TIRF) uses the properties of an evanescent wave close to the interface of two media (Fig. 4.16), for example, the region between the specimen and the glass coverslip. The technique relies on the fact that the intensity of the evanescent field falls off rapidly so that the excitation of any fluorophore is confined to a region of just 100 nm above the glass interface. This is thinner than the optical section thickness achieved using confocal methods and allows the imaging of single molecules at the interface.

4.4 IMAGING LIVING CELLS AND TISSUES

There are two basically different approaches to imaging biochemical events over time. One strategy is to collect images from a series of fixed and stained tissues at different developmental ages. Each animal represents a single time point in the experiment. Alternatively, the same tissue can be imaged in the living state. Here the events of interest are captured directly. The second approach, imaging living cells and tissues, is technically more challenging than the first approach.

4.4.1 Avoidance of artifacts

The only way to eliminate artifacts from specimen preparation is to view the specimen in the living state. Many living specimens are sensitive to light, and especially those labelled with fluorescent dyes. This is because the excitation of fluorophores can release cytotoxic free radicals into the cell. Moreover, some wavelengths are more deleterious than others. Generally, the shorter wavelengths are more harmful than the longer ones and near-infrared light rather than ultraviolet light is preferred for imaging (Fig. 4.5). The levels of light used for imaging must not compromise the cells. This is achieved using extremely low levels of light, using relatively bright fluorescent dyes and extremely sensitive photodetectors. Moreover, the viability of cells may also depend upon the cellular compartment that has been labelled with the fluorochrome. For example, imaging the nucleus with a dye that is excited with a short wavelength will cause more cellular damage than imaging in the cytoplasm with a dye that is excited in the far red.

Great care has to be observed in order to maintain the tissue in the living state on the microscope stage. A live cell chamber is usually required for mounting the specimen on the microscope stage. This is basically a modified slide and coverslip arrangement that allows access to the specimen by the objective and condenser lenses. It also supports the cells in a constant environment, and depending on the cell type of interest, the chamber may have to provide a constant temperature, humidity, pH, carbon dioxide and/or oxygen levels. Many chambers have the facility for introducing fluids or perfusing the preparation with drugs for experimental treatments.

4.4.2 Time-lapse imaging

Time-lapse imaging continues to be used for the study of cellular dynamics. Here images are collected at predetermined time intervals (Fig. 4.14). Usually a shutter arrangement is placed in the light path so that the shutter is only open when an image is collected in order to reduce the amount of light energy impacting the cells. When the images are played back in real time, a movie of the process of interest is produced, albeit speeded up from real time. Time-lapse is used to study cell behaviour in tissues and embryos and the dynamics of macromolecules within single cells. The event of interest and also the amount of light energy absorbed and tolerated by the cells govern the time interval used. For example, a cell in tissue culture moves relatively slowly and a time interval of 30 s between images might be used. Stability of the specimen and of the microscope is extremely important for successful time-lapse imaging. For example, the focus should not drift during the experiment.

Phase contrast was the traditional choice for imaging cell movement and behaviour of cells growing in tissue culture. DIC or fluorescence microscopy is generally chosen for imaging the development of eggs and embryos. Computer imaging methods can be used in conjunction with DIC to improve resolution. Here a background image is subtracted from each time-lapse frame and the contrast of the images is enhanced electronically. In this way microtubules assembled *in vitro* from tubulin in the presence of microtubule associated proteins can be visualised on glass. These images are below the resolution of the light microscope. Such preparations have formed the basis of motility assays for motor proteins, for example kinesin and dynein.

4.4.3 Fluorescent stains of living cells

Relatively few cells possess any inherent fluorescence (autofluorescence) although some endogenous molecules are fluorescent and can be used for imaging, for example, NAD(P)H. Relatively small fluorescent molecules are loaded into living cells using many different methods including diffusion, microinjection, bead loading or electroporation. Relatively larger fluorescently labelled proteins are usually injected into cells, and after time they are incorporated into the general protein pool of the cell for imaging.

Many reporter molecules are now available for recording the expression of specific genes in living cells using fluorescence microscopy including viewing whole transgenic animals using fluorescence stereomicroscopes (Table 4.3). The green fluorescent protein (GFP) is a very convenient reporter of gene expression because it is directly visible in the living cell using epifluorescence light microscopy with standard filter sets. The GFP gene can be linked to another gene of interest so that its expression is accompanied by GFP fluorescence in the living cell. No fixation, substrates or co-enzymes are required. The fluorescence of GFP is extremely bright and is not susceptible to photobleaching. Spectral variants of GFP and additional reporters such as DsRed are now available for multiple labelling of living cells. These probes have revolutionised the ability to image living cells and tissues using light microscopy (Fig. 4.17, see also colour section).

4.4.4 Multidimensional imaging

The collection of *Z*-series over time is called four-dimensional (4D) imaging where individual optical sections (*X* and *Y* dimensions) are collected at different depths in the specimen (*Z* dimension) at different times (the fourth dimension), i.e. one time and three space dimensions (Fig. 4.18). Moreover multiple wavelength images can also be collected over time. This approach has been called 5D imaging. Software is now available for the analysis and display of such 4D and 5D data sets. For example, the movement of a structure through the consecutive stacks of images can be traced, changes in volume of a structure can be measured, and the 4D data sets can be

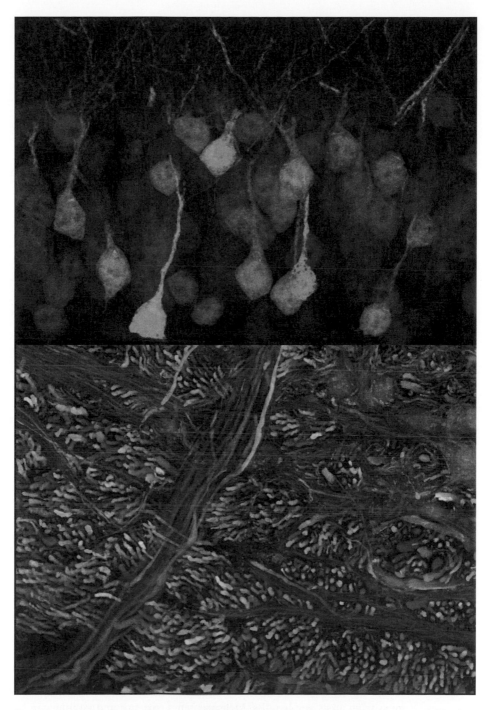

Fig. 4.17 Multiple labelling in a living mouse brain using the 'Brainbow' technique. Unique colour combinations in individual neurons are achieved by the relative levels of three or more fluorescent proteins (XFPs). The images are collected using a multi-channel laser scanning confocal microscope. Up to 90 different colours (neurons) can be distinguished using this technique. Top image, hippocampus; bottom image, brainstem. (Image courtesy of Jeff Lichtman, Harvard University, USA.) (See also colour plate.)

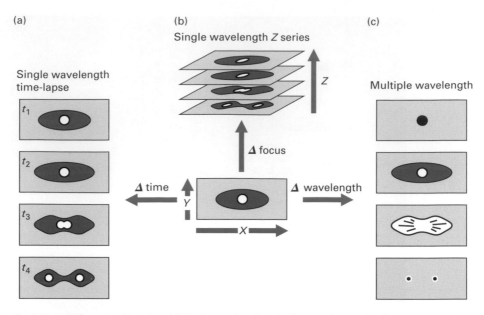

Fig. 4.18 Multidimensional imaging. (a) Single wavelength excitation over time or time-lapse X,Y imaging; (b) Z-series or X,Y,Z imaging. The combination of (a) and (b) is 4D imaging. (c) Multiple wavelength imaging. The combination of (a) and (b) and (c) is 5D imaging.

displayed as series of Z-projections or stereo movies. Multidimensional experiments can present problems for handling large amounts of data since gigabytes of information can be collected from a single 4D imaging experiment.

4.5 MEASURING CELLULAR DYNAMICS

Understanding the function of proteins within the context of the intact living cell is one of the main aims of contemporary biological research. The visualisation of specific cellular events has been greatly enhanced by modern microscopy. In addition to qualitatively viewing the images collected with a microscope, quantitative information can be gleaned from the images. The collection of meaningful measurements has been greatly facilitated by the advent of digital image processing. Subtle changes in intensity of probes of biochemical events can be detected with sensitive digital detectors. These technological advancements have allowed insight into the spatial aspects of molecular mechanisms.

Relatively simple measurements include counting features within a 2D image or measuring areas and lengths. Measurements of depth and volume can be made in 3D, 4D and 5D data sets. Images can be calibrated by collecting an image of a calibration grid at the same settings of the microscope as were used for collecting the images during the experiment. Many image processing systems allow for a calibration factor to be added into the program, and all subsequent measurements will then be comparable.

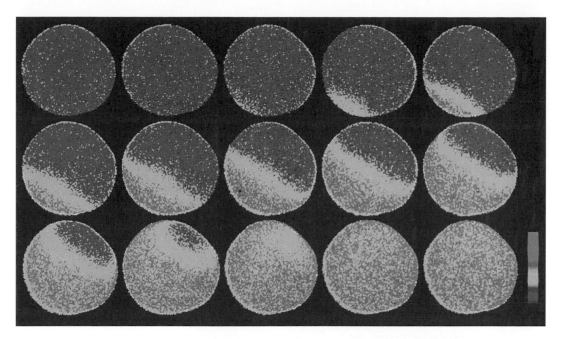

Fig. 4.19 Calcium imaging in living cells. A fertilisation-induced calcium wave in the egg of the starfish. The egg was microinjected with the calcium-sensitive fluorescent dye fluo-3 and subsequently fertilised by the addition of sperm during observation using time-lapse confocal microscopy with a 40× water immersion lens and a LSCM. An optical section located near the egg equator was collected every 4 s using the normal scan mode accumulated for two frames, and afterwards the images were corrected for offset and ratioed by linearly dividing the initial pre-fertilisation image into each successive frame of the time-lapse run. The ratioed images were then prepared as a montage and outputted with a pseudocolour look-up table in which blue regions represent low ratios and free calcium levels, and red areas depict high ratios and free calcium levels. Note that the wave sweeps through the entire ooplasm, rather than being cortically restricted. (Image kindly provided by Steve Stricker, University of New Mexico, USA.) (See also colour plate.)

The rapid development of fluorescence microscopy together with digital imaging and, above all, the development of new fluorescent probes of biological activity have brought a new level of sophistication into quantitative imaging. Most of the measurements are based on the ability to measure accurately the brightness of and the wavelength emitted from a fluorescent probe within a sample using a digital imaging system. This is also the basis of flow cytometry, which measures the brightness of each cell in a population of cells as they pass through a laser beam. Cells can be sorted into different populations using a related technique, fluorescence–activated cell sorting.

The brightness of the fluorescence from the probe can be calibrated to the amount of probe present at any given location in the cell at high resolution. For example, the concentration of calcium is measured in different regions of living embryos using calcium indicator dyes, for example fluo-3, whose fluorescence increases in proportion to the amount of free calcium in the cell (Fig. 4.19, see also colour section). Many probes have been developed for making such measurements in living tissues. Controls are a necessary part of such measurements since photobleaching and various dye artifacts during the experiment can obscure the true measurements. This can be achieved by staining the sample with two ion-sensitive dyes, and comparing their measured brightness during the experiment. These measurements are usually expressed as ratios (ratio imaging) and control for dye loading problems, photobleaching and instrument variation.

Fluorescently labelled proteins can be injected into cells where they incorporate into macromolecular structures over time. This makes the structures accessible to time-lapse imaging using fluorescence microscopy. Such methods can lead to high backgrounds, and can be difficult to interpret. In addition to optical sectioning methods several methods have been developed for avoiding high backgrounds for fluorescence measurements of biochemical events in cells.

Fluorescence recovery after photobleaching (FRAP) uses the high light flux from a laser to locally destroy fluorophores labelling the macromolecules to create a bleached zone (photobleaching). The observation and recording of the subsequent movement of undamaged fluorophores into the bleached zone gives a measure of molecular mobility. This enables biochemical analysis within the living cell. A second technique related to FRAP, photoactivation, uses a probe whose fluorescence can be induced by a flash of short wavelength (UV) light. The method depends upon 'caged' fluorescent probes that are locally activated (uncaged) by a pulse of UV light. Alternatively variants of GFP can be expressed in cells and selectively photoactivated. The activated probe is imaged using a longer wavelength of light. Here the signal-to-noise ratio of the images can be better than that for photobleaching experiments.

A third method, fluorescence speckle microscopy, was discovered as a chance observation while microinjecting fluorescently labelled proteins into living cells. Basically, when a really low concentration of fluorescently labelled protein is injected into cells, the protein of interest is not fully labelled inside the cell. When viewed in the microscope, structures inside cells that have been labelled in this way have a speckled appearance. The dark regions act as fiduciary marks for the observation of dynamics.

Fluorescence resonance energy transfer (FRET) is a fluorescence-based method that can take fluorescence microscopy past the theoretical resolution limit of the light microscope allowing the observation of protein–protein interactions *in vivo* (Fig. 4.20). FRET occurs between two fluorophores when the emission of the first one (the donor) serves as the excitation source for the second one (the acceptor). This will only occur when two fluorophore molecules are very close to one another, at a distance of 6 nm or less.

An example of a FRET experiment would be to use spectral variants of GFP (Fig. 4.20). Here the excitation of a cyan fluorescent protein (CFP)-tagged protein is used to monitor the emission of a yellow fluorescent protein (YFP)-tagged protein. YFP fluorescence will only be observed under the excitation conditions of CFP if the proteins are close together. Since this can be monitored over time, FRET can be used to measure direct binding of proteins or protein complexes.

A more complex technique, fluorescence lifetime imaging (FLIM) measures the amount of time a fluorophore is fluorescent after excitation with a 10 ns pulse of laser light. FLIM is a method used for detecting multiple fluorophores with different fluorescent lifetimes and overlapping emission spectra.

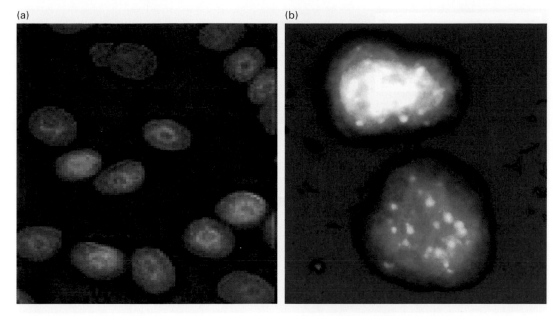

Fig. 2.4 Hoechst 33258 staining of mycoplasma in cells. (a) A Hoechst-negative stain, with the dye staining cellular DNA in the nucleus and thus showing nuclear fluorescence. (b) A Hoechst-positive stain, showing staining of mycoplasma DNA in the cytoplasm of the cells.

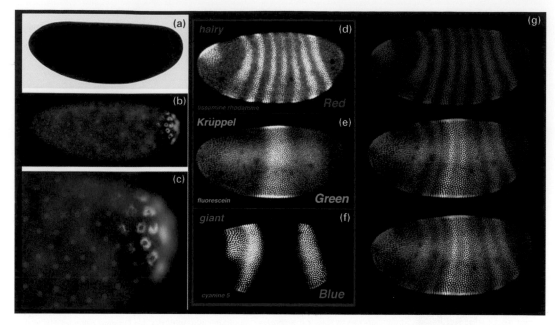

Fig. 4.13 Optical sectioning. Optical sections produced using laser scanning confocal microscopy. Comparison of alkaline phosphatase (a) and tyramide-amplified detection of mRNAs (b,c). Staining patterns obtained using DIG-labelled antisense probes directed against the CG14217 mRNAs, through conventional AP-based detection (a) or tyramide signal amplification (b), using tyramide–Alexa Fluor 488 (green fluorescence). Close-up images of tyramide-amplified samples are also shown (c). In (b) and (c), nuclei were labelled in red with propidium iodide. (d, e, f, g) Triple-labelled *Drosophila* embryo at the cellular blastoderm stage. The images were produced using an air-cooled 25 mW krypton argon laser which has three major lines at 488 nm (blue), 568 nm (yellow) and 647 nm (red). The three fluorochromes used were fluorescein (exc. 496 nm; em. 518 nm), lissamine rhodamine (exc. 572 nm; em. 590 nm) and cyanine 5 (exc. 649 nm; em. 666 nm). The images were collected simultaneously as single optical sections into the red, the green and the blue channels respectively, and merged as a three-colour (red/green/blue) image (Fig. 4.11). The image shows the expression of three genes: *hairy* (in red), *Krüppel* (in green) and *giant* (in blue). Regions of overlap of gene expression appear as an additive colour in the image, for example, the two yellow stripes of *hairy* expression in the *Krüppel* domain (g). (Images (a), (b) and (c) were kindly provided by Henry Krause, University of Toronto, Canada.)

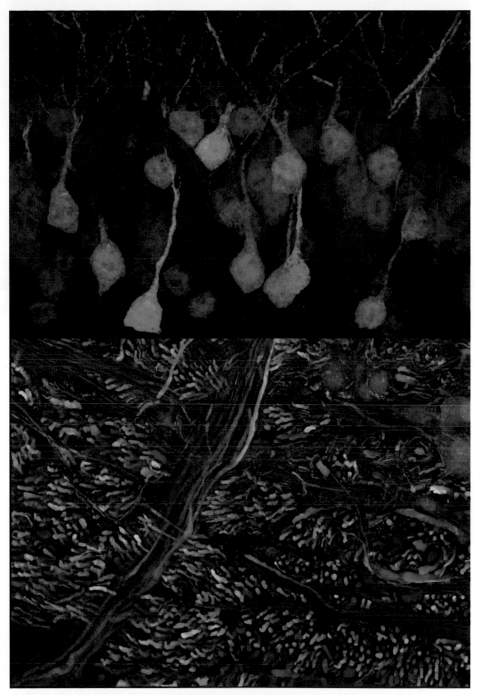

Fig. 4.17 Multiple labelling in a living mouse brain using the 'Brainbow' technique. Unique colour combinations in individual neurons are achieved by the relative levels of three or more fluorescent proteins (XFPs). The images are collected using a multi-channel laser scanning confocal microscope. Up to 90 different colours (neurons) can be distinguished using this technique. Top image, hippocampus; bottom image, brainstem. (Image courtesy of Jeff Lichtman, Harvard University, USA.)

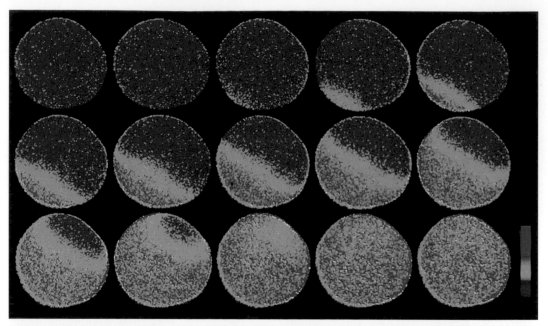

Fig. 4.19 Calcium imaging in living cells. A fertilisation-induced calcium wave in the egg of the starfish. The egg was microinjected with the calcium-sensitive fluorescent dye fluo-3 and subsequently fertilised by the addition of sperm during observation using time-lapse confocal microscopy with a 40× water immersion lens and a LSCM. An optical section located near the egg equator was collected every 4 s using the normal scan mode accumulated for two frames, and afterwards the images were corrected for offset and ratioed by linearly dividing the initial pre-fertilisation image into each successive frame of the time-lapse run. The ratioed images were then prepared as a montage and outputted with a pseudocolour look-up table in which blue regions represent low ratios and free calcium levels, and red areas depict high ratios and free calcium levels. Note that the wave sweeps through the entire ooplasm, rather than being cortically restricted. (Image kindly provided by Steve Stricker, University of New Mexico, USA.)

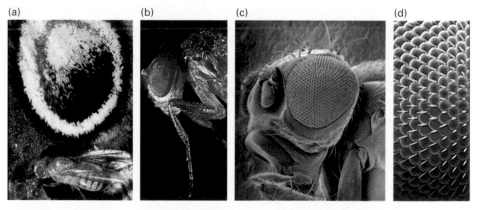

(a)   (b)   (c)   (d)

Fig. 4.22 Imaging surfaces using the light microscope (stereomicroscope) and the electron microscope (scanning electron microscope). Images produced using the stereomicroscope (a) and (b) and the scanning electron microscope (c) and (d). A stereomicroscope view of a fly (*Drosophila melanogaster*) on a butterfly wing (*Precis coenia*) (a) zoomed in to view the head region of the red-eyed fly (b). SEM image of a similar region of the fly's head (c) and zoomed more to view the individual ommatidia of the eye (d). Note that the stereomicroscope images can be viewed in real colour whereas those produced using the SEM are in greyscale. Colour can only be added to EM images digitally (d). (Images (b), (c) and (d) kindly provided by Georg Halder, MD Anderson Medical Centre, Houston, USA.)

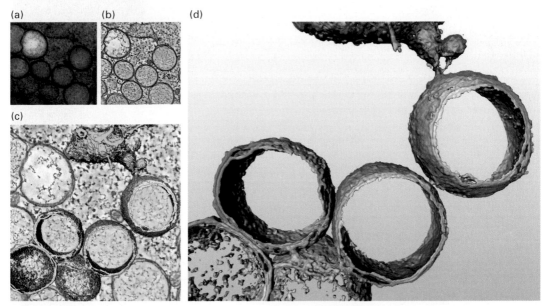

Fig. 4.24 Electron tomography revealing the interconnected nature of SARS–*Coronavirus*-induced double-membrane vesicles. Monkey kidney cells were infected with SARS–*Coronavirus* in a biosafety level-3 laboratory and pre-fixed using 3% paraformaldehyde at 7 h post-infection. Subsequently, the cells were rapidly frozen by plunge-freezing and freeze substitution was performed at low temperature, using osmium tetraoxide and uranyl acetate in acetone to optimally preserve cellular ultrastructure and gain maximal contrast. After washing with pure acetone at room temperature, the samples were embedded in an epoxy resin and polymerised at 60 °C for 2 days. Using an ultramicrotome, 200-nm thick sections were cut, placed on a 100 mesh EM grid, and used for electron tomography. To facilitate the image alignment that is required for the final 3D reconstruction, a suspension of 10 nm gold particles was layered on top of the sections as fiducial markers (a). Scale bar represents 100 nm. Images were recorded with an FEI T12 transmission electron microscope operating at an acceleration voltage of 120 kV. A tilt series consisted of 131 images recorded using 1° tilt increments between −65° and 65°. For dual-axis tomography, which improves resolution in the *X* and *Y* directions, the specimen was rotated 90° around the *Z*-axis and a second tilt series was recorded. To compute the final electron tomogram, the dual-axis tilt series were aligned by means of the fiducial markers using the IMOD software package. A single tomogram slice through the 3D reconstruction with a digital thickness of 1.2 nm is shown in (b). The 3D surface-rendered reconstruction of viral structures and adjacent cellular features (c) was made by thresholding and subsequent surface rendering using the AMIRA Visualization Package (TGS Europe). The final 3D surface-rendered model (d) shows interconnected double-membrane vesicles (outer membrane, gold; inner membrane, silver) and their connection to an endoplasmic reticulum stack (depicted in bronze). (Images kindly provided by Kevin Knoops and Eric Snijder, Leiden University, The Netherlands.)

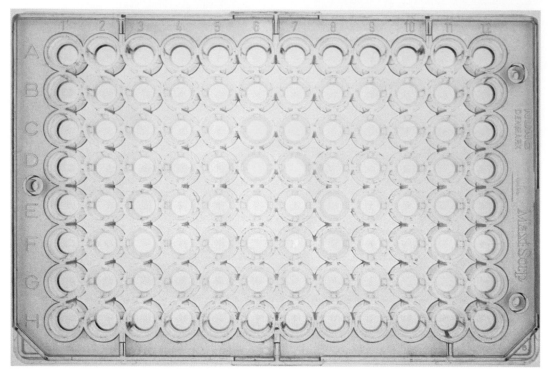
Fig. 7.12 A microtitre plate.

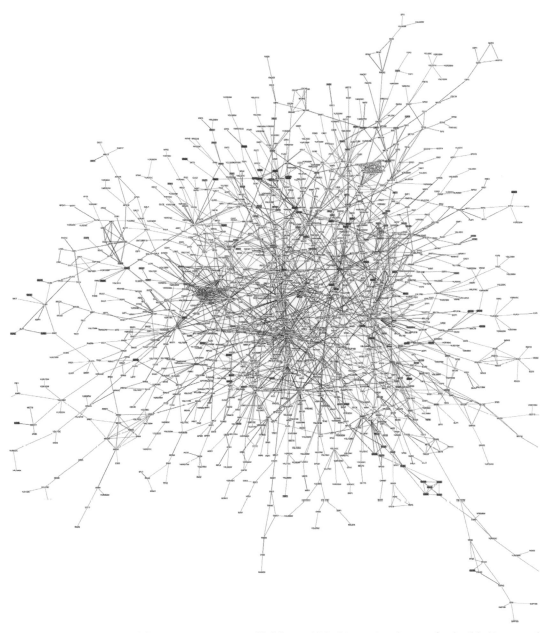

Fig. 8.9 An interaction map of the yeast proteome, assembled from published interactions (see text for details). (Courtesy of Benno Schwikowski, Peter Uetz and Stanley Fields. Reprinted with the permission of Nature Publishing Group.)

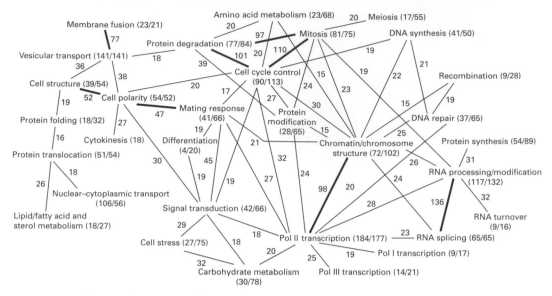

Fig. 8.10 A simplification of Fig. 8.9 identifying interactions between functional groups of proteins (see text for details). (Courtesy of Benno Schwikowski, Peter Uetz and Stanley Fields. Reprinted with the permission of Nature Publishing Group.)

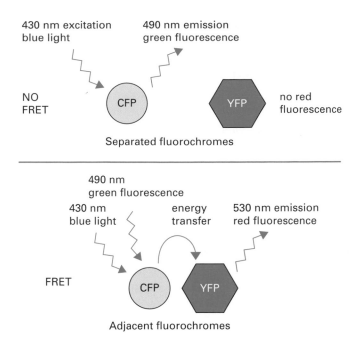

Fig. 4.20 Fluorescence resonance energy transfer (FRET). In the upper example (NO FRET) the cyan fluorescent protein (CFP) and the yellow fluorescent protein (YFP) are not close enough for FRET to occur (more than 60 nm separation). Here excitation with the 430 nm blue light results in the green 490 nm emission of the CFP only. In contrast, in the lower example (FRET), the CFP and YFP are close enough for 'energy transfer' or FRET to occur (closer than 6 nm). Here excitation with the 430 nm blue light results in fluorescence of the CFP (green) and of the YFP (red).

Example 1 **LOCATING AN UNKNOWN PROTEIN TO A SPECIFIC CELLULAR COMPARTMENT**

Question You have isolated and purified a novel protein from a biochemical preparation. How might you determine its subcellular distribution and possible function in the cell?

Answer Many fluorescent probes are available that label specific cellular compartments. For example, ToTo3 labels the nucleus and fluorescent phalloidins label cell outlines. An antibody to your protein could be raised and used to immunofluorescently label cells. Using a multiple-labelling approach and perhaps an optical sectioning technique such as laser scanning confocal microscopy the distribution of the protein in the cell relative to known distributions can be ascertained. For higher resolution immuno-EM or FRET studies could be performed.

4.6 THE ELECTRON MICROSCOPE (EM)

4.6.1 Principles

Electron microscopy is used when the greatest resolution is required, and when the living state can be ignored. The images produced in an electron microscope reveal the ultrastructure of cells. There are two different types of electron microscope – the transmission electron microscope (TEM) and the scanning electron microscope (SEM). In the TEM, electrons that pass through the specimen are imaged. In the SEM electrons that are reflected back from the specimen (secondary electrons) are collected, and the surfaces of specimens are imaged.

The equivalent of the light source in an electron microscope is the electron gun. When a high voltage of between 40 000 and 100 000 volts (the accelerating voltage) is passed between the cathode and the anode, a tungsten filament emits electrons (Fig. 4.1). The negatively charged electrons pass through a hole in the anode forming an electron beam. The beam of electrons passes through a stack of electromagnetic lenses (the column). Focussing of the electron beam is achieved by changing the voltage across the electromagnetic lenses. When the electron beam passes through the specimen some of the electrons are scattered while others are focussed by the projector lens onto a phosphorescent screen or recorded using photographic film or a digital camera. The electrons have limited penetration power which means that specimens must be thin (50–100 nm) to allow them to pass through.

Thicker specimens can be viewed by using a higher accelerating voltage, for example in the high-voltage electron microscope (HVEM) which uses 1 000 000 V accelerating voltage or in the intermediate voltage electron microscope (IVEM) which uses an accelerating voltage of around 400 000 V. Here stereo images are made by collecting two images at 8–10° tilt angles. Such images are useful in assessing the 3D relationships of organelles within cells when viewed in a stereoscope or with a digital stereo projection system.

4.6.2 Preparation of specimens

Contrast in the EM depends on atomic number; the higher the atomic number the greater the scattering and the contrast. Thus heavy metals are used to add contrast in the EM, for example uranium, lead and osmium. Labelled structures appear black or electron dense in the image (Fig. 4.21).

All of the water has to be removed from any biological specimen before it can be imaged in the EM. This is because the electron beam can only be produced and focussed in a vacuum. The major drawback of EM observation of biological specimens therefore is the non-physiological conditions necessary for their observation. Nevertheless, the improved resolution afforded by the EM has provided much information about biological structures and biochemical events within cells that could not have been collected using any other microscopical technique.

Extensive specimen preparation is required for EM analysis, and for this reason there can be issues of interpreting the images because of artifacts from specimen preparation. For example, specimens have been traditionally prepared for the TEM by fixation in glutaraldehyde to cross-link proteins followed by osmium tetroxide to fix and stain lipid membranes. This is followed by dehydration in a series of alcohols to remove the water, and then embedding in a plastic such as Epon for thin sectioning (Fig. 4.21).

Small pieces of the embedded tissue are mounted and sectioned on an ultramicro-tome using either a glass or a diamond knife. Ultrathin sections are cut to a thickness

(a)

(b)

(c)

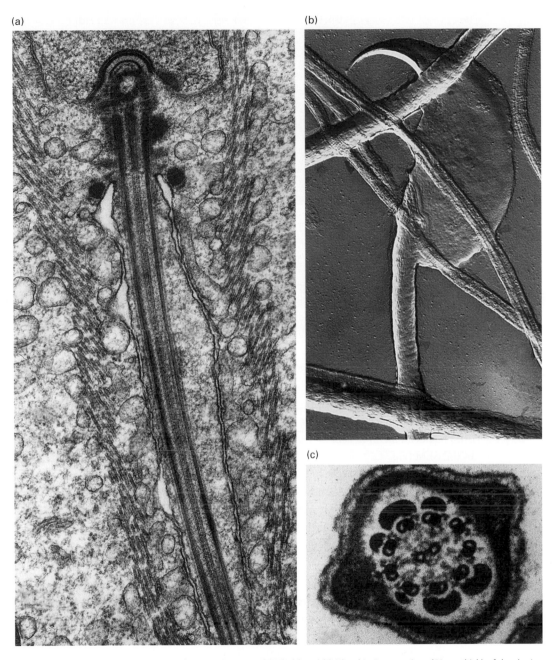

Fig. 4.21 Transmission electron microscopy (TEM). (a) and (c) Ultrathin Epon sections (60 nm thick) of developing rat sperm cells stained with uranyl acetate and lead citrate. (b) Carbon surface replica of a mouse sperm preparation.

of approximately 60 nm. The ribbons of sections are floated onto the surface of water and their interference colours are used to assess their thickness. The desired 60 nm section thickness has a silver/gold interference colour on the water surface. The sections are then mounted onto copper or gold EM grids, and are subsequently stained with heavy metals, for example uranyl acetate and lead citrate.

(a)   (b)   (c)   (d)

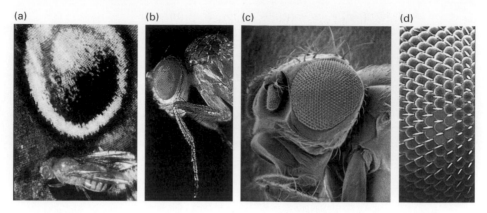

Fig. 4.22 Imaging surfaces using the light microscope (stereomicroscope) and the electron microscope (scanning electron microscope). Images produced using the stereomicroscope (a) and (b) and the scanning electron microscope (c) and (d). A stereomicroscope view of a fly (*Drosophila melanogaster*) on a butterfly wing (*Precis coenia*) (a) zoomed in to view the head region of the red-eyed fly (b). SEM image of a similar region of the fly's head (c) and zoomed more to view the individual ommatidia of the eye (d). Note that the stereomicroscope images can be viewed in real colour whereas those produced using the SEM are in greyscale. Colour can only be added to EM images digitally (d). (Images (b), (c) and (d) kindly provided by Georg Halder, MD Anderson Medical Centre, Houston, USA.) (See also colour plate.)

For the SEM, samples are fixed in glutaraldehyde, dehydrated through a series of solvents and dried completely either in air or by critical point drying. This method removes all of the water from the specimen instantly and avoids surface tension in the drying process thereby avoiding artifacts of drying. The specimens are then mounted onto a special metal holder or stub and coated with a thin layer of gold before viewing in the SEM (Fig. 4.22, see also colour section). Surfaces can also be viewed in the TEM using either negative stains or carbon replicas of air-dried specimens (Fig. 4.21).

Immuno-EM methods allow the localisation of molecules within the cellular microenvironment for TEM and on the cell surface for SEM (Fig. 4.23). Cells are prepared in a similar way to indirect immunofluorescence, with the exception that rather than a fluorescent probe bound to the secondary antibody, electron dense colloidal gold particles (10 nm) are used. Multiple labelling can be achieved using different sizes of gold particles attached to antibodies to the proteins of interest. The method depends upon the binding of protein A to the gold particles since protein A binds in turn to antibody fragments. Certain resins, for example Lowicryl and LR White, have been formulated to allow antibodies and gold particles to be attached to ultrathin sections for immunolabelling.

4.6.3 Electron tomography

New methods of fixation continue to be developed in an attempt to avoid the artifacts of specimen preparation and to observe the specimen more closely to its living state. Specimens are rapidly frozen in milliseconds by high-pressure freezing. Under these conditions the biochemical state of the cell is more likely to be preserved. Many of these frozen hydrated samples can be observed directly in the EM or they can be chemically

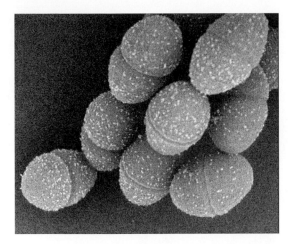

Fig. 4.23 Immunoelectron microscopy. Scanning electron microscope (SEM) image of microbes *Enterococcus faecalis* labelled with 10 nm collidal gold for the surface adhesion protein 'aggregation substance'. This protein facilitates exchange of DNA during conjugation. The gold labels appear as white dots on the surface of the bacteria. (Image kindly provided by the late Stan Erlandsen, University of Minnesota, USA.)

fixed using freeze substitution methods. Here fixatives are infused into the preparation at low temperature, after which the specimen is slowly warmed to room temperature.

Using cryo-electron tomography (Cryo-ET) the 3D structure of cells and macromolecules can be visualised at 5–8 nm resolution. Cells are typically rapidly frozen, fixed by freeze substitution and embedded in epoxy resin. Thick 200 nm sections are cut and imaged in the TEM equipped with a tilting stage. A typical tilt series of 100 or so images is collected in a digital form and exported to a computer reconstruction program for analysis. By using electron tomography, a 2D digital EM image is converted into a high-resolution 3D representation of the specimen (Fig. 4.24, see also colour section). The method is especially useful for imaging the fine connections within cells especially the cytoskeleton and nuclear pores and elucidating the surface structures of viruses.

4.6.4 Integrated microscopy

The same specimen can be viewed in the light microscope and subsequently in the EM. This approach is called integrated microscopy. The correlation of images of the same cell collected using the high temporal resolution of the light microscope and the high spatial resolution of the EM gives additional information to imaging using the two techniques separately (Fig. 4.25). The integrated approach also addresses the problem of artifacts. Probes are now available that are fluorescent in the light microscope and are electron dense in the EM.

4.7 IMAGE ARCHIVING

Most images produced by any kind of modern microscope are collected in a digital form. In addition to greatly speeding up the collection of the images (and experiment times), the use of digital imaging has allowed the use of digital image databases and

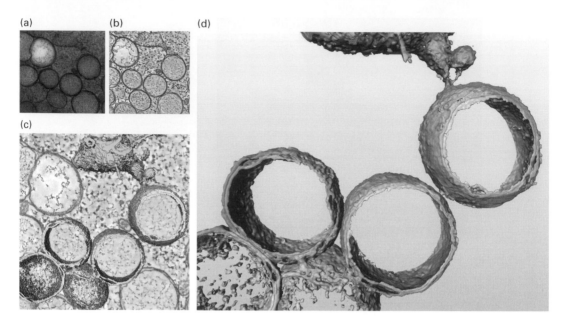

Fig. 4.24 Electron tomography revealing the interconnected nature of SARS–*Coronavirus*-induced double-membrane vesicles. Monkey kidney cells were infected with SARS–*Coronavirus* in a biosafety level-3 laboratory and pre-fixed using 3% paraformaldehyde at 7 h post-infection. Subsequently, the cells were rapidly frozen by plunge-freezing and freeze substitution was performed at low temperature, using osmium tetraoxide and uranyl acetate in acetone to optimally preserve cellular ultrastructure and gain maximal contrast. After washing with pure acetone at room temperature, the samples were embedded in an epoxy resin and polymerised at 60 °C for 2 days. Using an ultramicrotome, 200-nm thick sections were cut, placed on a 100 mesh EM grid, and used for electron tomography. To facilitate the image alignment that is required for the final 3D reconstruction, a suspension of 10 nm gold particles was layered on top of the sections as fiducial markers (a). Scale bar represents 100 nm. Images were recorded with an FEI T12 transmission electron microscope operating at an acceleration voltage of 120 kV. A tilt series consisted of 131 images recorded using 1° tilt increments between −65° and 65°. For dual-axis tomography, which improves resolution in the *X* and *Y* directions, the specimen was rotated 90° around the *Z*-axis and a second tilt series was recorded. To compute the final electron tomogram, the dual-axis tilt series were aligned by means of the fiducial markers using the IMOD software package. A single tomogram slice through the 3D reconstruction with a digital thickness of 1.2 nm is shown in (b). The 3D surface-rendered reconstruction of viral structures and adjacent cellular features (c) was made by thresholding and subsequent surface rendering using the AMIRA Visualization Package (TGS Europe). The final 3D surface-rendered model (d) shows interconnected double-membrane vesicles (outer membrane, gold; inner membrane, silver) and their connection to an endoplasmic reticulum stack (depicted in bronze). (Images kindly provided by Kevin Knoops and Eric Snijder, Leiden University, The Netherlands.) (See also colour plate.)

the rapid transfer of information between laboratories across the World Wide Web. Moreover there is no loss in resolution or colour balance from the images collected at the microscope as they pass between laboratories and journal web pages.

International image databases are under development for the storage and access of microscope image data from many different locations. One such effort is the Open Microscopy Environment (OME). There is a trend for modern microscopes to produce more and more data, especially when multi-dimensional datasets are generated. This trend is continuing with the need to develop automated methods of image analysis for large scale screening of gene expression data from genomic screens.

Table 4.4 **Websites of interest**

http://www.microscopyu.com/
http://www.microscopy.fsu.edu/
http://www.microscopy-analysis.com/
http://www.msa.microscopy.org
http://www.rms.org.uk/index.shtml
http://www.peachpit.com/articles/article.aspx?p=1221827
http://www.openmicroscopy.org
http://swehsc.pharmacy.arizona.edu/exppath/micro/index.html
http://www.itg.uiuc.edu/
http://rsb.info.nih.gov/ij/
http://www.openmicroscopy.org

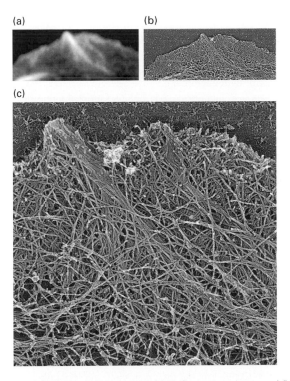

Fig. 4.25 Integrated microscopy. (a) Epifluorescence image and (b) and (c) whole mount TEM at different magnifications of the same cell. The fluorescence image is labelled with rhodamine phalloidin, which stains polymerised actin. A stress fibre at the periphery of the cell appears as a white line in the fluorescence image (a), and when viewed in the TEM the stress fibres appear as aligned densities of actin filaments. The TEM whole mount was prepared using detergent extraction, chemical fixation, critical point drying and platinum/carbon coating. (Image kindly provided by Tatyana Svitkina, University of Pennsylvania, USA.)

More detailed information on any of the microscopes and their applications in biochemistry and molecular biology can accessed on the World Wide Web. Several websites have been included as starting points for further study (Table 4.4). Should any of these listed websites become out of date, more information on any topic can be accessed using a web search engine. In addition, a comprehensive reference list has been provided for more detailed information (Section 4.8). The field of microscopy continues to be advanced but the basic principles and practices of light and electron microscopy remain unchanged.

4.8 SUGGESTIONS FOR FURTHER READING

Abramowitz, M. (2003). *Microscope Basics and Beyond.* Melville, NY: Olympus of America. (Good well-illustrated primer on all aspects of basic light microscopy, also available online as a pdf. file.)

Afzelius, B. A. and Maunsbach, A. B. (2004). Biological ultrastructure research: the first 50 years. *Tissue Cell*, **36**, 83–94. (Ageless review of the early history of electron microscopy.)

Alberts, B., Johnson, A., Lewis, J., Raff, M., Roberts, K. and Walter, P. (2007). *Molecular Biology of the Cell*, 5th edn. New York: Garland Science. (Basic introduction to all forms of microscopy and live cell imaging for the cell biologist.)

Andrews, P. D., Harper, I. S. and Swedlow, J. R. (2002). To 5D and beyond: quantitative fluorescence microscopy in the postgenomic era. *Traffic*, **3**, 29–36. (Review of multidimensional imaging, methods of coping with large data sets and international image databases.)

Baumeister, W. (2004). Mapping molecular landscapes inside cells. *Biological Chemistry*, **385**, 865–872. (Review of electron tomography.)

Cox, G. C. (2006). *Optical Imaging Techniques in Cell Biology.* Boca Raton, FL: CRC Press. (Overview of the entire field of light microscopy.)

Damle, S., Hanser, B., Davidson, E. H. and Fraser, S. E. (2006). Confocal quantification of *cis*-regulatory reporter gene expression in living sea urchins. *Developmental Biology*, **299**, 543–550. (Practical example of quantitative measurements in living cells.)

Darzacq, X. *et al.* (2009). Imaging transcription in living cells. *Annual Review of Biophysics*, **38**, 173–196.

Dunn, G. A. and Jones, G. E. (2004). Cell motility under the microscope: Vorsprung durch Technik. *Nature Reviews Molecular and Cell Biology*, **5**, 667–672. (Review of techniques used to study cell motility.)

Evanko, D., Heinrichs, A. and Karlsson-Rosenthal, C. (eds.) (2009). *Light Microscopy.* Nature Milestones. www.nature.com/milestones/light-microscopy (Well-produced and complete review of all aspects of contemporary light microscopy.)

Frankel, F. (2002). *Envisioning Science: The Design and Craft of the Science Image.* Cambridge, MA: MIT Press. (Popular work on imaging with some great tips and tricks for the stereomicroscope.)

Giepmans, B. N. G., Adams, S. R., Ellisman, M. H. and Tsien, R. Y. (2006). The fluorescent toolbox for assessing protein location and function. *Science*, **312**, 217–224. (A review of the characteristics and benefits of using fluorescent probes to study proteins.)

Hadjantonakis, A. K., Dickinson, M. E., Fraser, S. E. and Papaioannou, V. E. (2003). Technicolor transgenics: imaging tools for functional genomics in the mouse. *Nature Review Genetics*, **4**, 613–625.

Heath, J. P. (2005). *Dictionary of Microscopy.* Chichester, UK: John Wiley.

Hoenger, A. and McIntosh, J. R. (2009). Probing the macromolecular organisation of cells by electron tomography. *Current Opinion in Cell Biology*, **21**, 89–96.

Inoue, S. and Spring, K. (1997). *Video Microscopy: The Fundamentals*, 2nd edn. New York: Plenum Press. (The classic text on live cell imaging, video microscopy and general microscopy.)

Jaiswal, J. K. and Simon, S. M. (2007). Imaging single events at the cell membrane. *Nature Chemical Biology*, **3**, 92–98. (Overview of high resolution methods of light microscopy including TIRF.)

Keller, P. J., Schimdt, A. D., Wittbrodt, J. and Stelzer, E. H. K. (2008). Reconstruction of zebrafish early embryonic development by scanned light sheet microscopy. *Sciencexpress*, www.sciencexpress.org, 9 October 2008. (Application of scanning light microscopy to image living zebrafish embryos – stunning movies of zebrafish embryogenesis available online.)

Knoops, K., Kikkert, M., van den Worm, S. H. E., Zevenhoven-dobbe, J. C., van der Meer, Y., Koster, A. J., Mommaas, A. M. and Snijder, E. J. (2008). SARS–*Coronavirus* replication is supported by a reticulovesicular network of modified endoplasmic reticulum. *PLoS Biology*, **6**, 1957–1974. (Cryo-electron tomography in action.)

Lichtman, J. W. and Fraser, S. E. (2001). The neuronal naturalist: watching neurons in their native habitat. *Nature Neuroscience (Suppl.)*, **4**, 1215–1220.

Livet, J., Weissman, T. A., Kang, H., Draft, R. W., Lu, J., Bennis, R. A., Sanes, J. R. and Lichtman, J. W. (2007). Transgenic strategies for combinatorial expression of fluorescent proteins in the nervous system. *Nature*, **450**, 56–62. (Imaginative use of reporter gene technology to label multiple neurons in living brains.)

McGurk, L., Morrison, H., Keegan, L. P., Sharpe, J. and O'Connell, M. A. (2007). Three-dimensional imaging of *Drosophila melanogaster*. *PLoS ONE*, **2**, E834. (Methods of three-dimensional imaging including confocal and optical projection tomography.)

Sedgewick, J. (2008). *Scientific Imaging with PhotoShop: Methods, Measurement, and Output*. Berkeley, CA: Pearson Education, Peachpit Press. (Practical manual on the use of PhotoShop for measuring and preparing images for publication.)

Shapiro, H. M. (2003). *Practical Flow Cytometry*, 4th edn. New York: John Wiley. (Wonderfully written book on basic fluorescence and flow cytometry.)

Spector, D. L. and Goldman, R. D. (2006). *Basic Methods in Microscopy*. Plainview, NY: Cold Spring Harbor Laboratory Press. (A good introduction to contemporary methods of imaging both fixed and living cells at both the light and electron microscope level.)

Swedlow, J. R., Lewis, S. E. and Goldberg, I. G. (2006). Modeling data across labs, genomes, space and time. *Nature Cell Biology*, **8**, 1190–1194.

Swedlow, J. R., Goldberg, I. G., Eliceiri, K. W. and the OME Consortium (2009). Bioimage informatics for experimental biology. *Annual Review of Biophysics*, **38**, 327–346.

Tomancak, P., Berman, B. P., Beaton, A., Weiszmann, R., Kwan, E., Hartenstein, V., Celniker, S. E. and Rubin, G. M. (2007). Global analysis of gene expression during *Drosophila* embryogenesis. *Genome Biology*, **8**, R145.

Van Roessel, P. and Brand, A. H. (2002). Imaging into the future: visualizing gene expression and protein interaction with fluorescent proteins. *Nature Cell Biology*, **4**, E15–E20. (Good primer on GFP and FRET.)

Volpi, E. V. and Bridger, J. M. (2000). FISH glossary: an overview of the fluorescence *in situ* hybridization technique. *BioTechniques*, **45**, 385–409.

Wallace, W., Schaefer, L. H. and Swedlow, J. R. (2001). Workingperson's guide to deconvolution in light microscopy. *BioTechniques*, **31**, 1076–1097. (Comprehensive review of the deconvolution technique.)

Wilt, B. A., Burns, L. D., Tatt Wei Ho, E., Ghosh, K. K., Mukamel, E. A. and Schnitzer, M. J. (2009). Advances in light microscopy for neuroscience. *Annual Review of Neuroscience*, **32**, 435–506. (Complete coverage of all modern methods of imaging including super-resolution methods.)

Zhang, J., Campbell, R. E., Ting, A. Y. and Tsien, R. Y. (2002). Creating new fluorescent probes for cell biology. *Nature Reviews Molecular Cell Biology*, **3**, 906–918. (Review of the development of fluorescent probes of biological activity especially reporter molecules.)

5 Molecular biology, bioinformatics and basic techniques

R. RAPLEY

5.1 INTRODUCTION

The completion of the Human Genome Project (HGP) has been heralded as one of the major landmark events in science. The human genome contains the blueprint for human development and maintenance and may ultimately provide the means to understand human cellular and molecular processes in both health and disease. The genome is the full complement of DNA from an organism and carries all the information needed to specify the structure of every protein the cell can produce. The realisation that DNA lies behind all of the cell's activities led to the development of what is termed molecular biology. Rather than a discrete area of biosciences, molecular biology is now accepted as a very important means of understanding and describing complex biological processes. The development of methods and techniques for studying processes at the molecular level has led to new and powerful ways of isolating, analysing, manipulating and exploiting nucleic acids. Moreover, to keep pace with the explosion in biological information the discipline termed bioinformatics has evolved and provides a vital role in current biosciences. The completion of the human genome project and numerous other genome projects has allowed the continued development of new exciting areas of biological sciences such as biotechnology, genome mapping, molecular medicine and gene therapy.

In considering the potential utility of molecular biology techniques it is important to understand the basic structure of nucleic acids and gain an appreciation of how this dictates the function *in vivo* and *in vitro*. Indeed many techniques used in molecular biology mimic in some way the natural functions of nucleic acids such as replication and transcription. This chapter is therefore intended to provide an overview of the general features of nucleic acid structure and function and describe some of the basic methods used in its isolation and analysis.

5.2   STRUCTURE OF NUCLEIC ACIDS

5.2.1   Primary structure of nucleic acids

DNA and RNA are macromolecular structures composed of regular repeating polymers formed from nucleotides. These are the basic building blocks of nucleic acids and are derived from nucleosides which are composed of two elements: a five-membered pentose carbon sugar (2-deoxyribose in DNA and ribose in RNA), and a nitrogenous base. The carbon atoms of the sugar are designated 'prime' (l', 2', 3', etc.) to distinguish them from the carbons of nitrogenous bases of which there are two types, either a purine or a pyrimidine. A nucleotide, or nucleoside phosphate, is formed by the attachment of a phosphate to the 5' position of a nucleoside by an ester linkage (Fig. 5.1). Such nucleotides can be joined together by the formation of a second ester bond by reaction between the phosphate of one nucleotide and the 3' hydroxyl of another, thus generating a 5' to 3' phosphodiester bond between adjacent sugars; this process can be repeated indefinitely to give long polynucleotide molecules (Fig. 5.2). DNA has two such polynucleotide strands; however, since each strand has both a free 5' hydroxyl group at one end, and a free 3' hydroxyl at the other end, each strand has a polarity or directionality. The polarity of the two strands of the molecule is in opposite directions, and thus DNA is described as an antiparallel structure (Fig. 5.3).

The purine bases (composed of fused five- and six-membered rings), adenine (A) and guanine (G), are found in both RNA and DNA, as is the pyrimidine (a single six-membered ring) cytosine (C). The other pyrimidines are each restricted to one type of nucleic acid: uracil (U) occurs exclusively in RNA, whilst thymine (T) is limited to DNA. Thus it is possible to distinguish between RNA and DNA on the basis of the presence of ribose and uracil in RNA, and deoxyribose and thymine in DNA. However, it is the sequence of bases along a molecule that distinguishes one DNA (or RNA) from another. It is conventional to write a nucleic acid sequence starting at the 5' end of the molecule, using single capital letters to represent each of the bases, e.g. CGGATCT. Note that there is usually no point in including the sugar or phosphate groups, since these are identical throughout the length of the molecule. Terminal phosphate groups can, when necessary, be indicated by use of a 'p'; thus 5' pCGGATCT 3' indicates the presence of a phosphate on the 5' end of the molecule.

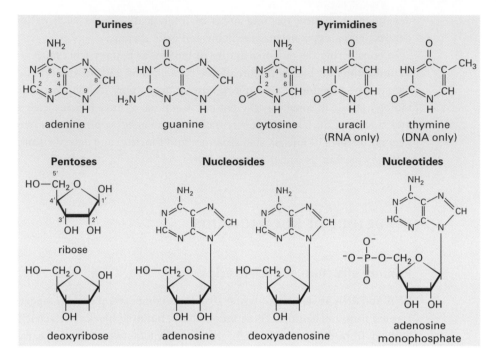

Fig. 5.1 Structure of bases, nucleosides and nucleotides.

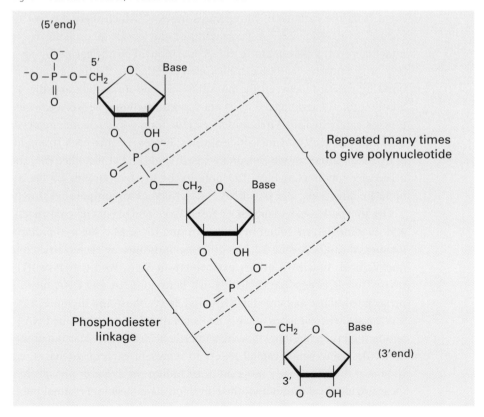

Fig. 5.2 Polynucleotide structure.

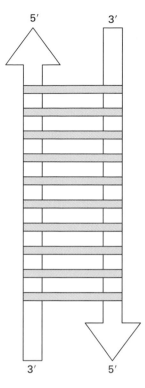

Fig. 5.3 The antiparallel nature of DNA. One strand in a double helix runs 5' to 3', whilst the other strand runs in the opposite direction 3' to 5'. The strands are held together by hydrogen bonds between the bases.

5.2.2 Secondary structure of nucleic acids

The two polynucleotide chains in DNA are usually found in the shape of a right-handed double helix, in which the bases of the two strands lie in the centre of the molecule, with the sugar–phosphate backbones on the outside. A crucial feature of this double-stranded structure is that it depends on the sequence of bases in one strand being complementary to that in the other. A purine base attached to a sugar residue on one strand is always hydrogen bonded to a pyrimidine base attached to a sugar residue on the other strand. Moreover, adenine (A) always pairs with thymine (T) or uracil (U) in RNA, via two hydrogen bonds, and guanine (G) always pairs with cytosine (C) by three hydrogen bonds (Fig. 5.4). When these conditions are met a stable double helical structure results in which the backbones of the two strands are, on average, a constant distance apart. Thus, if the sequence of one strand is known, that of the other strand can be deduced. The strands are designated as plus (+) and minus (−) and an RNA molecule complementary to the minus (−) strand is synthesised during transcription (Section 5.5.3). The base sequence may cause significant local variations in the shape of the DNA molecule and these variations are vital for specific interactions between the DNA and various proteins to take place. Although the three-dimensional structure of DNA may vary it generally adopts a double helical structure termed the B form or B-DNA *in vivo*. There are also other forms of right-handed DNA such as A and C, which are formed when DNA fibres are subjected to different relative humidities (Table 5.1).

Table 5.1 **The various forms of DNA**

DNA form	% humidity	Helix direction	Base/turn helix	Helix diameter (A)
B	92%	RH	10	19
A	75%	RH	11	23
C	66%	RH	9.3	19
Z	$(Pu-Py)_n$	LH	12	18

Notes: RH, right-handed helix; LH, left-handed helix; Pu, Purine; Py, Pyrimidine. Different forms of DNA may be obtained by subjecting DNA fibres to different relative humidities. The B form is the most common form of DNA whilst the A and C forms have been derived under laboratory conditions. The Z form may be produced with a DNA sequence made up from alternating purine and pyrimidine nucleotides.

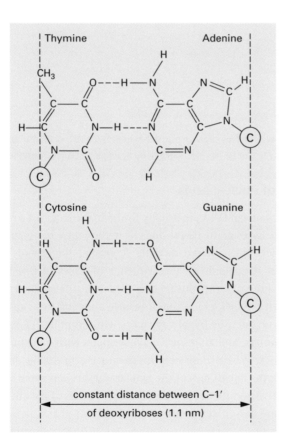

Fig. 5.4 Base-pairing in DNA. C in a circle represents carbon at the 1′ position of deoxyribose.

The major distinguishing feature of B-DNA is that it has approximately 10 bases for one turn of the double helix; furthermore a distinctive major and minor groove may be identified (Fig. 5.5). In certain circumstances where repeated DNA sequences or motifs are found the DNA may adopt a left-handed helical structure termed Z-DNA.

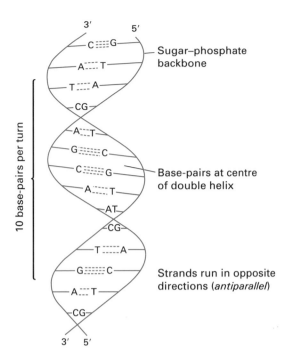

Fig. 5.5 The DNA double helix.

This form of DNA was first synthesised in the laboratory and is thought not to exist *in vivo*. The various forms of DNA serve to show that it is not a static molecule but dynamic and constantly in flux, and may be coiled, bent or distorted at certain times. Although RNA almost always exists as a single strand, it often contains sequences within the same strand that are self-complementary, and which can therefore base-pair if brought together by suitable folding of the molecule. A notable example is transfer RNA (tRNA) which folds up to give a clover-leaf secondary structure (Fig. 5.6)

5.2.3 Separation of double-stranded DNA

The two antiparallel strands of DNA are held together only by the weak forces of hydrogen bonding between complementary bases, and partly by hydrophobic inter-actions between adjacent, stacked base pairs, termed base-stacking. Little energy is needed to separate a few base pairs, and so, at any instant, a few short stretches of DNA will be opened up to the single-stranded conformation. However, such stretches immediately pair up again at room temperature, so the molecule as a whole remains predominantly double-stranded.

If, however, a DNA solution is heated to approximately 90 °C or above there will be enough kinetic energy to denature the DNA completely, causing it to separate into single strands. This is termed denaturation and can be followed spectrophotometrically by monitoring the absorbance of light at 260 nm. The stacked bases of double-stranded DNA are less able to absorb light than the less constrained bases of single-stranded molecules, and so the absorbance of DNA at 260 nm increases as the DNA becomes denatured, a phenomenon known as the hyperchromic effect.

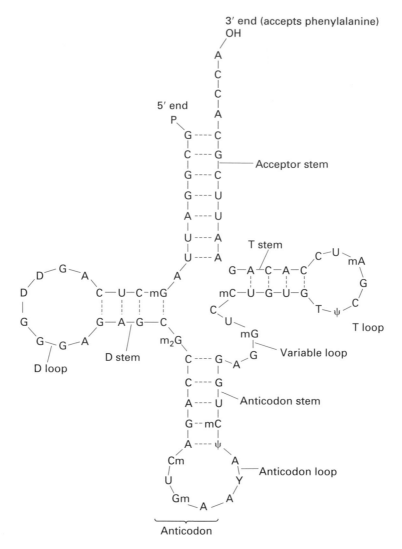

Fig. 5.6 Secondary structure of yeast tRNA^Phe. A single strand of 76 ribonucleotides forms four double-stranded 'stem' regions by base-pairing between complementary sequences. The anticodon will base-pair with UUU or UUC (both are codons for phenylalanine); phenylalanine is attached to the 3' end by a specific aminoacyl tRNA synthetase. Several 'unusual' bases are present: D, dihydrouridine; T, ribothymidine; ψ, pseudouridine; Y, very highly modified, unlike any 'normal' base. mX indicates methylation of base X (m_2X shows dimethylation); Xm indicates methylation of ribose on the 2' position.

The absorbance at 260 nm may be plotted against the temperature of a DNA solution which will indicate that little denaturation occurs below approximately 70 °C, but further increases in temperature result in a marked increase in the extent of denaturation. Eventually a temperature is reached at which the sample is totally denatured, or melted. The temperature at which 50% of the DNA is melted is termed the melting temperature or T_m, and this depends on the nature of the DNA (Fig. 5.7). If several different samples of DNA are melted, it is found that the T_m is highest for those DNAs which contain the highest proportion of cytosine and guanine, and T_m can actually be used to estimate the percentage (C + G) in a DNA sample. This relationship

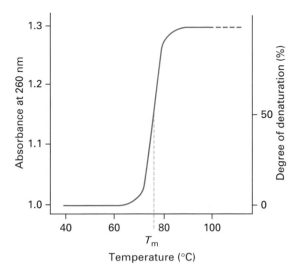

Fig. 5.7 Melting curve of DNA.

between T_m and $(C+G)$ content arises because cytosine and guanine form three hydrogen bonds when base-paired, whereas thymine and adenine form only two. Because of the differential numbers of hydrogen bonds between A–T and C–G pairs those sequences with a predominance of C–G pairs will require greater energy to separate or denature them. The conditions required to separate a particular nucleotide sequence are also dependent on environmental conditions such as salt concentration.

If melted DNA is cooled it is possible for the separated strands to reassociate, a process known as renaturation. However, a stable double-stranded molecule will only be formed if the complementary strands collide in such a way that their bases are paired precisely, and this is an unlikely event if the DNA is very long and complex (i.e. if it contains a large number of different genes). Measurements of the rate of renaturation can give information about the complexity of a DNA preparation.

Strands of RNA and DNA will associate with each other, if their sequences are complementary, to give double-stranded, hybrid molecules. Similarly, strands of radio-actively labelled RNA or DNA, when added to a denatured DNA preparation, will act as probes for DNA molecules to which they are complementary. This hybridisation of complementary strands of nucleic acids is very useful for isolating a specific fragment of DNA from a complex mixture. It is also possible for small single-stranded fragments of DNA (up to 40 bases in length) termed oligonucleotides to hybridise to a denatured sample of DNA. This type of hybridisation is termed annealing and again is dependent on the base sequence of the oligonucleotide and the salt concentration of the sample.

5.3 GENES AND GENOME COMPLEXITY

5.3.1 Gene complexity

Each region of DNA which codes for a single RNA or protein is called a gene, and the entire set of genes in a cell, organelle or virus forms its genome. Cells and organelles

Table 5.2 **Repetitive satellite sequences found in DNA, and their characteristics**

Types of repetitive DNA	Repeat unit size (bp)	Characteristics/motifs
Satellite DNA	5–200	Large repeat unit range (Mb) usually found at centromeres
Minisatellite DNA		
Telomere sequence	6	Found at the ends of chromosomes. Repeat unit may span up to 20 kb G-rich sequence
Hypervariable sequence	10–60	Repeat unit may span up to 20 kb
Microsatellite DNA	1–4	Mononucleotide repeat of adenine dinucleotide repeats common (CA). Usually known as VNTR (variable number tandem repeat)

Notes: bp, base-pairs; kb, kilobase-pairs.

may contain more than one copy of their genome. Genomic DNA from nearly all prokaryotic and eukaryotic organisms is also complexed with protein and termed chromosomal DNA. Each gene is located at a particular position along the chromosome, termed the locus, whilst the particular form of the gene is termed the allele. In mammalian DNA each gene is present in two allelic forms which may be identical (homozygous) or which may vary (heterozygous). It is thought that there are approximately 20 000 genes present in the human genome, although not all will be expressed in a given cell at the same time. However various processing events such as alternative splicing or RNA editing can increase the number of actual proteins found in the cell in relation to the number of genes to nearly 1 million. The occurrence of different alleles at the same site in the genome is termed polymorphism. In general the more complex an organism the larger its genome, although this is not always the case since many higher organisms have non-coding sequences some of which are repeated numerous times and termed repetitive DNA. In mammalian DNA repetitive sequences may be divided into low copy number and high copy number DNA. The latter is composed of repeat sequences that are dispersed throughout the genome and those that are clustered together. The repeat cluster DNA may be defined into so-called classical satellite DNA, minisatellite and microsatellite DNA, the latter being mainly composed of dinucleotide repeats (Table 5.2). These sequences are termed polymorphic, collectively termed polymorphisms, and vary between individuals; they also form the basis of genetic fingerprinting.

5.3.2 Single nucleotide polymorphisms (SNPs)

A further important source of polymorphic diversity known to be present in genomes is termed single nucleotide polymorphisms or SNPs (pronounced snips). SNPs are substitutions of one base at a precise location within the genome. Those that occur in coding regions are termed cSNPs. Estimates indicate that an SNP occurs every once in every 300 bases and there are thought to be approximately 10 million in the human genome. Interest in SNPs lies in the fact that these polymorphisms may account for the differences in disease susceptibility, drug metabolism and response to environmental factors between individuals. Indeed there are now a number of initiatives to identify SNPs and produce genomic SNP maps. One initiative is the international HapMap project. This will enable a haplotype map of common sources of variations from groups of associated SNPs to be produced. This will potentially allow a set of so-called tag SNPs to be identified and potentially provide an association between the haplotype and a disease.

5.3.3 Chromosomes and karyotypes

Higher organisms may be identified by using the size and shape of their genetic material at a particular point in the cell division cycle, termed metaphase. At this point DNA condenses to form a number of very distinct chromosome structures. Various morphological characteristics of chromosomes may be identified at this stage including the centromere and the telomere. The array of chromosomes from a given organism may also be stained with dyes such as giemsa stain and subsequently analysed by light microscopy. The complete array of chromosomes in an organism is termed the karyotype. In certain genetic disorders aberrations in the size, shape and number of chromosomes may occur and thus the karyotype may be used as an indicator of the disorder. Perhaps the most well known example of this is the correlation of Down syndrome, where three copies of chromosome 21 (trisomy 21) exist rather than two as in the normal state.

5.3.4 Renaturation kinetics and genome complexity

When preparations of double-stranded DNA are denatured and allowed to renature, measurement of the rate of renaturation can give valuable information about the complexity of the DNA, i.e. how much information it contains (measured in base-pairs). The complexity of a molecule may be much less than its total length if some sequences are repetitive, but complexity will equal total length if all sequences are unique, appearing only once in the genome. In practice, the DNA is first cut randomly into fragments about 1 kb in length (Section 5.9), and is then completely denatured by heating above its T_m (Section 5.2.3). Renaturation at a temperature about 10 °C below the T_m is monitored either by decrease in absorbance at 260 nm (the hypochromic effect), or by passing samples at intervals through a column of hydroxylapatite, which will adsorb only double-stranded DNA, and measuring how much of the sample is bound. The degree of renaturation after a given time will depend on C_0, the concentration (in nucleotides per unit volume) of double-stranded DNA prior to denaturation, and t, the duration of the renaturation in seconds.

For a given C_0, it should be evident that a preparation of bacteriophage λ DNA (genome size 49 kb) will contain many more copies of the same sequence per unit

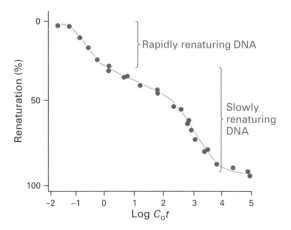

Fig. 5.8 Cot curve of human DNA. DNA was allowed to renature at 60 °C after being completely dissociated by heat. Samples were taken at intervals and passed through a hydroxylapatite column to determine the percentage of double-stranded DNA present. This percentage was plotted against log $C_{o}t$ (original concentration of DNA × 3 time of sampling).

volume than a preparation of human DNA (haploid genome size 3×10^6 kb), and will therefore renature far more rapidly, since there will be more molecules complementary to each other per unit volume in the case of λ DNA, and therefore more chance of two complementary strands colliding with each other. In order to compare the rates of renaturation of different DNA samples it is usual to measure C_o and the time taken for renaturation to proceed half way to completion, $t_{1/2}$, and to multiply these values together to give a $C_o t_{1/2}$ value. The larger the $C_o t_{1/2}$, the greater the complexity of the DNA; hence λ DNA has a far lower $C_o t_{1/2}$ than does human DNA.

In fact, the human genome does not renature in a uniform fashion. If the extent of renaturation is plotted against log $C_o t$ (this is known as a Cot curve), it is seen that part of the DNA renatures quite rapidly, whilst the remainder is very slow to renature (Fig. 5.8). This indicates that some sequences have a higher concentration than others; in other words, part of the genome consists of repetitive sequences. These repetitive sequences can be separated from the single-copy DNA by passing the renaturing sample through a hydroxylapatite column early in the renaturation process, at a time which gives a low value of $C_o t$. At this stage only the rapidly renaturing sequences will be double-stranded, and they will therefore be the only ones able to bind to the column.

5.3.5 The nature of the genetic code

DNA encodes the primary sequence of a protein by utilising sets of three nucleotides, termed a codon or triplet, to encode a particular amino acid. The four bases (A, C, G and T) present in DNA allow a possible 64 triplet combinations; however, since there are only 20 naturally occurring amino acids more than one codon may encode an amino acid. This phenomenon is termed the degeneracy of the genetic code. With the exception of a limited number of differences found in mitochondrial DNA and one or

First position (5′ end)	Second position				Third position (3′ end)
	T	C	A	G	
	Phe	Ser	Tyr	Cys	T
T	Phe	Ser	Tyr	Cys	C
	Leu	Ser	Stop	Stop	A
	Leu	Ser	Stop	Trp	G
	Leu	Pro	His	Arg	T
C	Leu	Pro	His	Arg	C
	Leu	Pro	Gln	Arg	A
	Leu	Pro	Gln	Arg	G
	Ile	Thr	Asn	Ser	T
A	Ile	Thr	Asn	Ser	C
	Ile	Thr	Lys	Arg	A
	Met	Thr	Lys	Arg	G
	Val	Ala	Asp	Gly	T
G	Val	Ala	Asp	Gly	C
	Val	Ala	Glu	Gly	A
	Val	Ala	Glu	Gly	G

Fig. 5.9 The genetic code. Note that the codons in blue represent the start codon (ATG) and the three stop codons.

two other species the genetic code appears to be universal. In addition to coding for amino acids particular triplet sequences also indicate the beginning (Start) and the end (Stop) of a particular gene. Only one start codon exists (ATG) which also codes for the amino acid methionine, whereas three dedicated stop codons are available (TAT, TAG and TGA) (Fig. 5.9). A sequence flanked by a start and a stop codon containing a number of codons that may be read in-frame to represent a continuous protein sequence is termed an open reading frame (ORF).

5.4 LOCATION AND PACKAGING OF NUCLEIC ACIDS

5.4.1 Cellular compartments

In general, DNA in eukaryotic cells is confined to the nucleus and organelles such as mitochondria or chloroplasts which contain their own genome. The predominant RNA species are however normally located within the cytoplasm. The genetic information of cells and most viruses is stored in the form of DNA. This information is used to

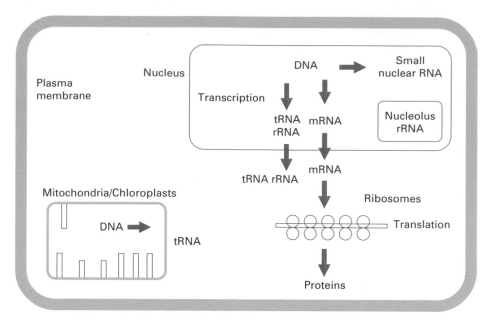

Fig. 5.10 Location of DNA and RNA molecules in eukaryotic cells and the flow of genetic information.

direct the synthesis of RNA molecules, which fall into three classes. Figure 5.10 indicates the locations of nucleic acids in prokaryotic and eukaryotic cells.

- *Messenger RNA* (mRNA) contains sequences of ribonucleotides which code for the amino acid sequences of proteins. A single mRNA codes for a single polypeptide chain in eukaryotes, but may code for several polypeptides in prokaryotes.
- *Ribosomal RNA* (rRNA) forms part of the structure of ribosomes, which are the sites of protein synthesis. Each ribosome contains only three or four different rRNA molecules, complexed with a total of between 55 and 75 proteins.
- *Transfer RNA* (tRNA) molecules carry amino acids to the ribosomes, and interact with the mRNA in such a way that their amino acids are joined together in the order specified by the mRNA. There is at least one type of tRNA for each amino acid.

In eukaryotic cells alone a further group of RNA molecules termed small nuclear RNA (snRNA) is present which function within the nucleus and promote the maturation of mRNA molecules. All RNA molecules are associated with their respective binding proteins and are essential for their cellular functions. Nucleic acids from prokaryotic cells are less well compartmentalised although they serve similar functions.

5.4.2 The packaging of DNA

The DNA in prokaryotic cells resides in the cytoplasm although it is associated with nucleoid proteins, where it is tightly coiled and supercoiled by topoisomerase enzymes to enable it to physically fit into the cell. By contrast eukaryotic cells have

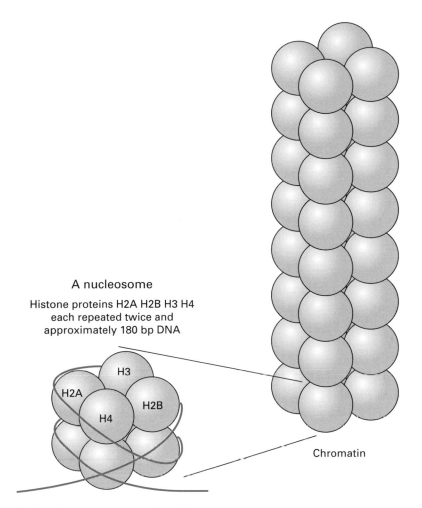

A nucleosome

Histone proteins H2A H2B H3 H4
each repeated twice and
approximately 180 bp DNA

H3

H2A

H4   H2B

Chromatin

Fig. 5.11 Structure and composition of the nucleosome and chromatin.

many levels of packaging of the DNA within the nucleus involving a variety of DNA binding proteins.

First-order packaging involves the winding of the DNA around a core complex of four small proteins repeated twice, termed histones (H2A, H2B, H3 and H4). These are rich in the basic amino acids lysine and arginine and form a barrel-shaped core octomer structure. Approximately 180 bp of DNA is wound twice around the structure which is termed a nucleosome. A further histone protein, H1, is found to associate with the outer surface of the nucleosome. The compacting effect of the nucleosome reduces the length of the DNA by a factor of six.

Nucleosomes also associate to form a second order of packaging termed the 30 nm chromatin fibre thus further reducing the length of the DNA by a factor of seven (Fig. 5.11). These structures may be further folded and looped through the interaction with other non-histone proteins and ultimately form chromosome structures.

DNA is found closely associated with the nuclear lamina matrix, which forms a protein scaffold within the nucleus. The DNA is attached at certain positions within the scaffold, usually coinciding with origins of replication. Many other DNA binding proteins are also present, such as high mobility group (HMG) proteins, which assist in promoting certain DNA conformations during processes such as replication or active gene expression.

5.5 FUNCTIONS OF NUCLEIC ACIDS

5.5.1 DNA replication

The double-stranded nature of DNA provides a means of replication during cell division since the separation of two DNA strands allows complementary strands to be synthesised upon them. Many enzymes and accessory proteins are required for *in vivo* replication, which in prokaryotes begins at a region of the DNA termed the origin of replication.

DNA has to be unwound before any of the proteins and enzymes needed for replication can act, and this involves separating the double-helical DNA into single strands. This process is carried out by the enzyme DNA helicase. Furthermore, in order to prevent the single strands from re-annealing small proteins termed single-stranded DNA binding proteins (SSBs) attach to the single DNA strands (Fig. 5.12).

On each exposed single strand a short, complementary RNA chain termed a primer is first produced, using the DNA as a template. The primer is synthesised by an RNA polymerase enzyme known as a primase which uses ribonucleoside triphosphates and itself requires no primer to function. Then DNA polymerase III (DNApolIII) also uses the original DNA as a template for synthesis of a DNA strand, using the RNA primer as a starting point. The primer is vital since it leaves an exposed 3′ hydroxyl group. This is necessary since DNA polymerase III can only add new nucleotides to the 3′ end and not the 5′ end of a nucleic acid. Synthesis of the DNA strand therefore occurs only in a 5′ to 3′ direction from the RNA primer. This DNA strand is usually termed the leading strand and provides the means for continuous DNA synthesis.

Since the two strands of double-helical DNA are antiparallel, only one can be synthesised in a continuous fashion. Synthesis of the other strand must take place in a more complex way. The precise mechanism was worked out by Reiji Okazaki in the 1960s. Here the strand, usually termed the lagging strand, is produced in relatively short stretches of 1–2 kb termed Okazaki fragments. This is still in a 5′ to 3′ direction, using many RNA primers for each individual stretch. Thus, discontinuous synthesis of DNA takes place and allows DNA polymerase III to work in the 5′ to 3′ direction. The RNA primers are then removed by DNA polI, which has a 5′ to 3′ exonuclease, and the gaps are filled by the same enzyme acting as a polymerase. The separate fragments are joined together by DNA ligase to give a newly formed strand of DNA on the lagging strand (Fig. 5.13).

The replication of eukaryotic DNA is less well characterised, involves multiple origins of replication and is certainly more complex than that of prokaryotes; however, in both cases the process involves 5′ to 3′ synthesis of new DNA strands. The net result of the replication is that the original DNA is replaced by two molecules, each

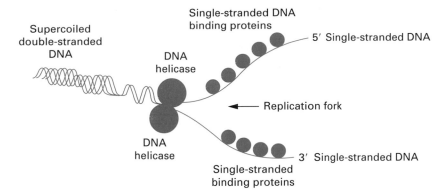

Fig. 5.12 Initial events at the replication fork involving DNA unwinding.

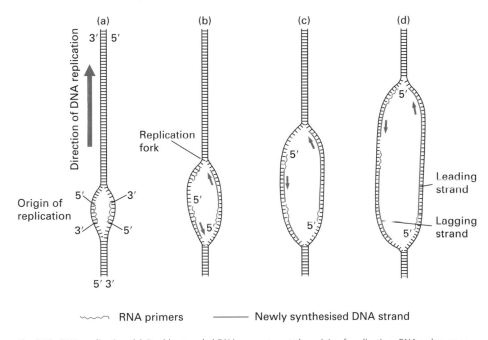

Fig. 5.13 DNA replication. (a) Double-stranded DNA separates at the origin of replication. RNA polymerase synthesises short DNA primer strands complementary to both DNA strands. (b) DNA polymerase III synthesises new DNA strands in a 5′ to 3′ direction, complementary to the exposed, old DNA strands, and continuing from the 3′end of each RNA primer. Consequently DNA synthesis is in the same direction as DNA replication for one strand (the leading strand) and in the opposite direction for the other (the lagging strand). RNA primer synthesis occurs repeatedly to allow the synthesis of fragments of the lagging strand. (c) As the replication fork moves away from the origin of replication, DNA polymerase III continues the synthesis of the leading strand, and synthesises DNA between RNA primers of the lagging strand. (d) DNA polymerase I removes RNA primers from the lagging strand and fills the resulting gaps with DNA. DNA ligase then joins the resulting fragments, producing a continuous DNA strand.

containing one 'old' and one 'new' strand; the process is therefore known as semi-conservative replication. The ideas behind DNA synthesis, replication and the enzymes involved in them have been adopted in many molecular biology techniques and form the basis of many manipulations in genetic engineering.

5.5.2 DNA protection and repair systems

Cellular growth and division require the correct and coordinated replication of DNA. Mechanisms that proofread replicated DNA sequences and maintain integrity of those sequences are, however, complex and are only beginning to be elucidated for prokaryotic systems. Bacterial protection is afforded by the use of a restriction modification system based on differential methylation of host DNA, so as to distinguish it from foreign DNA such as viruses. The most common is type II and consists of a host DNA methylase and restriction endonuclease that recognises short (4–6 bp) palindromic sequences and cleaves foreign unmethylated DNA at a particular target sequence. The enzymes involved in this process have been of enormous benefit for the manipulation and analysis of DNA, as indicated in Section 5.9.

Repair systems allow the recognition of altered, mispaired or missing bases in double-stranded DNA and invoke an excision repair process. The systems characterised for bacterial systems are based on the length of repairable DNA during either replication (dam system) or in general repair (urr system). In some cases damage to DNA activates a protein termed RecA to produce an SOS response that includes the activation of many enzymes and proteins; however, this has yet to be fully characterised. The recombination–repair systems in eukaryotic cells may share some common features with prokaryotes although the precise mechanism has yet to be established. Defects in DNA repair may result in the stable incorporation of errors into genomic sequences which may underscore several genetic-based diseases.

5.5.3 Transcription of DNA

Expression of genes is carried out initially by the process of transcription, whereby a complementary RNA strand is synthesised by an enzyme termed RNA polymerase from a DNA template encoding the gene. Most prokaryotic genes are made up of three regions. At the centre is the sequence which will be copied in the form of RNA, called the structural gene. To the 5′ side (upstream) of the strand which will be copied (the plus (+) strand) lies a region called the promoter, and downstream of the transcription unit is the terminator region. Transcription begins when DNA-dependent RNA polymerase binds to the promoter region and moves along the DNA to the transcription unit. At the start of the transcription unit the polymerase begins to synthesise an RNA molecule complementary to the minus (−) strand of the DNA, moving along this strand in a 3′ to 5′ direction, and synthesising RNA in a 5′ to 3′ direction, using ribonucleoside triphosphates. The RNA will therefore have the same sequence as the + strand of DNA, apart from the substitution of uracil for thymine. On reaching the stop site in the terminator region, transcription is stopped, and the RNA molecule is released. The numbering of bases in genes is a useful way of identifying key elements. Point or base +1 is the residue located at the transcription start site; positive numbers denote 3′ regions, whilst negative numbers denote 5′ regions (Fig. 5.14).

In eukaryotes, three different RNA polymerases exist, designated I, II and III. Messenger RNA is synthesised by RNA polymerase II, while RNA polymerase I and

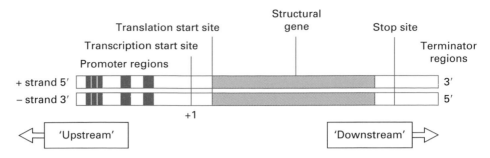

Fig. 5.14 Structure and nomenclature of a typical gene.

III catalyse the synthesis of rRNA (I), tRNA and snRNA (III). Many non-expressed genes tend to have residues that are methylated, usually the C of a GC dinucleotide, and in general active genes tend to be hypomethylated. This is especially prevalent at the 5′ flanking regions and is a useful means of discovering and identifying new genes.

5.5.4 Promoter and terminator sequences in DNA

Promoters are usually to the 5′ end or upstream of the structural gene and have been best characterised in prokaryotes such as *Escherichia coli*. They comprise two highly conserved sequence elements: the TATA box (consensus sequence 'TATATT') which is centred approximately 10 bp upstream from the transcription initiation site (−10 in the gene numbering system), and a 'GC-rich' sequence which is centred about −25 bp upstream from the TATA box. The GC element is thought to be important in the initial recognition and binding of RNA polymerase to the DNA, while the −10 sequence is involved in the formation of a transcription initiation complex (Fig. 5.15a).

The promoter elements serve as recognition sites for DNA binding proteins that control gene expression and these proteins are termed transcription factors or trans-acting factors. These proteins have a DNA binding domain for interaction with promoters and an activation domain to allow interaction with other transcription factors. A well-studied example of a transcription factor is TFIID which binds to the −35 promoter sequence in eukaryotic cells. Gene regulation occurs in most cases at the level of transcription, and primarily by the rate of transcription initiation, although control may also be by modulation of mRNA stability, or at other levels such as translation. Terminator sequences are less well characterised, but are thought to involve nucleotide sequences near the end of mRNA with the capacity to form a hairpin loop, followed by a run of U residues, which may constitute a termination signal for RNA polymerase.

In the case of eukaryotic genes numerous short sequences spanning several hundred bases may be important for transcription, compared to normally less than 100 bp for prokaryotic promoters. Particularly critical is the TATA box sequence, located approximately −35 bp upstream of the transcription initiation point in the majority of genes (Fig. 5.15b). This is analogous to the −10 sequence in prokaryotes. A number

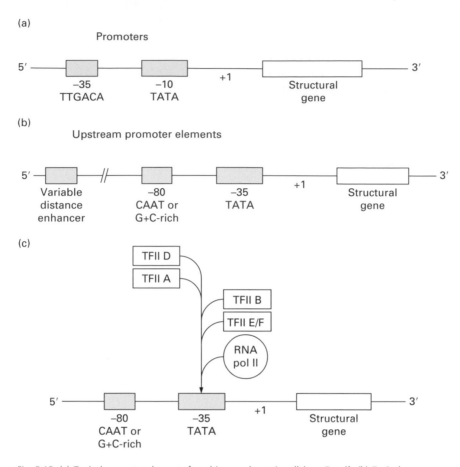

Fig. 5.15 (a) Typical promoter elements found in a prokaryotic cell (e.g. *E. coli*). (b) Typical promoter elements found in eukaryotic cells. (c) Generalised scheme of binding of transcription factors to the promoter regions of eukaryotic cells. Following the binding of the transcription factors IID, IIA, IIB, IIE and IIF a pre-initiation complex is formed. RNA polymerase II then binds to this complex and begins transcription from the start point + 1.

of other transcription factors also bind sequentially to form an initiation complex that includes RNA polymerase, subsequent to which transcription is initiated. In addition to the TATA box, a CAT box (consensus GGCCAATCT) is often located at about −80 bp, which is an important determinant of promoter efficiency. Many upstream promoter elements (UPEs) have been described that are either general in their action or tissue (or gene) specific. GC elements that contain the sequence GGGCG may be present at multiple sites and in either orientation and are often associated with housekeeping genes such as those encoding enzymes involved in general metabolism. Some promoter sequence elements, such as the TATA box, are common to most genes, while others may be specific to particular genes or classes of genes.

Of particular interest is a class of promoter first investigated in the virus SV40 and termed an enhancer. These sequences are distinguished from other promoter sequences by their unique ability to function over several kilobases either upstream or downstream of a particular gene in an orientation-independent manner. Even at such great distances from the transcription start point they may increase transcription by several hundred-fold. The precise interactions between transcription factors, RNA polymerase or other DNA binding proteins and the DNA sequences they bind to may be identified and characterised by the technique of DNA footprinting (Section 6.8.3). For transcription in eukaryotic cells to proceed a number of transcription factors need to interact with the promoters and with each other. This cascade mechanism is indicated in Fig. 5.15c and is termed a pre-initiation complex. Once this has been formed around the −35 TATA sequence RNA polymerase II is able to transcribe the structural gene and form a complementary RNA copy (Section 5.5.6).

5.5.5 Transcription in prokaryotes

Prokaryotic gene organisation differs from that found in eukaryotes in a number of ways. Prokaryotic genes are generally found as continuous coding sequences which are not interrupted. Moreover they are frequently found clustered into operons which contain genes that relate to a particular function such as the metabolism of a substrate or synthesis of a product. This is particularly evident in the best-known operon identified in *E. coli* termed the lactose operon where three genes *lacZ*, *lacY* and *lacA* share the same promoter and are therefore switched on and off at the same time. In this model the absence of lactose results in a repressor protein binding to an operator region upstream of the *Z*, *Y* and *A* gene and prevents RNA polymerase from transcribing the genes (Fig. 5.16a). However the presence of lactose requires the genes to be transcribed to allow its metabolism. Lactose binds to the repressor protein and causes a conformational change in its structure. This prevents it binding to the operator and allows RNA polymerase to bind and transcribe the three genes (Fig. 5.16b). Transcription and translation in prokaryotes is also closely linked or coupled whereas in eukaryotic cells the two processes are distinct and take place in different cell compartments.

5.5.6 Post-transcriptional processing

Transcription of a eukaryotic gene results in the production of a heterogeneous nuclear RNA transcript (hnRNA) which faithfully represents the entire structural gene (Fig. 5.17). Three processing events then take place. The first processing step involves the addition of a methylated guanosine residue (m7Gppp) termed a cap to the 5′ end of the hnRNA. This may be a signalling structure or aid in the stability of the molecule (Fig. 5.18). In addition, 150 to 300 adenosine residues termed a poly(A) tail are attached at the 3′ end of the hnRNA by the enzyme poly(A) polymerase. The poly(A) tail allows the specific isolation of eukaryotic mRNA from total RNA by affinity chromatography (Section 5.7.2); its presence is thought to confer stability on the transcript.

Unlike prokaryotic transcripts those from eukaryotes have their coding sequence (expressed regions or exons) interrupted by non-coding sequence (intervening

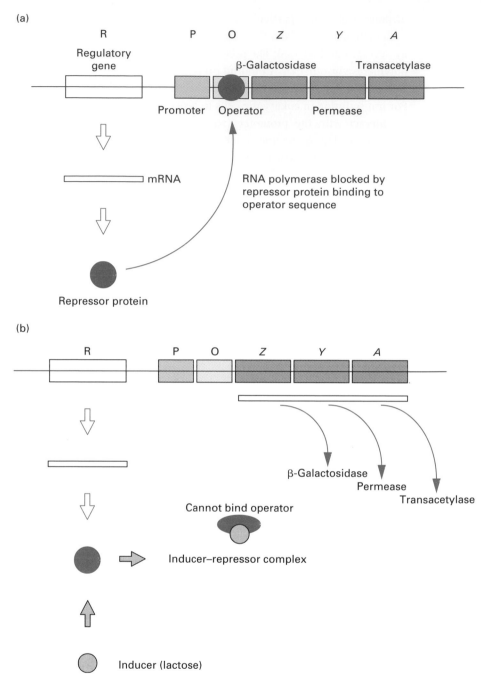

Fig. 5.16 Lactose operon (a) in a state of repression (no lactose present) and (b) following induction by lactose.

regions or introns). Intron–exon boundaries are generally determined by the sequence GU–AG and need to be removed or spliced before the mature mRNA is formed (Fig. 5.18). The process of intron splicing is mediated by small nuclear RNAs (snRNAs) which exist in the nucleus as ribonuclear protein particles. These are often found in

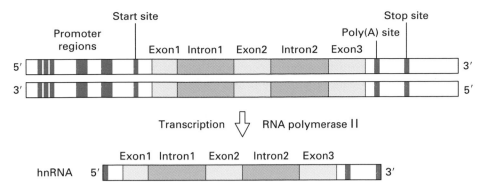

Fig. 5.17 Transcription of a typical eukaryotic gene to form heterogeneous nuclear RNA.

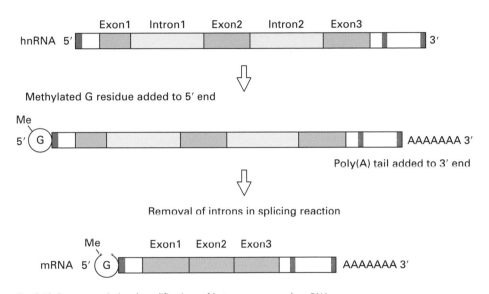

Fig. 5.18 Post-transcriptional modifications of heterogeneous nuclear RNA.

a large nuclear structure complex termed the spliceosome where splicing takes place. Introns are usually removed in a sequential manner from the 5′ to the 3′ end and their number varies between different genes. Some eukaryotic genes such as histone genes contain no introns whereas the gene for dystrophin, the gene responsible for muscular dystrophy, contains over 250 introns. In some cases, however, the same hnRNA transcript may be processed in different ways to produce different mRNAs coding for different proteins in a process known as alternative splicing. Thus a sequence that constitutes an exon for one RNA species may be part of an excised intron in another. The particular type or amount of mRNA synthesised from a cell or cell type may be analysed by a variety of molecular biology techniques (Section 6.8.1).

5.5.7 **Translation of mRNA**

Messenger RNA molecules are read and translated into protein by complex RNA–protein particles termed ribosomes. The ribosomes are termed 70S or 80S depending on their sedimentation coefficient. Prokaryotic cells have 70S ribosomes whilst those of the eukaryotic cytoplasm are 80S. Ribosomes are composed of two subunits that are held apart by ribosomal binding proteins until translation proceeds. There are sites on the ribosome for the binding of one mRNA and two tRNA molecules and the translation process is in three stages.

- *Initiation*: involving the assembly of the ribosome subunits and the binding of the mRNA.
- *Elongation*: where specific amino acids are used to form polypeptides, this being directed by the codon sequence in the mRNA.
- *Termination*: which involves the disassembly of the components of translation following the production of a polypeptide.

Transfer RNA molecules are also essential for translation. Each of these are covalently linked to a specific amino acid, forming an aminoacyl tRNA, and each has a triplet of bases exposed which is complementary to the codon for that amino acid. This exposed triplet is known as the anticodon, and allows the tRNA to act as an 'adapter' molecule, bringing together a codon and its corresponding amino acid. The process of linking an amino acid to its specific tRNA is termed charging and is carried out by the enzyme aminoacyl tRNA synthetase.

In prokaryotic cells the ribosome binds to the 5′ end of the mRNA at a sequence known as a ribosome binding site or sometimes termed the Shine–Dalgarno sequence after the discoverers of the sequence. In eukaryotes the situation is similar but involves a Kozak sequence located around the initiation codon. Following translation initiation the ribosome moves towards the 3′ end of the mRNA, allowing an aminoacyl tRNA molecule to base-pair with each successive codon, thereby carrying in amino acids in the correct order for protein synthesis. There are two sites for tRNA molecules in the ribosome, the A site and the P site, and when these sites are occupied, directed by the sequence of codons in the mRNA, the ribosome allows the formation of a peptide bond between the amino acids. The process is also under the control of an enzyme, peptidyl transferase. When the ribosome encounters a termination codon (UAA, UGA or UAG) a release factor binds to the complex and translation stops, the polypeptide and its corresponding mRNA are released and the ribosome divides into its two subunits (Fig. 5.19). A myriad of accessory initiation and elongation protein factors are involved in this process. In eukaryotic cells the polypeptide may then be subjected to post-translational modifications such as glycosylation and by virtue of specific amino acid signal sequences may be directed to specific cellular compartments or exported from the cell.

Since the mRNA base sequence is read in triplets, an error of one or two nucleotides in positioning of the ribosome will result in the synthesis of an incorrect polypeptide. Thus it is essential for the correct reading frame to be used during translation. This is ensured

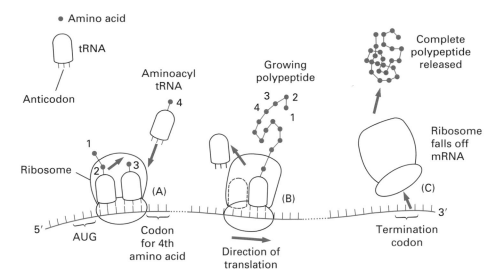

Fig. 5.19 Translation. Ribosome A has moved only a short way from the 5' end of the mRNA, and has built up a dipeptide (on one tRNA) that is about to be transferred onto the third amino acid (still attached to tRNA). Ribosome B has moved much further along the mRNA and has built up an oligopeptide that has just been transferred onto the most recent aminoacyl tRNA. The resulting free tRNA leaves the ribosome and will receive another amino acid. The ribosome moves towards the 3' end of the mRNA by a distance of three nucleotides, so that the next codon can be aligned with its corresponding aminoacyl tRNA on the ribosome. Ribosome C has reached a termination codon, has released the completed polypeptide, and has fallen off the mRNA.

in prokaryotes by base-pairing between the Shine–Dalgarno sequence (Kozak sequence in eukaryotes) and a complementary sequence of one of the ribosome's rRNAs, thus establishing the correct starting point for movement of the ribosome along the mRNA. However if a mutation such as a deletion/insertion takes place within the coding sequence it will also cause a shift of the reading frame and result in an aberrant polypeptide. Genetic mutations and polymorphisms are considered in more detail in Section 6.8.6.

5.5.8 Control of protein production – RNA interference

There are a number of mechanisms by which protein production is controlled; however the control may be either at the gene level or at the protein level. Typically this could include controlling levels of expression of mRNA, an increase or decrease in mRNA turnover, or controlling mRNA availability for translation. One recently discovered control mechanism that has also been adapted as a molecular biology technique to aid in the modulation of mRNA is termed *RNA interference* (RNAi). This involves the synthesis of short double-stranded RNA molecules which are cleaved into 21–23 nucleotide-long fragments to form an RNA-induced silencing complex (RISC). This complex potentially uses the short RNA molecules complementary to mRNA transcripts which, following hybridisation, allow an RNase to destroy the bound mRNA. The technique has important implications for medical conditions where, for example, increased levels of specific mRNA molecules in certain cancers and viral infections may be reduced using RNAi.

5.6 THE MANIPULATION OF NUCLEIC ACIDS – BASIC TOOLS AND TECHNIQUES

5.6.1 Enzymes used in molecular biology

The discovery and characterisation of a number of key enzymes has enabled the development of various techniques for the analysis and manipulation of DNA. In particular the enzymes termed type II restriction endonucleases have come to play a key role in all aspects of molecular biology. These enzymes recognise certain DNA sequences, usually 4–6 bp in length, and cleave them in a defined manner. The sequences recognised are palindromic or of an inverted repeat nature. That is they read the same in both directions on each strand. When cleaved they leave a flush-ended or staggered (also termed a cohesive-ended) fragment depending on the particular enzyme used (Fig. 5.20). An important property of staggered ends is that those produced from different molecules by the same enzyme are complementary (or 'sticky') and so will anneal to each other. The annealed strands are held together only by hydrogen bonding between complementary bases on opposite strands. Covalent joining of ends on each of the two strands may be brought about by the enzyme DNA ligase (Section 6.2.2). This is widely exploited in molecular biology to enable the construction of recombinant DNA, i.e. the joining of DNA fragments from different sources. Approximately 500 restriction

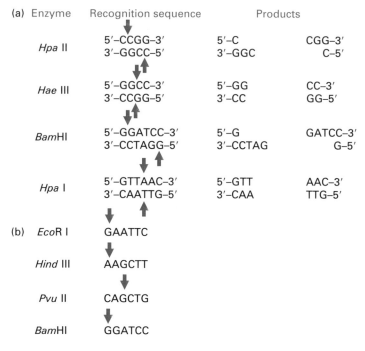

Fig. 5.20 Recognition sequences of some restriction enzymes showing (a) full descriptions and (b) conventional representations. Arrows indicate positions of cleavage. Note that all the information in (a) can be derived from knowledge of a single strand of the DNA, whereas in (b) only one strand is shown, drawn 5′ to 3′; this is the conventional way of representing restriction sites.

Table 5.3 **Types and examples of typical enzymes used in the manipulation of nucleic acids**

Enzyme	Specific example	Use in nucleic acid manipulation
	DNA pol I	DNA-dependent DNA polymerase $5'\rightarrow3'\rightarrow5'$ exonuclease activity
	Klenow	DNA pol I lacks $5'\rightarrow3'$ exonuclease activity
	T4 DNA pol	Lacks $5'\rightarrow3'$ exonuclease activity
DNA polymerases	Taq DNA pol	Thermostable DNA polymerase used in PCR
	Tth DNA pol	Thermostable DNA polymerase with RT activity
	T7 DNA pol	Used in DNA sequencing
	T7 RNA pol	DNA-dependent RNA polymerase
RNA polymerases	T3 RNA pol	DNA-dependent RNA polymerase
	Qß replicase	RNA-dependent RNA polymerase, used in RNA amplification
	DNase I	Non-specific endonuclease that cleaves DNA
	Exonuclease III	DNA-dependent $3'\rightarrow5'$ stepwise removal of nucleotides
Nucleases	RNase A	RNases used in mapping studies
	RNase H	Used in second strand cDNA synthesis
	S1 nuclease	Single-strand-specific nuclease
Reverse transcriptase	AMV-RT	RNA-dependent DNA polymerase, used in cDNA synthesis
Transferases	Terminal transferase (TdT)	Adds homopolymer tails to the $3'$ end of DNA
Ligases	T4 DNA ligase	Links $5'$-phosphate and $3'$-hydroxyl ends via phosphodiester bond
Kinases	T4 polynucleotide kinase (PNK)	Transfers terminal phosphate groups from ATP to $5'$-OH groups
Phosphatases	Alkaline phosphatase	Removes $5'$-phosphates from DNA and RNA
Transferases	Terminal transferase	Adds homopolymer tails to the $3'$ end of DNA
Methylases	EcoRI methylase	Methylates specific residues and protects from cleavage by restriction enzymes

Notes: PCR, polymerase chain reaction; RT, reverse transcriptase; cDNA, complementary DNA; AMV, avian myeloblastosis virus.

enzymes have been characterised that recognise over 100 different target sequences. A number of these, termed isoschizomers, recognise different target sequences but produce the same staggered ends or overhangs. A number of other enzymes have proved to be of value in the manipulation of DNA, as summarised in Table 5.3, and are indicated at appropriate points within the text.

5.7 ISOLATION AND SEPARATION OF NUCLEIC ACIDS

5.7.1 Isolation of DNA

The use of DNA for analysis or manipulation usually requires that it is isolated and purified to a certain extent. DNA is recovered from cells by the gentlest possible method of cell rupture to prevent the DNA from fragmenting by mechanical shearing. This is usually in the presence of EDTA which chelates the Mg^{2+} ions needed for enzymes that degrade DNA termed DNase. Ideally, cell walls, if present, should be digested enzymatically (e.g. lysozyme treatment of bacteria), and the cell membrane should be solubilised using detergent. If physical disruption is necessary, it should be kept to a minimum, and should involve cutting or squashing of cells, rather than the use of shear forces. Cell disruption (and most subsequent steps) should be performed at 4 °C, using glassware and solutions that have been autoclaved to destroy DNase activity.

After release of nucleic acids from the cells, RNA can be removed by treatment with ribonuclease (RNase) that has been heat-treated to inactivate any DNase contaminants; RNase is relatively stable to heat as a result of its disulphide bonds, which ensure rapid renaturation of the molecule on cooling. The other major contaminant, protein, is removed by shaking the solution gently with water-saturated phenol, or with a phenol/chloroform mixture, either of which will denature proteins but not nucleic acids. Centrifugation of the emulsion formed by this mixing produces a lower, organic phase, separated from the upper, aqueous phase by an interface of denatured protein. The aqueous solution is recovered and deproteinised repeatedly, until no more material is seen at the interface. Finally, the deproteinised DNA preparation is mixed with two volumes of absolute ethanol, and the DNA allowed to precipitate out of solution in a freezer. After centrifugation, the DNA pellet is redissolved in a buffer containing EDTA to inactivate any DNases present. This solution can be stored at 4 °C for at least a month. DNA solutions can be stored frozen although repeated freezing and thawing tends to damage long DNA molecules by shearing. The procedure described above is suitable for total cellular DNA. If the DNA from a specific organelle or viral particle is needed, it is best to isolate the organelle or virus before extracting its DNA, since the recovery of a particular type of DNA from a mixture is usually rather difficult. Where a high degree of purity is required DNA may be subjected to density gradient ultracentrifugation through caesium chloride which is particularly useful for the preparation of plasmid DNA. A flow chart of DNA extraction is indicated in Fig. 5.21.

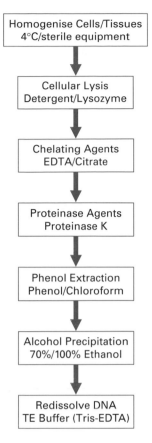

Fig. 5.21 General steps involved in extracting DNA from cells or tissues.

It is possible to check the integrity of the DNA by agarose gel electrophoresis and determine the concentration of the DNA by using the fact that 1 absorbance unit equates to $50\,\mu g\,ml^{-1}$ of DNA and so:

$$50 \times A_{260} = \text{concentration of DNA sample } (\mu g\,ml^{-1})$$

Contaminants may also be identified by scanning UV spectrophotometry from 200 nm to 300 nm. A ratio of 260 nm : 280 nm of approximately 1.8 indicates that the sample is free of protein contamination, which absorbs strongly at 280 nm.

5.7.2 Isolation of RNA

The methods used for RNA isolation are very similar to those described above for DNA; however, RNA molecules are relatively short, and therefore less easily damaged by shearing, so cell disruption can be rather more vigorous. RNA is, however, very vulnerable to digestion by RNases which are present endogenously in various concentrations in certain cell types and exogenously on fingers. Gloves should therefore

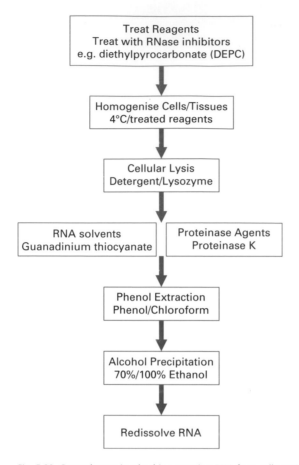

Fig. 5.22 General steps involved in extracting RNA from cells or tissues.

be worn, and a strong detergent should be included in the isolation medium to immediately denature any RNases. Subsequent deproteinisation should be particularly rigorous, since RNA is often tightly associated with proteins. DNase treatment can be used to remove DNA, and RNA can be precipitated by ethanol. One reagent in particular which is commonly used in RNA extraction is guanadinium thiocyanate which is both a strong inhibitor of RNase and a protein denaturant. A flow chart of RNA extraction is indicated in Fig. 5.22. It is possible to check the integrity of an RNA extract by analysing it by agarose gel electrophoresis. The most abundant RNA species, the rRNA molecules 23S and 16S for prokaryotes and 18S and 28S for eukaryotes, appear as discrete bands on the agarose gel and thus indicate that the other RNA components are likely to be intact. This is usually carried out under denaturing conditions to prevent secondary structure formation in the RNA. The concentration of the RNA may be estimated by using UV spectrophotometry. At 260 nm 1 absorbance unit equates to $40\,\mu\mathrm{g}\,\mathrm{ml}^{-1}$ of RNA and therefore:

$$40 \times A_{260} = \text{concentration of DNA sample } (\mu\mathrm{g}\,\mathrm{ml}^{-1})$$

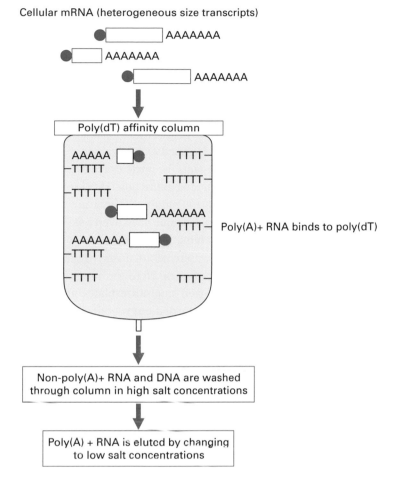

Cellular mRNA (heterogeneous size transcripts)

Poly(dT) affinity column

Poly(A)+ RNA binds to poly(dT)

Non-poly(A)+ RNA and DNA are washed through column in high salt concentrations

Poly(A) + RNA is eluted by changing to low salt concentrations

Fig. 5.23 Affinity chromatography of poly(A)+RNA.

Contaminants may also be identified in the same way as that for DNA by scanning UV spectrophotometry; however, in the case of RNA a 260 nm : 280 nm ratio of approximately 2 would be expected for a sample containing no protein (Section 5.8.1).

In many cases it is desirable to isolate eukaryotic mRNA which constitutes only 2–5% of cellular RNA from a mixture of total RNA molecules. This may be carried out by affinity chromatography on oligo(dT)-cellulose columns. At high salt concentrations, the mRNA containing poly(A) tails binds to the complementary oligo(dT) molecules of the affinity column, and so mRNA will be retained; all other RNA molecules can be washed through the column by further high salt solution. Finally, the bound mRNA can be eluted using a low concentration of salt (Fig. 5.23). Nucleic acid species may also be subfractionated by more physical means such as electrophoretic or chromatographic separations based on differences in nucleic acid fragment sizes or physicochemical characteristics. Nanodrop spectrophotometer systems have also aided the analysis of nucleic acids in recent years in allowing the full spectrum of information whilst requiring only a very small (microlitre) sample volume.

5.7.3 **Automated and kit-based extraction of nucleic acids**

Most of the current reagents used in molecular biology and the most common techniques can now be found in kit form or can be automated, and the extraction of nucleic acids by these means is no exception. The advantage of their use lies in the fact that the reagents are standardised and quality control tested providing a high degree of reliability. For example glass bead preparations for DNA purification have been used increasingly and with reliable results. Small compact column-type preparations such as QIAGEN columns are also used extensively in research and in routine DNA analysis. Essentially the same reagents for nucleic acid extraction may be used in a format that allows reliable and automated extraction. This is of particular use where a large number of DNA extractions are required. There are also many kit-based extraction methods for RNA; these in particular have overcome some of the problems of RNA extraction such as RNase contamination. A number of fully automated nucleic acid extraction machines are now employed in areas where high throughput is required, e.g. clinical diagnostic laboratories. Here the raw samples such as blood specimens are placed in 96- or 384-well microtitre plates and these follow a set computer-controlled processing pattern carried out robotically. Thus the samples are rapidly manipulated and extracted in approximately 45 min without any manual operations being undertaken.

5.7.4 **Electrophoresis of nucleic acids**

Electrophoresis in agarose or polyacrylamide gels is the most usual way to separate DNA molecules according to size. The technique can be used analytically or preparatively, and can be qualitative or quantitative. Large fragments of DNA such as chromosomes may also be separated by a modification of electrophoresis termed pulsed field gel electrophoresis (PFGE). The easiest and most widely applicable method is electrophoresis in horizontal agarose gels, followed by staining with ethidium bromide. This dye binds to DNA by insertion between stacked base pairs (intercalation), and it exhibits a strong orange/red fluorescence when illuminated with ultraviolet light (Fig. 5.24). Very often electrophoresis is used to check the purity and intactness of a DNA preparation or to assess the extent of an enzymatic reaction during for example the steps involved in the cloning of DNA. For such checks 'minigels' are particularly convenient, since they need little preparation, use small samples and give results quickly. Agarose gels can be used to separate molecules larger than about 100 bp. For higher resolution or for the effective separation of shorter DNA molecules polyacrylamide gels are the preferred method.

When electrophoresis is used preparatively, the piece of gel containing the desired DNA fragment is physically removed with a scalpel. The DNA may be recovered from the gel fragment in various ways. This may include crushing with a glass rod in a small volume of buffer, using agarase to digest the agarose leaving the DNA, or by the process of electroelution. In this method the piece of gel is sealed in a length of dialysis tubing containing buffer, and is then placed between two electrodes in a tank containing more buffer. Passage of an electrical current between the electrodes causes

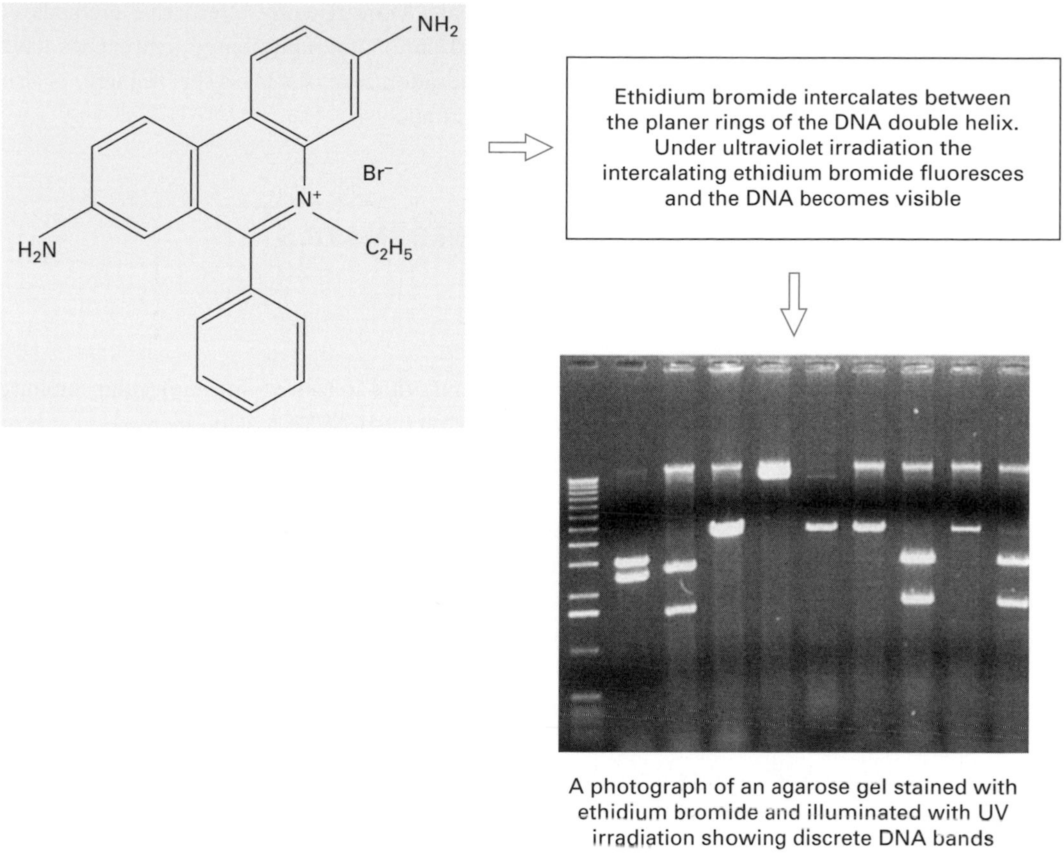

Ethidium bromide intercalates between the planer rings of the DNA double helix. Under ultraviolet irradiation the intercalating ethidium bromide fluoresces and the DNA becomes visible

A photograph of an agarose gel stained with ethidium bromide and illuminated with UV irradiation showing discrete DNA bands

Fig. 5.24 The use of ethidium bromide to detect DNA.

DNA to migrate out of the gel piece, but it remains trapped within the dialysis tubing, and can therefore be recovered easily.

5.7.5 Automated analysis of nucleic acid fragments

Gel electrophoresis remains the established method for the separation and analysis of nucleic acids. However a number of automated systems using pre-cast gels and standardised reagents are available that are now very popular. This is especially useful in situations where a large number of samples or high-throughput analysis is required. In addition technologies such as the Agilents' Lab-on-a-chip have been developed that obviate the need to prepare electrophoretic gels. These employ microfluidic circuits constructed on small cassette units that contain interconnected micro-reservoirs. The sample is applied in one area and driven through microchannels under computer-controlled electrophoresis. The channels lead to reservoirs allowing, for example, incubation with other reagents such as dyes for a specified time. Electrophoretic separation is thus carried out in a microscale format. The small sample size minimises sample and reagent consumption and the units, being computer controlled, allow data to be captured within a very short timescale. More recently alternative methods of analysis including high performance liquid chromatography based approaches have gained in popularity, especially for DNA mutation analysis. Mass spectrometry is also becoming increasingly used for nucleic acid analysis.

5.8   MOLECULAR BIOLOGY AND BIOINFORMATICS

5.8.1   Basic bioinformatics

Bioinformatics is now an established and vital resource for molecular biology research and is also a mainstay of routine analysis of DNA. This increase in use of bioinformatics has been driven by the increase in genetic sequence information and the need to store, analyse and manipulate the data. There are now a huge number of sequences stored in genetic databases from a variety of organisms, including the human genome. Indeed the genetic information from various organisms is now an indispensable starting point for molecular biology research. The main primary databases include GenBank at the National Institutes of Health (NIH) in the USA, EMBL at the European Bioinformatics Institute (EBI) at Cambridge, UK and the DNA Database of Japan (DDBJ) at Mishima in Japan. These databases contain the nucleotide sequences which are annotated to allow easy identification. There are also many other databases such as secondary databases that contain information relating to sequence motifs, such as core sequences found in cytochrome P450 domains, or DNA-binding domains. Importantly all of the databases may be freely accessed over the internet. A number of these important databases and internet resources are listed in Table 5.4. Consequently the new expanding and exciting areas of bioscience research are those that analyse genome and cDNA sequence databases (genomics) and also their protein counterparts (proteomics). This is sometimes referred to as *in silico* research.

5.8.2   Analysing information using bioinformatics

One of the most useful bioinformatics resources is termed BLAST (Basic Local Alignment Search Tool) located at the NCBI (www.ncbi.nlm.nih.gov). This allows a DNA sequence to be submitted via the internet in order to compare it to all the sequences contained within a DNA database. This is very useful since it is possible once a nucleotide sequence has been deduced by, for example, Sanger sequencing, to identify sequences of similarity. Indeed if human sequences are used and have already been mapped it is possible to locate their position to a particular chromosome using NCBI Map Viewer. Further resources such as ORF (open reading frame) finder allow a search to be undertaken for open reading frames, e.g. sequences beginning with a start codon (ATG) and continuing with a significant number of 'coding' triplets before a stop codon is reached. There are a number of other sequences that may be used to define coding sequences; these include ribosome binding sites, splice site junctions, poly(A) polymerase sequences and promoter sequences that lie outside the coding

Table 5.4 **Nucleic acid and protein database resources available on the World Wide Web**

Database or resource		URL (uniform resource locator)
General DNA sequence databases		
EMBL	European Bioinformatics Institute	\<http://www.ebi.ac.uk\>
GenBank	US genetic database resource	\<http://www.ncbi.nlm.nih.gov\>
DDBJ	Japanese genetic database	\<http://www.ddbj.nig.ac.jp\>
Protein sequence databases		
Swiss-Prot	European protein sequence database	\<http://www.expasy.org\>
UniProt TREMBL	European protein sequence database	\<http://www.ebi.ac.uk/trembl\>
Protein structure databases		
PDB	Protein structure database	\<http://www.rcsb.org\>
Genome project databases		
Human Genome Database, USA		\<http://gdbwww.gdb.org\>
dbEST	cDNA and partial sequences	\<http://www.ncbi.nih.gov/dbEST/index.html\>
Généthon	Genetic maps based on repeat markers	\<http://www.genethon.fr\>

regions. A number of bioinformatics resources such as GRAIL can be used to identify such features in a DNA sequence.

5.9 MOLECULAR ANALYSIS OF NUCLEIC ACID SEQUENCES

5.9.1 Restriction mapping of DNA fragments

Restriction mapping involves the size analysis of restriction fragments produced by several restriction enzymes individually and in combination (Section 5.6.1). The principle of this mapping is illustrated in Fig. 5.25, in which the restriction sites of two enzymes, A and B, are being mapped. Cleavage with A gives fragments 2 and 7 kb from a 9 kb molecule, hence we can position the single A site 2 kb from one end. Similarly, B gives fragments 3 and 6 kb, so it has a single site 3 kb from one end; but it is not possible at this stage to say if it is near to A's site, or at the opposite end of the DNA. This can be resolved by a double digestion. If the resultant fragments are 2, 3 and 4 kb, then A and B cut at opposite ends of the molecule; if they are 1, 2 and 6 kb, the sites are near each other. Not surprisingly, the mapping of real molecules is rarely

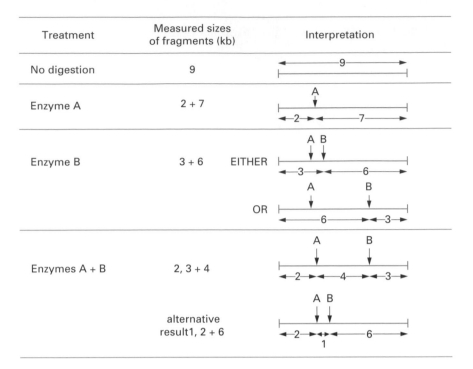

Treatment	Measured sizes of fragments (kb)	Interpretation
No digestion	9	
Enzyme A	2 + 7	
Enzyme B	3 + 6 EITHER	
	OR	
Enzymes A + B	2, 3 + 4	
alternative result1, 2 + 6		

Fig. 5.25 Restriction mapping of DNA. Note that each experimental result and its interpretation should be considered in sequence, thus building up an increasingly unambiguous map.

as simple as this, and bioinformatic analysis of the restriction fragment lengths is usually needed to construct a map.

5.9.2 Nucleic acid blotting methods

Electrophoresis of DNA restriction fragments allows separation based on size to be carried out, however it provides no indication as to the presence of a specific, desired fragment among the complex sample. This can be achieved by transferring the DNA from the intact gel onto a piece of nitrocellulose or nylon membrane placed in contact with it. This provides a more permanent record of the sample since DNA begins to diffuse out of a gel that is left for a few hours. First the gel is soaked in alkali to render the DNA single stranded. It is then transferred to the membrane so that the DNA becomes bound to it in exactly the same pattern as that originally on the gel. This transfer, named a Southern blot after its inventor Ed Southern, can be performed electrophoretically or by drawing large volumes of buffer through both gel and membrane, thus transferring DNA from one to the other by capillary action (Fig. 5.26). The point of this operation is that the membrane can now be treated with a labelled DNA molecule, for example a gene probe (Section 5.9.4). This single-stranded DNA probe will hybridise under the right conditions to complementary fragments immobilised onto the membrane. The conditions of hybridisation, including the temperature and salt concentration, are critical for this process to take place effectively. This is usually referred to as

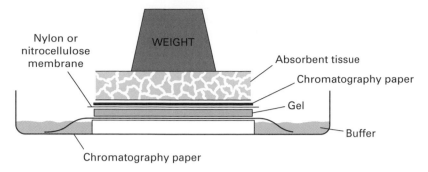

Fig. 5.26 Southern blot apparatus.

the stringency of the hybridisation and it is particular for each individual gene probe and for each sample of DNA. A series of washing steps with buffer is then carried out to remove any unbound probe and the membrane is developed after which the precise location of the probe and its target may be visualised. It is also possible to analyse DNA from different species or organisms by blotting the DNA and then using a gene probe representing a protein or enzyme from one of the organisms. In this way it is possible to search for related genes in different species. This technique is generally termed zoo blotting.

The same basic process of nucleic acid blotting can be used to transfer RNA from gels onto similar membranes. This allows the identification of specific mRNA sequences of a defined length by hybridisation to a labelled gene probe and is known as Northern blotting. It is possible with this technique to not only detect specific mRNA molecules but it may also be used to quantify the relative amounts of the specific mRNA. It is usual to separate the mRNA transcripts by gel electrophoresis under denaturing conditions since this improves resolution and allows a more accurate estimation of the sizes of the transcripts (Section 5.7.2). The format of the blotting may be altered from transfer from a gel to direct application to slots on a specific blotting apparatus containing the nylon membrane. This is termed slot or dot blotting and provides a convenient means of measuring the abundance of specific mRNA transcripts without the need for gel electrophoresis; it does not, however, provide information regarding the size of the fragments.

5.9.3 Design and production of gene probes

The availability of a gene probe is essential in many molecular biology techniques yet in many cases is one of the most difficult steps. The information needed to produce a gene probe may come from many sources; however, the availability of bioinformatics resources and genetic databases has ensured that this is the usual starting point for gene probe design.

In some cases it is possible to use related genes, that is from the same gene family, to gain information on the most useful DNA sequence to use as a probe. Similar proteins or DNA sequences but from different species may also provide a starting

Polypeptide		Phe	Met	Pro	Trp	His	
		T		T		T	
Corresponding nucleotide sequences	5′	TTC	ATC	CCC	TGG	CAC	3′
				A			
				G			

Fig. 5.27 Oligonucleotide probes. Note that only methionine and tryptophan have unique codons. It is impossible to predict which of the indicated codons for phenylalanine, proline and histidine will be present in the gene to be probed, so all possible combinations must be synthesised (16 in the example shown).

point with which to produce a so-called heterologous gene probe. Although in some cases probes are already produced and cloned it is possible, armed with a DNA sequence from a DNA database, to chemically synthesise a single-stranded oligo-nucleotide probe. This is usually undertaken by computer-controlled gene synthesisers which link dNTPs (deoxyribonucleoside triphosphates) together based on a desired sequence. It is essential to carry out certain checks before probe production to determine that the probe is unique, is not able to self-anneal or that it is self-complementary, all of which may compromise its use.

Where little DNA information is available to prepare a gene probe it is possible in some cases to use the knowledge gained from analysis of the corresponding protein. Thus it is possible to isolate and purify proteins and sequence part of the N-terminal end or an internal region of the protein. From our knowledge of the genetic code, it is possible to predict the various DNA sequences that could code for the protein, and then synthesise appropriate oligonucleotide sequences chemically. Due to the degen-eracy of the genetic code most amino acids are coded for by more than one codon, therefore there will be more than one possible nucleotide sequence that could code for a given polypeptide (Fig. 5.27). The longer the polypeptide, the greater the number of possible oligonucleotides that must be synthesised. Fortunately, there is no need to synthesise a sequence longer than about 20 bases, since this should hybridise effi-ciently with any complementary sequences, and should be specific for one gene. Ideally, a section of the protein should be chosen which contains as many tryptophan and methionine residues as possible, since these have unique codons, and there will therefore be fewer possible base sequences that could code for that part of the protein. The synthetic oligonucleotides can then be used as probes in a number of molecular biology methods.

5.9.4 Labelling DNA gene probe molecules

An essential feature of a gene probe is that it can be visualised or labelled by some means. This allows any complementary sequence that the probe binds to be flagged up or identified.

There are two main types of label used for gene probes: traditionally this has been carried out using radioactive labels, but gaining in popularity are non-radioactive labels.

Perhaps the most common radioactive label is 32-phosphorus (^{32}P), although for certain techniques 35-sulphur (^{35}S) and tritium (^{3}H) are used. These may be detected by the process of autoradiography where the labelled probe molecule, bound to sample DNA, located for example on a nylon membrane, is placed in contact with an X-ray-sensitive film. Following exposure the film is developed and fixed just as a black-and-white negative. The exposed film reveals the precise location of the labelled probe and therefore the DNA to which it has hybridised.

Non-radioactive labels are increasingly being used to label DNA gene probes. Until recently radioactive labels were more sensitive than their non-radioactive counterparts. However, recent developments have led to similar sensitivities which, when combined with their improved safety, have led to their greater acceptance.

The labelling systems are either termed direct or indirect. Direct labelling allows an enzyme reporter such as alkaline phosphatase to be coupled directly to the DNA. Although this may alter the characteristics of the DNA gene probe it offers the advantage of rapid analysis since no intermediate steps are needed. However indirect labelling is at present more popular. This relies on the incorporation of a nucleotide which has a label attached. At present three of the main labels in use are biotin, fluorescein and digoxygenin. These molecules are covalently linked to nucleotides using a carbon spacer arm of 7, 14 or 21 atoms. Specific binding proteins may then be used as a bridge between the nucleotide and a reporter protein such as an enzyme. For example, biotin incorporated into a DNA fragment is recognised with a very high affinity by the protein streptavidin. This may either be coupled or conjugated to a reporter enzyme molecule such as alkaline phosphatase. This is able to convert a colourless substrate p-nitrophenol phosphate (PNPP) into a yellow-coloured compound p-nitrophenol (PNP) and also offers a means of signal amplification. Alternatively labels such as digoxygenin incorporated into DNA sequences may be detected by monoclonal antibodies, again conjugated to reporter molecules such as alkaline phosphatase. Thus rather than the detection system relying on autoradiography which is necessary for radiolabels, a series of reactions resulting in the products of either a colour, light or the product of a chemiluminescence reaction take place. This has important practical implications since autoradiography may take 1–3 days whereas colour and chemiluminescent reactions take minutes.

5.9.5 End labelling of DNA molecules

The simplest form of labelling DNA is by 5′ or 3′ end-labelling. 5′ end labelling involves a phosphate transfer or exchange reaction where the 5′ phosphate of the DNA to be used as the probe is removed and in its place a labelled phosphate, usually ^{32}P, is added. This is usually carried out by using two enzymes; the first, alkaline phosphatase, is used to remove the existing phosphate group from the DNA. Following removal of the released phosphate from the DNA, a second enzyme, polynucleotide kinase, is added which catalyses the transfer of a phosphate group (^{32}P-labelled) to the 5′ end of the DNA. The newly labelled probe is then purified, usually by chromatography through a Sephadex column, and may be used directly (Fig. 5.28).

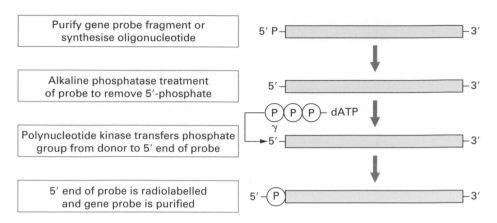

Fig. 5.28 End-labelling of a gene probe at the 5′ end with alkaline phosphatase and polynucleotide kinase.

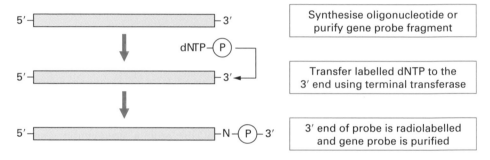

Fig. 5.29 End-labelling of a gene probe at the 3′ end using terminal transferase. Note that the addition of a labelled dNTP at the 3′ end alters the sequence of the gene probe.

Using the other end of the DNA molecule, the 3′ end, is slightly less complex. Here a new dNTP which is labelled (e.g. ^{32}P-αdATP or biotin-labelled dNTP) is added to the 3′ end of the DNA by the enzyme terminal transferase. Although this is a simpler reaction a potential problem exists because a new nucleotide is added to the existing sequence and so the complete sequence of the DNA is altered which may affect its hybridisation to its target sequence. End-labelling methods also suffer from the fact that only one label is added to the DNA so they are of a lower activity in comparison to methods which incorporate label along the length of the DNA (Fig. 5.29).

5.9.6 Random primer labelling and nick translation

The DNA to be labelled is first denatured and then placed under renaturing conditions in the presence of a mixture of many different random sequences of hexamers or hexanucleotides. These hexamers will, by chance, bind to the DNA sample wherever they encounter a complementary sequence and so the DNA will rapidly acquire an approximately random sprinkling of hexanucleotides annealed to it. Each of the hexamers can act as a primer for the synthesis of a fresh strand of DNA catalysed by DNA polymerase since it has an exposed 3′ hydroxyl group. The Klenow fragment of DNA polymerase is used for random primer labelling because it lacks a 5′ to

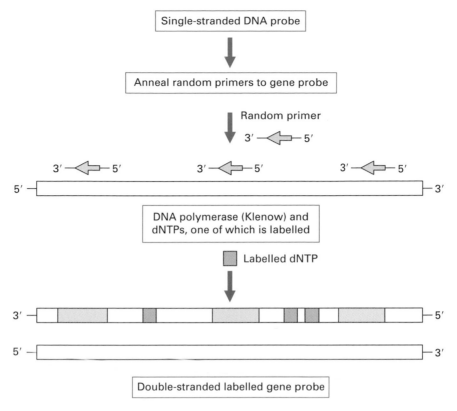

Fig 5.30 Random primer gene probe labelling. Random primers are incorporated and used as a start point for Klenow DNA polymerase to synthesise a complementary strand of DNA whilst incorporating a labelled dNTP at complementary sites.

3′ exonuclease activity. This is prepared by cleavage of DNA polymerase with subtilisin, giving a large enzyme fragment which has no 5′ to 3′ exonuclease activity, but which still acts as a 5′ to 3′ polymerase. Thus when the Klenow enzyme is mixed with the annealed DNA sample in the presence of dNTPs, including at least one which is labelled, many short stretches of labelled DNA will be generated (Fig. 5.30). In a similar way to random primer labelling the polymerase chain reaction may also be used to incorporate radioactive or non-radioactive labels (Section 5.11.4).

A further traditional method of labelling DNA is by the process of nick translation. Low concentrations of DNase I are used to make occasional single-strand nicks in the double-stranded DNA that is to be used as the gene probe. DNA polymerase then fills in the nicks, using an appropriate dNTP, at the same time making a new nick to the 3′ side of the previous one (Fig. 5.31). In this way the nick is translated along the DNA. If labelled dNTPs are added to the reaction mixture, they will be used to fill in the nicks, and so the DNA can be labelled to a very high specific activity.

5.9.7 Molecular-beacon-based probes

A more recent development in the design of labelled oligonucleotide hybridisation probes is that of molecular beacons. These probes contain a fluorophore at one end of the probe

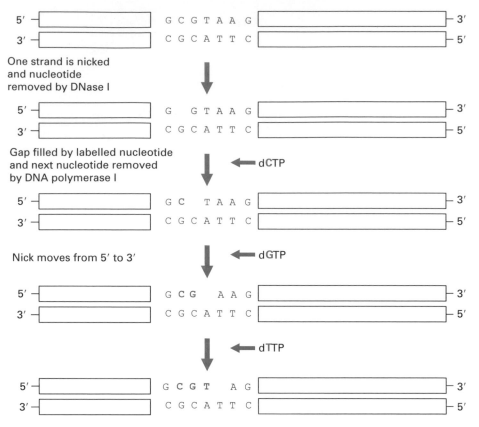

One strand is nicked
and nucleotide
removed by DNase I

Gap filled by labelled nucleotide
and next nucleotide removed
by DNA polymerase I

Nick moves from 5′ to 3′

Fig. 5.31 Nick translation. The removal of nucleotides and their subsequent replacement with labelled nucleotides by DNA polymerase I increase the label in the gene probe as nick translation proceeds.

and a quencher molecule at the other. The oligonucleotide has a stem–loop structure where the stems place the fluorophore and quencher in close proximity. The loop structure is designed to be complementary to the target sequence. When the stem–loop structure is formed the fluorophore is quenched by Förster or fluorescence resonance energy transfer (FRET), i.e. the energy is transferred from the fluorophore to the quencher and given off as heat. The elegance of these types of probe lies in the fact that upon hybridisation to a target sequence the stem and loop move apart, the quenching is then lost and emission of light occurs from the fluorophore upon excitation. These types of probe have also been used to detect nucleic acid amplification system products such as the polymerase chain reaction (PCR) and have the advantage that it is unnecessary to remove the unhybridised probes.

5.10 THE POLYMERASE CHAIN REACTION (PCR)

5.10.1 Basic concept of the PCR

The polymerase chain reaction or PCR is one of the mainstays of molecular biology. One of the reasons for the wide adoption of the PCR is the elegant simplicity of the

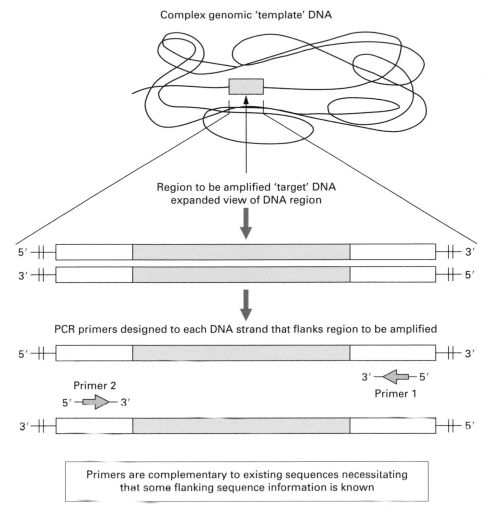

Fig. 5.32 The location of polymerase chain reaction (PCR) primers. PCR primers designed for sequences adjacent to the region to be amplified allow a region of DNA (e.g. a gene) to be amplified from a complex starting material of genomic template DNA.

reaction and relative ease of the practical manipulation steps. Indeed combined with the relevant bioinformatics resources for its design and for determination of the required experimental conditions it provides a rapid means for DNA identification and analysis. It has opened up the investigation of cellular and molecular processes to those outside the field of molecular biology.

The PCR is used to amplify a precise fragment of DNA from a complex mixture of starting material usually termed the template DNA and in many cases requires little DNA purification. It does require the knowledge of some DNA sequence information which flanks the fragment of DNA to be amplified (target DNA). From this information two oligonucleotide primers may be chemically synthesised each complementary to a stretch of DNA to the 3′ side of the target DNA, one oligonucleotide for each of the two DNA strands (Fig. 5.32). It may be thought of as a technique

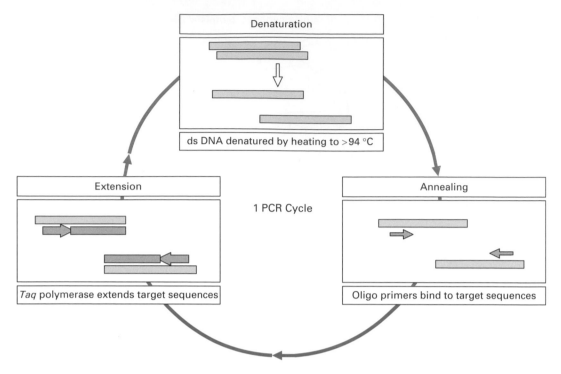

Fig. 5.33 A simplified scheme of one PCR cycle that involves denaturation, annealing and extension. ds, double-stranded.

analogous to the DNA replication process that takes place in cells since the outcome is the same: the generation of new complementary DNA stretches based upon the existing ones. It is also a technique that has replaced, in many cases, the traditional DNA cloning methods since it fulfils the same function, the production of large amounts of DNA from limited starting material; however, this is achieved in a fraction of the time needed to clone a DNA fragment (Chapter 6). Although not without its drawbacks the PCR is a remarkable development which is changing the approach of many scientists to the analysis of nucleic acids and continues to have a profound impact on core biosciences and biotechnology.

5.10.2 Stages in the PCR

The PCR consists of three defined sets of times and temperatures termed steps: (i) denaturation, (ii) annealing and (iii) extension. Each of these steps is repeated 30–40 times, termed cycles (Fig. 5.33). In the first cycle the double-stranded template DNA is (i) denatured by heating the reaction to above 90 °C. Within the complex DNA the region to be specifically amplified (target) is made accessible. The temperature is then cooled to 40–60 °C. The precise temperature is critical and each PCR system has to be defined and optimised. One useful technique for optimisation is touchdown PCR where a programmable cycler is used to incrementally decrease the annealing temperature until the optimum is derived. Reactions that are not optimised may give rise to other DNA products in addition to the specific target or may not produce any amplified products at all. The annealing step allows the hybridisation of the two oligonucleotide primers, which are present in excess, to bind to their complementary sites that flank the target DNA. The annealed oligonucleotides act as primers for DNA synthesis, since they provide a free 3′ hydroxyl group for DNA polymerase. The DNA synthesis step is termed extension and is carried out by a thermostable DNA polymerase, most commonly *Taq* DNA polymerase.

DNA synthesis proceeds from both of the primers until the new strands have been extended along and beyond the target DNA to be amplified. It is important to note that, since the new strands extend beyond the target DNA, they will contain a region near their 3′ ends that is complementary to the other primer. Thus, if another round of DNA synthesis is allowed to take place, not only the original strands will be used as templates but also the new strands. Most interestingly, the products obtained from the new strands will have a precise length, delimited exactly by the two regions complementary to the primers. As the system is taken through successive cycles of denaturation, annealing and extension all the new strands will act as templates and so there will be an exponential increase in the amount of DNA produced. The net effect is to selectively amplify the target DNA and the primer regions flanking it (Fig. 5.34).

One problem with early PCR reactions was that the temperature needed to denature the DNA also denatured the DNA polymerase. However the availability of a thermostable DNA polymerase enzyme isolated from the thermophilic bacterium *Thermus aquaticus* found in hot springs provided the means to automate the reaction. *Taq* DNA polymerase has a temperature optimum of 72 °C and survives prolonged exposure to temperatures as high as 96 °C and so is still active after each of the denaturation steps. The widespread utility of the technique is also due to the ability to automate the reaction and as such many thermal cyclers have been produced in which it is possible to program in the temperatures and times for a particular PCR reaction.

5.10.3 PCR primer design and bioinformatics

The specificity of the PCR lies in the design of the two oligonucleotide primers. These have to not only be complementary to sequences flanking the target DNA but also must not be self-complementary or bind each other to form dimers since both prevent DNA amplification. They also have to be matched in their GC content and have similar annealing temperatures. The increasing use of bioinformatics resources such as Oligo, Generunner and Genefisher in the design of primers makes the design and the selection of reaction conditions much more straightforward. These resources allow the sequences to be amplified, primer length, product size, GC content, etc. to be input and, following analysis, provide a choice of matched primer sequences. Indeed the initial selection and design of primers without the aid of bioinformatics would now be unnecessarily time-consuming.

It is also possible to design primers with additional sequences at their 5′ end such as restriction endonuclease target sites or promoter sequences. However modifications such as these require that the annealing conditions be altered to compensate for the areas of non-homology in the primers. A number of PCR methods have been developed where either one of the primers or both are random. This gives rise to

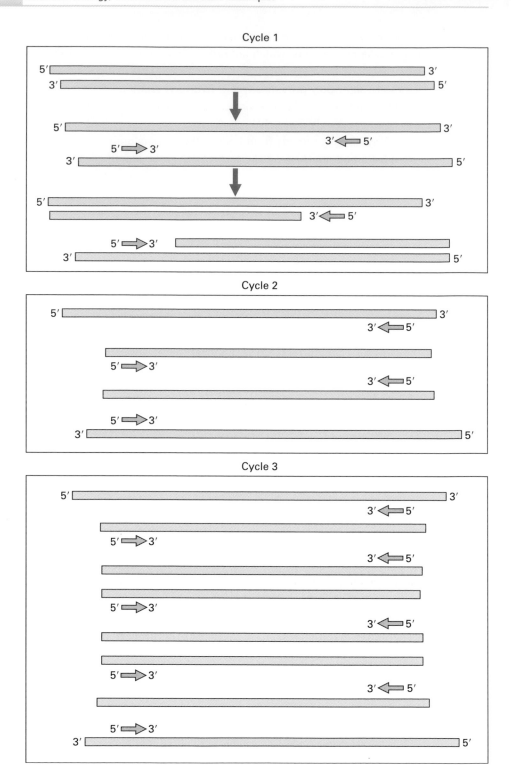

Fig. 5.34 Three cycles in the PCR. As the number of cycles in the PCR increases, the DNA strands that are synthesised and become available as templates are delimited by the ends of the primers. Thus specific amplification of the desired target sequence flanked by the primers is achieved. Primers are denoted as 5′ to 3′.

arbitrary priming in genomic templates but interestingly may give rise to discrete banding patterns when analysed by gel electrophoresis. In many cases this technique may be used reproducibly to identify a particular organism or species. This is sometimes referred to as random amplified polymorphic DNA (RAPD) and has been used successfully in the detection and differentiation of a number of pathogenic strains of bacteria. In addition primers can now be synthesised with a variety of labels such as fluorophores bound to them allowing easier detection and quantitation using techniques such as qPCR (Section 5.10.7).

5.10.4   **PCR amplification templates**

DNA from a variety of sources may be used as the initial source of amplification templates. It is also a highly sensitive technique and requires only one or two molecules for successful amplification. Unlike many manipulation methods used in current molecular biology the PCR technique is sensitive enough to require very little template preparation. The extraction from many prokaryotic and eukaryotic cells may involve a simple boiling step. Indeed the components of many extraction techniques such as SDS and proteinase K may adversely affect the PCR. The PCR may also be used to amplify RNA, a process termed RT–PCR (reverse transcriptase–PCR). Initially a reverse transcription reaction which converts the RNA to cDNA is carried out (Section 6.2.5). This reaction normally involves the use of the enzyme reverse transcriptase although some thermostable DNA polymerases used in the PCR such as *Tth* have a reverse transcriptase activity under certain buffer conditions. This allows mRNA transcription products to be effectively analysed. It may also be used to differentiate latent viruses (detected by standard PCR) or active viruses which replicate and thus produce transcription products and are thus detectable by RT–PCR (Fig. 5.35). In addition the PCR may be extended to determine relative amounts of a transcription product.

5.10.5   **Sensitivity of the PCR**

The enormous sensitivity of the PCR system is also one of its main drawbacks since the very large degree of amplification makes the system vulnerable to contamination. Even a trace of foreign DNA, such as that even contained in dust particles, may be amplified to significant levels and may give misleading results. Hence cleanliness is paramount when carrying out PCR, and dedicated equipment and in some cases dedicated laboratories are used. It is possible that amplified products may also contaminate the PCR although this may be overcome by UV irradiation to damage already amplified products so that they cannot be used as templates. A further interesting solution is to incorporate uracil into the PCR and then treat the products with the enzyme uracil *N*-glycosylase (UNG) which degrades any PCR amplicons with incorporated uracil rendering them useless as templates. In addition most PCRs are now undertaken using hotstart. Here the reaction mixture is physically separated from the template or the enzyme: when the reaction begins mixing occurs and thus avoids any mispriming that may have arisen.

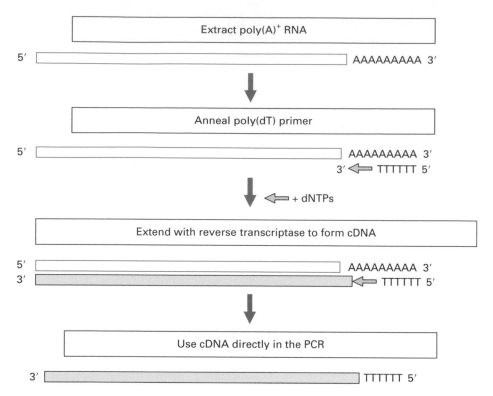

Fig. 5.35 Reverse transcriptase–PCR (RT–PCR): mRNA is converted to complementary DNA (cDNA) using the enzyme reverse transcriptase. The cDNA is then used directly in the PCR.

5.10.6 Applications of the PCR

Many traditional methods in molecular biology have now been superseded by the PCR and the applications for the technique appear to be unlimited. Some of the main techniques derived from the PCR are introduced in Chapter 6 while some of the main areas to which the PCR has been put to use are summarised in Table 5.5. The success of the PCR process has given impetus to the development of other amplification techniques that are based on either thermal cycling or non-thermal cycling (isothermal) methods. The most popular alternative to the PCR is termed the ligase chain reaction or LCR. This operates in a similar fashion to the PCR but a thermostable DNA ligase joins sets of primers together which are complementary to the target DNA. Following this a similar exponential amplification reaction takes place producing amounts of DNA that are similar to the PCR. A number of alternative amplification techniques are listed in Table 5.6.

5.10.7 Quantitative PCR (qPCR)

One of the most useful PCR applications is quantitative PCR or qPCR. This allows the PCR to be used as a means of identifying the initial concentrations of DNA or cDNA template used. Early qPCR methods involved the comparison of a standard or

Table 5.5 **Selected applications of the PCR. A number of the techniques are described in the text of Chapters 5 and 6**

Field or area of study	Application	Specific examples or uses
General molecular biology	DNA amplification	Screening gene libraries
Gene probe production	Production/labelling	Use with blots/hybridisations
RNA analysis	RT–PCR	Active latent viral infections
Forensic science	Scenes of crime	Analysis of DNA from blood
Infection/disease monitoring	Microbial detection	Strain typing/analysis RAPDs
Sequence analysis	DNA sequencing	Rapid sequencing possible
Genome mapping studies	Referencing points in genome	Sequence-tagged sites (STS)
Gene discovery	mRNA analysis	Expressed sequence tags (EST)
Genetic mutation analysis	Detection of known mutations	Screening for cystic fibrosis
Quantification analysis	Quantitative PCR	$5'$ Nuclease (TaqMan assay)
Genetic mutation analysis	Detection of unknown mutations	Gel-based PCR methods (DGGE)
Protein engineering	Production of novel proteins	PCR mutagenesis
Molecular archaeology	Retrospective studies	Dinosaur DNA analysis
Single-cell analysis	Sexing or cell mutation sites	Sex determination of unborn
In situ analysis	Studies on frozen sections	Localisation of DNA/RNA

Notes; RT, reverse transcriptase; RAPDs, rapid amplification polymorphic DNA; DGGE, denaturing gradient gel electrophoresis.

control DNA template amplified with separate primers at the same time as the specific target DNA. However these types of quantitation rely on the fact that all the reactions are identical and so any factors affecting this may also affect the result. The introduction of thermal cyclers that incorporate the ability to detect the accumulation of DNA through fluorescent dyes binding to the DNA has rapidly transformed this area.

In its simplist form a PCR is set up that includes a DNA-binding cyanine dye such as SYBR green. This dye binds to the major groove of double-stranded DNA but not single-stranded DNA and so as amplicons accumulate during the PCR process SYBR green binds the double-stranded DNA proportionally and fluorescence emission of the dye can be detected following excitation. Thus the accumulation of DNA amplicons can be followed in real time during the reaction run. In order to quantitate unknown DNA templates a standard dilution is prepared using DNA of known concentration. As the DNA accumulates during the early exponential phase of the reaction an arbitrary point is taken where each of the dilluted DNA samples cross. This is termed the crossing threshold on Ct value. From the various Ct values a log

Table 5.6 **Selected alternative amplification techniques to the PCR. Two broad methodologies exist that either amplify the target molecules such as DNA and RNA or detect the target and amplify a signal molecule bound to it**

Technique	Type of assay	Specific examples or uses
Target amplification methods		
Ligase chain reaction (LCR)	Non-isothermal, employs thermostable DNA ligase	Mutation detection
Nucleic acid sequence based amplification (NASBA)	Isothermal, involving use of RNA, RNase H/reverse transcriptase, and T7 DNA polymerase	Viral detection, e.g. HIV
Signal amplification methods		
Branched DNA amplification (b-DNA)	Isothermal microwell format using hybridisation or target/capture probe and signal amplification	Mutation detection

Note: HIV, human immunodeficiency virus.

graph is prepared from which an unknown concentration can be deduced. Since SYBR green and similar DNA-binding dyes are non-specific, in order to determine if a correctly sized PCR product is present most qPCR cyclers have a built-in melting curve function. This gradually increases the temperature of each tube until the double-stranded PCR product denatures or melts and allows a precise although not definitive determination of the product. Confirmation of the product is usually obtained by DNA sequencing.

5.10.8 The TaqMan system

In order to make qPCR specific a number of strategies may be employed that rely on specific hybridisation probes. One ingenious method is called the TaqMan assay or 5′ nuclease assay. Here the probe consists of an oligonucleotide labelled with a fluorescent reporter at one end of the molecule and quencher at the other end.

The PCR proceeds as normal and the oligonucleotide probe binds to the target sequence in the annealing step. As the *Taq* polymerase extends from the primer its 5′ exonuclease activity degrades the hybridisation probe and releases the reporter from the quencher. A signal is thus generated which increases in direct proportion to the number of starting molecules and fluorescence can be detected in real time as the PCR proceeds (Fig. 5.36). Although relatively expensive in comparison to other methods for determining expression levels it is simple, rapid and reliable and now in use in many research and clinical areas. Further developments in probe-based PCR systems have also been used and include scorpion probe systems, amplifluor and real-time LUX probes.

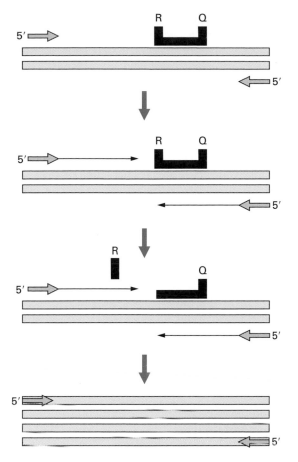

Fig. 5.36 5′ Nuclease assay (TaqMan assay). PCR is undertaken with RQ probe (reporter/quencher dye). As R–Q are in close proximity, fluorescence is quenched. During extension by *Taq* polymerase the probe is cleaved as a result of *Taq* having 5′ nuclease activity. This cleaves R–Q probe and the reporter is released. This results in detectable increase in fluorescence and allows real-time PCR detection.

5.11 NUCLEOTIDE SEQUENCING OF DNA

5.11.1 Concepts of nucleic acid sequencing

The determination of the order or sequence of bases along a length of DNA is one of the central techniques in molecular biology. Although it is now possible to derive amino acid sequence information with a degree of reliability it is frequently more convenient and rapid to analyse the DNA coding information. The precise usage of codons, information regarding mutations and polymorphisms and the identification of gene regulatory control sequences are also only possible by analysing DNA sequences. Two techniques have been developed for this, one based on an enzymatic method frequently termed Sanger sequencing after its developer, and a chemical method called Maxam and Gilbert, named for the same reason. At present Sanger sequencing is by far the most popular method and many commercial kits are available for its use. However, there are certain occasions such as the sequencing of short oligonucleotides where the Maxam and Gilbert method is more appropriate.

One absolute requirement for Sanger sequencing is that the DNA to be sequenced is in a single-stranded form. Traditionally this demanded that the DNA fragment of interest be inserted and cloned into a specialised bacteriophage vector termed *M13* which is naturally single-stranded (Section 6.3.3). Although M13 is still universally used the advent of the PCR has provided the means not only to amplify a region of any genome or cDNA but also very quickly generate the corresponding nucleotide sequence. This has led to an explosion in the accumulation of DNA sequence information and has provided much impetus for gene discovery and genome mapping (Section 6.9).

The Sanger method is simple and elegant and mimics in many ways the natural ability of DNA polymerase to extend a growing nucleotide chain based on an existing template. Initially the DNA to be sequenced is allowed to hybridise with an oligonucleotide primer, which is complementary to a sequence adjacent to the 3′ side of DNA within a vector such as M13 or in an amplicon. The oligonucleotide will then act as a primer for synthesis of a second strand of DNA, catalysed by DNA polymerase. Since the new strand is synthesised from its 5′ end, virtually the first DNA to be made will be complementary to the DNA to be sequenced. One of the dNTPs that must be provided for DNA synthesis is radioactively labelled with ^{32}P or ^{35}S, and so the newly synthesised strand will be labelled.

5.11.2 Dideoxynucleotide chain terminators

The reaction mixture is then divided into four aliquots, representing the four dNTPs, A, C, G and T. In addition to all of the dNTPs being present in the A tube an analogue of dATP is added (2′3′-dideoxyadenosine triphosphate (ddATP)) which is similar to A but has no 3′ hydroxyl group and so will terminate the growing chain since a 5′ to 3′ phosphodiester linkage cannot be formed without a 3′-hydroxyl group. The situation for tube C is identical except that ddCTP is added; similarly the G and T tubes contain ddGTP and ddTTP respectively (Fig. 5.37).

Since the incorporation of ddNTP rather than dNTP is a random event, the reaction will produce new molecules varying widely in length, but all terminating at the same type of base. Thus four sets of DNA sequence are generated, each terminating at a different type of base, but all having a common 5′ end (the primer). The four labelled and chain-terminated samples are then denatured by heating and loaded next to each other on a polyacrylamide gel for electrophoresis. Electrophoresis is performed at approximately 70 °C in the presence of urea, to prevent renaturation of the DNA, since even partial renaturation alters the rates of migration of DNA fragments. Very thin, long gels are used for maximum resolution over a wide range of fragment lengths. After electrophoresis, the positions of radioactive DNA bands on the gel are determined by autoradiography. Since every band in the track from the ddATP sample must contain molecules which terminate at adenine, and those in the ddCTP terminate

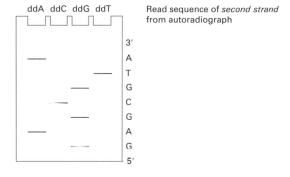

Fig. 5.37 Sanger sequencing of DNA.

at cytosine, etc., it is possible to read the sequence of the newly synthesised strand from the autoradiogram, provided that the gel can resolve differences in length equal to a single nucleotide (Fig. 5.38). Under ideal conditions, sequences up to about 300 bases in length can be read from one gel.

5.11.3 Direct PCR pyrosequencing

Rapid PCR sequencing has also been made possible by the use of pyrosequencing. This is a sequencing by synthesis whereby a PCR template is hybridised to an oligonucleotide and incubated with DNA polymerase, ATP sulphurylase, luciferase and apyrase. During the reaction the first of the four dNTPs are added and if incorporated release pyrophosphate (PP_i). The ATP sulphurylase converts the PP_i to ATP which drives the luciferase-mediated conversion of luciferin to oxyluciferin to generate light. Apyrase degrades the resulting component dNTPs and ATP. This is followed by another round of dNTP addition. A resulting pyrogram provides an output of the sequence. The method provides short reads very quickly and is especially useful for the determination of mutations or SNPs.

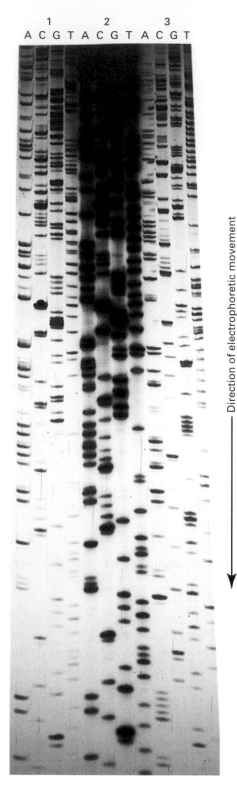

Fig. 5.38 Autoradiograph of a DNA sequencing gel. Samples were prepared using the Sanger dideoxy method of DNA sequencing. Each set of four samples was loaded into adjacent tracks, indicated by A, C, G and T, depending on the identity of the dideoxyribonucleotide used for that sample. Two sets of samples were labelled with ^{35}S (1 and 3) and one was labelled with ^{32}P (2). It is evident that ^{32}P generates darker but more diffuse bands than does ^{35}S, making the bands nearer the bottom of the autoradiograph easy to see. However, the broad bands produced by ^{32}P cannot be resolved near the top of the autoradiograph, making it impossible to read a sequence from this region. The much sharper bands produced by ^{35}S allow sequences to be read with confidence along most of the autoradiograph and so a longer sequence of DNA can be obtained from a single gel.

Direction of electrophoretic movement

It is also possible to undertake nucleotide sequencing from double-stranded molecules such as plasmid cloning vectors and PCR amplicons directly. The double-stranded DNA must be denatured prior to annealing with primer. In the case of plasmids an alkaline denaturation step is sufficient; however, for amplicons this is more problematic and a focus of much research. Unlike plasmids amplicons are short and reanneal rapidly, therefore preventing the reannealing process or biasing the amplification towards one strand by using a primer ratio of 100 : 1 overcomes this problem to a certain extent. Denaturants such as formamide or DMSO have also been used with some success in preventing the reannealing of PCR strands following their separation.

It is possible to physically separate and retain one PCR strand by incorporating a molecule such as biotin into one of the primers. Following PCR one strand with an affinity molecule may be removed by affinity chromatography with strepavidin, leaving the complementary PCR strand. This affinity purification provides single-stranded DNA derived from the PCR amplicon and although it is somewhat time-consuming does provide high-quality single-stranded DNA for sequencing.

5.11.4 PCR cycle sequencing

One of the most useful methods of sequencing PCR amplicons is termed PCR cycle sequencing. This is not strictly a PCR since it involves linear amplification with a single primer. Approximately 20 cycles of denaturation, annealing and extension take place. Radiolabelled or fluorescent-labelled dideoxynucleotides are then introduced in the final stages of the reaction to generate the chain-terminated extension products (Fig. 5.39). Automated direct PCR sequencing is increasingly being refined allowing greater lengths of DNA to be analysed in one sequencing run and provides a very rapid means of analysing DNA sequences.

5.11.5 Automated fluorescent DNA sequencing

Advances in fluorescent dye terminator and labelling chemistry have led to the development of high-throughput automated sequencing techniques. Essentially most systems involve the use of dideoxynucleotides labelled with different fluorochromes. Thus the label is incorporated into the ddNTP and this is used to carry out chain termination as in the standard reaction indicated in Section 5.11.1. The advantage of this modification is that since a different label is incorporated with each ddNTP it is unnecessary to perform four separate reactions. Therefore the four chain-terminated products are run on the same track of a denaturing electrophoresis gel. Each product with its base-specific dye is excited by a laser and the dye then emits light at its characteristic wavelength. A diffraction grating separates the emissions which are detected by a charge-coupled device (CCD) and the sequence is interpreted by a computer. The advantages of the technique include real-time detection of the sequence. In addition the lengths of sequence that may be analysed are in excess of 500 bp (Fig. 5.40). Capillary electrophoresis is increasingly being used for the detection of

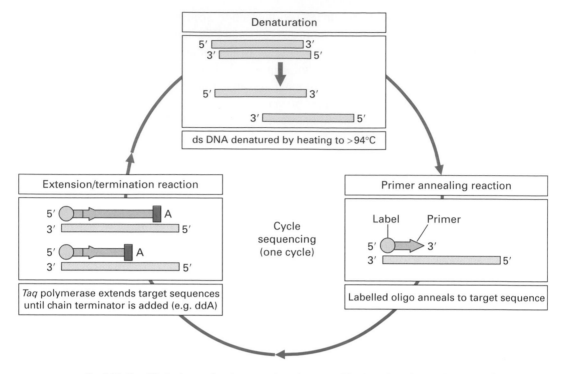

Fig. 5.39 Simplified scheme of cycle sequencing. Linear amplification takes place with the use of labelled primers. During the extension and termination reaction, the chain terminator dideoxynucleotides are incorporated into the growing chain. This takes place in four separate reactions (A, C, G and T). The products are then run on a polyacrylamide gel and the sequence analysed. The scheme indicates the events that take place in the A reaction only. ds, double-stranded.

sequencing products. This is where liquid polymers in thin capillary tubes are used obviating the need to pour sequencing gels and requiring little manual operation. This substantially reduces the electrophoresis run times and allows high throughput to be achieved. A number of large-scale sequence facilities are now fully automated using 96-well microtitre-based formats. The derived sequences can be downloaded automatically to databases and manipulated using a variety of bioinformatics resources.

5.11.6 Alternative DNA sequencing methods

Developments in the technology of DNA sequencing have made whole-genome sequencing projects a realistic proposition within achievable timescales; indeed the first diploid genome sequence to be completed was of Craig Venter who pioneered high-throughput sequencing. This makes studies on genome variation and evolution viable, as evidenced by the 1000 Genomes Project which is providing high-resolution sequence analysis of genomes. This has been made possible not only by refinements in traditional automated sequencing but also by new developments such as sequencing by synthesis and the development of sequencing by hybridisation arrays. These methods are changing the way genome analysis is undertaken and makes individual

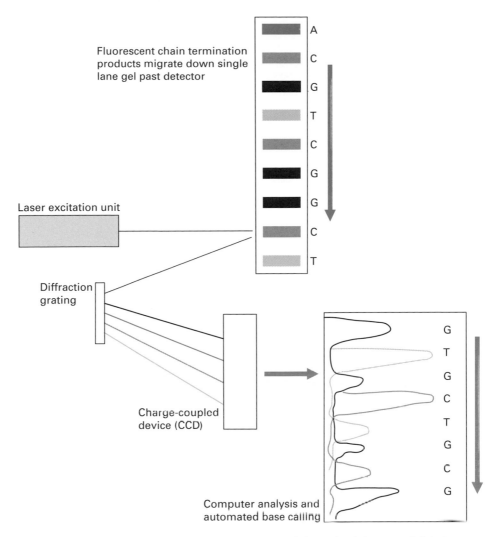

Fig. 5.40 Automated fluorescent sequencing detection using single-lane gel and charge-coupled device.

genome analysis a reality. Indeed more advanced methods using nanotechnology are in development and may provide an even more effective means of DNA sequencing.

5.11.7 Maxam and Gilbert sequencing

Sanger sequencing is by far the most popular technique for DNA sequencing; however, an alternative technique developed at the same time may also be used. The chemical cleavage method of DNA sequencing developed by Maxam and Gilbert is often used for sequencing small fragments of DNA such as oligonucleotides, where Sanger sequencing is problematic. A radioactive label is added to either the 3′ or the 5′ ends of a double-stranded DNA sample (Fig. 5.41). The strands are then

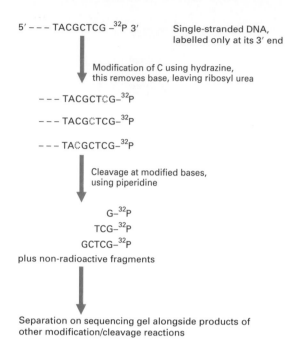

5' – – – TACGCTCG –^{32}P 3'   Single-stranded DNA,
labelled only at its 3' end

Modification of C using hydrazine,
this removes base, leaving ribosyl urea

– – – TACGCTCG–^{32}P

– – – TACGCTCG–^{32}P

– – – TACGCTCG–^{32}P

Cleavage at modified bases,
using piperidine

G–^{32}P
TCG–^{32}P
GCTCG–^{32}P
plus non-radioactive fragments

Separation on sequencing gel alongside products of
other modification/cleavage reactions

Fig. 5.41 Maxam and Gilbert sequencing of DNA. Only modification and cleavage of deoxycytidine is shown, but three more portions of the end-labelled DNA would be modified and cleaved at G, G+A, and T+C, and the products would be separated on the sequencing gel alongside those from the C reactions.

separated by electrophoresis under denaturing conditions, and analysed separately. DNA labelled at one end is divided into four aliquots and each is treated with chemicals which act on specific bases by methylation or removal of the base. Conditions are chosen so that, on average, each molecule is modified at only one position along its length; every base in the DNA strand has an equal chance of being modified. Following the modification reactions, the separate samples are cleaved by piperidine, which breaks phosphodiester bonds exclusively at the 5' side of nucleotides whose base has been modified. The result is similar to that produced by the Sanger method, since each sample now contains radioactively labelled molecules of various lengths, all with one end in common (the labelled end), and with the other end cut at the same type of base. Analysis of the reaction products by electrophoresis is as described for the Sanger method.

5.12   SUGGESTIONS FOR FURTHER READING

Augen, J. (2005). *Bioinformatics in the Post-Genomic Era*. Reading, MA: Addison-Wesley.
Brooker, R. J. (2005). *Genetics Analysis and Principles*, 2nd edn. New York: McGraw-Hill.
Hartwell, L. et al. (2008). *Genetics: From Genes to Genomes*, 3rd edn. New York: McGraw-Hill.
Lodish, H. et al. (2008). *Molecular Cell Biology*, 6th edn. San Francisco, CA: W. H. Freeman.
Lewin, B. (2007). *Genes IX*. Sudbury, MA: Jones & Bartlett.
Strachan, T. and Read, A. P (2004). *Human Molecular Genetics*, 3rd edn. Oxford, UK: Bios.
Walker, J. M. and Rapley, R. (2008). *Molecular Biomethods Handbook*, 2nd edn. Totowa, NJ: Humana Press.

6 Recombinant DNA and genetic analysis

R. RAPLEY

6.1 INTRODUCTION

The considerable advances made in microarray, sequencing technologies and bioinformatics analysis are now beginning to provide true insights into the development and maintenance of cells and tissues. Indeed areas of analysis such as metabolomics, transcriptomics and systems biology are now well established and allow analysis of vast numbers of samples simultaneously. This type of large-scale parallel analysis is now the main driving force of biological discovery and analysis. However, the techniques of molecular biology and genetic analysis have their foundations in methods developed a number of decades ago. One of the main cornerstones on which molecular biology analysis was developed was the discovery of restriction endonucleases in the early 1970s which not only led to the possibility of analysing DNA more effectively but also provided the ability to cut different DNA molecules so that they could later be joined together to create new recombinant DNA fragments. The newly created DNA molecules heralded a new era in the manipulation, analysis and exploitation of biological molecules. This process, termed gene cloning, has enabled numerous discoveries and insights into gene structure, function and regulation. Since their initial use the methods for the production of gene libraries have been steadily refined and developed. Although microarray analysis and the polymerase chain reaction (PCR) have provided short cuts to gene analysis there are still many cases where gene cloning methods are not only useful but are an absolute requirement. The following provides an account of the process of gene cloning and other methods based on recombinant DNA technology.

6.2 CONSTRUCTING GENE LIBRARIES

6.2.1 Digesting genomic DNA molecules

Following the isolation and purification of genomic DNA it is possible to specifically fragment it with enzymes termed restriction endonucleases. These enzymes are the key to molecular cloning because of the specificity they have for particular DNA sequences. It is important to note that every copy of a given DNA molecule from a specific organism will give the same set of fragments when digested with a particular enzyme. DNA from different organisms will, in general, give different sets of fragments when treated with the same enzyme. By digesting complex genomic DNA from an organism it is possible to reproducibly divide its genome into a large number of small fragments, each approximately the size of a single gene. Some enzymes cut straight across the DNA to give flush or blunt ends. Other restriction enzymes make staggered single-strand cuts, producing short single-stranded projections at each end of the digested DNA. These ends are not only identical, but complementary, and will base-pair with each other; they are therefore known as cohesive or sticky ends. In addition the 5′ end projection of the DNA always retains the phosphate groups.

Over 600 enzymes, recognising more than 200 different restriction sites, have been characterised. The choice of which enzyme to use depends on a number of factors. For example, the recognition sequence of 6 bp will occur, on average, every 4096 (46) bases assuming a random sequence of each of the four bases. This means that digesting genomic DNA with *Eco*R1, which recognises the sequence 5′-GAATTC-3′, will produce fragments each of which is on average just over 4 kb. Enzymes with 8 bp recognition sequences produce much longer fragments. Therefore very large genomes, such as human DNA, are usually digested with enzymes that produce long DNA fragments. This makes subsequent steps more manageable, since a smaller number of those fragments need to be cloned and subsequently analysed (Table 6.1).

6.2.2 Ligating DNA molecules

The DNA products resulting from restriction digestion to form sticky ends may be joined to any other DNA fragments treated with the same restriction enzyme. Thus, when the two sets of fragments are mixed, base-pairing between sticky ends will result in the annealing together of fragments that were derived from different starting DNA. There will, of course, also be pairing of fragments derived from the same starting DNA

Table 6.1 **Numbers of clones required for representation of DNA in a genome library**

Species	Genome size (kb)	No. of clones required	
		17 kb fragments	35 kb fragments
Bacteria (*E. coli*)	4 000	700	340
Yeast	20 000	3 500	1 700
Fruit fly	165 000	29 000	14 500
Man	3 000 000	535 000	258 250
Maize	15 000 000	2 700 000	1 350 000

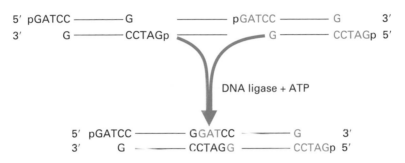

Fragments produced by cleavage with *Bam*HI

Fig. 6.1 Ligation molecules with cohesive ends. Complementary cohesive ends base-pair, forming a temporary link between two DNA fragments. This association of fragments is stabilised by the formation of 3′ to 5′ phosphodiester linkages between cohesive ends, a reaction catalysed by DNA ligase.

molecules, termed reannealing. All these pairings are transient, owing to the weakness of hydrogen bonding between the few bases in the sticky ends, but they can be stabilised by use of an enzyme termed DNA ligase in a process termed ligation. This enzyme, usually isolated from bacteriophage T4 and termed T4 DNA ligase, forms a covalent bond between the 5′ phosphate at the end of one strand and the 3′ hydroxyl of the adjacent strand (Fig. 6.1). The reaction, which is ATP dependent, is often carried out at 10 °C to lower the kinetic energy of molecules, and so reduce the chances of base-paired sticky ends parting before they have been stabilised by ligation. However, long reaction times are needed to compensate for the low activity of DNA ligase in the cold. It is also possible to join blunt ends of DNA molecules, although the efficiency of this reaction is much lower than sticky-ended ligations.

Since ligation reconstructs the site of cleavage, recombinant molecules produced by ligation of sticky ends can be cleaved again at the 'joins', using the same restriction enzyme that was used to generate the fragments initially. In order to propagate digested DNA from an organism it is necessary to join or ligate that DNA with

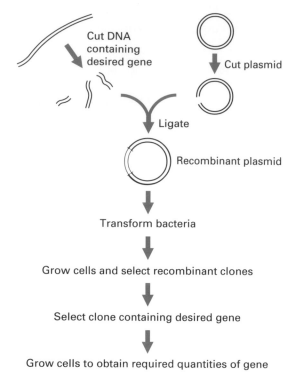

Fig. 6.2 Outline of gene cloning.

a specialised DNA carrier molecule termed a vector (Section 6.3). Thus each DNA fragment is inserted by ligation into the vector DNA molecule, which allows the whole recombined DNA to then be replicated indefinitely within microbial cells (Fig. 6.2). In this way a DNA fragment can be cloned to provide sufficient material for further detailed analysis, or for further manipulation. Thus, all of the DNA extracted from an organism and digested with a restriction enzyme will result in a collection of clones. This collection of clones is known as a gene library.

6.2.3 Aspects of gene libraries

There are two general types of gene library: a genomic library which consists of the total chromosomal DNA of an organism and a cDNA library which represents only the mRNA from a particular cell or tissue at a specific point in time (Fig. 6.3). The choice of the particular type of gene library depends on a number of factors, the most important being the final application of any DNA fragment derived from the library. If the ultimate aim is understanding the control of protein production for a particular gene or the analysis of its architecture, then genomic libraries must be used. However, if the goal is the production of new or modified proteins, or the determination of the tissue-specific expression and timing patterns, cDNA libraries are more appropriate. The main consideration in the construction of genomic or cDNA libraries is therefore

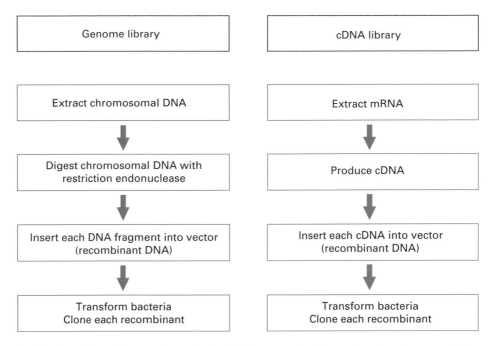

Fig. 6.3 Comparison of the general steps involved in the construction of genomic and complementary DNA (cDNA) libraries.

the nucleic acid starting material. Since the genome of an organism is fixed, chromosomal DNA may be isolated from almost any cell type in order to prepare genomic libraries. In contrast, however, cDNA libraries only represent the mRNA being produced from a specific cell type at a particular time. Thus, it is important to consider carefully the cell or tissue type from which the mRNA is to be derived in the construction of cDNA libraries.

There are a variety of cloning vectors available, many based on naturally occurring molecules such as bacterial plasmids or bacteria-infecting viruses. The choice of vector depends on whether a genomic library or cDNA library is constructed. The various types of vectors are explained in more detail in Section 6.3.

6.2.4 Genomic DNA libraries

Genomic libraries are constructed by isolating the complete chromosomal DNA from a cell, then digesting it into fragments of the desired average length with restriction endonucleases. This can be achieved by partial restriction digestion using an enzyme that recognises tetranucleotide sequences. Complete digestion with such an enzyme would produce a large number of very short fragments, but if the enzyme is allowed to cleave only a few of its potential restriction sites before the reaction is stopped, each DNA molecule will be cut into relatively large fragments. Average fragment size will depend on the relative concentrations of DNA and restriction enzyme, and in particular, on the conditions and duration of incubation (Fig. 6.4). It is also possible to produce fragments of DNA by physical shearing although the ends of the fragments

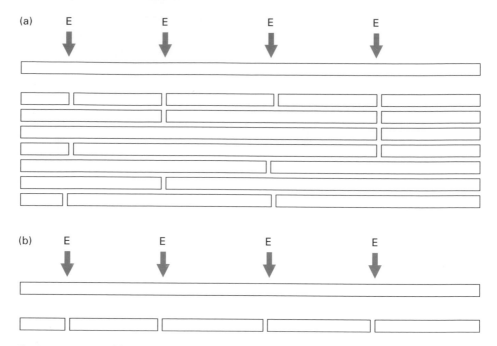

Fig. 6.4 Comparison of (a) partial and (b) complete digestion of DNA molecules at restriction enzyme sites (E).

may need to be repaired to make them flush-ended. This can be achieved by using a modified DNA polymerase termed Klenow polymerase. This is prepared by cleavage of DNA polymerase with subtilisin, giving a large enzyme fragment which has no 5′ to 3′ exonuclease activity, but which still acts as a 5′ to 3′ polymerase. This will fill in any recessed 3′ ends on the sheared DNA using the appropriate dNTPs.

The mixture of DNA fragments is then ligated with a vector, and subsequently cloned. If enough clones are produced there will be a very high chance that any particular DNA fragment such as a gene will be present in at least one of the clones. To keep the number of clones to a manageable size, fragments about 10 kb in length are needed for prokaryotic libraries, but the length must be increased to about 40 kb for mammalian libraries. It is possible to calculate the number of clones that must be present in a gene library to give a probability of obtaining a particular DNA sequence. This formula is:

$$N = \frac{\ln(1-P)}{\ln(1-f)}$$

where N is the number of recombinants, P is the probability and f is the fraction of the genome in one insert. Thus for the E. coli DNA chromosome of 5×10^6 bp and with an insert size of 20 kb the number of clones needed (N) would be 1×10^3 with a probability of 0.99.

6.2.5 cDNA libraries

There may be several thousand different proteins being produced in a cell at any one time, all of which have associated mRNA molecules. To identify any one of those

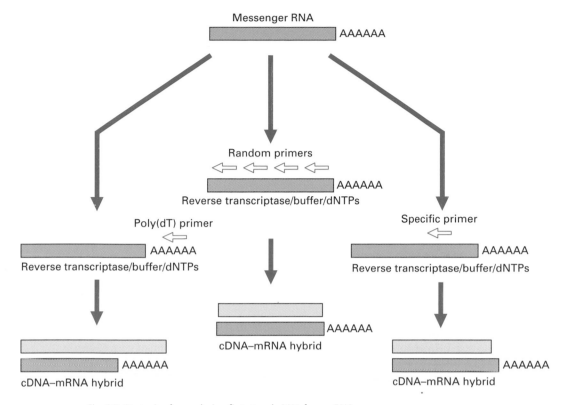

Fig. 6.5 Strategies for producing first-strand cDNA from mRNA.

mRNA molecules the clones of each individual mRNA have to be synthesised. Libraries that represent the mRNA in a particular cell or tissue are termed cDNA libraries. mRNA cannot be used directly in cloning since it is too unstable. However it is possible to synthesise complementary DNA molecules (cDNAs) to all the mRNAs from the selected tissue. The cDNA may be inserted into vectors and then cloned. The production of cDNA (complementary DNA) is carried out using an enzyme termed reverse transcriptase which is isolated from RNA-containing retroviruses.

Reverse transcriptase is an RNA-dependent DNA polymerase, and will synthesise a first-strand DNA complementary to an mRNA template, using a mixture of the four dNTPs. There is also a requirement (as with all polymerase enzymes) for a short oligonucleotide primer to be present (Fig. 6.5). With eukaryotic mRNA bearing a poly(A) tail, a complementary oligo(dT) primer may be used. Alternatively random hexamers may be used which randomly anneal to the mRNAs in the complex. Such primers provide a free 3′ hydroxyl group which is used as the starting point for the reverse transcriptase. Regardless of the method used to prepare the first-strand cDNA one absolute requirement is high-quality undegraded mRNA (Section 5.7.2). It is usual to check the integrity of the RNA by gel electrophoresis (Section 5.7.4). Alternatively a fraction of the extract may be used in a cell-free translation system, which, if intact mRNA is present, will direct the synthesis of proteins represented by the mRNA molecules in the sample (Section 6.7).

Following the synthesis of the first DNA strand, a poly(dC) tail is added to its 3′ end, using terminal transferase and dCTP. This will also, incidentally, put a poly(dC) tail on

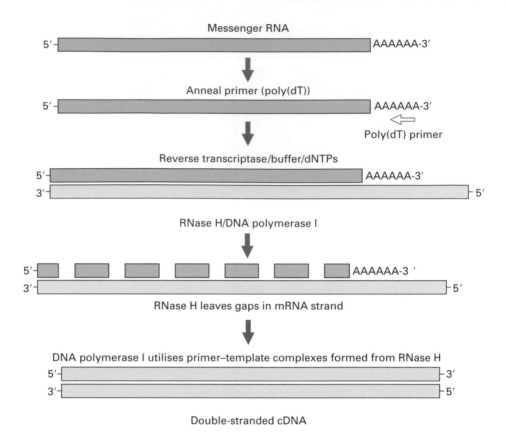

Fig. 6.6 Second-strand cDNA synthesis using the RNase H method.

the poly(A) of mRNA. Alkaline hydrolysis is then used to remove the RNA strand, leaving single-stranded DNA which can be used, like the mRNA, to direct the synthesis of a complementary DNA strand. The second-strand synthesis requires an oligo(dG) primer, base-paired with the poly(dC) tail, which is catalysed by the Klenow fragment of DNA polymerase I. The final product is double-stranded DNA, one of the strands being complementary to the mRNA. One further method of cDNA synthesis involves the use of RNase H. Here the first-strand cDNA is carried out as above with reverse transcriptase but the resulting mRNA–cDNA hybrid is retained. RNase H is then used at low concentrations to nick the RNA strand. The resulting nicks expose 3′ hydroxyl groups which are used by DNA polymerase as a primer to replace the RNA with a second strand of cDNA (Fig. 6.6).

6.2.6 Treatment of blunt cDNA ends

Ligation of blunt-ended DNA fragments is not as efficient as ligation of sticky ends, therefore with cDNA molecules additional procedures are undertaken before ligation with cloning vectors. One approach is to add small double-stranded molecules with one internal site for a restriction endonuclease, termed nucleic acid linkers, to the cDNA. Numerous linkers are commercially available with internal restriction sites for

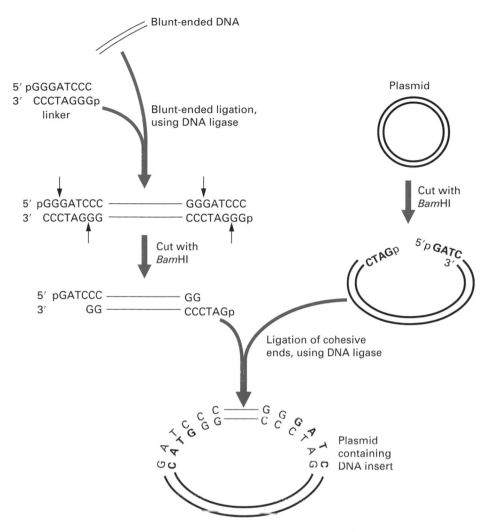

Fig. 6.7 Use of linkers. In this example, blunt-ended DNA is inserted into a specific restriction site on a plasmid, after ligation to a linker containing the same restriction site.

many of the most commonly used restriction enzymes. Linkers are blunt–end ligated to the cDNA but since they are added much in excess of the cDNA the ligation process is reasonably successful. Subsequently the linkers are digested with the appropriate restriction enzyme which provides the sticky ends for efficient ligation to a vector digested with the same enzyme. This process may be made easier by the addition of adaptors rather than linkers which are identical except that the sticky ends are preformed and so there is no need for restriction digestion following ligation (Fig. 6.7).

6.2.7 Enrichment methods for RNA

Frequently an attempt is made to isolate the mRNA transcribed from a desired gene within a particular cell or tissue that produces the protein in high amounts. Thus if the

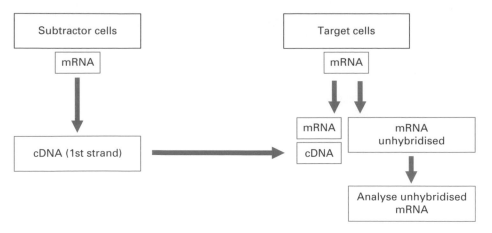

Fig. 6.8 Scheme of analysing specific mRNA molecules by subtractive hybridisation.

cell or tissue produces a major protein of the cell a large fraction of the total mRNA will code for the protein. An example of this are the B cells of the pancreas, which contain high levels of pro-insulin mRNA. In such cases it is possible to precipitate polysomes which are actively translating the mRNA, by using antibodies to the ribosomal proteins; mRNA can then be dissociated from the precipitated ribosomes. More usually the mRNA required is only a minor component of the total cellular mRNA. In such cases total mRNA may be fractionated by size using sucrose density gradient centrifugation. Then each fraction is used to direct the synthesis of proteins using an *in vitro* translation system (Section 6.7).

6.2.8 Subtractive hybridisation

It is often the case that genes are transcribed in a specific cell type or differentially activated during a particular stage of cellular growth, often at very low levels. It is possible to isolate those mRNA transcripts by subtractive hybridisation. Usually the the mRNA species common to the different cell types are removed, leaving the cell type or tissue-specific mRNAs for analysis (Fig. 6.8). This may be undertaken by isolating the mRNA from the so-called subtractor cells and producing a first-strand cDNA (Section 6.2.5). The original mRNA from the subtractor cells is then degraded and the mRNA from the target cells isolated and mixed with the cDNA. All the complementary mRNA–cDNA molecules common to both cell types will hybridise leaving the unbound mRNA which may be isolated and further analysed. A more rapid approach of analysing the differential expression of genes has been developed using the PCR. This technique, termed differential display, is explained in greater detail in Section 6.8.1.

6.2.9 Cloning PCR products

While PCR has to some extent replaced cloning as a method for the generation of large quantities of a desired DNA fragment there is, in certain circumstances, still

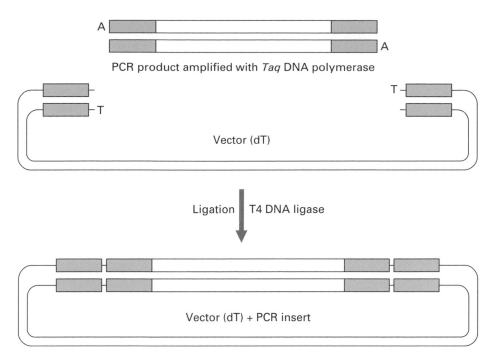

Fig. 6.9 Cloning of PCR products using dA : dT cloning.

a requirement for the cloning of PCR-amplified DNA. For example certain techniques such as *in vitro* protein synthesis are best achieved with the DNA fragment inserted into an appropriate plasmid or phage cloning vector (Section 6.7.1). Cloning methods for PCR follow closely the cloning of DNA fragments derived from the conventional manipulation of DNA. The techniques with which this may be achieved are through one of two ways, blunt-ended or cohesive-ended cloning. Certain thermostable DNA polymerases such as *Taq* DNA polymerase and *Tth* DNA polymerase give rise to PCR products having a 3′ overhanging A residue. It is possible to clone the PCR product into dT vectors termed dA : dT cloning. This makes use of the fact that the terminal additions of A residues may be successfully ligated to vectors prepared with T residue overhangs to allow efficient ligation of the PCR product (Fig. 6.9). The reaction is catalysed by DNA ligase as in conventional ligation reactions (Section 6.2.2).

It is also possible to carry out cohesive ended cloning with PCR products. In this case oligonucleotide primers are designed with a restriction endonuclease site incorporated into them. Since the complementarity of the primers needs to be absolute at the 3′ end the 5′ end of the primer is usually the region for the location of the restriction site. This needs to be designed with care since the efficiency of digestion with certain restriction endonuclease decreases if extra nucleotides, not involved in recognition, are absent at the 5′ end. In this case the digestion and ligation reactions are the same as those undertaken for conventional reactions (Section 6.2.1).

6.3 CLONING VECTORS

For the cloning of any molecule of DNA it is necessary for that DNA to be incorporated into a cloning vector. These are DNA elements that may be stably maintained and propagated in a host organism for which the vector has replication functions. A typical host organism is a bacterium such as *E. coli* which grows and divides rapidly. Thus any vector with a replication origin in *E. coli* will replicate (together with any incorporated DNA) efficiently. Thus, any DNA cloned into a vector will enable the amplification of the inserted foreign DNA fragment and also allow any subsequent analysis to be undertaken. In this way the cloning process resembles the PCR although there are some major differences between the two techniques. By cloning, it is possible to not only store a copy of any particular fragment of DNA, but also produce unlimited amounts of it (Fig. 6.10).

The vectors used for cloning vary in their complexity, their ease of manipulation, their selection and the amount of DNA sequence they can accommodate (the insert

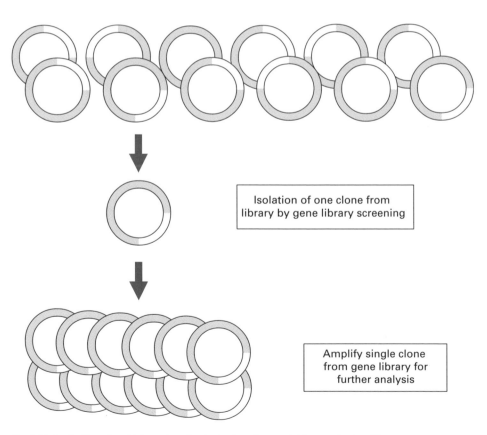

Stable gene bank (gene library), each vector containing a different foreign DNA fragment

Isolation of one clone from library by gene library screening

Amplify single clone from gene library for further analysis

Fig. 6.10 Production of multiple copies of a single clone from a stable gene bank or library.

Table 6.2 **Comparison of vectors generally available for cloning DNA fragments**

Vector	Host cell	Vector structure	Insert range (kb)
M13	*E. coli*	Circular virus	1–4
Plasmid	*E. coli*	Circular plasmid	1–5
Phage λ	*E. coli*	Linear virus	2–25
Cosmids	*E. coli*	Circular plasmid	35–45
BACs	*E. coli*	Circular plasmid	50–300
YACs	*S. cerevisiae*	Linear chromosome	100–2000

Notes: BAC, bacterial artificial chromosome; YAC, yeast artificial chromosome.

capacity). Vectors have in general been developed from naturally occurring molecules such as bacterial plasmids, bacteriophages or combinations of the elements that make them up, such as cosmids (Section 6.3.4). For gene library constructions there is a choice and trade-off between various vector types, usually related to ease of the manipulations needed to construct the library and the maximum size of foreign DNA insert of the vector (Table 6.2). Thus, vectors with the advantage of large insert capacities are usually more difficult to manipulate, although there are many more factors to be considered, which are indicated in the following treatment of vector systems.

6.3.1 Plasmids

Many bacteria contain an extrachromosomal element of DNA, termed a plasmid, which is a relatively small, covalently closed circular molecule, carrying genes for antibiotic resistance, conjugation or the metabolism of 'unusual' substrates. Some plasmids are replicated at a high rate by bacteria such as *E. coli* and so are excellent potential vectors. In the early 1970s a number of natural plasmids were artificially modified and constructed as cloning vectors, by a complex series of digestion and ligation reactions. One of the most notable plasmids, termed pBR322 after its developers Bolivar and Rodriguez (pBR), was widely adopted and illustrates the desirable features of a cloning vector as indicated below (Fig. 6.11).

- The plasmid is much smaller than a natural plasmid, which makes it more resistant to damage by shearing, and increases the efficiency of uptake by bacteria, a process termed transformation.
- A bacterial origin of DNA replication ensures that the plasmid will be replicated by the host cell. Some replication origins display stringent regulation of replication, in which rounds of replication are initiated at the same frequency as cell division. Most plasmids, including pBR322, have a relaxed origin of replication, whose activity is not tightly linked to cell division, and so plasmid replication will be

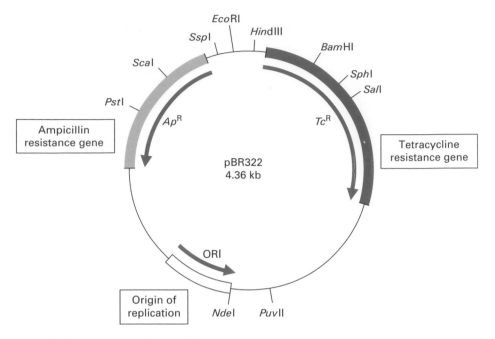

Fig. 6.11 Map and important features of pBR322.

initiated far more frequently than chromosomal replication. Hence a large number of plasmid molecules will be produced per cell.

- Two genes coding for resistance to antibiotics have been introduced. One of these allows the selection of cells which contain plasmid: if cells are plated on medium containing an appropriate antibiotic, only those that contain plasmid will grow to form colonies. The other resistance gene can be used, as described below, for detection of those plasmids that contain inserted DNA.
- There are single recognition sites for a number of restriction enzymes at various points around the plasmid, which can be used to open or linearise the circular plasmid. Linearising a plasmid allows a fragment of DNA to be inserted and the circle closed. The variety of sites not only makes it easier to find a restriction enzyme that is suitable for both the vector and the foreign DNA to be inserted, but, since some of the sites are placed within an antibiotic resistance gene, the presence of an insert can be detected by loss of resistance to that antibiotic. This is termed insertional inactivation.

Insertional inactivation is a useful selection method for identifying recombinant vectors with inserts. For example, a fragment of chromosomal DNA digested with *Bam*H1 would be isolated and purified. The plasmid pBR322 would also be digested at a single site, using *Bam*H1, and both samples would then be deproteinised to inactivate the restriction enzyme. *Bam*H1 cleaves to give sticky ends, and so it is possible to obtain ligation between the plasmid and digested DNA fragments in the presence of T4 DNA ligase. The products of this ligation will include plasmid containing a single fragment of the DNA as an insert, but there will also be unwanted products, such as

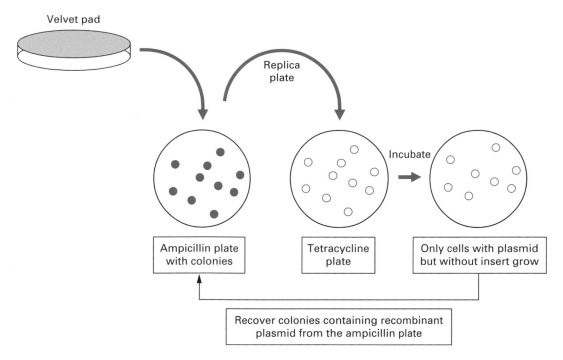

Fig. 6.12 Replica plating to detect recombinant plasmids. A sterile velvet pad is pressed onto the surface of an agar plate, picking up some cells from each colony growing on that plate. The pad is then pressed on to a fresh agar plate, thus inoculating it with cells in a pattern identical with that of the original colonies. Clones of cells that fail to grow on the second plate (e.g. owing to the loss of antibiotic resistance) can be recovered from their corresponding colonies on the first plate.

plasmid that has recircularised without an insert, dimers of plasmid, fragments joined to each other, and plasmid with an insert composed of more than one fragment. Most of these unwanted molecules can be eliminated during subsequent steps. The products of such reactions are usually identified by agarose gel electrophoresis (Section 5.7.4).

The ligated DNA must now be used to transform *E. coli*. Bacteria do not normally take up DNA from their surroundings, but can be induced to do so by prior treatment with Ca^{2+} at 4 °C; they are then termed competent, since DNA added to the suspension of competent cells will be taken up during a brief increase in temperature termed heat shock. Small, circular molecules are taken up most efficiently, whereas long, linear molecules will not enter the bacteria.

After a brief incubation to allow expression of the antibiotic resistance genes the cells are plated onto medium containing the antibiotic, e.g. ampicillin. Colonies that grow on these plates must be derived from cells that contain plasmid, since this carries the gene for resistance to ampicillin. It is not, at this stage, possible to distinguish between those colonies containing plasmids with inserts and those that simply contain recircularised plasmids. To do this, the colonies are replica plated, using a sterile velvet pad, onto plates containing tetracycline in their medium. Since the *Bam*HI site lies within the tetracycline resistance gene, this gene will be inactivated by the presence of insert, but will be intact in those plasmids that have merely recircularised (Fig. 6.12). Thus colonies that grow on ampicillin but not on tetracycline must contain

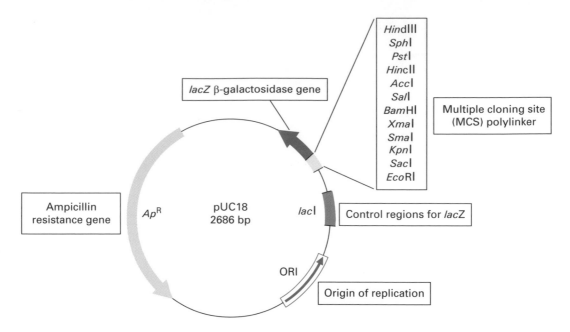

Fig. 6.13 Map and important features of pUC18.

plasmids with inserts. Since replica plating gives an identical pattern of colonies on both sets of plates, it is straightforward to recognise the colonies with inserts, and to recover them from the ampicillin plate for further growth. This illustrates the importance of a second gene for antibiotic resistance in a vector.

Although recircularised plasmid can be selected against, its presence decreases the yield of recombinant plasmid containing inserts. If the digested plasmid is treated with the enzyme alkaline phosphatase prior to ligation, recircularisation will be prevented, since this enzyme removes the 5′ phosphate groups that are essential for ligation. Links can still be made between the 5′ phosphate of insert and the 3′ hydroxyl of plasmid, so only recombinant plasmids and chains of linked DNA fragments will be formed. It does not matter that only one strand of the recombinant DNA is ligated, since the nick will be repaired by bacteria transformed with these molecules.

The valuable features of pBR322 have been enhanced by the construction of a series of plasmids termed pUC (produced at the University of California) (Fig. 6.13). There is an antibiotic resistance gene for tetracycline and origin of replication for *E. coli*. In addition the most popular restriction sites are concentrated into a region termed the multiple cloning site (MCS). In addition the MCS is part of a gene in its own right and codes for a portion of a polypeptide called β-galactosidase. When the pUC plasmid has been used to transform the host cell *E. coli* the gene may be switched on by adding the inducer IPTG (isopropyl-β-D-thiogalactopyranoside). Its presence causes the enzyme β-galactosidase to be produced (Section 5.5.5). The functional enzyme is able to hydrolyse a colourless substance called X-gal (5-bromo-4-chloro-3-indolyl-β-galactopyranoside) into a blue insoluble material (5,5′-dibromo-4,4′-dichloro indigo) (Fig. 6.14). However if the gene is disrupted by the insertion of a foreign

Non-recombinant vector (no insert)

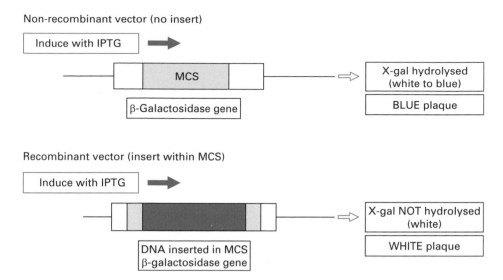

Recombinant vector (insert within MCS)

Fig. 6.14 Principle of blue/white selection for the detection of recombinant vectors.

fragment of DNA, a non-functional enzyme results which is unable to carry out hydrolysis of X-gal. Thus, a recombinant pUC plasmid may be easily detected since it is white or colourless in the presence of X-gal, whereas an intact non-recombinant pUC plasmid will be blue since its gene is fully functional and not disrupted. This elegant system, termed blue/white selection, allows the initial identification of recombinants to be undertaken very quickly and has been included in a number of subsequent vector systems. This selection method and insertional inactivation of antibiotic resistance genes do not, however, provide any information on the character of the DNA insert, just the status of the vector. To screen gene libraries for a desired insert hybridisation to gene probes is required and this is explained in Section 6.5.

6.3.2 Virus-based vectors

A useful feature of any cloning vector is the amount of DNA it may accept or have inserted before it becomes unviable. Inserts greater than 5 kb increase plasmid size to the point at which efficient transformation of bacterial cells decreases markedly, and so bacteriophages (bacterial viruses) have been adapted as vectors in order to propagate larger fragments of DNA in bacterial cells. Cloning vectors derived from λ bacteriophage are commonly used since they offer an approximately 16-fold advantage in cloning efficiency in comparison with the most efficient plasmid cloning vectors.

Phage λ is a linear double-stranded phage approximately 49 kb in length (Fig. 6.15). It infects *E. coli* with great efficiency by injecting its DNA through the cell membrane. In the wild-type phage λ the DNA follows one of two possible modes of replication. Firstly the DNA may either become stably integrated into the *E. coli* chromosome where it lies dormant until a signal triggers its excision. This is termed the lysogenic life cycle. Alternatively, it may follow a lytic life cycle where the DNA is replicated

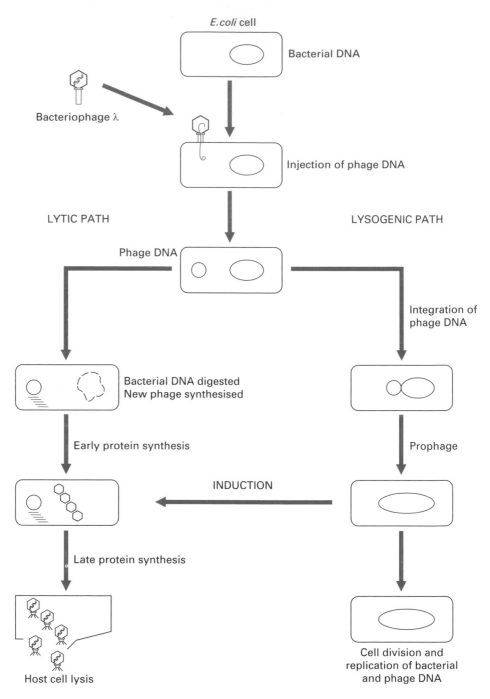

Fig. 6.15 The lysogenic and lytic cycles of bacteriophage λ.

upon entry to the cell, phage head and tail proteins synthesised rapidly and new functional phage assembled. The phage are subsequently released from the cell by lysing the cell membrane to infect further *E. coli* cells nearby. At the extreme ends of the phage λ are 12 bp sequences termed cos (cohesive) sites. Although they are

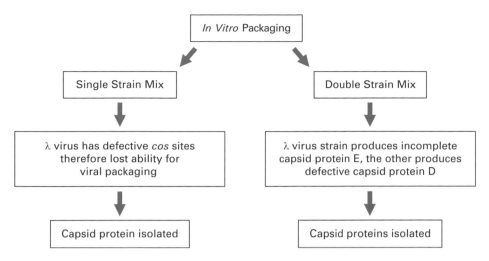

Fig. 6.16 Two strategies for producing *in vitro* packaging extracts for bacteriophage λ.

asymmetric they are similar to restriction sites and allow the phage DNA to be circularised. Phage may be replicated very efficiently in this way, the result of which are concatemers of many phage genomes which are cleaved at the cos sites and inserted into newly formed phage protein heads.

Much use of phage λ has been made in the production of gene libraries mainly because of its efficient entry into the *E. coli* cell and the fact that larger fragments of DNA may be stably integrated. For the cloning of long DNA fragments, up to approximately 25 kb, much of the non-essential λ DNA that codes for the lysogenic life cycle is removed and replaced by the foreign DNA insert. The recombinant phage is then assembled into pre-formed viral protein particles, a process termed *in vitro* packaging. These newly formed phage are used to infect bacterial cells that have been plated out on agar (Fig. 6.16).

Once inside the host cells, the recombinant viral DNA is replicated. All the genes needed for normal lytic growth are still present in the phage DNA, and so multiplication of the virus takes place by cycles of cell lysis and infection of surrounding cells, giving rise to plaques of lysed cells on a background, or lawn, of bacterial cells. The viral DNA including the cloned foreign DNA can be recovered from the viruses from these plaques and analysed further by restriction mapping (Section 5.9.1) and agarose gel electrophoresis (Section 5.7.4).

In general two types of λ phage vectors have been developed, λ insertion vectors and λ replacement vectors (Fig. 6.17). The λ insertion vectors accept less DNA than the replacement type since the foreign DNA is merely inserted into a region of the phage genome with appropriate restriction sites; common examples are λgt10 and λcharon16A. With a replacement vector a central region of DNA not essential for lytic growth is removed (a stuffer fragment) by a double digestion with for example *Eco*RI and *Bam*HI. This leaves two DNA fragments termed right and left arms. The central stuffer fragment is replaced by inserting foreign DNA between the arms to form a functional recombinant λ phage. The most notable examples of λ replacement vectors are λEMBL and λZap.

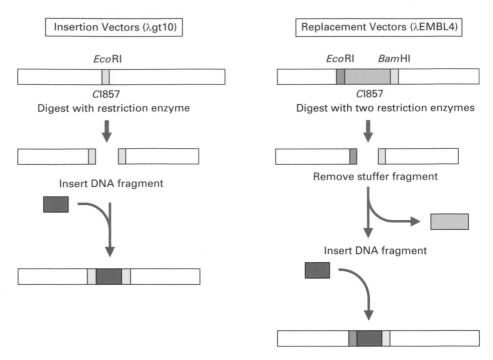

Fig. 6.17 General schemes used for cloning in λ insertion and λ replacement vectors. *Cl*857 is a temperature-sensitive mutation that promotes lysis at 42 °C after incubation at 37 °C.

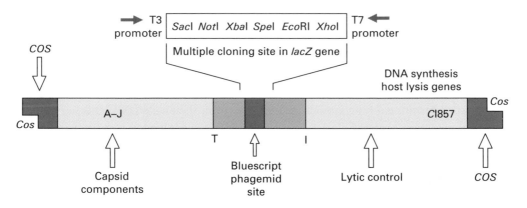

Fig. 6.18 General map of λZap cloning vector, indicating important areas of the vector. The multiple cloning site is based on the *lacZ* gene, providing blue/white selection based on the β-galactosidase gene. In between the initiator (I) site and terminator (T) site lie sequences encoding the phagemid Bluescript.

λZap is a commercially produced cloning vector that includes unique cloning sites clustered into a multiple cloning site (MCS) (Fig. 6.18). Furthermore the MCS is located within a *lacZ* region providing a blue/white screening system based on insertional inactivation. It is also possible to express foreign cloned DNA from this vector. This is a very useful feature of some λ vectors since it is then possible to screen for protein

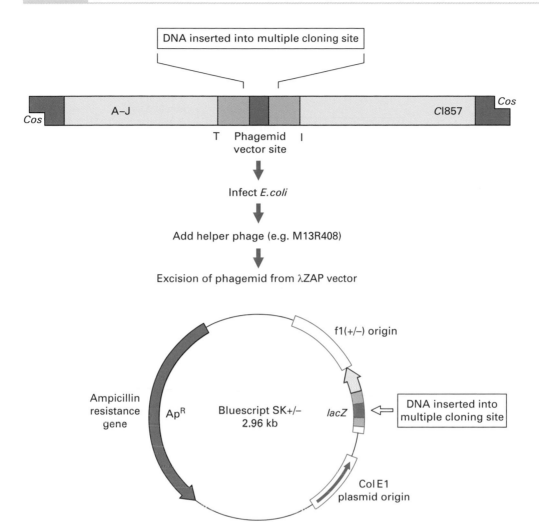

Fig. 6.19 Single-stranded DNA rescue of phagemid from λZap. The single-stranded phagemid pBluescript SK may be excised from λZap by addition of helper phage. This provides the necessary proteins and factors for transcription between the I and T sites in the parent phage to produce the phagemid with the DNA cloned into the parent vector.

product rather than the DNA inserted into the vector. This screening is therefore undertaken with antibody probes directed against the protein of interest (Section 6.5.4). Other features that make this a useful cloning vector are the ability to produce RNA transcripts termed cRNA or riboprobes. This is possible because two promoters for RNA polymerase enzymes exist in the vector, a T7 and a T3 promoter which flank the MCS (Section 6.4.2).

One of the most useful features of λZap is that it has been designed to allow automatic excision *in vivo* of a small 2.9 kb colony-producing vector termed a phagemid, pBluescript SK (Section 6.3.3). This technique is sometimes termed single-stranded DNA rescue and occurs as the result of a process termed superinfection where helper phage are added to the cells which are grown for an additional period of approximately 4 h (Fig. 6.19).

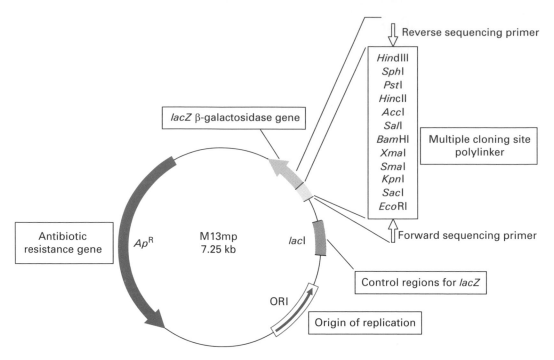

Fig. 6.20 Genetic map and important features of bacteriophage vector M13.

The helper phage displaces a strand within the λZap which contains the foreign DNA insert. This is circularised and packaged as a filamentous phage similar to M13 (Section 6.3.3). The packaged phagemid is secreted from the *E. coli* cell and may be recovered from the supernatant. Thus the λZap vector allows a number of diverse manipulations to be undertaken without the necessity of recloning or subcloning foreign DNA fragments. The process of subcloning is sometimes necessary when the manipulation of a gene fragment cloned in a general purpose vector needs to be inserted into a more specialised vector for the application of techniques such as *in vitro* mutagenesis or protein production (Section 6.6).

6.3.3 M13 and phagemid-based vectors

Much use has been made of single-stranded bacteriophage vectors such as M13 and vectors which have the combined properties of phage and plasmids, termed phagemids. M13 is a filamentous coliphage with a single-stranded circular DNA genome (Fig. 6.20). Upon infection of *E. coli*, the DNA replicates initially as a double-stranded molecule but subsequently produces single-stranded virions for infection of further bacterial cells (lytic growth). The nature of these vectors makes them ideal for techniques such as chain termination sequencing (Section 6.6.1) and *in vitro* mutagenesis (Section 6.6.3) since both require single-stranded DNA.

M13 or phagemids such as pBluescript SK infect *E. coli* harbouring a male-specific structure termed the F-pilus (Fig. 6.21). They enter the cell by adsorption to this structure and once inside the phage DNA is converted to a double-stranded replicative form or

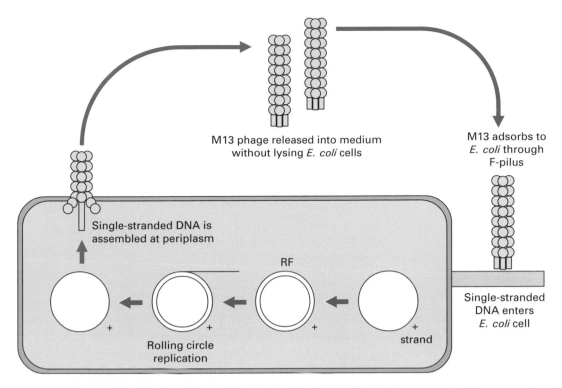

Fig. 6.21 Life cycle of bacteriophage M13. The bacteriophage virus enters the *E. coli* cell through the F-pilus. It then enters a stage where the circular single strands are converted to double strands. Rolling-circle replication then produces single strands, which are packaged and extruded through the *E. coli* cell membrane.

RF DNA. Replication then proceeds rapidly until some 100 RF molecules are produced within the *E. coli* cell. DNA synthesis then switches to the production of single strands and the DNA is assembled and packaged into the capsid at the bacterial periplasm. The bacteriophage DNA is then encapsulated by the major coat protein, gene VIII protein, of which there are approximately 2800 copies with three to six copies of the gene III protein at one end of the particle. The extrusion of the bacteriophage through the bacterial periplasm results in a decreased growth rate of the *E. coli* cell rather than host cell lysis and is visible on a bacterial lawn as an area of clearing. Approximately 1000 packaged phage particles may be released into the medium in one cell division.

In addition to producing single-stranded DNA the coliphage vectors have a number of other features that make them attractive as cloning vectors. Since the bacteriophage DNA is replicated as a double-stranded RF DNA intermediate a number of regular DNA manipulations may be performed such as restriction digestion, mapping and DNA ligation. RF DNA is prepared by lysing infected *E. coli* cells and purifying the super-coiled circular phage DNA with the same methods used for plasmid isolation. Intact single-stranded DNA packaged in the phage protein coat located in the supernatant may be precipitated with reagents such as polyethylene glycol, and the DNA purified with phenol/chloroform (Section 5.7.1). Thus the bacteriophage may act as a plasmid under certain circumstances and at other times produce DNA in the fashion of a virus. A family of vectors derived from M13 are currently widely used termed M13mp8/9,

M13 Multiple Cloning Site/Polylinker

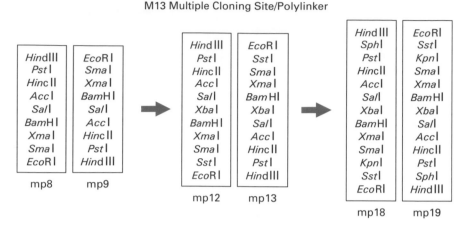

Fig. 6.22 Design and orientation of polylinkers in M13 series. Only the main restriction enzymes are indicated.

mp18/19, etc., all of which have a number of highly useful features. All contain a synthetic MCS, which is located in the *lacZ* gene without disruption of the reading frame of the gene. This allows efficient selection to be undertaken based on the technique of blue/white screening (Section 6.3.1). As the series of vectors were developed the number of restriction sites was increased in an asymmetric fashion. Thus M13mp8, mp12, mp18 and sister vectors which have the same MCS but in reverse orientation, M13mp9, mp13 and mp19 respectively have more restriction sites in the MCS making the vector more useful since greater choice of restriction enzymes is available (Fig. 6.22). However, one problem frequently encountered with M13 is the instability and spontaneous loss of inserts that are greater than 6 kb.

Phagemids are very similar to M13 and replicate in a similar fashion. One of the first phagemid vectors, pEMBL, was constructed by inserting a fragment of another phage termed f1 containing a phage origin of replication and elements for its morphogenesis into a pUC8 plasmid. Following superinfection with helper phage the f1 origin is activated allowing single-stranded DNA to be produced. The phage is assembled into a phage coat extruded through the periplasm and secreted into the culture medium in a similar way to M13. Without superinfection the phagemid replicates as a pUC type plasmid and in the replicative form (RF) the DNA isolated is double-stranded. This allows further manipulations such as restriction digestion, ligation and mapping analysis to be performed. The pBluescript SK vector is also a phagemid and can be used in its own right as a cloning vector and manipulated as if it were a plasmid. It may, like M13, be used in nucleotide sequencing and site-directed mutagenesis, and it is also possible to produce RNA transcripts that may be used in the production of labelled cRNA probes or riboprobes (Section 6.4.2).

6.3.4 Cosmid based vectors

The way in which the phage λ DNA is replicated is of particular interest in the development of larger insert cloning vectors termed cosmids (Fig. 6.23). These are

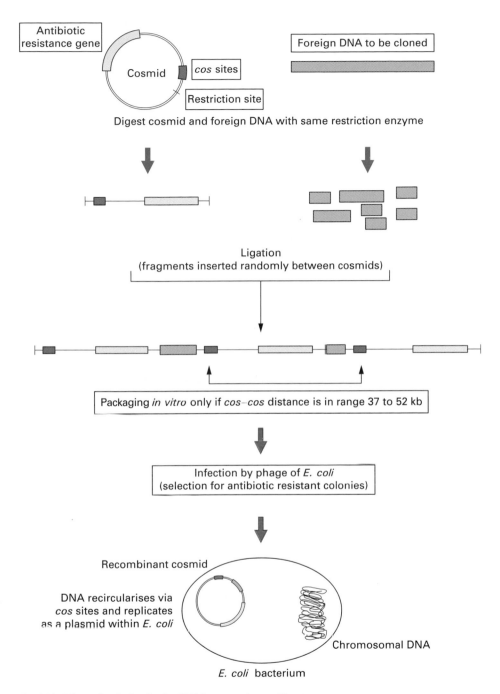

Fig. 6.23 Scheme for cloning foreign DNA fragments in cosmid vectors.

especially useful for the analysis of highly complex genomes and are an important part of various genome mapping projects (Section 6.9).

The upper limit of the insert capacity of phage λ is approximately 21 kb. This is because of the requirement for essential genes and the fact that the maximum length between the *cos* sites is 52 kb. Consequently cosmid vectors have been constructed that incorporate the *cos* sites from phage λ and also the essential features of a plasmid, such as the plasmid origin of replication, a gene for drug resistance, and several unique restriction sites for insertion of the DNA to be cloned. When a cosmid preparation is linearised by restriction digestion, and ligated to DNA for cloning, the products will include concatamers of alternating cosmid vector and insert. Thus the only requirement for a length of DNA to be packaged into viral heads is that it should contain *cos* sites spaced the correct distance apart; in practice this spacing can range between 37 and 52 kb. Such DNA can be packaged *in vitro* if phage head precursors, tails and packaging proteins are provided. Since the cosmid is very small, inserts of about 40 kb in length will be most readily packaged. Once inside the cell, the DNA recircularises through its *cos* sites, and from then onwards behaves exactly like a plasmid.

6.3.5 Large insert capacity vectors

The advantage of vectors that accept larger fragments of DNA than phage λ or cosmids is that fewer clones need to be screened when searching for the foreign DNA of interest. They have also had an enormous impact in the mapping of the genomes of organisms such as the mouse and are used extensively in the human genome mapping project (Section 6.9.3). Recent developments have allowed the production of large insert capacity vectors based on human artificial chromosomes, bacterial artificial chromosomes (BACs), mammalian artificial chromosomes (MACs) and on the virus P1 (PACs), P1 artificial chromosomes. However, perhaps the most significant development are vectors based on yeast artificial chromosomes (YACs).

6.3.6 Yeast artificial chromosome (YAC) vectors

Yeast artificial chromosomes (YACs) are linear molecules composed of a centromere, telomere and a replication origin termed an ARS element (autonomous replicating sequence). The YAC is digested with restriction enzymes at the SUP4 site (a suppressor tRNA gene marker) and *Bam*HI sites separating the telomere sequences (Fig. 6.24). This produces two arms and the foreign genomic DNA is ligated to produce a functional YAC construct. YACs are replicated in yeast cells; however, the external cell wall of the yeast needs to be removed to leave a spheroplast. These are osmotically unstable and also need to be embedded in a solid matrix such as agar. Once the yeast cells are transformed only correctly constructed YACs with associated selectable markers are replicated in the yeast strains. DNA fragments with repeat sequences are sometimes difficult to clone in bacterial-based vectors but may be successfully cloned in YAC systems. The main advantage of YAC-based vectors, however, is the ability to clone very large fragments of DNA. Thus the stable maintenance and replication of foreign DNA fragments of up to 2000 kb have been carried out in YAC vectors and they are the main vector of choice in the various genome mapping and sequencing projects (Section 6.9).

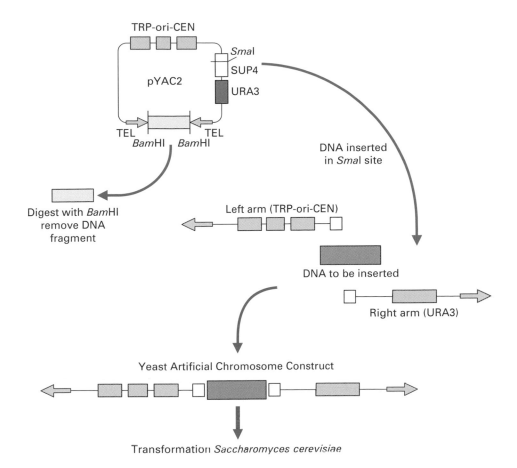

Fig. 6.24 Scheme for cloning large fragments of DNA into YAC vectors.

6.3.7 Vectors used in eukaryotes

The use of *E. coli* for general cloning and manipulation of DNA is well established; however, numerous developments have been made for cloning in eukaryotic cells. Plasmids used for cloning DNA in eukaryotic cells require a eukaryotic origin of replication and marker genes that will be expressed by eukaryotic cells. At present the two most important applications of plasmids to eukaryotic cells are for cloning in yeast and in plants.

Although yeast has a natural plasmid, called the 2μ circle, this is too large for use in cloning. Plasmids such as the yeast episomal plasmid (YEp) have been created by genetic manipulation using replication origins from the 2μ circle, and by incorporating a gene which will complement a defective gene in the host yeast cell. If, for example, a strain of yeast is used which has a defective gene for the biosynthesis of an amino acid, an active copy of that gene on a yeast plasmid can be used as a selectable marker for the presence of that plasmid. Yeast, like bacteria, can be grown rapidly, and it is therefore well suited for use in cloning. Of particular use has been the creation of shuttle vectors which have origins of replication for yeast and bacteria such as *E. coli*.

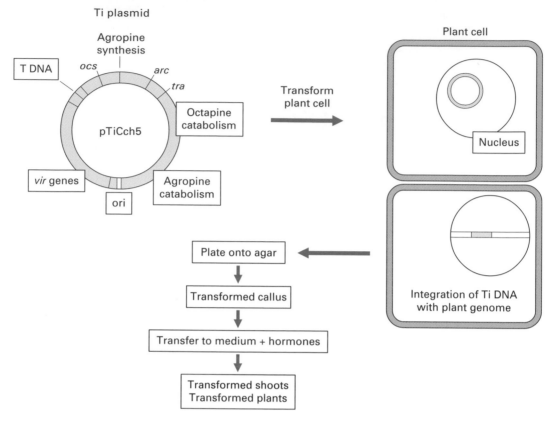

Fig. 6.25 Scheme for cloning in plant cells using the Ti plasmid.

This means that constructs may be prepared rapidly in the bacteria and delivered into yeast for expression studies.

The bacterium *Agrobacterium tumefaciens* infects plants that have been damaged near soil level, and this infection is often followed by the formation of plant tumours in the vicinity of the infected region. It is now known that *A. tumefaciens* contains a plasmid called the Ti plasmid, part of which is transferred into the nuclei of plant cells which are infected by the bacterium. Once in the nucleus, this DNA is maintained by integrating with the chromosomal DNA. The integrated DNA carries genes for the synthesis of opines (which are metabolised by the bacteria but not by the plants) and for tumour induction (hence 'Ti'). DNA inserted into the correct region of the Ti plasmid will be transferred to infected plant cells, and in this way it has been possible to clone and express foreign genes in plants (Fig. 6.25). This is an essential prerequisite for the genetic engineering of crops.

6.3.8 Delivery of vectors into eukaryotes

Following the production of a recombinant molecule, the so-called constructs are subsequently introduced into cells to enable it to be replicated a large number of times as the cells replicate. Initial recombinant DNA experiments were performed in bacterial cells, because of their ease of growth and short doubling time. Gram-negative bacteria such as *E. coli* can be made competent for the introduction of extraneous plasmid DNA into cells (Section 6.3.1). The natural ability of bacteriophage to introduce DNA into *E. coli* has also been well exploited and results in 10–100-fold higher efficiency for the introduction of recombinant DNA compared to transformation of competent bacteria with plasmids. These well-established and traditional approaches are the reason why so many cloning vectors have been developed for *E. coli*. The delivery of cloning vectors into eukaryotic cells is, however, not as straightforward as that for the bacterium *E. coli*.

It is possible to deliver recombinant molecules into animal cells by transfection. The efficiency of this process can be increased by first precipitating the DNA with Ca^{2+} or making the membrane permeable with divalent cations. High-molecular-weight polymers such as DEAE-dextran or polyethylene glycol (PEG) may also be used to maximise the uptake of DNA. The technique is rather inefficient although a selectable marker that provides resistance to a toxic compound such as neomycin can be used to monitor the success. Alternatively, DNA can be introduced into animal cells by electroporation. In this process the cells are subjected to pulses of a high-voltage gradient, causing many of them to take up DNA from the surrounding solution. This technique has proved to be useful with cells from a range of animal, plant and microbial sources. More recently the technique of lipofection has been used as the delivery method. The recombinant DNA is encapsulated by a core of lipid-coated particles which fuse with the lipid membrane of cells and thus release the DNA into the cell. Microinjection of DNA into cell nuclei of eggs or embryos has also been performed successfully in many mammalian cells.

The ability to deliver recombinant molecules into plant cells is not without its problems. Generally the outer cell wall of the plant must be stripped, usually by enzymatic digestion, to leave a protoplast. The cells are then able to take up recombinants from the supernatant. The cell wall can be regenerated by providing appropriate media. In cases where protoplasts have been generated transformation may also be achieved by electroporation. An even more dramatic transformation procedure involves propelling microscopically small titanium or gold pellet microprojectiles coated with the recombinant DNA molecule, into plant cells in intact tissues. This biolistic technique involves the detonation of an explosive charge which is used to propel the microprojectiles into the cells at a high velocity. The cells then appear to reseal themselves after the delivery of the recombinant molecule. This is a particularly promising technique for use with plants whose protoplasts will not regenerate whole plants.

6.4   HYBRIDISATION AND GENE PROBES

6.4.1   Cloned DNA probes

The increasing accumulation of DNA sequences in nucleic acid databases coupled with the availability of custom synthesis of oligonucleotides has provided a relatively straightforward means to design and produce gene probes and primers for PCR. Such

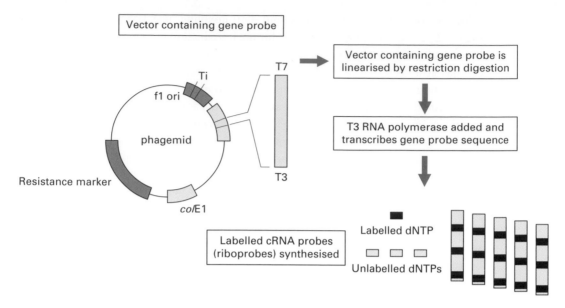

Fig. 6.26 Production of cRNA (riboprobes) using T3 RNA polymerase and phagemid vectors.

probes and primers are usually designed with bioinformatics software using sequence information from nucleic acid databases. Alternatively, gene family related sequences as indicated in Section 5.11 may also be successfully employed. However, there are many gene probes that have traditionally been derived from cDNA or from genomic sequences and which have been cloned into plasmid and phage vectors. These require manipulation before they may be labelled and used in hybridisation experiments. Gene probes may vary in length from 100 bp to a number of kilobases, although this is dependent on their origin. Many are short enough to be cloned into plasmid vectors and are useful in that they may be manipulated easily and are relatively stable both in transit and in the laboratory. The DNA sequences representing the gene probe are usually excised from the cloning vector by digestion with restriction enzymes and purified. In this way vector sequences which may hybridise non-specifically and cause high background signals in hybridisation experiments are removed. There are various ways of labelling DNA probes and these are described in Section 5.9.4.

6.4.2 RNA gene probes

It is also possible to prepare cRNA probes or riboprobes by *in vitro* transcription of gene probes cloned into a suitable vector. A good example of such a vector is the phagemid pBluescript SK since at each end of the multiple cloning site where the cloned DNA fragment resides are promoters for T3 or T7 RNA polymerase (Section 6.3.3). The vector is then made linear with a restriction enzyme and T3 or T7 RNA polymerase is used to transcribe the cloned DNA fragment. Provided a labelled NTP is added in the reaction a riboprobe labelled to a high specific activity will be produced (Fig. 6.26). One advantage of riboprobes is that they are single stranded and their sensitivity is generally regarded as superior to cloned double-stranded probes indicated in Section 6.4.1. They are used extensively in *in situ* hybridisation and for identifying and analysing mRNA and are described in more detail in Section 6.8.

6.5 SCREENING GENE LIBRARIES

6.5.1 Colony and plaque hybridisation

Once a cDNA or genomic library has been prepared the next task requires the identification of the specific fragment of interest. In many cases this may be more problematic than the library construction itself since many hundreds of thousands of clones may be in the library. One clone containing the desired fragment needs to be isolated from the library and therefore a number of techniques mainly based on hybridisation have been developed.

Colony hybridisation is one method used to identify a particular DNA fragment from a plasmid gene library (Fig. 6.27). A large number of clones are grown up to form colonies on one or more plates, and these are then replica plated onto nylon membranes placed on solid agar medium. Nutrients diffuse through the membranes and allow colonies to grow on them. The colonies are then lysed, and liberated DNA is denatured and bound to the membranes, so that the pattern of colonies is replaced by an identical pattern of bound DNA. The membranes are then incubated with a prehybridisation mix containing non-labelled non-specific DNA such as salmon sperm DNA to block non-specific sites. Following this denatured, labelled gene probe is added. Under hybridising conditions the probe will bind only to cloned fragments containing at least part of its corresponding gene (Section 5.9.3). The membranes are then washed to remove any unbound probe and the binding detected by autoradiography of the membranes. If non-radioactive labels have been used then alternative methods of detection must be employed (Section 5.9.4). By comparison of the patterns on the autoradiograph with the original plates of colonies, those that contain the desired gene (or part of it) can be identified and isolated for further analysis. A similar procedure is used to identify desired genes cloned into bacteriophage vectors. In this case the process is termed plaque hybridisation. It is the DNA contained in the bacteriophage particles found in each plaque that is immobilised on to the nylon membrane. This is then probed with an appropriately labelled complementary gene probe and the detection undertaken as for colony hybridisation.

6.5.2 PCR screening of gene libraries

In many cases it is now possible to use the PCR to screen cDNA or genomic libraries constructed in plasmids or bacteriophage vectors. This is usually undertaken with primers which anneal to the vector rather than the foreign DNA insert. The size of an amplified product may be used to characterise the cloned DNA and subsequent restriction mapping is then carried out (Fig. 6.28). The main advantage of the PCR over traditional hybridisation based screening is the rapidity of the technique, as PCR

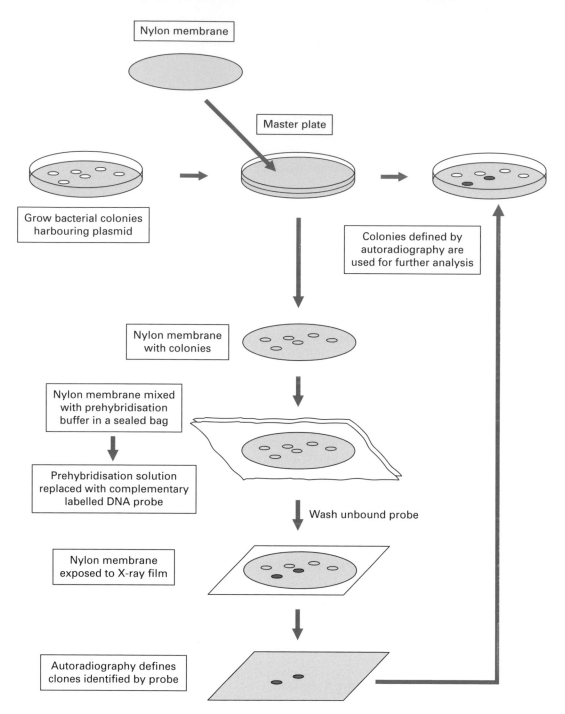

Fig. 6.27 Colony hybridisation technique for locating specific bacterial colonies harbouring recombinant plasmid vectors containing desired DNA fragments. This is achieved by hybridisation to a complementary labelled DNA probe and autoradiography.

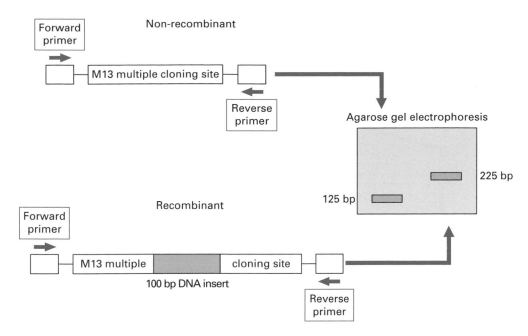

Fig. 6.28 PCR screening of recombinant vectors. In this figure, the M13 non-recombinant has no insert and so the PCR undertaken with forward and reverse sequencing primers gives rise to a product 125 bp in length. The M13 recombinant with an insert of 100 bp will give rise to a PCR product of 125 bp + 100 bp = 225 bp and thus may be distinguished from the non-recombinant by analysis on agarose gel electrophoresis.

screening may be undertaken in 3–4 h whereas it may be several days before detection by hybridisation is achieved. The PCR screening technique gives an indication of the size of the cloned insert rather than the sequence of the insert; however, PCR primers that are specific for a foreign DNA insert may also be used. This allows a more rigorous characterisation of clones from cDNA and genomic libraries.

6.5.3 Hybrid select/arrest translation

The difficulty of characterising clones and detecting a desired DNA fragment from a mixed cDNA library may be made simpler by two useful techniques termed hybrid select (release) translation or hybrid arrest translation. Following the preparation of a cDNA library in a plasmid vector the plasmid is extracted from part of each colony, and each preparation is then denatured and immobilised on a nylon membrane (Fig. 6.29). The membranes are soaked in total cellular mRNA, under stringent conditions (usually a temperature only a few degrees below T_m) in which hybridisation will occur only between complementary strands of nucleic acid. Hence each membrane will bind just one species of mRNA, since it has only one type of cDNA immobilised on it. Unbound mRNA is washed off the membranes, and then the bound mRNA is eluted and used to direct *in vitro* translation (Section 6.7). By immunoprecipitation or electrophoresis of the protein, the mRNA coding for a particular protein can be detected, and the clone containing its corresponding cDNA isolated. This technique is known as hybrid release translation. In a related method called

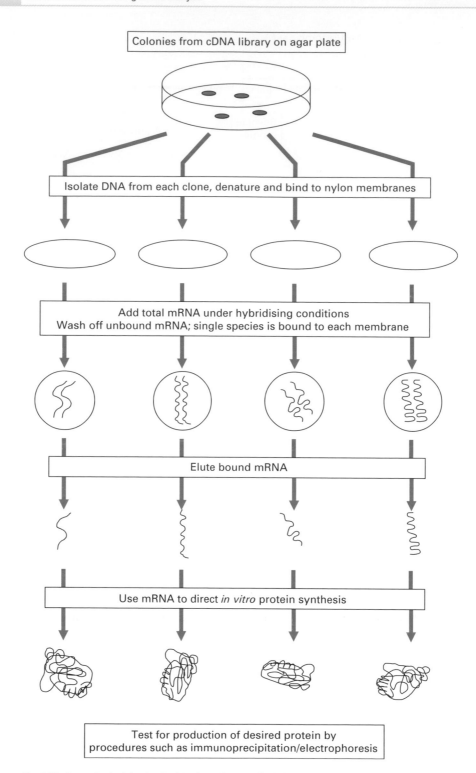

Fig. 6.29 General principles involved in the technique of a hybrid select translation.

hybrid arrest translation a positive result is indicated by the absence of a particular translation product when total mRNA is hybridised with excess cDNA. This is a consequence of the fact that mRNA cannot be translated when it is hybridised to another molecule.

6.5.4 Screening expression cDNA libraries

In some cases the protein for which the gene sequence is required is partially character-ised and in these cases it may be possible to produce antibodies to that protein. This allows immunological screening to be undertaken rather than gene hybridisation. Such anti-bodies are useful since they may be used as the probe if little or no gene sequence is available. In these cases it is possible to prepare a cDNA library in a specially adapted vector termed an expression vector which transcribes and translates any cDNA inserted into it. The protein is usually synthesised as a fusion with another protein such as β-galactosidase. Common examples of expression vectors are those based on bacterio-phage such as λgt11 and λZap or plasmids such as pEX. The precise requirements for such vectors are identical to vectors which are dedicated to producing proteins *in vitro* and are described in Section 6.7.1. In some cases expression vectors incorporate inducible promoters which may be activated by for example increasing the temperature allowing stringent control of expression of the cloned cDNA molecules (Fig. 6.30).

The cDNA library is plated out and nylon membrane filters prepared as for colony/plaque hybridisation. A solution containing the antibody to the desired protein is then added to the membrane. The membrane is then washed to remove any unbound protein and a further labelled antibody which is directed to the first antibody is applied. This allows visualisation of the plaque or colony that contains the cloned cDNA for that protein and this may then be picked from the agar plate and pure preparations grown for further analysis.

6.6 APPLICATIONS OF GENE CLONING

6.6.1 Sequencing cloned DNA

Most of the DNA sequencing now undertaken is based on the use of PCR products as the template; however, DNA fragments, including PCR products cloned into plasmid vectors, may be subjected to the chain termination sequencing (Section 5.9.5). How-ever, due to the double-stranded nature of plasmids further manipulation needs to be undertaken before this may be attempted. In these cases the plasmids are denatured usually by alkali treatment. Although the plasmids containing the foreign DNA inserts may reanneal the kinetics of the reaction is such that the strands are single-stranded for a long enough period of time to allow the sequencing method to succeed. It is also possible to include denaturants such as formamide in the reaction to further prevent reannealing. In general, however, superior results may be gained with sequencing single-stranded DNA from M13 or single-stranded phagemids which means that the cloned DNA of interest is usually subcloned into these vectors. A further modification

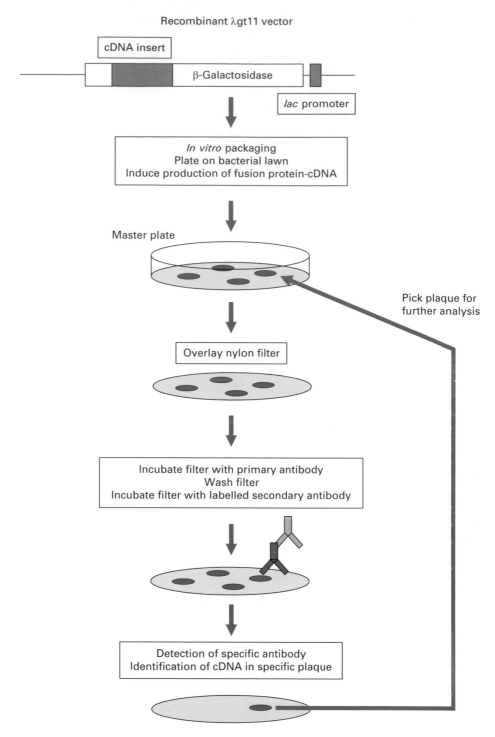

Fig. 6.30 Screening of cDNA libraries in expression vector λgt11. The cDNA inserted upstream from the gene for λβ-galactosidase will give rise to a fusion protein under induction (e.g. with IPTG). The plaques are then blotted onto a nylon membrane filter and probed with an antibody specific for the protein coded for by the cDNA. A secondary labelled antibody directed to the specific antibody can then be used to identify the location (plaque) of the cDNA.

that makes M13 useful in chain termination sequencing is the placement of universal priming sites at -20 or -40 bases from the start of the MCS. This allows any gene to be sequenced by using one universal primer since annealing of the primer prior to sequencing occurs outside the MCS and so is M13-specific rather than gene-specific. This obviates the need to synthesise new oligonucleotide primers for each new foreign DNA insert. A further, reverse priming site is also located at the opposite end of the polylinker allowing sequencing in the opposite orientation to be undertaken.

6.6.2 *In vitro* mutagenesis

One of the most powerful developments in molecular biology has been the ability to artificially create defined mutations in a gene and analyse the resulting protein following *in vitro* expression. Numerous methods are now available for producing site-directed mutations many of which now involve the PCR. Commonly termed protein engineering, this process involves a logical sequence of analytical and computational techniques centred around a design cycle. This includes the biochemical preparation and analysis of proteins, the subsequent identification of the gene encoding the protein and its modification. The production of the modified protein and its further biochemical analysis completes the concept of rational redesign to improve a protein's structure and function (Fig. 6.31).

The use of design cycles and rational design systems are exemplified by the study and manipulation of subtilisin. This is a serine protease of broad specificity and of considerable industrial importance being used in soap powder and in the food and leather industries. Protein engineering has been used to alter the specificity, pH profile and stability to oxidative, thermal and alkaline inactivation. Analysis of homologous thermophiles and their resistance to oxidation has also been improved. Engineered subtilisins of improved bleach resistance and wash performance are now used in many brands of washing powders. Furthermore mutagenesis has played an important role in the re-engineering of important therapeutic proteins such as the Herceptin antibody which has been used to successfully treat certain types of breast cancer.

6.6.3 Oligonucleotide-directed mutagenesis

The traditional method of site-directed mutagenesis demands that the gene be already cloned or subcloned into a single-stranded vector such as M13. Complete sequencing of the gene is essential to identify a potential region for mutation. Once the precise base change has been identified an oligonucleotide is designed that is complementary to part of the gene but has one base difference. This difference is designed to alter a particular codon, which, following translation, gives rise to a different amino acid and hence may alter the properties of the protein.

The oligonucleotide and the single-stranded DNA are annealed and DNA polymerase is added together with the dNTPs. The primer for the reaction is the $3'$ end of the oligonucleotide. The DNA polymerase produces a new complementary DNA strand to the existing one but which incorporates the oligonucleotide with the base mutation. The subsequent cloning of the recombinant produces multiple copies, half of which

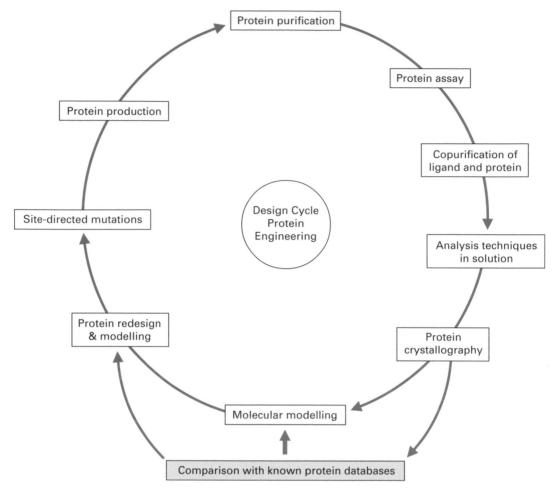

Fig. 6.31 Protein design cycle used in the rational redesign of proteins and enzymes.

contain a sequence with the mutation and half contain the wild-type sequence. Plaque hybridisation using the oligonucleotide as the probe is then used at a stringency that allows only those plaques containing a mutated sequence to be identified (Fig. 6.32). Further methods have also been developed which simplify the process of detecting the strands with the mutations.

6.6.4 PCR-based mutagenesis

The PCR has been adapted to allow mutagenesis to be undertaken and this relies on single bases mismatched between one of the PCR primers and the target DNA to become incorporated into the amplified product following thermal cycling.

The basic PCR mutagenesis system involves the use of two primary PCR reactions to produce two overlapping DNA fragments both bearing the same mutation in the overlap region. The technique is termed overlap extension PCR. The two separate PCR products are made single-stranded and the overlap in sequence allows the products

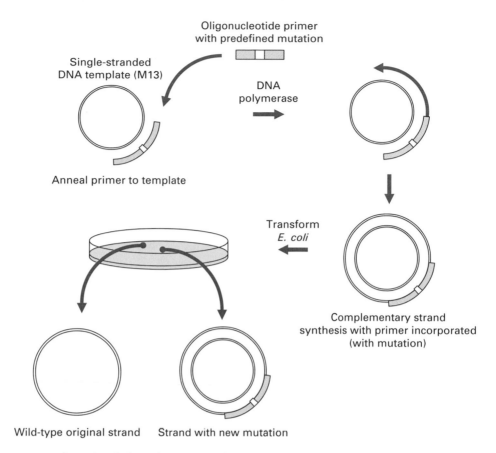

Fig. 6.32 Oligonucleotide-directed mutagenesis. This technique requires a knowledge of nucleotide sequence, since an oligonucleotide may then be synthesised with the base mutation. Annealing of the oligonucleotide to complementary (except for the mutation) single-stranded DNA provides a primer for DNA polymerase to produce a new strand and thus incorporates the primer with the mutation.

from each reaction to hybridise. Following this, one of the two hybrids bearing a free 3′ hydroxyl group is extended to produce a new duplex fragment. The other hybrid with a 5′ hydroxyl group cannot act as substrate in the reaction. Thus, the overlapped and extended product will now contain the directed mutation (Fig. 6.33). Deletions and insertions may also be created with this method although the requirements of four primers and three PCR reactions limits the general applicability of the technique. A modification of the overlap extension PCR may also be used to construct directed mutations; this is termed megaprimer PCR. This method utilises three oligonucleotide primers to perform two rounds of PCR. A complete PCR product, the megaprimer is made single-stranded and this is used as a large primer in a further PCR reaction with an additional primer.

The above are all methods for creating rational defined mutations as part of a design cycle system. However it is also possible to introduce random mutations into a gene and select for enhanced or new activities of the protein or enzyme it encodes. This accelerated form of artificial molecular evolution may be undertaken using

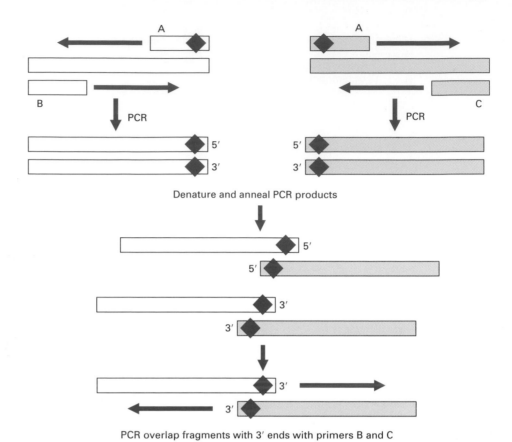

Fig. 6.33 Construction of a synthetic DNA fragment with a predefined mutation using overlap PCR mutagenesis.

error-prone PCR where deliberate and random mutations are introduced by a low-fidelity PCR amplification reaction. The resulting amplified gene is then translated and its activity assayed. This has already provided novel evolved enzymes such as a *p*-nitrobenzyl esterase which exhibits an unusual and surprising affinity for organic solvents. This accelerated evolutionary approach to protein engineering has been useful in the production of novel phage displayed antibodies and in the development of antibodies with enzymic activities (catalytic antibodies).

6.7 EXPRESSION OF FOREIGN GENES

One of the most useful applications of recombinant DNA technology is the ability to artificially synthesise large quantities of natural or modified proteins in a host cell such as bacteria or yeast. The benefits of these techniques have been enjoyed for many years since the first insulin molecules were cloned and expressed in 1982 (Table 6.3). Contamination of other proteins such as the blood product factor VIII with infectious agents has also increased the need to develop effective vectors for *in vitro* expression

Table 6.3 **A number of recombinant DNA-derived human therapeutic reagents**

Therapeutic area	Recombinant product
Drugs	Erythropoietin
	Insulin
	Growth hormone
	Coagulation factors (e.g. factor VIII)
	Plasminogen activator
Vaccines	Hepatitis B
Cytokines/growth factors	GM-CSF
	G-CSF
	Interleukins
	Interferons

Notes: GM-CSF, granulocyte–macrophage colony-stimulating factor; G-CSF, granulocyte colony-stimulating factor.

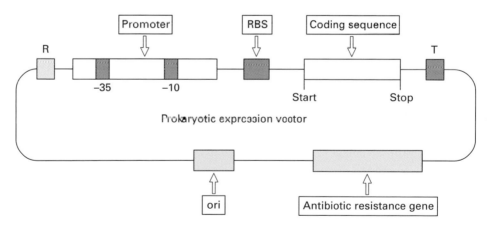

Fig. 6.34 Components of a typical prokaryotic expression vector. To produce a transcript (coding sequence) and translate it, a number of sequences in the vector are required. These include the promoter and ribosome-binding site (RBS). The activity of the promoter may be modulated by a regulatory gene (R), which acts in a way similar to that of the regulatory gene in the *lac* operon. T indicates a transcription terminator.

of foreign genes. In general the expression of foreign genes is carried out in specialised cloning vectors (Fig. 6.34). However it is possible to use cell-free transcription and translation systems that direct the synthesis of proteins without the need grow and maintain cells. *In vitro* translation is carried out with the appropriate amino acids, ribosomes, tRNA molecules and isolated mRNA fractions. Wheat germ extracts or rabbit reticulocyte lysates are usually the systems of choice for *in vitro* translation. The resulting

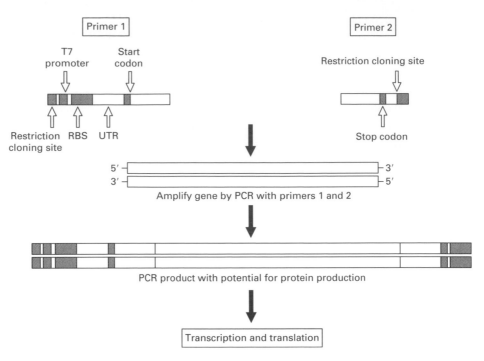

Fig. 6.35 Expression PCR (E-PCR). This technique amplifies a target sequence with one promoter that contains a transcriptional promoter, ribosome binding site (RBS), untranslated leader region (UTR) and start codon. The other primer contains a stop codon. The amplified PCR products may be used in transcription and translation to produce a protein.

proteins may be detected by polyacrylamide gel electrophoresis or by immunological detection using **western blotting**. Recently oligonucleotide PCR primers have been designed to incorporate a promoter for RNA polymerase and a ribosome-binding site. When the so-called **expression PCR** (E-PCR) is carried out the amplified products are denatured and transcribed by RNA polymerase after which they are translated *in vitro*. The advantage of this system is that large amounts of specific RNA are synthesised thus increasing the yield of specific proteins (Fig. 6.35).

6.7.1 Production of fusion proteins

For a foreign gene to be expressed in a bacterial cell, it must have particular promoter sequences upstream of the coding region, to which the RNA polymerase will bind prior to transcription of the gene. The choice of promoter is vital for correct and efficient transcription since the sequence and position of promoters are specific to a particular host such as *E. coli* (Section 5.5.4). It must also contain a ribosome-binding site, placed just before the coding region. Unless a cloned gene contains both of these sequences, it will not be expressed in a bacterial host cell. If the gene has been produced via cDNA from a eukaryotic cell, then it will certainly not have any such sequences. Consequently, expression vectors have been developed which contain promoter and ribosome-binding sites positioned just before one or more restriction sites for the insertion of foreign DNA. These regulatory sequences, such as that from the *lac* operon of *E. coli*, are usually derived from genes that, when induced, are strongly expressed in bacteria. Since the mRNA produced from the gene is read as triplet codons, the inserted sequence must be placed so that its reading frame is in phase with the regulatory sequence. This can be ensured by the use of three vectors which differ only in the number of bases between promoter and insertion site, the second and third vectors being respectively one and two bases longer than the first. If an insert is cloned in all three vectors then in general it will subsequently be in the correct reading frame in one of them. The resulting clones can be screened for the production of a functional foreign protein (Section 6.5.4).

In some cases the protein is expressed as a fusion with a general protein such as β-galactosidase or glutathione-S-transferase (GST) to facilitate its recovery. It may also be tagged with a moiety such as a polyhistidine (6×His-Tag) which binds strongly to a nickel-chelate-nitrilotriacetate (Ni-NTA) chromatography column. The usefulness of this method is that the binding is independent of the three-dimensional structure of the 6×His-tag and so recovery is efficient even under strong denaturing conditions, often required for membrane proteins and inclusion bodies (Fig. 6.36). The tags are subsequently removed by cleavage with a reagent such as cyanogen bromide and the protein of interest purified by protein biochemical methods such as chromatography and polyacrylamide gel electrophoresis.

It is not only possible, but usually essential, to use cDNA instead of a eukaryotic genomic DNA to direct the production of a functional protein by bacteria. This is because bacteria are not capable of processing RNA to remove introns, and so any foreign genes must be pre-processed as cDNA if they contain introns. A further problem arises if the protein must be glycosylated, by the addition of oligosaccharides at specific sites, in order to become functional. Although the use of bacterial expression systems is somewhat limited for eukaryotic systems there are a number of eukaryotic expression systems based on plant, mammalian, insect and yeast cells. These types of cells can perform such post-translational modifications, producing a correct glycosylation pattern and in some cases the correct removal of introns. It is also possible to include a signal or address sequence at the 5′ end of the mRNA which directs the protein to a particular cellular compartment or even out of the cell altogether into the supernatant. This makes the recovery of expressed recombinant proteins much easier since the supernatant may be drawn off while the cells are still producing protein.

One useful eukaryotic expression system is based on the monkey COS cell line. These cells each contain a region derived from a mammalian monkey virus termed simian virus 40 (SV40). A defective region of the SV40 genome has been stably integrated into the COS cell genome. This allows the expression of a protein termed the large T antigen which is required for viral replication. When a recombinant vector having the SV40 origin of replication and carrying foreign DNA is inserted into the COS cells viral replication takes place. This results in a high level expression of foreign proteins. The disadvantage of this system is the ultimate lysis of the COS cells and limited insert capacity of the vector. Much interest is also currently focussed on other modified viruses, vaccinia virus and baculovirus. These have been developed for high-level expression in mammalian cells and insect cells respectively. The vaccinia virus

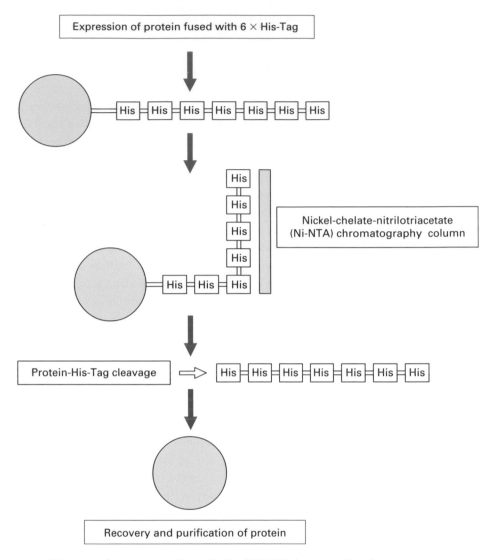

Fig. 6.36 Recovery of proteins using (6 × His-Tag) and (Ni-NTA) chromatography columns.

in particular has been used to correct the defective ion transport by introducing a wild-type cystic fibrosis gene into cells bearing a mutated cystic fibrosis (CFTR) gene. There is no doubt that the further development of these vector systems will enhance eukaryotic protein expression in the future.

6.7.2 Phage display techniques

As a result of the production of phagemid vectors and as a means of overcoming the problems of screening large numbers of clones generated from genomic libraries of antibody genes, a method for linking the phenotype or expressed protein with the genotype has been devised. This is termed phage display, since a functional protein is

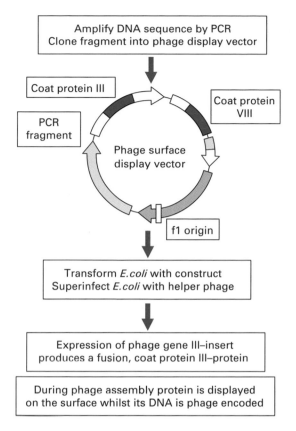

Amplify DNA sequence by PCR
Clone fragment into phage display vector

Coat protein III

PCR fragment

Coat protein VIII

Phage surface display vector

f1 origin

Transform *E. coli* with construct
Superinfect *E. coli* with helper phage

Expression of phage gene III–insert
produces a fusion, coat protein III–protein

During phage assembly protein is displayed
on the surface whilst its DNA is phage encoded

Fig. 6.37 Flow diagram indicating the main steps in the phage display technique.

linked to a major coat protein of a coliphage whilst the single-stranded gene encoding the protein is packaged within the virion. The initial steps of the method rely on the PCR to amplify gene fragments that represent functional domains or subunits of a protein such as an antibody. These are then cloned into a phage display vector which is an adapted phagemid vector (Section 6.3.3) and used to transform *E. coli*. A helper phage is then added to provide accessory proteins for new phage molecules to be constructed. The DNA fragments representing the protein or polypeptide of interest are also transcribed and translated, but linked to the major coat protein g III. Thus when the phage is assembled the protein or polypeptide of interest is incorporated into the coat of the phage and displayed, whilst the corresponding DNA is encapsulated (Fig. 6.37).

There are numerous applications for the display of proteins on the surface of bacteriophage viruses, bacteria and other organisms, and commercial organisations have been quick to exploit this technology. One major application is the analysis and production of engineered antibodies from which the technology was mainly developed. In general phage-based systems have a number of novel applications in terms of ease of selection rather than screening of antibody fragments, allowing analysis by methods such as affinity chromatography. In this way it is possible to generate large numbers of antibody heavy and light chain genes by PCR amplification and mix them in a random fashion. This recombinatorial library approach may allow new or novel partners to be formed as well as naturally existing ones. This strategy is not restricted to antibodies and vast libraries of peptides may be used in this combinatorial chemistry approach to identify novel compounds of use in biotechnology and medicine.

Phage-based cloning methods also offer the advantage of allowing mutagenesis to be performed with relative ease. This may allow the production of antibodies with affinities approaching that derived from the human or mouse immune system. This may be brought about by using an error prone DNA polymerase in the initial steps of constructing a phage display library. It is possible that these types of libraries may provide a route to high affinity recombinant antibody fragments that are difficult to produce by more conventional hybridoma fusion techniques. Surface display libraries have also been prepared for the selection of ligands, hormones and other polypeptides in addition to allowing studies on protein–protein or protein–DNA interactions or determining the precise binding domains in these receptor–ligand interactions.

6.7.3 Alternative display systems

A number of display systems have been developed based on the original phage display technique. One interesting method is ribosome display where a sequence or even a library of sequences are transcribed and translated *in vitro*. However in the DNA library the sequences are fused to spacer sequences lacking a stop codon. During translation at the ribosome the protein protrudes from the ribosome and is locked in with the mRNA. The complex can be stabilised by adding salt. In this way it is possible to select the appropriate protein through binding to its ligand. Thus a high-affinity protein-ligand can be isolated which has the mRNA that originally encoded it. The mRNA may then be reverse transcribed into cDNA and amplified by PCR to allow further methods such as mutagenesis to be undertaken. A related technique, mRNA display, is similar except the association between the protein and mRNA is through a more stable covalent puromycin link rather than the salt-induced link as in ribosome display. Further display systems, based on yeast or bacteria, have also been developed and provide powerful *in vitro* selection methods.

6.8 ANALYSING GENES AND GENE EXPRESSION

6.8.1 Identifying and analysing mRNA

The levels and expression patterns of mRNA dictate many cellular processes and therefore there is much interest in the ability to analyse and determine levels of a particular mRNA. Technologies such as real-time or quantitative PCR and microchip expression arrays are currently being employed and refined for high throughput analysis. A number of other informative techniques have been developed that allow the fine structure of a particular mRNA to be analysed and the relative amounts of an RNA quantitated by non-PCR-based methods. This is important not only for gene regulation studies but may also be used as a marker for certain clinical disorders. Traditionally the Northern blot has been used for

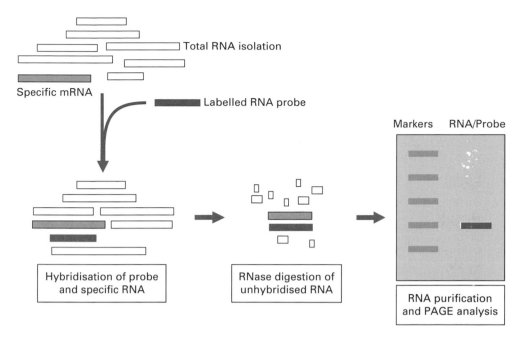

Total RNA isolation

Specific mRNA

Labelled RNA probe

Markers   RNA/Probe

Hybridisation of probe
and specific RNA

RNase digestion of
unhybridised RNA

RNA purification
and PAGE analysis

Fig. 6.38 Steps involved in the ribonuclease protection assay (RPA). PAGE, polyacrylamide gel electrophoresis.

detection of particular RNA transcripts by blotting extracted mRNA and immobilising it to a nylon membrane (Section 5.9.2). Subsequent hybridisation with labelled gene probes allows precise determination of the size and nature of a transcript. However, much use has been made of a number of nucleases that digest only single-stranded nucleic acids and not double-stranded molecules. In particular the ribonuclease protection assay (RPA) has allowed much information to be gained regarding the nature of mRNA transcripts (Fig. 6.38). In the RPA single-stranded mRNA is hybridised in solution to a labelled single-stranded RNA probe which is in excess. The hybridised part of the complex becomes protected whereas the unhybridised part of the probe made from RNA is digested with RNase A and RNase T1. The protected fragment may then be analysed on a high-resolution polyacrylamide gel. This method may give valuable information regarding the mRNA in terms of the precise structure of the transcript (transcription start site, intron/exon junctions, etc.). It is also quantitative and requires less RNA than a Northern blot. A related technique, S1 nuclease mapping, is similar although the unhybridised part of a DNA probe, rather than an RNA probe, is digested, this time with the enzyme S1 nuclease.

The PCR has also had an impact on the analysis of RNA via the development of a technique known as reverse transcriptase–PCR (RT–PCR). Here the RNA is isolated and a first strand cDNA synthesis undertaken with reverse transcriptase; the cDNA is then used in a conventional PCR (Section 6.2.5). Under certain circumstances a number of thermostable DNA polymerases have reverse transcriptase activity which obviates the need to separate the two reactions and allows the RT–PCR to be carried out in one tube. One of the main benefits of RT–PCR is the ability to identify rare or low levels of mRNA transcripts with great sensitivity. This is especially useful when

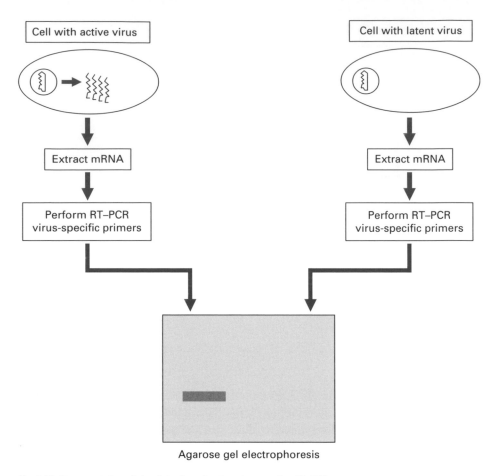

Fig. 6.39 Representation of the detection of active viruses using RT–PCR.

detecting, for example, viral gene expression and furthermore allows the means of differentiating between latent and active virus (Fig. 6.39). The level of mRNA production may also be determined by using a PCR-based method, termed quantitative PCR (Section 5.10.7).

In many cases the analysis of tissue-specific gene expression is required and again the PCR has been adapted provide a solution. This technique, termed differential display, is also an RT–PCR-based system requiring that isolated mRNA be first converted into cDNA. Following this, one of the PCR primers, designed to anneal to a general mRNA element such as the poly(A) tail in eukaryotic cells, is used in conjunction with a combination of arbitrary 6–7 bp primers which bind to the 5′ end of the transcripts. Consequently this results in the generation of multiple PCR products with reproducible patterns (Fig. 6.40). Comparative analysis by gel electrophoresis of PCR products generated from different cell types therefore allows the identification and isolation of those transcripts that are differentially expressed. As with many PCR-based techniques the time to identify such genes is dramatically reduced from the weeks that are required to construct and screen cDNA libraries to a few days.

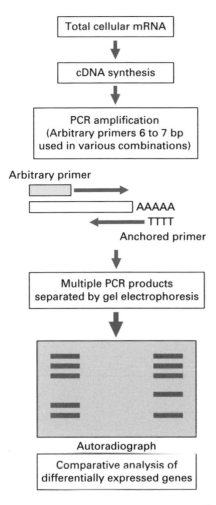

Fig. 6.40 Analysis of gene expression using differential display PCR.

6.8.2 **Analysing genes *in situ***

Gross chromosomal changes are often detectable by microscopic examination of the chromosomes within a karyotype (Section 5.3). Single or restricted numbers of base substitutions, deletions, rearrangements or insertions are far less easily detectable but may induce similarly profound effects on normal cellular biochemistry. *In situ* hybridisation makes it possible to determine the chromosomal location of a particular gene fragment or gene mutation. This is carried out by preparing a radiolabelled DNA or RNA probe and applying this to a tissue or chromosomal preparation fixed to a microscope slide. Any probe that does not hybridise to complementary sequences is washed off and an image of the distribution or location of the bound probe is viewed by autoradiography (Fig. 6.41). Using tissue or cells fixed to slides it is also possible to carry out *in situ* PCR and qPCR. This is a highly sensitive technique where PCR is carried out directly on the tissue slide with the standard PCR reagents. Specially adapted thermal cycling machines are required to hold the slide preparations and allow the PCR to proceed.

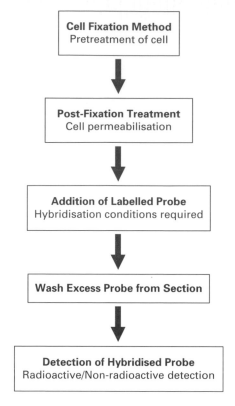

Fig. 6.41 General scheme for *in situ* hybridisation.

This allows the localisation and identification of, for example, single copies of intracellular viruses and in the case of qPCR the determination of initial concentrations of nucleic acid.

An alternative labelling strategy used in karyotyping and gene localisation is **fluorescent *in situ* hybridisation** (FISH). This method sometimes termed chromosome painting is based on *in situ* hybridisation but in which different gene probes are labelled with different fluorochromes, each specific for a particular chromosome. The advantage of this method is that separate gene regions may be identified and comparisons made within the same chromosome preparation. The technique is also likely to be highly useful in genome mapping for ordering DNA probes along a chromosomal segment (Section 6.9).

6.8.3 Analysing promoter–protein interactions

To determine potential transcriptional regulatory sequences genomic DNA fragments may be cloned into specially devised promoter probe vectors. These contain sites for insertion of foreign DNA which lies upstream of a reporter gene. A number of reporter genes are currently used, including the *lacZ* gene encoding β-galactosidase, the *CAT* gene encoding chloramphenicol acetyl transferase (CAT) and the *lux* gene which produces luciferase and is determined in a bioluminescent assay. Fragments of DNA potentially containing a promoter region are cloned into the vector and the constructs

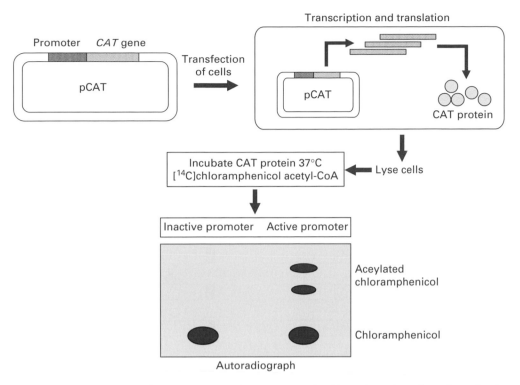

Fig. 6.42 Assay for promoters using the reporter gene for chloramphenicol acetyl transferase (CAT).

transfected into eukaryotic cells. Any expression of the reporter gene will be driven by the foreign DNA which must therefore contain promoter sequences (Fig. 6.42). These plasmids and other reporter genes such as those using **green fluorescent protein** (GFP) or the firefly luciferase gene allow quantitation of gene transcription in response to transcriptional activators.

The binding of a regulatory protein or transcription factor to a specific DNA site results in a complex that may be analysed by the technique termed **gel retardation**. Under gel electrophoresis the migration of a DNA fragment bound to a protein of a relatively large mass will be retarded in comparison to the DNA fragment alone. For gel retardation to be useful the region containing the promoter DNA element must be digested or mapped with a restriction endonuclease before it is complexed with the protein. The location of the promoter may then be defined by finding the position on the restriction map of the fragment that binds to the regulatory protein and therefore retards it during electrophoresis. One potential problem with gel retardation is the ability to define the precise nucleotide binding region of the protein, since this depends on the accuracy and detail of the restriction map and the convenience of the restriction sites. However it is a useful first step in determining the interaction of a regulatory protein with a DNA binding site.

DNA footprinting relies on the fact that the interaction of a DNA-binding protein with a regulatory DNA sequence will protect that DNA sequence from degradation by an enzyme such as DNase I. The DNA regulatory sequence is first labelled at one end with a radioactive label and then mixed with the DNA-binding protein

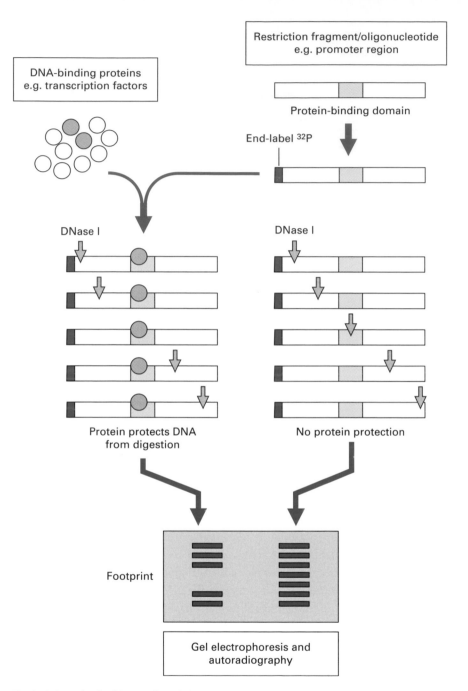

Fig. 6.43 Steps involved in DNA footprinting.

(Fig. 6.43). DNase I is added and conditions favouring a partial digestion are then carried out. This limited digestion ensures that a number of fragments are produced where the DNA is not protected by the DNA-binding protein. The region protected by the DNA-binding protein will remain undigested. All the fragments are then separated on a high-resolution polyacrylamide gel alongside a control digestion where no DNA-binding

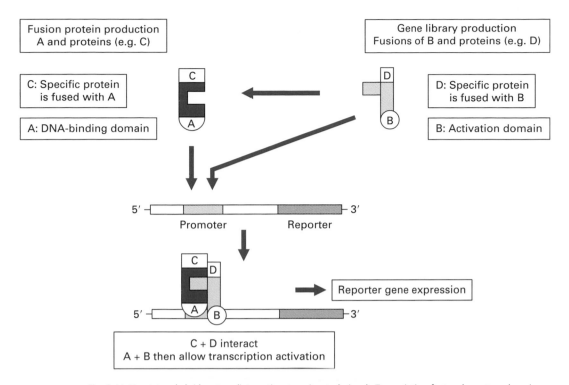

Fig. 6.44 Yeast two-hybrid system (interaction trapping technique). Transcription factors have two domains, one for DNA binding (A) and the other to allow binding to further proteins (B). Thus a recombinant molecule is formed from a protein (C) as a fusion with the DNA-binding domain. It cannot, however, activate transcription alone. Genes from a cDNA library (D) are expressed as a fusion with the activator domain (B) but also cannot initiate transcription alone. When the two fractions are mixed together, transcription is initiated if the domains are complementary and expression of a reporter gene takes place.

protein is present. The autoradiograph of a gel will contain a ladder of bands representing the partially digested fragments. Where DNA has been protected no bands appear; this region or hole is termed the DNA footprint. The position of the protein-binding sequence within the DNA may be elucidated from the size of the fragments either side of the footprint region. Footprinting is a more precise method of locating a DNA–protein interaction than gel retardation; however, it also is unable to give any information as to the precise interaction or the contribution of individual nucleotides.

In addition to the detection of DNA sequences that contribute to the regulation of gene expression an ingenious way of detecting the protein transcription factors has been developed. This is termed the yeast two–hybrid system. Transcription factors have two domains, one for DNA binding and the other to allow binding to further proteins (activation domain). These occur as part of the same molecule in natural transcription factors, for example TFIID (Section 5.5.4). However they may also be formed from two separate domains. Thus a recombinant molecule is formed encoding the protein under study as a fusion with the DNA-binding domain. It cannot however activate transcription. Genes from a cDNA library are expressed as a fusion with the activator domain; this also cannot initiate transcription. However, when the two fractions are mixed together transcription is initiated if the domains are complementary (Fig. 6.44). This is indicated

Table 6.4 **Use of transgenic mice for investigation of selected human disorders**

Gene/protein	Genetic lesion	Disorder in humans
Tyrosine kinase (TK)	Constitutive expression of gene	Cardiac hypertrophy
HIV transactivator	Expression of HIV *tat* gene	Kaposi's sarcoma
Angiotensinogen	Expression of rat angiotensinogen gene	Hypertension
Cholesterol ester transfer protein (CET protein)	Expression of *CET* gene	Atherosclerosis
Hypoxanthine-guanine phosphoribosyl transferase (HPRT)	Inactivation of *HPRT* gene	HPRT deficiency

by the transcription of a reporter gene such as the *CAT* gene. The technique is not just confined to transcription factors and may be applied to any protein system where interaction occurs.

6.8.4 Transgenics and gene targeting

In many cases it is desirable to analyse the effect of certain genes and proteins in an organism rather than in the laboratory. Furthermore the production of pharmaceutical products and therapeutic proteins is also desirable in a whole organism. This also has important consequences for the biotechnology and agricultural industry (Section 6.10) (Table 6.4). The introduction of foreign genes into germ line cells and the production of an altered organism is termed transgenics. There are two broad strategies for transgenesis. The first is direct transgenesis in mammals whereby recombinant DNA is injected directly into the male pronucleus of a recently fertilised egg. This is then raised in a foster mother animal resulting in an offspring that is all transgenic. Selective transgenesis is where the recombinant DNA is transferred into embryo stem (ES) cells. The cells are then cultured in the laboratory and those expressing the desired protein selected and incorporated into the inner cell mass of an early embryo. The resulting transgenic animal is raised in a foster mother but in this case the transgenic animal is a mosaic or chimeric since only a small proportion of the cells will be expressing the protein. The initial problem with both approaches is the random nature of the integration of the recombinant DNA into the genome of the egg or embryo stem cells. This may produce proteins in cells where it is not required or disrupt genes necessary for correct growth and development.

A refinement of this however is gene targeting which involves the production of an altered gene in an intact cell, a form of *in vivo* mutagenesis as opposed to *in vitro* mutagenesis (Section 6.6.2). The gene is inserted into the genome of, for example, an ES cell by specialised viral-based vectors. The insertion is non-random, however, since homologous sequences exist on the vector to the gene and on the gene to be targeted. Thus, homologous recombination may introduce a new genetic property to the cell, or inactivate an already existing one, termed gene knockout. Perhaps the most important aspect of these techniques is that they allow animal models of human diseases to be created. This is useful since the physiological and biochemical consequences of a disease are often complex and difficult to study impeding the development of diagnostic and therapeutic strategies.

6.8.5 Modulating gene expression by RNAi

There are a number of ways of experimentally changing the expression of genes. Traditionally methods have focussed on altering the levels of mRNA by manipulation of promoter sequences or levels of accessory proteins involved in control of expression. In addition post-mRNA production methods have also been employed such as antisense RNA, where a nucleic acid sequence complementary to an expressed mRNA is delivered into the cell. This antisense sequence binds to the mRNA and prevents its translation. A development of this theme and a process that is found in a variety of normal cellular processes is termed RNA interference (RNAi) and uses microRNA. Here a number of techniques have been developed that allow the modulation of gene expression in certain cells. This type of cellular-based gene expression modulation will no doubt extend to many organisms in the next few years.

6.8.6 Analysing genetic mutations

There are several types of mutations that can occur in nucleic acids, either transiently or those that are stably incorporated into the genome. During evolution, mutations may be inherited in one or both copies of a chromosome, resulting in polymorphisms within the population (Section 5.3). Mutations may potentially occur at any site within the genome; however, there are several instances whereby mutations occur in limited regions. This is particularly obvious in prokaryotes, where elements of the genome (termed hypervariable regions) undergo extensive mutations to generate large numbers of variants, by virtue of the high rate of replication of the organisms. Similar hypervariable sequences are generated in the normal antibody immune response in eukaryotes. Mutations may have several effects upon the structure and function of the genome. Some mutations may lead to undetectable effects upon normal cellular functions, termed conservative mutations. An example of these are mutations that occur in intron sequences and therefore play no part in the final structure and function of the protein or its regulation. Alternatively, mutations may result in profound effects upon normal cell function such as altered transcription rates or on the sequence of mRNAs necessary for normal cellular processes.

Mutations occurring within exons may alter the amino acid composition of the encoded protein by causing amino acid substitution or by changing the reading frame used during translation. These point mutations were traditionally detected by Southern blotting or, if a convenient restriction site was available, by restriction fragment length polymorphism (RFLP) (Section 5.9). However, the PCR has been used to great effect in mutation detection since it is possible to use allele-specific oligonucleotide PCR (ASO–PCR) where two competing primers and one general primer are used in the reaction (Fig. 6.45). One of the primers is directly complementary to the known point mutation whereas the other is a wild-type primer; that is, the primers are identical

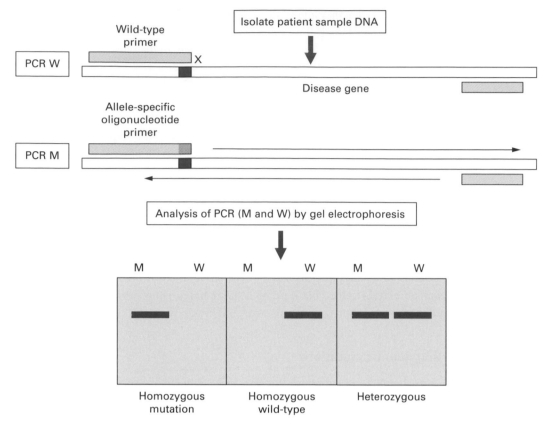

Fig. 6.45 Point mutation detection using allele-specific oligonucleotide PCR (ASO–PCR).

except for the terminal 3′ end base. Thus, if the DNA contains the point mutation only the primer with the complementary sequence will bind and be incorporated into the amplified DNA, whereas if the DNA is normal the wild-type primer is incorporated. The results of the PCR are analysed by agarose gel electrophoresis. A further modification of ASO–PCR has been developed where the primers are each labelled with a different fluorochrome. Since the primers are labelled differently a positive or negative result is produced directly without the need to examine the PCRs by gel electrophoresis.

Various modifications now allow more than one PCR to be carried out at a time (multiplex PCR), and hence the detection of more than one mutation is possible at the same time. Where the mutation is unknown it is also possible to use a PCR system with a gel-based detection method termed denaturing gradient gel electrophoresis (DGGE). In this technique a sample DNA heteroduplex containing a mutation is amplified by the PCR which is also used to attach a GC-rich sequence to one end of the heteroduplex. The mutated heteroduplex is identified by its altered melting properties through a polyacryl-amide gel which contains a gradient of denaturant such as urea. At a certain point in the gradient the heteroduplex will denature relative to a perfectly matched homoduplex and thus may be identified. The GC clamp maintains the integrity of the end of the duplex on passage through the gel (Fig. 6.46). The sensitivity of this and other mutation detection methods has been substantially increased by the use of PCR, and further mutation

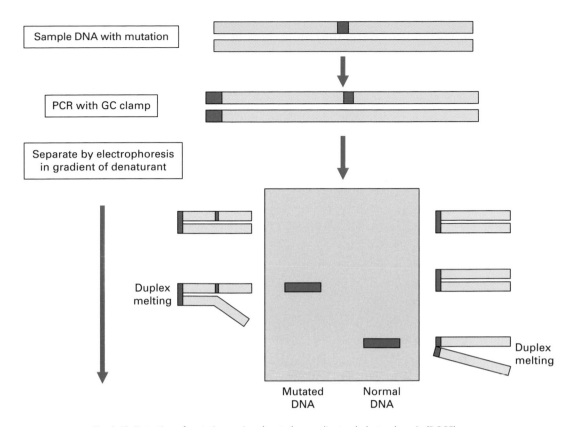

Fig. 6.46 Detection of mutations using denaturing gradient gel electrophoresis (DGGE).

techniques used to detect known or unknown mutations are indicated in Table 6.5. An extension of this principle is used in a number of detection methods employing denaturing high-performance liquid chromatography (dHPLC). Commonly known as wave technology the detection of denatured single strands containing mismatches is rapid allowing a high-throughput analysis of samples to be achieved.

6.8.7 Detecting DNA polymorphisms

Polymorphisms are particularly interesting elements of the human genome and as such may be used as the basis for differentiating between individuals. All humans carry repeats of sequences known as minisatellite DNA of which the number of repeats varies between unrelated individuals. Hybridisation of probes which anneal to these sequences using Southern blotting provides the means to type and identify those individuals (Section 5.3).

DNA fingerprinting is the collective term for two distinct genetic testing systems that use either 'multilocus' probes or 'single-locus' probes. Initially described DNA fingerprinting probes were multilocus probes and so termed because they detect hypervariable minisatellites throughout the genome, i.e. at multiple locations within the genome. In contrast, several single-locus probes were discovered which under

Table 6.5 **Main methods of detecting mutations in DNA samples**

Technique	Basis of method	Main characteristics of detection
Southern blotting	Gel based	Labelled probe hybridisation to DNA
Dot/slot blotting	Sample application	Labelled probe hybridisation to DNA
Allele-specific oligo-PCR (ASO–PCR)	PCR based	Oligonucleotide matching to DNA sample
Denaturing gradient gel electrophoresis (DGGE)	Gel/PCR based	Melting temperature of DNA strands
Single-stranded conformation polymorphism (SSCP)	Gel/PCR based	Conformation difference of DNA strands
Ligase chain reaction (LCR)	Gel/automated	Oligonucleotide matching to DNA sample
DNA sequencing	Gel based	Nucleotide sequence analysis of DNA
DNA microchips	Glass chip based	Sample DNA hybridisation to oligo arrays

specific conditions only detect the two alleles at a single locus and generate what have been termed DNA profiles because, unlike multilocus probes, the two-band pattern result is in itself insufficient to uniquely identify an individual.

Techniques based on the PCR have been coupled to the detection of minisatellite loci. The inherent larger size of such DNA regions was not best suited to PCR amplification; however, new PCR developments are beginning to allow this to take place. The discovery of polymorphisms within the repeating sequences of minisatellites has led to the development of a PCR-based method that distinguishes an individual on the basis of the random distribution of repeat types along the length of a person's two alleles for one such minisatellite. Known as **minisatellite variant repeat** (MVR) analysis or **digital DNA typing**, this technique can lead to a simple numerical coding of the repeat variation detected. Potentially this combines the advantages of PCR sensitivity and rapidity with the discriminating power of minisatellite alleles. Thus for the future there are a number of interesting identification systems under development and evaluation. Techniques for genetic detection of polymorphisms have been used in many cases of paternity testing and immigration control, and are becoming central factors in many criminal investigations. They are also valuable tools in plant biotechnology for cereal typing and in the field of pedigree analysis and animal breeding.

6.8.8 **Microarrays and DNA microchips**

One firmly established area under rapid development in molecular biology is the use of microarrays or **DNA microchips**. These provide a radically different approach to current laboratory molecular biology research strategies in that large-scale analysis and quantification of genes and gene expression is possible simultaneously. A microarray consists of an ordered arrangement of potentially hundreds of thousands of DNA sequences such as oligonucleotides or cDNAs deposited onto a solid surface. The solid support may be either glass or silicon and currently the arrays are synthesised on or off the chip. They require complex fabrication methods similar to that used in producing computer microchips. Most commercial productions employ robotic ultrafine microarray deposition instruments which dispense volumes in the picolitre range. Alternatively on-chip fabrication as used by Affymetrix builds up layers of nucleotides using a process borrowed from the computer industry termed photolithography. Here wafer-thin masks with holes allow photoactivation of specific dNTPs which are linked together at specific regions on the chip. The whole process allows layers of oligonucleotides to be built up with each nucleotide at each position being defined by computer.

The arrays themselves may represent a variety of nucleic acid material. This may be mRNA produced in a particular cell type, termed **cDNA expression arrays,** or may alternatively represent coding and regulatory regions of a particular gene or group of genes. A number of arrays are now available that may determine mutations in DNA, mRNA transcript levels or other polymorphisms such as SNPs. Sample DNA is placed on the array and any unhybridised DNA washed off. The array is then analysed and scanned for patterns of hybridisation by detection of fluorescence signals. Any mutations or genetic polymorphisms in relevant genes may be rapidly analysed by computer interpretation of the resulting hybridisation pattern and mutation, transcript level or polymorphism defined. Indeed the collation and manipulation of data from microarrays presents as big a problem as fabricating the chips in the first place. The potential of microarrays appears to be limitless and a number of arrays have been developed for the detection of various genetic mutations including the cystic fibrosis CFTR gene (cystic fibrosis transmembrane regulator), the breast cancer gene BRCA1 and in the study of the human immunodeficiency virus (HIV).

At present microarrays require DNA to be highly purified, which limits their applicability. However as DNA purification becomes automated and microarray technology develops it is not difficult to envisage numerous laboratory tests on a single DNA microchip. This could not only be used for analysing single genes but large numbers of genes or DNA representing microorganisms, viruses, etc. Since the potential for quantitation of gene transcription exists expression arrays could also be used in defining a particular disease status. This technique may be very significant since it will allow large amounts of sequence information to be gathered very rapidly and assist in many fields of molecular biology, especially in large genome sequencing projects or in so-called resequencing projects where gene regions such as those containing potentially important polymorphisms require analysis in a number of samples.

One current application of microarray technology is the generation of a catalogue of SNPs across the human genome. Estimates indicate that there are approximately 10 million SNPs and importantly 200 000 coding or cSNPs that lie within genes and may point to the development of certain diseases. SNP analysis is therefore clearly a candidate for microarray analysis and developments such as Affymetrix Genome Wide SNP array enables the simultaneous analysis of nearly 1 million SNPs on one gene chip. In order to simplify the problem of the vast numbers of SNPs that need to be analysed the HapMap project currently analyses SNPs that are inherited as a block, and in theory as few as 500 000 SNPs will be required to genotype an individual.

An extension of microarray technology may also be used to analyse tissue sections. This process, termed tissue microarrays (TMA), uses tissue cores or biopsies from conventional paraffin-embedded tissues. Thousands of tissue cores are sliced and placed on a solid support such as glass where they may all be subjected to the same immuno-histochemical staining process or analysis with gene probes using *in situ* hybridisation. As with DNA microarrays many samples may be analysed simultaneously, less tissue is required and greater standardisation is possible.

6.9 ANALYSING WHOLE GENOMES

Perhaps the most ambitious project in biosciences is the initiative to map and completely sequence a number of genomes from various organisms. The mapping and sequencing of a number of organisms indicated in Table 6.6. has been completed and many more are due for completion. A number have been completed already such as the bacterium *E. coli.* The demands of such large-scale mapping and sequencing have provided the impetus for the development and refinement of even the most standard of molecular biology tech-niques such as DNA sequencing. It has also led to new methods of identifying the important coding sequences that represent proteins and enzymes. The use of bioinfor-matics to collate, annotate and publish the information on the World Wide Web has also been an enormous undertaking. The availability of an informative map of the human genome that may be analysed and studied in detail chromosome by chromosome, such as the Map Viewer (NCBI), is just one of the rapid developments in the field of genome analysis and bioinformatics. Such is the power and ease of use of resources such as these that it is now inconceivable to work without these resources.

6.9.1 Physical genome mapping

In terms of genome mapping a physical map is the primary goal. Genetic linkage maps have also been produced by determining the recombination frequency between two particular loci. YAC-based vectors essential for large-scale cloning contain DNA inserts that are on average 300 000 bp in length, which is longer by a factor of ten than the longest inserts in the clones used in early mapping studies. The development of vectors with large insert capacity has enable the production of contigs. These are continuous overlapping cloned fragments that have been positioned relative to one another. Using these maps any cloned fragment may be identified and aligned to an area in one of the contig maps. In order to position cloned DNA fragments resulting from the construction of a library in a YAC or cosmid it is necessary to detect overlaps between the cloned DNA fragments. Overlaps are created because of the use of partial digestion conditions with a particular restriction endonuclease when constructing the libraries. This ensures that when each DNA fragment is cloned into a vector it has overlapping ends which theoretically may be identified and the clones positioned or ordered so that a physical map may be produced (Fig. 6.47).

In order to position the overlapping ends it is preferable to undertake DNA sequen-cing; however, due to the impracticality of this approach a fingerprint of each clone is

Table 6.6 **Current selected genome-sequencing projects**

Organism		Genome size (Mb)
Bacteria	*Escherichia coli*	4.6
Yeast	*Saccharomyces cerevisiae*	14
Roundworm	*Caenorhabditis elegans*	100
Fruit fly	*Drosophila melanogaster*	165
Puffer fish	*Fugu rubripes rubripes*	400
Mouse	*Mus musculus*	3000

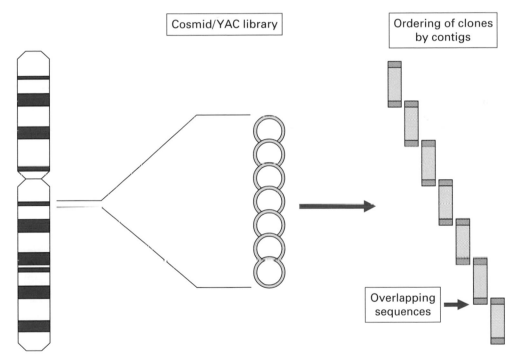

Fig. 6.47 Physical mapping using continuous overlapping cloned fragments (contigs). In order to assign the position of cloned DNA fragments resulting from the construction of a library in a YAC or cosmid vector, overlaps are detected between the clone fragments. These are created because of the use of partial digestion conditions when the libraries are constructed.

carried out by using restriction enzyme mapping. Although this is not an unambiguous method of ordering clones it is useful when also applying statistical probabilities of the overlap between clones. In order to link the contigs techniques such as *in situ* hybrid-isation may be used or a probe generated from one end of a contig in order to screen a different disconnected contig. This method of probe production and identification is termed walking, and has been used successfully in the production of physical maps

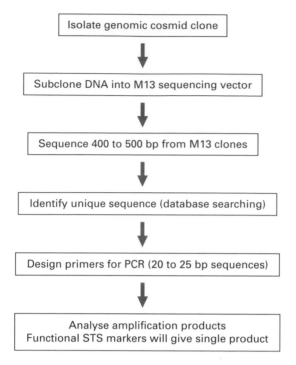

Fig. 6.48 General scheme of the production of a functional STS marker.

of *E. coli* and yeast genomes. This cycle of clone to fingerprint to contig is amenable to automation; however, the problem of closing the gaps between contigs remains very difficult.

In order to define a common way for all research laboratories to order clones and connect physical maps together an arbitrary molecular technique based on the polymerase chain reaction has been developed based on sequence-tagged sites (STS). This is a small unique sequence between 200–300 bp that is amplified by PCR (Fig. 6.48). The uniqueness of the STS is defined by the PCR primers that flank the STS. A PCR with those primers is performed and if the PCR results in selected amplification of target region it may be defined as a potential STS marker. In this way defining STS markers that lie approximately 100 000 bases apart along a contig map allows the ordering of those contigs. Thus, all groups working with clones have definable landmarks with which to order clones produced in their libraries.

An STS that occurs in two clones will overlap and thus may be used to order the clones in a contig. Clones containing the STS are usually detected by Southern blotting where the clones have been immobilised on a nylon membrane. Alternatively a library of clones may be divided into pools and and each pool PCR screened. This is usually a more rapid method of identifying an STS within a clone and further refinement of the PCR-based screening method allows the identification of a particular clone within a pool (Fig. 6.49). STS elements may also be generated from variable regions of the genome to produce a polymorphic marker that may be traced through families along with other DNA markers and located on a genetic linkage map. These

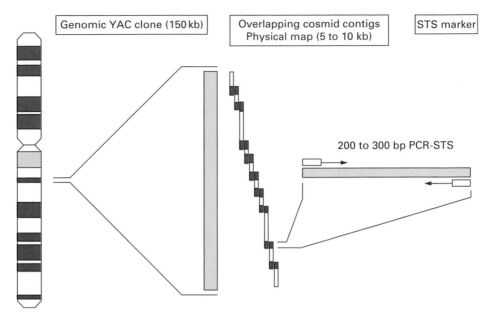

Fig. 6.49 The derivation of an STS marker. An STS is a small unique sequence of between 200 and 300 bp that is amplified by PCR and allows ordering along a contig map. Such sequences are definable landmarks with which to order clones produced in genome libraries and usually lie approximately 100 000 bp apart.

polymorphic STSs are useful since they may serve as markers on both a physical map and a genetic linkage map for each chromosome and therefore provide a useful marker for aligning the two types of map.

6.9.2 Gene discovery and localisation

A number of disease loci have been identified and located to certain chromosomes. This has been facilitated by the use of *in situ* mapping techniques such as FISH. In fact a number of genes have been identified and the protein determined where little was initially known about the gene except for its location. This method of gene discovery is known as positional cloning and was instrumental in the isolation of the *CFTR* gene responsible for the disorder cystic fibrosis (Fig. 6.50).

The genes that are actively expressed in a cell at any one time are estimated to be as little as 10% of the total. The remaining DNA is packaged and serves an as yet unknown function. Investigations have found that certain active genes may be identified by the presence of so-called HTF (*Hpa*II tiny fragments) islands often found at the 5′ end of genes. These are CpG-rich sequences that are not methylated and form tiny fragments on digestion with the restriction enzyme *Hpa*II. A further gene discovery method that has been used extensively in the past few years is a PCR-based technique giving rise to a product termed an **expressed sequence tag** (EST). This represents part of a putative gene for which a function has yet to be assigned. It is carried out on cDNA by using primers that bind to an anchor sequence such as a poly(A) tail and primers which bind to sequences at the 5′ end of the gene. Such PCRs may

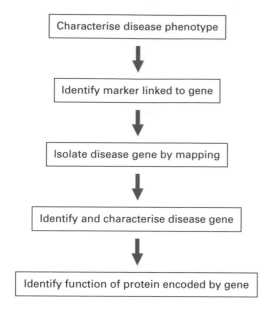

Fig. 6.50 The scheme of identification of a disease gene by positional cloning.

subsequently be used to map the putative gene to a chromosomal region or be used itself as a probe to search a genomic DNA library for the remaining parts of the gene. This type of information can be visualised using bioinformatics and useful information determined in a process termed data mining. Much interest currently lies in ESTs since they may represent a short cut to gene discovery.

A further gene isolation system that uses adapted vectors, termed exon trapping or exon amplification, may be used to identify exon sequences. Exon trapping requires the use of a specialised expression vector that will accept fragments of genomic DNA containing sequences for splicing reactions to take place. Following transfection of a eukaryotic cell line a transcript is produced that may be detected by using specific primers in a RT–PCR. This indicates the nature of the foreign DNA by virtue of the splicing sequences present. A list of further techniques that aid in the identification of a potential gene-encoding sequence is indicated in Table 6.7.

6.9.3 Genome mapping projects

As a result of the technological advances in large-scale DNA sequencing as indicated in Chapter 5 it is now possible not only to map genomes of various species but also to determine their sequence reliably and rapidly. The genomes of hundreds of species have been determined and this is increasing each month. Sequencing and mapping of the human genome was completed ahead of schedule and has provided many new insights into gene function and gene regulation. It was also a multi-collaboration effort that engaged many scientific research groups around the world and has given rise to many scientific, technical, financial and ethical debates. One interesting issue is the sequencing of the whole genome in relation to the coding sequences. Much of the human genome appears to be non-coding and composed of repetitive sequences.

Table 6.7 **Techniques used to determine putative gene-encoding sequences**

Identification method	Main details
Zoo blotting (cross-hybridisation)	Evolutionary conservation of DNA sequences that suggest functional significance
Homology searching	Gene database searching to gene family-related sequences
Identification of CpG islands	Regions of hypomethylated CpG frequently found 5′ to genes in vertebrate animals
Identification of open reading frames (ORF) promoters/splice sites/RBS	DNA sequences scanned for consensus sequences by computer
Northern blot hybridisation	mRNA detection by binding to labelled gene probes
Exon trapping technique	Artificial RNA splicing assay for exon identification
Expressed sequence tags (ESTs)	cDNAs amplified by PCR that represent part of a gene

Notes: RBS, ribosome binding site; cDNA, complementary DNA.

Estimates indicate that as little as 10% of the genome appears to encode enzymes and proteins. Current estimates equate this to approximately 20 000 genes which are important for human cellular development and maintenance. However it is the understanding of the complete function of many of the genes and their variants coupled with their interaction that now provides a major challenge. It also points to the fact that there is an extensive use of alternative splicing where exons are essentially mixed and matched to form different mRNA and thus different proteins. The study further aims to understand and possibly provide the eventual means of treating some of the 4000 genetic diseases in addition to other diseases whose inheritance is multifactorial. In this respect there are a number of specific genome projects such as the Cancer Genome Anatomy Project (CGAP) which aims to understand the part certain mutations play in the development of tumours.

6.10 PHARMACOGENOMICS

As a result of the developments in genomics new methods of providing targeted drug treatment are beginning to be developed. This area is linked to the proposal that it is possible to identify those people who react in a specific way to drug treatment by identifying their genetic make-up. In particular SNPs may provide a key marker of potential disease development and reaction to a particular treatment. A simple example that has been known for some time is the reaction to a drug used to treat a particular type of childhood leukaemia. Successful treatment of the majority of patients may be achieved with 6-mercaptopurine. A number of patients do not respond well, but in some cases it may be fatal to administer this drug. This is now known to be due to a mutation in the gene encoding the enzyme that metabolises the drug. Thus, it is possible to analyse patient DNA prior to administration of a drug to determine what the likely response will be. The technology to deduce a patient's genotype is already developed and indicated in Section 6.8.7. It is also now possible to analyse SNPs which may also correlate with certain disease processes in a microarray type format. This opens up the possibility that it may be possible to assign a **pharmacogenetic** profile at birth, in much the same way as blood typing for later treatment. A further possibility is the determination of likely susceptibility to a disease based on genetic information. A number of companies including the Icelandic genetics company deCode are able to provide personal genetic information based on modelling and analysis of disease genes in large population studies for certain conditions such as diabetes.

6.11 MOLECULAR BIOTECHNOLOGY AND APPLICATIONS

It is a relatively short period of time since the early 1970s when the first recombinant DNA experiments were carried out. However, huge strides have been made not only in the development of molecular biology techniques but also in their practical application. The molecular basis of disease and the new areas of genetic analysis and gene therapy hold great promise. In the past medical science relied on the measurement of protein and enzyme markers which reflected disease states. It is possible now not only to detect such abnormalities at an earlier stage using mRNA techniques but also in some cases to predict such states using genome analysis. The complete mapping and sequencing of the human genome and the development of techniques such as DNA microchips will certainly accelerate such events. Perhaps even more difficult is

Table 6.8 **General classification of oncogenes and their cellular and biochemical functions**

Oncogene	Example	Main details
G-proteins	H-K- and N-*ras*	GTP-binding protein/GTPase
Growth factors	*sis, nt-2, hst*	β-chain of platelet-derived growth factor (PDGF)
Growth factor receptors	*erbB*	Epidermal growth factor receptor (EGFR)
	fms	Colony-stimulating factor-1 receptor
Protein kinases	*abl, src*	Protein tyrosine kinases
	mos, ras	Protein serine kinases
Nucleus-located transcription factors	*mye*	DNA-binding protein
	myb	DNA-binding protein
	jun, fos	DNA-binding protein

Table 6.9 **A number of selected examples of targets for gene therapy**

Disorder	Defect	Gene target	Target cell
Emphysema	Deficiency (α1-AT)	α1-Antitrypsin (α1-AT)	Liver cells
Gaucher disease (storage disorder)	GC deficiency	Glucocerebrosidase	GC fibroblasts
Haemoglobinopathies	Thalassaemia	β-Globin	Fibroblasts
Lesch–Nyhan syndrome	Metabolic deficiency	Hypoxanthine guanine phosphoribosyl transferase (HPRT)	HPRT cells
Immune system disorder	Adenosine deaminase deficiency	Adenosine deaminase (ADA)	T and B cells

Table 6.10 **Current selected plant/crops modified by genetic manipulation**

Crop or plant	Genetic modification
Canola (oil seed rape)	Insect resistance, seed oil modification
Maize	Herbicide tolerance, resistance to insects
Rice	Modified seed storage protein, insect resistance
Soya bean	Tolerance to herbicide, modified seed storage protein
Tomato	Modified ripening, resistance to insects and viruses
Sunflower	Modified seed storage protein

the elucidation of diseases that are multifactorial and involve a significant contribution from environmental factors. One of the best-studied examples of this type of disease is cancer. Molecular genetic analysis has allowed a discrete set of cellular genes, termed **oncogenes,** to be defined which play key roles in such events. These genes and their proteins are also major points in the cell cycle and are intimately invloved in cell regulation. A number of these are indicated in Table 6.8. In a number of cancers well-defined molecular events have been correlated with mutations in these oncogenes and therefore in the corresponding protein. It is already possible to screen and predict the fate of some disease processes at an early stage, a point which itself raises significant ethical dilemmas. In addition to understanding cellular processes both in normal and disease states great promise is also held in drug discovery and molecular gene therapy. A number of genetically engineered therapeutic proteins and enzymes have been developed and are already having an impact on disease management. In addition the correction of disorders at the gene level (**gene therapy**) is also under way and perhaps is one of the most startling applications of molecular biology to date. A number of these developments are indicated in Table 6.9.

The production of modified crops and animals for farming and as producers of important therapeutic proteins is also one of the most exciting developments of molecular biology. This has allowed the production of modified crops, improving their resistance to environmental factors and their stability (Table 6.10). The production of transgenic animals also holds great promise for improved livestock quality, low-cost production of pharmaceuticals and disease-free or disease-resistant strains. In the future this may overcome such factors as contamination with agents such as BSE. There is no doubt that improved methods of producing livestock by whole-animal cloning will also be a major benefit. All of these developments do however require debate and the many ethical considerations that arise from them require careful consideration.

6.12 SUGGESTIONS FOR FURTHER READING

Augen, J. (2005). *Bioinformatics in the Post-Genomic Era*. Reading, MA: Addison-Wesley.

Brooker, R. J. (2005). *Genetics Analysis and Principles*, 2nd edn. McGraw-Hill.

Brown, T. A. (2006). *Gene Cloning and DNA Analysis*. Oxford, UK: Wiley–Blackwell.

Primrose, S. B. and Twyman, R. (2006). *Principles of Gene Manipulation and Genomics*. Oxford, UK: Wiley–Blackwell.

Strachan, T. and Read, A. P. (2004). *Human Molecular Genetics*, 3rd edn. Oxford, UK: Bios.

Walker, J. M. and Rapley, R. (2008). *Molecular Biomethods Handbook*, 2nd edn. Totowa, NJ: Humana Press.

Watson, J. D., Caudy, A. A., Myers, R. M. and Witkowski, J. A. (2007). *Recombinant DNA: Genes and Genomes*. San Francisco, CA: W. H. Freeman.

7 Immunochemical techniques

R. BURNS

7.1 INTRODUCTION

The immune system of mammals has evolved over millions of years and provides an incredibly elegant protection system which is capable of responding to infective challenges as they arise. The system is fluid-based and both the cells of immunity and their products are transported throughout the body, primarily in the blood and secondarily through fluid within the tissues and organs themselves. All areas of the body are protected by immunity apart from the central nervous system including the brain and eyes. There are several cell types involved in immune responses, each with a role to play and each controlled by chemical mediators known as cytokines. This control is essential as the immune system is such a powerful tool it needs careful management to ensure its effective operation. Both over- and under-activity could have fatal consequences.

All vertebrates have advanced immune systems which show the similarities that you would expect from our common evolutionary past. The more advanced the vertebrate the more complex the immune system. Fish and amphibians have fairly rudimentary immunity with the most sophisticated being found in mammals. The immune system is broadly additive; more complex animals have elements analogous to those found in primitive species but have extra features as well.

For the purposes of this chapter we will focus on the mammalian immune system although the use of birds for antibody production will be discussed in Section 7.1.2.

Immunity is monitored, delivered and controlled by specialised cells all derived from stem cells in the bone marrow. There are motile macrophages which move around the body removing debris and foreign materials, and two lineages of lymphocyte, B and T, which provide immediate killing potential but also provide the mechanism for the production of antibodies. There are also assorted other cells whose function is to rush to areas of the body where a breach of security has occurred and deliver potent chemicals capable of sterilising and neutralising any foreign bodies.

For mammalian immunity to function effectively it is vital that the cells of the system can recognise the difference between self and non-self. There are three ways in which this is achieved. Mammals have a pre-programmed ability to recognise and immediately act against substances derived from fungal and bacterial micro-organisms. This is mediated through a series of biological chemicals known as the complement system which are capable of adhering to and killing bacteria, fungi and some viruses. Secondly, the immune system is capable of recognising when a substance is close to but not quite the same as self. This is a response based on 'generic' circulating antibodies which are able to discriminate between self and non-self. Lastly, every individual has a unique 'signature' which is caused by a pattern of molecules on every cell surface. Cells of the immune system read this unique code and any cells differing from the authorised version are targeted and destroyed through a T–cell–mediated response. These three systems do not operate in isolation, they form a cohesive network of surveillance in which all of the cell types co-operate to provide the most appropriate response to any breach of security. In mammals the first line of defence against attack is the skin and any breaches of it are responded to by the cells of the immune system even though no foreign material is present. This response is mediated by cell messengers known as cytokines which can be released from damaged tissue or cells of the immune system near to the site of injury.

For the purposes of this chapter it is the antibody response in mammals that will be focussed on as these are the molecules that we are able to harness for our uses where a specific protein sequence or molecular structure has to be identified. As previously mentioned it is impossible to discuss one area of cellular immunity in isolation and so reference will be made to how the rest of the immune system contributes to the manufacture of antibodies by mammals.

An **antigen** is a substance capable of causing an immune response leading to the production of antibodies and they are also the targets to which antibodies will bind. Antibodies are antigen specific and will only bind to the antigen that initiated their production.

7.1.1 Development of the immune system

Mammalian embryos develop an immune system before birth which is capable of providing the newborn with immediate protection. Additional defences are acquired from maternal milk and this covers the period during which the juvenile immune system matures to deal with the requirements of the organism after weaning takes place. For an immune system to function effectively it must be organic in its ability to react to situations as they arise and the mammalian system has an extremely elegant way of dealing with this.

The cells of the mammalian immune system are descended from distinct lineages derived initially from stem cells, and those producing antibodies are known as B cells (also called B lymphocytes). These cells have the ability to produce antibodies which recognise specific molecular shapes. Cells of the immune system known as macrophages, dendritic cells and other antigen-presenting cells (APCs) have the ability to recognise 'foreign' substances (antigens) within the body and will attack and digest them when encountered. The majority of antigens found by the body are from viruses, bacteria, parasites and fungi, all of which may infect an individual. All of these organisms have proteins and other substances that will be antigenic (behave as an antigen) when encountered by the immune system. Organisms that infect or invade the body are known as pathogens and will have many antigens within their structures. The presenting cells process antigens into small fragments and present them to the B cells. The fragments contain epitopes which are typically about 15 amino acid residues in size. This size corresponds to the size that the antibody binding site can bind to. After ingesting the antigen fragments the B cells recruit 'help' in the form of cytokines from T cells which stimulates cell division and secretion. Each B cell that was capable of binding an antigen fragment and has ingested it will then model an antibody on the shape of an epitope and start to secrete it into the blood.

During embryonic development the immune system has to learn what is self and what is foreign. Failure to do this would lead to self-destruction or would lead to an inability to mount an immune response to foreign substances. During development this is achieved by selective clonal deletion (see Fig. 7.1) of self-recognising B cells. Early in the development of the immune system, B cell lineages randomly reassort the antibody-creating genes to produce a 'starter pack' of B cells that will respond to a huge number of molecular shapes. These cells have these randomly produced antibodies bound to their surface ready to bind should an antigen fit the antibody-binding site. These provide crude but instant protection to a large number of foreign substances immediately after birth. They also are the basis for the B cells that will provide protection for the rest of the animal's life. However, within the population of randomly produced B cells are a number which will be responsive to self-antigens, which are extremely dangerous as they could lead to the destruction of parts of the animal's body. Embryos are derived exclusively from cells derived from the fusion of egg and sperm. There are no cells derived from the mother within the embryonic sac in the uterus and so everything can be regarded as being immunologically part of 'self'. Any B cells that start to divide within the embryo prior to birth are responding to 'self' antigens and are destroyed as potentially dangerous. This selective clonal deletion is

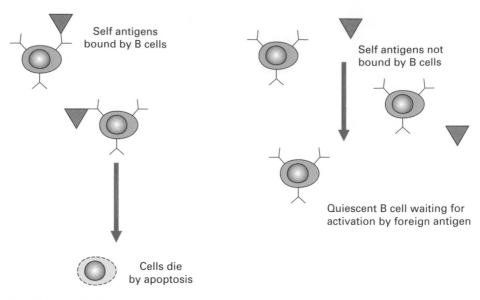

Self antigens
bound by B cells

Self antigens not
bound by B cells

Quiescent B cell waiting for
activation by foreign antigen

Cells die
by apoptosis

Fig. 7.1 Clonal deletion.

fundamental to the development of the immune system and without it the organism
could not continue to develop. The remainder of the B cells that have not undergone
cell division will only recognise non-self antigens and are retained within the bone
marrow of the animal as a quiescent cell population waiting for stimulation from
passing stimulated antigen-presenting cells. Stimulation of B cells requires both the
presence of macrophages and also T lymphocytes. T cells are descended from an
alternative lymphocyte lineage to the B cells and are responsible for 'helping' and
'suppressing' the immune response. T cells also undergo clonal selection during
development to ensure that they do not recognise self antigens. In addition they are
positively selected to ensure that they do recognise proteins of the major histocom-
batibility complex (MHC) found on cell surfaces. The balance of appropriate immune
response is governed by the interplay of T and B lymphocytes along with macrophages and other antigen-presenting cells to ensure that an individual is protected but
not endangered by inappropriate responses.

    After birth, exposure to foreign materials will cause an immediate response
resulting in antibody production and secretion by B cells. The antibody binds to the
target and marks it as foreign and it is then removed by the body. Macrophages
are responsible for much of the removal of foreign material which they ingest by
phagocytosis (an uptake system that some cells use to transport particles from outside
the surface membrane into the cell body). The material is then digested and exported
to the cell surface as small fragments (antigens) which are then presented to passing
B cells. Should a B cell carry an antibody that binds the antigen then it will take the
antigen from the macrophages and this causes a number of intracellular changes
known as B cell activation. B cell activation involves the recruitment of T cells which
stimulates cell growth and metabolism. B cell activation may also occur without the
presence of macrophages or other presenting cells when the lymphocyte is directly

exposed to antigens. The stimulation of the B cells leads to two major changes apart from antibody secretion. It leads to a larger population of cells being retained in the bone marrow that recognise the antigen. These are known as memory cells as they have the ability to recognise and rapidly respond should the antigen be encountered again. Binding antigen leads to cell division and antibody secretion but it also causes the cells to refine the quality of the antibody they produce. Avidity is the strength with which the antibody sticks to the antigen and affinity is the 'fit' of the antibody shape to the target. Both of these can be improved after the B cells have been stimulated but require more than one exposure to the antigen. The process is known as affinity maturation and is characterised by a change in antibody type from predo- minantly low-affinity pentameric (five molecules linked together) immunoglobulin M (IgM) to the high-affinity immunoglobulin G (IgG). Other antibody types may be produced in specific tissues and in response to particular antigens. For example parasites in the intestines often induce high levels of IgE in the gut mucosa (innermost layer of the gut which secretes large amounts of mucous). After several encounters with an antigen a background level of specific antibody will be found in the animal's blood along with a population of memory cells capable of rapidly responding to its presence by initiating high levels of antibody secretion. This status is known as immune and is the basis of both artificial immunisations for protection against disease and also for the production of antibodies for both diagnostic and therapeutic use.

7.1.2 Harnessing the immune system for antibody production

There are two major classes of antibodies used in immunochemistry: these are polyclonal and monoclonal. Polyclonal antibodies are produced in animals by injecting them with antigens. They are derived from the animal blood serum and their name means that they have been produced by many clones. This refers to the fact that the B cells that have made them will be producing antibodies to many different epitopes on the antigen and will involve the secretion of antibody by many B cell clones. Polyclonal antibodies are essentially a population of antibody molecules contributed by many B cell clones.

Monoclonal antibodies are produced by animal cells artificially in tissue culture, and as their name suggests the antibody produced comes from a single cell clone. The cells that make them are known as hybridomas and are produced from the fusion of a cancer cell line and B cells. Monoclonal antibodies are epitope specific whereas polyclonal antibodies are antigen specific. This difference is fundamental to the way in which they can be used for both diagnostics and therapeutics.

Mammals will produce antibodies to practically any foreign material that is introduced into their bodies providing it has a molecular weight greater than 5 000 Da. The only restriction to this is antigens that are closely related to substances found in the animal itself. Many mammalian proteins and other biochemical substances are highly conserved and are antigenically very similar in many species. This can lead to problems in producing antibodies for diagnostic and therapeutic use. The immune system is incapable of mounting a response to 'self' as discussed earlier and because of this, some antigens may not be able to produce an antibody response in some species. Providing that the antigen is large enough and that it does not resemble proteins in the host animal then antibodies can be produced to a huge number of substances which can be used in all branches of diagnostics and thera-peutics. There are three types of antibodies that can be produced: these are polyclonal, monoclonal and recombinant. Each of these antibody types has advantages but also limitations and should be viewed as complementary to each other as each has specific areas where they are particularly useful.

Polyclonal antibodies are produced in a number of animal species. Antibodies are generated by immunising the host with the substance of interest usually three or four times. Blood is collected on a number of occasions and the antibody fraction purified from the blood serum. The exception to this is chicken polyclonal antibodies which are harvested from eggs. Generally, larger animals are used since antibody is harvested from the blood of the animal and bigger volumes can be obtained from larger species. Historically, the first antibodies produced artificially for diagnostic purposes were polyclonal. They are the cheapest of antibodies to produce and have many uses in diagnostics. They have limited use in therapeutics, however, as there are problems in that they themselves can be antigenic when injected into other animals. There are exceptions to this and neutralising antibodies to snake venom and prophylactic (reducing risk of infection) antiviral injections fall into this category. Polyclonal antibodies are cheap to produce, robust but less specific than other antibodies and will have variable qualities depending on the batch and specific donor animal.

Monoclonal antibodies are secreted by mammalian cells grown in synthetic medium in tissue culture. The cells that produce them are known as hybridomas and are usually derived from donor mouse or rat lymphocytes. Human monoclonal antibodies are also available but they are produced by different methodologies to the rodent ones. The murine system was first described in 1976 when Kohler and Milstein published their work. Monoclonal antibodies have radically altered the possible uses for antibodies in both diagnostics and therapeutics. The basis of the technology is the creation of the hybridoma by fusing antibody-secreting B lymphocytes from a donor animal to a tumour cell line. B lymphocytes have a limited lifespan in tissue culture but the hybridoma has immortality conferred by the tumour parent and continues to produce antibody. Each hybridoma is derived from a single tumour cell and a single lymphocyte and this has to be ensured by cloning. Cell cloning is the process where single cells are grown into colonies, in isolation from each other so that they can be assessed and the best chosen for future development. Once cloned, the cell lines are reasonably stable and can be used to produce large quantities of antibody which they secrete into the tissue culture medium that they are grown in. The antibody they produce has the qualities that the parent lymphocyte had and it is this uniqueness that makes monoclonal antibodies so useful. During immuni-sation the B cells are presented with antigen fragments by macrophages and other antigen-presenting cells and each cell then produces a specific antibody to the fragment it has been presented with. The specific site that the antibody recognises is known as an epitope which is approximately 15 amino acids in size. There are thousands of potential epitopes on the antigen. The cell fusion process generates many hundreds of hybridoma clones, each making an individual antibody. The most important part of making hybridomas is the screening process that is used to select those of value. Monoclonal antibodies are epitope specific and so it is important that the screening process takes this into account to ensure that antibodies selected have the correct qualities needed for the final intended use. They can be used for human and veterinary therapeutics although they are antigenic if used unmodified. Monoclonal antibodies can be processed to modify antigenicity to make them more useful in therapeutics. Mouse hybridomas can also be engineered so that the antibodies that they make have human sequences in them. These humanised antibodies have been used very successfully for treating a range of human conditions including breast cancer, lymphoma and the rejection symptoms after organ transplantation. Monoclonal antibodies are more expensive to produce than their polyclonal counterparts but have qualities that can make them more valuable. They are highly specific and reasonably robust but may be less avid than polyclonal antibodies. They are produced from established cell lines in tissue culture and should show little in the way of batch variation.

Recombinant antibodies are produced by molecular methodologies and are expressed in a number of systems, both prokaryotic and eukaryotic. Attempts have also been made to express antibodies in plants and this has had some success. The idea of producing antibody in crop plant species such as potato is very attractive as the costs of growing are negligible and the amounts of antibody produced could be very large.

Two basic methods can be used to produce recombinant antibodies. Existing DNA libraries can be used to produce bacteriophage expressing antibody fragments on their surface. Useful antibodies can be identified by assay and the bacteriophage producing it then used to transfect the antibody DNA into a prokaryotic host cell type. The antibody can then be produced in culture by the recombinant cells. The antibodies produced are monoclonal but do not have the full structure of those expressed by animals or cell lines derived from them. They are less robust and as they are much smaller than native antibodies it may not be possible to modify them without losing binding function. The great advantage of using this system is the speed with which antibodies can be generated, generally in a matter of weeks. Typically the timescale for producing monoclonal antibodies from cell fusions is about 6 months.

Antibodies can also be generated from donor lymphocyte (B cell) DNA. The highest concentration of B cells is found in the spleen after immunisation and so this is the tissue usually used for DNA extraction. The antibody-coding genes are then selectively amplified by polymerase chain reaction (PCR) and then transfected (inserted into DNA) into a eukaryotic cell line. Usually a resistance gene is co-transfected so that only recombinant cells containing antibody genes will grow in culture. The cells chosen for this work are often those most easily grown in culture and may be rodent or other mammalian lines. Chinese hamster ovary (CHO) cells are often used for this and have become the industry standard amongst biotechnology companies. Yeasts, filamentous algae and insect cells have all also been used as recipients for antibody genes with varying degrees of success.

7.1.3 Antibody structure and function

Antibodies as they are found in nature are all based on a Y-shaped molecule consisting of four polypeptide chains held together by disulphide bonds (see Fig. 7.2). There are two pairs of chains, known as heavy (H) and light (L); each member of the pair is identical. Functionally the base of the Y is known as the constant region and the tips of the arms are the variable region. The amino acid structure in the constant region is fairly fixed in an individual but varies between animal species. The amino acid structure in the variable region is composed of between 110 and 130 amino acids and it is variations in these that forms the different binding sites of the antibodies. The ends of both the heavy and light chains are variable and the antigen-binding site is formed by a combination of the two. The variable part of the antibody contains two further areas, the framework and hypervariable regions. There are three hypervariable and four framework regions per binding site. The hypervariable regions are structurally supported by the framework regions and form the area of direct contact with the antigen.

Antibodies can be fragmented by enzymatic degradation and the subunits produced are sometimes used to describe portions of the antibody molecule. Treatment with the enzyme papain gives rise to three fragments: two antigen-binding fragments (Fab) and one constant fragment (Fc). The enzyme digests the molecule at the hinge region and the resulting Fab fragments retain their antigen-binding capability. The Fc fragment has no binding region and has no practical use. Fab fragments are sometimes prepared and used for some immunochemical applications. Their smaller size may mean that they can bind to antigens in certain situations where the larger native molecule would have difficulty binding.

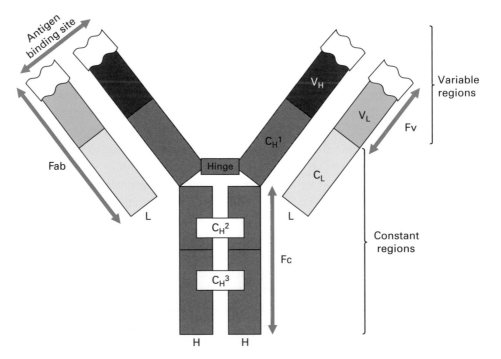

Fig. 7.2 Immunoglobulin G.

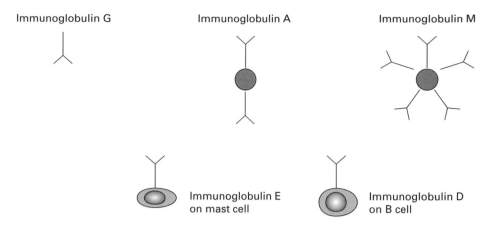

Fig. 7.3 Immunoglobulin classes.

Treatment with the enzyme papain gives rise to one double antigen-binding fragment (F(ab′)$_2$) and multiple Fc fragments. The enzyme digests the Fc until it reaches the hinge region of the antibody which is protected by a disulphide bond. F(ab′)$_2$ fragments have two binding sites and can be used in place of native antibody molecules.

There are five major classes of antibody molecule, also known as immunoglobulins (Ig); G is the commonest and it is characterised by its Y-shaped structure. The other classes of antibody are immunoglobulins M, A, D and E (Fig. 7.3).

Immunoglobulin M (IgM) is produced early in immune responses. It is produced by immature and newly activated B lymphocytes that have been exposed to an antigen for the first time. It is found on the surface of B cells frequently in association with IgD. Structurally it is formed from five immunoglobulin G molecules in a ring complexed by a mu chain. It may also be found as a hexamer without the mu chain. The molecule tends to have low affinity and poor avidity to antigen. It is much less specific and will react to a range of antigens without immunisation having taken place. It is known as 'natural antibody' as a result. Its production rises dramatically after first exposure to antigen and is characteristic of the primary immune response. It is generally only found in serum as its large size prevents it from crossing tissue boundaries. The pentameric form is particularly useful for complexing antigen such as bacteria into aggregates either for disposal or for further processing by the immune system. Cells secreting IgM can progress to IgG production, in time, if the animal is challenged again by the antigen. This progression to IgG production is known as affinity maturation and requires maturation of the cells to memory cell status. After several encounters with an antigen a background level of specific antibody will be found in the animal's blood along with a population of B memory cells capable of rapidly responding to its presence by initiating high levels of antibody secretion. This status is known as immune and is the basis of both artificial immunisations for protection against disease and also for the production of antibodies for both diagnostic and therapeutic use. A status of hyperimmunity may be reached after repeated exposure to an antigen leading to extremely high levels of circulating antibody. Hyperimmunity carries risk, as additional exposure to antigen can lead to anaphylactic shock due to the overwhelmingly large immune response. Conversely, a total loss of immune response to an antigen can occur after repeated immunisation as a state of immune tolerance to the antigen is reached. Acquired immune tolerance is a response to overstimulation by an antigen and is characterised by a loss of circulating B cells reactive to the antigen and also by a loss of T cell response to the antigen. This can be used therapeutically to protect individuals against allergic responses.

Immunoglobulin A (IgA) is a dimeric form of immunoglobulin essentially with two IgG molecules placed end to end with the binding sites facing outwards. They are complexed with a J chain. It is predominantly found in secretions from mucosa and is resistant to enzyme degradation due to its structure. It is primarily concerned with protection of the mucosal surface of the mouth, nose, eyes, digestive tract and genito-urinary system. It is produced by B cells resident in the mucosa and is directly secreted into the fluids associated with the individual tissues. It is of little use in immuno-chemistry as it cannot be purified easily and is prone to spontaneous aggregation. Occasionally a hybridoma is derived secreting antibodies with this isotype and it may be that this is the only source of a rare antibody. In this case an indirect assay may be developed using the tissue culture supernatant derived from the hybridoma along with a specific anti-IgA antibody–enzyme conjugate.

Immunoglobulin D (IgD) is an antibody resembling IgG and is found on the surface of immature B cells along with IgM. It is a cellular marker which indicates that an immature B cell is ready to mount an immune response and may be responsible for the migration of the cells from the spleen into the blood. It is used by the macrophages to identify cells to which they can present antigen fragments.

Immunoglobulin E (IgE) also resembles IgG structurally and is produced in response to allergens and parasites. It is secreted by B lymphocytes and attaches itself to the surface of specialised cells known as mast cells. Exposure to allergen and its subsequent binding to the IgE molecules on the cell surface cause the antibodies to cross-link and move together in the cell membrane. This cross-linking causes the cell to degranulate releasing histamine. Histamine is responsible for the symptoms suffered by individuals as a result of exposure to allergens.

Immunoglobulin G and to a lesser extent IgM are the only two antibodies that are of practical use in immunochemistry. IgG is the antibody of choice used for development of assays as it is easily purified from serum and tissue culture medium. It is very robust and can be modified by labelling with marker molecules (see Section 7.4) without losing function. It can be stored for extended periods of time at 4 °C or lower. Occasionally antigens will not generate IgG responses *in vivo* and instead IgM is produced. This is caused by the antigen being unable to activate the B cells fully and as a result no memory cells being produced. Such antigens are often highly glycosylated and it is the large number of sugar residues that block the full activation of the B cells. IgM can be used for assay development but is more difficult to work with. IgM molecules tend to be unstable and are difficult to label without losing function as the binding sites become blocked by the proximity of each other. This is known as stearic hindrance. They can be used directly from cell tissue culture supernatant in assays with an appropriate secondary anti-IgM enzyme conjugate.

7.2 MAKING ANTIBODIES

All methods used in immunochemistry rely on the antibody molecule or derivatives of it. Antibodies can be made in various ways and the choice of which method to use is very much dependent on the final assay format. For an antibody to be of use it has to have a defined specificity, affinity and avidity as these are the qualities that determine its usefulness in the method to be used. There are considerable cost differences in producing the various antibody types and it is important to remember that the most expensive product is not always the best.

7.2.1 Polyclonal antibody production

Polyclonal antibodies are raised in appropriate donor animals, generally rabbits for smaller amounts and sheep or goats for larger quantities. Occasionally rats or mice can be used for small research quantities of antibody. It is important that animals are sourced from reputable suppliers and that they are housed and managed according to domestic welfare legislation.

Usually antigens are mixed with an appropriate adjuvant prior to immunising the animals. Adjuvants are substances which increase the immunogenicity of the antigen and are used to reduce the amount of antigen required as well as stimulate specific immunity to it. Adjuvants may be chemicals such as detergents and oils or complex proprietary products containing bacterial cell walls or preparations of them. Pre-immune blood samples are taken to provide baseline IgG levels (Fig. 7.4). Immunisations are spaced at intervals to maximise antibody production usually at least 4–6 weeks apart although the first two may be given within 14 days. Blood samples are taken 10 days after the immunisation programme is complete and the serum tested for specific activity to antigen by a method such enzyme-linked immunosorbent assay (ELISA) (see Section 7.3). Usually a range of doubling serum dilutions are made (1/100–1/12 800) and tested against the antigen. Serum from a satisfactory course of immunisations will detect antigen at 1/6400 dilution indicating high levels of circulating antibody. Once a high level of circulating antibody is detected in test bleeds then donations can be taken. Animal welfare legislation governs permissible amounts and frequency of bleeds. Donations can be taken until the antibody titre begins to drop and if necessary the animal can be immunised again and a second round of donations taken.

Blood donations are allowed to clot and the serum collected. Individual bleeds may be kept separate or pooled to provide a larger volume of standard product. Serum can be stored at 4 °C or lower for longer periods.

It is also possible to produce antibodies in chicken eggs. Avian immunoglobulin is known as immunoglobulin Y (IgY) and chickens secrete it into eggs to provide protection for the developing embryo. This can be utilised for effective polyclonal antibody production. The chickens are immunised three or four times with the antigen and the immune status monitored by test bleeds. Eggs are collected and can yield up to 50 mg antibody per yolk. The antibody has to be purified from the egg yolks prior to use and a number of proprietary kits can be used to do this. Occasionally antigens that give a poor response in mammals can give much higher yields in chickens.

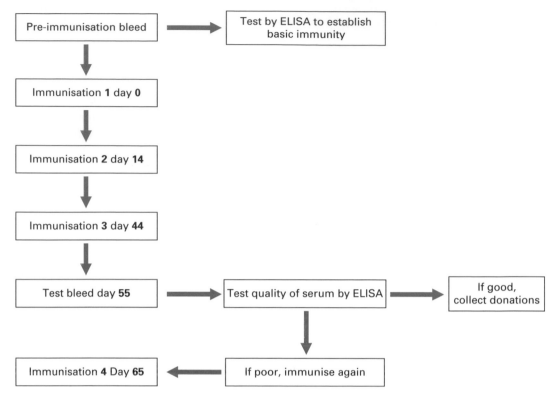

Fig. 7.4 Immunisation schedule.

Small quantities of very pure polyclonal antibodies can be produced in rats and mice in ascitic fluid. Ascites is a mammalian response to a tumour within the peritoneum (cavity containing the intestines). Fluid similar to plasma is secreted into the cavity of the animal and contains very high levels of the antibodies that the animal is currently secreting in its blood. Animals are immunised with the antigen of interest and once a high serum level is detected then ascitic fluid production is induced. Non-secretory myeloma cells such as NS-0 are introduced into the peritoneal cavity of the animal by injection and allowed to grow there. The presence of the tumour cells causes the animal to produce ascitic fluid which contains high levels of immunoglobulins to the original antigen. The fluid is removed by aspiration with a syringe and needle usually on three or four occasions over a month or so.

7.2.2 Monoclonal antibody production

Mice are usually the donor animal of choice for monoclonal antibody production although rats and other rodent species may be used. They are cheap to buy and house, and easy to manage and handle. The limitation on using other species is the availability of a suitable tumour partner for performing fusions. Balb/C is the usual mouse strain used for monoclonal antibody production and most of the tumour cell lines used for fusion are derived from this mouse. Females are usually used as they can be housed together without too much aggression.

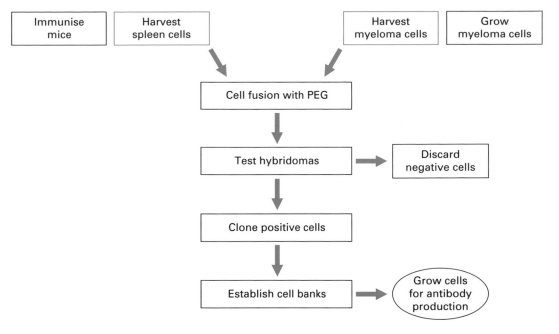

Fig. 7.5 Monoclonal antibody production.

Mice are immunised, usually three or four times over the course of 3–4 months, by the intraperitoneal route using antigen mixed with an appropriate adjuvant (Fig. 7.5). Test bleeds can be taken to monitor the immune status of the animals. Once the mice are sufficiently immune they are left for 2–3 months to 'rest'. This is important as the cells that will be used for the hybridoma production are memory B cells and require the rest period to become quiescent.

Mice are sacrificed and the spleens removed; a single spleen will provide sufficient cells for two or three cell fusions. Three days prior to cell fusion the partner cell line NS-0 is cultured to provide a log phase culture. If rat hybridomas are to be made then the fusion partner Y3 or its derivative Y0 can be used. Cell fusions can be carried out by a number of methods but one of the most commonly used is fusion by centrifugation in the presence of polyethylene glycol (PEG). Then 26×10^6 cells of spleen and fusion partner are mixed together in a centrifuge tube. A quantity of PEG is added to solubilise the cell membranes and the fusion carried out by gentle centrifugation. The PEG is removed from the cells by dilution with culture medium and the cells placed into 96-well tissue culture plates at a cell density of 10×10^5 per well. From experience, these cell numbers will produce only a single recombinant cell capable of growth in each well. Fusion partners are required to have a defective enzyme pathway to allow selection after cell fusion. NS-0 lacks the enzyme hypoxanthine-guanine phosphoribosyl transferase (HGPRT) which prevents it from using a nucleoside salvage pathway when the primary pathway is disabled by the use of the antibiotic aminopterin. The tissue culture additive HAT which contains hypoxanthine aminopterin and thymidine is used to select for hybridomas after cell fusion. They inherit an intact nucleoside salvage pathway from the spleen cell parent which allows them to grow in the presence of aminopterin. Unfused NS-0 cells are unable to assimilate nucleosides and die after a few days. Unfused spleen cells are unable to divide more than a few times in tissue culture and will die after a few weeks. Two weeks after the cell fusion the only cells surviving in tissue culture are hybridomas. The immunisation process ensures that many of the spleen cells that have fused will be secreting antibody to the antigen; however this cannot be relied upon and rigorous screening is required to ensure that the hybridomas selected are secreting an antibody of interest. Screening is often carried out by ELISA but other antibody assessing methods may be used. It is important that hybridomas are assessed more than once as they can lose the ability to secrete antibody after a few cell divisions. This occurs as chromosomes are lost during division to return the hybridoma to its normal chromosome quota.

Once hybridomas have been selected they have to be cloned to ensure that they are stable. Cloning involves the derivation of cell colonies from individual cells grown isolated from each other. In limiting dilution cloning, a cell count is carried out and dilutions of cells in media made. The aim is to ensure that only one cell is present in each well of the tissue culture plate. The plates are incubated for 7 days and cell growth assessed after this time. Colonies derived from single cells are then tested for antibody production by ELISA. It is essential to clone cell lines to ensure that they are truly monoclonal. It is desirable that a cell line should exhibit 100% cloning efficiency in terms of antibody secretion but some cell lines are inherently unstable and will always produce a small number of non-secretory clones. Providing such cell lines are not subcultured excessively then the problem may not be too great although it is usual to reclone these lines regularly to ensure that cultures are never too far from an authenticated clone.

It is very important to know the antibody isotype of the hybridomas as discussed previously and a number of commercial kits are available to do this. Most are based on lateral flow technology which will be discussed later in this chapter. Once the isotype of the antibody is established and it is clonally stable then cultures can be grown to provide both cell banks and antibody for use in testing or for reagent development.

Record-keeping is absolutely vital so that the pedigree of every cell line is known. It is also very important to be vigilant in handling and labelling flasks to prevent cross-contamination of cell lines. It is usual to name cell lines and use the clone and subclone number as part of the name. One such naming system used is: <fusion number>/<clone number>•<sub-clone number>•<additional sub-sub-clone numbers>. Other naming systems are used and it is up to the individual to find one that suits them best.

7.2.3 Freezing cells

Cell lines are frozen to provide a source of inoculum for future cultures. Cells cannot be grown indefinitely in culture as the required incubator space would be impractical in most tissue culture laboratories. Additionally, although established cell lines should be stable it is known that long-term culture leads to cellular instability and the increased risk of cellular change. Cells stored at the low temperatures achieved using liquid nitrogen vapour are stable for many years and can be resuscitated successfully after decades. Cells are transferred into a specialist medium prior to freezing to protect them both as the temperature is lowered and also as the temperature is raised when thawed. Serum containing 10% DMSO works well as a freezing medium although serum-free media can be used if required. Cells must be in perfect health and in log phase prior to freezing. A typical freezing should contain around 1×10^6 cells and this can be assessed by performing a cell count using a counting chamber. A confluent 25-cm^2 tissue culture T flask will contain approximately this many cells and for many applications it may not be necessary to carry out a cell count. The cells are harvested from the flask by tapping to dislodge them and pelleted by centrifugation to remove the culture medium. The cells are then resuspended in 1.0 ml freezing medium chilled to 4 °C, placed into a cryogenic vial and transferred to a cell freezing container. The freezing container contains butan-1-ol which when placed into a −70 °C freezer controls the rate of freezing to 1 °C per minute. The gradual freezing is necessary so that as the ice forms within the cells it does so as a glass and not as crystals which would expand and damage the cell structure. The cells are left for a minimum of 24 h and a maximum of 72 h prior to transfer into cryogenic storage. Transfer to liquid nitrogen storage must be rapid to prevent thawing of the cells. It is imperative that the vials are permanently marked and that the storage locations within the cryogenic vessel are noted for future retrieval.

7.2.4 Cell banking

Cell banks are established from known positive clones and are produced in a way that maximises reproducibility between frozen cell stocks and minimises the risk of cellular change (Fig. 7.6). A positive clone derived from a known positive clone is rapidly expanded in tissue culture until enough cells are present to produce 12 vials of frozen cells simultaneously. This is the master cell bank and is stored at −196 °C under liquid nitrogen vapour. The working cell bank is then derived from the master cell bank. One of the frozen vials from the working cell bank is thawed and rapidly grown until there are enough cells to produce 50 vials of frozen cells simultaneously. This is the working cell bank and it is also stored at −196 °C under liquid nitrogen vapour. This strategy ensures that all of the vials of the working cell bank are identical. All of the vials of the master cell bank are also identical and if a new working bank is required then it can be made from another vial from the master.

Cell banks work well if managed correctly but record-keeping is vital for their operation. A cell bank derived in this way will provide 550 working vials before the process of deriving a new master cell bank is required. If a new master cell bank is required this is produced by thawing and cloning from the last master cell bank vial and selecting a positive clone for expansion.

7.2.5 Antibodies to small molecules

The immune system will recognise foreign proteins and peptides providing that they have a molecular weight (mw) greater than about 2 000 Da (although above 5 000 Da is

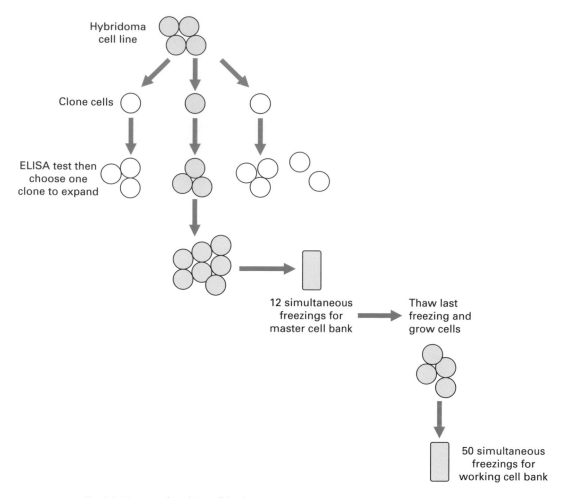

Fig. 7.6 Master and working cell banks.

optimal). The magnitude of the response will increase the greater the molecular weight. If an antibody is to be made to a molecule smaller than 2 000 Da then it has to be conjugated to a carrier molecule to effectively increase its size above the threshold for immune surveillance. These small molecules are known as haptens and may be peptides, organic molecules or other small chemicals. They are usually conjugated to a protein such as albumin, keyhole limpet haemocyanin or thyroglobulin and then used to immunise animals for antibody production (Fig. 7.7). If a polyclonal antibody is being made it is advisable to change the carrier protein at least once in the immunisation procedure as this favours more antibody being made to the hapten and less to each of the carrier proteins. If a monoclonal antibody is being made then the carrier protein can be the same throughout the immunisations. When screening hybridomas for monoclonal antibody production it is necessary to screen against the hapten and carrier separately. Any antibodies responding to both should be discarded as these will be recognising the junction between the hapten and carrier and will not recognise the native hapten.

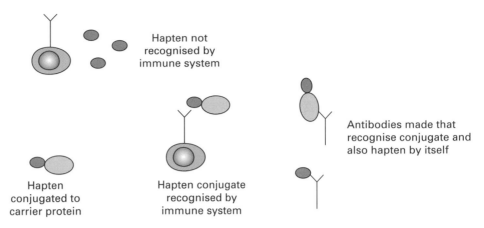

Fig. 7.7 Making antibodies to haptens.

7.2.6 Anti-idiotype antibodies

The binding site of an anti-idiotype antibody is a copy of an epitope. They are made by deriving a primary monoclonal antibody to the epitope of interest, usually a cell membrane receptor or other important binding site. These primary antibodies are then themselves used as antigen to produce secondary antibodies, some of which will recognise the binding site on the primary antibody. These are the anti-idiotype antibodies and they have the unique quality that their binding site structurally resembles the original epitope. They themselves can be used as vaccines as the immune response raised to them will cross-react with the native original epitope. Some human cancers have cell surface receptors that are unique to them and these can be used as a target for antibody therapy. Anti-idiotype antibodies raised to the cell receptors are used to immunise the patient. The resulting antibodies made by the patient bind to the receptors on the tumour cells allowing the immune system to recognise and destroy the tumour. The method has had some success in the treatment of ovarian and bowel tumours.

7.2.7 Phage display for development of antibody fragments

Bacteriophage or phage, as they are known, are viruses that infect and replicate within bacteria. They can be engineered by molecular methods to express proteins and providing the protein sequence is tagged to the coat protein gene then the foreign protein will be expressed on the virus surface. It is possible to isolate the variable (V) antibody coding genes from various sources and insert these into the phage resulting in single-chain antigen-binding (scFv) fragments. Whole antibodies are too large and complex to be expressed by this system but the scFv fragments can be used for diagnostic purposes. The DNA used in this process may come from immunised mouse B cells or from libraries derived from naive mouse (or other species). The V genes are cloned into the phage producing a library which is then assessed for specific activity. It is important to isolate clones that have the specific activity that is required and this can be done by immobilising the antigen onto a solid surface and then adding phage clones to the immobilised antigen. The clones that bind to the antigen are desirable and those that do not bind are washed away. This technique is known as panning and refers to the technique used by nineteenth-century gold prospectors who washed gravel from rivers, using shallow pans that retained gold fragments. Once clones have been selected for antibody expression they can be multiplied in their host bacterium in liquid culture. The scFv fragments can be harvested using proprietary extraction kits and used to develop ELISA and other immunoassays.

7.2.8   Growing hybridomas for antibody production

Cell growth and storage is carried out for the development of cell banks but hybridomas are primarily grown for their products, monoclonal antibodies. All monoclonal antibodies are secreted into the tissue culture media that the cells are growing in. There are a number of ways that cells can be grown to maximise antibody yield, reduce media costs and simplify purification of the product from tissue culture medium. The simplest method for antibody production is static bulk cultures of cells growing in T flasks. T flasks are designed for tissue culture and have various media capacities and cell culture surface areas. For most applications a production run is between 250 ml and 1000 ml medium. Most cell lines produce between 4 and 40 mg of antibody per litre so the size of the production run is based on requirement. The cells from a working cell bank vial are thawed rapidly into 15 ml medium containing 10% foetal bovine serum and placed in an incubator at 37 °C supplemented with 5% CO_2. Once cell division has started, the flask sizes are increased using medium supplemented with 5% foetal bovine serum until the desired volume is reached. Once the working volume has been achieved the cells are left to divide until all nutrients are utilised and cell death occurs. Usually the timespan for this is around 10 days. The cell debris can then be removed by centrifugation and the antibody harvested from the tissue culture medium. For some applications the antibody can be used in this form without further processing.

Monoclonal antibodies can also be produced in ascitic fluid in mice. As described previously, cells can be grown in the peritoneal cavities of mice. Nude mice have no T cells and because of this have poor immune systems. They are often used for ascitic fluid production as they do not mount an immune response to the implanted cells. The mice should not be immunised prior to use as it is important that the only antibody present in the ascitic fluid is derived from the implanted cells. Hybridoma cells are injected into the peritoneum of the mice and allowed to grow there. These cells are secretory and produce high levels of monoclonal antibody in the ascitic fluid. The fluid is harvested by aspiration with a syringe and needle.

A number of in vitro bioreactor systems have been developed to produce high yields of monoclonal antibody in small volumes of fluid which mimics ascitic fluid production (Fig. 7.8). All of them rely on physically separating the cells from the culture medium by semipermeable membrane which allows nutrient transfer but prevents monoclonal antibody from crossing. The culture medium can be changed to maximise cell growth and health, and fluid can be removed from around the cells to harvest antibody. Some are based on a rotating cylinder with a cell-growing compartment at

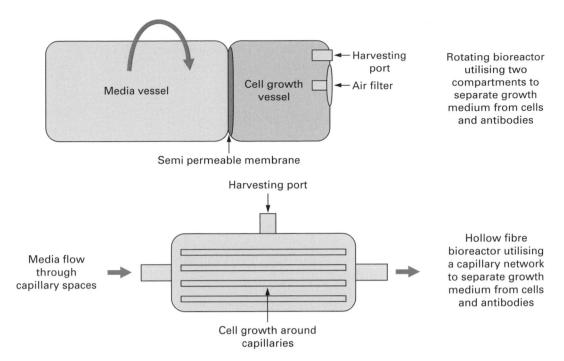

Fig. 7.8 Bioreactors for antibody production.

one end separated from the media container by a membrane. Others have capillary systems formed from membrane running through the cell culture compartment and in these the media is pumped through the cartridge to facilitate nutrient and gas exchange. These systems do produce high yields of antibody but can be problematical to set up and run. They are ideal where large quantities of monoclonal antibody are needed and space is at a premium. They are however prone to contamination by yeasts and great care must be exercised when handling them. Cells are grown in bioreactors for up to 6 weeks so the clone used must be stable and it is advisable to carry out studies on long-term culture prior to embarking on this form of culture. The major advantage of bioreactor culture is that the antibody is produced in high concentration without the presence of media components making it easy to purify. Total quantities per bioreactor run may be several hundred milligrams to gram quantities.

7.2.9 Antibody purification

The choice of method used for the purification of antibodies depends very much on the fluid that they are in. Antibody can be purified from serum by the addition of chaotropic ions in the form of saturated ammonium sulphate. This preferentially precipitates the antibody fraction at around 60% saturation and provides a rapid method for IgG purification. This method does not work well in tissue culture supernatant as media components such as ferritin are co-precipitated. Ammonium sulphate precipitation may be used as a preparatory method prior to further chromatographic purification.

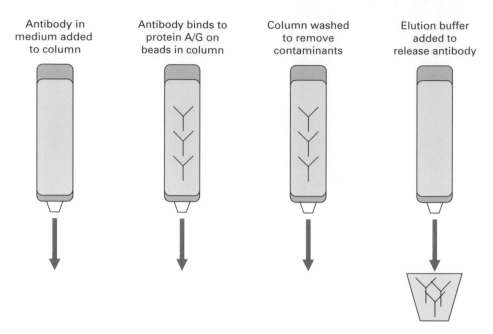

Fig. 7.9 Affinity chromatography.

Tissue culture supernatant is often concentrated before purification to reduce the volume of liquid. Tangential flow devices and centrifugal concentrators may be used to reduce the volume to 10% of the starting amount. This makes antibody purification by affinity chromatography much easier with the smaller volume of liquid (Fig. 7.9).

Antibodies from both polyclonal and monoclonal sources can be purified by similar means. In both cases the antibody type is IgG which allows purification by protein A/G affinity chromatography. Proteins A and G are derivatives of bacterial cells and have the ability to reversibly bind IgG molecules. Binding to the column occurs at neutral pH and the pure antibody fraction can be eluted at pH 2.0. Fractions are collected and neutralised back to pH 7.0. Antibody-containing fractions are identified by spectrophotometry using absorbance at 280 nm (specific wavelength for protein absorbance) and are pooled. A solution of protein at $1\,mg\,cm^{-3}$ will give an absorbance reading of 1.4 at 280 nm. This can be used to calculate the amount of antibody in specific aliquots after purification.

Purified antibody should be adjusted to $1\,mg\,cm^{-3}$ and kept at 4 °C, or −20 °C for long-term storage. It is usual to add 0.02% sodium azide to the antibody solution as this increases shelf-life by suppressing the growth of adventitious microorganisms. Antibodies can be stored for several years at 4 °C and for decades if kept below −20 °C without losing activity.

7.2.10 Antibody modification

Antibodies can be labelled for use in assays such as ELISA by the addition of marker enzyme such as horse radish peroxidise (HRP) or alkaline phosphatase (AP). Other enzymes such as urease have been used but HRP and AP are by far the most popular. Linkage is achieved by simple chemistry to provide stable antibody–enzyme conjugates. Glutaraldehyde is a cross-linking compound and conjugation to HRP is carried out in two stages. Firstly the glutaraldehyde is coupled to reactive amino groups on the enzyme. The HRP–glutaraldehyde is then purified by gel permeation chromatography and added to the antibody solution. The glutaraldehyde reacts with amino groups on the antibody forming a strong link between the antibody and HRP. Alkaline phosphatase can be linked to antibody by glutaraldehyde using a one-step conjugation. The linkage is achieved through amino groups on the antibody and on the enzyme coupled with the glutaraldehyde.

A number of proprietary labelling reagents are also available for making antibody enzyme conjugates.

Fluorescent labels can also be added for use in immunofluorescent assays; usually fluorescein is the molecule of choice. Fluorescein isothiocyanate (FITC) is often the derivative used to label antibodies. FITC is a fluorescein molecule with an isothiocyanate reactive group (–N=C=S) replacing one of the hydrogens. This derivative is reactive towards primary amines on proteins and will readily react with antibodies to produce fluorescent conjugates.

It is also possible to link antibodies to gold particles for use in immunosorbent electron microscopy (ISEM) and lateral flow devices. Gold particles are prepared by citrate reduction of auric acid. The size of particle is predictable and can be controlled by pH manipulation. The gold particles are reactive and will bind antibodies to their surface forming immunogold. The immunogold particles are stable and can be stored at 4 °C until required.

Rare earth lanthanides can also be used as labels and have the advantage that a single assay can be used to detect two or three different antibody bindings. The lanthanides are attached to the antibodies as a chelate. The commonest of the chelating compounds used is diethylenetriamine pentaacetate (DTPA). Each lanthanide fluoresces at a different light frequency and so multiple assays can be carried out and the individual reactions visualised by the use of a variable wavelength spectrophotometer. Antibodies can also be attached to latex particles either by passive absorption or to reactive groups or attachment molecules on the surface of the latex. These can be used either as the solid phase for an immunoassay or as markers in lateral flow devices. Magnetised latex particles are available allowing the easy separation of latex particle/antibody/antigen complex from a liquid phase. Latex and magnetic particles may be purchased which have protein A covalently attached to their surface. Protein A binds the Fc portion of the antibody which orientates the molecules with the binding sites facing outwards.

7.3   IMMUNOASSAY FORMATS

The first immunoassay formats described were methods based on the agglutination reaction (Fig. 7.10). The reaction between antibody and antigen can be observed when agglutination occurs and is characterised either by a gel formation in a liquid phase or

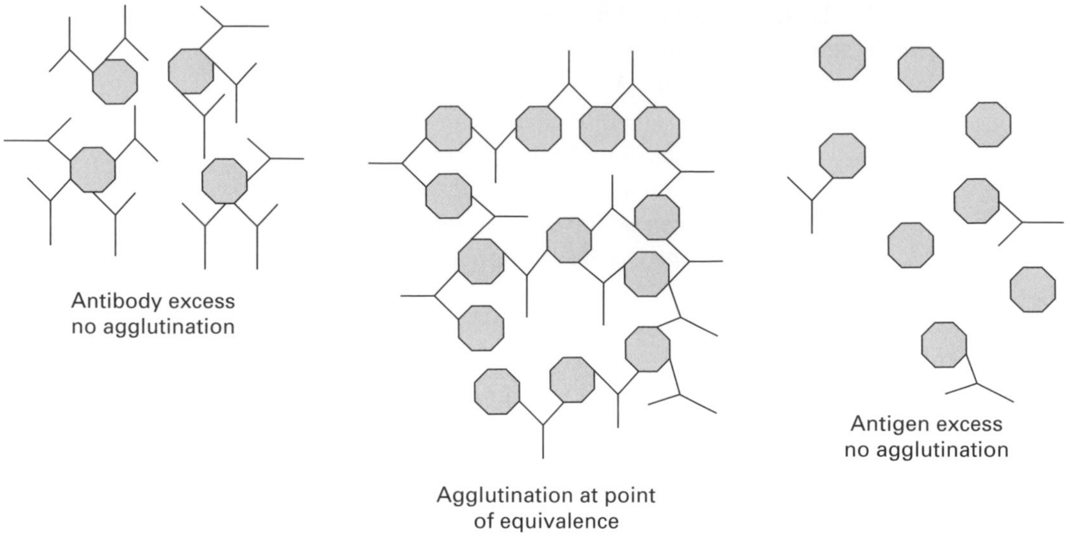

Antibody excess
no agglutination

Agglutination at point
of equivalence

Antigen excess
no agglutination

Fig. 7.10 Agglutination reaction.

as an opaque band in an agar plate assay. Agglutination only occurs when there is the right amount of antibody and antigen present. It relies on the fact that as an antibody has two binding sites then each of them can be bound to different antigen particles. As this happens bridges are formed created by the antibody molecules spanning two antigen molecules. The resulting lattice that is created forms a stable structure where antigen and antibody particles are suspended in solution by their attachment to each other. For this to take place there must be a precise amount of antigen and antibody present and this is known as equivalence. If too much antibody is present then each antigen molecule will bind multiple antibodies and the meshwork will not develop. If too much antigen is present then each antibody will bind only one antigen particle and no lattice will form. For this reason a dilution series of antibody is often made and a measure of antigen concentration can be made from the end point at which agglutination occurs.

Modifications of the agglutination reaction involve the use of antibody bound to red blood cells or latex particles which allow the reaction to be observed more easily in a liquid phase. Agglutination immunoassays are still used as they provide rapid results with the minimum of equipment. They are commonly used for the detection of viral antigens in blood serum. In commercial tests the antibody concentration in the reagent is provided at a working dilution known to produce a positive for the normal range of antigen concentration.

The Ouchterlony double diffusion agar plate method is the commonest gel-based assay system used (Fig. 7.11). Wells are cut in an agar plate which is then used to load samples and an antibody solution. The antigens and the antibody diffuse through the agar and if the antibody recognises antigen within the gel then a precipitin band is formed. A diffusion gradient is formed through the agar as the reagents progress and so there is no requirement for dilutions to be made to ensure that agglutination occurs.

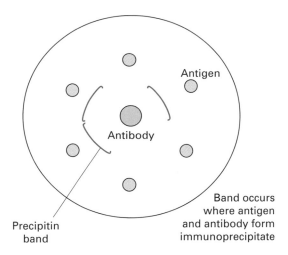

Fig. 7.11 Ouchterlony double diffusion plate.

Polyclonal antibodies can be made to different subspecies of bacteria and these will recognise surface epitopes on them. Because the subspecies are related then they will share some surface epitopes and the closer they are related the more they will share. Antibodies made to one organism will therefore react to a greater or lesser degree to a related organism depending on how many surface epitopes they share.

This has given rise to a systematic method of identification known as serotyping which is based on the reaction of microorganisms to antibodies. The system works well with closely related organisms but is not definitive, as it only assesses surface markers. The pattern of precipitin bands obtained to reference antibodies is specific and can be used to assign samples into serotypes. The method was used until very recently to characterise *Salmonella* strains as their pathogenicity can be assessed according to their relatedness to known strains. *Salmonellas* from food and water samples were tested by this method to establish if they had been the cause of food poisoning incidents.

Agglutination reactions using sensitised erythrocytes (red blood cells) or latex particles are carried out in liquid phase usually in small tubes or more recently in round-bottom microtitre plates. As discussed before, the agglutination reaction only occurs at the point of equivalence. A positive agglutination test appears cloudy to the eye as the erythrocytes or latex particles are suspended in solution. A negative result is characterised by a 'button' at the bottom of the reaction vessel which is formed from non-reacted particles. A negative result may be obtained from excess antigen or antibody, as the binding reaction favours the production of small aggregates of antigen/marker particles rather than the agglutination gel.

7.3.1 Enzyme immunosorbent assays

By far, the vast majority of immunoassays carried out fall into the category of enzyme immunosorbent assays. These are routinely used for the diagnosis of infectious agents

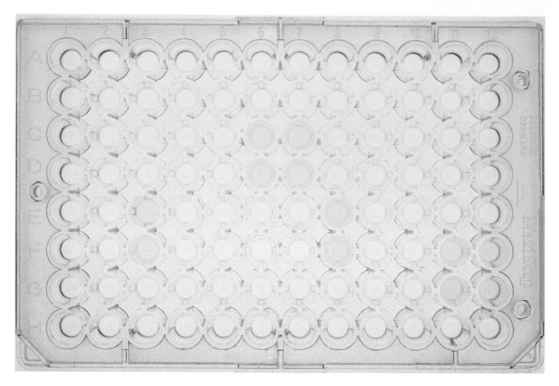

Fig. 7.12 A microtitre plate. (See also colour plate.)

such as viruses, and other substances in blood. The antigen is the substance or agent to be measured. In this technique the antigen is immobilised on to a solid phase, either the reaction vessel or a bead. The most commonly used solid phase is the enzyme linked immunosorbent assay (ELISA) plate. Immobilisation is achieved by the use of a coating antibody which actively traps antigen to the solid phase. A second antibody (antibody enzyme conjugate) which is labelled with a reporter enzyme is allowed to bind to the immobilised antigen. The enzyme substrate is then added to the antigen/antibody/enzyme complex and a reaction, usually involving a colour change, is seen (Fig. 7.12, see also colour section). There are many permutations of this method but all of them rely on the antibody–antigen complex being formed and the presence of it being confirmed by the reactions of the reporter enzyme. These assays rely on a stepwise addition of layers with each one being linked to the one before. The antigen is central to the assay as it provides the bridge between the solid phase and the signal-generating molecule. Without antigen, the antibody enzyme conjugate cannot be bound to the solid phase and no signal can be generated. The coating antibody also concentrates the antigen from the sample as it binds the antigen irreversibly and so the coating layer has the ability to concentrate available antigen until saturation has been reached. This is particularly useful when testing for low levels of antigens in fluids such as blood serum.

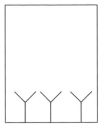

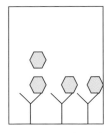

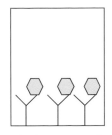

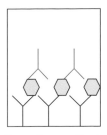

ELISA plate coated
with antibody

Sample incubated
on plate

Antigen trapped
by antibody

Secondary antibody
incubated on plate

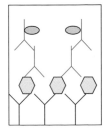

Anti-species
antibody
conjugate
incubated
on plate

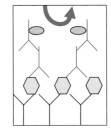

Substrate added
to plate causing
colour change in
positive wells

Fig. 7.13 TAS ELISA.

7.3.2 Triple antibody sandwich ELISA (TAS ELISA)

Triple antibody sandwich (TAS) ELISA, also known as indirect ELISA, is a widely used method (Fig. 7.13). It is often used to identify antibodies in patient blood which may be there as the result of infection. As with other immunoassays, layers of reagents are built up, each dependent on the binding of the previous one. The system is used to test patient blood for the presence of hepatitis B virus (HBV) antibodies as a diagnostic test for this disease. In this test HBV coating antibody is bound to the wells of a microtitre plate and HBV coat protein added to them. The live virus is not used as antigen as this would be too dangerous to use in the laboratory. HBV coat protein is made synthetically specifically for use as antigen in this type of test. After incubation and washing, patient serum is added which if it contains antibodies reacts to the antigen. Anti-human antibody conjugated to an enzyme marker is then added which will bind to the patient antibodies. Substrate is then added to identify samples which were positive. The test works well for the diagnosis of HBV infection and is also used to ensure that blood donations given for transfusion are free from this virus.

7.3.3 Double antibody sandwich ELISA (DAS ELISA)

Double antibody sandwich (DAS) ELISA is probably the most widely used immunochemical technique in diagnostics (Fig. 7.14). It is rapid, robust, and reliable and can be performed and the results interpreted with minimal training. The principle is the same as other ELISA techniques in that the antigen is immobilised to a solid phase by a primary antibody and detected with a second antibody which has been labelled with a marker enzyme. The antigen creates a bridge between the two antibodies and the

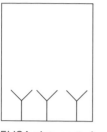

ELISA plate coated
with antibody

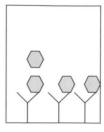

Sample incubated
on plate

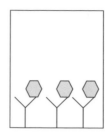

Antigen trapped
by antibody

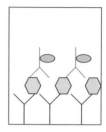

Antibody/enzyme
conjugate
incubated on plate

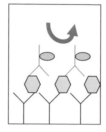

Substrate added
to plate causing
colour change in
positive wells

Fig. 7.14 DAS ELISA.

presence of the enzyme causes a colour change in the chromogenic (colour-producing) substrate. The marker enzyme used is usually either horseradish peroxidase (HRP) or alkaline phosphatase (AP). Other enzymes have been used and claims have been made for increased sensitivity but this is at the expense of more complex substrates and buffers. In some systems the enzyme is replaced with a radioactive label and this format is known as the immunoradiometric assay (IRMA). DAS ELISA is used extensively in horticulture and agriculture to ensure that plant material is free of virus. Potato tubers that are to be used as seed for growing new crops have to be free of potato viruses and screening for this is carried out by DAS ELISA. There are many potato viruses but potato leafroll virus (PLRV) in particular causes considerable problems. PLRV antibodies are coated onto the wells of ELISA plates and then the sap to be tested is added. After incubation, the plates are washed and PLRV antibody conjugated to alkaline phosphatase is added. The plates are incubated and after washing, substrate is used to identify the positive wells. The system again requires the presence of the antigen (PLRV) for the sandwich of antibodies to be built up.

7.3.4 Enhanced ELISA systems

The maximum sensitivity of ELISA is in the picomole range and there have been many attempts to increase the detection threshold for assays beyond this. The physical limitations are based on the dynamics of the double binding event and the subsequent generation of signal above the background substrate value. Most workers have concentrated their efforts on the amplification of signal. Antibody binding cannot be improved as it is primarily a random event modified by the individual avidities of the antibodies themselves. Some improvement in some assays can be made by temperature modification as some antibodies may perform better at specific temperatures.

There are two basic ways that signal can be amplified in ELISA. More enzyme can be bound by using multivalent attachment molecules. Systems using the avidin–biotin binding system allow amplification through this route. Both avidin and biotin are tetravalent (i.e. they have four binding sites) and it is this property that produces the amplification. The detection antibody is labelled with biotin and the reporter enzyme with avidin. The high affinity and multivalency of the reagents allows larger complexes of enzyme to be linked to the detection antibody, producing an increase in substrate conversion and improved colour development in positive samples. The alternative amplification step is by enhancing the substrate reaction usually by using a double enzyme system. The primary enzyme bound to the antigen catalyses a change in the second enzyme which then generates signal. Both of these methods will increase the signal generated but may also increase the background reaction. Alkaline phosphatase conjugated secondary enzyme can be used to drive a secondary reaction involving NADP dephosphorylation to NAD which is further reduced to NADH by alcohol dehydrogenase. This in turn creates a loop in which a tetrazolium salt is oxidised as the NADH returns to NAD. The tetrazolium salt is chromogenic when in the oxidised state. The cyclic nature of the reaction causes the amplification and increases the observed colour development. Some claims have been made for 'supersubstrates' which work directly with enzyme–antibody conjugates but there is usually little in the way of true gain if standard curves are calculated for the various substrate types.

7.3.5 Competitive ELISA

Competitive ELISA is used in assays for small molecules such as hormones in blood samples where often only a single epitope is present on the antigen (Fig. 7.15). It is quantitative when used in conjunction with a standard curve. The principle is based on competition between the natural antigen (hormone) to be tested for and an enzyme-conjugated form of the antigen which is the detection reagent. The test sample and a defined amount of enzyme-conjugated antigen are mixed together and placed into the coated wells of a microtitre plate. The antigen and conjugated form of it compete for the available spaces on the coating antibody layer. The more natural antigen present the more it will displace (compete out) the conjugated form leading to a reduction in enzyme bound to the plate. The relationship of substrate colour development is therefore inverted; the more natural antigen bound the lower the signal generated. This form of ELISA is routinely used for testing blood samples for thyroxin. Thyroxin is a hormone that is responsible for regulating metabolic rate and deficiencies (hypothyroidism) and excesses (hyperthyroidism) of it will slow or speed up the metabolism. Patients can be given additional thyroxin if required if they are deficient but it is important to establish the baseline level before treating the condition. Competitive ELISA is used for this as an accurate measure of the circulating level of the hormone can be made from a standard curve of known dilutions.

In some assays the enzyme is replaced with a radioactive label and this form of competitive ELISA is known as the radioimmunoassay (RIA).

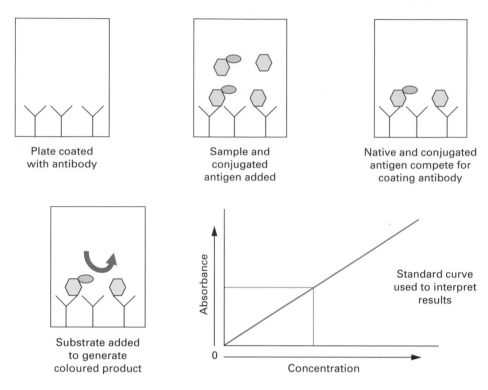

Fig. 7.15 Competitive ELISA.

Competitive ELISA using conjugated antibody can also be used to quantify levels of circulating antibody in test serum. The solid phase has the antigen to which the antibody will attach directly coated on to it. The test serum and the conjugated form of antibody are mixed together and added to the reaction wells. The conjugated and test antibody then compete to bind to the antigen. The level of antibody can again be determined by the reduction in signal observed by addition of the substrate.

7.3.6 Dissociation enhanced lanthanide fluorescence immunoassay (DELFIA)

DELFIA is a time-resolved fluorometric assay which relies on the unique properties of lanthanide chelate antibody labels. The lanthanides will generate a fluorescent signal when stimulated with light of a specific wavelength. The light signal generated has a long decay which enhances the negative to positive ratio of the assay. DELFIA offers a signal enhancement greater than that possible from conventional enzyme-linked assays. The lanthanide chelates are conjugated onto the secondary antibody and as there are a number of lanthanides each with a unique signal which can be used, multiplexing (more than one test carried out in the same reaction vessel) is possible. The assay is carried out similarly to standard ELISA and may be competitive or non-competitive. The assay is concluded by adding an enhancement solution which causes dissociation of the lanthanide from the antibody molecules. The signal is generated by stimulating the lanthanide with light of a specific wavelength and measuring the resulting fluorescence. If europium is used the stimulatory wavelength is 340 nm, and the fluorescence generated is 615 nm.

7.4   IMMUNO MICROSCOPY

7.4.1   Immunofluorescent (IF) microscopy

Immunofluorescent (IF) microscopy uses antibodies conjugated to fluorescent markers to locate specific structures on specimens and allows them to be visualised by illuminating them with ultraviolet light. Fluoroscein and rhodamine are the usual labels used but alternative markers are available. Fluoroscein produces a green fluorescence and rhodamine is red. Microscopes equipped to carry out IF have dual sources of light allowing the operator to view the specimen under white light before illuminating with ultraviolet to look for specific fluorescence. The technique is particularly useful for looking at surface markers on eukaryotic cells but is also used as a whole-cell staining technique in bacteriology.

Membrane studies on whole mammalian cells can be undertaken and the migration, endocytosis (uptake of membrane-bound particles by cells) and fate of labelled receptors studied in real time. Bound receptors in cell membranes frequently migrate to one end of the cell prior to being endocytosed. This phenomenon is known as capping and is easily viewed in living cells using antibodies specific to cell membrane receptors labelled with a fluorescent marker.

7.4.2   Immunosorbent electron microscopy

Immunosorbent electron microscopy (ISEM) is a diagnostic technique used primarily in virology. Virus-specific antibodies conjugated to gold particles are used to visualise virus particles on electron microscopes. The gold is electron-dense and is seen as a dark shadow against the light background of the specimen field. The technique can be used for both transmission or scanning systems. If gold-labelled primary antibodies are not available then anti-IgG–gold conjugated antibodies can be used with the primary antibody in a double antibody system. Both monoclonal and polyclonal antibodies can be used for ISEM depending on the required specificity.

7.5   LATERAL FLOW DEVICES

Lateral flow devices (LFD) are used as rapid diagnostic platforms allowing almost instant results from fluid samples (Fig. 7.16). They are simple to use and contain all of the required components within the strip itself. They are usually supplied as a plastic cassette with a port for applying the sample and an observation window for viewing the result. The technology is based on a solid phase consisting of a nitrocellulose or polycarbonate membrane which has a detection zone which is coated with a trapping

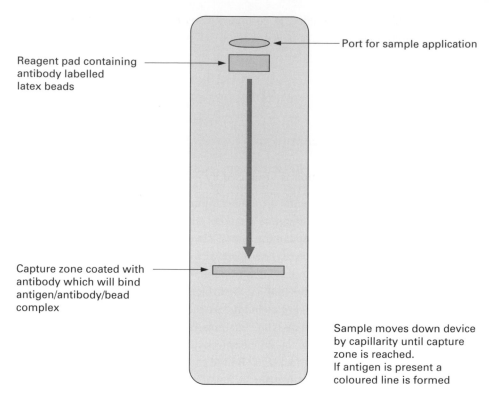

Port for sample application

Reagent pad containing
antibody labelled
latex beads

Capture zone coated with
antibody which will bind
antigen/antibody/bead
complex

Sample moves down device
by capillarity until capture
zone is reached.
If antigen is present a
coloured line is formed

Fig. 7.16 Lateral flow device.

antibody. The detection antibody is conjugated to a solid coloured marker, usually latex or colloidal gold, and is stored in a fibre pad which acts as a reservoir. The solid phase has a layer of transparent plastic overlaying it leaving a very narrow gap which will draw liquid by capillarity. The sample is applied to the reservoir pad through the sample port where it can react with the conjugated antibody if the specific antigen is present. The liquid then leaves the reservoir and travels up the solid phase pad to the location of the trapping antibody. If the sample contains the specific antigen then it will react to both the conjugated and trapping antibody. This results in a coloured line if the sample is positive. If the sample is negative then no coloured line develops. The system lends itself to multiplexing and up to three different antigens can be tested for simultaneously with appropriate trapping antibodies and different coloured marker particles. The technology has been applied to home pregnancy testing and various other 'self-diagnostic' kits. Lateral flow devices are also used by police forces and regulatory authorities for the rapid identification of recreational drugs.

7.6 EPITOPE MAPPING

Epitope mapping is carried out to establish where on the target protein the antibody binds. The method works well with new monoclonal antibodies where it may be necessary to know the precise epitope to which binding occurs. This however can only be performed where the epitope is linear. A linear epitope is formed by amino acids lying adjacent to each other and the antibody binds to the structure that they form. Non-linear epitopes are formed from non-adjacent amino acids when they interact with each other in space, as is found in helical or hairpin structures. To carry out epitope mapping the amino acid sequence of the target protein must be known. The sequence is then used to design and make synthetic peptides each containing around 15 amino acid residues in length and overlapping with the previous one by about five residues. The synthetic peptides are then coated on to the wells of microtitre plates or onto nitrocellulose membranes and reacted with the antibody of interest. The reaction is visualised by using a secondary antibody enzyme conjugate and substrate. From the reaction to the peptides and the position of the sequence in the native protein it is possible to predict where the epitope lies and also what its sequence is.

7.7 IMMUNOBLOTTING

This technique is also known as western blotting and is used to identify proteins from samples after electrophoresis. The sample may be tissue homogenate in origin or an extract of cells or other biological source. The sample may be electrophoresed under reducing or non-reducing conditions until separation is achieved. This is usually visualised by staining with a general protein stain. The separated proteins are transferred onto a nitrocellulose or polyvinyl membrane either passively or by using an electroblotter. The membrane is treated with a protein-blocking solution to prevent non-specific binding of antibody to the membrane itself. Popular blocking compounds are dried milk or bovine serum albumin. Either direct or indirect antibody systems can be used but often indirect methods are used for reasons of cost. Directly conjugating primary antibodies may be expensive and so very often anti-species enzyme conjugate is used. For indirect labelling the membrane is incubated in antibody solution and after washing, it is treated with a solution of a secondary antibody–enzyme conjugate. Both peroxidase and alkaline phosphatase have substrates that will produce a solid colour reaction on the blot where the antibodies have bound. The substrate reaction can be stopped after optimum colour development, dried and the blots stored for reference. Further details can be found in Section 10.3.8.

The method is particularly useful during development of new antibodies as part of epitope mapping studies.

7.8 FLUORESCENT ACTIVATED CELL SORTING (FACS)

Fluorescent activated cell sorting (FACS) machines are devices that are capable of separating populations of cells into groups of cells with similar characteristics based on antibody binding (Fig. 7.17). The technique is used on live cells and allows recovery and subsequent culture of the cells after separation. Many cell markers are known which identify subsets of cell types and specific antibodies to them are available.

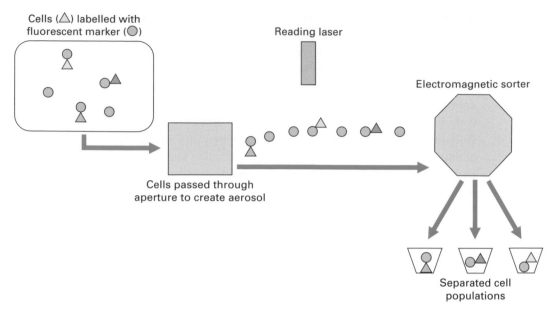

Fig. 7.17 Fluorescent activated cell sorting.

The method can be used quite successfully to separate normal from abnormal cells in bone marrow samples from patients with leukaemia. This can be used as a method of cleaning the marrow prior to autologous (from the patient themselves) marrow transplantation.

The technique is also used for diagnostic tests where the numbers of cell subtypes need to be known. This is of particular use when looking at bloodborne cells such as lymphocytes where the ratios of cell types can be of diagnostic significance. For example, in HIV infection the numbers of specific T cell subtypes are of great diagnostic significance in the progress of the infection to AIDS.

The cells are labelled with the antibodies to specific cell markers labelled with a fluorescent label and then they are passed through a narrow gauge needle to produce an aerosol. The droplet size is adjusted so that each one should contain only one cell. The aerosol is then passed through a scanning Laser which allows detection of the fluorescent label. The droplets have a surface charge and can be deflected by an electron magnet based on their fluorescent label status. The system relies on computer control to effectively sort the cells into labelled and non-labelled populations. The desired cell population can then be recovered and subsequently counted or cultured if desired. More than one label can be used simultaneously so that multiple sorting can be undertaken.

7.9 CELL AND TISSUE STAINING TECHNIQUES

There are many antibodies available that recognise receptors on and structural proteins in cells and tissues and these can be of use diagnostically. Generally immunostaining is carried out on fixed tissues but this is not always the case as it may be important to observe a dynamic event only seen in living cells. Different antibodies may be required for living and fixed tissues for the same protein, as fixation may destroy the structure of the epitopes in some cases. Fixed tissues are prepared by standard histological methods. The tissue is fixed with a preservative which kills the cells but maintains structure and makes the cell membranes permeable. The sample is embedded in wax or epoxy resin and fine slices are taken using a microtome and they are then mounted onto microscope slides. The antibodies that are used for immuno-pathology may carry enzyme, fluorescent markers or labels such as gold particles. They may also be unconjugated and in this case would require a secondary antibody conjugate and solid substrate to visualise them. It is important to remember that enzymes such as alkaline phosphatase may be endogenous (found naturally) in mammalian tissue samples and their activity is not easily blocked. Often horseradish peroxidase is used as an alternative. Any endogenous peroxidase activity in the sample can be blocked by treating the sample with hydrogen peroxide. Antibodies may recognise structural proteins within the cells and can access them in fixed tissues through the permeabilised membranes. More than one antibody can be used to produce a composite stain with more than one colour of marker being used. Combi-nations of fluorescent and enzyme staining may also be used but this has to be carried out sequentially. Fluorescent stains can also be used in conjunction with standard histological stains viewed with a microscope equipped with both white and ultraviolet light. Fluorescence will decay in time and although anti-quench products can be used the specimens should not be considered to be permanent. Photographs can be taken of slides through the microscope and kept as a permanent record.

7.10 IMMUNOCAPTURE POLYMERASE CHAIN REACTION (PCR)

Immunocapture PCR is a hybrid method which uses the specificity of antibodies to capture antigen from the sample and the diagnostic power of PCR to provide a result. The method is particularly of use in diagnostic virology where the technique allows the capture of virus from test samples and subsequent diagnosis by PCR. It is useful where levels of virus are low such as in water samples and other non-biological sources. The technique can be carried out in standard PCR microtitre plates or in PCR tubes. The antibody is bound passively to the plastic of the plate or tube and the sample incubated afterwards (see Fig. 7.18). After washing to remove excess sample material the PCR reagents can be added and thermocycling carried out on the bound viral nucleic acid. RNA viruses will require an additional reverse transcription step prior to PCR (see also Section 6.8.1).

7.11 IMMUNOAFFINITY CHROMATOGRAPHY (IAC)

Immunoaffinity chromatography can be used for a number of applications. The principle is based on the immobilisation of antibody onto a matrix, normally beads, which are then placed into a chromatography column. Antibody may be permanently

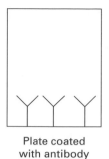
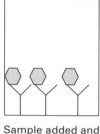

| Plate coated with antibody | Sample added and antigen captured | Sample lysed to release nucleic acid followed by PCR with specific primers |

Fig. 7.18 Immunocapture PCR.

linked to the beads by covalent linkage to reactive sites on a resin or bound using protein A or G. Usually, antibody is permanently bound to column beads for most applications as this is a more stable linkage and allows repeated use of the columns following regeneration. IAC may be used as a clean-up method in analytical chemistry to extract small quantities of chemical residues such as pesticides from wastewater and other sources. The method also works well for the extraction of biological compounds such as hormones from patient samples. The columns are made by reacting highly purified antibody (monoclonal or polyclonal) with the chromatography beads to form the affinity matrix. Harsh conditions have to be avoided as denaturation of the antibody molecules could occur. A number of proprietary resins are available which have reactive sites suitable for antibody immobilisation. The affinity matrix is loaded into chromatography columns prior to use. Antibody binding of antigen generally occurs best at around pH 7.4 but individual monoclonal antibodies may vary considerably from this pH. Once the sample has been loaded onto the column it should be washed to remove contaminating material from sample fluid. Conditions for elution vary according to individual antibodies and antigens but pH 2.0 buffer, methanol and 10% acetonitrile have all been used successfully. The column can be regenerated after elution by incubating with pH 7.4 buffer. The technique works extremely well for clean-up and concentration of sample from dilute sources prior to additional analysis. Samples eluted from IAC columns may be tested further by high-performance liquid chromatography, ELISA or other analytical techniques.

7.12 ANTIBODY-BASED BIOSENSORS

A biosensor is a device that is composed of a biological element and a physicochemical transduction part which converts signal reception by the biological entity into an electrical impulse. A number of biosensor devices are available that use enzymes as the biological part of the device. The enzyme is used to catalyse a chemical reaction which generates an electrical charge at an electrode. Antibodies have the potential to be excellent biological molecules to use for this technology as they can be developed to detect virtually any molecule. The main problem with developing this technology with antibodies has been the lack of adequate physicochemical transduction systems. Three methods have been developed that will provide a signal from antibody binding and these are likely to produce a new generation of biosensors in the future. Antibodies may be bound onto thin layers of gold which in turn are coated onto refractive glass slides. If the slides are illuminated at a precise angle with fixed-wavelength Laser light then electron waves are produced on the surface of the gold. This is known as surface plasmon resonance and only occurs if the incident angle and wavelength of light are precisely right. If the antibody binds antigen then the surface plasmon resonance pattern is changed and a measurable change in emitted energy is observed.

Fibre optic sensors have also been developed which rely on the natural ability of biological materials to fluoresce with light at defined frequency. The reaction vessel is coated with antibody and the fibre optic sensor used to illuminate and read light scatter from the vessel. The sample is then applied and the sample vessel washed. The fibre optic sensor is again used to illuminate and read backscatter from the vessel. Changes in the fluorescence will give a change in the observed returned light.

A third approach relies on changes in crystals as a result of surface molecules bound to them. Piezoelectric crystals generate a characteristic signature resonance when stimulated with an alternating current. The crystals are elastic and changes to their surface will produce a change in the signature resonance. The binding of antigen to antibody located on the surface of the crystal can be sufficient to alter the signature and therefore induce a signal indicating that antigen has been detected by antibody.

7.13 THERAPEUTIC ANTIBODIES

Therapeutic antibodies fall into a number of different classes but are all designed to bind to specific structures or molecules to alter cellular or systemic responses *in vivo*. The simplest of these are the inhibitory systemic (found throughout the body) antibodies that will bind to substances to render them ineffective. At their crudest, they consist of hyperimmune serum and are used to alleviate the symptoms of bites and stings from a number of poisonous animals. Antivenom produced in horses for treatment of snake bite is a good example of this. Hyperimmune serum derived from human patients who have had the disease has also been used prophylactically (reduce the risk of disease) after exposure to pathogenic viruses. Hyperimmune serum is available to help to treat a number of pathogenic viral conditions such as West Nile Fever, AIDS and hepatitis B. These are used after exposure to the pathogen, for example by needle-stick injury, and help to reduce the risk of infection occurring. The next class of therapeutic antibodies are those that bind bioactive molecules and reduce their effects *in vivo*. They are all monoclonal and have a number of targets which help to alleviate the symptoms of a number of human diseases. One of the major targets for this approach are systemic cytokines which have been implicated in the progression of diseases such as arthritis; results using antibody therapy have been encouraging. Monoclonal antibodies can also be used to reduce the numbers of specific cell types *in vivo* by binding to surface markers on them. The binding of the antibody to the cells alerts the immune system and causes the cells to be cleared from circulation. Chimeric (formed from two sources) mouse/human monoclonal antibodies consisting of mouse variable regions and human constant regions which are specific to the B cell marker CD20 have been used successfully for the treatment of systemic lupus erythematosus. This disease is characterised by the development of aberrant B cells secreting autoantibodies which cause a number of immune phenomena. The decrease in circulating B cells reduces the number producing the autoantibodies and alleviates some of the symptoms.

Agonistic (causing upregulation of a biological system) monoclonal antibodies are therapeutic antibodies which have the ability to influence living cells *in vivo*. They upregulate cellular systems by binding to surface receptor molecules. Normally, cell receptors are stimulated briefly by their ligand (substance that binds to them) and the resulting upregulation is also brief. Agonistic monoclonal antibodies bind to the receptor molecule and mimic their ligand, but have the capacity to remain in place for much longer than the natural molecule. This is due to the fact that the cell finds it much more difficult to clear the antibody than it would the natural ligand. The action of agonistic antibodies is incredibly powerful as the internal system cascades that can be generated are potentially catastrophic for both the cell and the organism. Their use has been mainly restricted to induction of apoptosis (programmed cell death) in cancer cells and only where a known unique cellular receptor is being stimulated.

There are a number of therapeutic inhibitory antibodies available and all of them downregulate cellular systems by blocking the binding of antigen to receptor. They behave as competitive analogues to the inhibitor and have a long dwell time (the time they remain bound) on the receptor increasing their potency. They may block the binding of hormones, cytokines and other cellular messengers. They have been used successfully for the management of some hormone-dependent tumours such as breast cancer and also for the downregulation of the immune system to help prevent rejection after organ transplantation.

These therapeutic antibody types need to be carefully engineered to make them effective as treatment agents. The avidity and affinity of the antibodies is critical to their therapeutic efficacy as their specific binding ability is critical to their length and specificity of action. Additionally, they must not appear as 'foreign' to the immune system or they will be rapidly cleared by the body. Often, the original monoclonal antibody will have been derived using a mouse system and as a result is a murine antibody. These antibodies can be humanised by engineering the cells, retaining the murine binding site and replacing the constant region genes with human ones. The resulting antibody escapes immune surveillance but retain their effective binding capacity. Natural antibodies may remain in the circulation for up to 6 weeks but engineered antibodies survive a much shorter time. The shortened survival time is due to the humanisation which still leaves a degree of murine antibody visible to the immune system. Each engineered, therapeutic antibody has a different half-life *in vivo* and this factor is of great importance when baseline dosage is being established. All of the currently used therapeutic antibodies may cause side effects in patients and so this line of therapy has only been exploited where the benefits outweigh the problems that may be encountered. Great success has been seen in the treatment of prostate cancer in men and breast cancer in women using humanised monoclonal antibodies which bind to hormone receptors on the tumour cells and inhibit their growth as a result.

7.14 THE FUTURE USES OF ANTIBODY TECHNOLOGY

Antibodies are incredibly useful molecules which can be designed to detect an almost limitless number of antigens. They are adaptable and will operate in many conditions. They can be used in both diagnostic and therapeutic scenarios. In the future there will be a rise in the availability of therapeutic antibodies both for the up- and downregulation of cellular and systemic responses. Cancer therapy and immune modulation of autoimmune phenomena are probably the two areas where greatest developments will take place.

Biosensors for the detection of disease will become increasingly available as will multiple lab on a chip (LOC) formats. LOC devices are miniaturised devices that are capable of handling microscopic amounts of liquids and perform a number of laboratory assays in miniature. They are frequently fully automated and can give rapid results without the equipment normally required for laboratory assays. They have the added advantage that they can be used in field situations as they are becoming increasingly portable.

The use of non-animal systems for antibody generation will be exploited more fully with more use of phage display and other DNA library based systems.

7.15 SUGGESTIONS FOR FURTHER READING

Burns, R. (ed.) (2005). *Immunochemical Protocols*, 3rd edn. Totowa, NJ: Humana Press.

Coligan, J. (2005). *Short Protocols in Immunology*. New York: John Wiley. (A good background book which gives detail of immunological protocols and how they can be used to investigate the immune system.)

Cruse, J. and Lewis, R. (2002). *Illustrated Dictionary of Immunology*, 2nd edn. Boca Raton, FL: CRC Press. (An excellent book which describes in detail immunological processes and how they interact. A good balance of text and graphics.)

Howard, G. and Kaser, M. (2007). *Making and Using Antibodies: A Practical Handbook*. Boca Raton, FL: CRC Press. (An excellent book which describes in detail methods for producing, validating, purifying, modifying and storing antibodies.)

Subramanian, G. (ed.) (2004). *Antibodies,* Volume 1, *Production and Purification*. Dordrecht: Kluwer Academic. (This book gives good coverage of methods for antibody production, purification, modification and storage.)

Wild, D. (2005). *Immunoassay Handbook*, 3rd edn. New York: Elsevier. (This book describes in detail background to many clinical immunoassays and how to design and validate them.)

8 Protein structure, purification, characterisation and function analysis

J. WALKER

8.1 IONIC PROPERTIES OF AMINO ACIDS AND PROTEINS

Twenty amino acids varying in size, shape, charge and chemical reactivity are found in proteins and each has at least one codon in the genetic code (Section 5.3.5). Nineteen of the amino acids are α-amino acids (i.e. the amino and carboxyl groups are attached to the carbon atom that is adjacent to the carboxyl group) with the general formula $RCH(NH_2)COOH$, where R is an aliphatic, aromatic or heterocyclic group. The only exception to this general formula is proline, which is an imino acid in which the $-NH_2$ group is incorporated into a five-membered ring. With the exception of the simplest amino acid glycine (R = H), all the amino acids found in proteins contain one asymmetric carbon atom and hence are optically active and have been found to have the L configuration.

For convenience, each amino acid found in proteins is designated by either a three-letter abbreviation, generally based on the first three letters of their name, or a one-letter symbol, some of which are the first letter of the name. Details are given in Table 8.1.

Since they possess both an amino group and a carboxyl group, amino acids are ionised at all pH values, i.e. a neutral species represented by the general formula does not exist in solution irrespective of the pH. This can be seen as follows:

Table 8.1 **Abbreviations for amino acids**

Amino acid	Three-letter code	One-letter code
Alanine	Ala	A
Arginine	Arg	R
Asparagine	Asn	N
Aspartic acid	Asp	D
Asparagine or aspartic acid	Asx	B
Cysteine	Cys	C
Glutamine	Gln	Q
Glutamic acid	Glu	E
Glutamine or glutamic acid	Glx	Z
Glycine	Gly	G
Histidine	His	H
Isoleucine	Ile	I
Leucine	Leu	L
Lysine	Lys	K
Methionine	Met	M
Phenylalanine	Phe	F
Proline	Pro	P
Serine	Ser	S
Threonine	Thr	T
Tryptophan	Trp	W

Table 8.1 (*cont.*)

Amino acid	Three-letter code	One-letter code
Tyrosine	Tyr	Y
Valine	Val	V

Thus at low pH values an amino acid exists as a cation and at high pH values as an anion. At a particular intermediate pH the amino acid carries no net charge, although it is still ionised, and is called a zwitterion. It has been shown that, in the crystalline state and in solution in water, amino acids exist predominantly as this zwitterionic form. This confers upon them physical properties characteristic of ionic compounds, i.e. high melting point and boiling point, water solubility and low solubility in organic solvents such as ether and chloroform. The pH at which the zwitterion predominates in aqueous solution is referred to as the isoionic point, because it is the pH at which the number of negative charges on the molecule produced by ionisation of the carboxyl group is equal to the number of positive charges acquired by proton acceptance by the amino group. In the case of amino acids this is equal to the isoelectric point (pI), since the molecule carries no net charge and is therefore electrophoretically immobile. The numerical value of this pH for a given amino acid is related to its acid strength (pK_a values) by the equation:

$$pI = \frac{pK_{a1} + pK_{a2}}{2} \tag{8.1}$$

where pK_{a1} and pK_{a2} are equal to the negative logarithm of the acid dissociation constants, K_{a1} and K_{a2} (Section 1.3.2).

In the case of glycine, pK_{a1} and pK_{a2} are 2, 3 and 9.6, respectively, so that the isoionic point is 6.0. At pH values below this, the cation and zwitterion will coexist in equilibrium in a ratio determined by the Henderson–Hasselbalch equation (Section 1.3.3), whereas at higher pH values the zwitterion and anion will coexist in equilibrium.

For acidic amino acids such as aspartic acid, the ionisation pattern is different owing to the presence of a second carboxyl group:

```
COOH            COOH            COO⁻            COO⁻
 |               |               |               |
CH₂             CH₂             CH₂             CH₂
 |      pKa₁      |      pKa₂      |      pKa₃      |
CH — NH₃⁺ ⇌ CH — NH₃⁺ ⇌ CH — NH₃⁺ ⇌ CH — NH₂
 |      2.1       |      3.9       |      9.8       |
COOH            COO⁻            COO⁻            COO⁻

Cation          Zwitterion      Anion           Anion
(1 net          pH 3.0          (1 net          (2 net
positive        (isoionic       negative        negative
charge)         point)          charge)         charges)
```

In this case, the zwitterion will predominate in aqueous solution at a pH determined by pK_{a_1} and pK_{a_2}, and the isoelectric point is the mean of pK_{a_1} and pK_{a_2}.

In the case of lysine, which is a basic amino acid, the ionisation pattern is different again and its isoionic point is the mean of pK_{a_2} and pK_{a_3}:

$$
\overset{+}{N}H_3 \qquad \overset{+}{N}H_3 \qquad \overset{+}{N}H_3 \qquad \overset{+}{N}H_2
$$

$\overset{+}{N}H_3$	$\overset{+}{N}H_3$	$\overset{+}{N}H_3$	$\overset{+}{N}H_2$
$\|$	$\|$	$\|$	$\|$
$(CH_2)_4$	$(CH_2)_4$	$(CH_2)_4$	$(CH_2)_4$
$\|$ $\quad pK_{a_1}$	$\|$ $\quad pK_{a_2}$	$\|$ $\quad pK_{a_3}$	$\|$
$CH - \overset{+}{N}H_3 \rightleftharpoons CH - \overset{+}{N}H_3 \rightleftharpoons CH - NH_2 \rightleftharpoons CH - NH_2$			
$\|$ $\quad\quad 2.2$	$\|$ $\quad\quad 9.0$	$\|$ $\quad\quad 10.5$	$\|$
$COOH$	COO^-	COO^-	COO^-
Cation	Cation	Zwitterion	Anion
(2 net	(1 net	pH 3.0	(1 net
positive	positive	(isoionic	negative
charges)	charge)	point)	charge)

As an alternative to possessing a second amino or carboxyl group, an amino acid side chain may contain in the R of the general formula a quite different chemical group that is also capable of ionising at a characteristic pH. Such groups include a phenolic group (tyrosine), guanidino group (arginine), imidazolyl group (histidine) and sulphydryl group (cysteine) (Table 8.2). It is clear that the state of ionisation of the main groups of amino acids (acidic, basic, neutral) will be grossly different at a particular pH. Moreover, even within a given group there will be minor differences due to the precise nature of the R group. These differences are exploited in the electrophoretic and ion-exchange chromatographic separation of mixtures of amino acids such as those present in a protein hydrolysate (Section 8.4.2).

Proteins are formed by the condensation of the α-amino group of one amino acid with the α-carboxyl of the adjacent amino acid (Section 8.2). With the exception of the two terminal amino acids, therefore, the α-amino and carboxyl groups are all involved in peptide bonds and are no longer ionisable in the protein. Amino, carboxyl, imidazolyl, guanidino, phenolic and sulphydryl groups in the side chains are, however, free to ionise and of course there will be many of these. Proteins fold in such a manner that the majority of these ionisable groups are on the outside of the molecule, where they can interact with the surrounding aqueous medium. Some of these groups are located within the structure and may be involved in electrostatic attractions that help to stabilise the three-dimensional structure of the protein molecule. The relative numbers of positive and negative groups in a protein molecule influence aspects of its physical behaviour, such as solubility and electrophoretic mobility.

The isoionic point of a protein and its isoelectric point, unlike that of an amino acid, are generally not identical. This is because, by definition, the isoionic point is the pH at which the protein molecule possesses an equal number of positive and negative groups formed by the association of basic groups with protons and dissociation of acidic groups, respectively. In contrast, the isoelectric point is the pH at which the protein is electrophoretically immobile. In order to determine electrophoretic mobility experimentally, the protein must be dissolved in a buffered medium containing anions and cations, of low relative molecular mass, that are capable of binding to the multi-ionised protein. Hence the observed balance of charges at the isoelectric point could be due in

Table 8.2 **Ionisable groups found in proteins**

Amino acid group	pH-dependent ionisation	Approx. pK_a
N-terminal α-amino	$-NH_3 \rightleftharpoons NH_2 + H^+$	8.0
C-terminal α-carboxyl	$-COOH \rightleftharpoons COO^- + H^+$	3.0
Asp-β-carboxyl	$-CH_2COOH \rightleftharpoons CH_2COO^- + H^+$	3.9
Glu-γ-carboxyl	$-(CH_2)_2COOH \rightleftharpoons (CH_2)_2COO^- + H^+$	4.1
His-imidazolyl		6.0
Cys-sulphydryl	$-CH_2SH \rightleftharpoons -CH_2S^- + H^+$	8.4
Tyr-phenolic		10.1
Lys-ε-amino	$-(CH_2)_4\overset{+}{N}H_3 \rightleftharpoons -(CH_2)_4NH_2 + H^+$	10.3
Arg-guanidino	$-NH-\underset{\underset{+NH_2}{\|\|}}{C}-NH_2 \rightleftharpoons -NH-\underset{\underset{NH}{\|\|}}{C}-NH_2 + H^+$	12.5

part to there being more bound mobile anions (or cations) than bound cations (anions) at this pH. This could mask an imbalance of charges on the actual protein.

In practice, protein molecules are always studied in buffered solutions, so it is the isoelectric point that is important. It is the pH at which, for example, the protein has minimum solubility, since it is the point at which there is the greatest opportunity for attraction between oppositely charged groups of neighbouring molecules and consequent aggregation and easy precipitation.

8.2 PROTEIN STRUCTURE

Proteins are formed by condensing the α-amino group of one amino acid or the imino group of proline with the α-carboxyl group of another, with the concomitant loss of a molecule of water and the formation of a peptide bond.

The progressive condensation of many molecules of amino acids gives rise to an unbranched polypeptide chain. By convention, the N-terminal amino acid is taken as the beginning of the chain and the C-terminal amino acid as the end of the chain (proteins are biosynthesised in this direction). Polypeptide chains contain between 20 and 2 000 amino acid residues and hence have a relative molecular mass ranging between about 2 000 and 2 00 000. Many proteins have a relative molecular mass in the range 20 000 to 1 00 000. The distinction between a large peptide and a small protein is not clear. Generally, chains of amino acids containing fewer than 50 residues are referred to as peptides, and those with more than 50 are referred to as proteins. Most proteins contain many hundreds of amino acids (ribonuclease is an extremely small protein with only 103 amino acid residues) and many biologically active peptides contain 20 or fewer amino acids, for example oxytocin (9 amino acid residues), vasopressin (9), enkephalins (5), gastrin (17), somatostatin (14) and lutenising hormone (10).

The primary structure of a protein defines the sequence of the amino acid residues and is dictated by the base sequence of the corresponding gene(s). Indirectly, the primary structure also defines the amino acid composition (which of the possible 20 amino acids are actually present) and content (the relative proportions of the amino acids present).

The peptide bonds linking the individual amino acid residues in a protein are both rigid and planar, with no opportunity for rotation about the carbon–nitrogen bond, as it has considerable double bond character due to the delocalisation of the lone pair of electrons on the nitrogen atom; this, coupled with the tetrahedral geometry around each α-carbon atom, profoundly influences the three-dimensional arrangement which the polypeptide chain adopts.

Secondary structure defines the localised folding of a polypeptide chain due to hydrogen bonding. It includes structures such as the α-helix and β-pleated sheet. Certain of the 20 amino acids found in proteins, including proline, isoleucine, tryptophan and asparagine, disrupt α-helical structures. Some proteins have up to 70% secondary structure but others have none.

Tertiary structure defines the overall folding of a polypeptide chain. It is stabilised by electrostatic attractions between oppositely charged ionic groups ($-\overset{+}{N}H_3$, COO^-), by weak van der Waals forces, by hydrogen bonding, hydrophobic interactions and, in some proteins, by disulphide (-S – S-) bridges formed by the oxidation of spatially adjacent sulphydryl groups (-SH) of cysteine residues (Fig. 8.1). The three-dimensional folding of polypeptide chains is such that the interior consists predominantly of non-polar, hydrophobic amino acid residues such as valine, leucine and phenylalanine. The polar, ionised, hydrophilic residues are found on the outside of the molecule, where they are compatible with the aqueous environment. However, some proteins also have hydrophobic residues on their outside and the presence of these residues is important in the processes of ammonium sulphate fractionation and hydrophobic interaction chromatography (Section 8.3.4).

Quaternary structure is restricted to oligomeric proteins, which consist of the association of two or more polypeptide chains held together by electrostatic attractions, hydrogen bonding, van der Waals forces and occasionally disulphide bridges. Thus disulphide bridges may exist within a given polypeptide chain (intra-chain) or

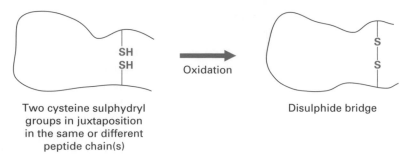

Two cysteine sulphydryl
groups in juxtaposition
in the same or different
peptide chain(s)

Oxidation

Disulphide bridge

Fig. 8.1 The formation of a disulphide bridge.

linking different chains (inter-chain). An individual polypeptide chain in an oligomeric protein is referred to as a subunit. The subunits in a protein may be identical or different: for example, haemoglobin consists of two α- and two β-chains, and lactate dehydrogenase of four (virtually) identical chains.

Traditionally, proteins are classified into two groups – globular and fibrous. The former are approximately spherical in shape, are generally water soluble and may contain a mixture of α-helix, β-pleated sheet and random structures. Globular proteins include enzymes, transport proteins and immunoglobulins. Fibrous proteins are structural proteins, generally insoluble in water, consisting of long cable-like structures built entirely of either helical or sheet arrangements. Examples include hair keratin, silk fibroin and collagen. The native state of a protein is its biologically active form.

The process of protein denaturation results in the loss of biological activity, decreased aqueous solubility and increased susceptibility to proteolytic degradation. It can be brought about by heat and by treatment with reagents such as acids and alkalis, detergents, organic solvents and heavy-metal cations such as mercury and lead. It is associated with the loss of organised (tertiary) three-dimensional structure and exposure to the aqueous environment of numerous hydrophobic groups previously located within the folded structure.

In enzymes, the specific three-dimensional folding of the polypeptide chain(s) results in the juxtaposition of certain amino acid residues that constitute the active site or catalytic site. Oligomeric enzymes may possess several such sites. Many enzymes also possess one or more regulatory site(s). X-ray crystallography studies have revealed that the active site is often located in a cleft that is lined with hydrophobic amino acid residues but which contains some polar residues. The binding of the substrate at the catalytic site and the subsequent conversion of substrate to product involves different amino acid residues.

Some oligomeric enzymes exist in multiple forms called isoenzymes or isozymes (Section 15.1.2). Their existence relies on the presence of two genes that give similar but not identical subunits. One of the best-known examples of isoenzymes is lactate dehydrogenase, which reversibly interconverts pyruvate and lactate. It is a tetramer and exists in five forms (LDH1 to 5) corresponding to the five permutations of arranging the two types of subunits (H and M), which differ only in a single amino acid substitution, into a tetramer:

H$_4$	LDH1
H$_3$M	LDH2
H$_2$M$_2$	LDH3
HM$_3$	LDH4
M$_4$	LDH5

Each isoenzyme promotes the same reaction but has different kinetic constants (K_m, V_{max}), thermal stability and electrophoretic mobility. The tissue distribution of isoenzymes within an organism is frequently different, for example, in humans LDH1 is the dominant isoenzyme in heart muscle but LDH5 is the most abundant form in liver and muscle. These differences are exploited in diagnostic enzymology to identify specific organ damage, for example following myocardial infarction, and thereby aiding clinical diagnosis and prognosis.

8.2.1 Post-translational modifications

Proteins are synthesised at the ribosome and as the growing polypeptide chain emerges from the ribosome it folds up into its native three-dimensional structure. However, this is often not the final active form of the protein. Many proteins undergo modifications once they leave the ribosome, where one or more amino acid side chains are modified by the addition of a further chemical group; this is referred to as post-translational modification. Such changes include extensive modifications of the protein structure, for example the addition of chains of carbohydrates to form glycoproteins (see Section 8.4.4), where in some cases the final protein consists of as much as over 40% carbohydrate. Less dramatic, but equally important modifications include the addition of a hydroxyl group to proline to produce hydroxyproline (found in the structure of collagen), or the phosphorylation of one or more amino acids (tyrosine, serine and threonine residues are all capable of being phosphorylated). Many cases are known, for example, where the addition of a single phosphate group (by enzymes known as kinases) can activate a protein molecule, and the subsequent removal of the phosphate group (by a phosphatase) can inactivate the molecule; protein phosphorylation reactions are a central part of intracellular signalling. Another example can be found in the post-translational modification of proline residues in the transcription factor HIF (the α subunit of the hypoxia-inducible factor), which is a key oxygen-sensing mechanism in cells. Many proteins therefore are not in their final active, biological form until post-translational modifications have taken place. Over 200 different post-translational modifications have been reported for proteins from microbial, plant and animal sources. Mass spectrometry is used to determine such modifications (see Section 9.5.5).

8.3 PROTEIN PURIFICATION

8.3.1 Introduction

At first sight, the purification of *one* protein from a cell or tissue homogenate that will typically contain 10 000–20 000 different proteins, seems a daunting task. However, in practice, on average, only four different fractionation steps are needed to purify a given protein. Indeed, in exceptional circumstances proteins have been purified in a single chromatographic step. Since the reason for purifying a protein is normally to provide material for structural or functional studies, the final degree of purity required depends on the purposes for which the protein will be used, i.e. you may not need a protein sample that is 100% pure for your studies. Indeed, to define what is meant by a 'a pure protein' is not easy. Theoretically, a protein is pure when a sample contains only a single protein species, although in practice it is more or less impossible to achieve 100% purity. Fortunately, many studies on proteins can be carried out on samples that contain as much as 5–10% or more contamination with other proteins. This is an important point, since each purification step necessarily involves loss of some of the protein you are trying to purify. An extra (and unnecessary) purification step that increases the purity of your sample from, say, 90% to 98% may mean that you now have a more pure protein, but insufficient protein for your studies. Better to have studied the sample that was 90% pure and have enough to work on!

For example, a 90% pure protein is sufficient for amino acid sequence determination studies as long as the sequence is analysed quantitatively to ensure that the deduced sequence does not arise from a contaminant protein. Similarly, immunisation of a rodent to provide spleen cells for monoclonal antibody production (Section 7.2.2) can be carried out with a sample that is considerably less than 50% pure. As long as your protein of interest raises an immune response it matters not at all that antibodies are also produced against the contaminating proteins. For kinetic studies on an enzyme, a relatively impure sample can be used provided it does not contain any competing activities. On the other hand, if you are raising a monospecific polyclonal antibody in an animal (see Section 7.2.1), it is necessary to have a highly purified protein as antigen, otherwise immunogenic contaminating proteins will give rise to additional antibodies. Equally, proteins that are to have a therapeutic use must be extremely pure to satisfy regulatory (safety) requirements. Clearly, therefore, the degree of purity required depends on the purpose for which the protein is needed.

8.3.2 The determination of protein concentration

The need to determine protein concentration in solution is a routine requirement during protein purification. The only truly accurate method for determining protein concentration is to acid hydrolyse a portion of the sample and then carry out amino acid analysis on the hydrolysate (see Section 8.4.2). However, this is relatively time-consuming, particularly if multiple samples are to be analysed. Fortunately, in practice, one rarely needs decimal place accuracy and other, quicker methods that give a reasonably accurate assessment of protein concentrations of a solution are acceptable. Most of these (see below) are colorimetric methods, where a portion of the protein solution is reacted with a reagent that produces a coloured product. The amount of this coloured product is then measured spectrophotometrically and the amount of colour related to the amount of protein present by appropriate calibration. However, none of these methods is absolute, since, as will be seen below, the development of colour is often at least partly dependent on the amino acid composition of the protein(s). The presence of prosthetic groups (e.g. carbohydrate) also influences colorimetric assays. Many workers prepare a standard calibration curve using bovine serum albumin (BSA), chosen because of its low cost, high purity and ready availability. However, it should be understood that, since the amino acid composition of BSA will differ from the composition of the sample being tested, any concentration values deduced from the calibration graph can only be approximate.

Ultraviolet absorption

The aromatic amino acid residues tyrosine and tryptophan in a protein exhibit an absorption maximum at a wavelength of 280 nm. Since the proportions of these aromatic amino acids in proteins vary, so too do extinction coefficients for individual proteins. However, for most proteins the extinction coefficient lies in the range 0.4–1.5; so for a complex mixture of proteins it is a fair approximation to say that a solution with an absorbance at 280 nm (A_{280}) of 1.0, using a 1 cm pathlength, has a protein concentration of approximately $1 \, \text{mg cm}^{-3}$. The method is relatively sensitive, being able to measure protein concentrations as low as $10 \, \mu\text{g cm}^{-3}$, and, unlike colorimetric methods, is non-destructive, i.e. having made the measurement, the sample in the cuvette can be recovered and used further. This is particularly useful when one is working with small amounts of protein and cannot afford to waste any. However, the method is subject to interference by the presence of other compounds that absorb at 280 nm. Nucleic acids fall into this category having an absorbance as much as 10 times that of protein at this wavelength. Hence the presence of only a small percentage of nucleic acid can greatly influence the absorbance at this wavelength. However, if the absorbances (A) at 280 and 260 nm wavelengths are measured it is possible to apply a correction factor:

$$\text{Protein} \, (\text{mg cm}^{-3}) = 1.55 \, A_{280} - 0.76 A_{260}$$

The great advantage of this protein assay is that it is non-destructive and can be measured continuously, for example in chromatographic column effluents.

Even greater sensitivity can be obtained by measuring the absorbance of ultraviolet light by peptide bonds. The peptide bond absorbs strongly in the far ultraviolet, with a maximum at about 190 nm. However, because of the difficulties caused by the absorption by oxygen and the low output of conventional spectro-photometers at this wavelength, measurements are usually made at 205 or 210 nm. Most proteins have an extinction coefficient for a $1 \, \mu\text{g cm}^{-3}$ solution of about 30 at 205 nm and about 20 at 210 nm. Clearly therefore measuring at these wavelengths is 20 to 30 times more sensitive than measuring at 280 nm, and protein concentration can be measured to less than $1 \, \mu\text{g cm}^{-3}$. However, one disadvantage of working at these lower wavelengths is that a number of buffers and other buffer components commonly used in protein studies also absorb strongly at this wavelength, so it is not always practical to work at this lower wavelength.

Nowadays all purpose-built column chromatography systems (e.g. fast protein liquid chromatography and high-performance liquid chromatography (HPLC)) have in-line variable wavelength ultraviolet light detectors that monitor protein elution from columns.

Lowry (Folin–Ciocalteau) method

In the past this has been the most commonly used method for determining protein concentration, although it is tending to be replaced by the more sensitive methods described below. The Lowry method is reasonably sensitive, detecting down to $10\,\mu g$ cm^{-3} of protein, and the sensitivity is moderately constant from one protein to another. When the Folin reagent (a mixture of sodium tungstate, molybdate and phosphate), together with a copper sulphate solution, is mixed with a protein solution, a blue-purple colour is produced which can be quantified by its absorbance at 660 nm. As with most colorimetric assays, care must be taken that other compounds that interfere with the assay are not present. For the Lowry method this includes Tris, zwitterionic buffers such as Pipes and Hepes, and EDTA. The method is based on both the Biuret reaction, where the peptide bonds of proteins react with Cu^{2+} under alkaline conditions producing Cu^{+}, which reacts with the Folin reagent, and the Folin–Ciocalteau reaction, which is poorly understood but essentially involves the reduction of phosphomolybdotungstate to hetero-polymolybdenum blue by the copper-catalysed oxidation of aromatic amino acids. The resultant strong blue colour is therefore partly dependent on the tyrosine and tryptophan content of the protein sample.

The bicinchoninic acid method

This method is similar to the Lowry method in that it also depends on the conversion of Cu^{2+} to Cu^{+} under alkaline conditions. The Cu^{+} is then detected by reaction with bicinchoninic acid (BCA) to give an intense purple colour with an absorbance maximum at 562 nm. The method is more sensitive than the Lowry method, being able to detect down to $0.5\,\mu g$ protein cm^{-3}, but perhaps more importantly it is generally more tolerant of the presence of compounds that interfere with the Lowry assay, hence the increasing popularity of the method.

The Bradford method

This method relies on the binding of the dye Coomassie Brilliant Blue to protein. At low pH the free dye has absorption maxima at 470 and 650 nm, but when bound to protein has an absorption maximum at 595 nm. The practical advantages of the method are that the reagent is simple to prepare and that the colour develops rapidly and is stable. Although it is sensitive down to $20\,\mu g$ protein cm^{-3}, it is only a relative method, as the amount of dye binding appears to vary with the content of the basic amino acids arginine and lysine in the protein. This makes the choice of a standard difficult. In addition, many proteins will not dissolve properly in the acidic reaction medium.

Kjeldahl analysis

This is a general chemical method for determining the nitrogen content of any compound. It is not normally used for the analysis of purified proteins or for monitoring column fractions but is frequently used for analysing complex solid samples and microbiological samples for protein content. The sample is digested by boiling

Example 1 **PROTEIN ASSAY**

Question A series of dilutions of bovine serum albumin (BSA) was prepared and 0.1 cm^3 of
each solution subjected to a Bradford assay. The increase in absorbance at 595 nm
relative to an appropriate blank was determined in each case, and the results are
shown in the table.

Concentration of BSA (mg cm^{-3})	A_{595}
1.5	1.40
1.0	0.97
0.8	0.79
0.6	0.59
0.4	0.37
0.2	0.17

A sample (0.1 cm^3) of a protein extract from *E. coli* gave an A_{595} of 0.84 in the same
assay. What was the concentration of protein in the *E. coli* extract?

Answer If a graph of BSA concentration against A_{595} is plotted it is seen to be linear.
From the graph, at an A_{595} of 0.84 it can be seen that the protein concentration of the
E. coli extracted is 0.85 mg cm^{-3}.

with concentrated sulphuric acid in the presence of sodium sulphate (to raise the
boiling point) and a copper and/or selenium catalyst. The digestion converts all the
organic nitrogen to ammonia, which is trapped as ammonium sulphate. Completion of
the digestion stage is generally recognised by the formation of a clear solution. The
ammonia is released by the addition of excess sodium hydroxide and removed by
steam distillation in a Markham still. It is collected in boric acid and titrated with
standard hydrochloric acid using methyl red–methylene blue as indicator. It is possible to carry out the analysis automatically in an autokjeldahl apparatus. Alternatively, a selective ammonium ion electrode may be used to directly determine the
content of ammonium ion in the digest. Although Kjeldahl analysis is a precise and
reproducible method for the determination of nitrogen, the determination of the
protein content of the original sample is complicated by the variation of the nitrogen
content of individual proteins and by the presence of nitrogen in contaminants such
as DNA. In practice, the nitrogen content of proteins is generally assumed to be 16%
by weight.

8.3.3 Cell disruption and production of initial crude extract

The initial step of any purification procedure must, of course, be to disrupt the starting
tissue to release proteins from within the cell. The means of disrupting the tissue will
depend on the cell type (see Cell disruption, below), but thought must first be given to
the composition of the buffer used to extract the proteins.

Extraction buffer

Normally extraction buffers are at an ionic strength (0.1–0.2 M) and pH (7.0–8.0) that is considered to be compatible with that found inside the cell. Tris or phosphate buffers are most commonly used. However, in addition a range of other reagents may be included in the buffer for specific purposes. These include:

- *An anti-oxidant*: Within the cell the protein is in a highly reducing environment, but when released into the buffer it is exposed to a more oxidising environment. Since most proteins contain a number of free sulphydryl groups (from the amino acid cysteine) these can undergo oxidation to give inter- and intramolecular disulphide bridges. To prevent this, reducing agents such as dithiothreitol, β-mercaptoethanol, cysteine or reduced glutathione are often included in the buffer.

- *Enzyme inhibitors*: Once the cell is disrupted the organisational integrity of the cell is lost, and proteolytic enzymes that were carefully packaged and controlled within the intact cells are released, for example from lysosomes. Such enzymes will of course start to degrade proteins in the extract, including the protein of interest. To slow down unwanted proteolysis, all extraction and purification steps are carried out at 4 °C, and in addition a range of protease inhibitors is included in the buffer. Each inhibitor is specific for a particular type of protease, for example serine proteases, thiol proteases, aspartic proteases and metalloproteases. Common examples of inhibitors include: di-isopropylphosphofluoridate (DFP), phenylmethyl sulphonylfluoride (PMSF) and tosylphenylalanyl-chloromethylketone (TPCK) (all serine protease inhibitors); iodoacetate and cystatin (thiol protease inhibitors); pepstatin (aspartic protease inhibitor); EDTA and 1,10-phenanthroline (metalloprotease inhibitors); and amastatin and bestatin (exopeptidase inhibitors).

- *Enzyme substrate and cofactors*: Low levels of substrate are often included in extraction buffers when an enzyme is purified, since binding of substrate to the enzyme active site can stabilise the enzyme during purification processes. Where relevant, cofactors that otherwise might be lost during purification are also included to maintain enzyme activity so that activity can be detected when column fractions, etc. are screened.

- *EDTA*: This can be present to remove divalent metal ions that can react with thiol groups in proteins giving *mercaptids*.

$$R - SH + Me^{2+} \rightarrow R - S - Me^+ + H^+$$

- *Polyvinylpyrrolidone (PVP)*: This is often added to extraction buffers for plant tissue. Plant tissues contain considerable amounts of phenolic compounds (both monomeric, such as *p*-hydroxybenzoic acid, and polymeric, such as tannins) that can bind to enzymes and other proteins by non-covalent forces, including hydrophobic, ionic and hydrogen bonds, causing protein precipitation. These phenolic compounds are also easily oxidised, predominantly by endogenous phenol oxidases, to form quinones, which are highly reactive and can combine with reactive groups in proteins causing cross-linking, and further aggregation and precipitation. Insoluble PVP (which mimics the polypeptide backbone) is therefore added to adsorb the phenolic compounds which can then be removed by centrifugation. Thiol compounds (reducing agents) are also added to minimise the activity of phenol oxidases, and thus prevent the formation of quinones.

- *Sodium azide*: For buffers that are going to be stored for long periods of time, antibacterial and/or antifungal agents are sometimes added at low concentrations. Sodium azide is frequently used as a bacteriostatic agent.

Membrane proteins

Membrane-bound proteins (normally glycoproteins) require special conditions for extraction as they are not released by simple cell disruption procedures alone. Two classes of membrane proteins are identified. Extrinsic (or peripheral) membrane proteins are bound only to the surface of the cell, normally via electrostatic and hydrogen bonds. These proteins are predominantly hydrophilic in nature and are relatively easily extracted either by raising the ionic concentration of the extraction buffer (e.g. to 1 M NaCl) or by changes of pH (e.g. to pH 3–5 or pH 9–12). Once extracted, they can be purified by conventional chromatographic procedures. Intrinsic membrane proteins are those that are embedded in the membrane (integrated membrane proteins). These invariably have significant regions of hydrophobic amino acids (those regions of the protein that are embedded in the membrane, and associated with lipids) and have low solubility in aqueous buffer systems. Hence, once extracted into an aqueous polar environment, appropriate conditions must be used to retain their solubility. Intrinsic proteins are usually extracted with buffer containing detergents. The choice of detergent is mainly one of trial and error but can include ionic detergents such as sodium dodecyl sulphate (SDS), sodium deoxycholate, cetyl trimethylammonium bromide (CTAB) and CHAPS, and non-ionic detergents such as Triton X-100 and Nonidet P-40.

Once extracted, intrinsic membrane proteins can be purified using conventional chromatographic techniques such as gel filtration, ion-exchange chromatography or affinity chromatography (using lectins). However, in each case it is necessary to include detergent in all buffers to maintain protein solubility. The level of detergent used is normally 10- to 100-fold less than that used to extract the protein, in order to minimise any interference of the detergent with the chromatographic process.

Cell disruption

Unless one is isolating proteins from extracellular fluids such as blood, protein purification procedures necessarily start with the disruption of cells or tissue to release the protein content of the cells into an appropriate buffer. This initial extract is therefore the starting point for protein purification. Clearly one chooses, where possible, a starting material that has a high level of the protein of interest. Depending on the protein being isolated one might therefore start with a microbial culture, plant tissue, or mammalian tissue. The last of these has generally been the tissue of choice where possible, owing to the relatively large amounts of starting material available. However, the ability to clone and overexpress genes for proteins from any source, in both bacteria and yeast, means that nowadays more and more protein purification protocols are starting with a microbial lysate. The different methods available for

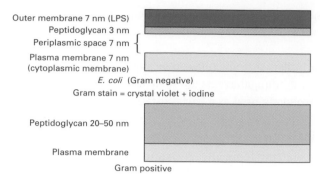

Outer membrane 7 nm (LPS)
Peptidoglycan 3 nm
Periplasmic space 7 nm
Plasma membrane 7 nm
(cytoplasmic membrane)
E. coli (Gram negative)
Gram stain = crystal violet + iodine

Peptidoglycan 20–50 nm

Plasma membrane

Gram positive

Fig. 8.2 The structure of the cell wall of Gram-positive and of Gram-negative bacteria. LPS, lipopolysaccharide.

disrupting cells are described below. Which method one uses depends on the nature of the cell wall/membrane being disrupted.

Mammalian cells

Mammalian cells are of the order of 10 μm in diameter and enclosed by a plasma membrane, weakly supported by a cytoskeleton. These cells therefore lack any great rigidity and are easy to disrupt by shear forces.

Plant cells

Plant cells are of the order of 100 μm in diameter and have a fairly rigid cell wall, comprising carbohydrate complexes and lignin or wax that surround the plasma membrane. Although the plasma membrane is protected by this outer layer, the large size of the cell still makes it susceptible to shear forces.

Bacteria

Bacteria have cell diameters of the order of 1 to 4 μm and generally have extremely rigid cell walls. Bacteria can be classified as either Gram positive or Gram negative depending on whether or not they are stained by the Gram stain (crystal violet and iodine). In Gram-positive bacteria (Fig. 8.2) the plasma membrane is surrounded by a thick shell of peptidoglycan (20–50 nm), which stains with the Gram stain. In Gram-negative bacteria (e.g. *Escherichia coli*) the plasma membrane is surrounded by a thin (2–3 nm) layer of peptidoglycan but this is compensated for by having a second outer membrane of lipopolysaccharide. The negatively charged lipopolysaccharide polymers interact laterally, being linked by divalent cations such as Mg^{2+}. A number of Gram-negative bacteria secrete proteins into the periplasmic space.

Fungi and yeast

Filamentous fungi and yeasts have a rigid cell wall that is composed mainly of polysaccharide (80–90%). In lower fungi and yeast the polysaccharides are mannan and glucan. In filamentous fungi it is chitin cross-linked with glucans. Yeasts also have a small percentage of glycoprotein in the cell wall, and there is a periplasmic space between the cell wall and cell membrane. If the cell wall is removed the cell content, surrounded by a membrane, is referred to as a spheroplast.

Cell disruption methods

Blenders

These are commercially available, although a typical domestic kitchen blender will suffice. This method is ideal for disrupting mammalian or plant tissue by shear force. Tissue is cut into small pieces and blended, in the presence of buffer, for about 1 min to disrupt the tissue, and then centrifuged to remove debris. This method is inappropriate for bacteria and yeast, but a blender can be used for these microorganisms if small glass beads are introduced to produce a bead mill. Cells are trapped between colliding beads and physically disrupted by shear forces.

Grinding with abrasives

Grinding in a pestle and mortar, in the presence of sand or alumina and a small amount of buffer, is a useful method for disrupting bacterial or plant cells; cell walls are physically ripped off by the abrasive. However, the method is appropriate for handling only relatively small samples. The Dynomill is a large-scale mechanical version of this approach. The Dynomill comprises a chamber containing glass beads and a number of rotating impeller discs. Cells are ruptured when caught between colliding beads. A $600 \, \text{cm}^3$ laboratory scale model can process 5 kg of bacteria per hour.

Presses

The use of a press such as a French Press, or the Manton–Gaulin Press, which is a larger-scale version, is an excellent means for disrupting microbial cells. A cell suspension ($\sim 50 \, \text{cm}^3$) is forced by a piston-type pump, under high pressure ($10\,000 \, \text{PSI} = \text{lbfin.}^{-2} \approx 1450 \, \text{kPa}$) through a small orifice. Breakage occurs due to shear forces as the cells are forced through the small orifice, and also by the rapid drop in pressure as the cells emerge from the orifice, which allows the previously compressed cells to expand rapidly and effectively burst. Multiple passes are usually needed to lyse all the cells, but under carefully controlled conditions it can be possible to selectively release proteins from the periplasmic space. The X-Press and Hughes Press are variations on this method; the cells are forced through the orifice as a frozen paste, often mixed with an abrasive. Both the ice crystal and abrasive aid in disrupting the cell walls.

Enzymatic methods

The enzyme lysozyme, isolated from hen egg whites, cleaves peptidoglycan. The peptidoglycan cell wall can therefore be removed from Gram-positive bacteria (see Fig. 8.2) by treatment with lysozyme, and if carried out in a suitable buffer, once the cell wall has been digested the cell membrane will rupture owing to the osmotic effect of the suspending buffer.

Gram-negative bacteria can similarly be disrupted by lysozyme but treatment with EDTA (to remove Ca^{2+}, thus destabilising the outer lipopolysaccharide layer) and the inclusion of a non-ionic detergent to solubilise the cell membrane are also needed. This effectively permeabilises the outer membrane, allowing access of the lysozyme to the peptidoglycan layer. If carried out in an isotonic medium so that the cell membrane is not ruptured, it is possible to selectively release proteins from the periplasmic space.

Yeast can be similarly disrupted using enzymes to degrade the cell wall and either osmotic shock or mild physical force to disrupt the cell membrane. Enzyme digestion alone allows the selective release of proteins from the periplasmic space. The two most commonly used enzyme preparations for yeast are zymolyase or lyticase, both of which have β-1, 3-glucanase activity as their major activity, together with a proteolytic activity specific for the yeast cell wall. Chitinase is commonly used to disrupt filamentous fungi. Enzymic methods tend to be used for laboratory-scale work, since for large-scale work their use is limited by cost.

Sonication

This method is ideal for a suspension of cultured cells or microbial cells. A sonicator probe is lowered into the suspension of cells and high frequency sound waves (<20 kHz) generated for 30–60 s. These sound waves cause disruption of cells by shear force and cavitation. Cavitation refers to areas where there is alternate compression and rarefaction, which rapidly interchange. The gas bubbles in the buffer are initially under pressure but, as they decompress, shock waves are released and disrupt the cells. This method is suitable for relatively small volumes (50–100 cm^3). Since considerable heat is generated by this method, samples must be kept on ice during treatment.

8.3.4 Fractionation methods

Monitoring protein purification

As will be seen below, the purification of a protein invariably involves the application of one or more column chromatographic steps, each of which generates a relatively large number of test tubes (fractions) containing buffer and protein eluted from the column. It is necessary to determine how much protein is present in each tube so that an elution profile (a plot of protein concentration versus tube number) can be produced. Appropriate methods for detecting and quantifying protein in solution are described in Section 8.3.2. A method is also required for determining which tubes contain the protein of interest so that their contents can be pooled and the pooled sample progressed to the next purification step. If one is purifying an enzyme, this is relatively easy as each tube simply has to be assayed for the presence of enzyme activity.

For proteins that have no easily measured biological activity, other approaches have to be used. If an antibody to the protein of interest is available then samples from each tube can be dried onto nitrocellulose and the antibody used to detect the protein-containing fractions using the dot blot method (Section 5.9.2). Alternatively, an immunoassay such as ELISA or radioimmunoassay (Section 7.3.1) can be used to detect the protein. If an antibody is not available, then portions from each fraction can be run on a sodium dodecyl sulphate–polyacrylamide gel and the protein-containing fraction identified from the appearance of the protein band of interest on the gel (Section 10.3.1).

An alternative approach that can be used for cloned genes that are expressed in cells is to express the protein as a fusion protein, i.e. one that is linked via a short peptide sequence to a second protein. This can have advantages for protein purification (see Section 8.3.5). However, it can also prove extremely useful for monitoring the purification of a protein that has no easily measurable activity. If the second protein is an enzyme that can be easily assayed (e.g. using a simple colorimetric assay), such as β-galactosidase, then the presence of the protein of interest can be detected by the presence of the linked β-galactosidase activity.

A successful fractionation step is recognised by an increase in the specific activity of the sample, where the specific activity of the enzyme relates its total activity to the total amount of protein present in the preparation:

$$\text{specific activity} = \frac{\text{total units of enzyme in fraction}}{\text{total amount of protein in fraction}}$$

The measurement of units of an enzyme relies on an appreciation of certain basic kinetic concepts and upon the availability of a suitable analytical procedure. These are discussed in Section 15.2.2.

The amount of enzyme present in a particular fraction is expressed conventionally not in terms of units of mass or moles but in terms of units based upon the rate of the reaction that the enzyme promotes. The international unit (IU) of an enzyme is defined as the amount of enzyme that will convert 1 μmole of substrate to product in 1 minute under defined conditions (generally 25 or 30 °C at the optimum pH). The SI unit of enzyme activity is defined as the amount of enzyme that will convert 1 mole of substrate to product in 1 second. It has units of katal (kat) such that 1 kat $= 6 \times 10^7$ IU and 1 IU $= 1.7 \times 10^{-8}$ kat. For some enzymes, especially those where the substrate is a macromolecule of unknown relative molecular mass (e.g. amylase, pepsin, RNase, DNase), it is not possible to define either of these units. In such cases arbitrary units are used generally that are based upon some observable change in a chemical or physical property of the substrate.

For a purification step to be successful, therefore, the specific activity of the protein must be greater after the purification step than it was before. This increase is best represented as the fold purification:

$$\text{fold purification} = \frac{\text{specific activity of fraction}}{\text{original specific activity}}$$

A significant increase in specific activity is clearly necessary for a successful purification step. However, another important factor is the yield of the step. It is no use having an increased specific activity if you lose 95% of the protein you are trying to purify. Yield is defined as follows:

$$\text{yield} = \frac{\text{units of enzyme in fraction}}{\text{units of enzyme in original preparation}}$$

A yield of 70% or more in any purification step would normally be considered as acceptable. Table 8.3 shows how yield and specific activity vary during a purification schedule.

Preliminary purification steps

The initial extract, produced by the disruption of cells and tissue, and referred to at this stage as a homogenate, will invariably contain insoluble matter. For example, for mammalian tissue there will be incompletely homogenised connective and/or vascular tissue, and small fragments of non-homogenised tissue. This is most easily removed by filtering through a double layer of cheesecloth or by low speed (5 000 g)

Example 2 ENZYME FRACTIONATION

Question A tissue homogenate was prepared from pig heart tissue as the first step in the preparation of the enzyme aspartate aminotransferase (AAT). Cell debris was removed by filtration and nucleic acids removed by treatment with polyethyleneimine, leaving a total extract (solution A) of 2 dm^3. A sample of this extract (50 mm^3) was added to 3 cm^3 of buffer in a 1 cm pathlength cuvette and the absorbance at 280 nm shown to be 1.7.

(i) Determine the approximate protein concentration in the extract, and hence the total protein content of the extract.

(ii) One unit of AAT enzyme activity is defined as the amount of enzyme in 3 cm^3 of substrate solution that causes an absorbance change at 260 nm of 0.1 min^{-1}. To determine enzyme activity, 100 mm^3 of extract was added to 3 cm^3 of substrate solution and an absorbance change of 0.08 min^{-1} was recorded. Determine the number of units of AAT actively present per cm^3 of extract A, and hence the total number of enzyme units in the extract.

(iii) The initial extract (solution A) was then subjected to ammonium sulphate fractionation. The fraction precipitating betweeen 50% and 70% saturation was collected and redissolved in 120 cm^3 of buffer (solution B). Solution B (5 mm^3 (0.005 cm^3)) was added to 3 cm^3 of buffer and the absorbance at 280 nm determined to be 0.89 using a 1 cm pathlength cuvette. Determine the protein concentration, and hence total protein content, of solution B.

(iv) Solution B 20 mm^3 was used to assay for AAT activities and an absorbance change of 0.21 per min at 260 nm was recorded. Determine the number of AAT units cm^{-3} in solution B and hence the total number of enzyme units in solution B.

(v) From your answers to (i) to (iv), determine the specific activity of AAT in both solutions A and B.

(vi) From your answers to question (v), determine the fold purification achieved by the ammonium sulphate fractionation step.

(vii) Finally, determine the yield of AAT following the ammonium sulphate fractionation step.

Answer (i) Assuming the approximation that a 1 mg protein cm^{-3} solution has an absorbance of 1.0 at 280 nm using a 1 cm pathlength cell, then we can deduce that the protein concentration *in the cuvette* is approximately 1.7 mg cm^{-3}. Since 50 μl (0.05 cm^3) of the solution A was added to 3.0 cm^3 then the solution A sample had been diluted by a factor of 3.05/0.05 = 61.

Therefore the protein concentration of solution A is 61 × 1.7 mg cm^{-3} = ~104 mg cm^{-3}. Since there is 2 dm^3 (2000 cm^3) of solution A, the *total* amount of protein in solution A is 2000 × 104 = 208 000 mg or 208 g.

(ii) Since one enzyme unit causes an absorbance change of 0.1 per minute, there was 0.08/0.1 = 0.8 enzyme units in the cuvette. These 0.8 enzyme units came from the 100 mm^3 of solution A that was added to the cuvette.

Example 2 (*cont.*)

Therefore in 100 mm^3 of solution A there is 0.8 enzyme unit.

Therefore in 1 cm^3 of solution A there are 8.0 enzyme units.

Since we have 2000 cm^3 of solution A there is a total of $2000 \times 8.0 = 16\,000$ enzyme units in solution A.

(iii) Using the same approach as in Example 2(i), the protein concentration of solution B is $3.005/0.005 \times 0.89 = 601 \times 0.89 = 535$ mg cm^{-3}.

Therefore the total protein present in solution B $= 120 \times 535 = 64\,200$ mg.

(iv) Using the same approach as in Example 2(ii), there are $0.21/0.1 = 2.1$ units of enzyme activity in the cuvette. These units came from the 20 mm^3 that was added to the cell.

Therefore, 20 mm^3 (0.020 cm^3) of solution B contains 2.1 enzyme units.

Thus, 1 cm^3 of solution B contains $1.0/0.02 \times 2.1 = 105$ units. Therefore, solution B has 105 units cm^{-3}.

Since there are 120 cm^3 of solution B, total units in solution B $= 120 \times 105 = 12\,600$.

(v) For solution A, specific activity $= 16\,000/208\,000 = 0.077$ units mg^{-1}.

For solution B, specific activity $= 12\,600/64\,200 = 0.197$ units mg^{-1}.

(vi) Fold purification $= 0.197/0.077 = 2.6$ (approx.).

(vii) Yield $= (12\,600/16\,000) \times 100\% = 79\%$.

---

centrifugation. Any fat floating on the surface can be removed by coarse filtration through glass wool or cheesecloth. However, the solution will still be cloudy with organelles and membrane fragments that are too small to be conveniently removed by filtration or low speed centrifugation. These may not be much of a problem as they will often be lost in the preliminary stages of protein purification, for example during salt fractionation. However, if necessary they can be removed first by precipitation using materials such as Celite (a diatomaceous earth that provides a large surface area to trap the particles), Cell Debris Remover (CDR, a cellulose-based absorber), or any number of flocculants such as starch, gums, tannins or polyamines, the resultant precipitate being removed by centrifugation or filtration.

It is tempting to assume that the cell extract contains only protein, but of course a range of other molecules is present such as DNA, RNA, carbohydrate and lipid as well as any number of small molecular weight metabolites. Small molecules tend to be removed later on during dialysis steps or steps that involve fractionation based on size (e.g. gel filtration) and therefore are of little concern. However, specific attention has to be paid at this stage to macromolecules such as nucleic acids and polysaccharides. This is particularly true for bacterial extracts, which are particularly viscous owing to the presence of chromosomal DNA. Indeed microbial extracts can be extremely difficult to centrifuge to produce a supernatant extract. Some workers include DNase I in the extraction buffer to reduce viscosity, the small DNA fragments generated being removed at later dialysis/gel filtration steps. Likewise RNA can be removed by treatment with RNase. DNA and RNA can also be removed by precipitation with protamine

Table 8.3 **Example of a protein purification schedule**

Fraction	Volume (cm^3)	Protein concentration (mg U cm^{-3})	Total protein (mg)	Activity[a] (mg U cm^{-3})	Total activity (U)	Specific activity (U mg^{-1})	Purification factor[b]	Overall yield[c] (%)
Homogenate	8 500	40	340 000	1.8	15 300	0.045	1	100
45%–70%(NH$_4$)$_2$SO$_4$	530	194	103 000	23.3	12 350	0.12	2.7	81
CM-cellulose	420	19.5	8 190	25	10 500	1.28	28.4	69
Affinity chromotography	48	2.2	105.6	198	9 500	88.4	1 964	62
DEAE-Sepharose	12	2.3	27.6	633	7 600	275	6 110	50

Notes: [a]The unit of enzyme activity (U) is defined as that amount which produces 1 μmole of product per minute under standard assay conditions.
[b]Defined as: purification factor = (specific activity of fraction/specific activity of homogenate).
[c]Defined as: overall yield = (total activity of fraction/total activity of homogenate).
Reproduced with permission from *Methods in Molecular Biology*, 59, *Protein Purification Protocols*, ed. S. Doonan (1996), Humana Press Inc, Totowa, NJ.

sulphate. Protamine sulphate is a mixture of small, highly basic (i.e. positively charged) proteins, whose natural role is to bind to DNA in the sperm head. (Protamines are usually extracted from fish organs, which are obtained as a waste product at canning factories.) These positively charged proteins bind to negatively charged phosphate groups on nucleic acids, thus masking the charged groups on the nucleic acids and rendering them insoluble. The addition of a solution of protamine sulphate to the extract therefore precipitates most of the DNA and RNA, which can subsequently be removed by centrifugation. An alternative is to use polyethyleneimine, a synthetic long chain cationic (i.e. positively charged) polymer (molecular mass 24 kDa). This also binds to the phosphate groups in nucleic acids, and is very effective, precipitating DNA and RNA almost instantly. For bacterial extracts, carbohydrate capsular gum can also be a problem as this can interfere with protein precipitation methods. This is best removed by totally precipitating the protein with ammonium sulphate (see below) leaving the gum in solution. The protein can then be recovered by centrifugation and redissolved in buffer. However, if lysozyme (plus detergent) is used to lyse the cells (see Section 8.3.3) capsular gum will not be a problem as it is digested by the lysozyme.

The clarified extract is now ready for protein fractionation steps to be carried out. The concentration of the protein in this initial extract is normally quite low, and in fact the major contaminant at this stage is water! The initial purification step is frequently based on solubility methods. These methods have a high capacity, can therefore be easily applied to large volumes of initial extracts and also have the advantage of concentrating the protein sample. Essentially, proteins that differ considerably in their physical characteristics from the protein of interest are removed at this stage, leaving a more concentrated solution of proteins that have more closely similar physical characteristics. The next stages, therefore, involve higher resolution techniques that can separate proteins with similar characteristics. Invariably these high resolution techniques are chromatographic. Which technique to use, and in which order, is more often than not a matter of trial and error. The final research paper that describes in four pages a three-step, four-day protein purification procedure invariably belies the months of hard work that went into developing the final 'simple' purification protocol!

All purification techniques are based on exploiting those properties by which proteins differ from one another. These different properties, and the techniques that exploit these differences, are as follows.

Stability

Denaturation fractionation exploits differences in the heat sensitivity of proteins. The three-dimensional (tertiary) structure of proteins is maintained by a number of forces, mainly hydrophobic interactions, hydrogen bonds and sometimes disulphide bridges. When we say that a protein is denatured we mean that these bonds have by some means been disrupted and that the protein chain has unfolded to give the insoluble, 'denatured' protein. One of the easiest ways to denature proteins in solution is to heat them. However, different proteins will denature at different temperatures, depending on their different thermal stabilities; this, in turn, is a measure of the number of bonds holding the tertiary structure together. If the protein of interest is particularly heat stable, then heating the extract to a temperature at which the protein is stable yet other proteins denature can be a very useful preliminary step. The temperature at which the protein being purified is denatured is first determined by a small-scale experiment. Once this temperature is known, it is possible to remove more thermolabile contaminating proteins by heating the mixture to a temperature 5–10 °C below this critical temperature for a period of 15–30 min. The denatured, unwanted protein is then removed by centrifugation. The presence of the substrate, product or a competitive inhibitor of an enzyme often stabilises it and allows an even higher heat denaturation temperature to be employed. In a similar way, proteins differ in the ease with which they are denatured by extremes of pH (<3 and >10). The sensitivity of the protein under investigation to extreme pH is determined by a small-scale trial. The whole protein extract is then adjusted to a pH not less than 1 pH unit within that at which the test protein is precipitated. More sensitive proteins will precipitate and are removed by centrifugation.

Solubility

Proteins differ in the balance of charged, polar and hydrophobic amino acids that they display on their surfaces. Charged and polar groups on the surface are solvated by water molecules, thus making the protein molecule soluble, whereas hydrophobic residues are masked by water molecules that are necessarily found adjacent to these regions. Since solubility is a consequence of solvation of charged and polar groups on the surfaces of the protein, it follows that, under a particular set of conditions, proteins will differ in their solubilities. In particular, one exploits the fact that proteins precipitate differentially from solution on the addition of species such as neutral salts or organic solvents. It should be stressed here that these methods precipitate native (i.e. active) protein that has become insoluble by aggregation; we have not denatured the protein.

Salt fractionation is frequently carried out using ammonium sulphate. As increasing salt is added to a protein solution, so the salt ions are solvated by water molecules in the solution. As the salt concentration increases, freely available water molecules that can solvate the ions become scarce. At this stage those water molecules that have been forced into contact with hydrophobic groups on the surface of the protein are the next most freely available water molecules (rather than those involved in solvating polar groups on the protein surface, which are bound by electrostatic interactions and are far less easily given up) and these are therefore removed to solvate the salt molecules, thus leaving the hydrophobic patches exposed. As the ammonium sulphate concentration increases, the hydrophobic surfaces on the protein are progressively exposed. Thus revealed, these hydrophobic patches cause proteins to aggregate by hydrophobic interaction, resulting in precipitation. The first proteins to aggregate are therefore those with the most hydrophobic residues on the surface, followed by those with less hydrophobic residues. Clearly the aggregates formed are made of mixtures of more than one protein. Individual identical molecules do not seek out each other, but simply bind to another adjacent molecule with an exposed hydrophobic patch. However, many proteins are precipitated from solution over a narrow range of salt concentrations, making this a suitably simple procedure for enriching the proteins of interest.

Organic solvent fractionation is based on differences in the solubility of proteins in aqueous solutions containing water-miscible organic solvents such as ethanol, acetone and butanol. The addition of organic solvent effectively 'dilutes out' the water present (reduces the dielectric constant) and at the same time water molecules are used up in hydrating the organic solvent molecules. Water of solvation is therefore removed from the charged and polar groups on the surface of proteins, thus exposing their charged groups. Aggregation of proteins therefore occurs by charge (ionic) interactions between molecules. Proteins consequently precipitate in decreasing order of the number of charged groups on their surface as the organic solvent concentration is increased.

Organic polymers can also be used for the fractional precipitation of proteins. This method resembles organic solvent fractionation in its mechanism of action but requires lower concentrations to cause protein precipitation and is less likely to cause protein denaturation. The most commonly used polymer is polyethylene glycol (PEG), with a relative molecular mass in the range 6000–20 000.

The fractionation of a protein mixture using ammonium sulphate is given here as a practical example of fractional precipitation. As explained above, as increasing amounts of ammonium sulphate are dissolved in a protein solution, certain proteins start to aggregate and precipitate out of solution. Increasing the salt strength results in further, different proteins precipitating out. By carrying out a controlled pilot experiment where the percentage of ammonium sulphate is increased stepwise say from 10% to 20% to 30% etc., the resultant precipitate at each step being recovered by centrifugation, redissolved in buffer and analysed for the protein of interest, it is possible to determine a fractionation procedure that will give a significantly purified sample. In the example shown in Table 8.3, the original homogenate was made in 45% ammonium sulphate and the precipitate recovered and discarded. The supernatant was then made in 70% ammonium sulphate, the precipitate collected, redissolved in buffer, and kept, with the supernatant being discarded. This produced a purification factor of 2.7. As can be seen, a significant amount of protein has been removed at this step (237 000 mg of protein) while 81% of the total enzyme present was recovered, i.e. the yield was good. This step has clearly produced an enrichment of the protein of interest from a large volume of extract and at the same time has concentrated the sample.

Isoelectric precipitation fractionation is based upon the observations that proteins have their minimum solubility at their isoelectric point. At this pH there are equal numbers of positive and negative charges on the protein molecule; intermolecular repulsions are therefore minimised and protein molecules can approach each other. This therefore allows opposite charges on different molecules to interact, resulting in the formation of insoluble aggregates. The principle can be exploited either to remove unwanted protein, by adjusting the pH of the protein extract so as to cause the precipitation of these proteins but not that of the test protein, or to remove the test protein, by adjusting the pH of the extract to its pI. In practice, the former alternative is preferable, since some denaturation of the precipitation protein inevitably occurs.

Finally, an unusual solubility phenomenon can be utilised in some cases for protein purification from *E. coli*. Early workers who were overexpressing heterologous proteins in *E. coli* at high levels were alarmed to discover that, although their protein was expressed in high yield (up to 40% of the total cell protein), the protein aggregated to form insoluble particles that became known as inclusion bodies. Initially this was seen as a major impediment to the production of proteins in *E. coli*, the inclusion bodies effectively being a mixture of monomeric and polymeric denatured proteins formed by partial or incorrect folding, probably due to the reducing environment of the *E. coli* cytoplasm. However, it was soon realised that this phenomenon could be used to advantage in protein purification. The inclusion bodies can be separated from a large proportion of the bacterial cytoplasmic protein by centrifugation, giving an effective purification step. The recovered inclusion bodies must then be solubilised and denatured and subsequently allowed to refold slowly to their active, native configuration. This is normally achieved by heating in 6 M guanidinium hydrochloride (to denature the protein) in the presence of a reducing agent (to disrupt any disulphide bridges). The denatured protein is then either diluted in buffer or dialysed against buffer, at which time the protein slowly refolds. Although the refolding method is not always 100% successful, this approach can often produce protein that is 50% or more pure.

Having carried out an initial fractionation step such as that described above, one would then move towards using higher resolution chromatographic methods. Chromatographic techniques for purifying proteins are summarised in Table 8.4, and some of the more commonly used methods are outlined below. The precise practical details of each technique are discussed in Chapter 11.

Charge

Proteins differ from one another in the proportions of the charged amino acids (aspartic and glutamic acids, lysine, arginine and histidine) that they contain. Hence proteins will differ in net charge at a particular pH. This difference is exploited in ion-exchange chromatography (Section 11.6), where the protein of interest is bound onto a solid support material bearing charged groups of the opposite sign (ion-exchange resin). Proteins with the same charge as the resin pass through the column to waste, after which bound proteins, containing the protein of interest, are selectively released from the column by gradually increasing the strength of salt ions in the buffer passing through the column or by gradually changing the pH of the eluting buffer. These ions compete with the protein for binding to the resin, the more weakly charged protein being eluted at the lower salt strength and the more strongly charged protein being eluted at higher salt strengths.

Size

Differences between proteins can be exploited in molecular exclusion (also known as gel filtration) chromatography. The gel filtration medium consists of a range of beads with slighly differing amounts of cross-linking and therefore slightly different pore sizes. The separation process depends on the different abilities of the various proteins to enter some, all or none of the beads, which in turn relates to the size of this protein (Section 11.7). The method has limited resolving power, but can be used to obtain a separation between large and small protein molecules and therefore be useful when the protein of interest is either particularly large or particularly small. This method can also be used to determine the relative molecular mass of a protein (Section 11.7.2) and for concentrating or desalting a protein solution (Section 11.7.2).

Affinity

Certain proteins bind strongly to specific small molecules. One can take advantage of this by developing an affinity chromatography system where the small molecule

Table 8.4 **Summary of chromatographic techniques commonly used in protein purification**

Technique	Property exploited	Capacity	Resolution	Practical points	Further details
Hydrophobic interaction	Hydrophobicity	High	Medium	Can cope with high ionic strength samples, e.g. ammonium sulphate precipitates. Fractions are of varying pH and/or ionic strength. Medium yield. Commonly used in early stages of purification protocol. Unpredictable	Section 11.4.3
Ion exchange	Charge	High	Medium	Sample ionic strength must be low. Fractions are of varying pH and/or ionic strength. Medium yield. Commonly used in early stages of purification protocol	Section 11.6
Affinity	Biological function	Medium (cost limited)	High	Limited by availability of immobilised ligand. Elution may denature protein. Yield medium–low. Commonly used towards end of purification protocol	Section 11.8
Dye affinity	Structure and hydrophobicity		High	Necessary to carry out initial screening of a wide range of dye–ligand supports	Section 11.8.5
Covalent	Thiol groups	Medium–low	High	Specific for thiol-containing proteins. Limited by high cost and long (3 h) regeneration time	Section 11.8.6
Metal chelate	Imidazole, thiol, tryptophan groups	Medium–low	High	Expensive	Section 11.8.4
Exclusion	Molecular size	Medium	Low	Can give information about protein molecular weight. Good for desalting protein samples	Section 11.7

(ligand) is bound to an insoluble support. When a crude mixture of proteins containing the protein of interest is passed through the column, the ligand binds the protein to the matrix whilst all other proteins pass through the column. The bound protein can then be eluted from the column by changing the pH, increasing salt strength or passing through a high concentration of unbound free ligand. For example, the protein concanavalin A (con A) binds strongly to glucose. An affinity column using glucose as the ligand can therefore be used to bind con A to the matrix, and the con A can be recovered by passing a high concentration of glucose through the column. Affinity chromatography is covered in detail in Section 11.8.

Hydrophobicity

Proteins differ in the amount of hydrophobic amino acids that are present on their surface. This difference can be exploited in salt fractionation (see above) but can also be used in a higher resolution method using hydrophobic interaction chromatography (HIC) (Section 11.4.3). A typical column material would be phenyl-Sepharose, where phenyl groups are bonded to the insoluble support Sepharose. The protein mixture is loaded on the column in high salt (to ensure hydrophobic patches are exposed) where hydrophobic interaction will occur between the phenyl groups on the resin and hydrophobic regions on the proteins. Proteins are then eluted by applying a decreasing salt gradient to the column and should emerge from the column in order of increasing hydrophobicity. However, some highly hydrophobic proteins may not even be eluted in the total absence of salt. In this case it is necessary to add a small amount of water-miscible organic solvent such as propanol or ethylene glycol to the column buffer solution. This will compete with the proteins for binding to the hydrophobic matrix and will elute any remaining proteins.

8.3.5   Engineering proteins for purification

With the ability to clone and overexpress genes for proteins using genetic engineering methodology has also come the ability to aid considerably the purification process by manipulation of the gene of interest prior to expression. These manipulations are carried out either to ensure secretion of the proteins from the cell or to aid protein purification.

Ensuring secretion from the cell

For cloned genes that are being expressed in microbial or eukaryotic cells, there are a number of advantages in manipulating the gene to ensure that the protein product is secreted from the cell:

- *To facilitate purification*: Clearly if the protein is secreted into the growth medium, there will be far fewer contaminating proteins present than if the cells had to be ruptured to release the protein, when all the other intracellular proteins would also be present.
- *Prevention of intracellular degradation of the cloned protein*: Many cloned proteins are recognised as 'foreign' by the cell in which they are produced and are therefore degraded by intracellular proteases. Secretion of the protein into the culture medium should minimise this degradation.

- *Reduction of the intracellular concentration of toxic proteins*: Some cloned proteins are toxic to the cell in which they are produced and there is therefore a limit to the amount of protein the cell will produce before it dies. Protein secretion should prevent cell death and result in continued production of protein.

- *To allow post-translational modification of proteins*: Most post-translational modifications of proteins occur as part of the secretory pathway, and these modifications, for example glycosylation (see Section 8.4.4), are a necessary process in producing the final protein structure. Since prokaryotic cells do not glycosylate their proteins, this explains why many proteins have to be expressed in eukaryotic cells (e.g. yeast) rather than in bacteria. The entry of a protein into a secretory pathway and its ultimate destination is determined by a short amino acid sequence (signal sequence) that is usually at the N terminus of the protein. For proteins going to the membrane or outside the cell the route is via the endoplasmic reticulum and Golgi apparatus, the signal sequence being cleaved-off by a protease prior to secretion. For example, human γ-interferon has been secreted from the yeast *Pichia pastoris* using the protein's native signal sequence. Also there are a number of well-characterised yeast signal sequences (e.g. the α-factor signal sequence) that can be used to ensure secretion of proteins cloned into yeast.

Fusion proteins to aid protein purification

This approach requires an additional gene to be joined to the gene of the protein of interest such that the protein is produced as a fusion protein (i.e. linked to this second protein, or tag). As will be seen below, the purpose of this tag is to provide a means whereby the fusion protein can be selectively removed from the cell extract. The fusion protein can then be cleaved to release the protein of interest from the tag protein. Clearly the amino acid sequence of the peptide linkage between tag and protein has to be carefully designed to allow chemical or enzymatic cleavage of this sequence. The following are just a few examples of many different types of fusion proteins that have been used to aid protein purification.

Flag™

This is a short hydrophilic amino acid sequence that is attached to the N-terminal end of the protein, and is designed for purification by immunoaffinity chromatography.

Asp-Tyr-Lys-Asp-Asp-Asp-Asp-Lys-Protein

A monoclonal antibody against this Flag sequence is available on an immobilised support for use in affinity chromatography. The cell extract, which includes the Flag-labelled protein, is passed through the column where the antibody binds to the Flag-labelled protein, allowing all other proteins to pass through. This is carried out in the presence of Ca^{2+}, since the binding of the Flag sequence to the monoclonal antibody is Ca^{2+} dependent. Once all unbound protein has been eluted from the column, the Flag-linked protein is released by passing EDTA through the column, which chelates the Ca^{2+}. Finally the Flag sequence is removed by the enzyme enterokinase, which recognises the following amino acid sequence and cleaves the C-terminal to the lysine residue:

N–Asp–Asp–Asp–Lys–C.

Using this approach, granulocyte-macrophage colony-stimulating factor (GMCSF) was cloned in and secreted from yeast, and purified in a single step. GMCSF was produced in the cell as signal peptide-Flag-gene. The signal sequence used was the signal sequence for the outer membrane protein OmpA. The Flag-gene protein was thus secreted into the periplasm, the fusion protein purified, and finally the Flag sequence removed, as described above.

Glutathione affinity agarose

In this method the protein of interest is expressed as a fusion protein with the enzyme glutathione *S*-transferase. The cell extract is passed through a column of glutathione-linked agarose beads, where the enzyme binds to the glutathione. Once all unbound protein has been washed through the column, the fusion protein is eluted by passing reduced glutathione through the column. Finally, cleavage of the fusion protein is achieved using human thrombin, which recognises a specific amino acid sequence in the linker region.

Protein A

Protein A binds to the Fc region of the immunoglobulin G (IgG) molecule. The protein of interest is cloned fused to the protein A gene, and the fusion protein purified by affinity chromatography on a column of IgG-Sepharose. The bound fusion protein is then eluted using either high salt or low pH, to disrupt the binding between the IgG molecule and the protein A–protein fusion product. Protein A is then finally removed by treatment with 70% (v/v) formic acid for 2 days, which cleaves an acid-labile Asp-Pro bond in the linker region.

Poly(arginine)

This method requires the addition of a series of arginine residues to the C terminus of the protein to be purified. This makes the protein highly basic (positively charged at neutral pH). The cell extract can therefore be fractionated using cation-exchange chromatography. Bound proteins are sequentially released from the column by applying a salt gradient, with the poly(Arg)-containing protein, because of its high overall positive charge, being the last to be eluted. The poly(Arg) tail is then removed by incubation with the enzyme carboxypeptidase B. Carboxypeptidase B is an exoprotease that sequentially removes arginine or lysine residues from the C terminus of proteins. The arginine residues are therefore sequentially removed from the C terminus, the removal of amino acid residues stopping when the 'normal' (i.e. non-arginine) C-terminal amino acid residue of the protein is reached.

8.4 PROTEIN STRUCTURE DETERMINATION

8.4.1 Relative molecular mass

There are three methods available for determining protein relative molecular mass, M_r, frequently referred to as molecular weight. The first two described here are quick and easy methods that will give a value to $\pm$ 5–10%. For many purposes one simply needs a rough estimate of size and these methods are sufficient. The third method, mass spectrometry, requires expensive specialist instruments and can give accuracy to $\pm 0.001\%$. This kind of accuracy is invaluable in detecting postsynthetic modification of proteins.

SDS-polyacrylamide gel electrophoresis (SDS–PAGE)

This form of electrophoresis, described in Section 10.3.1, separates proteins on the basis of their shape (size), which in turn relates to their relative molecular masses. A series of proteins of known molecular mass (molecular weight markers) are run on a gel on a track adjacent to the protein of unknown molecular mass. The distance each marker protein moves through the gel is measured and a calibration curve of log M_r versus distance moved is plotted. The distance migrated by the protein of unknown M_r is also measured, and from the graph its log M_r and hence M_r is calculated. The method is suitable for proteins covering a large M_r range (10 000–300 000). The method is easy to perform and requires very little material. If silver staining (Section 10.3.7) is used, as little as 1 ng of protein is required. In practice SDS–PAGE is the most commonly used method for determining protein M_r values.

Molecular exclusion (gel filtration) chromatography

The elution volume of a protein from a molecular exclusion chromatography column having an appropriate fractionation range is determined largely by the size of the protein such that there is a logarithmic relationship between protein relative molecular mass and elution volume (Section 11.7.1). By calibrating the column with a range of proteins of known M_r, the M_r of a test protein can be calculated. The method is carried out on HPLC columns ($\sim 1 \times 30$ cm) packed with porous silica beads. Flow rates are about $1\,\text{cm}^3\,\text{min}^{-1}$, giving a run time of about 12 min, producing sharp, well-resolved peaks. A linear calibration line is obtained by plotting a graph of log M_r versus K_d for the calibrating proteins. K_d is calculated from the following equation:

$$K_d = \frac{(V_e - V_o)}{(V_t - V_o)}$$

where V_o is the volume in which molecules that are wholly excluded from the column material emerge (the excluded volume), V_t is the volume in which small molecules that can enter all the pores emerge (the included volume) and V_e is the volume in which the marker protein elutes. This method gives values that are accurate to $\pm 10\%$.

Mass spectrometry

Using either electrospray ionisation (ESI) (Section 9.2.4) or matrix-assisted laser desorption ionisation (MALDI) (Section 9.3.8) intact molecular ions can be produced for proteins and hence their masses accurately measured by mass spectrometry. ESI produces molecular ions from molecules with molecular masses up to and in excess of 100 kDa, whereas MALDI produces ions from intact proteins up to and in excess of 200 kDa. In either case, only low picomole quantities of protein are needed. For example, $\alpha\beta_2$ crystallin gave a molecular mass value (20 200 $\pm$ 0.9), in excellent agreement with the deduced mass of 20 201. However, in addition about 10% of the analysed material produced an ion of mass 20 072.2. This showed that some of the purified protein molecules had lost their N-terminal amino acid (lysine). The deduced mass with the loss of N-terminal lysine was 20 072.8. Clearly mass spectrometry has the ability to provide highly accurate molecular mass measurements for proteins and peptides, which in turn can be used to deduce small changes made to the basic protein structure.

8.4.2 Amino acid analysis

The determination of which of the 20 possible amino acids are present in a particular protein, and in what relative amounts, is achieved by hydrolysing the protein to yield its component amino acids and identifying and quantifying them chromatographically. Hydrolysis is achieved by heating the protein with 6 M hydrochloride acid for 14 h at 110 °C *in vacuo*. Unfortunately, the hydrolysis procedure destroys or chemically modifies the asparagine, glutamine and tryptophan residues. Asparagine and glutamine are converted to their corresponding acids (Asp and Glu) and are quantified with them. Tryptophan is completely destroyed and is best determined spectrophotometrically on the unhydrolysed protein.

The amino acids in the protein hydrolysate are then separated chromatographically. Nowadays this is normally done using the method of precolumn derivatisation, followed by separation by reverse-phase HPLC. In this approach the amino acid hydrolysate is first treated with a molecule that (i) reacts with amino groups in amino acids, (ii) is hydrophobic, thus allowing separation of derivatised amino acids by reversed-phase HPLC and (iii) is easily detected by its ultraviolet absorbance or fluorescence. Reagents routinely used for precolumn derivatisation include *o*-phthalaldehyde and 6-aminoquinolyl-*N*-hydroxysuccinimidyl carbamate (AQC), which both produce fluorescent derivatives, and phenylisothiocyanate, which produces a phenylthiocarbamyl derivative that is detected by its absorbance at 254 nm. Analysis times can be as little as 20 min, and sensitivity is down to 1 pmole or less of amino acid.

8.4.3 Primary structure determination

For many years the amino acid sequence of a protein was determined from studies made on the purified protein alone. This in turn meant that sequence data available were limited to those proteins that could be purified in sufficiently large amounts. Knowledge of the complete primary structure of the protein was (and still is) a prerequisite for the determination of the three-dimensional structure of the protein, and hence an understanding of how that protein functions. However, nowadays the protein biochemist is normally satisfied with data from just a relatively short length of sequence either from the N terminus of the protein or from an internal sequence, obtained by sequencing peptides produced by cleavage of the native protein. The sequence data will then most likely be used for one of three purposes:

- To search sequence databases to see whether the protein of interest has already been isolated, and hence can therefore be identified. For this type of search extremely short lengths of sequence (three to five residues), known as sequence tags, need to be used. Examples of this type of data search are given in Sections 8.5.1 and 9.5.2.

- To search for sequence homology using computerised databases in order to identify the function of the protein. For example, the search may show significant sequence identity with the amino acid sequence of some known protein tyrosine kinases, strongly suggesting that the protein is also a tyrosine kinase.
- The sequence will be used to design an oligonucleotide probe for selecting appropriate clones from complementary DNA libraries. In this way the DNA coding for the protein can be isolated and the DNA sequence, and hence the protein sequence, determined. Obtaining a protein sequence in this way is far less laborious and time-consuming than having to determine the total protein sequence by analysis of the protein.

A further use of protein sequence data is in quality control in the biopharmaceutical industry. Many pharmaceutical companies produce products that are proteins, for example peptide hormones, antibodies, therapeutic enzymes, etc., and synthetic peptides also require analysis to confirm their identities. Sequence analysis, especially to determine sites and nature of postsynthetic modifications such as glycosylation, is necessary to confirm the structural integrity of these products.

Edman degradation

In 1950, Per Edman published a chemical method for the stepwise removal of amino acid residues from the N terminus of a peptide or protein. This series of reactions came to be known as the Edman degradation, and the method still remains the most effective chemical means for removing amino acid residues in a stepwise fashion from a polypeptide chain and thus determining the order of amino acids at the N-terminus of a protein or peptide.

However, the method is only infrequently used nowadays and will not be described in any detail here. Developments in the use of mass spectrometry over the past 20 years has led to mass spectrometry being the method of choice nowadays for determining protein sequences, and is discussed in more detail below and in Chapter 9.

Protein cleavage and peptide production

When studying proteins there are many occasions when one might wish to cleave a protein into peptide fragments (see, for example, peptide mass fingerprinting, Section 8.5.1). Peptides can be produced by either chemical or enzymatic cleavage of the native protein (see Table 8.5). Chemical methods tend to produce large fragments, as they cleave at the less common amino acids (often giving as few as two or three large peptides). Enzymatic methods tend to cleave adjacent to the more common amino acids (e.g. trypsin cleaves at every arginine and lysine residue in a protein), thus often producing as many as 50 or more peptides from a protein. Throughout this and other chapters, you will come across examples of where it is necessary to study peptide fragments of a protein.

Mass spectrometry

Because of the absolute requirement to produce ions in the gas phase for the analysis of any sample by mass spectrometry (MS), for many years MS analysis was applicable only to small, non-polar molecules ($< 500\,M_r$). However, in the early 1980s the

Table 8.5 **Specific cleavage of polypeptide**

Reagent	Specificity
Enzymic cleavage	
Chymotrypsin	C-terminal side of hydrophobic amino acid residues, e.g. Phe, Try, Tyr, Leu
Endoproteinase Arg-C	C-terminal side of arginine
Endoproteinase Asp-N	Peptide bonds N-terminal to aspartate or cysteine residues
Endoproteinase Glu-C	C-terminal side of glutamate residues and some aspartate residues
Endoproteinase Lys-C	C-terminal side of lysine
Thermolysin	N-terminal side of hydrophobic amino acid residues excluding Trp
Trypsin	C-terminal side of arginine and lysine residues but Arg-Pro and Lys-Pro poorly cleaved
Chemical cleavage	
BNPS skatole	
N-Bromosuccinimide	C-terminal side of tryptophan residues
o-Iodosobenzoate	
Cyanogen bromide	C-terminal side of methionine residues
Hydroxylamine	Asparagine–glycine bonds
2-Nitro-5-thiocyanobenzoate	N-terminal side of cysteine residues

introduction of fast atom bombardment (FAB) MS allowed the analysis of large, charged molecules such as proteins and peptides to be achieved for the first time. The further development of more sophisticated methods such as electrospray ionisation (ESI) and matrix-assisted laser desorption ionisation (MALDI) (see Chapter 9) has led to the analysis of protein by mass spectrometry becoming routine. Although the Edman degradation still has occasional applications in protein structure analysis, mass spectrometry is now the method of choice for determining amino acid sequence data. When peptides are fragmented by MS it is fortunate that cleavage occurs predominantly at the peptide bond (although it must be noted that other fragmentations, such as internal cleavages, secondary fragmentations, etc. do occur, thus complicating the mass spectrum). This means that the peptide fragments produced each differ sequentially by the mass of one amino acid residue. The amino acid sequence can thus be readily deduced. In particular, if side-chain modifications occur, these can also be observed due to the corresponding increase in mass difference. The use of mass spectrometry to obtain sequence data from proteins and peptides is described more fully in Section 9.5. Tandem mass spectrometry (MS/MS or MS^2) is also increasingly being used to obtain sequence data. A digest of the protein (e.g. with trypsin) is separated by MS. The ion corresponding to one peptide is selected in the first analyser and collided with argon gas in a collision cell to generate fragment ions. The fragment ions thus generated are then separated, according to mass, in a second analyser, identified, and the sequence determined as described in Section 9.5.2.

A further method, ladder sequencing, has been developed, and combines the Edman chemistry with MS. Edman sequencing is carried out using a mixture of PITC and phenylisocyanate (PIC) (at about 5% of the concentration of PITC). N-terminal amino groups that react with PIC are effectively blocked as they are not cleaved at the acid cleavage step. Consequently, at each cycle, approximately 5% of the protein molecules are blocked. Thus, after 20 to 30 cycles of Edman degradation, a nested set of peptides is produced, each differing by the loss of one amino acid. Analysis of the mass of each of these polypeptides using ESI or MALDI allows the determination of the molecular mass of each polypeptide and the difference in mass between each molecule identifies the lost amino acid residue.

Detection of disulphide linkages

For proteins that contain more than one cysteine residue it is important to determine whether, and if so how many, cysteine residues are joined by disulphide bridges. The most commonly used method involves the use of MS (Section 9.5.5). The native protein (i.e. with disulphide bridges intact) is cleaved with a proteolytic enzyme (e.g. trypsin) to produce a number of small peptides. The same experiment is also carried out on proteins treated with dithiothreitol (DTT) which reduces (cleaves) the disulphide bridges. MALDI spectra of the tryptic digest before and after reduction with DTT allows identification of disulphide-linked peptides. Linked peptides from the native protein will disappear from the spectrum of the reduced protein and reappear as *two* peptides of lower mass. Knowledge of the exact mass of each of the two peptides, and knowledge of the cleavage site of the enzyme used, will allow easy identification of the two peptides from the known protein sequence. Thus, if the mass of two disulphide-linked peptides is M, and this is reduced to two separate chains of masses A and B, respectively, then $A + B = M + 2$. The extra two mass units derive from the fact that reduction of the disulphide bond results in an increase of mass of $+1$ for both cysteine residues.

$$-S - S- \xrightarrow{2H} - SH + HS-$$

Hydrophobicity profile

Having determined the amino acid sequence of a protein, analysis of the distribution of hydrophobic groups along the linear sequence can be used in a predictive manner. This requires the products of a hydrophobicity profile for the protein, which graphs the average hydrophobicity per residue against the sequence number. Averaging is achieved by evaluating, using a predictive algorithm, the mean hydrophobicity within a moving window that is stepped along the sequence from each residue to the next. In this way, a graph comprising a series of curves is produced and reveals areas of minima and maxima in hydrophobicity along the linear polypeptide chain. For membrane proteins, such profiles allow the identification of potential membrane-spanning segments. For example, an analysis of a thylakoid membrane protein revealed seven general regions of the protein sequence that contained spans of 20–28 amino acid residues, each of which contained predominantly hydrophobic residues flanked on either side by hydrophilic residues. These regions represent the seven membrane-spanning helical regions of the protein.

For membrane proteins defining aqueous channels, hydrophilic residues are also present in the transmembrane section. Pores comprise amphipathic α-helices, the polar sides of which line the channel, whereas the hydrophobic sides interact with the membrane lipids. More advanced algorithms are used to detect these sequences, since such helices would not necessarily be revealed by simple hydrophobicity analysis.

8.4.4 Glycoproteins

Glycoproteins result from the covalent attachment of carbohydrate chains (glycans), both linear and branched in structure, to various sites on the polypeptide backbone of a protein. These post-translational modifications are carried out by cytoplasmic enzymes within the endoplasmic reticulum and Golgi apparatus. The amount of polysaccharide attached to a given glycoprotein can vary enormously, from as little as a few per cent to more than 60% by weight. Glycoproteins tend to be found in the serum and in cell membranes. The precise role played by the carbohydrate moiety of glycoproteins includes stabilisation of the protein structure, protection of the protein from degradation by proteases, control of protein half-life in blood, the physical maintenance of tissue structure and integrity, a role in cellular adhesion and cell–cell interaction, and as an important determinant in receptor–ligand binding.

The major types of protein glycoconjugates are:

- N-linked;
- O-linked;
- glycosylphosphatidylinositol (GPI)-linked.

N-linked glycans are always linked to an asparagine residue side-chain (Fig. 8.3) at a consensus sequence Asn-X-Ser/Thr where X is any amino acid except proline. O-linked glycosylation occurs where carbohydrate is attached to the hydroxyl group of a serine or threonine residue (Fig. 8.3). However, there is no consensus sequence similar to that found for N-linked oligosaccharides. GPI membrane anchors are a more recently discovered modification of proteins. They are complex glycophospholipids that are covalently attached to a variety of externally expressed plasma membrane proteins. The role of this anchor is to provide a stable association of protein with the membrane lipid bilayer, and will not be discussed further here.

There is considerable interest in the determination of the structure of O- and N-linked oligosaccharides, since glycosylation can affect both the half-life and func- tion of a protein. This is particularly important of course when producing therapeutic glycoproteins by recombinant methods as it is necessary to ensure that the correct carbohydrate structure is produced. It should be noted that prokaryotic cells do not produce glycoproteins, so cloned genes for glycoproteins need to be expressed in eukaryotic cells. The glycosylation of proteins is a complex subject. From one glyco- protein to another there are variations in the sites of glycosylation (e.g. only about

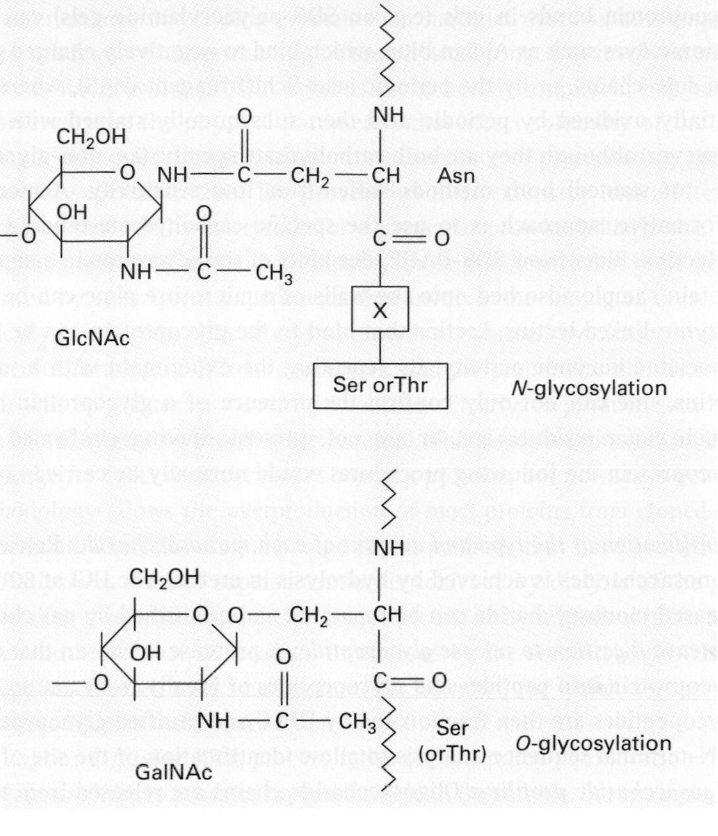

Fig. 8.3 The two types of oligosaccharide linkages found in glycoproteins.

30% of consensus sequences for N-linked attachments are occupied by polysacchar-ide; the nature of the secondary structure at this position also seems to play a role in deciding whether glycosylation takes place), variations in the type of amino acid–carbohydrate bond, variations in the composition of the sugar chains, and variations in the particular carbohydrate sequences and linkages in each chain. There are eight monosaccharide units commonly found in mammalian glycoproteins, although other less common units are also known to occur. These eight are *N*-acetyl neuraminic acid (NeuNAc), *N*-glycolyl neuraminic acid (NeuGc), D-galactose (Gal), *N*-acetyl-D-glucosamine (GlcNac), *N*-acetyl-D-galactosamine (GalNAc), D-mannose (Man), L-fucose (Fuc) and D-xylose (Xyl). To further complicate the issue, within any population of molecules in a purified glycoprotein there can be considerable heterogeneity in the carbohydrate structure (glycoforms). This can include some molecules showing increased branching of sugar side-chains, reduced chain length and further addition of single carbohydrate units to the same polypeptide chain. The complete determination of the glycosylation status of a molecule clearly requires considerable effort. However, the steps involved are fairly straight forward and the following therefore provides a generalised (and idealised) description of the overall procedures used.

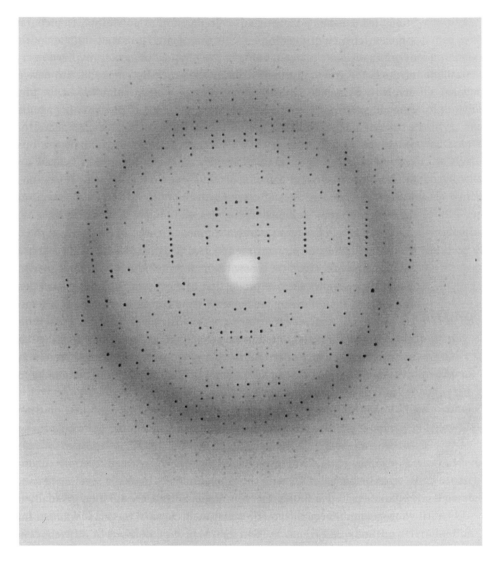

Fig. 8.4 X-ray diffraction frame of data from a crystal of herpes simplex virus type 1 thymidine kinase, complexed with substrate deoxythymidine, at 2 Å resolution. (Picture provided by John N. Champness, Matthew S. Bennett and Mark R. Sanderson of King's College London.)

- Unfortunately the diffraction pattern alone is insufficient to determine the crystal structure. Each diffraction maximum has both an amplitude and a phase associated with it, and both need to be determined. But the phases are not directly measurable in a diffraction experiment and must be estimated from further experiments. This is usually done by the method of isomorphous replacement (MIR). The MIR method requires at least two further crystals of the protein (derivatives), each being crystallised in the presence of a different heavy-metal ion (e.g. Hg^{2+}, Cu^{2+}, Mn^{2+}). Comparison of the diffraction patterns from the crystalline protein and the crystalline heavy-metal atom derivative allows phases to be estimated. A more recent approach to producing a heavy-metal derivative is to clone the protein of interest into a methionine

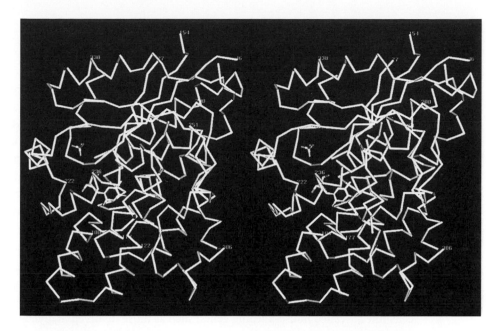

Fig. 8.5 (*Relaxed-eye stereo pair*): A C$^\alpha$-trace of herpes simplex virus type 1 thymidine kinase from a crystallographic study of a complex of the enzyme with one of its substrates, deoxythymidine. The enzyme is an α–β protein, having a five-stranded parallel β-sheet surrounded by 14 α-helices. The active site, occupied by deoxythymidine, is a volume surrounded by four of the helices, the C-terminal edge of the β-sheet and a short 'flap' segment; a sulphate ion occupies the site of the β-phosphate of the absent co-substrate ATP. (Short missing regions of chain indicate where electron density calculated from the X-ray data could not be interpreted.) (Picture provided by John N. Champness, Matthew S. Bennett and Mark R. Sanderson of King's College London.)

auxotroph, and then grow this strain in the presence of selenomethionine (a selenium-containing analogue of methionine). Selenomethionine is therefore incorporated into the protein in the place of methionine, and the final purified and crystallised protein has the selenium heavy metal conveniently included in its structure.

- Diffraction data and phase information having been collected, these data are processed by computer to construct an electron density map. The known sequence of the protein is then fitted into the electron density map using computer graphics, to produce a three-dimensional model of the protein (Fig. 8.5.). In the past there had been concern that the three-dimensional structure determined from the rigid molecules found in a crystal may differ from the true, more flexible, structure found in free solution. These concerns have been effectively resolved by, for example, diffusing substrate into an enzyme crystal and showing that the substrate is converted into product by the crystalline enzyme (there is sufficient mother liquor within the crystal to maintain the substrate in solution). In a more recent development, it is now becoming possible to determine the solution structure of protein using NMR. At present the method is capable of determining the structure of a protein up to about 20 000 kDa but will no doubt be developed to study larger proteins. Although the time-consuming step of producing a crystal is obviated, the methodology and data analysis involved are at present no less time-consuming and complex than that for X-ray crystallography.

8.5 PROTEOMICS AND PROTEIN FUNCTION

In order to completely understand how a cell works, it is necessary to understand the function (role) of every single protein in that cell. The analysis of any specific disease (e.g. cancer) will also require us to understand what changes have taken place in the protein component of the cell, so that we can use this information to understand the molecular basis of the disease, and thus design appropriate drug therapies and develop diagnostic methods. (Just about every therapeutic drug that is currently in use has a protein as its target.) The completion of the Human Genome Project might suggest that it is not now necessary to study proteins directly, since the amino acid sequence of each protein can be deduced from the DNA sequence. This is not true for the following reasons:

- First, although the DNA in each cell type in the body is the same, different sets of genes are expressed in different tissues, and hence the protein component of a cell varies from cell type to cell type. For example, some proteins are found in nearly all cells (the so-called house-keeping genes) such as those involved in glycolysis, whereas specific cell types such as kidney, liver, brain, etc. contain specific proteins unique to that tissue and necessary for the functioning of that particular tissue/organ. It is therefore only by studying the protein component of a cell directly that we can identify which proteins are actually present.
- Secondly, it is now appreciated that a single DNA sequence (gene) can encode multiple proteins. This can occur in a number of ways:
  (i) Alternative splicing of the mRNA transcript.
  (ii) Variation in the translation 'stop' or 'start' sites.
  (iii) Frameshifting, where a different set of triplet codons is translated, to give a totally different amino acid sequence.
  (iv) Post-translational modifications. The genome sequence defines the amino acid sequence of a protein, but tells us nothing of any post-translational modifications (Sections 8.2.1 and 9.5.5) that can occur once the polypeptide chain is synthesised at the ribosome. Up to 10 different forms (variants) of a single polypeptide chain can be produced by phosphorylation, glycosylation, etc.

The consequence of the above is that the total protein content of the human body is an order of magnitude more complex than the genome. The human genome sequence suggests there may be 30 000–40 000 genes (and hence proteins) whereas estimates of the actual number of proteins in human cells suggests possibly as many as 200 000 or even more. The dogma that one gene codes for one protein has been truly demolished!

From the above, I hope it is easy to appreciate the need to directly analyse the protein component of the cell, and the need for an understanding of the function of each individual protein in the cell. In recent years, development of new techniques (discussed below) has enhanced our ability to study the protein component of the cell and has led to the introduction of the terms proteome and proteomics. The total DNA composition of a cell is referred to as the genome, and the study of the structure and function of this DNA is called genomics. By analogy, the proteome is defined as the total protein component of a cell, and the study of the structure and function of these proteins is called proteomics. The ultimate aim of proteomics is to catalogue the identity and amount of every protein in a cell, and determine the function of each protein.

Earlier sections of this chapter and Chapter 11 describe the traditional, but still very valid approach to studying proteins, where individual proteins are extracted from tissue and purified so that studies can be made of the structure and function of the purified proteins. The subject of proteomics has developed from a different approach, where modern techniques allow us to view and analyse much of the total protein content of the cell in a single step. The development of these newer techniques has gone hand-in-hand with the development of techniques for the analysis of proteins by mass spectrometry, which has revolutionised the subject of protein chemistry. The cornerstone of proteomics has been two-dimensional (2-D) PAGE (described in Section 10.3) and the applications of this technique in proteomics are described below. However, although 2-D PAGE remains central to proteomics, the study of proteomics has stimulated the development of further methods for studying proteins and these will also be described below.

8.5.1   2-D PAGE

2-D PAGE has found extensive use in detecting changes in gene expressions between two different biological states, for example comparing normal and diseased tissue. In this case, a 2-D gel pattern would be produced of an extract from a diseased tissue such as a liver tumour and compared with the 2-D gel patterns of an extract from normal liver tissue. The two gel patterns are then compared to see whether there are any differences in the two patterns. If it is found that a protein is present (or is absent) only in the liver tumour sample, then by identifying this protein we are directed to the gene for this protein and can thus try to understand why this gene is expressed (or not) in the diseased state. In this way it is possible to obtain an understanding of the molecular basis of diseases. This approach can be taken to study *any* disease process where normal and diseased tissue can be compared, for example arthritis, kidney disease, or heart valve disease.

Under favourable circumstances up to 5000 protein spots can be identified on a large format 2-D gel. Thus with 2-D PAGE we now have the ability to follow changes in the expression of a significant proportion of the proteins in a cell or tissue type, rather than just one or two, which has been the situation in the past. The potential applications of proteome analysis are vast. Initially one must produce a 2-D map of the proteins expressed by an organism, tissue or cell under 'normal' conditions. This 2-D reference map and database can then be used to compare similar information from 'abnormal' or treated organisms, tissues or cells. For example, as well as comparing normal tissue with diseased tissue (as described above), we can:

- analyse the effects of drug treatment or toxins on cells;
- observe the changing protein component of the cell at different stages of tissue development;

- observe the response to extracellular stimuli such as hormones or cytokines;
- compare pathogenic and non-pathogenic bacterial strains;
- compare serum protein profiles from healthy individuals and Alzheimer or cancer patients to detect proteins, produced in the serum of patients, which can then be developed as diagnostic markers for diseases (e.g. by setting up an enzyme-linked immunosorbent assay (ELISA) to measure the specific protein).

As a typical example, a research group studying the toxic effect of drugs on the liver can compare the 2-D gel patterns from their 'damaged' livers with the normal liver 2-D reference map, thus identifying protein changes that occur as a result of drug treatment.

The sheer complexity and amount of data available from 2-D gel patterns is daunting, but fortunately there is a range of commercial 2-D gel analysis software, compatible with personal computer workstations, which can provide both qualitative and quantitative information from gel patterns, and can also compare patterns between two different 2-D gels (see below). This has allowed the construction of a range of databases of quantitative protein expression in a range of tissue and cell types. For example, an extensive series of 2-DE databases, known as SWISS-2D PAGE, is maintained at Geneva University Hospital and is accessible via the World Wide Web at <http://au.expasy.org/ch2d/>. This facility therefore allows an individual laboratory to compare their own 2-D protein database with that in another laboratory.

The comparison of two gel patterns is made by using any one of a number of software packages designed for this purpose. One of the more interesting approaches to comparing gel patterns is the use of the Flicker program, which is available on the Web at <http://open2dprot.sourceforge.net/Flicker>. This program superimposes the two 2-D patterns to be compared and then alternately, and rapidly, displays one pattern and then the other. Spots that appear on both gel patterns (the majority) will be seen as fixed spots, but a spot that appears on one gel and not the other will seen to be flashing (hence 'flicker'). When one has compared two 2-DE patterns and identified any proteins spot(s) of interest, it is then necessary to identify each specific protein. In the majority of cases this is done by peptide mass-fingerprinting. The spot of interest is cut out of the gel and incubated in a solution of the proteolytic enzyme trypsin, which cleaves the protein C-terminal to each arginine and lysine residue. In this way the protein is reduced to a set of peptides. This collection of peptides is then analysed by MALDI-MS (see Section 9.3.8) to give an accurate mass measurement for each of the peptides in the sample. This set of masses, derived from the tryptic digestion of the protein, is highly diagnostic for this protein, as no other protein would give the same set of peptide masses (fingerprint). Using Web-based programs such as Mascot or Protein Prospector this experimentally derived peptide mass-fingerprint is compared with databases of tryptic peptide mass-fingerprints generated from sequences of known proteins (or predicted sequences deduced from nucleotide sequences). If a match is found with a fingerprint from the database then the protein will be identified.

However, sometimes results from peptide mass-fingerprinting can be ambiguous. In this case it is necessary to obtain some partial amino acid sequence data from one of the peptides. This is done by tandem mass spectrometry (MS/MS; Section 9.5),

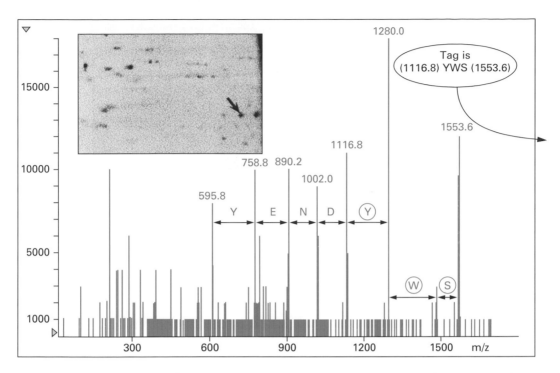

Fig. 8.6 Nano-ESI MS^2 spectrum of *m/z* 890 from RBL spot 2 showing construction of a sequence tag. The *y*-axis shows relative intensity. (Courtesy of Glaxo SmithKline, Stevenage, UK.)

where one of the peptides separated for mass-fingerprinting is further fragmented in a second analyser, and from the fragmentation pattern sequence data can be deduced (mass spectrometry conveniently fragments peptides at the peptide bond, such that the difference in the mass of fragments produced can be related to the loss of specific amino acids; Section 9.5.2). This partial sequence data is then used to search the protein sequence databases for sequence identity. Universal databases are available that store information on all types of protein from all biological species. These databases can be divided into two categories: (i) databases that are a simple repository of sequence data, mostly deduced directly from DNA sequences, for example the Tr EMBL database; and (ii) annotated databases where information in addition to the sequence is extracted by the biologist (the annotator) from the literature, review article, etc., for example the SWISS-PROT database.

An example of how sequence data can be produced is shown in Fig. 8.6. A lysate of 2×10^6 rat basophil leukaemic (RBL) cells were separated by 2-D electrophoresis and spot 2 chosen for analysis. This spot was digested *in situ* using trypsin and the resultant peptides extracted. This sample was then analysed by tandem MS using a triple quadrupole instrument (ESI-MS^2). MS of the peptide mixture showed a number of molecular ions relating to peptides. One of these (*m/z* 890) was selected for further analysis, being further fragmented in a quadrupole mass spectrometer to give fragment ions ranging from *m/z* 595.8 to 1553.6 (Fig. 8.6). The ions at *m/z* 1002.0, 1116.8, 1280.0, 1466.2 and 1553.6 are likely to be part of a Y ion series (see Fig. 8.6) as they appear at higher *m/z* than the precursor at *m/z* 890. The gap between adjacent Y ions is

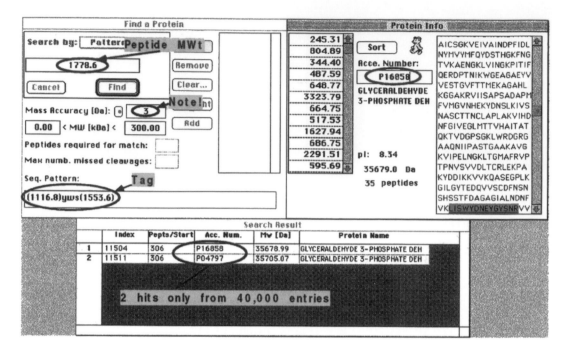

Fig. 8.7 The PeptideSearch™ input form and search result based on data obtained from nano-ESI MS2 of m/s 890 from RBL Spot 2. (Courtesy of Glaxo SmithKline, Stevenage, UK.)

related directly to an amino acid residue because the two flanking Y ions result from cleavage of two adjacent amide bonds. Therefore, with a knowledge of the relative molecular masses of each of the 20 naturally occurring amino acids, it is possible to determine the presence of a particular residue at any point within the peptide. The position of the assigned amino acid is deduced by virtue of the m/z ratio of the two ions. By reading several amino acids it was possible to assemble a sequence of amino acids, in this case (using the one-letter code) YWS. Database searching was then possible using the peptide 1778 Da, the position of the lower m/z Y ion (1116.8), the proposed amino acid sequence (YWS) and the higher Y ion at m/z 1553.6. This provides a sequence tag, which is written as (1116.8) YWS (1553.6).

A search of the SWISS-PROT database (Fig. 8.7), showed just two 'hits' from 40 000 entries, suggesting the protein is glyceraldehyde-3-phosphate dehydrogenase. The full sequence of this peptide is LISWYDNEYGYSNR and the MS/MS fragmentation data give a perfect match. Other peptides in the sample can also be analysed in the same manner, confirming the identity of the protein.

A further development of 2-D PAGE has been the introduction of difference gel electrophoresis (DIGE). This again allows the comparison of protein components of similar mixtures, but has the advantage that only one 2-D gel has to be run rather than two. In this method the two samples to be compared are each treated with one of two different, yet structurally very similar, fluorescent dyes (cy3 and cy5). Each dye reacts with amino groups, so that each protein is fluorescently labelled by the dye binding to lysine residues and the N-terminal amino groups. The two protein solutions to be compared are then mixed and run on a *single* 2-D gel. Thus every protein in one sample superimposes with its differentially labelled identical counterpart in the other sample. Scanning of the gel at two different wavelengths that excite the two dye molecules reveals whether any individual spot is associated with only one dye molecule rather than two. Most spots will, of course, fluoresce at both wavelengths, but if a spot is associated with only one dye molecule then this tells us that that protein can have been present in only one of the extracts, and the wavelength at which it fluoresces tells you which extract it was originally in.

8.5.2 Isotope-coded affinity tags (ICAT)

Isotope-coded affinity tags (ICAT) uses mass spectrometry (rather than 2-D gels) to identify differences in the protein content of two complex mixtures. For example, the method can be used to identify protein differences between tumour and normal tissue, in the same way that 2-D PAGE can be used to address the same question (Section 8.5.1). This method uses two protein 'tags' that, whilst being in every other respect identical, differ slightly in molecular mass; hence one is 'heavy' and one is 'light'. Both contain (a) a chemical group that reacts with the amino acid cysteine, and (b) a biotin group. In both molecules these groups are joined by a linker region, but in one case the linker contains eight hydrogen atoms, in the other, eight deuterium atoms; one molecule (tag) is thus heavier than the other by 8 Da (see Fig. 9.26). One cell extract (e.g. from cancer cells) is thus treated with one tag (which binds to cysteine residues in all the proteins in the extract) and the second tag is used to treat the second extract (e.g. from normal cells). Both extracts are then treated with trypsin to produce mixtures of peptides, those peptides that contain cysteine having been 'tagged'. The two extracts are then combined and an avidin column used to affinity-purify the labelled peptides by binding to the biotin moiety. When released from the column this mixture of labelled peptides will contain pairs of identical peptides (derived from identical proteins) from the two cell extracts, each pair differing by a mass of 8 Da.

Analysis of this peptide mixture by liquid chromatography–MS will then reveal a series of peptide mass signals, each one existing as a 'pair' of signals separated by eight mass units. These data will reveal the relative abundance of each peptide in the pair. Since most proteins present in the two samples originally being compared will be present at much the same levels, most peptide pairs will have equal signal strengths. However, for proteins that exist in greater or lesser amounts in one of the extracts, different signal strengths will be observed for each of the peptides in the pair, reflecting the relative abundance of this protein in the two samples. Further analysis of either of these pairs via tandem mass spectrometry will provide some sequence data that should allow the protein to be identified. ICAT is discussed in more detail in Section 9.6.2.

8.5.3 Determining the function of a protein

Successfully applied, the methods described in the preceding section will have provided the amino acid sequence (or partial sequence) of a protein of interest. The next step is to identify the function and role of this protein. The first step is invariably to search the databases of existing protein sequences to find a protein or proteins that have sequence homology with the protein of interest (the homology method). This is done using programs such as BLAST and PSI-BLAST. If sequence homology is found with a protein of known function, either from the same or different species, then this invariably identifies the function of the protein. However, this approach does not always work. For example, when the genome of the yeast *Saccharomyces cerevisiae* was completely sequenced in 1996, 6000 genes were identified. Of these, approximately 2000 coded for proteins that were already known to exist in yeast (i.e. had been purified and studied in previous years), 2000 had homology with known sequences and hence their function could be deduced by the homology method but 2000 could not be matched to any known genes, i.e. they were 'new', previously undiscovered genes. In these cases, there are a number of other computational methods that can be used to help to identify the protein's function. These include:

- *Phylogenic profile method:* This method aims to identify any other protein(s) that has the same phylogenic profile (i.e. the same pattern of presence or absence) as the unknown protein, in all known genomes. If such proteins are found it is inferred that the unknown protein is involved in the same cellular process as these other protein(s) (i.e. they are said to have a functional link) and will give a strong clue as to the function of the unknown protein. This method is based on the premise that two proteins would not always both be inherited into a new species (or neither inherited) unless the two proteins have a functional link. At the time of writing there are over 100 published genome sequences that can be surveyed with this method. Fig. 8.8 shows a simple, hypothetical example, where just five genomes are analysed.
- *Method of correlated gene neighbours:* If two genes are found to be neighbours in several different genomes, a functional linkage may be inferred between the two proteins. The central assumption of this approach is based on the observation that functionally related genes in prokaryotes tend to be linked to form operons (e.g. the *lac* operon). Although operons are rare in eukaryotic species, it does appear that proteins involved in the same biological process/pathway within the cell have their genes situated in close proximity (e.g. within 500 bp) in the genome. Thus, if two genes are found to be in close proximity across a number of genomes, it can be inferred that the protein products of these genes have a functional linkage. This method is most robust for microbial genomics but works to some extent in human cells where operon-like clusters are also observed. As an example, this method correctly identified a functional link between eight enzymes in the biosynthetic pathway for the amino acid arginine in *Mycobacterium tuberculosis*.
- *Analysis of fusion:* This method is based on the observation that two genes may exist separately in one organism, whereas the genes are fused into a single multifunctional gene in another organism. The existence of the protein product of the fused gene, in which the two functions of the protein clearly interact (being part of the same protein molecule), suggests that in the first organism the two separate proteins also interact. It has been suggested that gene fusion events occur to reduce the regulational load of multiple interacting gene products.
- *Protein–protein interactions:* A further clue to identifying protein function can come from identifying protein–protein interactions, and methods to identify these are described in the next section.

	A	B	C	D	E
P1	1	1	1	0	0
P2	0	0	1	1	1
P3	1	0	1	1	0
P4	0	1	1	0	1
P5	1	1	0	0	1
P6	0	1	1	0	1
P7	1	0	0	1	0
P8	1	0	1	1	0

Fig. 8.8 Phylogenic profile method. Five genomes, A–E, are shown (e.g. *E. coli*, *S. cerevisiae*, etc.). The presence (1) or absence (0) of eight proteins (P1–P8) in each of these genomes is shown. It can be seen that proteins P3 and P8 have the same phylogenic profile and therefore may have a functional linkage. P4 and P6 are similarly linked.

8.5.4 Protein–protein interactions

Given the complex network of pathways that exist in the cell (signalling pathways, biosynthetic pathways, etc.), it is clear that all proteins must interact with other molecules to fulfil their role. Indeed, it is now apparent that proteins do not exist in isolation in the cell; proteins involved in a common pathway appear to exist in a loose interaction, sometimes referred to as a biomodule. Therefore, if one can identify an interaction between our unknown protein and a well-characterised protein, it can be inferred that the former has a function somehow related to the latter. For example, if the unknown protein is shown to interact with one or more proteins involved in the biosynthetic pathways for arginine, then this strongly suggests that the unknown protein is also involved in this pathway. Using this approach networks of interacting proteins are being identified in individual organisms. This has led to the development of the Database of Interacting Proteins (DIP), which can be found at <http://dip.doe-mbi.ucla.edu>. Given the current fad for inventing new words ending in 'ome', some refer to these maps of protein interactions as the interactome.

One of the most widely used, and successful, methods for investigating protein–protein interaction is the yeast two-hybrid (Y2H) system, which exploits the modular architecture of transcription factors. A transcription factor gene (GAL4) is split into the coding regions for two domains, a DNA-binding domain and a *trans*-activation domain. Both these domains are expressed, each linked to a different protein (one being the unknown protein, the other a protein with which it may interact), in separate yeast cells, which are then mated to produce diploid cells (the two proteins being studied are often referred to as the bait and prey). If, in this diploid cell, the bait and prey proteins bind to each other, they will bring together the two domains of the transcription factor, which will then be active and will bind to the promoter of a reporter gene (e.g. the *his* gene), inducing its expression. Identification of cells expressing the reporter gene product is evidence that the bait and prey proteins interact. In practice, following mating, diploids are selected on deficient medium (in this case, medium deficient in histidine), thus only yeast cells expressing interacting proteins survive (as they are capable of synthesising histidine). Once such a positive interaction is identified, the two interacting open reading frames (ORFs) are simply identified by sequencing a small part of the protein gene.

Using this approach, all 6000 ORFs from *S. cerevisiae* were individually cloned as both bait and prey. When the pool of 6000 prey clones was screened against each of the 6000 bait clones, 691 interactions were identified, only 88 of which were previously known. This therefore gave an indication of the function of over 600 proteins whose function was previously unknown. On a much larger scale, the same approach was used to identify protein–protein interactions in the fruit fly, *Drosophila melanogaster*. All 14 000 predicted *D. melanogaster* ORFs were amplified using the polymerase chain reaction (PCR) and each cloned into two-hybrid bait and prey vectors. A total of 45 417 two-hybrid positive colonies were obtained, from which 10 021 protein interactions involving 4500 proteins were obtained. The yeast two-hybrid system is described in greater detail in Section 6.8.3.

8.5.5 Protein arrays

A newly developing area for studying protein–protein interactions is the use of protein arrays (chips). Although the basic principle for screening and identifying interacting molecules is much the same as for DNA arrays (Section 6.8.8), the production of protein arrays is more technically demanding owing mainly to the difficulty of binding proteins to a surface and ensuring that the protein is not denatured at any stage of the assay procedure.

In a protein array, proteins are immobilised as small spots (150–200 μm) onto a solid support (typically glass or a nitrocellulose membrane), using high precision contact printing (not unlike a dot-matrix printer) at a spot density of the order of 1500 spots cm^{-2}. A solution of the protein of unknown function is then incubated on the array surface for a period of time, then washed off, and the position(s) where the protein has bound, identified (see below). Since it is known which protein was immobilised in each position of the chip, each pair of interacting proteins can be identified.

Saccharomyces cerevisiae again provides a good example of the successful use of this technology where a protein array was used to identify yeast proteins that bind to the protein calmodulin (an important protein involved in calcium regulation). Five thousand eight hundred yeast ORFs were cloned into a yeast high copy expression vector, and each of the expressed proteins purified. Each protein was then spotted at high density onto nickel-coated glass microscope slides. Since each protein also contained a $(His)_6$-Tag (which binds to nickel) introduced at the C terminus, proteins were attached to the surface in an orientated manner, the C terminus being linked to the nickel-coated glass through the $(His)_6$ sequence, while the rest of the molecule was therefore suitably orientated away from the surface of the array to be available for interaction with another protein. The array was then incubated in a solution of calmodulin that had been labelled with biotin. The calmodulin was then washed off and the positions where calmodulin had bound to the array were identified by incubating the array with a solution of fluorescently labelled avidin (the protein avidin binds strongly to the small-molecular-mass vitamin biotin: see Section 10.3.8). The use of ultraviolet light thus identified fluorescence where the screening molecules had bound. In total, 33 new proteins that bind calmodulin were discovered in this way.

Figure 8.9 (see also colour section) shows an interaction map of the yeast proteome. The authors constructed the map from published data on protein–protein interactions in yeast. The map contains 1584 proteins and 2358 interactions. Proteins are coloured according to their functional role, e.g. proteins involved in membrane fusion (blue), lipid metabolism (yellow), cell structure (green), etc. If one views the electronic version of this publication it is possible for the reader to zoom in and search for protein names and to read interactions more clearly.

Figure 8.10 (see also colour section) is a summary of Fig. 8.9 showing the number of interactions of proteins from each functional group with proteins of their own and other groups. The word function means the cellular role of the protein. Numbers in parentheses indicate, first, the number of interactions within a group and, secondly, the number of proteins within a group. Numbers on connecting lines indicate the numbers of interactions between proteins of the two connected groups. For example, in the upper left-hand corner, there are 77 interactions between the 21 proteins involved in membrane fusion and 141 proteins involved in vesicular transport. Looking at the bottom right of the diagram it can be seen that some proteins involved in RNA processing/modification not surprisingly also interact with proteins involved in RNA turnover, RNA splicing, RNA transcription and protein synthesis.

8.5.6 Systems biology

It can be seen from the section on proteomics that the study of proteins is moving away from methods that involve the purification and study of individual proteins. Nowadays proteins are more likely to be studied as a stained spot on a complex 2-D gel pattern, often present in as little as nanogram amounts, more often than not using

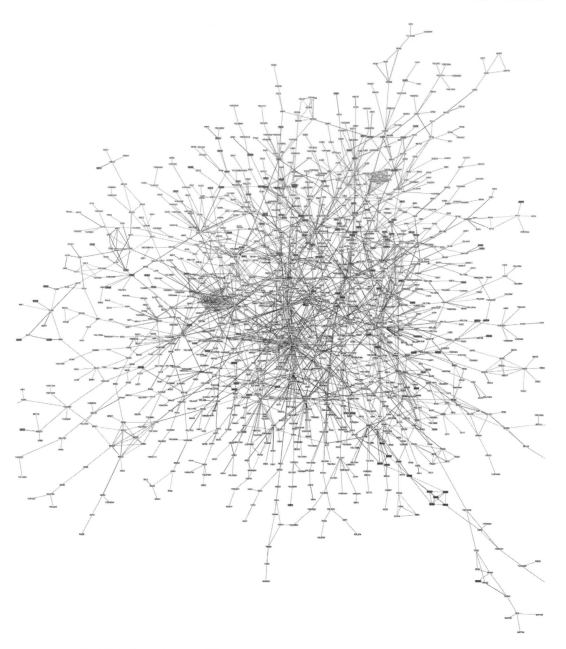

Fig. 8.9 An interaction map of the yeast proteome, assembled from published interactions (see text for details). (Courtesy of Benno Schwikowski, Peter Uetz and Stanley Fields. Reprinted with the permission of Nature Publishing Group.) (See also colour plate.)

analytical techniques such as mass spectrometry (see Chapter 9) and invariably requiring the interrogation of protein and genome sequence data on the Web (bioinformatics, Section 5.8). It is then necessary to determine which other proteins interact with the protein being studied. Proteomics is thus moving us away from studying proteins in isolation and encouraging us to consider the proteins in the cell as part

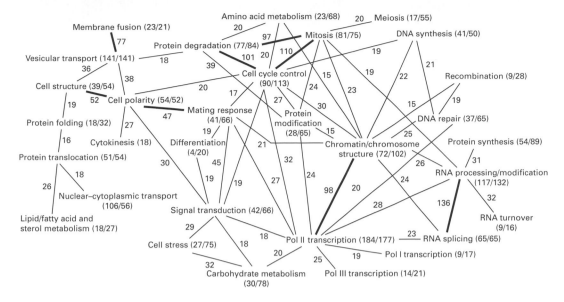

Fig. 8.10 A simplification of Fig. 8.9 identifying interactions between functional groups of proteins (see text for details). (Courtesy of Benno Schwikowski, Peter Uetz and Stanley Fields. Reprinted with the permission of Nature Publishing Group.) (See also colour plate.)

of a dynamic interacting system. This has led to the development of the concept of systems biology, which can be defined as the study of living organisms in terms of their underlying network structure rather than just their individual molecular components. Since systems biology requires a study of all interacting components in the cell the new high throughput and quantitative.

8.6 SUGGESTIONS FOR FURTHER READING

Cutler, P. (2004). *Protein Purification Protocols*. Totowa, NJ: Humana Press. (Detailed theory and practical procedures for a range of protein purification techniques.)

Walker, J. M. (2005). *Proteomics Protocols Handbook*. Totowa, NJ: Humana Press. (Theory and techniques of a spectrum of methods applied to proteomics.)

Nedelkov, D. (2006). *New and Emerging Proteomics Techniques*. New York: Humana Press. (In-depth details of a range of proteomics techniques.)

Thompson, J. D. (2008). *Functional Proteomics*. New York: Humana Press. (Comprehensive coverage of functional proteomics including protein analysis and mass spectrometry.)

Simpson, R. J., Adams, P. D. and Golemis, E. A. (2008). *Basic Methods in Protein Purification and Analysis: A Laboratory Manual*. New York: CSH Press. (A comprehensive collection of protein purification methods.)

9 Mass spectrometric techniques

A. AITKEN

9.1   INTRODUCTION

9.1.1   General

Mass spectrometry (MS) is an extremely valuable analytical technique in which the molecules in a test sample are converted to gaseous ions that are subsequently separated in a mass spectrometer according to their mass-to-charge (m/z) ratio and detected. The mass spectrum is a plot of the (relative) abundance of the ions at each m/z ratio. Note that it is the mass to charge ratios of ions (m/z) and not the actual mass that is measured. If for example, a biomolecule is ionised by the addition of one or more protons (H^+ ions) the instrument measures the m/z after addition of 1 Da for each proton if the instrument is measuring positive ions or m/z minus 1 Da for each proton lost if measuring negative ions.

The development of two ionisation techniques, electrospray (ESI) and matrix-assisted laser desorption/ionisation (MALDI), has enabled the accurate mass determination of high-molecular-mass compounds as well as low-molecular-mass molecules and has revolutionised the applicability of mass spectrometry to almost any biological molecule. Applications include the new science of proteomics as well as in drug discovery. The latter includes combinatorial chemistry where a large number of similar molecules (combinatorial libraries) are produced and analysed to find the most effective compounds from a group of related organic chemicals.

M_r is sometimes used to designate relative molar mass. Molecular weight (which is a force not a mass) is also frequently and incorrectly used. M_r is a relative measure and has no units. However, M_r is numerically equivalent to the mass, M, which does have units and the Dalton is frequently used (see Section 1.2.2).

The essential features of all mass spectrometers are therefore:

- production of ions in the gas phase;
- acceleration of the ions to a specific velocity in an electric field;
- separation of the ions in a mass analyser; and
- detection of each species of a particular m/z ratio.

The instruments are calibrated with standard compounds of accurately known M_r values. In mass spectrometry the carbon scale is used with $^{12}C = 12.000000$. This level of accuracy is achievable in high-resolution magnetic sector double-focussing, accelerator mass spectrometers and Fourier transform mass spectrometers (Sections 9.3.5, 9.3.6 and 9.3.13).

The mass analyser may separate ions either by use of a magnetic or an electrical field. Alternatively the time taken for ions of different masses to travel a given distance in space is measured accurately in the time-of-flight (TOF) mass spectrometer (Section 9.3.8).

Any material that can be ionised and whose ions can exist in the gas phase can be investigated by MS, remembering that very low pressures, i.e. high vacuum, in the region of 10^{-6} Torr are required (Torr is measure of pressure which equals 1 mm of mercury (133.3 Pa; atmospheric pressure is 760 Torr)). The majority of biological MS investigations on proteins, oligosaccharides and nucleic acids is carried out with quadrupole, quadrupole–ion trap and TOF mass spectrometers. In the organic chemistry/biochemistry area of analysis, the well-established magnetic sector mass spectrometers still find wide application and their main principles will also be described.

The treatment of mass spectrometry in this chapter will be strictly non-mathematical and non-technical. However, the intention is to give an overview of the types of instrumentation that will be employed, the main uses of each, complementary techniques and advantages/disadvantages of the different instruments and particular applications most suited to each type. Data analysis and sample preparation to obtain the best sensitivity for a particular type of compound will also be covered.

9.1.2 Components of a mass spectrometer

All mass spectrometers are basically similar (Fig. 9.1). They consist of the following:

- *A high vacuum system (10^{-6} torr or 1 μtorr)*: These include turbomolecular pumps, diffusion pumps and rotary vane pumps.
- *A sample inlet*: This comprises a sample or target plate; a high-performance liquid chromatography (HPLC), gas chromatography (GC) or capillary electrophoresis system; solids probe; electron impact or direct chemical ionisation chamber.
- *An ion source* (to convert molecules into gas-phase ions): This can be MALDI; ESI; fast atom bombardment (FAB); electron impact or direct chemical ionisation.
- *A mass filter/analyser*: This can be: TOF; quadrupole; ion trap; magnetic sector or ion cyclotron Fourier transform (the last is also actually a detector).
- *A detector*: This can be a conversion dynode, electron multiplier, microchannel plate or array detector.

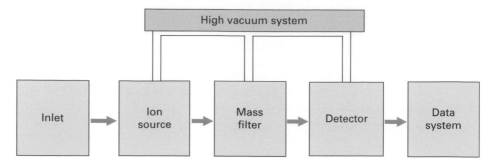

Fig. 9.1 Basic components of mass spectrometers.

9.1.3 Vacuum system

All mass analysers operate under vacuum in order to minimise collisions between ions and air molecules. Without a high vacuum, the ions produced in the source will not reach the detector. At atmospheric pressure, the mean free path of a typical ion is around 52 nm; at 1 mtorr, it is 40 mm; and at 1 μtorr, it is 40 m.

In most instruments, two vacuum pump types are used, e.g. a rotary vane pump (to produce the main reduction in pressure) followed by a turbomolecular pump or diffusion pump to produce the high vacuum.

The rotary vane pump can be an oil pump to provide initial vacuum (approximately 1 torr), while the turbomolecular pump provides working high vacuum (1 mtorr to 1 ntorr). This is a high-speed gas turbine with interspersed rotors (moving blades) and stators (i.e. fixed or stationary blades) whose rotation forces molecules through the blade system.

9.2 IONISATION

Ions may be produced from a neutral molecule by removing an electron to produce a positively charged cation, or by adding an electron to form an anion. Both positive- and negative-ion mass spectrometry may be carried out but the methods of analysis in the following sections will be described mainly for positive-ion MS, since this is more common and the principles of separation and detection are essentially the same for both types of ion.

9.2.1 Electron impact ionisation (EI)

Electron impact ionisation (EI) is widely used for the analysis of metabolites, pollutants and pharmaceutical compounds, for example in drug testing programmes. Electron impact (EI) has major applications as a mass detector for gas chromatography (GC/MS, Section 11.9.3). A stream of electrons from a heated metal filament is accelerated to 70 eV potential (the electron volt, eV, is a measure of energy). Sample ionisation occurs when the electrons stream across a high vacuum chamber into which molecules of the substance to be analysed (analyte) are allowed to diffuse (Fig. 9.2). Interaction with the analyte results in either loss of an electron from the

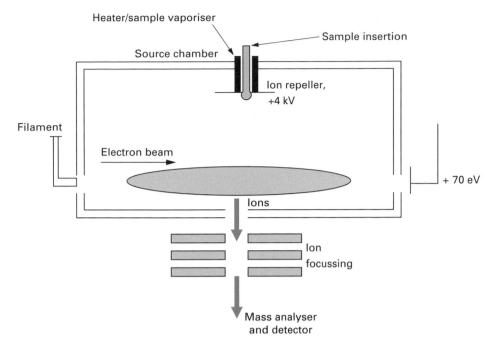

Fig. 9.2 Electron impact source. Electrons are produced by thermionic emission from a filament of tungsten or rhenium. The filament current is typically 0.1 mA. Electrons are accelerated toward the ion source chamber (held at a positive potential equal to the accelerating voltage) and acquire an energy equal to the voltage between the filament and the source chamber, typically 70 eV. The electron trap is held at a fixed positive potential with respect to the source chamber. Gaseous analyte molecules are introduced into the path of the electron beam where they are ionised. Owing to the positive ion repeller voltage and the negative excitation voltage that produce an electric field in the source chamber, the ions leave the source through the ion exit slit and are analysed.

substance (to produce a cation) or electron capture (to produce an anion). The analyte must be in the vapour state in the electron impact source, which limits the applicability to biological materials below ca. 400 Da. Before the advent of electrospray and MALDI, the method did have some applicability to peptides, for example, whose volatility could be increased by chemical modification. A large amount of fragmentation of the sample is common, which may or may not be desirable depending on the information required.

Chemical bonds in organic molecules are formed by the pairing of electrons. Ionisation resulting in a cation requires loss of an electron from one of these bonds (effectively knocked out by the bombarding electrons), but it leaves a bond with a single unpaired electron. This is a radical as well as being a cation and hence the representation as $M^{\cdot+}$, the $(^+)$ sign indicating the ionic state and the $(\cdot)$ a radical. Conversely, electron capture results both in an anion but also the addition of an unpaired electron and therefore a negatively charged radical, hence the symbol $M^{\cdot-}$. Such radical ions are termed molecular ions, parent ions or precursor ions and under the conditions of electron bombardment are relatively unstable. Their energy in excess of that required for ionisation has to be dissipated. This latter process results in the precursor ion disintegrating into a number of smaller fragment ions that may be relatively unstable and further fragmentation may occur. This gives rise to a series of daughter ions or product ions, which are recorded as the mass spectrum.

For the production of a radical cation, as it is not known where either the positive charge or the unpaired electron actually reside in the molecule, it has been the practice to place the dot signs outside the abbreviated bracket sign, $]^{\cdot}$. The recent recommendation by IUPAC for mass spectrometry notation is to write the sign first followed by the superscripted dot, i.e. $M^{+\cdot}$ or $M^{-\cdot}$.

When the precursor ion fragments, one of the products carries the charge and the other the unpaired electron, i.e. it splits into a radical and an ion. The product ions are therefore true ions and not radical ions. The radicals produced in the fragmentation process are neutral species and therefore do not take any further part in the mass spectrometry but are pumped away by the vacuum system. Only the charged species are accelerated out of the source and into the mass analyser. It is also important to recognise that almost all possible bond breakages can occur and any given fragment will arise both as an ion and a radical. The distribution of charge and unpaired electron, however, is by no means equal. The distribution depends entirely on the thermodynamic stability of the products of fragmentation. Furthermore, any fragment ion may break down further (until single atoms are obtained) and hence not many ions of a particular type may survive, resulting in a low signal being recorded.

A simple example is given by n-butane ($CH_3CH_2CH_2CH_3$) and some of the major fragmentations are shown Fig. 9.3.a. The resultant EI spectrum is shown in Fig. 9.3b.

9.2.2 Chemical ionisation

Chemical ionisation (CI) is used for a range of samples similar to those for EI. It is particularly useful for the determination of molecular masses, as high intensity molecular ions are produced due to less fragmentation. CI therefore gives rise to much cleaner spectra. The source is essentially the same as the EI source but it contains a suitable reagent gas such as methane (CH_4) or ammonia (NH_3) that is initially ionised by EI. The high gas pressure in the source results in ion–molecule reactions between reagent gas ions (such as NH_3^+ and CH_4^+) some of which react with the analyte to produce analyte ions. The mass differences from the neutral parent compounds therefore correspond to these adducts.

9.2.3 Fast atom bombardment (FAB)

At the time of its development in the early 1980s, fast atom bombardment (FAB) revolutionised MS for the biologist. The important advance was that this soft ionisation technique, which leads to the formation of ions with low internal energies and little consequent fragmentation, permitted analysis of biomolecules in solution without prior derivatisation. The sample is mixed with a relatively involatile, viscous matrix such as glycerol, thioglycerol or m-nitrobenzyl alcohol. The mixture, placed on a probe, is introduced into the source housing and bombarded with an ionising beam of neutral

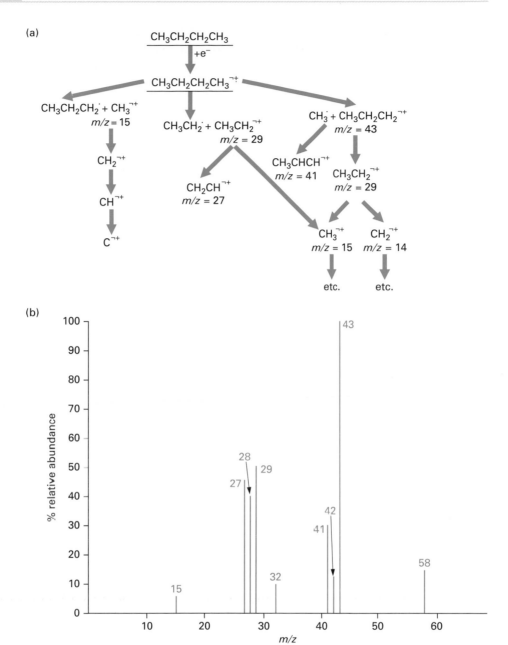

Fig. 9.3 Fragmentation pathways in *n*-butane and the EI spectrum. The pathway for fragmentation of *n*-butane is shown in (a) and the EI spectrum in (b). In the spectrum, the relative abundance is plotted from 0 to 100% where the largest peak is set at 100% (base peak). Spectra represented in this way are said to be **normalised**.

atoms (such as Ar, He, Xe) of high velocity. A later development was the use of a beam of caesium (Cs$^+$) ions and the term liquid secondary ion mass spectrometry (LSIMS) was introduced to distinguish this from FAB–MS. Pseudomolecular ion species arise as either protonated or deprotonated entities (M + H)$^+$ and (M − H)$^-$ respectively, which allows positive and negative ion mass spectra to be determined. The term pseudomolecular implies the mass of the ion formed from a substance of a given mass by the gain or loss of one or more protons. Other charged adducts can also be formed such as $(M + Na)^+$ and $(M + K)^+$.

9.2.4 Electrospray ionisation (ESI)

This involves the production of ions by spraying a solution of the analyte into an electrical field. This is a soft ionisation technique and enables the analysis of large intact (underivatised) biomolecules, such as proteins and DNA. The electrospray (ES) creates very small droplets of solvent-containing analyte. The essential principle in ES is that a spray of charged liquid droplets is produced by atomisation or nebulisation. Solvent (typically 50 : 50 water and organic solvent) is removed as the droplets enter the mass spectrometer. ESI is the result of the strong electric field (around 4 keV at the end of the capillary and 1 keV at the counter electrode) acting on the surface of the sample solution. As the solvent evaporates in the high-vacuum region, the droplet size decreases and eventually charged analyte (free of solvent) remains. Ionisation can occur at atmospheric pressure and this method is also sometimes referred to as atmospheric pressure ionisation (API).

The concentration of sample is usually around $1–10 \, \text{pmol mm}^{-3}$. Typical solvents are 50/50 acetonitrile (or methanol)/H_2O with 1% acetic acid or 0.1% formic acid. Ammonium hydroxide or trifluoroacetic acid (TFA, 0.02%) in 50/50 acetonitrile (or methanol)/H_2O can also be used. The organic acid (or the NH_4OH) aids ionisation of the analyte. At low pH, basic groups will be ionised. In the example of peptides these are the side groups of Lys, His, Arg and the N-terminal amino group. At alkaline pH the carboxylic acid side chains as well as stronger anions such as phosphate and sulphate groups will be ionised. The presence of organic solvent assists in formation of small droplets and facilitates evaporation.

The flow rate into the source is normally around a few $\text{mm}^3 \, \text{min}^{-1}$ although higher flow rates can be tolerated (up to $1 \, \text{cm}^3$) if the solution is an eluant from on-line HPLC for example.

Smaller molecules usually produce singly charged ions but multiply charged ions are frequently formed from larger biomolecules, in contrast to MALDI, resulting in m/z ratios that are sufficiently small to be observed in the quadrupole analyser. Thus masses of large intact proteins, DNA and organic polymers can also be accurately measured in electrospray MS although the m/z limit of measurement is normally 2000 or 3000 Da. For example, proteins are normally analysed in the positive ion mode where charges are introduced by addition of protons. The number of basic amino acids in the protein (mainly lysine and arginine) determines the maximum number of charges carried by the molecule. The distribution of basic residues in most proteins is such that the multiple peaks (one for each $M + nH)^{n+}$ ion, are centred on m/z about 1000. In Fig. 9.6 a large protein with a mass of over 100 000 Da behaves as if it were multiple mass species around 1020 Da. For the species with 100 protons (H^+) i.e. with 100 charges, $z = 100$, $m/z = 1027.6$ therefore $(M + 100H)^{100+} = 1027.6$. When the computer processes the data for the multiple peaks, the average for each set of peaks gives a mass determination

Example 1   **PROTEIN MASS DETERMINATION BY ESI**

Question   A protein was isolated from human tissue and subjected to a variety of investigations. Relative molecular mass determinations gave values of approximately 12 000 by size exclusion chromatography and 13 000 by gel electrophoresis. After purification, a sample was subjected to electrospray ionisation mass spectrometry and the following data obtained.

m/z	773.9	825.5	884.3	952.3	1031.3
Abundance (%)	59	88	100	66	37

Given that $n_2 = (m_1 - 1)/(m_2 - m_1)$ and $M = n_2 (m_2 - 1)$ and assuming that the only ions in the mixture arise by protonation, deduce an average molecular mass for the protein by this method.

Answer M_r by exclusion chromatography $= 12\,000$

M_r by gel electrophoresis $= 13\,000$
Taking ESI peaks in pairs:

$m_1 - 1$	$m_2 - m_1$	n_2	$m_2 - 1$	M (Da)	z
951.3	79.0	12.041	1030.3	12406.6	12
883.3	68.0	12.989	951.3	12357.1	13
824.5	58.8	14.022	883.3	12385.7	14
772.9	51.6	14.978	824.5	12349.9	15

$\Sigma M = 49\,499.3$ Da
Mean $M = 12\,374.8$ Da

Note: Relative abundance values are not required for the determination of the mass.

to high accuracy. The peaks can be deconvoluted and presented as a single peak representing the M_r (in this example $M = 102\,658$).

A diagrammatic representation of the ESI source is shown in Fig. 9.4. A curtain or sheath gas (usually nitrogen) around the spray needle at a slow flow rate may be used to assist evaporation of the solvent at or below room temperature. This may be an advantage for thermally labile compounds.

9.3   MASS ANALYSERS

9.3.1   Introduction

Once ions are created and leave the ion source, they pass into a mass analyser, the function of which is to separate the ions and to measure their masses. (Remember, what

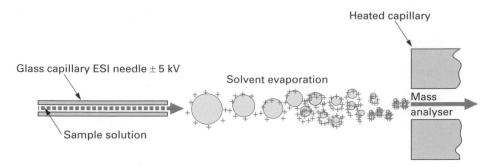

Fig. 9.4 Electrospray ionisation source. The ESI creates very small droplets of solvent-containing analyte by atomisation or **nebulisation** as the sample is introduced into the source through the fine glass (or other material) hollow needle capillary. The solvent evaporates in the high-vacuum region as the spray of droplets enters the source. As the result of the strong electric field acting on the surface of the sample droplets, and electrostatic repulsion, their size decreases and eventually single species of charged analyte (free of solvent) remain. These may have multiple charges depending on the availability of ionisable groups.

is really measured is the mass-to-charge ratio (m/z) for each ion.) At any given moment, ions of a particular mass are allowed to pass through the analyser where they are counted by the detector. Subsequently, ions of a different mass are allowed to pass through the analyser and again the detector counts the number of ions. In this way, the analyser scans through a large range of masses.

In the majority of instruments, a particular type of ionisation is coupled to a particular mass analyser that operates by a particular principle. That is, EI, CI and FAB are combined with magnetic sector instruments; ESI and its derivatives with quadrupole (or its variant ion-trap) and MALDI is coupled to TOF detection.

9.3.2 Quadrupole mass spectrometry

The quadrupole analyser consists of four parallel cylindrical rods (Fig. 9.5). A direct current (DC) voltage and a superimposed radio frequency (RF) voltage are applied to each rod, creating a continuously varying electric field along the length of the analyser. Once in this field, ions are accelerated down the analyser towards the detector. The varying electric field is precisely controlled so that during each stage of a scan, ions of one particular mass-to-charge ratio pass down the length of the analyser. Ions with any other mass-to-charge value impact on the quadrupole rods and are not detected. By changing the electric field (scanning), the ions of different m/z successively arrive at the detector.

Quadrupoles can routinely analyse up to m/z 3000, which is extremely useful for biological MS since, as we have seen, proteins and other biomolecules normally give a charge distribution of m/z that is centred below this value (see Fig. 9.6).

Note that hexapole and octapole devices are also used, to direct a beam into the next section of a triple quadrupole or into the ion trap for example, but the principle is the same.

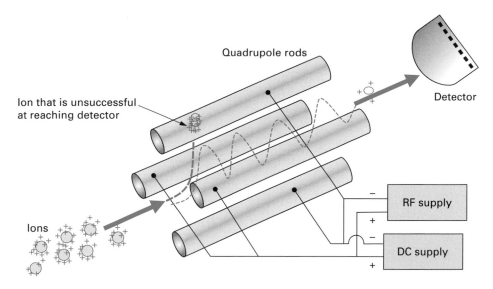

Fig. 9.5 Quadrupole analyser. The fixed (DC) and oscillating (RF) fields cause the ions to undergo complicated trajectories through the quadrupole filter. For a given set of fields, only certain trajectories are stable, which only allows ions of specific *m/z* to travel through to the detector. The efficiency of the quadrupole is impaired after a build-up of ions that do not reach the detector. Therefore a set of pre-filters is added to the quadrupole to remove the ions that would otherwise affect the main quadrupole.

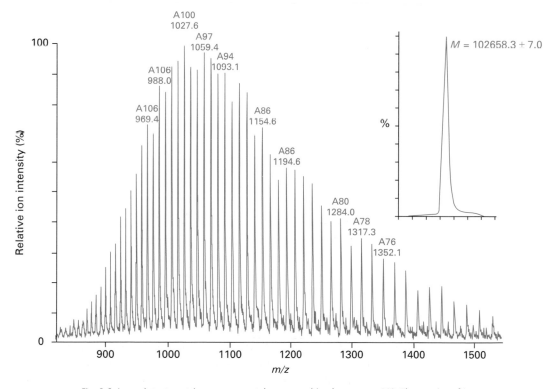

Fig. 9.6 Large intact protein mass accurately measured in electrospray MS. The species of ions are annotated by the charge state, e.g. with 99, 100, 101 charges, etc., and the associated *m/z* value. The inset shows the 'deconvoluted spectrum'.

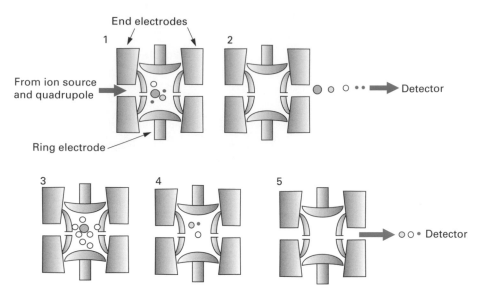

Fig. 9.7 Diagram of an ion trap. The ion trap contains three hyperbolic electrodes which form a cavity in a cylindrical device of around 5 cm diameter in which the ions are trapped (stored) and subsequently analysed. Each end-cap electrode has a small hole in the centre. Ions produced from the source enter the trap through the quadrupole and the entrance end-cap electrode. Potentials are applied to the electrodes to trap the ions (diagrams 1 and 2). The ring electrode has an alternating potential of constant radio frequency but variable amplitude. This results in a three-dimensional electrical field within the cavity. The ions are trapped in stable oscillating trajectories that depend on the potentials and the m/z of the ions. To detect these ions, the potentials are varied, resulting in the ion trajectories becoming unstable and the ions are ejected in the axial direction out of the trap in order of increasing m/z into the detector. A very low pressure of helium is maintained in the trap, which 'cools' the ions into the centre of the trap by low-speed collisions that normally do not result in fragmentation. These collisions merely slow the ions down so that during scanning, the ions leave quickly in a compact packet, producing narrower peaks with better resolution. In sequencing, all the ions are ejected except those of a particular m/z ratio that has been selected for fragmentation (see diagrams 3, 4 and 5). The steps are: (3) selection of precursor ion, (4) collision-induced dissociation of this ion, and (5) ejection and detection of the fragment ions.

9.3.3 Ion trap mass spectrometry

Ion trap mass spectrometers use ESI to produce ions, all of which are transferred into and subsequently measured almost simultaneously (within milliseconds) in a device called an ion trap (Fig. 9.7). The trap must then be refilled with the ions that are arriving from the source. Therefore, although the trap does not measure 100% of all ions produced (it depends on the cycle time to refill the trap then analyse the ions) this results nevertheless in a great improvement in sensitivity relative to quadrupole mass spectrometers where at any given moment only ions of one particular m/z are detected. ESI–ion trap mass spectrometers have found wide application for analysis of peptides and small biomolecules such as in protein identification by tandem MS; liquid chromatography/mass spectrometry (LC/MS); combinatorial libraries and rapid analysis in drug discovery and drug development.

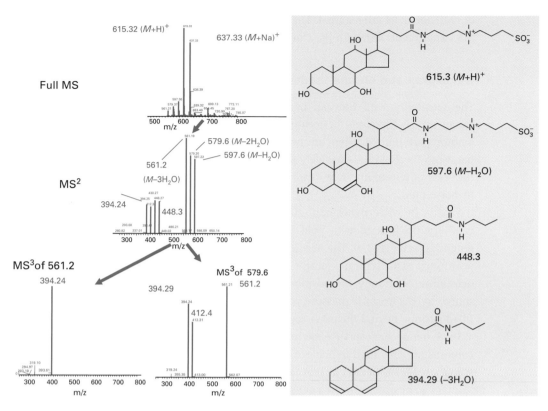

Fig. 9.8 Structural analysis, MS^n in an ion trap. In this example, of a steroid-related compound, the structure can be analysed when the (M + H)$^+$ ions at 615.3 are selected to be retained in the ion trap. These ions are subjected to collision-induced dissociation (CID) resulting in loss of the aliphatic sulphonate from the quaternary ammonium group and partial loss of some hydroxyl groups in the tandem MS (MS^2) experiment. The major fragment ions (561.2 and 579.6) are further selected for CID (MS^3) resulting in subsequent losses of more hydroxyl groups from specific parts of the steroid ring.

Ion trap MS permits structural information to be readily obtained (and sequence information in the case of polypeptides). Not only can MS–MS analysis be carried out but also due to the high efficiency of each stage, further fragmentation of selected ions may be carried out to MS to the power n (MS^n) (Fig. 9.8). The instrument still allows accurate molecular mass determination to over 100 000 Da at greater than 0.01% mass accuracy.

The MS^n procedure in an ion trap involves ejecting all ions that are stored in the trap, except those corresponding to the selected m/z value. To perform tandem MS (MS^2) a collision gas is introduced (a low pressure of helium) and collision–induced dissociation (CID) occurs (Fig. 9.7). The fragment ions are then ejected in turn and the fragment spectrum determined. The process can be repeated successively where all the fragment ions stored in the trap except those fragment ions corresponding to another selected m/z value are ejected. This fragment ion can then be further fragmented to obtain more structural information, as illustrated for the example shown in Fig. 9.8. This technique has a big advantage since no additional mass spectrometers or collision cells are required. The limitation is sensitivity, which decreases with each MS experiment, although the claimed record in an ion trap is currently MS^{14}.

9.3.4 Nanospray and on-line tandem mass spectrometry

The sensitivity with ESI can be greatly improved with a reduction in flow rate. Nanospray is therefore the technique of choice for ultimate sensitivity when sample amounts are limited. There are two ways of achieving this. Both static and dynamic nanospray techniques are widely used. Flow rates in both nanospray techniques are in the order of tens of $nm^3 \, min^{-1}$, which leads to low sample consumption and low signal-to-noise ratio.

Firstly, in static nanospray, glass needles are used with a very finely drawn out capillary tip (coated with gold to allow the needle to be held at the correct kV potential; see Fig. 9.4). The needles are filled with $1–2 \, mm^3$ of sample and accurately positioned at the entrance to the source. Closed-circuit television (CCTV) is used to determine accurately the position of the capillary. The solution is drawn into the source by electrostatic pressure, although a low pressure may be applied with an air-filled syringe behind the other (open) end of the needle if necessary.

In dynamic nanospray experiments, small-diameter microbore HPLC or capillary columns are also used to achieve separation at low flow rates. This can be combined with a stream splitter device that can further reduce flow rate (Section 11.9.3). The stream splitter can be used to divert a percentage of the solvent flow from the pump, say 99% to 99.9% to waste and allow the remainder to pass through the column. This allows for much more accurate flow rates since it is extremely difficult to directly and accurately pump at $0.5 \, mm^3$ or even $50 \, nm^3 \, min^{-1}$ with a high-pressure pump. Therefore one can use a pump that functions more efficiently at flow rates of 50 to $500 \, mm^3 \, min^{-1}$ to pass $0.5 \, nm^3 \, min^{-1}$ or less into the micro column.

Nanospray sources are used in triple quadrupole, ion trap and hybrid MALDI instruments.

Computer programs can be set up to perform tandem MS during the chromatographic separation on each component as it elutes from the column, if it gives a signal above a threshold that is set by the operator.

9.3.5 Magnetic sector analyser

A magnetic sector analyser is shown diagrammatically in Fig. 9.9. The ions are accelerated by an electric field. The electric sector acts as a kinetic energy filter and allows only ions of a particular kinetic energy to pass, irrespective of the m/z. This greatly increases the resolution since the ions emerge from the electrostatic analyser (ESA) with the whole range of masses but the same velocity. A given ion with the appropriate velocity then enters the magnetic sector analyser. It will travel in a curved trajectory in the magnetic field with a radius depending on the m/z and the velocity of the ion (the latter has already been selected). Thus only ions of a particular m/z will be detected at a particular magnetic field strength. The trajectory

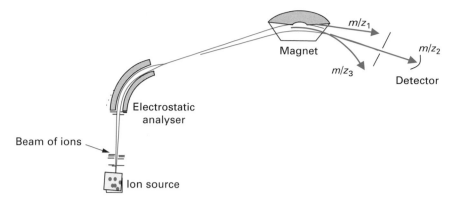

Fig. 9.9 Double-focussing magnetic sector mass spectrometer. The figure shows the 'forward geometry' arrangement where the electrostatic analyser is before the magnetic sector (known as EB; E for electric, B for magnetic). Similar results may be obtained if the reverse geometry (BE) type is used. The radial path followed by each ion is shown by scanning the magnetic field, B, and each ion of a particular m/z can be brought into the detector slit in turn.

of the ions is through a sector of the circular poles of the magnet, hence the term magnetic sector.

Figure 9.9 shows several possible trajectories for a given ion in the magnetic field. Only one set of ions will be focussed on the detector. If the field is changed, these ions will be defocussed because they will not be deflected to the correct extent. A new set of ions will be deflected and collected at the detector. By starting at either end of the magnet range the ions can be scanned from high to low mass or from low to high mass. This magnetic scanning is the most commonly used type of analysis in this instrument. Alternatively, the mass spectrum can be scanned electrically by varying the voltage, V, while holding the magnetic field, B, constant. This type of instrument is called a two-sector or double-focussing mass spectrometer and resolving power to parts per million may be obtained.

9.3.6 Accelerator mass spectrometry

Accelerator mass spectrometry (AMS) has proved to be extremely useful for quantifying rare isotopes and has had a major impact in archaeology (to measure ^{14}C) and geochronology. AMS can also measure radioisotopes such as ^{3}H, ^{10}Be, ^{26}Al, ^{36}Cl and ^{41}Ca with attomole (10^{-18}) to zeptomole (10^{-21}) levels of sensitivity and very high precision. AMS has found increasing application in human microdosing studies in drug development. This enables metabolites to be measured in human plasma or urine after administration of low, pharmacologically relevant doses of labelled drugs. Among the many applications of AMS are long-term pharmacokinetic studies to determine low-dose and chronic effects and the analysis of molecular targets of neurotoxins (see Section 18.3.1).

9.3.7 Plasma desorption ionisation

Plasma desorption ionisation mass spectrometry (PDMS) was the first mass spectrometer to be able to analyse proteins and other large biomolecules (although only those of relatively low M_r, less than 35 k). The technique and instruments developed are now obsolete and clearly overtaken by the much more powerful, sensitive and accurate instruments described elsewhere in this chapter. PDMS instruments are however still in use in some laboratories and research publications still appear with mass spectra obtained on this instrument. A basic understanding of the principle is therefore worth including. The source of the plasma (atomic nuclei stripped of electrons) was radioactive californium, ^{252}Cf, and two typical emission nuclei were the 100 MeV Ba^{20+} and Tc^{18+}, formed by the decay of the Cf, which are ejected in opposite directions, almost collinearly and with equal velocity. This is a pulsed technique, i.e. particles are emitted at discrete time intervals and require a TOF mass detector. The plasma particle emitted in the opposite direction to that passing through the sample triggers a time counter and the desorbed sample ions are accelerated electrically and detected as for other TOF analysers (Section 9.3.8).

9.3.8 MALDI, TOF mass spectrometry, MALDI-TOF

Matrix-assisted laser desorption ionisation (MALDI) produces gas phase protonated ions by excitation of the sample molecules from the energy of a laser transferred via a UV light-absorbing matrix. The matrix is a conjugated organic compound (normally a weak organic acid such as a derivative of cinnamic acid and dihydroxybenzoic acid) that is intimately mixed with the sample. Examples of MALDI matrix compounds and their application for particular biomolecules are shown in Table 9.1. These are designed to maximally absorb light at the wavelength of the laser, typically a nitrogen laser of 337 nm or a neodymium/yttrium-aluminium-garnet (Nd-YAG) at 355 nm.

The sample (1–10 pmol mm^{-3}) is mixed with an excess of the matrix and dried on to the target plate, where they co-crystallise on drying. Pulses of laser light of a few nanoseconds duration cause rapid excitation and vaporisation of the crystalline matrix and the subsequent ejection of matrix and analyte ions into the gas phase (Fig. 9.10). This generates a plume of matrix and analyte ions that are analysed in a TOF mass analyser.

The particular advantage of MALDI is the ability to produce large mass ions, with high sensitivity. MALDI is a very soft ionisation method that does not produce abundant amounts of fragmentation compared with some other ionisation methods. Since the molecular ions are produced with little fragmentation, it is a valuable technique for examining mixtures (see Fig. 9.14 and compare this to the more complex spectrum in Fig. 9.6).

TOF is the best type of mass analyser to couple to MALDI, as this technique has a virtually unlimited mass range. Proteins and other macromolecules of M_r greater than 400 000 have been accurately measured. The principle of TOF is illustrated in Fig. 9.11 and the main components of the instrument are shown in Fig. 9.12.

Table 9.1 **Examples of MALDI matrix compounds**

Compound	Structure	Application
α-Cyano-4-hydroxycinnamic acid (CHCA)		Peptides <10 kDa (glycopeptides)
Sinapinic acid (3,5-dimethoxy-4-hydoxycinnamic acid) (SA)		Proteins >10 kDa
'Super DHB', mixture of 10% 5-methoxysalycilic acid (2-hydroxy-5-methoxybenzoic acid) with DHB		Proteins, glycosylated proteins
2,5-Dihydroxybenzoic acid (DHB) (gentisic acid)		Neutral carbohydrates, synthetic polymers (oligos)
3-Hydroxypicolinic acid		Oligonucleotides
2,-(4-hydroxy-phenlyazo)-Benzoic acid (HABA)		Oligosaccharides, proteins

Sample concentration for MALDI

Maximum sensitivity is achieved in MALDI–TOF if samples are diluted to a particular concentration range. If the sample concentration is unknown a dilution series may be needed to produce a satisfactory sample/matrix spot of suitable concentration on the MALDI plate.

Peptides and proteins seem to give best spectra at around 0.1 to 10 pmol mm^{-3} (Figs. 9.13, 9.14). Some proteins, particularly glycoproteins, may yield better results at concentrations up to 10 pmol mm^{-3}. Oligonucleotides give better spectra at around 10 to 100 pmol mm^{-3} while polymers require a concentration around 100 pmol mm^{3}. (Note: 1 pmol nm^{-3} = 10^{-6} mol dm^{-3}.)

(a)

(b)

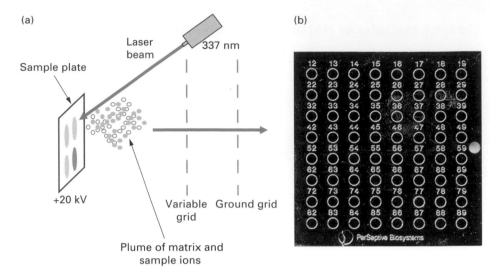

Fig. 9.10 MALDI ionisation mechanism and MALDI–TOF sample plate. (a) The sample is mixed, in solution, with a 'matrix' – the organic acid in excess of the analyte (in a ratio between 1000 : 1 to 10 000 : 1) and transferred to the MALDI plate. An ultraviolet laser is directed to the sample (with a beam diameter of a few micrometres) for desorption. The laser radiation of a few nanoseconds' duration is absorbed by the matrix molecules, causing rapid heating of the region around the area of laser impact and electronic excitation of the matrix. The immediate region of the sample explodes into the high vacuum of the mass spectrometer, creating gas phase protonated molecules of both the acid and the analyte. The laser flash ionises matrix molecules: neutrals (M) and matrix ions $(MH)^+$, $(M - H)^-$ and sample neutral fragments (A). Sample molecules are ionised by gas phase proton transfer from the matrix:

$MH^+ + A- > M + AH^+.$
$(M - H)^- + A- > (A - H)^- + M.$

The matrix serves as an absorbing medium for the ultraviolet light converting the incident laser energy into molecular electronic energy, both for desorption and ionisation and as a source of H^+ ions to transfer to, and ionise, the analyte molecule. (b) A MALDI sample plate.

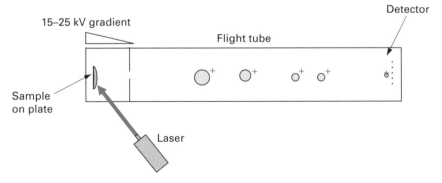

Fig. 9.11 Principle of time-of-flight (TOF). The ions enter the flight tube, where the lighter ions travel faster than the heavier ions to the detector. If the ions are accelerated with the same potential at a fixed point and a fixed initial time, the ions will separate according to their mass to charge ratios. This time of flight can be converted to mass. Typically a few 100 pulses of laser light are used, each of around a few nanoseconds' duration and the information is accumulated to build up a good spectrum. With the benefit of a camera that is used to follow the laser flashes one can move or 'track' the laser beam around the MALDI plate to find so called **sweet spots** where the composition of co-crystallised matrix and sample is optimal for good sensitivity.

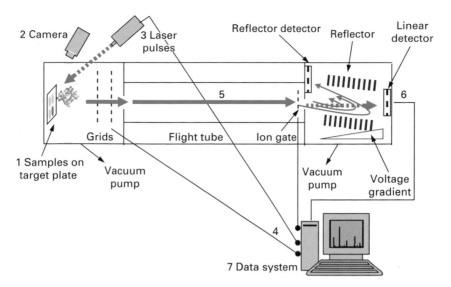

Fig. 9.12 MALDI–TOF instrument components. (1) Sample mixed with matrix is dried on the target plate which is introduced into high-vacuum chamber. (2) The camera allows viewing of the position of the laser beam which can be tracked to optimise the signal. (3) The sample/matrix is irradiated with laser pulses. (4) The clock is started to measure time-of-flight. (5) Ions are accelerated by the electric field to the same kinetic energy and are separated according to mass as they fly through the flight tube. (6) Ions strike the detector either in linear (dashed arrow) or reflectron (full arrows) mode at different times, depending on their m/z ratio. (7) A data system controls instrument parameters, acquires signal versus time and processes the data.

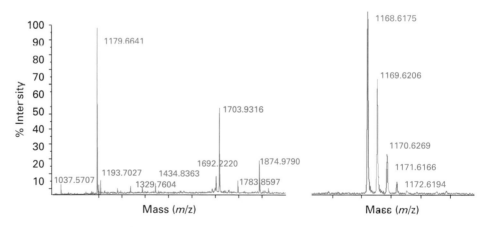

Fig. 9.13 Two examples of MALDI–TOF peptide spectra. The left-hand spectrum is from a protein digest mixture and the right-hand image is an expanded one of a small part of a spectrum showing ^{13}C-containing forms (see Section 9.5.4).

9.3.9 Delayed extraction

In the first MALDI–TOF instruments, the ions in the plume of material generated by the laser pulse were continuously extracted by a high electrostatic field. Since this plume of material occupies a small but finite volume of space, ions arising at different

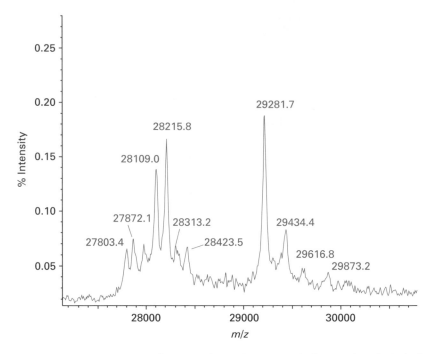

Fig. 9.14 MALDI–TOF spectrum of protein isoforms. The spectrum is almost exclusively singly charged ions representing the molecular ion species of the constituent proteins. Compare this spectrum with the electrospray spectrum of another protein (Fig 9.6) where the multiply charged ions result in multiple peaks which would make it harder to interpret masses of mixtures. (I acknowledge the assistance of Bruker Daltonics who carried out the analysis.)

places could have different energies. This energy spread (and fragmentation occurring during this initial extraction period) usually broadens the peak corresponding to any particular ion which leads to lower mass accuracy. However, if extraction is delayed until all ions have formed, this spread is minimised. The procedure is known as delayed extraction (DE), whereby the ions are formed in either a weak field or no field during a predetermined time delay, and then extracted by the application of a high-voltage pulse. The degree of fragmentation of ions (Section 9.3.10) can also be controlled, to some extent, by the length of the time delay. Delayed extraction is illustrated in Fig. 9.15.

9.3.10 **Post-source decay**

Post-source decay (PSD) is the process of fragmentation that may occur after an ion (the precursor ion) has been extracted from the source. Many biological molecules, particularly peptides, give rise to ions that dissociate over a timespan of microseconds and most precursor ions will have been extracted before this dissociation is complete. The fragment ions generated will have the same velocity as the precursor and cause peak broadening and loss of resolution in a linear TOF analyser (Fig. 9.16). The problem is overcome by the use of a reflector.

Example 2 **PEPTIDE MASS DETERMINATION (I)**

Question A peptide metabolite and an enzyme digest of it were analysed by a combination of mass spectrometric techniques giving the data listed below:

(i) The peptide showed two signals at 3841.5 and 1741 in the MALDI–TOF.
(ii) Five signals could be discerned when the peptide was introduced into a mass spectrometer via an electrospray ionisation source:

m/z	498.2	581.1	697.1	871.2	1161.2

(iii) HPLC-MS of the digest indicated *four* components, the $(M + H)^+$ data for the components being $m/z = 176, 625, 1229$ and 1508. The ions corresponding to the MS of the '625' component appeared at $m/z = 521, 406, 293, 130$ and 113.
(iv) HPLC-MS–MS of the $m/z = 406$ ion of the '625' component identified two ions at $m/z = 378$ and 336, and that of the $m/z = 113$ ion gave $m/z = 85$ and 57, in the product ion spectra.

Use the above data to compare and contrast the different ionisation methods, deduce a molecular mass for the peptide and determine a sequence for the '625' component. Use the amino acid residue mass values in Table 9.2.

Answer The data in (i) are $m/z = 3481.5$ and $m/z = 1741$. These data could represent either of the following possibilities:

(a) $m/z = 3481.5 = (M + H)^+$
    when $m/z = 1741 \equiv (M + H)^{2+}$, giving $M = 3480.5$
(b) $m/z = 3481.5 \equiv (2M + H)^+$
    when $m/z = 1741 \equiv (M + H)^+$, giving $M = 1740$

Consideration of the data in (ii) allows a choice to be made between these two alternatives, using $n_2 = (m_1 - 1)/(m_2 - m_1)$ and $M = n_2(m_2 - 1)$.

$m_1 - 1$	$m_2 - m_1$	n_2	$m_2 - 1$	M (Da)	z
870.2	290	3.0006	1160.2	3481.2	3
696.1	174.1	3.9982	870.2	3479.3	4
580.1	116	5.0000	696.1	3481.1	5
497.2	82.9	5.9975	580.1	3479.2	6

$\Sigma M = 13920.8$ Da
Mean $M = 3480.2$ Da

The mean M result confirms set (a) of the conclusions above concerning the data obtained from the MALDI experiments.

Example 2 (*cont.*)

The data in (iii) indicate that four products arise from the enzymatic digest of the original peptide. As these products arise directly from the original, the sum of these masses will be related to the M of the pepide.

Therefore

$$176 + 625 + 1229 + 1508 = 3538 \, \text{Da}$$

The difference between this mass and the M determined above is

$$3538 - 3480.2 = 57.8 \approx 58 \, \text{Da}$$

The difference of 58 mass units is explained as follows.

Each of the enzyme digest products is protonated (to be 'seen' in the mass spectrometer). Hence this accounts for 4 units. The remaining 54 unit increase arises from the enzymic hydrolysis. From a linear peptide, four products arise from three cleavage points (three cuts in a piece of string give four pieces). Each cleavage point requires the input of one water molecule (hydrolysis, $H_2O = 18$). Three cleavage points require $3 \times 18 = 54$.

The $m/z = 625$, $(M + H)^+$, peak was subjected to further mass spectrometry and sequence ions were observed.

m/z	624		521		406		293		130		113
Δ		103		115		113		163		17	
aa		Cys		Asp		Ile/Leu		Tyr		Ile/Leu	

The loss of 113 from the $m/z = 406$ ion indicates either Ile or Leu. MS2 shows consecutive losses of 28 (CO) and 42 ($CH_2 = CH = CH_3$) which is indicative of Leu. The loss of 17 (not a sequence ion) from 130 confirms this as the C-terminal amino acid.

The predicted sequence from the N-terminal end is
Cys - Asp - Leu - Tyr - Ile

The reflector

A reflector (or reflectron) is a type of ion mirror that provides higher resolution in MALDI–TOF. The reflector increases the overall path length for an ion and it corrects for minor variation in the energy spread of ions of the same mass. Both effects improve resolution. The device has a gradient electric field and the depth to which ions will penetrate this field, before reversal of direction of travel, depends upon their energy. Higher-energy ions will travel further and lower-energy ions a shorter distance. The flight times thus become focussed, while neutral fragments are unaffected by the deflection. Figure 9.16 shows a diagrammatic representation of a MALDI–TOF

Example 3 **PEPTIDE MASS DETERMINATION (II)**

Question Consider the following mass spectrometric data obtained for a peptide metabolite.

(i) The MALDI spectrum showed two signals at $m/z = 1609$ and 805.

(ii) There were two significant signals in positive ion trap MS mass spectrum at $m/z = 805$ and 827, the latter signal being enhanced on addition of sodium chloride.

(iii) Signals at $m/z = 161.8, 202.0, 269.0$ and 403.0 were observed when the sample was introduced into the mass spectrometer via an electrospray ionisation source.

Use these data to give an account of the ionisation methods used. Discuss the significance of the data and deduce a relative molecular mass for the metabolite. Use the amino acid residue mass values in Table 9.2.

Answer (i) Signals in the MALDI spectrum were observed at $m/z = 1609$ and 805. These data could represent the following possibilities:

(a) $m/z = 1609 \equiv (M + H)^+$
    when $m/z = 805 \equiv (M + 2H)^{2+}$
    and $m/z = 403 \equiv (M + 4H)^{4+}$, giving $M - 1608\,\text{Da}$

(b) $m/z = 1609 \equiv (2M + H)^+$
    when $m/z = 805 \equiv (M + H)^+$
    and $m/z - 403 \equiv (M + 2H)^{2+}$, giving $M = 804\,\text{Da}$

(ii) The distinction between the above options can be made by considering the ion trap data. This mode of ionisation gave peaks at $m/z = 805$ and 827, the latter being enhanced on the addition of sodium chloride. This evidence suggests:

$m/z = 805 = (M + H)^+$
$m/z = 827 \equiv (M + Na)^+$

giving $M = 804$ Da and supports option (b) from the MALDI data.

(iii) The multiply charged ions observed in the electrospray ionisation method allow an average M to be calculated. Using the standard formula:

$m_1 - 1$	$m_2 - m_1$	n_2	$m_2 - 1$	M (Da)	z
268.0	134	2.0	402.0	804	2
201.0	67	3.0	268.0	804	3
160.8	40.2	4.0	201.0	804	4

The molecular mass is clearly 804 Da, confirming the above conclusions.

instrument that includes the facility for both linear and reflectron modes of ion collection. The reflectron improves resolution and mass accuracy and also allows structure and sequence information (in the case of peptides) to be obtained by PSD analysis.

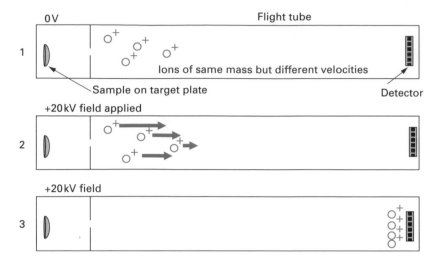

Fig. 9.15 Delayed extraction (DE). (1) No applied electric field. The ions spread out. (2) Field applied. The potential gradient accelerates slow ions more than fast ones. (3) Slow ions catch up with faster ones at the detector.

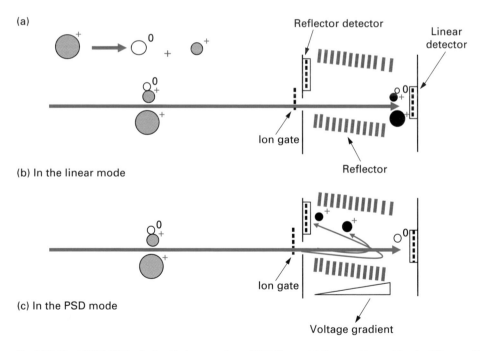

Fig. 9.16 The MALDI–TOF reflector. Post-source decay (PSD) theory. (a) Fragment ions arising by PSD as well as the neutral fragments and the precursor ions have the same velocity and reach the detector simultaneously. This prevents a distinction between precursor and PSD fragment. (b) In the linear mode the charged fragments are not separated. (c) In the reflector mode, the fragment that does not retain the charge (neutral, denoted by $0°$) is not deflected in the reflector but the charged fragments ($•^+$) are deflected according to their m/z and a spectrum of the fragment (daughter) ions is recorded, albeit of a limited m/z range for each setting of the reflector voltages.

Sequencing peptides by PSD analysis in MALDI–TOF is less straightforward (and in a large percentage of experiments is unsuccessful) than tandem MS on a quadrupole ESI or ion trap instrument. At any given setting of the reflector/ion mirror, charged fragments of a particular range of m/z are focussed in the reflector (Fig. 9.16). Fragment ions of m/z above and below this narrow range are poorly focussed. Therefore, since only fragment ions of a limited mass range are focussed for a given mirror ratio in the reflector, a number of spectra are run at different settings and stitched together to generate a composite spectrum.

Types of MALDI sample plates

MALDI sample plate types that are available include 100-well stainless steel flat plates. These are good for multiple sample analysis where close external calibration is used, that is the use of a compound or compounds of known molecular mass placed on an adjacent spot to calibrate the instrument. It is also easier to see crystallisation of the matrix on this type of surface.

Four-hundred-spot Teflon-coated plates have particular application for concentrating sample for increased sensitivity. Due to the very small diameter of the spots, it is difficult to spot accurately manually but these plates are good for automated sample spotting. Only in the centre of each spot is the surface of the plate exposed therefore the sample does not 'wet' over the whole surface but concentrates itself into the centre of each spot as it dries. Gold-coated plates with wells (2 mm diameter, see Fig. 9.10b) are good surfaces on which to contain the spread of sample and matrix when used with highly organic solvents, e.g. tetrahydrofuran (THF) preparations for polymers. They also allow on-plate reactions within the well with thiol-containing reagents that bind to the gold surface.

9.3.11   Novel hybrid instruments

There are a number of commercial developments of hybrid MS instruments that involve coupling an electrospray, ion trap or a MALDI ion source with a hybrid quadrupole orthogonal acceleration time-of-flight mass spectrometer (Fig. 9.17). This potentially leads to improved tandem MS performance from MALDI phase samples.

The intention of the development of these instruments is to combine the best features of both types of ion source with the best features of all types of analyser in order to improve tandem MS capability and increase sensitivity. Hybrid magnetic sector instruments are also manufactured where the first mass spectrometer is a two-sector device and the second mass spectrometer is a quadrupole.

9.3.12   Fourier-transform ion cyclotron resonance MS

The recent development of Fourier-transform ion cyclotron resonance (FT ICR) mass spectrometry has great potential in analysis of a wide range of biomolecules. It is potentially the most sensitive mass spectrometric technique and has very high

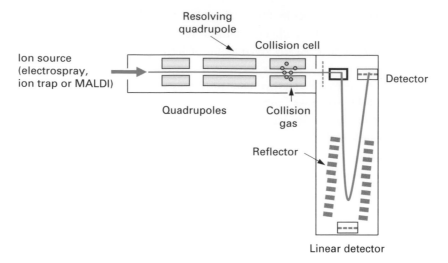

Fig. 9.17 Diagram of a hybrid quadrupole TOF MS. The diagram shown here does not represent any specific instrument from a particular manufacturer. The source may be an ion trap device, an electrospray or even a MALDI source (such as in the 'MALDI Q-TOF' from Micromass). Other hybrid instruments include the Bruker Daltonics 'BioTOF III, ESI-Q-q-TOF System' and the 'QSTAR' Hybrid LC/MS/MS from Applied Biosystems with an electrospray, nanospray or an optional MALDI source. The Shimadzu Biotech 'AXIMA MALDI QIT TOF' combines a MALDI source with an ion trap and reflectron TOF mass analyser.

mass resolution; $>10^6$ is observable with most instruments. The instrument also allows tandem MS to be carried out. The ions can be generated by a variety of techniques, such as an ESI or a MALDI source. FT-ICR MS is based on the principle of ions, which while orbiting in a magnetic field, are excited by radio frequency (RF) signals. As a result, the ions produce a detectable image current on the cell in which they are trapped. The time-dependent image current is Fourier transformed to obtain the component frequencies of the different ions, which correspond to their m/z (Fig. 9.18).

9.3.13 Orbitrap mass spectrometer

The resolving power of FTICR-MS is proportional to the strength of the magnetic field therefore superconducting magnets (3–12 tesla) are required, which makes for high maintenance cost. The Orbitrap mass analyser is a lower-cost alternative to the high magnetic field FTICR mass spectrometer. Ions are trapped in the orbitrap, where they undergo harmonic ion oscillations, along the axis of an electric field (see Fig. 9.19). Their m/z values are measured from the frequency of the ions, measured non-destructively, using Fourier transforms to obtain the mass spectrum. The instrument has high mass resolution (up to 150 000), high mass accuracy (2–5 p.p.m.), an m/z range of >6000 and a dynamic range greater than 10^3 with sub-femtomol sensitivity.

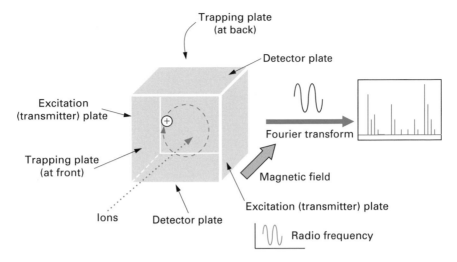

Fig. 9.18 Schematic diagram of the Fourier-transform ion cyclotron resonance (FT-ICR) instrument. The technique involves trapping, excitation and detection of ions to produce a mass spectrum. The **trapping plates** to maintain the ions in orbit are at the front and back in the schematic. The **excitation** or **transmitter plates** where the radio frequency (RF) pulse is given to the ions are shown at each side and the **detector plates** that detect the **image current** which is Fourier transformed are shown at the top and bottom. The sample source is normally electrospray (described in Section 9.2.4) or MALDI (see Section 9.3.8 and Fig. 9.10). The ions are focussed and transferred into the **analyser cell** under high vacuum. The analyser cell is a type of ion trap in a spatially uniform strong magnetic field which constrains the ions in a circular orbit, the frequency of which is determined by the mass, charge and velocity of the ion. While the ions are in these stable orbits between the detector electrodes they will not give a measurable signal. In order to achieve this, ions of a given *m/z* are excited to a wider orbit by applying a RF signal of a few milliseconds' duration. One frequency excites ions of one particular *m/z* which results in the ions producing a detectable **image current**. This time-dependent image current is Fourier transformed to obtain the component frequencies which correspond to the *m/z* of the different ions. The angular frequency measurements produce values for *m/z*. Therefore the mass spectrum is determined to a very high mass resolution since frequency can be measured more accurately than any other physical property. After excitation, the ions relax back to their previous orbits and high sensitivity can be achieved by repeating this process many times.

9.4 DETECTORS

9.4.1 Introduction

The ions from the mass analyser impinge on a surface of a detector where the charge is neutralised, either by collection or donation of electrons. An electric current flows that is amplified and ultimately converted into a signal that is processed by a computer. The total ion current (TIC) is the sum of the current carried by all the ions being detected at any given moment and is a very useful parameter to measure during on-line MS. A plot of ion current versus time complements the ultraviolet trace that is also normally recorded during the chromatography run. Unlike the ultraviolet trace which depends on the absorbance of each component at the particular wavelength(s) set on the ultraviolet detector, the TIC is of course independent of the light-absorbing properties of a substance and depends only on its ionisability in the instrument.

(a)

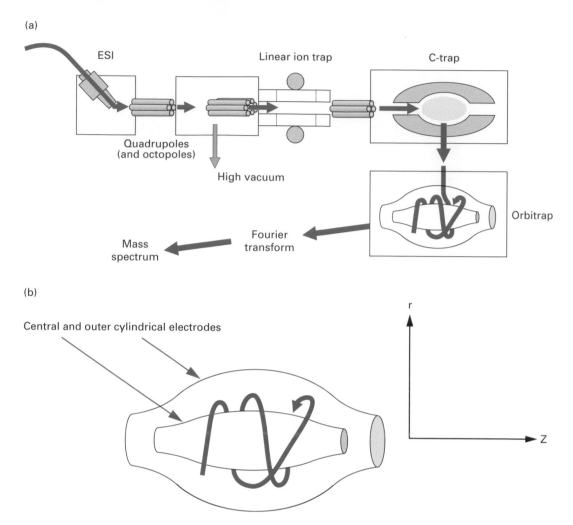

(b)

Fig. 9.19 (a) Simplified schematic of an Orbitrap mass spectrometer with an electrospray ionisation (ESI) source. Ions are transferred from an ESI source (described in Section 9.2.4) through three stages of differential vacuum pumping using RF guide quadrupoles and octopoles which focus and guide the ions through the various parts of the instrument. The ions are stored in the linear ion trap then axially ejected to the C-trap where they are squeezed into a small cloud. 'Bunches' of ions are then injected into the Orbitrap analyser. The third quadrupole, which is pressurised to less than 10^{-3} torr with collision gas, acts as an ion accumulator where ion/neutral collisions slow the ions which pool in an axial potential well at the end of the quadrupole (the C-trap). The linear ion trap operates on a similar principle to the ion trap described in Section 9.3.3. and Fig. 9.7. MALDI (see Section 9.3.8. and Fig. 9.10) is an alternative ion source. (b) Detail of the Orbitrap analyser. In the Orbitrap, the ions are trapped in a radial electric field between a central and an outer cylindrical electrode. They orbit around the central electrode with axial oscillations. The superimposed harmonic oscillations in the Z direction are detected by measuring the image current at the outer electrode. The frequency of oscillation is proportional to the m/z ratio and is detected and processed by fast Fourier transform, as in FT-ICR MS.

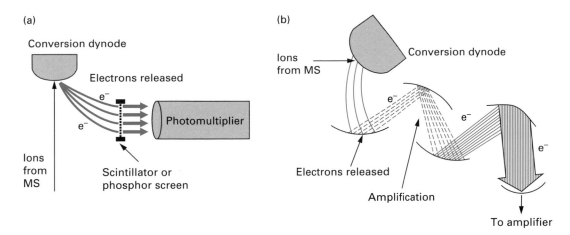

Fig. 9.20 Conversion dynode and electron multiplier. (a and b) Each ion strikes the conversion dynode (which converts ions to electrons) which emits a number of electrons that travel to the next, higher-voltage dynode. The secondary electrons from the conversion dynode are accelerated and focussed onto a second dynode, which itself emits secondary electrons. Each electron then produces several more electrons. Amplification is achieved through the 'cascading effect' of secondary electrons from dynode to dynode that finally results in a measurable current at the end of the electron multiplier. The cascade of electrons continues until a sufficiently large current for normal amplification is obtained. A series of up to 10–20 dynodes (set at different potentials) provides an amplification gain of 10^6 or 10^7.

9.4.2 Electron multiplier and conversion dynode

Electron multipliers are used as detectors for many types of mass spectrometers. These are frequently combined with a conversion dynode which is a device to increase sensitivity. The ion beam from the mass analyser is focussed onto the conversion dynode, which emits electrons in direct proportion to the number of bombarding ions. A positive ion or a negative ion hits the conversion dynode, causing the emission of secondary particles containing secondary ions, electrons and neutral particles (see Fig. 9.20). These secondary particles are accelerated into the dynodes of the electron multiplier. They strike the dynodes with sufficient energy to dislodge electrons, which pass further into the electron multiplier, colliding with the dynodes, producing more and more electrons.

9.5 STRUCTURAL INFORMATION BY TANDEM MASS SPECTROMETRY

9.5.1 Introduction

As mentioned above, the newer ionisation techniques ESI and MALDI are soft ionisation techniques (as is FAB and its derivative techniques). In contrast to EI, they do not produce significant amounts of fragment ions. Therefore in order to obtain structural information on biomolecules and sequence information (in the case of proteins and peptides), tandem MS has been developed. The technique can also be applied to obtain sequence information on oligosaccharides (see Sections 9.5.5 and 9.5.6) and oligonucleotides. Although it is unlikely that this method will ever replace DNA sequencing

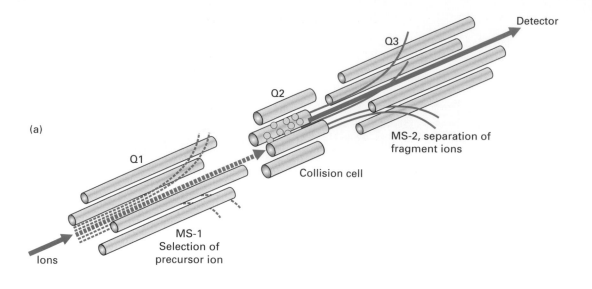

(a)

(b)

Fig. 9.21 Quadrupole MS sequencing. An ion of a particular *m/z* value is selected in the first quadrupole, Q1, as in Fig. 9.5, but instead of being detected, it passes through the second quadrupole, Q2, where it is subjected to collision with the collision gas. The Q3 acts like a second quadrupole mass spectrometer, MS-2, to scan *m/z* to obtain a spectrum of the fragment ions. The collision cell, Q2, is frequently a radio frequency (RF)-only quadrupole containing the appropriate collision gas. No mass filtering occurs here, the RF merely constrains the ions to allow a greater number of collisions to occur. The fragmentation depicted here is at the peptide bond and one of the fragments will retain the charge, resulting in either a y-series or a b-series ion (see Fig. 9.22).

gels, it can be used to identify positions of modified or labelled bases that might not be picked up by the Sanger dideoxy sequencing method.

Structural information can be obtained on almost any type of organic molecule, on an instrument that is suitable for that type of sample. This includes investigation of organic compounds on a magnet sector MS where two double-focussing magnetic sector machines can be combined into a four-sector device coupled through a collision cell.

The general procedure is that a mixture of ions is generated in the ion source of the mass spectrometer as normal and the ions are allowed to pass through the first mass analyser where an ion of a particular m/z is selected (but not detected). This ion then enters the collision cell and collides with an inert collision gas such as helium or argon. The kinetic energy of this ion is converted to vibrational energy and the ion fragments. This is known as collision-induced dissociation (CID) or collision-activated dissociation (CAD). The m/z values of the fragment ions are then determined in a second mass spectrometer (see Fig. 9.21 for an illustration of the principle in a quadrupole mass spectrometer). Collision cells may be placed in any of the field-free regions, leading to a wide variety of experimental methodologies for many different applications.

For example, as well as in the triple quadrupole MS this can be done in a hybrid instrument such as the Q-TOF (described in Section 9.3.11).

Since the principles of tandem MS are similar for most instrument configurations, further discussion will focus on electrospray tandem MS.

The procedure for obtaining structural and sequence information on polypeptides in ion trap MS has been described above (Section 9.3.3).

9.5.2 Sequencing of proteins and peptides

The identification of proteins involves protease cleavage, mostly by trypsin. Owing to the specificity of this protease, tryptic peptides usually have basic groups at the N- and C-terminis. Trypsin cleaves after lysine and arginine residues, both of which have basic side chains (an amino and a guanidino group respectively). This results in a large proportion of high-energy doubly charged positive ions that are more easily fragmented.

The digestion of the protein into peptides is followed by identification of the peptides by mass charge ratio (m/z) either as very accurate masses alone or by using a second fragmentation that gives ladders of fragments cleaved at the peptide bonds.

Although a wide variety of fragmentations may occur, there is a predominance of peptide bond cleavage which gives rise to peaks in the spectrum that differ sequentially by the residue mass. The mass differences are thus used to reconstruct the amino acid sequence (primary structure) of the peptide (Table 9.2).

Different series of ions, a, b, c and x, y, z, may be recognised, depending on which fragment carries the charge. Ions x, y and z arise by retention of charge on the C-terminal fragment of the peptide. For example, the z_1 ion is the first C-terminal residue; y_1 also contains the NH group (15 atomic mass units greater) and x_1 includes the carbonyl group; y_2 comprises the first two C-terminal residues, and so on. The a, b, and c ion series arise from the N-terminal end of the peptide, when the fragmentation results in retention of charge on these fragments.

Figure 9.22a shows an idealised peptide subjected to fragmentation. Particular series will generally predominate so that the peptide may be sequenced from both ends by obtaining complementary data (Fig. 9.22b). In addition, ions can arise from side chain fragmentation, which enables a distinction to be made between isomeric amino acids such as leucine and isoleucine.

The protein is identified by searching databases of expected masses from all known peptides from every protein (or translations from DNA) and theoretical masses from fragmented peptides. Sensitivity of tandem MS has been claimed down to zeptomole level.

Table 9.2 **Symbols and residue masses of the protein amino acids**

Name	Symbol	Residue mass^a	Side chain
Alanine	A, Ala	71.079	CH_3-
Arginine	R, Arg	156.188	$HN=C(NH_2)-NH-(CH_2)_3-$
Asparagine	N, Asn	114.104	$H_2N-CO-CH_2-$
Aspartic acid	D, Asp	115.089	$HOOC-CH_2-$
Cysteine	C, Cys	103.145	$HS-CH_2-$
Glutamine	Q, Gln	**128.131**	$H_2N-CO-(CH_2)_2-$
Glutamic acid	E, Glu	129.116	$HOOC-(CH_2)_2-$
Glycine	G, Gly	57.052	$H-$
Histidine	H, His	137.141	$Imidazole-CH_2-$
Isoleucine	I, Ile	**113.160**	$CH_3-CH_2-CH(CH_3)-$
Leucine	L, Leu	**113.160**	$(CH_3)_2-CH-CH_2-$
Lysine	K, Lys	**128.17**	$H_2N-(CH_2)_4-$
Methionine	M, Met	131.199	$CH_3-S-(CH_2)_2-$
Metsulphoxide	Met.SO	**147.199**	$CH_3-S(O)-(CD_2)_2-$
Phenylalanine	F, Phe	**147.177**	$Phenyl-CH_2-$
Proline	P, Pro	97.117	$Pyrrolidone-CH-$
Serine	S, Ser	87.078	$HO-CH_2-$
Threonine	T, Thr	101.105	$CH_3-CH(OH)-$
Tryptophan	W, Trp	186.213	$Indole-NH-CH=C-CH_2-$
Tyrosine	Y, Tyr	163.176	$4-OH-Phenyl-CH_2-$
Valine	V, Val	99.133	$CH_3-CH(CH_2)-$

Note: ^{a}Residue mass is the mass in a peptide bond, i.e. after loss of H_2O when the peptide bond is formed. The numbers in bold in the residue mass column indicate amino acids that may be ambiguous in a sequence determined by tandem MS due to close similarity or identity in mass.

9.5.3 Comparison of MS and Edman sequencing

Edman degradation (Section 8.4.3) to obtain the complete sequence of a protein is uncommon nowadays since genomes are available to search with fragmentary sequences. Most intact proteins, if they are not processed from a secretory or pro-peptide form, are blocked at the N-terminus, most commonly with an acetyl group. Other amino terminal blocking includes fatty acylation, most commonly with a myristoyl, C_{12} fatty acid, attached through a glycine residue but the presence of many shorter-chain fatty acids is known to occur. Cyclisation of glutamine to a pyroglutamyl

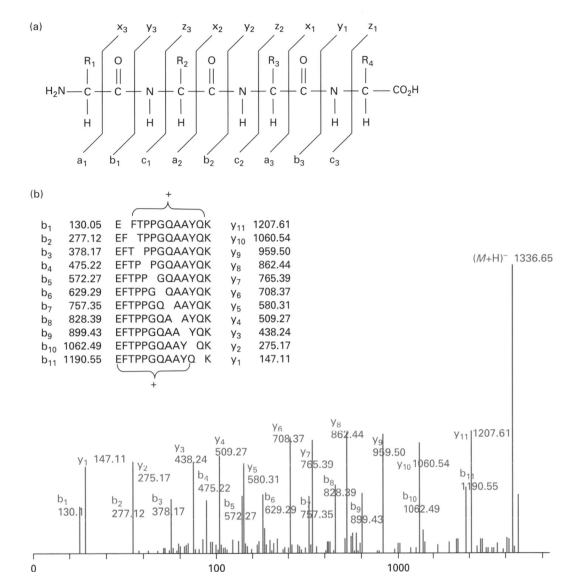

Fig. 9.22 Peptide fragment ion nomenclature and tandem MS spectrum of a peptide. (a) Charge may be retained by either the N- or C-terminal fragment, resulting in the a, b and c series of ions or x, y and z series respectively. Ions in the b and y series frequently predominate. Corresponding neutral fragments are of course not detected. (b) The sequence of the peptide from a mutant haemoglobin is: EFTPPGQAAYQK. The figure shows the tandem mass spectrum from collision-induced dissociation of the doubly charged $(M + 2H)^{2+}$ precursor, $m/z = 668.3$. Cleavage at each peptide bond results in the b or y ions when the positive charge is retained by the fragment containing the N- or C-terminus of the peptide respectively (see inset).

residue and post-translational modification to N-terminal trimethylalanine and dimethylproline also occur. In the case of recombinant proteins over-expressed in *E. coli*, the initiator residue *N*-formyl methionine is often incompletely removed. All these modifications leave the N-terminal residue without a free proton on the alpha nitrogen and Edman chemistry cannot proceed. Mass spectrometry has therefore been essential for their correct structural identification. The protein sequencing instruments are still important for solid phase sequencing to identify post-translational modifications; in particular, sites of phosphorylation and a combination of micro-sequencing and mass spectrometry techniques are now commonly employed for complete covalent structure determination of proteins.

Example 4   **PEPTIDE SEQUENCING (I)**

Question   An oligopeptide obtained by tryptic digestion was investigated by ESI–MS and ion trap MS–MS both in positive mode, and gave the following m/z data:

ESI	223.2	297.3								
Ion trap	146	203	260	357	444	591	648	705	802	890

  (i) Predict the sequence of the oligopeptide. Use the amino acid residual mass values in Table 9.2.
 (ii) Determine the average molecular mass.
(iii) Identify the peaks in the ESI spectrum.

Note: Trypsin cleaves on the C-terminal side of arginine and lysine.

Answer   (i) The highest mass peak in the ion trap MS spectrum is $m/z = 890$, which represents $(M + H)^+$.

Hence $M = 889\,\text{Da}$.

m/z	146	203	260	357	444	591	648	705	802	889
Δ		57	57	97	87	147	57	57	97	87
aa		Gly	Gly	Pro	Ser	Phe	Gly	Gly	Pro	Ser

The mass differences (Δ), between sequence ions, represent the amino acid (aa) residue masses. The lowest mass sequence ion, $m/z = 146$, is too low for arginine and must therefore represent Lys + OH. The sequence in conventional order from the N-terminal end would be:

Ser-Pro-Gly-Gly-Phe-Ser-Pro-Gly-Gly-Lys

 (ii) The summation of the residues = 889 Da, which is a check on the mass spectrometry value for M.
(iii) The m/z values in the ESI spectrum represent multiply charged species and may be identified as follows:
$m/z = 223.2 \equiv (M + 4H)^{4+}$ from $889/223.2 = 3.98$
$m/z = 297.3 \equiv (M + 3H)^{3+}$ from $889/297.3 = 2.99$

Remember that z must be an integer and hence values need to be rounded to the nearest whole number.

Table 9.3 **Mass differences due to isotopes in multiply charged peptides**

Charge on peptide	Apparent mass	Mass difference between isotope peaks
Single charge	$[(M + H)/1]$	1 Da
Double charge	$[(M + 2H)/2]$	0.5 Da
Triple charge	$[(M + 3H)/3]$	0.33 Da
n charges	$[(M + nH)/n]$	$1/n$ Da

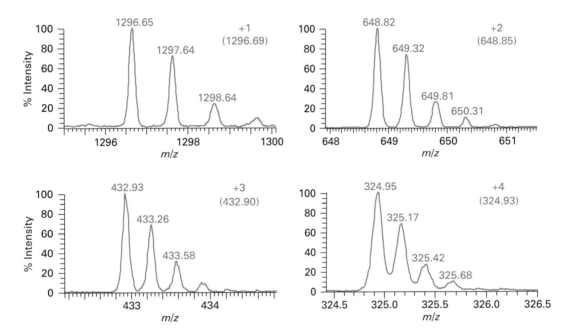

Fig. 9.23 Spectra of a multiply charged peptide. Finding the charge state of a peptide involves **zooming in** on a particular part of the mass spectrum to obtain a detailed image of the mass differences between different peaks that arise from the same biomolecule, due to isotopic abundance. This is mainly due to ^{12}C and its ^{13}C isotope, as described in the text.

9.5.4 Carbon isotopes and finding the charge state of a peptide

Since the mass detector operates on the basis of mass-to-charge ratio (m/z), mass assignment is normally made assuming a single charge per ion (i.e. $m/z = m + 1$ in positive ion mode). However, since there is around 1.1% ^{13}C natural abundance, with increasing size, peptides will have a greater chance of containing at least one ^{13}C and two ^{13}C, etc. A peptide of 20 residues has approximately equal peak heights of the 'all ^{12}C peptide' and of the peptide with one ^{13}C.

A singly charged peptide will show adjacent peaks differing in one mass unit; a doubly charged peptide will show adjacent peaks differing in half a mass unit and so on (Fig. 9.23 and Table 9.3). In the example illustrated, the peptide has a mass calculated from its sequence as 1295.69. The experimentally derived values are, for the singly charged ion, $[(M + H)/1] = 1296.65$ and for the doubly charged ion, $[(M + 2H)/2] = 648.82$.

For elements such as chlorine, the isotopic abundance is approximately $3:1$, $^{35}Cl:$ ^{37}Cl. If a compound contains a single chlorine atom, two ion species will be observed, with peak intensities in an approximate ratio of $3:1$. If a compound contains two chlorine atoms then three peaks will be seen.

The technique is particularly useful for determining which are the high-energy doubly charged tryptic peptides, for tandem MS.

Example 5 **PEPTIDE SEQUENCING (II)**

Question Determine the primary structure of the oligopeptide that gave the following, positive mode, MS–MS data:

m/z	149	305	442	529	617

Use the amino acid residual mass values in Table 9.2.

Answer $m/z = 617 \equiv (M + H)^+$

m/z	149	305	442	529	616
Δ		156	137	87	87
aa		Arg	His	Ser	Ser

Conventional order for the sequence would be:

Ser-Ser-His-Arg-?

It is important to note that no assignment has been given for the remaining $m/z = 149$. It may not in fact be a sequence ion and more information would be required, such as an accurate molecular mass of the oligopeptide, in order to proceed further. It is, however, possible to speculate as to the nature of this ion. If the $m/z = 149$ ion is the C-terminal amino acid then it would end in -OH and be 17 mass units greater than the corresponding residue mass. The difference between 149 and 17 is 132, which is extremely close to methionine, so this amino acid remains a possibility to end the chain.

9.5.5 **Post-translational modification of proteins**

Many chemically distinct types of post-translational modification of proteins are known to occur. These include the wide variety of acylations at the N-terminus of proteins (mentioned above) as well as acylations at the C-terminus and at internal sites. In this section, examples of the application of MS techniques employed for analysis of glycosylation, phosphorylation and disulphide bonds are given.

An up-to-date list of the broad chemical diversity of known modifications and the side chains of the amino acids to which they are attached is on the website

Example 6 PEPTIDE MASS DETERMINATION (III)

Question An unknown peptide and an enzymatic digest of it were analysed by mass spectrometric and chromatographic methods as follows:

(i) MALDI–TOF mass spectrometry of the peptide gave two signals at $m/z = 3569$ and 1785;

(ii) MALDI–TOF of the hydrolysate showed signals at $m/z = 766, 891, 953$ and 1016;

(iii) the data obtained from analysis of the peptide using coupled HPLC–MS operating through an electrospray ionisation source were $m/z = 510.7, 595.7, 714.6, 893.0$ and 1190.3;

(iv) when the hydrolysate was analysed by HPLC, four distinct components could be discerned.

Explain what information is available from these observations and determine a molecular mass, using the amino acid residue mass values in Table 9.2, for the unknown peptide.

Answer (i) Signals from MALDI–TOF were observed at $m/z = 3569$ and 1785. These data could represent either of the following possibilities:

(a) $m/z = 3569 \equiv (M + H)^+$
  when $m/z = 1785 \equiv (M + 2H)^{2+}$, giving $M = 3568$

(b) $m/z = 3569 \equiv (2M + H)^+$,
  when $m/z = 1785 \equiv (M + H)^+$, giving $M = 1784$

(ii) It is possible to distinguish between these two options by considering the MALDI–TOF of the products of hydrolysis. Four m/z values were obtained: 766, 891, 953 and 1016.

Each is a protonated species and the sum of these masses, 3626, will be of the order of the M of the original peptide. The value of this sum supports option (a) in (i) above

(iii) Electrospray ionisation data represent multiply charged ions. Using the standard formula the mean M may be obtained.

$m_1 - 1$	$m_2 - m_1$	n_2	$m_2 - 1$	M (Da)	z
892.0	297.3	3.0003	1189.3	3568.3	3
713.6	178.4	4.0000	892.0	3568.0	4
594.7	110.9	5.0016	713.6	3569.2	5
509.7	85.0	5.9964	594.7	3566.1	6

$\Sigma M = 14271.6$ Da
Mean $M = 3567.9$ Da

This more precise value confirms the conclusions found above. For an explanation of the mass difference between M_r and the sum of the hydrolysate products, refer to the answer to Example 2.

The data in (iv) are confirmatory chromatographic evidence that only four hydrolysis products were obtained.

'Delta Mass', which is a database of protein post-translational modifications that can be found at http://www.abrf.org/index.cfm/dm.home. There are hyperlinks to references to the modifications.

Protein phosphorylation and identification of phosphopeptides

Phosphate is reversibly covalently attached to eukaryotic proteins in order to regulate activity (Section 15.5.4). The modified residues are O-phosphoserine, O-phosphothreonine and O-phosphotyrosine but many other amino acids in proteins can be phosphorylated: O-phospho-Asp; S-phospho-Cys; N-phospho-Arg; N-phospho-His and N-phospho-Lys. Analysis of modified peptides by mass spectrometry is essential to confirm the exact location and number of phosphorylated residues, especially if no ^{32}P or other radiolabel is present. Identification of either positive or negative ions may yield more information, depending on the mode of ionisation and fragmentation of an individual peptide. Phosphopeptides may give better spectra in the negative ion mode since they have a strong negative charge due to the phosphate group. Phosphopeptides may not run well on MALDI–TOF and methods have been successfully developed for this type of instrument that employ examination of spectra before and after dephosphorylation of the peptide mixture with phosphatases.

Mass spectrometry of glycosylation sites and structures of the sugars

The attachment points of N-linked (through asparagine) and O-linked (through serine) glycosylation sites and the structures of the complex carbohydrates can be determined by MS. The loss of each monosaccharide unit of distinct mass can be interpreted to reconstruct the glycosylation pattern (see example in Fig. 9.24).

The 'GlycoMod' website, part of the ExPASy suite, provides valuable assistance in interpretation of the spectra. GlycoMod is a tool that can predict the possible oligosaccharide structures that occur on proteins from their experimentally determined masses. The program can be used for free or derivatised oligosaccharides and for glycopeptides. Another algorithm, GlycanMass, also part of the ExPASy suite, can be used to calculate the mass of an oligosaccharide structure from its oligosaccharide composition. GlyocoMod and GlycanMass are found at http://us.expasy.org/tools/glycomod/ and http://us.expasy.org/tools/glycomod/glycanmass.html respectively.

Identification of disulphide linkages by mass spectrometry

Mass spectrometry is also used in the location of disulphide bonds in a protein. Identification of the position of the disulphide linkages involves the fragmentation of proteins into peptides under low pH conditions to minimise disulphide exchange. Proteases with active site thiols should be avoided (e.g. papain, bromelain). Pepsin and cyanogen bromide are particularly useful. The disulphide-linked peptide fragments are separated and identified under mild oxidising conditions by HPLC–MS. The separation is repeated after reduction with reagents such as mercaptoethanol and dithiothreitol (DTT) to cleave –S–S– bonds and the products reanalysed as before. Peptides that were disulphide linked disappear from the spectrum and reappear at the appropriate positions for the individual components.

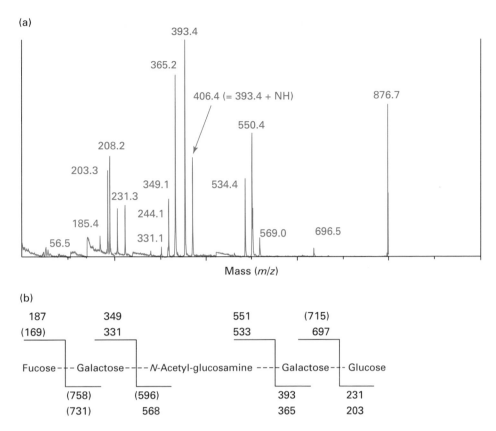

Fig. 9.24 MALDI–TOF PSD MS of carbohydrates. (a) PSD MS spectrum of the carbohydrate Fuc1–2Gal1–3GlucNAc1–3Gal1–4Glc using 2,5-dihydroxybenzoic acid (DHB) as matrix. On careful inspection of the spectrum one can observe a number of abrupt changes in baseline corresponding to where the PSD spectra have been stitched together. The peak at 876.7 Da is due to the mass of the intact molecule as a sodium adduct, i.e. the parent ion at 876.7 − $[M + Na]^+$ ion. (Spectrum courtesy of Dr Andrew Cronshaw.) (b) Interpretation of the spectrum. Experimentally derived fragment masses are mainly within 1 Da of the theoretical. The masses in parentheses were not seen in this experiment.

9.5.6 Selected ion monitoring

Selected ion monitoring (SIM) is typically used to look for ions that are characteristic of a target compound or family of compounds. This technique has particular application for on-line chromatography/MS where the instruments can be set up to monitor selected ion masses as the components elute successively from the capillary LC or reverse-phase HPLC column for example (Sections 11.3.3 and 11.9.3). Detection programmes or algorithms that are set up to carry out tandem MS on each component as it elutes from a chromatography column can be adapted to enable selective detection of many types of post-translationally modified peptides. This technique can selectively detect low-mass fragment ions that are characteristic markers that identify the presence of post-translational modifications such as phosphorylation, glycosylation, sulphation and acylation in any particular peptide. For example, phosphopeptides can be identified by production of phosphate-specific fragment ions of 63 Da (PO_2^-) and 79 Da (PO_3^-) by collision-induced dissociation during negative ion

HPLC–ES MS. Glycopeptides can be identified by characteristic fragment ions including hexose$^+$ (163 Da) and N-acetyl hexosamine$^+$ (204 Da).

Phosphoserine- and phosphothreonine-containing peptides can also be identified by a process known as neutral loss scanning where these peptides show loss of 98 Da by β-elimination of H_3PO_4 (Fig. 9.25).

9.6 ANALYSING PROTEIN COMPLEXES

Mass spectrometry is frequently used to identify partner proteins that interact with a particular protein of interest. Interacting proteins can be isolated by a number of methods including immunoprecipitation of tagged proteins from cell transfection; affinity chromatography and surface plasmon resonance. Surface plasmon resonance (SPR) (Section 13.3) technology has widespread application for biomolecular interaction analysis and during characterisation of protein–ligand and protein–protein interactions, direct analysis by MALDI–TOF MS of samples bound to the Biacore chips is now possible (where interaction kinetic data is also obtained; see Sections 13.3 and 17.3.2). Direct analysis of protein complexes by mass spectrometry is also possible. As well as accurate molecular weight of large biopolymers such as proteins of mass greater than 400 kDa, intact virus particles of M_r 40×10^6 (40 MDa) have been analysed using ESI–TOF. An icosahedral virus consisting of a single-stranded RNA surrounded by a homogeneous protein shell with a total mass of 6.5×10^6 Da and a rod-shaped RNA virus with a total mass of 40.5×10^6 Da were studied on a ESI–TOF hybrid mass spectrometer.

9.6.1 Sample preparation and handling

Mass analysis by ES–MS and MALDI–TOF is affected, seriously in some cases, by the presence of particular salts, buffers and detergents. Keratin contamination from flakes of skin and hair can be a major problem particularly when handling gels and slices; therefore gloves and laboratory coats must be worn. Work on a clean surface in a hood with air filter if possible and use a dedicated box of clean polypropylene microcentrifuge tubes tested to confirm that they do not leach out polymers, mould release agents, plasticisers, etc. Sample clean-up to remove or reduce levels of buffer salts, EDTA, DMSO, non-ionic and ionic detergents (e.g. SDS) etc. can be achieved by dilution, washing, drop dialysis and ion exchange resins. If one is analysing samples by MALDI–TOF, on-plate washing can remove buffers and salts. Sample clean-up can also be achieved by pipette tip chromatography (Section 11.2.5). This consists of a miniature C_{18} reverse-phase chromatography column, packed in a 10 nm^3 pipette tip. The sample, in low or zero organic solvent-containing buffer, is loaded into the tip with a few up- and-down movements of the pipette piston to ensure complete binding of the sample. Since most contaminants described above will not bind, the sample is trapped on the reverse phase material and eluted with a solvent containing high organic solvent (typically 50–75% acetonitrile). This is particularly applicable for clean-up of samples after in-gel digestion of protein bands separated on SDS-PAGE. Coomassie Brilliant Blue dye is also removed by this procedure. The technique can be used to concentrate samples and fractionate a mixture. Purification

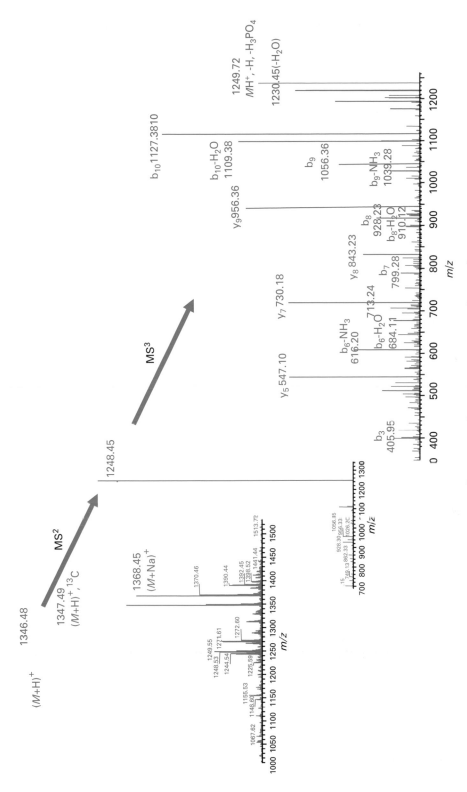

Fig. 9.25 MS identification of phosphopeptides. Sequence is YEILNS^{P}EKAC where S^P is phosphoserine. The MS^2 and MS^3 spectra are shown. The first tandem MS experiment mainly results in loss of H_3PO_4, 98 Da. Particular problems may also be associated with electrospray mass spectrometry of phosphopeptides, where a high level of Na+ and K+ adducts is regularly seen.

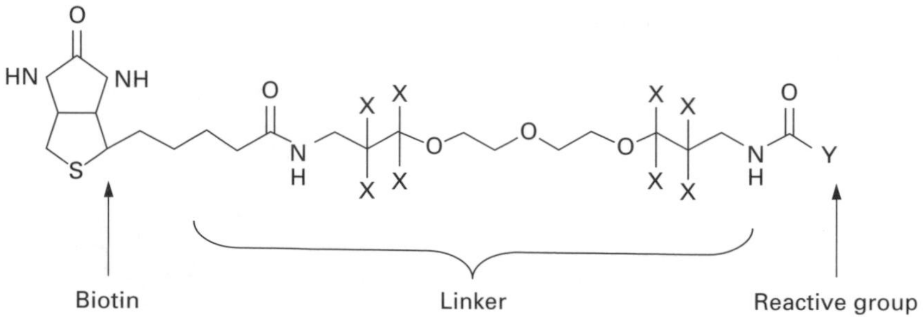

Fig. 9.26 Structure of the ICAT reagent. The ICAT reagent is in two forms, heavy (eight deuterium atoms) and light (no deuterium). The reagent has three elements: an affinity tag (biotin), to isolate ICAT-labelled peptides; a linker in two forms that has stable isotopes incorporated; and a reactive group (Y) with specificity toward thiol groups or other functional groups in proteins (e.g. SH, NH₂, COOH). The heavy reagent is D8-ICAT (where X is deuterium) and light reagent is D0-ICAT (where X is hydrogen). Two protein mixtures representing two different cell states are treated with the isotopically light and heavy ICAT reagents; an ICAT reagent is covalently attached to each cysteine residue in every protein. The protein mixtures are combined; proteolysed and ICAT-labelled peptides are isolated on an avidin column utilising the biotin tag. Peptides are separated by microbore HPLC. Since each pair of ICAT-labelled peptides is chemically identical they are easily visualised because they co-elute, with an 8 Da mass difference. The ratios of the original amounts of proteins from the two cell states are strictly maintained in the peptide fragments. The relative quantification is determined by the ratio of the peptide pairs. The protein is identified by database searching with the sequence information from tandem MS analysis by selecting peptides that show differential expression between samples.

can also be carried out to specifically bind one particular component in a mixture. Immobilised metal ion affinity columns are used to enrich phosphopeptides.

9.6.2 Quantitative analysis of complex protein mixtures by mass spectrometry

Proteome analysis (described in Section 8.5) involves the following basic steps:

- run a gel (one-dimensional (1D) or two dimensional (2D)),
- stain,
- scan to identify spots of interest,
- excise gel spots,
- extract and digest proteins,
- mass analyse the resulting peptides,
- search database.

The initial separation of proteins currently relies on gel electrophoresis which has a number of limitations including the difficulty in analysing all the proteins expressed due to huge differences in expression levels. Although thousands of proteins can be reproducibly separated on one 2D gel from approximately 1 mg of tissue/biopsy or biological fluid, the dynamic range of protein expression can be as high as nine orders of magnitude. One development that has helped to overcome some of the problems is the isotope–coded affinity tag (ICAT) strategy for quantifying differential protein expression.

The heavy and light forms of the sulphydryl (thiol-)-specific ICAT reagent (whose structure is illustrated in Fig. 9.26) are used to derivatise proteins in respective samples isolated from cells or tissues in different states. The two samples are combined and proteolysed, normally with trypsin, for reasons explained above. The labelled peptides are purified by affinity chromatography utilising the biotin group on the ICAT reagent then analysed by MS on either LC–MS MS (including ion trap) or MALDI–TOF instruments. The relative intensities of the ions from the two isotopically tagged forms of each specific peptide indicate their relative abundance. These pairs of peptides are easily detected because they co-elute from reverse-phase microcapillary liquid chromatography (RP–μLC) and contain eight mass units of difference due to the two forms of the ICAT tag. An initial MS scan identifies the peptides from proteins that show differential expression by measuring relative signal intensities of each ICAT-labelled peptide pair. Peptides of interest are then selected for sequencing by tandem MS and the particular protein from which a peptide originated can be identified by database searching the tandem MS spectral data.

9.6.3 iTRAQ

An alternative method for quantitative analysis of protein mixtures is to use the iTRAQ reagents from Applied Biosystems. These are a set of four isobaric (same mass) reagents which are amine-specific and yield-labelled peptides identical in mass and hence also identical in single MS mode, but which produce strong, diagnostic, low-mass tandem MS signature ions allowing quantitation of up to four different samples (or triplicate analyses plus control of the same sample) simultaneously (Fig. 9.27). Information on post-translational modifications is not lost and since all peptides are tagged, proteome coverage is expanded, and since multiple peptides per protein are analysed, this improves confidence and quantitation.

As a consequence of mixing the multiple proteome samples together, the complexity of both MS and tandem MS data is not increased and since there is no signal splitting in either mode, low-level analysis is enhanced as a result of the signal amplification.

The protocol involves reduction, alkylation and digestion with trypsin of the protein samples in parallel, in an amine-free buffer system (Fig. 9.28). The resulting peptides are labelled with the iTRAQ reagents. The samples are then combined and depending on sample complexity, they are may be directly analysed by LC–MS MS after one-step elution from a cation exchange column to remove reagent by-products. Alternatively, to reduce overall peptide complexity of the sample mixture, fractionation can be carried out on the cation-exchange column by stepwise elution of part of the complex mixture. Very recently the company has launched a kit with eight different isobaric tagging reagents (the principle is the same).

9.6.4 Stable isotope labelling with amino acids in cell culture

Alternatives to the above include stable isotope labelling with amino acids in cell culture (SILAC) which is useful in investigation of signalling pathways and protein interactions. As the name implies, this technique involves metabolically labelling protein samples in cell cultures that are grown with different stable isotopically labelled amino acids such as ^{13}C and/or ^{15}N lysine and arginine. It is also possible to use ^{12}C and ^{13}C leucine.

Fig. 9.27 iTRAQ reagent structure. The iTRAQ reagents consist of a charged reporter group, a peptide reactive group (an NHS, *n*-hydroxysuccinimide) which reacts with amino groups and a neutral balance group. The last part maintains an overall mass of 145. The term 'isobaric' is defined as two or more species that have the same mass. The peptide reactive group covalently links an iTRAQ reagent isobaric tag with each lysine side chain and N-terminus of a polypeptide, labelling all peptides in a given sample digest. By combining multiple iTRAQ-reagent-labelled digests into one sample mixture, the MS resembles that of an individual sample (assuming the same peptides are present). The balance group ensures that an iTRAQ-reagent-labelled peptide displays the same mass, whether labelled with iTRAQ reagent 114, 115, 116, or 117. The reporter group retains the charge after fragmentation and the balance group undergoes neutral loss. In the case of '8plex' iTRAQ reagents, additional isotopes are employed to provide reporter groups of 113 to 121 Da (with corresponding masses of balance groups of 192 to 184 (the chemical structure is changed to permit a greater variety of stable isotopes). A reporter group of 120 Da (and of course the balance group of 185) is not used as this may be confused with the ammonium ion for phenylalanine.

9.7 COMPUTING AND DATABASE ANALYSIS

9.7.1 Organic compound databases

Mass spectrometry organic compound databases are available to identify the compound(s) in analyte. The spectra in the databases are obtained by electron impact ionisation. Two such databases are:

- Integrated Spectral Data Base System for Organic Compounds (SDBS) from the National Institute of Advanced Industrial Science and Technology NIMC Spectral Database System. Data on specific compounds can be searched with compound name, molecular formula, number of atoms (CHNO) and molecular mass. The database contains 24 000 electron impact mass spectra as well as over 51 000 Fourier-transform infrared (FT-IR) spectra, 14 700 ^{1}H-NMR spectra and 13 000 ^{13}C-NMR spectra. The URL is http://riodb01.ibase.aist.go.jp/sdbs/cgi-bin/cre_index.cgi?lang=eng

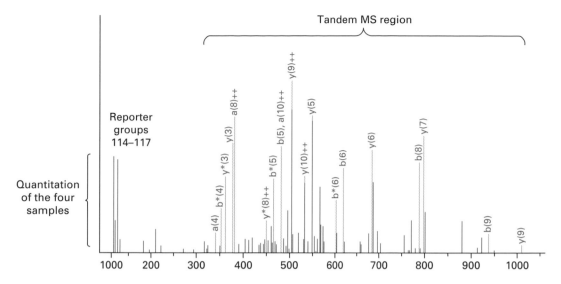

Fig. 9.28 Identification and quantitation of proteins by iTRAQ reagents. During MS/MS (along with the usual peptide fragmentation) the isobaric tag cleaves at the sites indicated. As result of fragmentation, there is neutral loss of the balance group, and the reporter groups are generated, displaying diagnostic ions in the low-mass region at between m/z values 114 to 117. Because this region is free of other common ions, quantification of the peak areas of these resultant ions represents the relative amount of a given peptide in the respective sample.

- NIST Chemistry WebBook. NIST is the National Institute for Standards and Technology. Data on specific compounds in the Chemistry WebBook can be searched by name, chemical formula, CAS (Chemical Abstracts Service) registry number, molecular mass or selected ion energetics and spectral properties. This site comprises electron impact mass spectra for over 15 000 compounds as well as thermochemical data for over 7000 organic and small inorganic compounds and IR spectra for over 16 000 compounds. The URL is: http://webbook.nist.gov/chemistry/.

9.7.2 Identification of proteins

Database searches to identify a particular protein that has been analysed by mass spectrometry are particularly important. This section gives an overview of websites for proteomic identification.

Identification of proteins can be carried out by using many websites, for example: 'Mascot' from Matrix Science (http://www.matrixscience.com/search_form_select.html) and 'Protein Prospector' (http://prospector.ucsf.edu/). The search can be limited by searching a particular species or taxon, e.g. mammalia only, thus increasing the speed. However, when looking for a homologous sequence the species should not be defined. The modification of cysteine residues, if any, should be included otherwise the number of peptides matched to the theoretical list will be decreased, producing a worse hit. If no cysteine modification has been carried out, and if the protein originates from a gel sample, then much of this residue will have been converted to acrylamide-modified Cys.

Unmatched masses should be re-searched since sometimes two or more proteins run together on electrophoresis. Note the delta p.p.m. (the difference between the theoretical and the experimental mass of a particular peptide) which should be low and consistent. This gives an indication of whether the result is genuine. If an internal calibration has been performed the mass accuracy parameter can be set to 20 p.p.m. For a close external calibration this should be set to 50 p.p.m. If a hit is not found with the first search this parameter can be increased.

Different protein databases can be searched:

- MSDB is a non-identical database maintained by the Proteomics Department at Imperial College, London.
- NCBInr is a non-identical database maintained by NCBI for use with their search tools BLAST and Entrez. The entries have been compiled from GenBank CDS translations, PIR, SWISS-PROT, PRF and PDB.
- SwissProt is a database that is ideal for peptide mass fingerprint searches in which sequences are non-redundant, rather than non-identical; therefore there may be fewer matches for a tandem MS search than from a comprehensive database, such as MSDB or NCBInr.
- dbEST is the division of GenBank that contains 'single-pass' cDNA sequences, or expressed sequence tags, from a number of organisms.

NCBInr is the largest database while Swiss Prot is smaller. However Swiss Prot provides the most information with the protein hits. During a Mascot search, the nucleic acid sequences are translated in all six reading frames. dbEST is very large and is divided into three sections: EST_human, EST_mouse and EST_others. Nevertheless, searches of these databases take far longer than a search of one of the non-redundant protein databases. An EST database should only be searched if a protein database search has failed to find a match. If it is known that the protein is not larger than e.g. 100 kDa then the mass range should be limited to prevent false hits. Although the search will be refined by limiting to a particular mass range of the intact protein, the possibility of subunits or fragments must be considered. Some information on the isoelectric point of a protein will also be known for a 2D gel sample but this should also be treated cautiously.

If a number of larger size peptides are seen in the digest then the missed cleavages parameter should be increased. Typically this is set to 1 or 2.

If the possibility of post-translational or other modification is uncertain then the top three options should be selected, i.e. acetylation of the N-terminus, oxidation of Met, and conversion of Glu to pyro-Glu. If phosphorylation of S, T or Y is selected when not suspected this may lead to false hits. More than one amino acid can usually be listed in the box (e.g. STY 80 to select any phosphorylation).

The list of peptide masses should be input to four decimal places if possible. In the initial search, use masses from the higher signal intensity peaks and set the minimum number of peptides low compared to number of masses in the peptide list. To increase the specificity of the search this number can be increased. If no hits are found then this number can be decreased in subsequent passes. Be sure to select whether the fragment and precursor ions have been calculated from monoisotopic or average masses.

Deisotoping software is available to artificially remove the ^{13}C peaks arising from chemically identical peptides but which arise from the presence of the ^{13}C isotopic form of carbon. This simplifies the spectrum but more importantly this will ensure that the search algorithm will not be not confused and attempt to find two or more distinct peptides that each differ by 1 Da. This is particularly valuable when analysing peptide mixtures since overlapping isotope clusters are thus identified correctly and only the genuine ^{12}C peaks are reported.

If the resolution of the mass spectrum is not sufficient to resolve individual isotope peaks then the average mass is often reported. This is still the case with larger polypeptides and proteins (see Fig. 9.14) but in modern instruments, the all ^{12}C, one ^{13}C, two ^{13}C (etc.) peptide forms can be resolved (see Fig. 9.13).

Various software (including commercial software packages such as SEQUEST) is available to use the information on the fragment ions obtained from a tandem MS experiment to search protein (and DNA translation) databases to identify the sequence and the protein from which it is derived. Once the protein has been identified one can view the full protein summary and link to protein structure, Swiss 2D PAGE, nucleic acid databases, etc.

9.8 SUGGESTIONS FOR FURTHER READING

Aebersold, R. and Mann, M. (2003). Mass spectrometry-based proteomics. *Nature*, **422**, 198–207. (There are also a number of other, very informative proteomics reviews in this issue between pages 193 and 225.)

Breitling, R., Pitt, A.R. and Barrett M.P. (2006). Precision mapping of the metabolome. *Trends in Biotechnology*, **24**, 543–548. (Review of the study of metabolic networks in complex mixtures using the high resolving power of FT-ICR and Orbitrap MS to discriminate between metabolites of near identical mass that differ perhaps by only 0.004 Da.)

Glish, G. L. and Burinsky, D. J. (2008). Hybrid mass spectrometers for tandem mass spectrometry. *Journal of the American Society for Mass Spectrometry*, **19**, 161–172. (Review of the development of hybrid mass spectrometers and a comparison of the particular applications of each type.)

Hah, S. S. (2009). Recent advances in biomedical applications of accelerator mass spectrometry. *Journal of Biomedical Science*, **16**, 54. (This is an Open Access article which reviews all aspects of this subject.)

Han, X., Aslanian, A. and Yates, J.R. 3rd (2008). Mass spectrometry for proteomics. *Current Opinion in Chemical Biology*, **12**, 1–8. (An excellent introductory review of the new instruments with a comparative table of their performance characteristics, such as mass resolution, mass accuracy, sensitivity, *m/z* range and main applications. Also covers quantitative proteomics and description of the latest jargon such as shotgun and top–down proteomics.)

Nita-Lazar, A., Saito Benz, H. and White, F. M. (2008). Quantitative phosphoproteomics by mass spectrometry: past, present, and future. *Proteomics*, **8**, 4433–4443. (This reviews methods for identification of this important and widespread post-translational modification. This volume, no. 8 issue 21, 4367–4612 is a Special Issue on signal transduction proteomics which includes other informative articles on this subject.)

Websites
The ExPASy (Expert Protein Analysis System) server of the Swiss Institute of Bioinformatics (SIB) contains a large suite of programs for the analysis of protein sequences, structures and proteomics as well as 2D PAGE analysis (2D gel documentation and 2D gel image analysis programmes). The ExPASy suite of programmes is at http://us.expasy.org/tools/
Glycomod and GlycanMass are found at http://us.expasy.org/tools/glycomod/ and http://us.expasy.org/tools/glycomod/glycanmass.html respectively.

Deltamass is a database of protein post-translational modifications at http://www.abrf.org/index.cfm/dm.home which can be accessed to determine whether post-translational modifications are present. There are hyperlinks to references to the modifications.

There is also a prediction program 'findmod' for finding potential protein post-translational modifications in the ExPASy suite at http://expasy.org/tools/findmod/

Information and protocols for sample clean-up are found at http://www.millipore.com/catalogue/module/c5737 and http://www.nestgrp.com/protocols/protocol.shtml#massspec

Products for Phosphorylated Peptide and Protein Enrichment and Detection IMAC columns and chromatography at http://www.piercenet.com/files/phosphor.pdf and GelCode Phosphoprotein Staining Kit. http://www.piercenet.com/

10 Electrophoretic techniques

J. WALKER

10.1 GENERAL PRINCIPLES

The term electrophoresis describes the migration of a charged particle under the influence of an electric field. Many important biological molecules, such as amino acids, peptides, proteins, nucleotides and nucleic acids, possess ionisable groups and, therefore, at any given pH, exist in solution as electrically charged species either as cations (+) or anions (−). Under the influence of an electric field these charged particles will migrate either to the cathode or to the anode, depending on the nature of their net charge.

The equipment required for electrophoresis consists basically of two items, a power pack and an electrophoresis unit. Electrophoresis units are available for running either vertical or horizontal gel systems. Vertical slab gel units are commercially available and routinely used to separate proteins in acrylamide gels (Section 10.2). The gel is formed between two glass plates that are clamped together but held apart by plastic spacers. The most commonly used units are the so-called minigel apparatus (Fig. 10.1). Gel dimensions are typically 8.5 cm wide × 5 cm high, with a thickness of 0.5−1 mm. A plastic comb is placed in the gel solution and is removed after polymerisation to provide loading wells for up to 10 samples. When the apparatus is assembled, the lower electrophoresis tank buffer surrounds the gel plates and affords some cooling of the gel plates. A typical horizontal gel system is shown in Fig. 10.2. The gel is cast on a glass or plastic sheet and placed on a cooling plate (an insulated surface through which cooling water is passed to conduct away generated heat). Connection between the gel and electrode buffer is made using a thick wad of wetted filter paper (Fig. 10.2); note, however, that agarose gels for DNA electrophoresis are run submerged in the buffer (Section 10.4.1). The power pack supplies a direct current between the electrodes in the

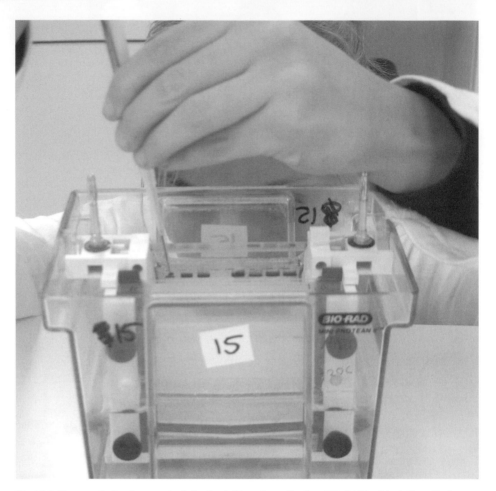

Fig. 10.1 Photograph showing samples being loaded into the wells of an SDS–PAGE minigel. Six wells that have been loaded can be identified by the blue dye (bromophenol blue) that is incorporated into the loading buffer.

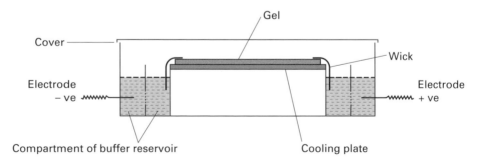

Fig. 10.2 A typical horizontal apparatus, such as that used for immunoelectrophoresis, isoelectric focussing and the electrophoresis of DNA and RNA in agarose gels.

electrophoresis unit. All electrophoresis is carried out in an appropriate buffer, which is essential to maintain a constant state of ionisation of the molecules being separated. Any variation in pH would alter the overall charge and hence the mobilities (rate of migration in the applied field) of the molecules being separated.

In order to understand fully how charged species separate it is necessary to look at some simple equations relating to electrophoresis. When a potential difference (voltage) is applied across the electrodes, it generates a potential gradient, E, which is the applied voltage, V, divided by the distance, d, between the electrodes. When this potential gradient E is applied, the force on a molecule bearing a charge of q coulombs is Eq newtons. It is this force that drives a charged molecule towards an electrode. However, there is also a frictional resistance that retards the movement of this charged molecule. This frictional force is a measure of the hydrodynamic size of the molecule, the shape of the molecule, the pore size of the medium in which electrophoresis is taking place and the viscosity of the buffer. The velocity, v, of a charged molecule in an electric field is therefore given by the equation:

$$\nu = \frac{Eq}{f} \qquad (10.1)$$

where f is the frictional coefficient.

More commonly the term electrophoretic mobility (μ) of an ion is used, which is the ratio of the velocity of the ion to field strength (v/E). When a potential difference is applied, therefore, molecules with different overall charges will begin to separate owing to their different electrophoretic mobilities. Even molecules with similar charges will begin to separate if they have different molecular sizes, since they will experience different frictional forces. As will be seen below, some forms of electrophoresis rely almost totally on the different charges on molecules to effect separation, whilst other methods exploit differences in molecular size and therefore encourage frictional effects to bring about separation.

Provided the electric field is removed before the molecules in the sample reach the electrodes, the components will have been separated according to their electrophoretic mobility. Electrophoresis is thus an incomplete form of electrolysis. The separated samples are then located by staining with an appropriate dye or by autoradiography (Section 14.3.3) if the sample is radiolabelled.

The current in the solution between the electrodes is conducted mainly by the buffer ions, a small proportion being conducted by the sample ions. Ohm's law expresses the relationship between current (I), voltage (V) and resistance (R):

$$\frac{V}{I} = R \qquad (10.2)$$

It therefore appears that it is possible to accelerate an electrophoretic separation by increasing the applied voltage, which would result in a corresponding increase in the current flowing. The distance migrated by the ions will be proportional to both current and time. However, this would ignore one of the major problems for most forms of electrophoresis, namely the generation of heat.

During electrophoresis the power (W, watts) generated in the supporting medium is given by

$$W = I^2 R \qquad (10.3)$$

Most of this power generated is dissipated as heat. Heating of the electrophoretic medium has the following effects:

- An increased rate of diffusion of sample and buffer ions leading to broadening of the separated samples.
- The formation of convection currents, which leads to mixing of separated samples.
- Thermal instability of samples that are rather sensitive to heat. This may include denaturation of proteins (and thus the loss of enzyme activity).
- A decrease of buffer viscosity, and hence a reduction in the resistance of the medium.

If a constant voltage is applied, the current increases during electrophoresis owing to the decrease in resistance (see Ohm's law, equation 10.2) and the rise in current increases the heat output still further. For this reason, workers often use a stabilised power supply, which provides constant power and thus eliminates fluctuations in heating.

Constant heat generation is, however, a problem. The answer might appear to be to run the electrophoresis at very low power (low current) to overcome any heating problem, but this can lead to poor separations as a result of the increased amount of diffusion resulting from long separation times. Compromise conditions, therefore, have to be found with reasonable power settings, to give acceptable separation times, and an appropriate cooling system, to remove liberated heat. While such systems work fairly well, the effects of heating are not always totally eliminated. For example, for electrophoresis carried out in cylindrical tubes or in slab gels, although heat is generated uniformly through the medium, heat is removed only from the edges, resulting in a temperature gradient within the gel, the temperature at the centre of the gel being higher than that at the edges. Since the warmer fluid at the centre is less viscous, electrophoretic mobilities are therefore greater in the central region (electrophoretic mobilities increase by about 2% for each $1\,^{\circ}\text{C}$ rise in temperature), and electrophoretic zones develop a bowed shape, with the zone centre migrating faster than the edges.

A final factor that can effect electrophoretic separation is the phenomenon of electroendosmosis (also known as electroosmotic flow), which is due to the presence of charged groups on the surface of the support medium. For example, paper has some carboxyl groups present, agarose (depending on the purity grade) contains sulphate groups and the surface of glass walls used in capillary electrophoresis (Section 10.5) contains silanol (Si–OH) groups. Figure 10.3 demonstrates how electroendosmosis occurs in a capillary tube, although the principle is the same for any support medium that has charged groups on it. In a fused-silica capillary tube, above a pH value of about 3, silanol groups on the silica capillary wall will ionise, generating negatively charged sites. It is these charges that generate electroendosmosis. The ionised silanol groups create an electrical double layer, or region of charge separation, at the capillary wall/electrolyte interface. When a voltage is applied, cations in the electrolyte near the capillary wall migrate towards the cathode, pulling electrolyte solution with them. This creates a net electroosmotic flow towards the cathode.

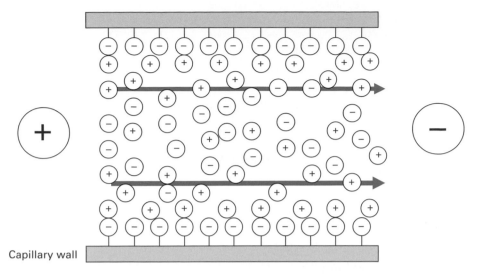

Capillary wall

• Acidic silanol groups impart negative charge on wall
• Counter ions migrate toward cathode, dragging solvent along

Fig. 10.3 Electroosmotic flow through a glass capillary. Electrolyte cations are attracted to the capillary walls, forming an electrical double layer. When a voltage is applied, the net movement of electrolyte solution towards the cathode is known as electroendosmotic flow.

10.2 SUPPORT MEDIA

The pioneering work on protein electrophoresis by Arne Tiselius (for which he received the Nobel Prize in Chemistry in 1948) was performed in free solution. However, it was soon realised that many of the problems associated with this approach, particularly the adverse effects of diffusion and convection currents, could be minimised by stabilising the medium. This was achieved by carrying out electrophoresis on a porous mechanical support, which was wetted in electrophoresis buffer and in which electrophoresis of buffer ions and samples could occur. The support medium cuts down convection currents and diffusion so that the separated components remain as sharp zones. The earliest supports used were filter paper or cellulose acetate strips, wetted in electrophoresis buffer. Nowadays these media are infrequently used, although cellulose acetate still has its uses (see Section 10.3.6). In particular, for many years small molecules such as amino acids, peptides and carbohydrates were routinely separated and analysed by electrophoresis on supports such as paper or thin-layer plates of cellulose, silica or alumina. Although occasionally still used nowadays, such molecules are now more likely to be analysed by more modern and sensitive techniques such as high-performance liquid chromatography (Section 11.3). While paper or thin-layer supports are fine for resolving small molecules, the separation of macromolecules such as proteins and nucleic acids on such supports is poor.

However, the introduction of the use of gels as a support medium led to a rapid improvement in methods for analysing macromolecules. The earliest gel system to be used was the starch gel and, although this still has some uses, the vast majority

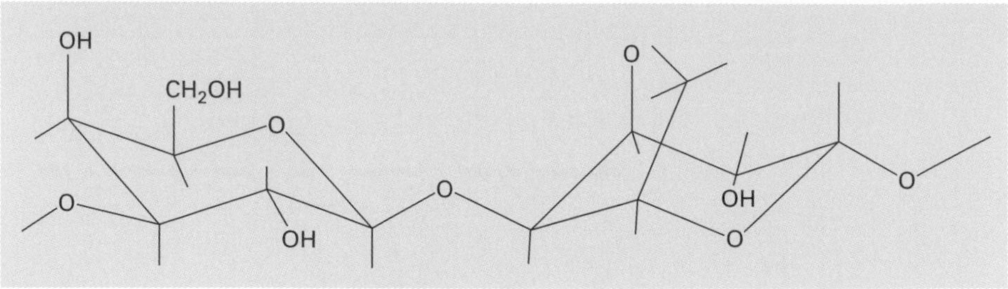

Fig. 10.4 Agarobiose, the repeating unit of agarose.

of electrophoretic techniques used nowadays involve either agarose gels or polyacryl-amide gels.

10.2.1 Agarose gels

Agarose is a linear polysaccharide (average relative molecular mass about 12 000) made up of the basic repeat unit agarobiose, which comprises alternating units of galactose and 3,6-anhydrogalactose (Fig. 10.4). Agarose is one of the components of agar that is a mixture of polysaccharides isolated from certain seaweeds. Agarose is usually used at concentrations of between 1% and 3%. Agarose gels are formed by suspending dry agarose in aqueous buffer, then boiling the mixture until a clear solution forms. This is poured and allowed to cool to room temperature to form a rigid gel. The gelling properties are attributed to both inter- and intramolecular hydrogen bonding within and between the long agarose chains. This cross-linked structure gives the gel good anticonvectional properties. The pore size in the gel is controlled by the initial concentration of agarose; large pore sizes are formed from low concentrations and smaller pore sizes are formed from the higher concentrations. Although essentially free from charge, substitution of the alternating sugar residues with carboxyl, methyoxyl, pyruvate and especially sulphate groups occurs to varying degrees. This substitution can result in electro-endosmosis during electrophoresis and ionic interactions between the gel and sample in all uses, both unwanted effects. Agarose is therefore sold in different purity grades, based on the sulphate concentration – the lower the sulphate content, the higher the purity.

Agarose gels are used for the electrophoresis of both proteins and nucleic acids. For proteins, the pore sizes of a 1% agarose gel are large relative to the sizes of proteins. Agarose gels are therefore used in techniques such as flat-bed isoelectric focussing (Section 10.3.4), where the proteins are required to move unhindered in the gel matrix according to their native charge. Such large pore gels are also used to separate much larger molecules such as DNA or RNA, because the pore sizes in the gel are still large enough for DNA or RNA molecules to pass through the gel. Now, however, the pore size and molecule size are more comparable and fractional effects begin to play a role in the separation of these molecules (Section 10.4). A further advantage of using agarose is the availability of low melting temperature agarose (62−65 °C). As the name suggests, these gels can be reliquified by heating to 65 °C and

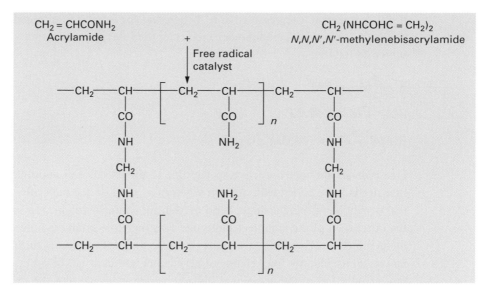

Fig. 10.5 The formation of a polyacrylamide gel from acrylamide and bis-acrylamide.

thus, for example, DNA samples separated in a gel can be cut out of the gel, returned to solution and recovered.

Owing to the poor elasticity of agarose gels and the consequent problems of removing them from small tubes, the gel rod system sometimes used for acrylamide gels is not used. Horizontal slab gels are invariably used for isoelectric focussing or immunoelectrophoresis in agarose. Horizontal gels are also used routinely for DNA and RNA gels (Section 10.4), although vertical systems have been used by some workers.

10.2.2 Polyacrylamide gels

Electrophoresis in acrylamide gels is frequently referred to as PAGE, being an abbreviation for *poly*acrylamide *g*el *e*lectrophoresis.

Cross-linked polyacrylamide gels are formed from the polymerisation of acrylamide monomer in the presence of smaller amounts of N,N'-methylene-bisacrylamide (normally referred to as 'bis'-acrylamide) (Fig. 10.5). Note that bisacrylamide is essentially two acrylamide molecules linked by a methylene group, and is used as a cross-linking agent. Acrylamide monomer is polymerised in a head-to-tail fashion into long chains and occasionally a bis-acrylamide molecule is built into the growing chain, thus introducing a second site for chain extension. Proceeding in this way a cross-linked matrix of fairly well-defined structure is formed (Fig. 10.5). The polymerisation of acrylamide is an example of free-radical catalysis, and is initiated by the addition of ammonium persulphate and the base N,N,N',N'-tetramethylenediamine (TEMED). TEMED catalyses the decomposition of the persulphate ion to give a free radical (i.e. a molecule with an unpaired electron):

$$S_2O_8^{2-} + e^- \rightarrow SO_4^{2-} + SO_4^{-\bullet}$$

If this free radical is represented as $R^\bullet$ (where the dot represents an unpaired electron) and M as an acrylamide monomer molecule, then the polymerisation can be represented as follows:

$$R^\bullet + M \rightarrow RM^\bullet$$

$$RM^\bullet + M \rightarrow RMM^\bullet$$

$$RMM^\bullet + M \rightarrow RMMM^\bullet \text{ etc.}$$

Free radicals are highly reactive species due to the presence of an unpaired electron that needs to be paired with another electron to stabilise the molecule. $R\bullet$ therefore reacts with M, forming a single bond by sharing its unpaired electron with one from the outer shell of the monomer molecule. This therefore produces a new free radical molecule $R - M^\bullet$, which is equally reactive and will attack a further monomer molecule. In this way long chains of acrylamide are built up, being cross-linked by the introduction of the occasional bis-acrylamide molecule into the growing chain. Oxygen mops up free radicals and therefore all gel solutions are normally degassed (the solutions are briefly placed under vacuum to remove loosely dissolved air) prior to use. The degassing of the gel solution also serves a second purpose. The polymerisation of acrylamide is an exothermic reaction (i.e. heat is liberated) and the warming up of the gel solution as it sets can liberate air bubbles that become trapped in the polymerised gel. The degassing step prevents this possibility.

Photopolymerisation is an alternative method that can be used to polymerise acrylamide gels. The ammonium persulphate and TEMED are replaced by riboflavin and when the gel is poured it is placed in front of a bright light for 2–3 h. Photodecomposition of riboflavin generates a free radical that initiates polymerisation.

Acrylamide gels are defined in terms of the total percentage of acrylamide present, and the pore size in the gel can be varied by changing the concentrations of both the acrylamide and bis-acrylamide. Acrylamide gels can be made with a content of between 3% and 30% acrylamide. Thus low percentage gels (e.g. 4%) have large pore sizes and are used, for example, in the electrophoresis of proteins, where free movement of the proteins by electrophoresis is required without any noticeable frictional effect, for example in flat-bed isoelectric focusing (Section 10.3.4) or the stacking gel system of an SDS–polyacrylamide gel (Section 10.3.1). Low percentage acrylamide gels are also used to separate DNA (Section 10.4). Gels of between 10% and 20% acrylamide are used in techniques such as SDS–gel electrophoresis, where the smaller pore size now introduces a sieving effect that contributes to the separation of proteins according to their size (Section 10.3.1).

Proteins were originally separated on polyacrylamide gels that were polymerised in glass tubes, approximately 7 mm in diameter and about 10 cm in length. The tubes were easy to load and run, with minimum apparatus requirements. However, only one sample could be run per tube and, because conditions of separation could vary from tube to tube, comparison between different samples was not always accurate. The later introduction of vertical gel slabs allowed running of up to 20 samples under identical conditions in a single run. Vertical slabs are now used routinely both for the analysis of proteins (Section 10.3) and for the separation of DNA fragments during DNA sequence analysis (Section 10.4). Although some workers prepare their own acrylamide gels, others purchase commercially available ready-made gels for techniques such as SDS–PAGE, native gels and isoelectric focusing (IEF) (see below).

10.3 ELECTROPHORESIS OF PROTEINS

10.3.1 Sodium dodecyl sulphate (SDS)–polyacrylamide gel electrophoresis

SDS–polyacrylamide gel electrophoresis (SDS–PAGE) is the most widely used method for analysing protein mixtures qualitatively. It is particularly useful for monitoring protein purification and, because the method is based on the separation of proteins according to size, it can also be used to determine the relative molecular mass of proteins. $SDS(CH_3 - (CH_2)_{10} - CH_2OSO_3^- Na^+)$ is an anionic detergent. Samples to be run on SDS–PAGE are firstly boiled for 5 min in sample buffer containing β-mercaptoethanol and SDS. The mercaptoethanol reduces any disulphide bridges present that are holding together the protein tertiary structure, and the SDS binds strongly to, and denatures, the protein. Each protein in the mixture is therefore fully denatured by this treatment and opens up into a rod-shaped structure with a series of negatively charged SDS molecules along the polypeptide chain. On average, one SDS molecule binds for every two amino acid residues. The original native charge on the molecule is therefore completely swamped by the negatively charged SDS molecules. The rod-like structure remains, as any rotation that tends to fold up the protein chain would result in repulsion between negative charges on different parts of the protein chain, returning the conformation back to the rod shape. The sample buffer also contains an ionisable tracking dye, usually bromophenol blue, that allows the electrophoretic run to be monitored, and sucrose or glycerol, which gives the sample solution density thus allowing the sample to settle easily through the electrophoresis buffer to the bottom when injected into the loading well (see Fig. 10.1). Once the samples are all loaded, a current is passed through the gel. The samples to be separated are not in fact loaded directly into the main separating gel. When the main separating gel (normally about 5 cm long) has been poured between the glass plates and allowed to set, a shorter (approximately 0.8 cm) stacking gel is poured on top of the separating gel and it is into this gel that the wells are formed and the proteins loaded. The purpose of this stacking gel is to concentrate the protein sample into a sharp band before it enters the main separating gel. This is achieved by utilising differences in ionic strength and pH between the electrophoresis buffer and the stacking gel buffer and involves a phenomenon known as isotachophoresis. The stacking gel has a very large pore size (4% acrylamide), which allows the proteins to move freely and concentrate, or stack, under the effect of the electric field. The band-sharpening effect relies on the fact that negatively charged glycinate ions (in the electrophoresis buffer) have a lower electrophoretic mobility than do the protein–SDS complexes, which, in turn, have lower mobility than the chloride ions (Cl^-) of the loading buffer and the stacking gel buffer. When the current is switched on, all the ionic species have to migrate at the same speed otherwise there would be a break in the electrical circuit. The glycinate ions can move at the same speed as Cl^- only if they are in a region of higher field strength. Field strength is inversely proportional to conductivity, which is proportional to concentration. The result is that the three species of interest adjust their concentrations so that $[Cl^-]$ > [protein–SDS] > [glycinate]. There is only a small quantity of protein–SDS complexes, so they concentrate in a very tight band between glycinate and Cl^- boundaries. Once the glycinate reaches the separating gel it becomes more fully ionised in the higher pH environment and its mobility increases. (The pH of the stacking gel is 6.8, that of the separating gel is 8.8.) Thus, the interface between glycinate and Cl^- leaves behind the protein–SDS complexes, which are left to electrophorese at their own rates. The negatively charged protein–SDS complexes now continue to move towards the anode, and, because they have the same charge per unit length, they travel into the separating gel under the applied electric field with the same mobility. However, as they pass through the separating gel the proteins separate, owing to the molecular sieving properties of the gel. Quite simply, the smaller the protein the more easily it can pass through the pores of the gel, whereas large proteins are successively retarded by frictional resistance due to the sieving effect of the gels. Being a small molecule, the bromophenol blue dye is totally unretarded and therefore indicates the electrophoresis front. When the dye reaches the bottom of the gel, the current is turned off, and the gel is removed from between the glass plates and shaken in an appropriate stain solution (usually Coomassie Brilliant Blue, see Section 10.3.7) and then washed in destain solution. The destain solution removes unbound background dye from the gel, leaving stained proteins visible as blue bands on a clear background. A typical minigel would take about 1 h to prepare and set, 40 min to run at 200 V and have a 1 h staining time with Coomassie Brilliant Blue. Upon destaining, strong protein bands would be seen in the gel within 10–20 min, but overnight destaining is needed to completely remove all background stain. Vertical slab gels are invariably run, since this allows up to 10 different samples to be loaded onto a single gel. A typical SDS–polyacrylamide gel is shown in Fig. 10.6.

Typically, the separating gel used is a 15% polyacrylamide gel. This gives a gel of a certain pore size in which proteins of relative molecular mass (M_r) 10 000 move through the gel relatively unhindered, whereas proteins of M_r 100 000 can only just enter the pores of this gel. Gels of 15% polyacrylamide are therefore useful for separating proteins in the range M_r 100 000 to 10 000. However, a protein of M_r 150 000, for example, would be unable to enter a 15% gel. In this case a larger-pored gel (e.g. a 10% or even 7.5% gel) would be used so that the protein could now enter the gel and be stained and identified. It is obvious, therefore, that the choice of gel to be used depends on the size of the protein being studied. The fractionation range of different percentage acrylamide gels is shown in Table 10.1. This shows, for example, that in a 10% polyacrylamide gel proteins greater than 200 kDa in mass cannot enter the gel, whereas proteins with relative molecular mass (M_r) in the range 200 000 to 15 000 will separate. Proteins of M_r 15 000 or less are too small to experience the sieving effect of the gel matrix, and all run together as a single band at the electrophoresis front.

Table 10.1 **The relationship between acrylamide gel concentration and protein fractionation range**

Acrylamide concentration (%)	Protein fractionation range ($M_r \times 10^{-3}$)
5	60–350
10	15–200
15	10–100

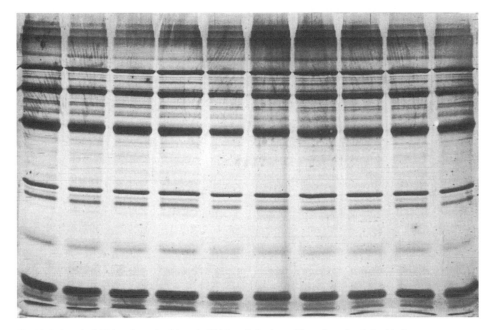

Fig. 10.6 A typical SDS–polyacrylamide gel. All 10 wells in the gel have been loaded with the same complex mixture of proteins. (Courtesy of Bio-Rad Laboratories.)

The M_r of a protein can be determined by comparing its mobility with those of a number of standard proteins of known M_r that are run on the same gel. By plotting a graph of distance moved against log M_r for each of the standard proteins, a calibration curve can be constructed. The distance moved by the protein of unknown M_r is then measured, and then its log M_r and hence M_r can be determined from the calibration curve.

SDS–gel electrophoresis is often used after each step of a purification protocol to assess the purity or otherwise of the sample. A pure protein should give a single band on an SDS–polyacrylamide gel, unless the molecule is made up of two unequal subunits. In the latter case two bands, corresponding to the two subunits, will be seen. Since only submicrogram amounts of protein are needed for the gel, very little material is used in this form of purity assessment and at the same time a value for the

---

Example 1 **MOLECULAR MASS DETERMINATION BY ELECTROPHORESIS**

Question The following table shows the distance moved in an SDS–polyacrylamide gel by a series of marker proteins of known relative molecular mass (M_r). A newly purified protein (X) run on the same gel showed a single band that had moved a distance of 45 mm. What was the M_r of protein X?

Protein	M_r	Distance moved (mm)
Transferrin	78 000	6.0
Bovine serum albumin	66 000	12.5
Ovalbumin (egg albumin)	45 000	32.0
Glyceraldehyde-3-phosphate dehydrogenase	36 000	38.0
Carbonic anhydrase	29 000	50.0
Trypsinogen	24 000	54.0
Soyabean trypsin inhibitor	20 100	61.0
β-Lactoglobulin	18 400[a]	69.0
Myoglobin	17 800	69.0
Lysozyme	14 300	79.0
Cytochrome c	12 400	86.5

Note: [a]β-lactoglobulin has a relative molecular mass of 36 800 but is a dimer of two identical subunits of 18 400 relative molecular mass. Under the reducing conditions of the sample buffer the disulphide bridges linking the subunits are reduced and thus the monomer chains are seen on the gel.

Answer Construct a calibration graph by plotting log M_r versus distance moved for each of the marker proteins. From a graph of log M_r versus the distance moved by each protein you can determine a relative molecular mass for protein X of approximately 31 000. Note that this method is accurate to $\pm$ 10%, so your answer is 31 000 $\pm$ 3100.

---

relative molecular mass of the protein can be determined on the same gel run (as described above), with no more material being used.

10.3.2 Native (buffer) gels

While SDS–PAGE is the most frequently used gel system for studying proteins, the method is of no use if one is aiming to detect a particular protein (often an enzyme) on the basis of its biological activity, because the protein (enzyme) is denatured by the SDS–PAGE procedure. In this case it is necessary to use non-denaturing conditions. In native or buffer gels, polyacrylamide gels are again used (normally a 7.5% gel) but the SDS is absent and the proteins are *not* denatured prior to loading. Since all the proteins in the sample being analysed carry their native charge at the pH of the gel (normally pH 8.7), proteins separate according to their different electrophoretic mobilities *and* the sieving effects of the gel. It is therefore not possible to predict the behaviour of a given protein in a buffer gel but, because of the range of different charges and sizes of proteins in a given protein mixture, good resolution is achieved. The enzyme of interest can be identified by incubating the gel in an appropriate substrate solution such that a coloured product is produced at the site of the enzyme. An alternative method for enzyme detection is to include the substrate in an agarose gel that is poured over the acrylamide gel and allowed to set. Diffusion and interaction of enzyme and substrate between the two gels results in colour formation at the site of the enzyme. Often, duplicate samples will be run on a gel, the gel cut in half and one half stained for activity, the other for total protein. In this way the total protein content of the sample can be analysed and the particular band corresponding to the enzyme identified by reference to the activity stain gel.

10.3.3   Gradient gels

This is again a polyacrylamide gel system, but instead of running a slab gel of uniform pore size throughout (e.g. a 15% gel) a gradient gel is formed, where the acrylamide concentration varies uniformly from, typically, 5% at the top of the gel to 25% acrylamide at the bottom of the gel. The gradient is formed via a gradient mixer and run down between the glass plates of a slab gel. The higher percentage acrylamide (e.g. 25%) is poured between the glass plates first and a continuous gradient of decreasing acrylamide concentration follows. Therefore at the top of the gel there is a large pore size (5% acrylamide) but as the sample moves down through the gel the acrylamide concentration slowly increases and the pore size correspondingly decreases. Gradient gels are normally run as SDS gels with a stacking gel. There are two advantages to running gradient gels. First, a much greater range of protein M_r values can be separated than on a fixed-percentage gel. In a complex mixture, very low molecular weight proteins travel freely through the gel to begin with, and start to resolve when they reach the smaller pore sizes towards the lower part of the gel. Much larger proteins, on the other hand, can still enter the gel but start to separate immediately due to the sieving effect of the gel. The second advantage of gradient gels is that proteins with very similar M_r values may be resolved, although they cannot otherwise be resolved in fixed percentage gels. As each protein moves through the gel the pore sizes become smaller until the protein reaches its pore size limit. The pore size in the gel is now too small to allow passage of the protein, and the protein sample stacks up at this point as a sharp band. A similar-sized protein but with slightly lower M_r will be able to travel a little further through the gel before reaching its pore size limit, at which point it will form a sharp band. These two proteins, of slightly different M_r values, therefore separate as two, close, sharp bands.

10.3.4   Isoelectric focussing gels

This method is ideal for the separation of amphoteric substances such as proteins because it is based on the separation of molecules according to their different isoelectric points (Section 8.1). The method has high resolution, being able to separate proteins that differ in their isoelectric points by as little as 0.01 of a pH unit. The most widely used system for IEF utilises horizontal gels on glass plates or plastic sheets.

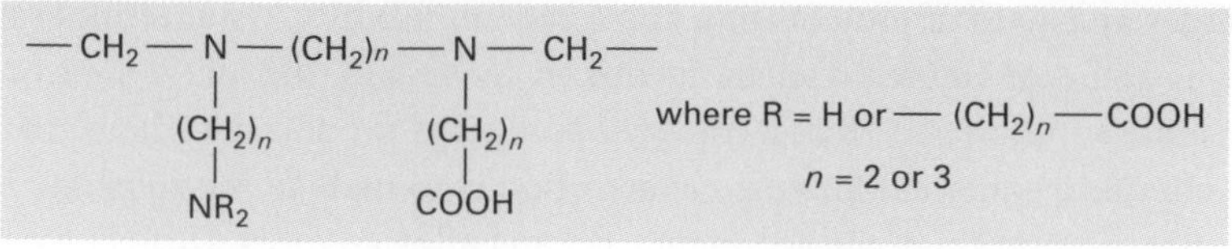

Fig. 10.7 The general formula for ampholytes.

Separation is achieved by applying a potential difference across a gel that contains a pH gradient. The pH gradient is formed by the introduction into the gel of compounds known as ampholytes, which are complex mixtures of synthetic polyamino-polycarboxylic acids (Fig. 10.7). Ampholytes can be purchased in different pH ranges covering either a wide band (e.g. pH 3−10) or various narrow bands (e.g. pH 7−8), and a pH range is chosen such that the samples being separated will have their isoelectric points (pI values) within this range. Commercially available ampholytes include Bio-Lyte and Pharmalyte.

Traditionally 1−2 mm thick IEF gels have been used by research workers, but the relatively high cost of ampholytes makes this a fairly expensive procedure if a number of gels are to be run. However, the introduction of thin-layer IEF gels, which are only 0.15 mm thick and which are prepared using a layer of electrical insulation tape as the spacer between the gel plates, has considerably reduced the cost of preparing IEF gels, and such gels are now commonly used. Since this method requires the proteins to move freely according to their charge under the electric field, IEF is carried out in low percentage gels to avoid any sieving effect within the gel. Polyacrylamide gels (4%) are commonly used, but agarose is also used, especially for the study of high M_r proteins that may undergo some sieving even in a low percentage acrylamide gel.

To prepare a thin-layer IEF gel, carrier ampholytes, covering a suitable pH range, and riboflavin are mixed with the acrylamide solution, and the mixture is then poured over a glass plate (typically 25 cm × 10 cm), which contains the spacer. The second glass plate is then placed on top of the first to form the gel cassette, and the gel polymerised by photopolymerisation by placing the gel in front of a bright light. The photodecomposition of the riboflavin generates a free radical, which initiates polymerisation (Section 10.2.2). This takes 2−3 h. Once the gel has set, the glass plates are prised apart to reveal the gel stuck to one of the glass sheets. Electrode wicks, which are thick (3 mm) strips of wetted filter paper (the anode is phosphoric acid, the cathode sodium hydroxide) are laid along the long length of each side of the gel and a potential difference applied. Under the effect of this potential difference, the ampholytes form a pH gradient between the anode and cathode. The power is then turned off and samples applied by laying on the gel small squares of filter paper soaked in the sample. A voltage is again applied for about 30 min to allow the sample to electrophorese off the paper and into the gel, at which time the paper squares can be removed from the gel. Depending on which point on the pH gradient the sample has been loaded, proteins that are initially at a pH region below their isoelectric point will be positively charged and will initially migrate towards the cathode. As they proceed, however, the surrounding pH will be steadily increasing, and therefore the positive charge on the protein will decrease correspondingly until eventually the protein arrives at a point where the pH is equal to its isoelectric point. The protein will now be in the zwitterion form with no net charge, so further movement will cease. Likewise, substances that are initially at pH regions above their isoelectric points will be negatively charged and will migrate towards the anode until they reach their isoelectric points and become stationary. It can be seen that as the samples will always move towards their isoelectric points it is not critical where on the gel they are applied. To achieve rapid separations (2−3 h) relatively high voltages (up to 2500 V) are used. As considerable heat is produced, gels are run on cooling plates (10 °C) and power packs used to stabilise the power output and thus to minimise thermal fluctuations. Following electrophoresis, the gel must be stained to detect the proteins. However, this cannot be done directly, because the ampholytes will stain too, giving a totally blue gel. The gel is therefore first washed with fixing solution (e.g. 10% (v/v) trichloroacetic acid). This precipitates the proteins in the gel and allows the much smaller ampholytes to be washed out. The gel is stained with Coomassie Brilliant Blue and then destained (Section 10.3.7). A typical IEF gel is shown in Fig. 10.8.

The pI of a particular protein may be determined conveniently by running a mixture of proteins of known isoelectric point on the same gel. A number of mixtures of proteins with differing pI values are commercially available, covering the pH range 3.5−10. After staining, the distance of each band from one electrode is measured and a graph of distance for each protein against its pI (effectively the pH at that point) plotted. By means of this calibration line, the pI of an unknown protein can be determined from its position on the gel.

IEF is a highly sensitive analytical technique and is particularly useful for studying microheterogeneity in a protein. For example, a protein may show a single band on an SDS gel, but may show three bands on an IEF gel. This may occur, for example, when a protein exists in mono-, di- and tri-phosphorylated forms. The difference of a couple of phosphate groups has no significant effect on the overall relative molecular mass of the protein, hence a single band on SDS gels, but the small charge difference introduced on each molecule can be detected by IEF.

The method is particularly useful for separating isoenzymes (Section 8.2), which are different forms of the same enzyme often differing by only one or two amino acid residues. Since the proteins are in their native form, enzymes can be detected in the gel either by washing the unfixed and unstained gel in an appropriate substrate or by overlayering with agarose containing the substrate. The approach has found particular use in forensic science, where traces of blood or other biological fluids can be analysed and compared according to the composition of certain isoenzymes.

Although IEF is used mainly for analytical separations, it can also be used for preparative purposes. In vertical column IEF, a water-cooled vertical glass column is used, filled with a mixture of ampholytes dissolved in a sucrose solution containing a density gradient to prevent diffusion. When the separation is complete, the current is switched off and the sample components run out through a valve in the base of the column. Alternatively, preparative IEF can be carried out in beds of granulated gel, such as Sephadex G-75 (Section 11.7).

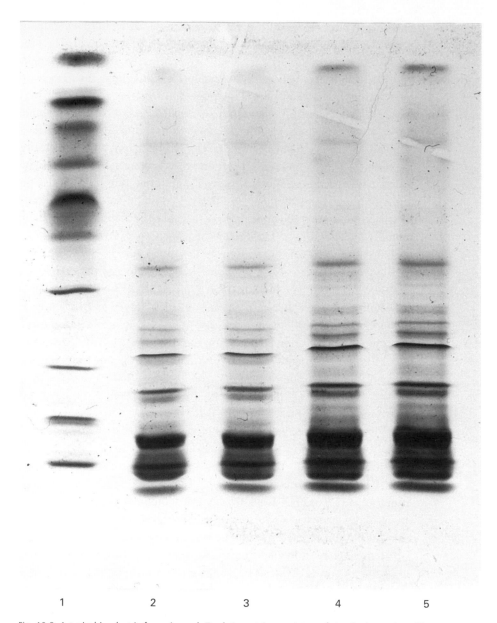

Fig. 10.8 A typical isoelectric focussing gel. Track 1 contains a mixture of standard proteins of known isoelectric points. Tracks 2–5 show increasing loadings of venom from the Japanese water moccasin snake. (Courtesy of Bio-Rad Laboratories Ltd.)

10.3.5 Two-dimensional polyacrylamide gel electrophoresis

This technique combines the technique of IEF (first dimension), which separates proteins in a mixture according to charge (pI), with the size separation technique of SDS–PAGE (second dimension). The combination of these two techniques to give two-dimensional (2-D) PAGE provides a highly sophisticated analytical method for analysing protein mixtures. To maximise separation, most workers use large format

2-D gels (20 cm × 20 cm), although the minigel system can be used to provide useful separation in some cases. For large-format gels, the first dimension (isoelectric focussing) is carried out in an acrylamide gel that has been cast on a plastic strip (18 cm × 3 mm wide). The gel contains ampholytes (for forming the pH gradient) together with 8 M urea and a non-ionic detergent, both of which denature and maintain the solubility of the proteins being analysed. The denatured proteins therefore separate in this gel according to their isoelectric points. The IEF strip is then incubated in a sample buffer containing SDS (thus binding SDS to the denatured proteins) and then placed between the glass plates of, and on top of, a previously prepared 10% SDS–PAGE gel. Electrophoresis is commenced and the SDS-bound proteins run into the gel and separate according to size, as described in Section 10.3.1. The IEF gels are provided as dried strips and need rehydrating overnight. The first dimension IEF run takes 6−8 h, the equilibration step with SDS sample buffer takes about 15 min, and then the SDS–PAGE step takes about 5 h. A typical 2-D gel is shown in Fig. 10.9. Using this method one can routinely resolve between 1000 and 3000 proteins from a cell or tissue extract and in some cases workers have reported the separation of between 5000 and 10 000 proteins. The applications of 2-D PAGE, and a description of the method's central role in proteomics is described in Section 8.5.1.

10.3.6 Cellulose acetate electrophoresis

Although one of the older methods, cellulose acetate electrophoresis still has a number of applications. In particular it has retained a use in the clinical analysis of serum samples. Cellulose acetate has the advantage over paper in that it is a much more homogeneous medium, with uniform pore size, and does not adsorb proteins in the way that paper does. There is therefore much less trailing of protein bands and resolution is better, although nothing like as good as that achieved with polyacrylamide gels. The method is, however, far simpler to set up and run. Single samples are normally run on cellulose acetate strips (2.5 cm × 12 cm), although multiple samples are frequently run on wider sheets. The cellulose acetate is first wetted in electrophoresis buffer (pH 8.6 for serum samples) and the sample (1−2 mm³) loaded as a 1 cm wide strip about one-third of the way along the strip. The ends of the strip make contact with the electrophoresis buffer tanks via a filter paper wick that overlaps the end of the cellulose acetate strip, and electrophoresis is conducted at 6–8 V cm⁻¹ for about 3 h. Following electrophoresis, the strip is stained for protein (see Section 10.3.7), destained, and the bands visualised. A typical serum protein separation shows about six major bands. However, in many disease states, this serum protein profile changes and a clinician can obtain information concerning the disease state of a patient from the altered pattern. Although still frequently used for serum analysis, electrophoresis on cellulose acetate is being replaced by the use of agarose gels, which give similar but somewhat better resolution. A typical example of the analysis of serum on an agarose gel is shown in Fig. 10.10. Similar patterns are obtained when cellulose acetate is used.

Enzymes can easily be detected, in samples electrophoresed on cellulose acetate, by using the zymogram technique. The cellulose strip is laid on a strip of filter paper

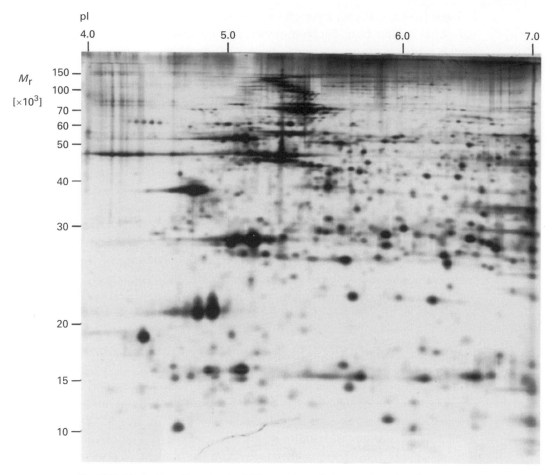

Fig. 10.9 A typical two-dimensional gel. The sample applied was 100 μg of total protein extracted from a normal dog heart ventricle. The first dimension was carried out using a pH 4–7 isoelectric focussing gel. The second dimension was a 12% SDS–PAGE vertical slab gel. The pattern was visualised by silver staining. (Courtesy of Monique Heinke and Dr Mike Dunn, Division of Cardiothoracic Surgery, Imperial College School of Medicine, Heart Science Centre, Harefield, UK.)

soaked in buffer and substrate. After an appropriate incubation period, the strips are peeled apart and the paper zymogram treated accordingly to detect enzyme product; hence, it is possible to identify the position of the enzyme activity on the original strip. An alternative approach to detecting and semiquantifying *any* particular protein on a strip is to treat the strip as the equivalent of a protein blot and to probe for the given protein using primary antibody and then enzyme-linked secondary antibody (Section 10.3.8). Substrate colour development indicates the presence of the particular protein and the amount of colour developed in a given time is a semiquantitative measure of the amount of protein. Thus, for example, large numbers of serum samples can be run on a wide sheet, the sheet probed using antibodies, and elevated levels of a particular protein identified in certain samples by increased levels of colour development in these samples.

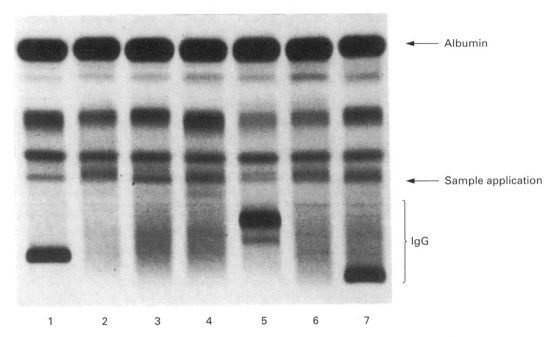

Fig. 10.10 Electrophoresis of human serum samples on an agarose gel. Tracks 2, 3, 4 and 6 show normal serum protein profiles. Tracks 1, 5 and 7 show myeloma patients, who are identified by the excessive production of a particular monoclonal antibody seen in the IgG fraction. (Courtesy of Charles Andrews and Nicholas Cundy, Edgware General Hospital, London.)

10.3.7 Detection, estimation and recovery of proteins in gels

The most commonly used general protein stain for detecting protein on gels is the sulphated trimethylamine dye Coomassie Brilliant Blue R-250 (CBB). Staining is usually carried out using 0.1% (w/v) CBB in methanol:water:glacial acetic acid (45:45:10, by vol.). This acid–methanol mixture acts as a denaturant to precipitate or fix the protein in the gel, which prevents the protein from being washed out whilst it is being stained. Staining of most gels is accomplished in about 2 h and destaining, usually overnight, is achieved by gentle agitation in the same acid–methanol solution but in the absence of the dye. The Coomassie stain is highly sensitive; a very weakly staining band on a polyacrylamide gel would correspond to about 0.1 μg (100 ng) of protein. The CBB stain is not used for staining cellulose acetate (or indeed protein blots) because it binds quite strongly to the paper. In this case, proteins are first denatured by brief immersion of the strip in 10% (v/v) trichloroacetic acid, and then immersed in a solution of a dye that does not stain the support material, for example Procion blue, Amido black or Procion S.

Although the Coomassie stain is highly sensitive, many workers require greater sensitivity such as that provided by silver staining. Silver stains are based either on techniques developed for histology or on methods based on the photographic process. In either case, silver ions (Ag$^+$) are reduced to metallic silver on the protein, where the silver is deposited to give a black or brown band. Silver stains can be used immediately after electrophoresis, or, alternatively, after staining with CBB. With the latter approach, the major bands on the gel can be identified with CBB and then minor bands, not detected with CBB, resolved using the silver stain. The silver stain is at least 100 times more sensitive than CBB, detecting proteins down to 1 ng amounts. Other stains with similar sensitivity include the fluorescent stains Sypro Orange (30 ng) and Sypro Ruby (10 ng).

Glycoproteins have traditionally been detected on protein gels by use of the periodic acid–Schiff (PAS) stain. This allows components of a mixture of glycoproteins to be distinguished. However, the PAS stain is not very sensitive and often gives very weak, red-pink bands, difficult to observe on a gel. A far more sensitive method used nowadays is to blot the gel (Section 10.3.8) and use lectins to detect the glycoproteins. Lectins are protein molecules that bind carbohydrates, and different lectins have been found that have different specificities for different types of carbohydrate. For example, certain lectins recognise mannose, fucose, or terminal glucosamine of the carbohydrate side-chains of glycoproteins. The sample to be analysed is run on a number of tracks of an SDS–polyacrylamide gel. Coloured bands appear at the point where the lectins bind if each blotted track is incubated with a different lectin, washed, incubated with a horseradish peroxidase-linked antibody to the lectin, and then peroxidase substrate added. In this way, by testing a protein sample against a series of lectins, it is possible to determine not only that a protein is a *glyco*protein, but to obtain information about the type of glycosylation.

Quantitative analysis (i.e. measurements of the relative amounts of different proteins in a sample) can be achieved by scanning densitometry. A number of commercial scanning densitometers are available, and work by passing the stained gel track over a beam of light (laser) and measuring the transmitted light. A graphic presentation of protein zones (peaks of absorbance) against migration distance is produced, and peak areas can be calculated to obtain quantitative data. However, such data must be interpreted with caution because there is only a limited range of protein concentrations over which there is a linear relationship between absorbance and concentration. Also, equal amounts of different proteins do not always stain equally with a given stain, so any data comparing the relative amounts of protein can only be semiquantitative. An alternative and much cheaper way of obtaining such data is to cut out the stained bands of interest, elute the dye by shaking overnight in a known volume of 50% pyridine, and then to measure spectrophotometrically the amount of colour released. More recently gel documentation systems have been developed, which are replacing scanning densitometers. Such benchtop systems comprise a video imaging unit (computer linked) attached to a small 'darkroom' unit that is fitted with a choice of white or ultraviolet light (transilluminator). Gel images can be stored on the computer, enhanced accordingly and printed as required on a thermal printer, thus eliminating the need for wet developing in a purpose built darkroom, as is the case for traditional photography.

Although gel electrophoresis is used generally as an analytical tool, it can be utilised to separate proteins in a gel to achieve protein purification. Protein bands can be cut out of protein blots and sequence data obtained by placing the blot in a protein sequencer (see Section 8.4.3). Stained protein bands can be cut out of protein gels and the protein recovered by electrophoresis of the protein out of the gel piece

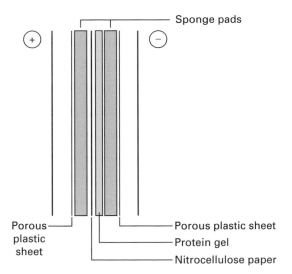

Fig. 10.11 Diagrammatic representation of electroblotting. The gel to be blotted is placed on top of a sponge pad saturated in buffer. The nitrocellulose sheet is then placed on top of the gel, followed by a second sponge pad. This sandwich is supported between two rigid porous plastic sheets and held together with two elastic bands. The sandwich is then placed between parallel electrodes in a buffer reservoir and an electric current passed. The sandwich must be placed such that the immobilising medium is between the gel and the anode for SDS–polyacrylamide gels, because all the proteins carry a negative charge.

(electroelution). A number of different designs of electroelution cells are commercially available, but perhaps the easiest method is to seal the gel piece in buffer in a dialysis sac and place the sac in buffer between two electrodes. Protein will electrophorese out of the gel piece towards the appropriate electrode but will be retained by the dialysis sac. After electroelution, the current is reversed for a few seconds to drive off any protein that has adsorbed to the wall of the dialysis sac and then the protein solution within the sac is recovered.

10.3.8 Protein (western) blotting

Although essentially an analytical technique, PAGE does of course achieve fractionation of a protein mixture during the electrophoresis process. It is possible to make use of this fractionation to examine further individual separated proteins. The first step is to transfer or blot the pattern of separated proteins from the gel onto a sheet of nitrocellulose paper. The method is known as protein blotting, or western blotting by analogy with Southern blotting (Section 5.9.2), the equivalent method used to recover DNA samples from an agarose gel. Transfer of the proteins from the gel to nitrocellulose is achieved by a technique known as electroblotting. In this method a sandwich of gel and nitrocellulose is compressed in a cassette and immersed, in buffer, between two parallel electrodes (Fig. 10.11). A current is passed at right angles to the gel, which causes the separated proteins to electrophorese out of the gel and into the nitrocellulose sheet. The nitrocellulose with its transferred protein is referred to as a blot. Once transferred onto nitrocellulose, the separated proteins can be examined further. This involves probing the blot, usually using an antibody to detect a specific

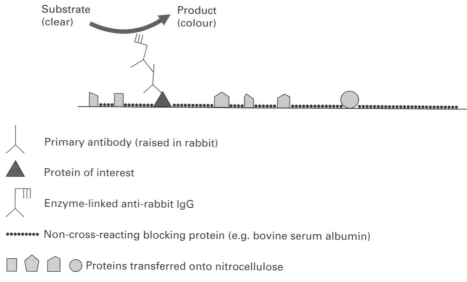

Primary antibody (raised in rabbit)

Protein of interest

Enzyme-linked anti-rabbit IgG

•••••••• Non-cross-reacting blocking protein (e.g. bovine serum albumin)

▢ ⬠ ⬠ ◯ Proteins transferred onto nitrocellulose

Fig. 10.12 The use of enzyme-linked second antibodies in immunodetection of protein blots. First, the primary antibody (e.g. raised in a rabbit) detects the protein of interest on the blot. Second, enzyme-linked anti-rabbit IgG detects the primary antibody. Third, addition of enzyme substrate results in coloured product deposited at the site of protein of interest on the blot.

protein. The blot is first incubated in a protein solution, for example 10% (w/v) bovine serum albumin, or 5% (w/v) non-fat dried milk (the so-called blotto technique), which will block all remaining hydrophobic binding sites on the nitrocellulose sheet. The blot is then incubated in a dilution of an antiserum (primary antibody) directed against the protein of interest. This IgG molecule will bind to the blot if it detects its antigen, thus identifying the protein of interest. In order to visualise this interaction the blot is incubated further in a solution of a secondary antibody, which is directed against the IgG of the species that provided the primary antibody. For example, if the primary antibody was raised in a rabbit then the secondary antibody would be anti-rabbit IgG. This secondary antibody is appropriately labelled so that the interaction of the secondary antibody with the primary antibody can be visualised on the blot. Anti-species IgG molecules are readily available commercially, with a choice of different labels attached. One of the most common detection methods is to use an enzyme-linked secondary antibody (Fig. 10.12). In this case, following treatment with enzyme-labelled secondary antibody, the blot is incubated in enzyme–substrate solution, when the enzyme converts the substrate into an insoluble coloured product that is precipitated onto the nitrocellulose. The presence of a coloured band therefore indicates the position of the protein of interest. By careful comparisons of the blot with a stained gel of the same sample, the protein of interest can be identified. The enzyme used in enzyme-linked antibodies is usually either alkaline phosphatase, which converts colourless 5-bromo-4-chloro-indolylphosphate (BCIP) substrate into a blue product, or horseradish peroxidase, which, with H_2O_2 as a substrate, oxidises either 3-amino-9-ethylcarbazole into an insoluble brown product, or 4-chloro-l-naphthol into an insoluble blue product. An alternative approach to the detection of

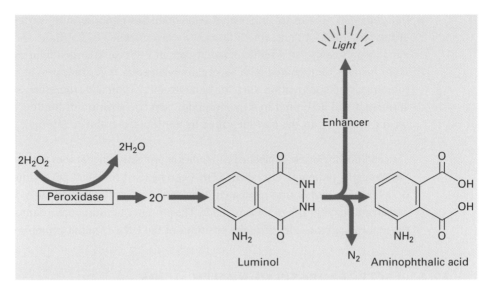

Fig. 10.13 The use of enhanced chemiluminescence to detect horseradish peroxidase.

horseradish peroxidase is to use the method of enhanced chemiluminescence (ECL). In the presence of hydrogen peroxide and the chemiluminescent substrate luminol (Fig. 10.13) horseradish peroxidase oxidises the luminol with concomitant production of light, the intensity of which is increased 1000-fold by the presence of a chemical enhancer. The light emission can be detected by exposing the blot to a photographic film. Corresponding ECL substrates are available for use with alkaline-phosphatase-labelled antibodies. The principle behind the use of enzyme-linked antibodies to detect antigens in blots is highly analogous to that used in enzyme-linked immuno-sorbent assays (Section 7.3.1).

Although enzymes are commonly used as markers for second antibodies, other markers can also be used. These include:

- *^{125}I-labelled secondary antibody:* Binding to the blot is detected by autoradiography (Section 14.3.3).
- *Fluorescein-labelled secondary antibody:* The fluorescent label is detected by exposing the blot to ultraviolet light.
- *^{125}I-labelled protein A:* Protein A is purified from *Staphylococcus aureus* and specifically binds to the Fc region of IgG molecules. ^{125}I-labelled protein A is therefore used instead of a second antibody, and binding to the blot is detected by autoradiography.
- *Biotinylated secondary antibodies:* Biotin is a small molecular weight vitamin that binds strongly to the egg protein avidin ($K_d = 10^{-15}$ M). The blot is incubated with biotinylated secondary antibody, then incubated further with enzyme-conjugated avidin. Since multiple biotin molecules can be linked to a single antibody molecule, many enzyme-linked avidin molecules can bind to a single biotinylated antibody molecule, thus providing an enhancement of the signal. The enzyme used is usually alkaline phosphatase or horseradish peroxidase.

- *Quantum dots*: These are engineered semiconductor nanoparticles, with diameters of the order of 2−10 nm, which fluoresce when exposed to UV light. Quantum dot nanocrystals comprise a semiconductor core of CdSe surrounded by a shell of ZnS. This crystal is then coated with an organic molecular layer that provides water solubility, and conjugation sites for biomolecules. Typically, therefore, second antibodies will be bound to a quantum dot, and the position of binding of the second antibody on the blot identified by exposing the blot to UV light.

In addition to the use of labelled antibodies or proteins, other probes are sometimes used. For example, radioactively labelled DNA can be used to detect DNA-binding proteins on a blot. The blot is first incubated in a solution of radiolabelled DNA, then washed, and an autoradiograph of the blot made. The presence of radioactive bands, detected on the autoradiograph, identifies the positions of the DNA-binding proteins on the blot.

10.4   ELECTROPHORESIS OF NUCLEIC ACIDS

10.4.1   Agarose gel electrophoresis of DNA

For the majority of DNA samples, electrophoretic separation is carried out in agarose gels. This is because most DNA molecules and their fragments that are analysed routinely are considerably larger than proteins and therefore, because most DNA fragments would be unable to enter a polyacrylamide gel, the larger pore size of an agarose gel is required. For example, the commonly used plasmid pBR322 has an M_r of 2.4×10^6. However, rather than use such large numbers it is more convenient to refer to DNA size in terms of the number of base-pairs. Although, originally, DNA size was referred to in terms of base-pairs (bp) or kilobase-pairs (kbp), it has now become the accepted nomenclature to abbreviate kbp to simply kb when referring to double-stranded DNA. pBR322 is therefore 4.36 kb. Even a small restriction fragment of 1 kb has an M_r of 620 000. When talking about single-stranded DNA it is common to refer to size in terms of nucleotides (nt). Since the charge per unit length (owing to the phosphate groups) in any given fragment of DNA is the same, all DNA samples should move towards the anode with the same mobility under an applied electrical field. However, separation in agarose gels is achieved because of resistance to their movement caused by the gel matrix. The largest molecules will have the most difficulty passing through the gel pores (very large molecules may even be blocked completely), whereas the smallest molecules will be relatively unhindered. Consequently the mobility of DNA molecules during gel electrophoresis will depend on size, the smallest molecules moving fastest. This is analogous to the separation of proteins in SDS–polyacrylamide gels (Section 10.3.1), although the analogy is not perfect, as double-stranded DNA molecules form relatively stiff rods and while it is not completely understood how they pass through the gel, it is probable that long DNA molecules pass through the gel pores end-on. While passing through the pores, a DNA molecule will experience drag; so the longer the molecule, the more it will be retarded by each pore. Sideways movement may become more important for very small double-stranded DNA and for the more flexible single-stranded DNA. It will be obvious from the above that gel concentrations must be chosen to suit the size range of the molecules to be separated. Gels containing 0.3% agarose will separate double-stranded DNA molecules of between 5 and 60 kb size, whereas 2% gels are used for samples of between 0.1 and 3 kb. Many laboratories routinely use 0.8% gels, which are suitable for separating DNA molecules in the range 0.5–10 kb. Since agarose gels separate DNA according to size, the M_r of a DNA fragment may be determined from its electrophoretic mobility by running a number of standard DNA markers of known M_r on the same gel. This is most conveniently achieved by running a sample of bacteriophage λ DNA (49 kb) that has been cleaved with a restriction enzyme such as *Eco*RI. Since the base sequence of λ DNA is known, and the cleavage sites for *Eco*RI are known, this generates fragments of accurately known size (Fig. 10.14).

DNA gels are invariably run as horizontal, submarine or submerged gels; so named because such a gel is totally immersed in buffer. Agarose, dissolved in gel buffer by boiling, is poured onto a glass or plastic plate, surrounded by a wall of adhesive tape or a plastic frame to provide a gel about 3 mm in depth. Loading wells are formed by placing a plastic well-forming template or comb in the poured gel solution, and removing this comb once the gel has set. The gel is placed in the electrophoresis tank, covered with buffer, and samples loaded by directly injecting the sample into the wells. Samples are prepared by dissolving them in a buffer solution that contains sucrose, glycerol or Ficoll, which makes the solution dense and allows it to sink to the bottom of the well. A dye such as bromophenol blue is also included in the sample solvent; it makes it easier to see the sample that is being loaded and also acts as a marker of the electrophoresis front. No stacking gel (Section 10.3.1) is needed for the electrophoresis of DNA because the mobilities of DNA molecules are much greater in the well than in the gel, and therefore all the molecules in the well pile up against the gel within a few minutes of the current being turned on, forming a tight band at the start of the run. General purpose gels are approximately 25 cm long and 12 cm wide, and are run at a voltage gradient of about $1.5\,\text{V cm}^{-1}$ overnight. A higher voltage would cause excessive heating. For rapid analyses that do not need extensive separation of DNA molecules, it is common to use mini-gels that are less than 10 cm long. In this way information can be obtained in 2–3 h.

Once the system has been run, the DNA in the gel needs to be stained and visualised. The reagent most widely used is the fluorescent dye ethidium bromide. The gel is rinsed gently in a solution of ethidium bromide ($0.5\,\mu\text{g cm}^{-3}$) and then viewed under ultraviolet light (300 nm wavelength). Ethidium bromide is a cyclic planar molecule that binds between the stacked base-pairs of DNA (i.e. it intercalates) (Section 5.7.4). The ethidium bromide concentration therefore builds up at the site of the DNA bands and under ultraviolet light the DNA bands fluoresce orange-red. As little as 10 ng of DNA can be visualised as a 1 cm wide band. It should be noted that extensive viewing of the DNA with ultraviolet light can result in damage of the DNA by nicking and base-pair dimerisation. This is of no consequence if a gel is only to be viewed, but obviously viewing of the gel should be kept to a minimum if the DNA is to be recovered (see below). It is essential to protect one's eyes by wearing goggles when ultraviolet light is used. If viewing of gels under ultraviolet is carried out for long periods, a plastic mask that covers the whole face should be used to avoid 'sunburn'.

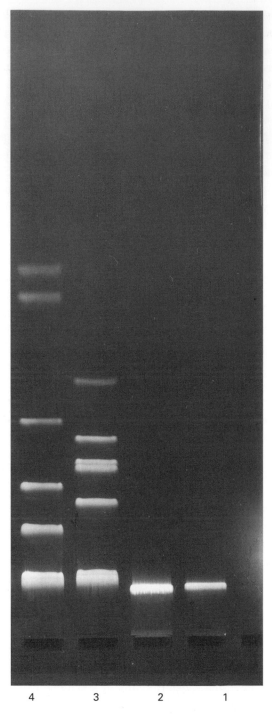

Fig. 10.14 Photograph showing four tracks from a 0.8% agarose submarine gel. The gel was run at 40 V in Tris/borate/EDTA buffer for 16 h, stained with ethidium bromide and viewed under ultraviolet light. Sample loadings were about 0.5 μg of DNA per track. Tracks 1 and 2, λ DNA (49 kb). Track 3, λ DNA cleaved with the enzyme EcoRI to generate fragments of the following size (in order from the origin): 21.80 kb, 7.52 kb, 5.93 kb, 5.54 kb, 4.80 kb, 3.41 kb. Track 4, λ DNA cleaved with the enzyme HindIII to generate fragments of the following size (in order from the origin): 23.70 kb, 9.46 kb, 6.75 kb, 4.26 kb, 2.26 kb, 1.98 kb. (Courtesy of Stephen Boffey, University of Hertfordshire.)

10.4.2   DNA sequencing gels

Although agarose gel electrophoresis of DNA is a 'workhorse' technique for the molecular biologist, a different form of electrophoresis has to be used when DNA sequences are to be determined. Whichever DNA sequencing method is used (Section 5.11), the final analysis usually involves separating single-stranded DNA molecules shorter than about 1000 nt and differing in size by only 1 nt. To achieve this it is necessary to have a small-pored gel and so acrylamide gels are used instead of agarose. For example, 3.5% polyacrylamide gels are used to separate DNA in the range 80–1000 nt and 12% gels to resolve fragments of between 20 and 100 nt. If a wide range of sizes is being analysed it is often convenient to run a gradient gel, for example from 3.5% to 7.5%. Sequencing gels are run in the presence of denaturing agents, urea and formamide. Since it is necessary to separate DNA molecules that are very similar in size, DNA sequencing gels tend to be very long (100 cm) to maximise the separation achieved. A typical DNA sequencing gel is shown in Fig. 5.38.

As mentioned above, electrophoresis in agarose can be used as a preparative method for DNA. The DNA bands of interest can be cut out of the gel and the DNA recovered by: (a) electroelution, (b) macerating the gel piece in buffer, centrifuging and collecting the supernatant; or (c), if low melting point agarose is used, melting the gel piece and diluting with buffer. In each case, the DNA is finally recovered by precipitation of the supernatant with ethanol.

10.4.3   Pulsed-field gel electrophoresis

The agarose gel methods for DNA described above can fractionate DNA of 60 kb or less. The introduction of pulsed-field gel electrophoresis (PFGE) and the further development of variations on the basic technique now means that DNA fragments up to 2×10^3 kb can be separated. This therefore allows the separation of whole chromosomes by electrophoresis. The method basically involves electrophoresis in agarose where two electric fields are applied alternately at different angles for defined time periods (e.g. 60 s). Activation of the first electric field causes the coiled molecules to be stretched in the horizontal plane and start to move through the gel. Interruption of this field and application of the second field force the molecule to move in the new direction. Since there is a length-dependent relaxation behaviour when a long-chain molecule undergoes conformational change in an electric field, the smaller a molecule, the quicker it realigns itself with the new field and is able to continue moving through the gel. Larger molecules take longer to realign. In this way, with continual reversing of the field, smaller molecules draw ahead of larger molecules and separate according to size. PFGE has proved particularly useful in identifying the course of outbreaks of bacterial foodborne illness (e.g. *Salmonella* infections). Having isolated the bacterial pathogen responsible for the illness from an individual, the DNA is isolated and cleaved into large fragments which are separated by PFGE. For example, DNA from *Salmonella* species, when digested with the restriction enzyme *Xba*1, gives around 15 fragments ranging from 25 kb to 680 kb. This pattern of fragments, or 'fingerprint', is unique to that strain. If the same fingerprint is found from bacteria

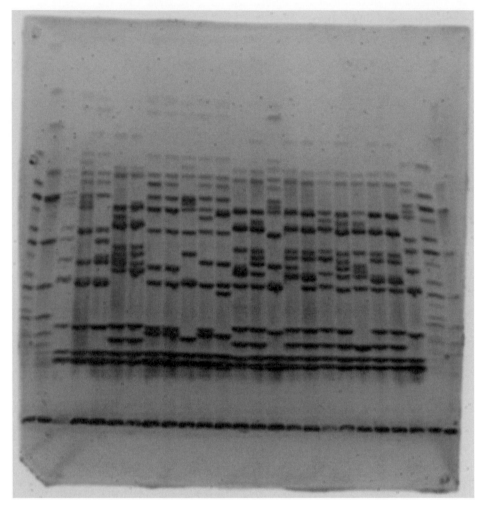

Fig. 10.15 PFGE separation of the digestion pattern produced with the restriction enzyme Nhe1, of 21 strains of *Neisseria meningitidis*. There are two molecular weight marker tracks at either end of the gel. (Courtesy of Dr Giovanna Morelli, Max-Planck Institute for Molecular Genetics, Berlin, Germany.)

from other infected people, then it can be assumed that they were all infected from a common source. Thus by comparing their eating habits, food sources, etc. the source of the infection can be traced to a restaurant, food item, etc. Figure 10.15 shows the restriction patterns from different strains of *Neisseria meningitidis*.

10.4.4 Electrophoresis of RNA

Like that of DNA, electrophoresis of RNA is usually carried out in agarose gels, and the principle of the separation, based on size, is the same. Often one requires a rapid method for checking the integrity of RNA immediately following extraction but before deciding whether to process it further. This can be achieved easily by electrophoresis in a 2% agarose gel in about 1 h. Ribosomal RNAs (18 S and 28 S) are clearly

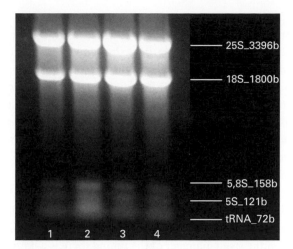

Fig. 10.16 Separation of yeast RNA on a 1.5% agarose gel. (Courtesy of Dr Tomas Masek, Department of Genetics and Microbiology, Charles University, Prague, Czech Republic.)

resolved and any degradation (seen as a smear) or DNA contamination is seen easily. This can be achieved on a 2.5–5% acrylamide gradient gel with an overnight run. Both these methods involve running native RNA. There will almost certainly be some secondary structure within the RNA molecule owing to intramolecular hydrogen bonding (see e.g. the clover leaf structure of tRNA, Fig. 5.6). For this reason native RNA run on gels can be stained and visualised with ethidium bromide. However, if the study objective is to determine RNA size by gel electrophoresis, then full denaturation of the RNA is needed to prevent hydrogen bond formation within or even between polynucleotides that will otherwise affect the electrophoretic mobility. There are three denaturing agents (formaldehyde, glyoxal and methylmercuric hydroxide) that are compatible with both RNA and agarose. Either one of these may be incorporated into the agarose gel and electrophoresis buffer, and the sample is heat denatured in the presence of the denaturant prior to electrophoresis. After heat denaturation, each of these agents forms adducts with the amino groups of guanine and uracil, thereby preventing hydrogen bond reformation at room temperature during electrophoresis. It is also necessary to run denaturing gels if the RNA is to be blotted (northern blots, Section 5.9.2) and probed, to ensure that the base sequence is available to the probe. Denatured RNA stains only very weakly with ethidium bromide, so acridine orange is commonly used to visualise RNA on denaturing gels. However, it should be noted that many workers will be using radiolabelled RNA and will therefore identify bands by autoradiography. An example of the electrophoresis of RNA is shown in Fig. 10.16.

10.5 CAPILLARY ELECTROPHORESIS

The technique has variously been referred to as high performance capillary electrophoresis (HPCE), capillary zone electrophoresis (CZE), free solution capillary electrophoresis (FSCE) and capillary electrophoresis (CE), but the term CE is the one most common nowadays. The microscale nature of the capillary used, where only microlitres of reagent are consumed by analysis and only nanolitres of sample needed for analysis, together with the ability for on-line detection down to femtomole (10^{-15} moles) sensitivity in some cases has for many years made capillary electrophoresis the method of choice for many biomedical and clinical analyses. Capillary electrophoresis can be used to separate a wide spectrum of biological molecules including amino acids, peptides, proteins, DNA fragments (e.g. synthetic oligonucleotides) and nucleic acids, as well as any number of small organic molecules such as drugs or even metal ions (see below). The method has also been applied successfully to the problem of chiral separations (Section 11.5.5).

As the name suggests, capillary electrophoresis involves electrophoresis of samples in very narrow-bore tubes (typically 50 µm internal diameter, 300 µm external diameter). One advantage of using capillaries is that they reduce problems resulting from heating effects. Because of the small diameter of the tubing there is a large surface-to-volume ratio, which gives enhanced heat dissipation. This helps to eliminate both convection currents and zone broadening owing to increased diffusion caused by heating. It is therefore not necessary to include a stabilising medium in the tube and allows free-flow electrophoresis.

Theoretical considerations of CE generate two important equations:

$$t = \frac{L^2}{\mu V} \tag{10.4}$$

where t is the migration time for a solute, L is the tube length, μ is the electrophoretic mobility of the solute, and V is the applied voltage.

The separation efficiency, in terms of the total number of theoretical plates, N, is given by

$$N = \frac{\mu V}{2D} \tag{10.5}$$

where D is the solute's diffusion coefficient.

From these equations it can be seen, first, that the column length plays no role in separation efficiency, but that it has an important influence on migration time and hence analysis time, and, secondly, high separation efficiencies are best achieved through the use of high voltages (μ and D are dictated by the solute and are not easily manipulated).

It therefore appears that the ideal situation is to apply as high a voltage as possible to as short a capillary as possible. However, there are practical limits to this approach. As the capillary length is reduced, the amount of heat that must be dissipated increases owing to the decreasing electrical resistance of the capillary. At the same time the surface area available for heat dissipation is decreasing. Therefore at some point a significant thermal effect will occur, placing a practical limit on how short a tube can be used. Also the higher the voltage that is applied, the greater the current, and therefore the heat generated. In practical terms a compromise between voltage used and capillary length is required. Voltages of 10–50 kV with capillaries of 50–100 cm are commonly used.

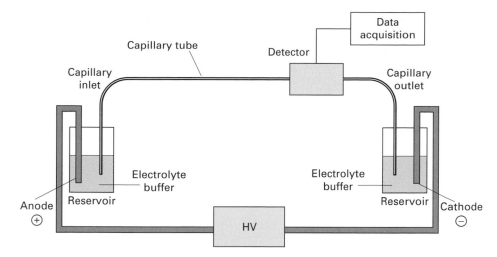

Fig. 10.17 Diagrammatic representation of a typical capillary electrophoresis apparatus.

The basic apparatus for CE is shown diagrammatically in Fig. 10.17. A small plug of sample solution (typically $5-30\,\mu m^3$) is introduced into the anode end of a fused silica capillary tube containing an appropriate buffer. Sample application is carried out in one of two ways: by high voltage injection or by pressure injection.

- *High voltage injection.* With the high voltage switched off, the buffer reservoir at the positive electrode is replaced by a reservoir containing the sample, and a plug of sample (e.g. $5-30\,\mu m^3$ of a $1\,mg\,cm^{-3}$ solution) is introduced into the capillary by briefly applying high voltage. The sample reservoir is then removed, the buffer reservoir replaced, voltage again applied and the separation is then commenced.
- *Pressure injection.* The capillary is removed from the anodic buffer reservoir and inserted through an air-tight seal into the sample solution. A second tube provides pressure to the sample solution, which forces the sample into the capillary. The capillary is then removed, replaced in the anodic buffer and a voltage applied to initiate electrophoresis.

A high voltage (up to $50\,kV$) is then put across the capillary tube and component molecules in the injected sample migrate at different rates along the length of the capillary tube. Electrophoretic migration causes the movement of charged molecules in solution towards an electrode of opposite charge. Owing to this electrophoretic migration, positive and negative sample molecules migrate at different rates. However, although analytes are separated by electrophoretic migration, they are all drawn towards the cathode by electroendosmosis (Section 10.1). Since this flow is quite strong, the rate of electroendosmotic flow usually being much greater than the electrophoretic velocity of the analytes, all ions, regardless of charge sign, and neutral species are carried towards the cathode. Positively charged molecules reach the cathode first because the combination of electrophoretic migration and electroosmotic flow causes them to move fastest. As the separated molecules approach the cathode, they pass through a viewing window where they are detected by an ultraviolet monitor that

Peptide	
1	Lys-Arg-Pro-Pro-Gly-Phe-Ser-Pro-Phe-Arg
2	Met-Lys-Arg-Pro-Pro-Gly-Phe-Ser-Pro-Phe-Arg
3	Arg-Pro-Pro-Gly-Phe-Ser-Pro-Phe-Arg
4	Ser-Arg-Pro-Pro-Gly-Phe-Ser-Pro-Phe-Arg
5	Ile-Ser-Arg-Pro-Pro-Gly-Phe-Ser-Pro-Phe-Arg

Source: Courtesy of Patrick Camilleri and George Okafo, GSK Ltd.

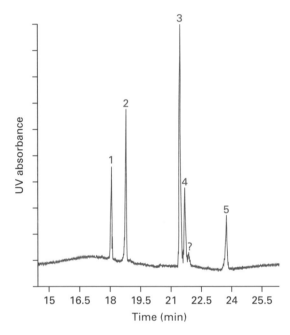

Fig. 10.18 Capillary electrophoresis of five structurally related peptides. Column length was 100 cm and the separation voltage 50 kV. Peptides were detected by their ultraviolet absorbance at 200 nm.

transmits a signal to a recorder, integrator or computer. Typical run times are between 10 and 30 min. A typical capillary electrophoretograph is shown in Fig. 10.18.

This free solution method is the simplest and most widely practised mode of capillary electrophoresis. However, while the generation of ionised groups on the capillary wall is advantageous via the introduction of electroendosmotic flow, it can also sometimes be a disadvantage. For example, protein adsorption to the capillary wall can occur with cationic groups on protein surfaces binding to the ionised silanols. This can lead to smearing of the protein as it passes through the capillary (recognised as peak broadening) or, worse, complete loss of protein due to total adsorption on the walls. Some workers therefore use coated tubes where a neutral coating group has been used to block the silanol groups. This of course eliminates electroendosmotic flow. Therefore, during electrophoresis in coated capillaries, neutral species are immobile while acid species migrate to the anode and basic species to the cathode. Since detection normally takes place at only one end of the capillary, only one class of species can be detected at a time in an analysis using a coated capillary.

A range of variations on this basic technique also exist. For example, as seen above, in normal CE neutral molecules do not separate but rather travel as a single band. However, separation of neutral molecules can be achieved by including a surfactant such as SDS with the buffer. Above a certain concentration some surfactant molecules agglomerate and form micelles, which, under the influence of an applied electric field, will migrate towards the appropriate electrode. Solutes will interact and partition with the moving micelles. If a solute interacts strongly it will reach the detector later than one which partitions to a lesser degree. This method is known as micellular electrokinetic capillary electrophoresis (MECC). Since ionic solutes will also migrate under the applied field, separation by MECC is due to a combination of both electrophoresis and chromatography.

Original developments in CE concentrated on the separation of peptides and proteins, but in recent years CE has been successfully applied to the separation of a range of other biological molecules. The following provides a few examples.

- In the past, peptide analysis has been performed routinely using reversed-phase HPLC, achieving separation based on hydrophobicity differences between peptides. Peptide separation by CE is now also routinely carried out, and is particularly useful, for example as a means of quality (purity) control for peptides and proteins produced by preparative HPLC. Fig. 10.18 shows the impressive separation that can be achieved for peptides with very similar structures.
- High purity synthetic oligodeoxyribonucleotides are necessary for a range of applications including use as hybridisation probes in diagnostic and gene cloning experiments, use as primers for DNA sequencing and the polymerase chain reaction (PCR), use in site-directed mutagenesis and use as antisense therapeutics. CE can provide a rapid method for analysing the purity of such samples. For example, analysis of an 18-mer antisense oligonucleotide containing contaminant fragments (8-mer to 17-mer) can be achieved in only 5 min.
- Point mutations in DNA, such as occur in a range of human diseases, can be identified by CE.
- CE can be used to quantify DNA. For example, CE analysis of PCR products from HIV-I allowed the identification of between 200 000 and 500 000 viral particles per cubic centimetre of serum.
- Chiral compounds can be resolved using CE. Most work has been carried out in free solution using cyclodextrins as chiral selectors.
- A range of small molecules, drugs and metabolites can be measured in physiological solutions such as urine and serum. These include amino acids (over 50 are found in urine), nucleotides, nucleosides, bases, anions such as chloride and sulphate (NO_2^- and NO_3^- can be separated in human plasma) and cations such as Ca^{2+} and Fe^{3+}.

10.6 MICROCHIP ELECTROPHORESIS

The further miniaturisation of electrophoretic systems has led to the development of microchip electrophoresis, which has many advantages over conventional electrophoresis methods, allowing very high speed analyses at very low sample sizes. For example, microchip analysis can often be completed in tens of seconds whereas capillary electrophoresis (CE) can take 20 min and conventional gel electrophoresis at least 2 h. Using new detection systems, such as laser-induced fluorescence, picomole to attomole (10^{-18} moles) sensitivity can be achieved, which is at least two orders of magnitude greater than for conventional CE. Detection systems for molecules that do not fluoresce include electrochemical detectors (Section 11.3.3), pulsed amperometric detection (PAD), and sinusoidal voltometry. All these detection techniques offer high sensitivity, are ideally suited to miniaturisation, are very low cost, and all are highly compatible with advanced micromachining and microfabrication (see below) technologies. Finally, the applied voltage required is only a few volts, which eliminates the need for the high voltages used by CE.

The manufacturing process that produces microchips is called microfabrication. The process etches precise and reproducible capillary-like channels (typically, 50 µm wide and 10 µm deep; slightly smaller than a strand of human hair) on the surface of sheets of quartz, glass or plastic. A second sheet is then fused on top of the first sheet, turning the etched channels into closed microfluidic channels. The end of each channel connects to a reservoir through which fluids are introduced/removed. Typically, the size of chips can be as small as 2 cm^2. Basically the microchip provides an electrophoretic system similar to CE but with more flexibility.

Current developments of this technology are based on integrating functions other than just separation into the chip. For example, sample extraction, pre-concentration of samples prior to separation, PCR amplification of DNA samples using infrared-mediated thermocycling for rapid on-chip amplification, and the extraction of separated molecules using microchamber-bound solid phases are all examples of where further functions have been built into a microchip electrophoresis system. An interface has also been developed for microchip electrophoresis–mass spectrometry (MCE–MS) where drugs have been separated by MCE and then identified by MS.

10.7 SUGGESTIONS FOR FURTHER READING

Walker, J. M. (2009). *The Protein Protocols Handbook*, 3rd edn. New York: Humana Press. (Detailed theory and laboratory protocols for a range of electrophoretic techniques and blotting procedures.)

Hames, B. D. and Rickwood, D. (2002). *Gel Electrophoresis of Proteins: A Practical Approach*, 3rd edn. Oxford: Oxford University Press. (Detailed theory and practical procedures for the electrophoresis of proteins.)

11 Chromatographic techniques

K. WILSON

11.1 PRINCIPLES OF CHROMATOGRAPHY

11.1.1 Distribution coefficients

The basis of all forms of chromatography is the distribution or partition coefficient (K_d), which describes the way in which a compound (the analyte) distributes between two immiscible phases. For two such phases A and B, the value for this coefficient is a constant at a given temperature and is given by the expression:

$$\frac{\text{concentration in phase A}}{\text{concentration in phase B}} = K_d \qquad (11.1)$$

The term effective distribution coefficient is defined as the total amount, as distinct from the concentration, of analyte present in one phase divided by the total amount present in the other phase. It is in fact the distribution coefficient multiplied by the ratio of the volumes of the two phases present. If the distribution coefficient of an analyte between two phases A and B is 1, and if this analyte is distributed between $10\,cm^3$ of A and $1\,cm^3$ of B, the concentration in the two phases will be the same, but the total amount of the analyte in phase A will be 10 times the amount in phase B.

All chromatographic systems consist of the stationary phase, which may be a solid, gel, liquid or a solid/liquid mixture that is immobilised, and the mobile phase, which may be liquid or gaseous, and which is passed over or through the stationary phase after the mixture of analytes to be separated has been applied to the stationary phase. During the chromatographic separation the analytes continuously pass back and forth between the two phases so that differences in their distribution coefficients result in their separation.

11.1.2 Column chromatography

In column chromatography the stationary phase is packed into a glass or metal column. The mixture of analytes is then applied and the mobile phase, commonly referred to as the eluent, is passed through the column either by use of a pumping system or applied gas pressure. The stationary phase is either coated onto discrete small particles (the matrix) and packed into the column or applied as a thin film to the inside wall of the column. As the eluent flows through the column the analytes separate on the basis of their distribution coefficients and emerge individually in the eluate as it leaves the column.

Basic column chromatographic components

A typical column chromatographic system using a gas or liquid mobile phase consists of the following components:

- *A stationary phase*: Chosen to be appropriate for the analytes to be separated.
- *A column*: In liquid chromatography these are generally 25−50 cm long and 4 mm internal diameter and made of stainless steel whereas in gas chromatography they are 1−3 m long and 2−4 mm internal diameter and made of either glass or stainless steel. They may be either of the conventional type filled with the stationary phase, or of the microbore type in which the stationary phase is coated directly on the inside wall of the column.
- *A mobile phase and delivery system*: Chosen to complement the stationary phase and hence to discriminate between the sample analytes and to deliver a constant rate of flow into the column.
- *An injector system*: To deliver test samples to the top of the column in a reproducible manner.
- *A detector and chart recorder*: To give a continuous record of the presence of the analytes in the eluate as it emerges from the column. Detection is usually based on the measurement of a physical parameter such as visible or ultraviolet absorption or fluorescence. A peak on the chart recorder represents each separated analyte.
- *A fraction collector*: For collecting the separated analytes for further biochemical studies.

The two forms of column chromatography to be discussed in this chapter are liquid chromatography (LC), mainly high-performance liquid chromatography (HPLC), and gas chromatography (GC).

11.1.3 Analyte development and elution

Analyte development and elution relates to the separation of the mixture of analytes applied to the stationary phase by the mobile phase and their elution from the column. Column chromatographic techniques can be subdivided on the basis of the development and elution modes.

- In zonal development, the analytes in the sample are separated on the basis of their distribution coefficients between the stationary and mobile phases. The sample is dissolved in a suitable solvent and applied to the stationary phase as a narrow, discrete band. The mobile phase is then allowed to flow continuously over the stationary phase, resulting in the progressive separation and elution of the sample analytes. If the composition of the mobile phase is constant as in GC and some forms of HPLC, the process is said to be isocratic elution. To facilitate separation however, the composition of the mobile phase may be gradually changed, for example with respect to pH, salt concentration or polarity. This is referred to as gradient elution. The composition of the mobile phase may be changed continuously or in a stepwise manner.
- In displacement or affinity development that is confined to some forms of HPLC the analytes in the sample are separated on the basis of their affinity for the stationary phase. The sample of analytes dissolved in a suitable solvent is applied to the stationary phase as a discrete band. The analytes bind to the stationary phase with a strength determined by their affinity constant for the phase. The analytes are then selectively eluted by using a mobile phase containing a specific solute that has a higher affinity for the stationary phase than have the analytes in the sample. Thus, as the mobile phase is added, this agent displaces the analytes from the stationary phase in a competitive fashion, resulting in their repetitive binding and displacement along the stationary phase and eventual elution from the column in the order of their affinity for the stationary phase, the one with the lowest affinity being eluted first.

11.2 CHROMATOGRAPHIC PERFORMANCE PARAMETERS

11.2.1 Introduction

The successful chromatographic separation of analytes in a mixture depends upon the selection of the most appropriate process of chromatography followed by the optimisation of the experimental conditions associated with the separation. Optimisation requires an understanding of the processes that are occurring during the development and elution, and of the calculation of a number of experimental parameters characterising the behaviour of each analyte in the mixture.

In any chromatographic separation two processes occur concurrently to affect the behaviour of each analyte and hence the success of the separation of the analytes from each other. The first involves the basic mechanisms defining the chromatographic

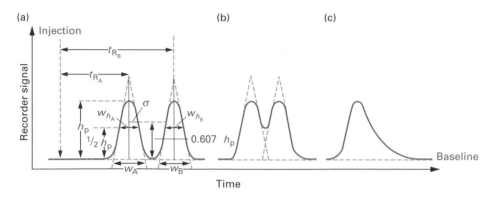

Fig. 11.1 (a) Chromatogram of two analytes showing complete resolution and the calculation of retention times; (b) chromatogram of two analytes showing incomplete resolution (fused peaks); (c) chromatogram of an analyte showing excessive tailing.

process such as adsorption, partition, ion exchange, ion pairing and molecular exclusion. These mechanisms involve the unique kinetic and thermodynamic processes that characterise the interaction of each analyte with the stationary phase. The second general process defines the other processes, such as diffusion, which tend to oppose the separation and which result in non-ideal behaviour of each analyte. These processes are manifest as a broadening and tailing of each analyte band. The analytical challenge is to minimise these secondary processes.

11.2.2 Retention time

A chromatogram is a pictorial record of the detector response as a function of elution volume or retention time. It consists of a series of peaks or bands, ideally symmetrical in shape, representing the elution of individual analytes, as shown in Fig. 11.1. The retention time t_R for each analyte has two components. The first is the time it takes the analyte molecules to pass through the free spaces between the particles of the matrix coated with the stationary phase. This time is referred to as the dead time, t_M. The volume of the free space is referred to as the column void volume, V_0. The value of t_M will be the same for all analytes and can be measured by using an analyte that does not interact with the stationary phase but simply spends all of the elution time in the mobile phase travelling through the void volume. The second component is the time the stationary phase retains the analyte, referred to as the adjusted retention time, t'_R. This time is characteristic of the analyte and is the difference between the observed retention time and the dead time:

$$t'_R = t_R - t_M \tag{11.2}$$

It is common practice to relate the retention time t_R or t'_R for an analyte to a reference internal standard (Section 11.2.5). In such cases the relative retention time is often calculated. It is simply the retention time for the analyte divided by that for the standard.

11.2.3 **Retention factor**

One of the most important parameters in chromatography is the retention factor, k (previously called capacity factor and represented by the symbol k'). It is simply the additional time that the analyte takes to elute from the column relative to an unretained or excluded analyte that does not interact with the stationary phase and which, by definition, has a k value of 0. Thus:

$$k = \frac{t_R - t_M}{t_M} = \frac{t_R'}{t_M} \tag{11.3}$$

It is apparent from this equation that if the analyte spends an equal time in the stationary and mobile phases, its t_R would equal $2t_M$ and its k would be 1, whilst if it spent four times as long in the stationary phase as the mobile phase t_R would equal $5t_M$ so that k would equal $5t_M - t_M /t_M = 4$. Note that k has no units.

If an analyte has a k of 4, it follows that there will be four times the amount of analyte in the stationary phase than in the mobile phase at any point in the column at any time. It is evident, therefore, that k is related to the distribution coefficient of the analyte (equation 11.1), which was defined as the relative concentrations of the analyte between the two phases. Since amount and concentration are related by volume, we can write:

$$k = \frac{t_R'}{t_M} = \frac{M_S}{M_M} = K_d \times \frac{V_S}{V_M} \tag{11.4}$$

where M_S is the mass of analyte in the stationary phase, M_M is the mass of analyte in the mobile phase, V_S is the volume of stationary phase and V_M is the volume of mobile phase. The ratio V_S/V_M is referred to as the volumetric phase ratio, β. Hence:

$$k = K_d\beta \tag{11.5}$$

Thus the retention factor for an analyte will increase with both the distribution coefficient between the two phases and the volume of the stationary phase. Values of k normally range from 1 to 10. Retention factors are important because they are independent of the physical dimensions of the column and the rate of flow of mobile phase through it. They can therefore be used to compare the behaviour of an analyte in different chromatographic systems. They are also a reflection of the selectivity of the system that in turn is a measure of its inherent ability to discriminate between two analytes. Such selectivity is expressed by the selectivity or separation factor, α, which can also be viewed as simply the relative retention ratio for the two analytes:

$$\alpha = \frac{k_A}{k_B} = \frac{K_{d_A}}{K_{d_B}} = \frac{t_{R_A}'}{t_{R_B}'} \tag{11.6}$$

The selectivity factor is influenced by the chemical nature of the stationary and mobile phases. Some chromatographic mechanisms are inherently highly selective. Good examples are affinity chromatography (Section 11.8) and chiral chromatography (Section 11.5.5).

11.2.4 **Plate height and resolution**

Plate height

Chromatography columns are considered to consist of a number of adjacent zones in each of which there is sufficient space for an analyte to completely equilibrate between the two phases. Each zone is called a theoretical plate (of which there are N in total in the column). The length of column containing one theoretical plate is referred to as the plate height, H, which has units of length normally in micrometres. The smaller the value of H and the greater the value of N, the more efficient is the column in separating a mixture of analytes. The numerical value of both N and H for a particular column is expressed by reference to a particular analyte. Plate height is simply related to the width of the analyte peak, expressed in terms of its standard deviation σ (Fig. 11.1), and the distance it travelled within the column, x. Specifically:

$$H = \frac{\sigma^2}{x} \tag{11.7}$$

For symmetrical Gaussian peaks, the base width is equal to 4σ and the peak width at the point of inflection, w_i, is equal to 2σ. Hence the value of H can be calculated from the chromatogram by measuring the peak width. The number of theoretical plates in the whole column of length L is equal to L divided by the plate height:

$$N = \frac{L}{H} = \frac{Lx}{\sigma^2} \tag{11.8}$$

If the position of a peak emerging from the column is such that $x = L$, from knowledge of the fact that the width of the peak at its base, w, obtained from tangents drawn to the two steepest parts of the peak, is equal to 4σ (this is a basic property of all Gaussian peaks) hence $\sigma = w/4$ and equation 11.8 can therefore be converted to:

$$N = \frac{L^2}{\sigma^2} = \frac{16L^2}{w^2} \tag{11.9}$$

If both L and w are measured in units of time rather than length, then equation 11.9 becomes:

$$N = 16(t_R/w)^2 \tag{11.10a}$$

Rather than expressing N in terms of the peak base width, it is possible to express it in terms of the peak width at half height ($w_{1/2}$) and this has the practical advantage that this is more easily measured:

$$N = 5.54(t_R/w_{1/2})^2 \tag{11.10b}$$

Equations 11.9 and 11.10 represent alternative ways to calculate the column efficiency in theoretical plates. The value of N, which has no units, can be as high as 50 000 to 100 000 per metre for efficient columns and the corresponding value of H can be as little as a few micrometres. The smaller the plate height (the larger the value of N), the narrower is the analyte peak (Fig. 11.2).

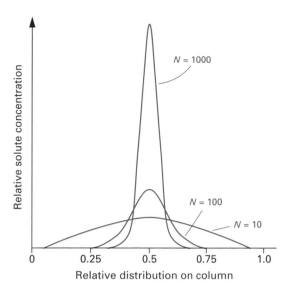

Fig. 11.2 Relationship between the number of theoretical plates (*N*) and the shape of the analyte peak.

Peak broadening

A number of processes oppose the formation of a narrow analyte peak thereby increasing the plate height:

- *Application of the sample to the column*: It takes a finite time to apply the analyte mixture to the column, so that the part of the sample applied first will already be moving along the column by the time the final part is applied. The part of the sample applied first will elute at the front of the peak.
- *Longitudinal diffusion*: Fick's law of diffusion states that an analyte will diffuse from a region of high concentration to one of low concentration at a rate determined by the concentration gradient between the two regions and the diffusion coefficient (*P*) of the analyte. Thus the analyte within a narrow band will tend to diffuse outwards from the centre of the band, resulting in band broadening.
- *Multiple pathways*: The random packing of the particles in the column results in the availability of many routes between the particles for both mobile phase and analytes. These pathways will vary slightly in length and hence elution time. The smaller the particle size the less serious is this problem and in open tubular columns the phenomenon is totally absent, which is one of the reasons why they give shorter elution times and better resolution than packed columns.
- *Equilibration time between the two phases*: It takes a finite time for each analyte in the test sample to equilibrate between the stationary and mobile phases as it passes down the column. As a direct consequence of the distribution coefficient, K_d, some of each analyte is retained by the stationary phase whilst the remainder stays in the mobile phase and continues its passage down the column. This partitioning automatically results in some spreading of each analyte band. Equilibration time, and hence band broadening, is also influenced by the particle size of the stationary phase. The smaller the size, the less time it takes to establish equilibration.

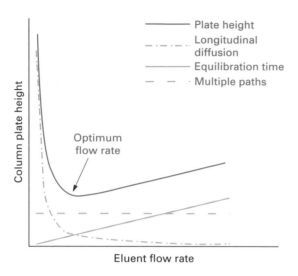

Fig. 11.3 Example of a van Deemter plot. The plot shows that the optimum flow rate for a given column is the net result of the influence of flow rate on longitudinal diffusion, equilibration time and multiple pathways.

Two of these four factors promoting the broadening of the analyte band are influenced by the flow rate of the eluent through the column. Longitudinal diffusion, defined by Fick's law, is inversely proportional to flow rate whilst equilibration time due to the partitioning of the analyte is directly proportional to flow rate. These two factors together with that of the multiple pathways factor determine the value of the plate height for a particular column and, as previously stated, plate height determines the width of the analyte peak. The precise relationship between the three factors and plate height is expressed by the van Deemter equation (equation 11.11) which is shown graphically in Fig. 11.3:

$$H = A + \frac{B}{u_x} + Cu_x \tag{11.11}$$

where u_x is the flow rate of the eluent and A, B and C are constants for a particular column and stationary phase relating to multiple paths, longitudinal diffusion and equilibration time respectively.

Figure 11.3 gives a clear demonstration of the importance of establishing the optimum flow rate for a particular column. Longitudinal diffusion is much faster in a gas than in a liquid and as a consequence flow rates are higher in gas chromatography than in liquid chromatography.

As previously stated, the width of an analyte peak is expressed in terms of the standard deviation σ, which is half the peak width at the point of inflexion ($0.607h_p$, where h_p is the peak height, Fig. 11.1). It can be shown that $\sigma = \sqrt{2Pt_R}$ where P is the diffusion coefficient of the analyte that is a measure of the rate at which the analyte moves randomly in the mobile phase from a region of high concentration to one of lower concentration. It has units of $m^2\ s^{-1}$. Since the value of σ is proportional to the square root of t_R it follows that if the elution time increases by a factor of four the width of the peak will double. Thus the longer it takes a given analyte to elute, the wider will be its peak. For this reason, increasing the column length is not the preferred way to improve resolution.

Asymmetric peaks

In some chromatographic separations, the ideal Gaussian-shaped peaks are not obtained, but rather asymmetrical peaks are produced. In cases where there is a gradual rise at the front of the peak and a sharp fall after the peak, the phenomenon is known as fronting. The most common cause of fronting is overloading the column so that reducing the amount of mixture applied to the column often resolves the problem. In cases where the rise in the peak is normal but the tail is protracted, the phenomenon is known as tailing (Fig. 11.1). The probable explanation for tailing is the retention of analyte by a few active sites on the stationary phase, commonly on the inert support matrix. Such sites strongly adsorb molecules of the analyte and only slowly release them. This problem can be overcome by chemically removing the sites, frequently hydroxyl groups, by treating the matrix with a silanising reagent such as hexamethyl-disilazine. This process is sometimes referred to as capping. Peak asymmetry is usually expressed as the ratio of the width of the peak from the centre of the peak at $0.1\,h_{\mathrm{p}}$.

Resolution

The success of a chromatographic separation is judged by the ability of the system to resolve one analyte peak from another. Resolution (R_S) is defined as the ratio of the difference in retention time (Δt_R) between the two peaks (t_{R_A} and t_{R_B}) to the mean (w_{av}) of their base widths (w_A and w_B):

$$R_S = \frac{\Delta t_R}{w_{av}} = \frac{2(t_{R_A} - t_{R_B})}{w_A + w_B} \tag{11.12}$$

When $R_S = 1.0$, the separation of the two peaks is 97.7% complete (thus the overlap is 2.3%). When $R_S = 1.5$ the overlap is reduced to 0.2%. Unresolved peaks are referred to as fused peaks (Fig. 11.1). Provided the overlap is not excessive, the analysis of the individual peaks can be made on the assumption that their characteristics are not affected by the incomplete resolution.

Resolution is influenced by column efficiency, selectivity factors and retention factors according to equation 11.13:

$$R_S = \frac{\sqrt{N}}{4} \left(\frac{\alpha - 1}{\alpha} \right) \left(\frac{k_2}{1 + k_{av}} \right) \tag{11.13}$$

where k_2 is the retention factor for the longest retained peak and k_{av} is the mean retention factor for the two analytes. Equation 11.13 is one of the most important in chromatography as it enables a rational approach to be taken to the improvement of the resolution between the analytes. For example, it can be seen that resolution increases with $\sqrt{N}$. Since N is linked to the length of the column, doubling the length of the column will increase resolution by $\sqrt{2}$, i.e. by a factor of 1.4 and in general is not the preferred way to improve resolution. Since both retention factors and selectivity factors are linked to retention times and retention volumes, altering the nature of the two phases or their relative volumes will impact on resolution. Retention factors are also dependent upon distribution coefficients, which in turn are temperature dependent; hence altering the column temperature may improve resolution.

The capacity of a particular chromatographic separation is a measure of the amount of material that can be resolved into its components without causing peak overlap or fronting. Ion-exchange chromatography (Section 11.6) has a high capacity, which is why it is often used in the earlier stages of a purification process.

11.2.5   Qualitative and quantitative analysis

Chromatographic analysis can be carried out on either a qualitative or quantitative basis.

Qualitative analysis

The objective of this approach is to confirm the presence of a specific analyte in a test sample. This is achieved on the evidence of:

- A comparison of the retention time of the peaks in the chromatograph with that of an authentic reference sample of the test analyte obtained under identical chromatographic conditions. Confirmation of the presence of the analyte in the sample can be obtained by spiking a second sample of the test sample with a known amount of the authentic compound. This should result in a single peak with the predicted increase in area.
- The use of either a mass spectrometer or nuclear magnetic resonance (NMR) spectrometer as a detector so that structural evidence for the identity of the analyte responsible for the peak can be obtained.

Quantitative analysis

The objective of this approach is to confirm the presence of a specific analyte in a test sample and to quantify its amount. Quantification is achieved on the basis of peak area coupled with an appropriate calibration graph. The area of each peak in a chromatogram can be shown to be proportional to the amount of the analyte producing the peak. The area of the peak may be determined by measuring the height of the peak (h_P) and its width at half the height (w_h) (Fig 11.1). The product of these dimensions is taken to be equal to the area of the peak. This procedure is time consuming when complex and/or a large number of analyses are involved and dedicated integrators or microcomputers best perform the calculations. These can be programmed to compute retention time and peak area and to relate them to those of a reference standard enabling relative retention ratios and relative peak area ratios to be calculated. These may be used to identify a particular analyte and to quantify it using previously obtained and stored calibration data. The data system can also be used to correct problems inherent in the chromatographic system. Such problems can arise either from the characteristics of the detector or from the efficiency of the separation process. Problems that are attributable to the detector are baseline drift, where the detector signal gradually changes with time, and baseline noise, which is a series of rapid minor fluctuations in detector signal, commonly the result of the operator using too high a detector sensitivity or possibly an electronic fault.

Example 1   CALCULATION OF THE RESOLUTION OF TWO ANALYTES

Question   Two analytes A and B were separated on a 25 cm long column. The observed retention times were 7 min 20 s and 8 min 20 s, respectively. The base peak width for analyte B was 10 s. When a reference compound, which was completely excluded from the stationary phase under the same elution conditions, was studied, its retention time was 1 min 20 s. What was the resolution of the two analytes?

Answer   In order to calculate the required resolution, it is first necessary to calculate other chromatographic parameters.

(i) The adjusted retention time for A and B based on equation 11.2: $t'_R = t_R - t_M$

For analyte A $t'_R = 440 - 80 = 360$ s
For analyte B $t'_R = 500 - 80 = 420$ s

(ii) The retention factor for A and B based on equation 11.3: $k = t_R/t_M$

For analyte A $k_A = 360/80 = 4.5$
For analyte B $k_B = 420/80 = 5.25$

(iii) The selectivity factor for the two analytes based on equation 11.5:

$\alpha = k_B/k_A$
$\alpha = 5.25/4.5 = 1.167$

(iv) The number of theoretical plates in the column; based on equation 11.10:
$N = (t_R/\omega)^2$ for analyte B

$N = (420/10)^2 = 1764$

(v) The resolution of the two analytes based on equation 11.13:

$R_S = (\sqrt{N}/4)[(\alpha - 1)/\alpha)](k'_B/(1 + k_{av})$

gives

$R_s = (\sqrt{1764}/4)(0.167/1.167)(5.25/1 + 4.875) = 1.34$

Discussion   From the earlier discussion on resolution, it is evident that a resolution of 1.34 is such as to give a peak separation of greater than 99%. If there were an analytical need to increase this separation it would be possible to calculate the length of column required to double the resolution. Since resolution is proportional to the square root of N, to double the resolution the number of theoretical plates in the column must be increased four-fold, i.e. to 4×1764 = 7056. The plate height in the column $H = L/N$, i.e. $250/1764 = 0.14$ mm. Hence to get 7056 plates in the column, its length must be increased to $0.14 \times 7056 = 987.84$ mm or 98.78 cm.

Quantification of a given analyte is based on the construction of a calibration curve obtained using a pure, authentic sample of the analyte. The construction of the calibration curve is carried out using the general principles discussed in Section 1.4.6. Most commonly the calibration curve is based on the use of relative peak areas obtained using an internal standard that has been subject to any preliminary extraction procedures adopted for the test samples. The standard must be carefully chosen to have similar physical and structural characteristics to those of the test analyte, and in practice is frequently an isomer or structural analogue of the analyte. Ideally, it should have a retention time close to that of the analyte but such that the resolution is greater than 99.5%.

11.2.6 Sample clean-up

Whilst chromatographic techniques are designed to separate mixtures of analytes, this does not mean that attention need not be paid to the preliminary purification (clean-up) of the test sample before it is applied to the column. On the contrary, it is clear that, for quantitative work using HPLC techniques in particular, such preliminary action is essential, particularly if the test analyte(s) is in a complex matrix such as plasma, urine, cell homogenate or microbiological culture medium. The extraction and purification of the components from a cell homogenate is often a complex multi-stage process. For some forms of analysis, for example the analysis of low-molecular-weight organic drugs in biological fluids, sample preparation is relatively easy. Solvent extraction is based on the extraction of the analytes from aqueous mixtures using a low-boiling-point water-immiscible solvent such as diethyl ether or dichloromethane. The extract is dried to remove traces of water before it is evaporated to dryness (often under nitrogen or *in vacuo*), the residue dissolved in the minimum volume of an appropriate solvent such as methanol or acetonitrile, and a sample applied to the column. This solvent extraction procedure tends to lack selectivity and is unsatisfactory for protein clean-up and for the HPLC analysis of analytes in the ng cm^{-3} or less range.

The alternative to solvent extraction is solid phase extraction, which unlike solvent extraction can be applied to the pre-purification of proteins. The test solution is passed through a small (few millimetres in length) disposable column (cartridge) packed with an appropriate stationary phase similar to those used for HPLC (Section 11.3). These selectively adsorb the analyte(s) under investigation and ideally allow interfering compounds to pass through. For example, for the purification of a mixture of proteins a reversed phase, affinity or ion exchange chromatography cartridge would be selected as these are ideal for desalting or concentrating protein mixtures containing in the range of femto- to picomoles of protein. Once the test solution has been passed through the column, either by simple gravity feed or by the application of a slight vacuum to the receiver vessel, the column is washed to remove traces of contaminants and the adsorbed analyte(s) recovered by elution with a suitable eluent. Complex protein and peptide mixtures can be partially fractionated at this stage by gradient elution. Several commercial forms of this solid phase extraction technique are available, for example Millipore Ziptips™, and the term pipette tip chromatography has been coined to describe it.

Table 11.1 **Examples of derivatising agents**

Analyte	Reagent
Pre-column	
Ultraviolet detection	
Alcohols, amines, phenols	3,5-Dinitrobenzoyl chloride
Amino acids, peptides	Phenylisothiocyanate, dansyl chloride
Carbohydrates	Benzoyl chloride
Carboxylic acids	1-*p*-Nitrobenzyl-*N*, *N′*-diisopropylisourea
Fatty acids, phospholipids	Phenacyl bromide, naphthacyl bromide
Electrochemical detection	
Aldehydes, ketones	2,4-Dinitrophenylhydrazine
Amines, amino acids	*o*-Phthalaldehyde, fluorodinitrobenzene
Carboxylic acids	*p*-Aminophenol
Fluorescent detection	
Amino acids, amines, peptides	Dansyl chloride, dabsyl chloride, fluoroescamine, *o*-phthalaldehyde
Carboxylic acids	4-Bromomethyl-7-methoxycoumarin
Carbonyl compounds	Dansylhydrazine
Post-column	
Ultraviolet detection	
Amino acids	Phenylisothiocyanate
Carbohydrates	Orcinol and sulphuric acid
Penicillins	Imidazole and mercuric chloride
Fluorescent detection	
Amino acids	*o*-Phthalaldehyde, fluorescamine, 6-aminoquinolyl-*N*-hydroxysuccinimidyl carbamate

Sample derivatisation

Some functional groups, especially hydroxyl, present in a test analyte may compromise the quality of its behaviour in a chromatographic system. The technique of analyte pre- or post-column derivatisation may facilitate better chromatographic separation and detection by masking these functional groups. Common derivatisation reagents are shown in Table 11.1.

11.3 HIGH-PERFORMANCE LIQUID CHROMATOGRAPHY

11.3.1 Principle

It is evident from equations 11.1 to 11.13 that the resolving power of a chromatographic column is determined by a number of factors that are embedded in equation 11.13. This shows that the resolution increases with:

- the number of theoretical plates (N) in the column and hence plate height (H). The value of N increases with column length but there are practical limits to the length of a column owing to the problem of peak broadening (Section 11.2.4);
- the selectivity of the column, α; and
- the retentivity of the column as determined by the retention factor, k.

As the number of theoretical plates in the column is related to the surface area of the stationary phase, it follows that the smaller the particle size of the stationary phase, the greater the value of N, i.e. N is inversely proportional to particle size. Unfortunately, the smaller the particle size, the greater is the resistance to the flow of the mobile phase for a given flow rate. This resistance creates a backpressure in the column that is directly proportional to both the flow rate and the column length and inversely proportional to the square of the particle size. The back-pressure may be sufficient to cause the structure of the matrix to collapse, thereby actually further reducing eluent flow and impairing resolution. This problem has been solved by the development of small particle size stationary phases, generally in the region of $5-10\,\mu m$ diameter with a narrow range of particle sizes, which can withstand pressures up to $40\,MPa$. This development, which is the basis of HPLC that was originally and incorrectly referred to as high-pressure liquid chromatography, explains why it has emerged as the most popular, powerful and versatile form of chromatography. Larger particle size phases are available commercially and form the basis of low-pressure liquid chromatography in which flow of the eluaent through the column is either gravity-fed or pumped by a low pressure pump, often a peristaltic pump. It is cheaper to run than HPLC but lacks the high resolution that is the characteristic of HPLC. Many commercially available HPLC systems are available and most are microprocessor-controlled to allow dedicated, continuous chromatographic separations.

Columns

The components of a typical HPLC system are shown in Fig. 11.4. Conventional columns used for HPLC are generally made of stainless steel and are manufactured so that they can withstand pressures of up to $50\,MPa$. The columns are generally 3–25 cm long and approximately 4.6 mm internal diameter to give typical flow rates of $1-3\,cm^3\ min^{-1}$. Microbore or open tubular columns have an internal diameter of 1–2 mm and are generally 25–50 cm long. They can sustain flow rates of $5-20\,mm^3\ min^{-1}$. Microbore columns have three important advantages over conventional columns:

- reduced eluent consumption due to the slower flow rates;
- ideal for interfacing with a mass spectrometer due to the reduced flow rate; and
- increased sensitivity due to the higher concentration of analytes that can be used.

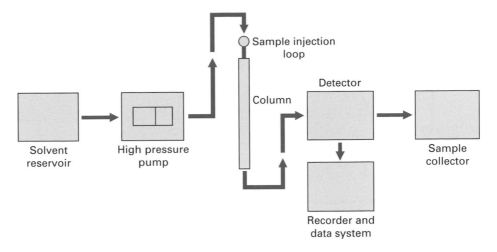

Fig. 11.4 Components of an isocratic HPLC system. For gradient elution two reservoirs and two pumps are used with liquid-phase mixing before entry to the sample injection loop.

Matrices and stationary phases

Two main forms of matrix/stationary phase material are available, based on a rigid solid structure. Both forms involve approximately spherical particles of a uniform size to minimise space for diffusion and hence band broadening to occur. They are made of chemically modified silica or styrene/divinylbenzene copolymers. The two forms are:

- *Microporous supports*: In which micropores ramify through the particles that are generally $5-10\,\mu m$ in diameter.
- *Bonded phases*: In which the stationary phase is chemically bonded onto an inert support such as silica.

11.3.2 Application of sample

The application of the sample onto an HPLC column in the correct way is a particularly important factor in achieving successful separations. The most common method of sample introduction is by use of a loop injector (Fig. 11.5). This consists of a metal loop, of fixed small volume, that can be filled with the sample. The eluent from the pump is then channelled through the loop by means of a valve switching system and the sample flushed onto the column via the loop outlet without interruption of the flow of eluent to the column.

Repeated application of highly impure samples such as sera, urine, plasma or whole blood, which have preferably been deproteinated, may eventually cause the column to lose its resolving power. To prevent this occurrence, a guard column is often installed between the injector and the column. This guard column is a short (1−2 cm) column of the same internal diameter and packed with material similar to that present in the analytical column. The packing in the guard column preferentially retains contaminating material and can be replaced at regular intervals.

(a) Loading position
(b) Injecting position

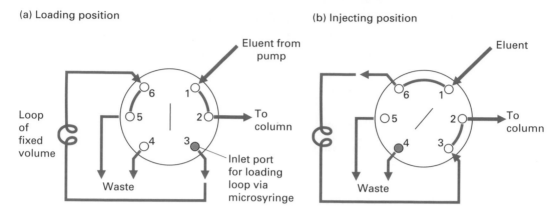

Fig. 11.5 HPLC loop injector. (a) The loop is loaded via port 3 with excess sample going to waste via port 5. In this position the eluent from the pump passes to the column via ports 1 and 2. (b) In the injecting position eluent flow is directed through the loop via ports 1 and 6 and then onto the column.

Mobile phases

The choice of mobile phase to be used in any separation depends on the type of separation to be achieved. Isocratic elution may be made with a single pump, using a single eluent or two or more eluents premixed in fixed proportions. Gradient elution generally uses separate pumps to deliver two eluents in proportions predetermined by a gradient programmer. All eluents for use in HPLC systems must be specially purified because traces of impurities can affect the column and interfere with the detection system. This is particularly the case if the detection system is based on the measurement of absorbance changes below 200 nm. Pure eluents for use in HPLC systems are available commercially, but even with these a $1-5$ mm microfilter is generally introduced into the system prior to the pump. It is also essential that all eluents be degassed before use otherwise gassing (the presence of air bubbles in the eluent) tends to occur in most pumps. Gassing, which tends to be particularly bad for eluents containing aqueous methanol and ethanol, can alter column resolution and interfere with the continuous monitoring of the eluate. Degassing of the eluent may be carried out in several ways – by warming, by stirring vigorously with a magnetic stirrer, by applying a vacuum, by ultrasonication, and by bubbling helium gas through the eluent reservoir.

Pumps

Pumping systems for delivery of the eluent are one of the most important features of HPLC systems. The main features of a good pumping system are that it is capable of outputs of at least 50 MPa and ideally there must be no pulses (i.e. cyclical variations in pressure) as this may affect the detector response. There must be a flow capability of at least 10 cm^3 min^{-1} and up to 100 cm^3 min^{-1} for preparative separations. Constant displacement pumps maintain a constant flow rate through the column irrespective of changing conditions within the column. The reciprocating pump is the most commonly used form of constant displacement pump. Such pumps produce small pulses of flow and pulse dampeners are usually incorporated into the system to minimise this pulsing effect. All constant displacement pumps have inbuilt safety cut-out mechanisms so that if the pressure within the column changes from pre-set limits the pump is inactivated automatically.

11.3.3 Detectors

Since the quantity of material applied to an HPLC column is normally very small, it is imperative that the sensitivity of the detector system is sufficiently high and stable to respond to the low concentrations of each analyte in the eluate. The most commonly used detectors are:

- *Variable wavelength detectors*: These are based upon ultraviolet–visible spectrophotometry. These types of detector are capable of measuring absorbances down to 190 nm and can give full-scale deflection (AUFS) for as little as 0.001 absorbance units. They have a detection sensitivity of the order of 5×10^{-10} g cm^{-3} and a linear range of 10^5. All spectrophotometric detectors use continuous flow cells with a small internal volume (typically 8 mm^3) and optical path length of 10 mm which allow the continuous monitoring of the column eluate.
- *Scanning wavelength detectors*: These have the facility to record the complete absorption spectrum of each analyte, thus aiding identification. Such opportunities are possible either by temporarily stopping the eluent flow or by the use of diode array techniques, which allow a scan of the complete spectrum of the eluate within 0.01 s and its display as a 3D plot on a VDU screen in real time (Fig. 11.6).
- *Fluorescence detectors*: These are extremely valuable for HPLC because of their greater sensitivity (10^{-12} g cm^{-3}) than UV detectors but they have a slightly reduced linear range (10^4). However, the technique is limited by the fact that relatively few analytes fluoresce. Pre-derivatisation of the test sample can broaden the applications of the technique.
- *Electrochemical detectors*: These are selective for electroactive analytes and are potentially highly sensitive. Two types are available, amperometric and coulometric, the principles of which are similar. A flow cell is fitted with two electrodes, a stable counter electrode and a working electrode. A constant potential is applied to the working electrode at such a value that, as an analyte flows through the flow cell, molecules of the analyte at the electrode surface undergo either an oxidation or a reduction, resulting in a current flow between the two electrodes. The size of the current is recorded to give the chromatogram. The potential applied to the counter electrode is sufficient to ensure that the current detected gives a full-scale deflection on the recorder within the working analyte range. The two types of detector differ in the extent of conversion of the analyte at the detector surface and on balance amperometric detectors are preferred since they have a higher sensitivity (10^{-12} g cm^{-3} as opposed to 10^{-8} g cm^{-3}) and greater linear range (10^5 as opposed to 10^4).

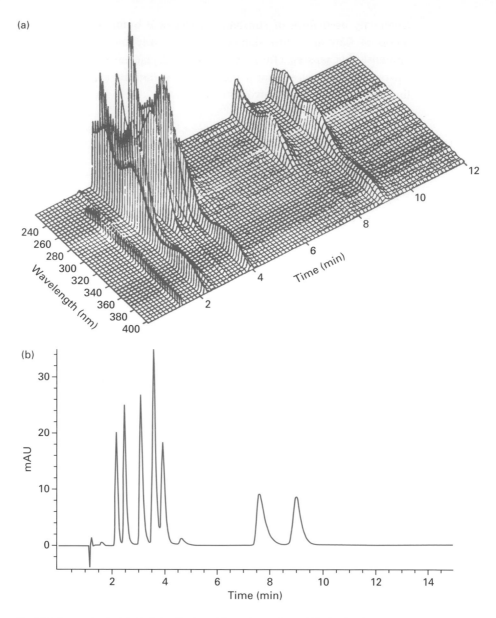

Fig. 11.6 Separation by HPLC of the dihydropyridine calcium channel blocker lacidipine and its metabolites. Column: ODS Hypercil. Eluent: methanol/acetonitrile/water (66%, 5%, 29% by volume) acidified to pH 3.5 with 1% formic acid. Flow rate: 1 cm^3 min^{-1}. Column temperature: 40 °C. (a) As recorded by a diode array detector and (b) by an ultraviolet detector. (Reproduced by permission of GlaxoSmithKline, UK.)

For reduction reactions the working electrode is normally mercury, and for oxidative reactions carbon or a carbon composite.

- *Mass spectrometer detectors*: These enable the analyte to be detected and its structure determined simultaneously. The technical problems associated with the logistics of removing the bulk of the mobile phase before the sample is introduced into the mass spectrometer have been resolved in a number of ways that are discussed in detail in Chapter 9. Analytes may be detected by total ion current (TIC) (Section 9.4.1) or selected ion monitoring (SIM) (Section 9.5.6). An advantage of mass spectrometry detection is that it affords a mechanism for the identification of overlapping peaks. If there is a suspicion that a large peak is masking a smaller peak then the presence of a minor analyte can be confirmed by selected ion monitoring provided that the minor and major analytes have a unique molecular ion or fragment ion.

- *NMR spectrometer detectors*: These give structural information about the analyte that is complementary to that obtained via HPLC–MS.
- *Refractive index detectors*: These rely on a change in the refractive index of the eluate as analytes emerge from the column. The great advantage is that they will respond to any analyte in any eluent, changes in refractive index being either positive or negative. Their limitation is the relatively modest sensitivity (10^{-7} g cm^{-3}) but they are commonly used in the analysis of carbohydrates.
- *Evaporative light-scattering detectors* (ELSD): These rely on the vaporisation of the eluate, evaporation of the eluent and the quantification of the analyte by light scattering. The eluate emerging from the column is combined with a flow of air or nitrogen to form an aerosol; the eluent is then evaporated from the aerosol by passage through an evaporator and the emerging dry particles of analyte irradiated with a light source and the scattered light detected by a photodiode. The intensity of the scattered light is determined by the quantity of analyte present and its particle size. It is independent of the analyte's spectroscopic properties and hence does not require the presence of a chromophoric group or any prior derivatisation of the analyte. It can quantify analytes in flow rates of up to 5 cm^3 min^{-1}. Appropriate calibration gives good, stable quantification of the analyte with no baseline drift. It is an attractive method for the detection of fatty acids, lipids and carbohydrates.

The sensitivity of ultraviolet absorption, fluorescence and electrochemical detectors can often be increased significantly by the process of derivatisation, whereby the analyte is converted pre- or post-column to a chemical derivative. Examples are given in Table 11.1.

Ultra-performance liquid chromatography (UPLC)

As was pointed out earlier, the resolution of a mixture of analytes increases as the particle size of the stationary phases decreases, but such a decrease leads to a high back-pressure from eluent flow. The technological solution to this problem represented by HPLC has recently been advanced by the development of new stationary phases of less than 2 μm diameter by the Waters Corporation. The particles of 1.7 μm diameter are made of 'Bridged Ethylsiloxane Silica Hybrid' (BEH)™ and are available in a range of forms suitable for various applications (Fig. 11.7). Back-pressures of up to 150 MPa are generated and this necessitated the development of special pumps, columns and detectors capable of operating in a pulse-free way at these high pressures. The instrumentation available under the trade name of ACQUITY™

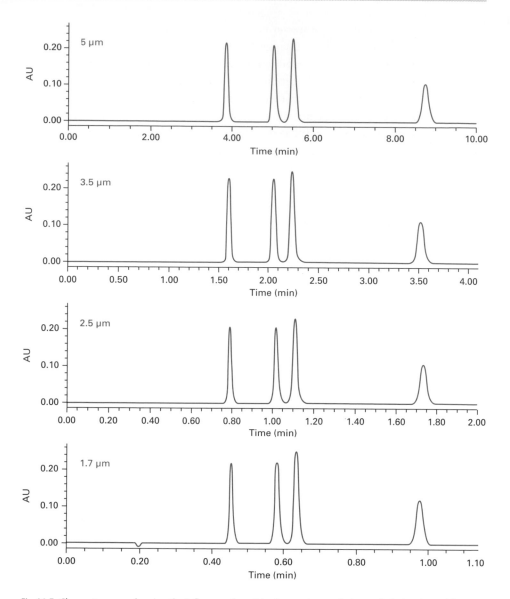

Fig 11.7 Chromatograms showing the influence of particle size on the resolution and elution time of four analytes in an HPLC system demonstrating the time advantages of the pressure-tolerant sub-2 μm ACQUITY UPLC system. (Reproduced by permission of Waters Corporation, UK.)

and the term UPLC are both registered to Waters. The system operates up to 10 times faster than conventional HPLC and chromatograms are routinely complete in less than 5 min. Whereas conventional HPLC peaks last up to 10 s, UPLC peaks may last 1 s so detectors have to respond ultra-fast to define the peak. The technology is now widely used in drug discovery and metabolite identification by the pharmaceutical industry.

11.3.4 Perfusion chromatography

The high resolution achieved by HPLC is based on the use of small diameter particles for the stationary phase. However, this high resolution is achieved at the cost of the generation of high pressures, relatively low flow rates and the constraints the high pressure imposes on the instrumentation. Perfusion chromatography overcomes some of these limitations by the use of small particles (10–50 µm diameter) that have channels of approximately 1 µm diameter running through them that allow the use of high flow rates without the generation of high pressures. The high flow rates result in small plate heights (Section 11.2.4) and hence high resolution in very short separation times. The particles are made of polystyrene-divinylbenzene and are available under the trade name POROS. Two types of pore are available: through pores that are long (up to 8000 Å) and diffusive pores that are shorter (up to 1000 Å). The stationary phase is coated onto the particles, including the surface of the pores. The eluent perfuses through the pores allowing the analyte to equilibrate rapidly with the stationary phase. By comparison, the microporous particles used in HPLC have a much smaller diameter pore, hence the greater back-pressure. All the forms of stationary phase used for the various forms of chromatography are available for perfusion chromatography. The technique uses the same type of instrumentation as HPLC. Protein separations in as short a time as 1 min can be achieved.

11.4 ADSORPTION CHROMATOGRAPHY

11.4.1 Principle

This is the classic form of chromatography, which is based upon the principle that certain solid materials, collectively known as adsorbents, have the ability to hold molecules at their surface. This adsorption process, which involves weak, non-ionic attractive forces of the van der Waals' and hydrogen-bonding type, occur at specific adsorption sites. These sites have the ability to discriminate between types of molecules and are occupied by molecules of either the eluent or of the analytes in proportions determined by their relative strength of interaction. As eluent is constantly passed down the column, differences in these binding strengths eventually lead to the separation of the analytes.

Silica is a typical adsorbent. It has silanol (Si-OH) groups on its surface, which are slightly acidic, and can interact with polar functional groups of the analyte or eluent. The topology (arrangement) of these silanol groups in different commercial preparations of silica explains their different separation properties. Other commonly used adsorbents are alumina and carbon.

In general, an eluent with a polarity comparable to that of the most polar analyte in the mixture is chosen. Thus, alcohols would be selected if the analytes contained hydroxyl groups, acetone or esters would be selected for analytes containing carbonyl groups, and hydrocarbons such as hexane, heptane and toluene for analytes that are predominantly non-polar. Mixtures of solvents are commonly used in the context of gradient elution.

11.4.2 **Hydroxylapatite chromatography**

Crystalline hydroxylapatite ($Ca_{10}(PO_4)_6(OH)_2$) is an adsorbent used to separate mixtures of proteins or nucleic acids. One of the most important applications of hydroxylapatite chromatography is the separation of single-stranded DNA from double-stranded DNA. Both forms of DNA bind at low phosphate buffer concentrations but as the buffer concentration is increased single-stranded DNA is selectively desorbed. As the buffer concentration is increased further, double-stranded DNA is released. This behaviour is exploited in the technique of Cot analysis (Section 5.3.4). The affinity of double-stranded DNA for hydroxylapatite is so high that it can be selectively removed from RNA and proteins in cell extracts by use of this type of chromatography.

11.4.3 **Hydrophobic interaction chromatography**

Hydrophobic interaction chromatography (HIC) was developed to purify proteins by exploiting their surface hydrophobicity which is a measure of their dislike of binding molecules of water. Groups of hydrophilic amino acid residues are scattered over the surface of proteins in a way that gives characteristic properties to each protein. In aqueous solution, these hydrophilic regions on the protein are covered with an ordered layer of water molecules that effectively mask the hydrophobic groups of the proteins the majority of which are in the interior of the folded molecule. These hydrophobic groups can, however, be exposed by the addition of salt ions, which preferentially take up the ordered water molecules. The exposed hydrophobic regions can then interact with each other by weak van der Waals' forces causing protein–protein aggregation. In HIC, the presence of hydrophobic groups such as butyl, octyl and phenyl attached to a matrix facilitates protein–matrix interaction rather than facilitating protein–protein interaction. Commercial materials include Phenyl Sepharose and Phenyl SPW, both for low-pressure HIC, and Poly PROPYL Asparta-mide, Bio-Gel TSK Phenyl and Spherogel TSK Phenyl for HPLC HIC.

Since HIC requires the presence of salting-out compounds such as ammonium sulphate to facilitate the exposure of the hydrophobic regions on the protein molecule, it is commonly used immediately after fractionation of protein mixtures with ammonium sulphate. To maximise the process, it is advantageous to adjust the pH of the protein sample to that of its isoelectric point. Once the proteins have been adsorbed to the stationary phase, selective elution can be achieved in a number of ways, including the use of an eluent of gradually decreasing ionic strength or of increasing pH (this increases the hydrophilicity of the protein) (Fig. 11.8) or by selective displacement by a displacer that has a stronger affinity for the stationary phase than has the protein. Examples include non-ionic detergents such as Tween 20 and Triton X-100, aliphatic alcohols such as 1-butanol and ethylene glycol, and aliphatic amines such as 1-aminobutane. HIC has many similarities with reversed-phase HPLC (RPC) but has two advantages over it. The first is that it uses aqueous elution conditions that minimise protein denaturation whereas RPC requires non-polar solvents for elution. The second is that it has a higher capacity.

The technique of **immobilised artificial membrane chromatography** (IAM) resembles HIP and uses phosphotidylcholine-based stationary phases. It is widely used in

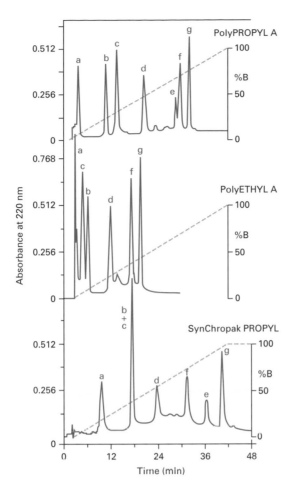

Fig. 11.8 Chromatogram of a mixture of proteins separated by hydrophobic interaction chromatography using different stationary phases. A linear gradient elution program was used changing from 0 to 100% mobile phase B in 40 min. Mobile phase A: 1.8 M ammonium sulphate + 0.1 M potassium phosphate pH 7.0. Mobile phase B: 0.1 M potassium phosphate pH 7.0. Elution was monitored at 220 nm. (Reproduced with permission from K. Benedek (2003) High-performance interaction chromatography, in *HPLC of Peptides and Proteins: Methods and Protocols*, Methods in Molecular Biology No. 251, M.-I. Aguilar (ed.), Humana Press, Totowa, NJ.)

the pharmaceutical industry to predict the ability of candidate drugs to be absorbed and distributed since they are good models of cell membranes. The technique uses a phospholipid-coated filter disc and is configured in 96-well format with an ultraviolet plate reader as a detection system.

11.5 PARTITION CHROMATOGRAPHY

11.5.1 Principle

Like other forms of chromatography, partition chromatography is based on differences in retention factor, k, and distribution coefficients, K_d, of the analytes using liquid stationary and mobile phases. It can be subdivided into liquid–liquid chromatography, in which the liquid stationary phase is attached to a supporting matrix by purely physical means, and bonded-phase liquid chromatography, in which the stationary phase is covalently attached to the matrix. An example of liquid–liquid chromatography is one in which a water stationary phase is supported by a cellulose, starch or silica matrix, all of which have the ability to physically bind as much as 50% (w/v) water and remain free-flowing powders. The advantages of this form of chromatography are that it is cheap, has a high capacity and has broad selectivity. Its disadvantage is that the elution process may gradually remove the stationary phase, thereby altering the chromatographic conditions. This problem is overcome by the use of bonded phases and this explains their more widespread use. Most bonded phases use silica as the matrix, which is derivatised to immobilise the stationary phase by reaction with an organochlorosilane.

11.5.2 Normal-phase liquid chromatography

In this form of partition chromatography, the stationary phase is polar and the mobile phase relatively non-polar. The most popular stationary phase is an alkylamine bonded to silica. The mobile phase is generally an organic solvent such as hexane, heptane, dichloromethane or ethyl acetate. These solvents form an elutropic series based on their polarity. Such a series in order of increasing polarity is as follows:

n-hexane $<$ cyclohexane $<$ trichloromethane $<$ dichloromethane $<$ tetrahydrofuran $<$ acetonitrile $<$ ethanol $<$ methanol $<$ ethanoic acid $<$ water

The order of elution of analytes is such that the least polar is eluted first and the most polar last. Indeed, polar analytes generally require gradient elution with a mobile phase of increasing polarity, generally achieved by the use of methanol or dioxane. The main applications of normal-phase liquid chromatography are its use to separate analytes that have low water solubility and those that are not amenable to reversed-phase liquid chromatography.

11.5.3 Reversed-phase liquid chromatography

In this form of liquid chromatography, which has many similarities with hydrophobic interaction chromatography, the stationary phase is non-polar and the mobile phase relatively polar, hence the name reversed-phase. By far the most commonly used type is the bonded-phase form, in which alkylsilane groups are chemically attached to silica. Butyl (C_4), octyl (C_8) and octadecyl (C_{18}) silane groups are most commonly used (Table 11.2). The mobile phase is commonly water or aqueous buffers, methanol, acetonitrile or tetrahydrofuran, or mixtures of them. The organic solvent is referred to as an organic modifier. Reversed-phase liquid chromatography differs from most other forms of chromatography in that the stationary phase is essentially inert and only non-polar (hydrophobic) interactions are possible with analytes.

Table 11.2 **Examples of silica bonded phases for reversed-phase HPLC**

Product	Particle size	Pore size (Å)
μBondapak octadecyl	10 μm	70
μBondapak phenyl	10 μm	125
μBondapak CN	10 μm	125
μBondapak NH_2	10 μm	80
Zorbax octadecyl	6 μm	70
Zorbax octyl	6 μm	70
Zorbax NH_2	6 μm	70
Discovery octyl	5 μm	180
Supelcosil LC-octadecyl	5 μm	120
Supelcosil LC-301 methyl	5 μm	300
Supelcosil LC-308 octyl	5 μm	300

Note: 1 Å = 0.1 nm.

Reversed-phase HPLC is probably the most widely used form of chromatography mainly because of its flexibility and high resolution. It is widely used to analyse drugs and their metabolites, insecticide and pesticide residues, and amino acids, peptides and proteins. Octadecylsilane (ODS) phases bind proteins more tightly than do octyl- or methylsilane phases and are therefore more likely to cause protein denaturation because of the more extreme conditions required for the elution of the protein.

11.5.4 Ion-pair reversed-phase liquid chromatography

Although the separation of some highly polar analytes, such as amino acids, peptides, organic acids and the catecholamines, is not possible by reversed-phase chromatography, it is sometimes possible to achieve such separations by one of two approaches:

- *Ion suppression*: The ionisation of the analytes is suppressed by using a mobile phase with an appropriately high or low pH thus giving the molecules greater hydrophobic character. For weak acid analytes, for example, an acidified mobile phase would be used.
- *Ion-pairing*: A counter ion that has a charge opposite to that of the analytes to be separated is added to the mobile phase so that the resulting ion-pair has sufficient hydrophobic, lipophilic character to be retained by the non-polar stationary phase of a reversed-phase system. Thus, to aid the separation of acidic analytes, which would be present as their conjugate anions, a quaternary alkylamine ion such as tetrabutylammonium would be used as the counter ion, whereas for the separation

of bases, which would be present as cations, an alkyl sulphonate such as sodium heptanesulphonate would be used:

$$RCOO^- \quad + \quad R'_4N^+ \quad \rightleftharpoons \quad [RCOO^-N^+R'_4]$$

carboxylic   counter cation   ion-pair
acid anion

$$RN^+H_3 \quad + \quad R'SO_3^- \quad \rightleftharpoons \quad [RH_3N^{+-}O_3SR']$$

conjugate acid   counter anion   ion-pair
of weak base

The technique can also be applied to proteins. The addition of trifluoroacetic acid (TFA) suppresses the ionisation of exposed groups on the protein surface giving it greater hydrophobic character. In practice, the success of the ion-pairing approach is variable and somewhat empirical. The size of the counter ion, its concentration and the pH of the solution are all factors that may profoundly influence the outcome of the separation.

Octyl- and octadecylsilane-bonded phases are used most commonly in conjunction with a water/methanol or water/acetonitrile mobile phase. One of the advantages of ion-pair reversed-phase chromatography is that if the sample to be resolved contains a mixture of non-ionic and ionic analytes, the two groups can be separated simultaneously because the ion-pair reagent does not affect the chromatography of the non-ionic species.

11.5.5 Chiral chromatography

This form of chromatography allows mixtures of enantiomers (mirror image forms, denoted either as D or L or as S or R) to be resolved. One of these techniques is based on the fact that diastereoisomers, which are optical isomers that do not have an object–image relationship, have different physical properties even though they contain identical functional groups. They can therefore be separated by conventional chromatographic techniques, most commonly reversed-phase chromatography. The diastereoisomer approach requires that the enantiomers contain a function group that can be derivatised by a chemically and optically pure chiral derivatising agent (CDA) that converts them to a mixture of diastereoisomers:

$$(R + S) \quad + \quad R' \quad \rightleftharpoons \quad RR' + SR'$$

mixture of   chiral   mixture of
enantiomers   derivatising   diastereoisomers
     agent

Examples of CDAs include the R or S form of the following:

For amines	*N*-trifluoroacetyl-1-prolylchloride, α-phenylbutyric anhydride
For alcohols	2-Phenylpropionyl chloride, 1-phenylethylisothiocyanate
For ketones	2,2,2-Trifluoro-1-pentylethylhydrazine
For aliphatic and alicyclic acids	1-Menthol, desoxyephedrine

An alternative approach to the resolution of enantiomers is to use a chiral mobile phase. In this technique a transient diastereomeric complex is formed between the enantiomers and the chiral mobile phase agent. Examples of chiral mobile phase agents include albumin, α_1-acid glycoprotein, α, β- and γ-cyclodextrins, camphor-10-sulphonic acid and N-benzoxycarbonylglycyl-L-proline, all of which are used with a reversed-phase chromatographic system. For example, this technique has been used to show that cannabidiol, one of the main components of marijuana, consists of (+) and (−) forms only one of which is physiologically active.

The most successful approach to chiral chromatography, however, has been the use of a chiral stationary phase. This is based upon the principle that the need for a three-point interaction between the stationary phase (working as a chiral discriminator) and the enantiomer would allow the resolution of racemic mixtures due to the different spatial arrangement of the functional groups at the chiral centre in the enantiomers. The cyclodextrins are cyclic oligosaccharides that have an open truncated conical structure 6 to 8 Å (0.6 to 0.8 nm) wide at their base. Their inner surface is predominantly hydrophobic, but secondary hydroxyl groups are located around the wide rim of the cone. β-Cyclodextrin has seven glucopyranose units and contains 35 chiral centres and α-cyclodextrin has six glucopyranose units, 30 chiral centres and is smaller than β-cyclodextrin. Collectively they are referred to as chiral cavity phases because they rely on the ability of the enantiomer to enter the three-dimensional cyclodextrin cage while at the same time presenting functional groups and hence the chiral centre for interaction with hydroxyl groups on the cone rim.

11.6 ION-EXCHANGE CHROMATOGRAPHY

11.6.1 Principle

This form of chromatography relies on the attraction between oppositely charged stationary phase, known as an ion exchanger, and analyte. It is frequently chosen for the separation and purification of proteins, peptides, nucleic acids, polynucleotides and other charged molecules, mainly because of its high resolving power and high capacity. There are two types of ion exchanger, namely cation and anion exchangers. Cation exchangers possess negatively charged groups and these will attract positively charged cations. These exchangers are also called acidic ion exchangers because their negative charges result from the ionisation of acidic groups. Anion exchangers have positively charged groups that will attract negatively charged anions. The term basic ion exchangers is also used to describe these exchangers, as positive charges generally result from the association of protons with basic groups.

11.6.2 Materials and applications

Matrices used include polystyrene, cellulose and agarose. Functional ionic groups include sulphonate ($-SO_3^-$) and quaternary ammonium ($-N^+R_3$), both of which are

Table 11.3 **Examples of commonly used ion exchangers**

Type	Functional groups	Functional group name	Matrices
Weakly acidic (cation exchanger)	$-COO^-$	Carboxy	Agarose
	$-CH_2COO^-$	Carboxymethyl	Cellulose
			Dextran
			Polyacrylate
Strongly acidic (cation exchanger)	$-SO_3^-$	Sulpho	Cellulose
	$-CH_2SO_3^-$	Sulphomethyl	Dextran
	$-CH_2CH_2CH_2SO_3^-$	Sulphopropyl	Polystyrene
			Polyacrylate
Weakly basic (anion exchanger)	$-CH_2CH_2N^+H_3$	Aminoethyl	Agarose
	$-CH_2CH_2N^+H$ $\mid$ $(CH_2CH_3)_2$	Diethylaminoethyl	Cellulose
			Dextran
			Polystyrene
			Polyacrylate
Strongly basic (anion exchanger)	$-CH_2N^+(CH_3)_3$	Trimethylaminomethyl	Cellulose
	$-CH_2CH_2N^+(CH_2CH_3)_3$	Triethylaminoethyl	Dextran
	$-CH_2N^+(CH_3)_2$ $\mid$ CH_2CH_2OH	Dimethyl-2-hydroxyethyl-aminomethyl	Polystyrene
	$-CH_2CH_2N^+(CH_2CH_3)_2$ $\mid$ $CH_2CH(OH)CH_3$	Diethyl-2-hydroxypropyl-aminoethyl	

strong exchangers because they are totally ionised at all normal working pH values, and carboxylate ($-COO^-$) and diethylammonium ($-HN^+(CH_2CH_3)_2$), both of which are termed weak exchangers because they are ionised over only a narrow range of pH values. Examples are given in Table 11.3. Bonded phase ion exchangers suitable for HPLC, containing a wide range of ionic groups, are commercially available. Porous varieties are based on polystyrene, porous silica or hydrophilic polyethers, and are particularly valuable for the separation of proteins. They have a particle diameter of 5–25 μm. Most HPLC ion exchangers are stable up to 60 °C and separations are often carried out at this temperature, owing to the fact that the raised temperature decreases the viscosity of the mobile phase and thereby increases the efficiency of the separation.

Choice of exchanger

The choice of the ion exchanger depends upon the stability of the test analytes, their relative molecular mass and the specific requirements of the separation. Many biological analytes, especially proteins, are stable within only a fairly narrow pH range so the exchanger selected must operate within this range. Generally, if an analyte is most stable below its isoionic point (giving it a net positive charge) a cation exchanger should be used, whereas if it is most stable above its isoionic point (giving it a net negative charge) an anion exchanger should be used. Either type of exchanger may be used to separate analytes that are stable over a wide range of pH values. The choice between a strong and weak exchanger also depends on analyte stability and the effect of pH on analyte charge. Weak electrolytes requiring a very low or high pH for ionisation can be separated only on strong exchangers, as they only operate over a wide pH range. In contrast, for strong electrolytes, weak exchangers are advantageous for a number of reasons, including a reduced tendency to cause protein denaturation, their inability to bind weakly charged impurities and their enhanced elution characteristics. Although the degree of cross-linking of an exchanger does not influence the ion-exchange mechanism, it does influence its capacity. The relative molecular mass and hence size of the proteins in the sample therefore determines which exchanger should be used.

Eluent pH

The pH of the buffer selected as eluent should be at least one pH unit above or below the isoionic point of the analytes. In general, cationic buffers such as Tris, pyridine and alkylamines are used in conjunction with anion exchangers, and anionic buffers such as acetate, barbiturate and phosphate are used with cation exchangers. The precise initial buffer pH and ionic strength should be such as just to allow the binding of the analytes to the exchanger. Equally, a buffer of the lowest ionic strength that effects elution should initially be used for the subsequent elution of the analytes. This ensures that initially the minimum numbers of contaminants bind to the exchanger and that subsequently the maximum number of these impurities remains on the column. If, however, gradient elution is to be used, the initial conditions chosen are such that the exchanger binds all the test analytes at the top of the column.

Elution

Gradient elution is far more common than isocratic elution. Continuous or stepwise pH and ionic strength gradients may be employed but continuous gradients tend to give better resolution with less peak tailing. Generally with an anion exchanger, the pH gradient decreases and the ionic strength increases, whereas for cation exchangers both the pH and ionic gradients increase during the elution.

11.7 MOLECULAR (SIZE) EXCLUSION CHROMATOGRAPHY

11.7.1 Principle

This chromatographic technique for the separation of molecules on the basis of their molecular size and shape exploits the molecular sieve properties of a variety of porous materials. The terms exclusion or permeation chromatography or gel filtration describe all molecular separation processes using molecular sieves. The general principle of exclusion chromatography is quite simple. A column of microparticulate cross-linked copolymers generally of either styrene or divinylbenzene and with a narrow range of pore sizes is in equilibrium with a suitable mobile phase for the analytes to be separated. Large analytes that are completely excluded from the pores will pass through the interstitial spaces between the particles and will appear first in the eluate. Smaller analytes will be distributed between the mobile phase inside and outside the particles and will therefore pass through the column at a slower rate, hence appearing last in the eluate (Fig. 11.9).

The mobile phase trapped by a particle is available to an analyte to an extent that is dependent upon the porosity of the particle and the size of the analyte molecule. Thus, the distribution of an analyte in a column of cross-linked particles is determined solely by the total volume of mobile phase, both inside and outside the particles, that is available to it. For a given type of particle, the distribution coefficient, K_d, of a particular analyte between the inner and outer mobile phase is a function of its molecular size. If the analyte is large and completely excluded from the mobile phase within the particle, $K_d = 0$, whereas, if the analyte is sufficiently small to gain complete access to the inner mobile phase, $K_d = 1$. Due to variation in pore size between individual particles, there is some inner mobile phase that will be available and some that will not be available to analytes of intermediate size; hence K_d values vary between 0 and 1. It is this complete variation of K_d between these two limits that makes it possible to separate analytes within a narrow molecular size range on a given particle type.

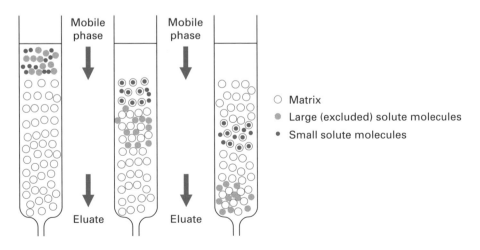

Fig. 11.9 Separation of different size molecules by exclusion chromatography. Large excluded molecules are eluted first in the void volume.

Example 2 ESTIMATION OF RELATIVE MOLECULAR MASS

Question The relative molecular mass (M_r) of a protein was investigated by exclusion chromatography using a Sephacryl S300 column and using aldolase, catalase, ferritin, thyroglobulin and Blue Dextran as standard. The following elution data were obtained.

	Retention volume	
	M_r	V_r (cm^3)
Aldolase	158 000	22.5
Catalase	210 000	21.4
Ferritin	444 000	18.2
Thyroglobulin	669 000	16.4
Blue Dextran	2 000 000	13.6
Unknown		19.5

What is the approximate M_r of the unknown protein?

Answer A plot of the logarithm of the relative molecular mass of individual proteins versus their retention volume has a linear section from which it can be deduced that the unknown protein with a retention volume of 19.5 cm^3 must have a relative molecular mass of 330 000.

For two analytes of different relative molecular mass and K_d values, K'_d and K''_d, the difference in their elution volumes, V_S, can be shown to be:

$$V_S = (K'_d - K''_d)V_i \tag{11.14}$$

where V_i is the inner volume within the particle available to a compound whose $K_d = 1$.

In practice, deviations from ideal behaviour, for example owing to poor packing of the column, make it advisable to reduce the sample volume below the value of V_S because the ratio between sample volume and inside particle volume affects both the sharpness of the separation and the degree of dilution of the sample.

The stationary phases for exclusion separations are generally based on silica, polymethacrylate or polyvinyl acetate or chloride or on cross-linked dextran or agarose (Table 11.4). All are available in a range of pore sizes. They are generally used where the eluent is an organic system. The supports for affinity separations are similar to those for exclusion separations.

11.7.2 Applications

Purification

The main application of exclusion chromatography is in the purification of biological macromolecules by facilitating their separation from larger and smaller molecules.

Table 11.4 **Stationary phases commonly used for exclusion chromatography**

Polymer	Trade name	Fractionation range[a] ($M_r \times 10^{-3}$)
Low-pressure liquid chromatography		
Dextran	Sephadex	
	G-10	<0.7
	G-25	1–5
	G-50	1.5–30
	G-100	4–150
	G-200	5–600
Dextran, cross-linked	Sephacryl	
	S-100	1–100
	S-200	5–250
	S-300	10–1500
	S-400	20–8000
Agarose	Sepharose	
	6B	10–4000
	4B	60–20 000
	2B	70–40 000
Polyacrylamide	Bio-Gel	
	P-2	0.1–1.8
	P-6	1–6
	P-30	2.5–40
	P-100	5–100
	P-300	60–400
High-performance liquid chromatography		
Polyvinyl chloride	Fractogel	
	TSK HW-40	0.1–10
	TSK HW-55	1–700
	TSK HW-65	50–5000
	TSK HW-75	500–50 000

Table 11.4 (*cont.*)

Polymer	Trade name	Fractionation range[a] $(M_r \times 10^{-3})$
Dextran linked to cross-linked agarose	Superdex	
	75	3–70
	200	10–600

Note: [a]Determined for globular proteins. The range is approximately the same for single-stranded nucleic acids and smaller for fibrous proteins and double-stranded DNA.

Viruses, enzymes, hormones, antibodies, nucleic acids and polysaccharides have all been separated and purified by use of appropriate gels or glass granules.

Relative molecular mass determination

The elution volumes of globular proteins are determined largely by their relative molecular mass (M_r). It has been shown that, over a considerable range of relative molecular masses, the elution volume or K_d is an approximately linear function of the logarithm of M_r. Hence the construction of a calibration curve, with proteins of a similar shape and known M_r, enables the M_r values of other proteins, even in crude preparations, to be estimated (See Example 2, p. 463).

Solution concentration

Solutions of high M_r substances can be concentrated by the addition of dry Sephadex G-25 (coarse). The swelling gel absorbs water and low M_r substances, whereas the high M_r substances remain in solution. After 10 min the gel is removed by centrifugation, leaving the high M_r material in a solution whose concentration has increased but whose pH and ionic strength are unaltered.

Desalting

By use of a column of, for example, Sephadex G-25, solutions of high M_r compounds may be desalted, i.e. removed from contaminants such as salts, detergents, lipids and chaotropic agents. The high M_r compounds move with the void volume, whereas the low M_r compounds are distributed between the mobile and stationary phases and hence move slowly. This method of desalting is faster and more efficient than dialysis. Applications include removal of phenol from nucleic acid preparations, ammonium sulphate from protein preparations and salt from samples eluted from ion-exchange chromatography columns.

11.8   AFFINITY CHROMATOGRAPHY

11.8.1   Principle

Separation and purification of analytes by affinity chromatography is unlike most other forms of chromatography and such techniques as electrophoresis and

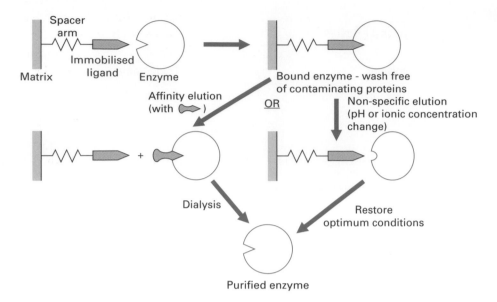

Fig. 11.10 Principle of purification of an enzyme by affinity chromatography.

centrifugation in that it does not rely on differences in the physical properties of the analytes. Instead, it exploits the unique property of extremely specific biological interactions to achieve separation and purification. As a consequence, affinity chromatography is theoretically capable of giving absolute purification, even from complex mixtures, in a single process. The technique was originally developed for the purification of enzymes, but it has since been extended to nucleotides, nucleic acids, immunoglobulins, membrane receptors and even to whole cells and cell fragments.

The technique requires that the material to be isolated is capable of binding reversibly to a specific ligand that is attached to an insoluble matrix:

$$\begin{array}{ccccc} & & & k_{+1} & \\ \text{M} & + & \text{L} & \overset{}{\underset{k_{-1}}{\rightleftharpoons}} & \text{ML} \\ \text{macromolecule} & & \text{ligand} & & \text{complex} \\ & & \text{(attached} & & \\ & & \text{to matrix)} & & \end{array}$$

Under the correct experimental conditions, when a complex mixture containing the specific compound to be purified is added to the immobilised ligand, generally contained in a conventional chromatography column, only that compound will bind to the ligand. All other compounds can therefore be washed away and the compound subsequently recovered by displacement from the ligand (Fig. 11.10). The method requires a detailed preliminary knowledge of the structure and biological specificity of the compound to be purified so that the separation conditions that are most likely to be successful may be carefully planned. In the case of an enzyme, the ligand may be the substrate, a competitive reversible inhibitor or an allosteric modifier. The conditions chosen would normally be those that are optimal for enzyme–ligand binding. Since the success of the method relies on the reversible formation of the complex and on the numerical values of the first-order rate constants k_{+1} and k_{-1}, as the enzyme is added progressively to the insolubilised ligand in a column, the enzyme molecules will be stimulated to bind and a dynamic situation develops in which the concentration of the complex and the strength of the binding increase.

11.8.2 Materials and applications

Matrix

An ideal matrix for affinity chromatography must have the following characteristics:

- possess suitable and sufficient chemical groups to which the ligand may be covalently coupled, and be stable under the conditions of the attachment;
- be stable during binding of the macromolecule and its subsequent elution;
- interact only weakly with other macromolecules to minimise non-specific adsorption;
- exhibit good flow properties.

In practice, particles that are uniform, spherical and rigid are used. The most common ones are the cross-linked dextrans and agarose, polyacrylamide, polymethacrylate, polystyrene, cellulose and porous glass and silica.

Ligand

The chemical nature of a ligand is dictated by the biological specificity of the compound to be purified. In practice it is sometimes possible to select a ligand that displays absolute specificity in that it will bind exclusively to one particular compound. More commonly, it is possible to select a ligand that displays group selectivity in that it will bind to a closely related group of compounds that possess a similar in-built chemical specificity. An example of the latter type of ligand is 5'AMP, which can bind reversibly to many NAD^+-dependent dehydrogenases because it is structurally similar to part of the NAD^+ molecule. It is essential that the ligand possesses a suitable chemical group that will not be involved in the reversible binding of the ligand to the macromolecule, but which can be used to attach the ligand to the matrix. The most common of such groups are $-NH_2$, $-COOH$, $-SH$ and $-OH$ (phenolic and alcoholic).

To prevent the attachment of the ligand to the matrix interfering with its ability to bind the macromolecule, it is generally advantageous to interpose a spacer arm between the ligand and the matrix. The optimum length of this spacer arm is six to ten carbon atoms or their equivalent. In some cases, the chemical nature of this spacer is critical to the success of separation. Some spacers are purely hydrophobic, most commonly consisting of methylene (CH_2) groups; others are hydrophilic, possessing carbonyl (CO) or imido (NH) groups. Spacers are most important for small immobilised ligands but generally are not necessary for macromolecular ligands (e.g. in immunoaffinity chromatography, Section 11.8.3) as their binding site for the mobile macromolecule is well displaced from the matrix. Several supports of the agarose, dextran and polyacrylamide type are commercially available with a variety of spacer arms and ligands pre-attached ready for immediate use. Examples of ligands are given in Table 11.5. Glutathione Sepharose High Performance is an agarose support used

Table 11.5 **Examples of group-specific ligands commonly used in affinity chromatography**

Ligand	Affinity
Nucleotides	
5′-AMP	NAD^+-dependent dehydrogenases, some kinases
2′5′-ADP	$NADP^+$-dependent dehydrogenases
Calmodulin	Calmodulin-binding enzymes
Avidin	Biotin-containing enzymes
Fatty acids	Fatty-acid-binding proteins
Heparin	Lipoproteins, lipases, coagulation factors, DNA polymerases, steroid receptor proteins, growth factors, serine protease inhibitors
Proteins A and G	Immunoglobulins
Concanavalin A	Glycoproteins containing α-D-mannopyranosyl and α-D-glucopyranosyl residues
Soybean lectin	Glycoproteins containing N-acetyl-α-(or β)-D-galactopyranosyl residues
Phenylboronate	Glycoproteins
Poly(A)	RNA containing poly(U) sequences, some RNA-specific proteins
Lysine	rRNA
Cibacron Blue F3G-A	Nucleotide-requiring enzymes, coagulation factors

for the isolation any GST-tagged cloned protein. It is possible to remove the GST tag in a one-step process by adding PreScission™ Protease to the matrix, as it will be bound to the column and remove the tag as the protein is eluted.

Practical procedure

The procedure for affinity chromatography is similar to that used in other forms of liquid chromatography. The buffer used must contain any cofactors, such as metal ions, necessary for ligand–macromolecule interaction. Once the sample has been applied and the macromolecule bound, the column is eluted with more buffer to remove non-specifically bound contaminants. The purified compound is recovered from the ligand by either specific or non-specific elution. Non-specific elution may be achieved by a change in either pH or ionic strength. pH shift elution using dilute acetic acid or ammonium hydroxide results from a change in the state of ionisation of groups in the ligand and/or the macromolecule that are critical to ligand–macromolecule binding. A change in ionic strength, not necessarily with a concomitant change in pH, also causes elution due to a disruption of the ligand–macromolecule interaction; 1M NaCl is frequently used for this purpose.

Applications

Many enzymes and other proteins, including receptor proteins and immunoglobulins, have been purified by affinity chromatography. The application of the technique is limited only by the availability of immobilised ligands. The principles have been extended to nucleic acids and have made a considerable contribution to developments in molecular biology. Messenger RNA, for example, is routinely isolated by selective hybridisation on poly(U)-Sepharose 4B by exploiting its poly(A) tail. Immobilised single-stranded DNA can be used to isolate complementary RNA and DNA. Whilst this separation can be achieved on columns, it is usually performed using single-stranded DNA immobilised on nitrocellulose filters. Immobilised nucleotides are useful for the isolation of proteins involved in nucleic acid metabolism.

11.8.3 Immunoaffinity chromatography

The use of antibodies as the immobilised ligand has been exploited in the isolation and purification of a range of proteins including membrane proteins of viral origin. Monoclonal antibodies may be linked to agarose matrices by the cyanogen bromide technique. Protein binding to the immobilised antibody is achieved in neutral buffer solution containing moderate salt concentrations. Elution of the bound protein quite often requires forceful conditions because of the need to disrupt the very tight ionic or hydrophobic binding with the antibody ($K_d = 10^{-8}$ to 10^{-12} M) and this may lead to protein denaturation. Examples of elution procedures include the use of high salt concentrations with or without the use of detergent and the use of urea or guanidine hydrochloride, both of which cause protein denaturation. The use of some other chaotropic agents (ions or small molecules that increase the water solubility of non-polar substances) such as thiocyanate, perchlorate and trifluoroacetate or lowering the pH to about 3 may avoid denaturation. Organic solvents such as acetonitrile can also be used to disrupt the hydrophobic interaction.

11.8.4 Metal chelate chromatography (immobilised metal affinity chromatography)

This is a special form of affinity chromatography in which an immobilised metal ion such as Cu^{2+}, Zn^{2+}, Hg^{2+} or Cd^{2+} or a transition metal ion such as Co^{2+}, Ni^{2+} or Mn^{2+} is used to bind proteins selectively by reaction with imidazole groups of histidine residues, thiol groups in cysteine residues and indole groups in tryptophan residues sterically available on the surface of the proteins. The immobilisation of the protein involves the formation of a coordinate bond that must be sufficiently stable to allow protein attachment and retention during the elution of non-binding contaminating material. The subsequent release of the protein can be achieved either by simply lowering the pH or by the use of complexing agents such as EDTA. Most commonly the metal atom is immobilised by attachment to an iminodiacetate- or tris(carboxymethyl)-ethylenediamine-substituted agarose. Nickel or cobalt immobilised metal affinity chromatography is commonly used to isolate and purify His-tagged proteins

(Section 6.7.1). The cobalt commercial product called Dynabeads Myone TALON™ has the practical advantage that it only binds adjacent histidines or histidines in certain arrangements and is commonly used for the isolation of recombinant proteins.

11.8.5 Dye–ligand chromatography

A number of triazine dyes that contain both conjugated rings and ionic groups fortuitously have the ability to bind to some proteins. The term pseudo-ligand has therefore been used to describe the dyes. It is not possible to predict whether a particular protein will bind to a given dye as the interaction is not specific but is thought to involve interaction with ligand-binding domains via both ionic and hydrophobic forces. Dye binding to proteins enhances their binding to materials such as Sepharose 4B and this is exploited in the purification process. The attraction of the technique is that the dyes are cheap, readily coupled to conventional matrices and are very stable. The most widely used dye is Cibacron Blue F3G-A. Dye selection for a particular protein purification is empirical and is made on a trial-and-error basis. Attachment of the protein to the immobilised dye is generally achieved at pH 7 to 8.5. Elution is most commonly brought about either by a salt gradient or by affinity (displacement) elution.

11.8.6 Covalent chromatography

This form of chromatography has been developed specifically to separate thiol(–SH)-containing proteins by exploiting their interaction with an immobilised ligand containing a disulphide group. The principle is illustrated in Fig. 11.11. The most commonly used ligand is a disulphide 2′-pyridyl group attached to an agarose matrix such as Sepharose 4B. On reaction with the thiol-containing protein, pyridine-2-thione is released. This process can be monitored spectrophotometrically at 343 nm, thereby allowing the adsorption of the protein to be followed. Once the protein has been attached covalently to the matrix, non-thiol-containing contaminants are eluted and unreacted thiopyridyl groups removed by use of 4 mM dithiothreitol or mercaptoethanol. The protein is then released by displacement with a thiol-containing compound such as 20–50 mM dithiothreitol, reduced glutathione or cysteine. The matrix is regenerated by reaction with 2,2′-dipyridyldisulphide. The method has been used successfully for many proteins but its use is limited by its cost and the rather difficult regeneration stage. It can, however, be applied to very impure protein preparations.

11.9 GAS CHROMATOGRAPHY

11.9.1 Principle

The principles of gas chromatography (GC) are similar to those of HPLC but the apparatus is significantly different. It exploits differences in the partition coefficients

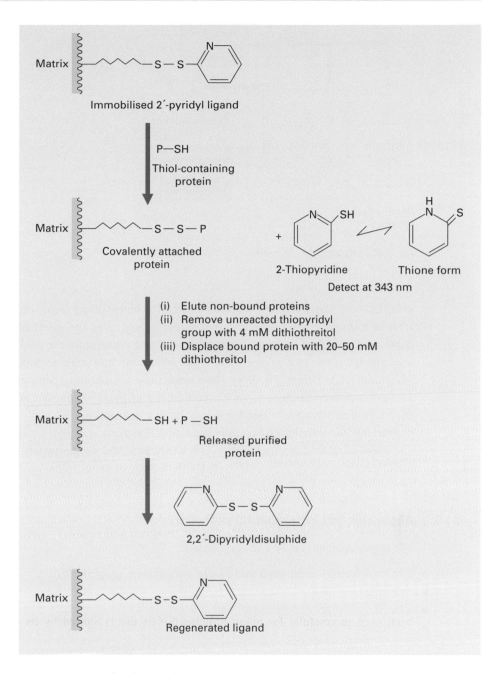

Fig. 11.11 Principle of purification of a protein (P-SH) by covalent chromatography.

between a stationary liquid phase and a mobile gas phase of the volatilised analytes as they are carried through the column by the mobile gas phase. Its use is therefore confined to analytes that are volatile but thermally stable. The partition coefficients are inversely proportional to the volatility of the analytes so that the most volatile elute first. The temperature of the column is raised to $50-300\,^{\circ}C$ to facilitate analyte

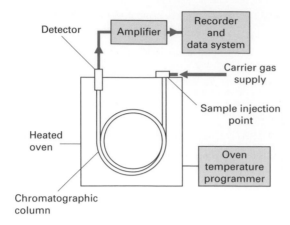

Fig. 11.12 Components of a GC system.

volatilisation. The stationary phase consists of a high-boiling-point liquid material such as silicone grease or wax that is either coated onto the internal wall of the column or supported on an inert granular solid and packed into the column. There is an optimum flow rate of the mobile gas phase for maximum column efficiency (minimum plate height, H). Very high resolutions are obtained (equations 11.8 to 11.12) hence the technique is very useful for the analysis of complex mixtures. Gas chromatography is widely used for the qualitative and quantitative analysis of a large number of low-polarity compounds because it has high sensitivity, reproducibility and speed of resolution. Analytically, it is a very powerful technique when coupled to mass spectrometry.

11.9.2 Apparatus and experimental procedure

The major components of a GC system are:

- a column housed in an oven that can be temperature programmed;
- a sample inlet point;
- a carrier gas supply and control; and
- a detector, amplifier and data recorder system (Fig. 11.12).

Columns

These are of two types:

- *Packed conventional columns*: These consist of a coiled glass or stainless steel column 1–3 m long and 2–4 mm internal diameter. They are packed with stationary phase coated on an inert silica support. Commonly used stationary phases include the polyethylene glycols (Carbowax 20M, very polar), methylphenyl- and methylvinylsilicone gums (OV17 and OV101, medium and non-polar respectively),

Apiezon L (non-polar) and esters of adipic, succinic and phthalic acids. β-Cyclodextrin-based phases are available for chiral separations (Section 11.5.5). The most commonly used support is Celite (diatomaceous silica), which because of the problem of support–sample interaction is often treated so that the hydroxyl groups that occur in the Celite are modified. This is normally achieved by silanisation of the support with such compounds as hexamethyldisilazane. The support particles have a large surface area and an even size, which, for the majority of practical applications, ranges from 60–80 mesh (0.25 mm) to 100–120 mesh (0.125 mm) (Section 11.3.1). The smaller the particle size and the thinner the coating the less band spreading occurs.

- *Capillary (open tubular) columns*: These are made of high-quality fused quartz and are 10–100 m long and 0.1–1.0 mm internal diameter. They are of two types known as wall-coated open tubular (WCOT) and support-coated open tubular (SCOT), also known as porous layer open tubular (PLOT) columns, for adsorption work. In WCOT columns the stationary phase is thinly coated (0.1–5 μm) directly onto the walls of the capillary whilst in SCOT columns the support matrix is bonded to the walls of the capillary column and the stationary phase coated onto the support. Commonly used stationary phases include polyethylene glycol (CP wax and DB wax, very polar) and methyl and phenyl-polysiloxanes (BP1, non-polar; BP10, medium polar). They are coated onto the supporting matrix to give a 1% to 25% loading, depending upon the analysis. The capacity of SCOT columns is considerably higher than that of WCOT columns.

The operating temperature for all types of column must be compatible with the stationary phase chosen for use. Too high a temperature results in excessive column bleed owing to the phase being volatilised off, contaminating the detector and giving an unstable recorder baseline. The working temperature range is chosen to give a balance between peak retention time and resolution. Column temperature is controlled to ±0.1 °C. Analyte partition coefficients are particularly sensitive to temperature so that analysis times may be regulated by adjustment of the column oven, which can be operated in one of two modes:

- *Isothermal analysis*: Here a constant temperature is employed.
- *Temperature programming*: The temperature is gradually increased to facilitate the separation of compounds of widely differing polarity or M_r. This, however, sometimes results in excessive bleed of the stationary phase as the temperature is raised, giving rise to baseline variation. Consequently some instruments have two identical columns and detectors, one set of which is used as a reference. The currents from the two detectors are opposed, hence, assuming equal bleed from both columns, the resulting current gives a steady baseline as the column temperature is raised. The choice of phase for analysis depends on the analytes under investigation and is best chosen after reference to the literature.

Application of sample

The majority of non- and low-polar compounds are directly amenable to GC, but other compounds possessing such polar groups as −OH, −NH₂ and −COOH are generally retained on the column for excessive periods of time if they are applied directly. Poor resolution and peak tailing usually accompany this excessive retention (Section 11.2.4). This problem can be overcome by derivatisation of the polar groups. This increases the volatility and effective distribution coefficients of the compounds. Methylation, silanisation and perfluoracylation are common derivatisation methods for fatty acids, carbohydrates and amino acids.

The test sample is dissolved in a suitable solvent such as acetone, heptane or methanol. Chlorinated organic solvents are generally avoided as they contaminate the detector. For packed and SCOT columns the sample is injected onto the column using a microsyringe through a septum in the injection port attached to the top of the column. Normally 0.1 to 10 mm^3 of solution is injected. As there is only a small amount of stationary phase present in WCOT columns, only very small amounts of sample may be applied to the column. Consequently a splitter system has to be used at the sample injection port so that only a small fraction of the injected sample reaches the column. The remainder of the sample is vented to waste. The design of the splitter is critical in quantitative analyses in order to ensure that the ratio of sample applied to the column to sample vented is always the same. It is common practice to maintain the injection region of the column at a slightly higher temperature ($+20$ to $50\,^\circ$C) than the column itself as this helps to ensure rapid and complete volatilisation of the sample. Sample injection is automated in many commercial instruments as this improves the precision of the analysis.

Mobile phase

The mobile phase consists of an inert gas such as nitrogen for packed columns or helium or argon for capillary columns. The gas from a cylinder is pre-purified by passing through a variety of molecular sieves to remove oxygen, hydrocarbons and water vapour. It is then passed through the chromatography column at a flow rate of $40-80\,\text{cm}^3\,\text{min}^{-1}$. A gas-flow controller is used to ensure a constant flow irrespective of the back-pressure and temperature of the column.

11.9.3 Detectors

Several types of detector are in common use in conjunction with GC:

- *Flame ionisation detector* (FID): This responds to almost all organic compounds. It has a minimum detection quantity of the order of $5 \times 10^{-12}\,\text{g}\,\text{s}^{-1}$, a linear range of 10^7 and an upper temperature limit of $400\,^\circ$C. A mixture of hydrogen and air is introduced into the detector to give a flame, the jet of which forms one electrode, whilst the other electrode is a brass or platinum wire mounted near the tip of the flame (Fig. 11.13). When the sample analytes emerge from the column they are ionised in the flame, resulting in an increased signal being passed to the recorder. The carrier gas passing through the column and the detector gives a small background signal, which can be offset electronically to give a stable baseline.

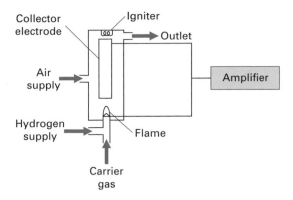

Fig. 11.13 GC flame ionisation detector. The tip of the flame forms the anode and the collector electrode the cathode.

$$\text{analyte} + H_2 + O_2 \longrightarrow \text{combustion products} + H_2O + \text{ions} + \text{radicals} + \text{electrons}$$

$$\sum(\text{ions})^- + \sum(\text{electrons})^- \longrightarrow \text{current}$$

- *Nitrogen–phosphorus detector* (NPD) (also called a thermionic detector): This is similar in design to an FID but has a crystal of a sodium salt fused onto the electrode system, or a burner tip embedded in a ceramic tube containing a sodium salt or a rubidium chloride tip. The NPD has excellent selectivity towards nitrogen- and phosphorus-containing analytes and shows a poor response to analytes possessing neither of these two elements. Its linearity (10^5) and upper temperature limit ($300\,°C$) are not quite as good as an FID but its detection limit (10^{-14} g s^{-1}) is better. It is widely used in organophosphorus pesticide residue analysis.
- *Electron capture detector* (ECD): This responds only to analytes that capture electrons, particularly halogen-containing compounds. This detector is widely used in the analysis of polychlorinated compounds, such as the pesticides DDT, dieldrin and aldrin. It has a very high sensitivity (10^{-13} g s^{-1}) and an upper temperature limit of $300\,°C$ but its linear range (10^2 to 10^4) is much lower than that of the FID. The detector works by means of a radioactive source (^{63}Ni) ionising the carrier gas and releasing an electron that gives a current across the electrodes when a suitable voltage is applied. When an electron-capturing analyte (generally one containing a halogen atom) emerges from the column, the ionised electrons are captured, the current drops and this change in current is recorded. The carrier gas most commonly used in conjunction with an ECD is nitrogen or an argon +5% methane mixture.
- *Flame photometric detector*: This exploits the fact the P- and S-containing analytes emit light when they are burned in a FID-type detector. This light is detected and quantified. The detection limit is of the order of 1.0 pg for P-containing compounds and 20 pg for S-containing compounds.
- *Rapid scanning Fourier transform infrared detector*: This records the infrared spectrum of the emerging analytes and can give structural as well as quantitative information about the analyte. Any analyte with an infrared spectrum can be detected with a detection limit of about 1 ng.

- *Mass spectrometer detector*: This is a universal detector that gives a mass spectrum of the analyte and therefore gives both structural and quantitative data. Its detection limit is less than 1 ng per scan. Analytes may be detected by a total ion current (TIC) (Section 9.4.1) trace that is non-selective, or by selected ion monitoring (SIM) (Section 9.5.11) that can be specific for a selected analyte. In cases where authentic samples of the test compounds are not available for calibration purposes or in cases where the identity of the analytes is not known, a mass spectrometer is the best means of detecting and identifying the analyte. Special separators are available for removing the bulk of the carrier gas from the sample emerging from the column prior to its introduction in the mass spectrometer (Section 9.3).

Modern GC systems are controlled by dedicated microcomputers capable of automating and optimising the experimental conditions, recording the calibration and test retention data and carrying out statistical analysis of it and displaying the outputs in colour graphics in real time. They are capable of carrying out both qualitative and quantitative analysis on a similar basis to that of LC.

11.10 SUGGESTIONS FOR FURTHER READING

Niessen, W. M. A. (2007). *Liquid Chromatography–Mass Spectrometry*, 3rd edn. Boca Raton, FL: CRC Press. (A definitive guide with particular emphasis on applications in proteomics, drug discovery, food safety and environmental monitoring.)

Pyell, U. (ed.) (2006). *Electrokinetic Chromatography: Theory, Instrumentation and Applications*. New York: John Wiley. (A comprehensive coverage of the techniques and its most recent applications.)

Zachariou, M. (ed.) (2007). *Affinity Chromatography: Methods and Protocols*. Totowa, NJ: Humana Press. (A detailed account of recent applications of this important method of protein purification.)

12 Spectroscopic techniques: I Spectrophotometric techniques

A. HOFMANN

12.1 INTRODUCTION

Spectroscopic techniques employ light to interact with matter and thus probe certain features of a sample to learn about its consistency or structure. Light is electromagnetic radiation, a phenomenon exhibiting different energies, and dependent on that energy, different molecular features can be probed. The basic principles of interaction of electromagnetic radiation with matter are treated in this chapter. There is no obvious logical dividing point to split the applications of electromagnetic radiation into parts treated separately. The justification for the split presented in this text is purely pragmatic and based on 'common practice'. The applications considered in this chapter use visible or UV light to probe consistency and conformational structure of biological molecules. Usually, these methods are the first analytical procedures used by a biochemical scientist. The applications covered in Chapter 13 present a higher level of complexity in undertaking and are employed at a later stage in biochemical or biophysical characterisation.

An understanding of the properties of electromagnetic radiation and its interaction with matter leads to an appreciation of the variety of types of spectra and, consequently, different spectroscopic techniques and their applications to the solution of biological problems.

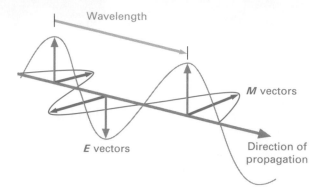

Fig. 12.1 Light is electromagnetic radiation and can be described as a wave propagating transversally in space and time. The electric (**E**) and magnetic (**M**) field vectors are directed perpendicular to each other. For UV/Vis, circular dichroism and fluorescence spectroscopy, the electric field vector is of most importance. For electron paramagnetic and nuclear magnetic resonance, the emphasis is on the magnetic field vector.

12.1.1 Properties of electromagnetic radiation

The interaction of electromagnetic radiation with matter is a quantum phenomenon and dependent upon both the properties of the radiation and the appropriate structural parts of the samples involved. This is not surprising, since the origin of electromagnetic radiation is due to energy changes within matter itself. The transitions which occur within matter are quantum phenomena and the spectra which arise from such transitions are principally predictable.

Electromagnetic radiation (Fig. 12.1) is composed of an electric and a perpendicular magnetic vector, each one oscillating in plane at right angles to the direction of propagation. The wavelength λ is the spatial distance between two consecutive peaks (one cycle) in the sinusoidal waveform and is measured in submultiples of metre, usually in nanometres (nm). The maximum length of the vector is called the amplitude. The frequency ν of the electromagnetic radiation is the number of oscillations made by the wave within the timeframe of 1 s. It therefore has the units of $1\ \mathrm{s}^{-1} = 1$ Hz. The frequency is related to the wavelength via the speed of light c ($c = 2.998 \times 10^8\ \mathrm{m\ s}^{-1}$ *in vacuo*) by $\nu = c\ \lambda^{-1}$. A historical parameter in this context is the wavenumber $\bar{\nu}$ which describes the number of completed wave cycles per distance and is typically measured in $1\ \mathrm{cm}^{-1}$.

12.1.2 Interaction with matter

Figure 12.2 shows the spectrum of electromagnetic radiation organised by increasing wavelength, and thus decreasing energy, from left to right. Also annotated are the types of radiation and the various interactions with matter and the resulting spectroscopic applications, as well as the interdependent parameters of frequency and wavenumber.

Electromagnetic phenomena are explained in terms of quantum mechanics. The photon is the elementary particle responsible for electromagnetic phenomena. It carries the electromagnetic radiation and has properties of a wave, as well as of

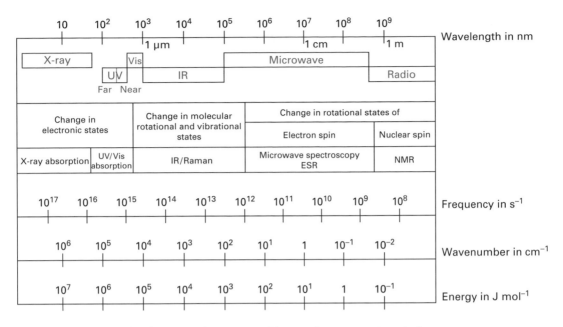

Fig. 12.2 The electromagnetic spectrum and its usage for spectroscopic methods.

a particle, albeit having a mass of zero. As a particle, it interacts with matter by transferring its energy E:

$$E = \frac{hc}{\lambda} = h\nu \tag{12.1}$$

where h is the Planck constant ($h = 6.63 \times 10^{-34}$ Js) and ν is the frequency of the radiation as introduced above.

When considering a diatomic molecule (see Fig. 12.3), rotational and vibrational levels possess discrete energies that only merge into a continuum at very high energy. Each electronic state of a molecule possesses its own set of rotational and vibrational levels. Since the kind of schematics shown in Fig. 12.3 is rather complex, the Jablonski diagram is used instead, where electronic and vibrational states are schematically drawn as horizontal lines, and vertical lines depict possible transitions (see Fig. 12.8 below).

In order for a transition to occur in the system, energy must be absorbed. The energy change ΔE needed is defined in quantum terms by the difference in absolute energies between the final and the starting state as $\Delta E = E_{final} - E_{start} = h\nu$.

Electrons in either atoms or molecules may be distributed between several energy levels but principally reside in the lowest levels (ground state). In order for an electron to be promoted to a higher level (excited state), energy must be put into the system. If this energy $E = h\nu$ is derived from electromagnetic radiation, this gives rise to an absorption spectrum, and an electron is transferred from the electronic ground state (S_0) into the first electronic excited state (S_1). The molecule will also be in an excited vibrational and rotational state. Subsequent relaxation of the molecule into the vibrational ground state of the first electronic excited state will occur. The electron can then revert back to the electronic ground state. For non-fluorescent molecules, this is accompanied by the emission of heat (ΔH).

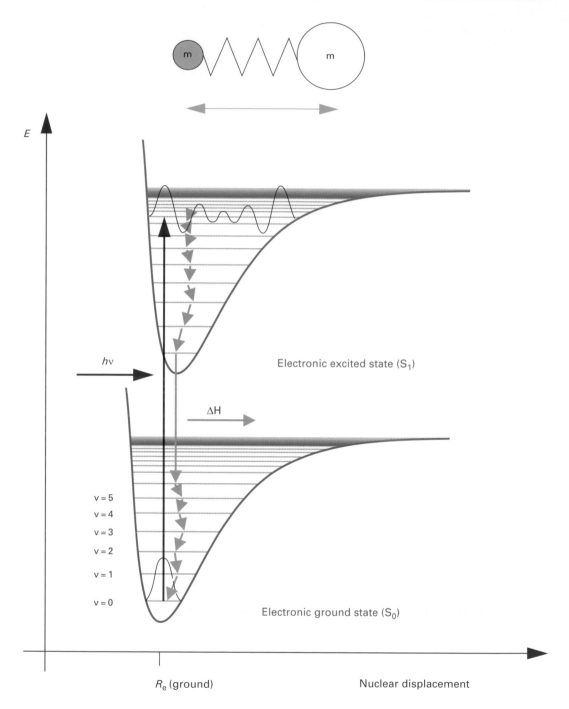

Fig. 12.3 Energy diagram for a diatomic molecule exhibiting rotation, vibration as well as an electronic structure. The distance between two masses m_1 and m_2 (nuclear displacement) is described as a Lennard–Jones potential curve with different equilibrium distances (R_e) for each electronic state. Energetically lower states always have lower equilibrium distances. The vibrational levels (horizontal lines) are superimposed on the electronic levels. Rotational levels are superimposed on the vibrational levels and not shown for reasons of clarity.

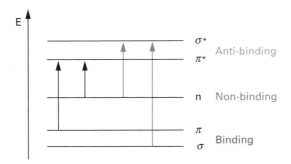

Fig. 12.4 Energy scheme for molecular orbitals (not to scale). Arrows indicate possible electronic transitions. The length of the arrows indicates the energy required to be put into the system in order to enable the transition. Black arrows depict transitions possible with energies from the UV/Vis spectrum for some biological molecules. The transitions shown by grey arrows require higher energies (e.g. X-rays).

The plot of absorption probability against wavelength is called absorption spectrum. In the simpler case of single atoms (as opposed to multi-atom molecules), electronic transitions lead to the occurrence of line spectra (see Section 12.7). Because of the existence of more different kinds of energy levels, molecular spectra are usually observed as band spectra (for example Fig. 12.7 below) which are molecule-specific due to the unique vibration states.

A commonly used classification of absorption transitions uses the spin states of electrons. Quantum mechanically, the electronic states of atoms and molecules are described by orbitals which define the different states of electrons by two parameters: a geometrical function defining the space and a probability function. The combination of both functions describes the localisation of an electron.

Electrons in binding orbitals are usually paired with antiparallel spin orientation (Fig. 12.8). The total spin S is calculated from the individual electron spins. The multiplicity M is obtained by $M = 2 \times S + 1$. For paired electrons in one orbital this yields:

$$S - \text{spin(electron 1)} + \text{spin(electron 2)} = (+1/2) + (-1/2) = 0$$

The multiplicity is thus $M = 2 \times 0 + 1 = 1$. Such a state is thus called a singlet state and denotated as 'S'. Usually, the ground state of a molecule is a singlet state, S_0.

In case the spins of both electrons are oriented in a parallel fashion, the resulting state is characterised by a total spin of $S = 1$, and a multiplicity of $M = 3$. Such a state is called a triplet state and usually exists only as one of the excited states of a molecule, e.g. T_1.

According to quantum mechanical transition rules, the multiplicity M and the total spin S must not change during a transition. Thus, the $S_0 \rightarrow S_1$ transition is allowed and possesses a high transition probability. In contrast, the $S_0 \rightarrow T_1$ is not allowed and has a small transition probability. Note that the transition probability is proportional to the intensity of the respective absorption bands.

Most biologically relevant molecules possess more than two atoms and, therefore, the energy diagrams become more complex than the ones shown in Fig. 12.3. Different orbitals combine to yield molecular orbitals that generally fall into one of five different classes (Fig. 12.4): s orbitals combine to the binding σ and the anti-binding σ^* orbitals. Some p orbitals combine to the binding π and the anti-binding π^*

orbitals. Other p orbitals combine to form non-binding n orbitals. The population of binding orbitals strengthens a chemical bond, and, vice versa, the population of anti-binding orbitals weakens a chemical bond.

12.1.3 Lasers

Laser is an acronym for light amplification by stimulated emission of radiation. A detailed explanation of the theory of lasers is beyond the scope of this textbook. A simplified description starts with the use of photons of a defined energy to excite an absorbing material. This results in elevation of an electron to a higher energy level. If, whilst the electron is in the excited state, another photon of precisely that energy arrives, then, instead of the electron being promoted to an even higher level, it can return to the original ground state. However, this transition is accompanied by the emission of two photons with the same wavelength and exactly in phase (coherent photons). Multiplication of this process will produce coherent light with extremely narrow spectral bandwidth. In order to produce an ample supply of suitable photons, the absorbing material is surrounded by a rapidly flashing light of high intensity (pumping).

Lasers are indispensable tools in many areas of science, including biochemistry and biophysics. Several modern spectroscopic techniques utilise laser light sources, due to their high intensity and accurately defined spectral properties. One of the probably most revolutionising applications in the life sciences, the use of lasers in DNA sequencing with fluorescence labels (see Sections 5.11.5, 5.11.6 and 12.3.3), enabled the breakthrough in whole-genome sequencing.

12.2 ULTRAVIOLET AND VISIBLE LIGHT SPECTROSCOPY

These regions of the electromagnetic spectrum and their associated techniques are probably the most widely used for analytical work and research into biological problems.

The electronic transitions in molecules can be classified according to the participating molecular orbitals (See Fig. 12.4). From the four possible transitions ($n{\to}\pi^*$, $\pi{\to}\pi^*$, $n{\to}\sigma^*$, $\sigma{\to}\sigma^*$), only two can be elicited with light from the UV/Vis spectrum for some biological molecules: $n{\to}\pi^*$ and $\pi{\to}\pi^*$. The $n{\to}\sigma^*$ and $\sigma{\to}\sigma^*$ transitions are energetically not within the range of UV/Vis spectroscopy and require higher energies.

Molecular (sub-)structures responsible for interaction with electromagnetic radiation are called chromophores. In proteins, there are three types of chromophores relevant for UV/Vis spectroscopy:

- peptide bonds (amide bond);
- certain amino acid side chains (mainly tryptophan and tyrosine); and
- certain prosthetic groups and coenzymes (e.g. porphyrine groups such as in haem).

The presence of several conjugated double bonds in organic molecules results in an extended π-system of electrons which lowers the energy of the π^* orbital through electron delocalisation. In many cases, such systems possess $\pi{\to}\pi^*$ transitions in the UV/Vis range of the electromagnetic spectrum. Such molecules are very useful tools in colorimetric applications (see Table 12.1).

Table 12.1 **Common colorimetric and UV absorption assays**

Substance	Reagent	Wavelength (nm)
Amino acids	(a) Ninhydrin	570 (proline : 420)
	(b) Cupric salts	620
Cysteine residues, thiolates	Ellman reagent (di-sodium-bis-(3-carboxy-4-nitrophenyl)-disulphide)	412
Protein	(a) Folin (phosphomolybdate, phosphotungstate, cupric salt)	660
	(b) Biuret (reacts with peptide bonds)	540
	(c) BCA reagent (bicinchoninic acid)	562
	(d) Coomassie Brilliant Blue	595
	(e) Direct	Tyr, Trp: 278, peptide bond : 190
Coenzymes	Direct	FAD: 438, NADH: 340, NAD^+ : 260
Carotenoids	Direct	420, 450, 480
Porphyrins	Direct	400 (Soret band)
Carbohydrate	(a) Phenol, H_2SO_4	Glucose: 490, xylose: 480
	(b) Anthrone (anthrone, H_2SO_4)	620 or 625
Reducing sugars	Dinitrosalicylate, alkaline tartrate buffer	540
Pentoses	(a) Bial (orcinol, ethanol, $FeCl_3$, HCl)	665
	(b) Cysteine, H_2SO_4	380–415
Hexoses	(a) Carbazole, ethanol, H_2SO_4	540 or 440
	(b) Cysteine, H_2SO_4	380–415
	(c) Arsenomolybdate	500–570
Glucose	Glucose oxidase, peroxidase, o-dianisidine, phosphate buffer	420
Ketohexose	(a) Resorcinol, thiourea, ethanoic acid, HCl	520
	(b) Carbazole, ethanol, cysteine, H_2SO_4	560
	(c) Diphenylamine, ethanol, ethanoic acid, HCl	635
Hexosamines	Ehrlich (dimethylaminobenzaldehyde, ethanol, HCl)	530

Table 12.1 (*cont.*)

Substance	Reagent	Wavelength (nm)
DNA	(a) Diphenylamine	595
	(b) Direct	260
RNA	Bial (orcinol, ethanol, $FeCl_3$, HCl)	665
Sterols and steroids	Liebermann–Burchardt reagent (acetic anhydride, H_2SO_4, chloroform)	425, 625
Cholesterol	Cholesterol oxidase, peroxidase, 4-amino-antipyrine, phenol	500
ATPase assay	Coupled enzyme assay with ATPase, pyruvate kinase, lactate dehydrogenase: ATP → ADP (consumes ATP) phosphoenolpyruvate → pyruvate (consumes ADP) pyruvate → lactate (consumes NADH)	NADH: 340

12.2.1   Chromophores in proteins

The electronic transitions of the peptide bond occur in the far UV. The intense peak at 190 nm, and the weaker one at 210–220 nm is due to the $\pi \rightarrow \pi^*$ and $n \rightarrow \pi^*$ transitions. A number of amino acids (Asp, Glu, Asn, Gln, Arg and His) have weak electronic transitions at around 210 nm. Usually, these cannot be observed in proteins because they are masked by the more intense peptide bond absorption. The most useful range for proteins is above 230 nm, where there are absorptions from aromatic side chains. While a very weak absorption maximum of phenylalanine occurs at 257 nm, tyrosine and tryptophan dominate the typical protein spectrum with their absorption maxima at 274 nm and 280 nm, respectively (Fig. 12.5). *In praxi*, the presence of these two aromatic side chains gives rise to a band at ∼278 nm. Cystine (Cys_2) possesses a weak absorption maximum of similar strength as phenylalanine at 250 nm. This band can play a role in rare cases in protein optical activity or protein fluorescence.

Proteins that contain prosthetic groups (e.g. haem, flavin, carotenoid) and some metal–protein complexes, may have strong absorption bands in the UV/Vis range. These bands are usually sensitive to local environment and can be used for physical studies of enzyme action. Carotenoids, for instance, are a large class of red, yellow and orange plant pigments composed of long carbon chains with many conjugated double bonds. They contain three maxima in the visible region of the electromagnetic spectrum (∼420 nm, 450 nm, 480 nm).

Porphyrins are the prosthetic groups of haemoglobin, myoglobin, catalase and cytochromes. Electron delocalisation extends throughout the cyclic tetrapyrrole ring of porphyrins and gives rise to an intense transition at ∼400 nm called the Soret band. The spectrum of haemoglobin is very sensitive to changes in the iron-bound ligand. These changes can be used for structure–function studies of haem proteins.

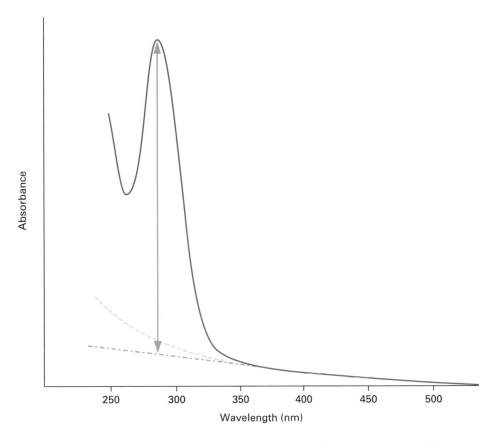

Fig. 12.5 The presence of larger aggregates in biological samples gives rise to Rayleigh scatter visible by a considerable slope in the region from 500 to 350 nm. The dashed line shows the correction to be applied to spectra with Rayleigh scatter which increases with λ^{-4}. Practically, linear extrapolation of the region from 500 to 350 nm is performed to correct for the scatter. The corrected absorbance is indicated by the double arrow.

Molecules such as FAD (flavin adenine dinucleotide), NADH and NAD^+ are important coenzymes of proteins involved in electron transfer reactions (RedOx reactions). They can be conveniently assayed by using their UV/Vis absorption: 438 nm (FAD), 340 nm (NADH) and 260 nm (NAD^+).

Chromophores in genetic material

The absorption of UV light by nucleic acids arises from $n \rightarrow \pi^*$ and $\pi \rightarrow \pi^*$ transitions of the purine (adenine, guanine) and pyrimidine (cytosine, thymine, uracil) bases that occur between 260 nm and 275 nm. The absorption spectra of the bases in polymers are sensitive to pH and greatly influenced by electronic interactions between bases.

12.2.2 Principles

Quantification of light absorption

The chance for a photon to be absorbed by matter is given by an extinction coefficient which itself is dependent on the wavelength λ of the photon. If light with the intensity I_0 passes through a sample with appropriate transparency and the path length (thickness) d, the intensity I drops along the pathway in an exponential manner. The characteristic absorption parameter for the sample is the extinction coefficient α, yielding the correlation $I = I_0\, e^{-\alpha d}$. The ratio $T = I/I_0$ is called transmission.

Biochemical samples usually comprise aqueous solutions, where the substance of interest is present at a molar concentration c. Algebraic transformation of the exponential correlation into an expression based on the decadic logarithm yields the law of Beer–Lambert:

$$\lg \frac{I_0}{I} = \lg \frac{1}{T} = \varepsilon \times c \times d = A \tag{12.2}$$

where $[d] = 1$ cm, $[c] = 1$ mol dm^{-3}, and $[\varepsilon] = 1$ dm^3 mol^{-1} cm^{-1}. ε is the molar absorption coefficient (also molar extinction coefficient) ($\alpha = 2.303 \times c \times \varepsilon$). A is the absorbance of the sample, which is displayed on the spectrophotometer.

The Beer–Lambert law is valid for low concentrations only. Higher concentrations might lead to association of molecules and therefore cause deviations from the ideal behaviour. Absorbance and extinction coefficients are additive parameters, which complicates determination of concentrations in samples with more than one absorbing species. Note that in dispersive samples or suspensions scattering effects increase the absorbance, since the scattered light is not reaching the detector for read-out. The absorbance recorded by the spectrophotometer is thus overestimated and needs to be corrected (Fig. 12.5).

Deviations from the Beer–Lambert law

According to the Beer–Lambert law, absorbance is linearly proportional to the concentration of chromophores. This might not be the case any more in samples with high absorbance. Every spectrophotometer has a certain amount of stray light, which is light received at the detector but not anticipated in the spectral band isolated by the monochromator. In order to obtain reasonable signal-to-noise ratios, the intensity of light at the chosen wavelength (I_λ) should be 10 times higher than the intensity of the stray light (I_{stray}). If the stray light gains in intensity, the effects measured at the detector have nothing or little to do with chromophore concentration. Secondly, molecular events might lead to deviations from the Beer–Lambert law. For instance, chromophores might dimerise at high concentrations and, as a result, might possess different spectroscopic parameters.

Absorption or light scattering – optical density

In some applications, for example measurement of turbidity of cell cultures (determination of biomass concentration), it is not the absorption but the scattering of light (see Section 12.6) that is actually measured with a spectrophotometer. Extremely turbid samples like bacterial cultures do not absorb the incoming light. Instead, the light is scattered and thus, the spectrometer will record an apparent absorbance (sometimes also called attenuance). In this case, the observed parameter is called optical density (OD). Instruments specifically designed to measure turbid samples are nephelometers or Klett meters; however, most biochemical laboratories use the general UV/Vis spectrometer for determination of optical densities of cell cultures.

Factors affecting UV/Vis absorption

Biochemical samples are usually buffered aqueous solutions, which has two major advantages. Firstly, proteins and peptides are comfortable in water as a solvent, which is also the 'native' solvent. Secondly, in the wavelength interval of UV/Vis (700–200 nm) the water spectrum does not show any absorption bands and thus acts as a silent component of the sample.

The absorption spectrum of a chromophore is only partly determined by its chemical structure. The environment also affects the observed spectrum, which mainly can be described by three parameters:

- protonation/deprotonation (pH, RedOx);
- solvent polarity (dielectric constant of the solvent); and
- orientation effects.

Vice versa, the immediate environment of chromophores can be probed by assessing their absorption, which makes chromophores ideal reporter molecules for environmental factors. Four effects, two each for wavelength and absorption changes, have to be considered:

- a wavelength shift to higher values is called red shift or bathochromic effect;
- similarly, a shift to lower wavelengths is called blue shift or hypsochromic effect;
- an increase in absorption is called hyperchromicity ('more colour'),
- while a decrease in absorption is called hypochromicity ('less colour').

Protonation/deprotonation arises either from changes in pH or oxidation/reduction reactions, which makes chromophores pH- and RedOx-sensitive reporters. As a rule of thumb, λ_{max} and ε increase, i.e. the sample displays a batho- and hyperchromic shift, if a titratable group becomes charged.

Furthermore, solvent polarity affects the difference between the ground and excited states. Generally, when shifting to a less polar environment one observes a batho- and hyperchromic effect. Conversely, a solvent with higher polarity elicits a hypso- and hypochromic effect.

Lastly, orientation effects, such as an increase in order of nucleic acids from single-stranded to double-stranded DNA, lead to different absorption behaviour. A sample of free nucleotides exhibits a higher absorption than a sample with identical amounts of nucleotides but assembled into a single-stranded polynucleotide. Accordingly, double-stranded polynucleotides exhibit an even smaller absorption than two single-stranded polynucleotides. This phenomenon is called the hypochromicity of polynucleotides. The increased exposure (and thus stronger absorption) of the individual nucleotides in the less ordered states provides a simplified explanation for this behaviour.

12.2.3 Instrumentation

UV/Vis spectrophotometers are usually dual-beam spectrometers where the first channel contains the sample and the second channel holds the control (buffer) for correction.

Alternatively, one can record the control spectrum first and use this as internal reference for the sample spectrum. The latter approach has become very popular as many spectrometers in the laboratories are computer-controlled, and baseline correction can be carried out using the software by simply subtracting the control from the sample spectrum.

The light source is a tungsten filament bulb for the visible part of the spectrum, and a deuterium bulb for the UV region. Since the emitted light consists of many different wavelengths, a monochromator, consisting of either a prism or a rotating metal grid of high precision called grating, is placed between the light source and the sample. Wavelength selection can also be achieved by using coloured filters as monochromators that absorb all but a certain limited range of wavelengths. This limited range is called the bandwidth of the filter. Filter-based wavelength selection is used in colorimetry, a method with moderate accuracy, but best suited for specific colorimetric assays where only certain wavelengths are of interest. If wavelengths are selected by prisms or gratings, the technique is called spectrophotometry (Fig. 12.6).

Example 1 ESTIMATION OF MOLAR EXTINCTION COEFFICIENTS

In order to determine the concentration of a solution of the peptide MAMVSEFLKQ AWFIENEEQE YVQTVKSSKG GPGSAVSPYP TFNPSS in water, the molar absorption coefficient needs to be estimated.

The molar extinction coefficient ε is a characteristic parameter of a molecule and varies with the wavelength of incident light. Because of useful applications of the law of Beer–Lambert, the value of ε needs be known for a lot of molecules being used in biochemical experiments.

Very frequently in biochemical research, the molar extinction coefficient of proteins is estimated using incremental ε_i values for each absorbing protein residue (chromophore). Summation over all residues yields a reasonable estimation for the extinction coefficient. The simplest increment system is based on values of Gill and von Hippel.[1] The determination of protein concentration using this formula only requires an absorption value at $\lambda = 280$ nm. Increments ε_i are used to calculate a molar extinction coefficient at 280 nm for the entire protein or peptide by summation over all relevant residues in the protein:

Residue	Gill and von Hippel ε_i (280 nm) in $dm^3\,mol^{-1}\,cm^{-1}$
Cys_2	120
Trp	5690
Tyr	1280

For the peptide above, one obtains $\varepsilon = (1 \times 5690 + 2 \times 1280)\,dm^3\,mol^{-1}\,cm^{-1} = 8250\,dm^3\,mol^{-1}\,cm^{-1}$

[1] Gill, S. C. and von Hippel, P. H. (1989). Calculation of protein extinction coefficients from amino acid sequence data. *Analytical Biochemistry*, **182**, 319–326. Erratum: *Analytical Biochemistry* (1990), **189**, 283.

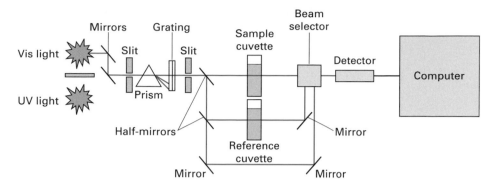

Fig. 12.6 Optical arrangements in a dual-beam spectrophotometer. Either a prism or a grating constitutes the monochromator of the instrument. Optical paths are shown as green lines.

A prism splits the incoming light into its components by refraction. Refraction occurs because radiation of different wavelengths travels along different paths in medium of higher density. In order to maintain the principle of velocity conservation, light of shorter wavelength (higher speed) must travel a longer distance (i.e. blue sky effect). At a grating, the splitting of wavelengths is achieved by diffraction. Diffraction is a reflection phenomenon that occurs at a grid surface, in this case a series of engraved fine lines. The distance between the lines has to be of the same order of magnitude as the wavelength of the diffracted radiation. By varying the distance between the lines, different wavelengths are selected. This is achieved by rotating the grating perpendicular to the optical axis. The resolution achieved by gratings is much higher than the one available by prisms. Nowadays instruments almost exclusively contain gratings as monochromators as they can be reproducibly made in high quality by photoreproduction.

The bandwidth of a colorimeter is determined by the filter used as monochromator. A filter that appears red to the human eye is transmitting red light and absorbs almost any other (visual) wavelength. This filter would be used to examine blue solutions, as these would absorb red light. The filter used for a specific colorimetric assay is thus made of a colour complementary to that of the solution being tested. Theoretically, a single wavelength is selected by the monochromator in spectrophotometers, and the emergent light is a parallel beam. Here, the bandwidth is defined as twice the half-intensity bandwidth. The bandwidth is a function of the optical slit width. The narrower the slit width the more reproducible are measured absorbance values. In contrast, the sensitivity becomes less as the slit narrows, because less radiation passes through to the detector.

In a dual-beam instrument, the incoming light beam is split into two parts by a half-mirror. One beam passes through the sample, the other through a control (blank, reference). This approach obviates any problems of variation in light intensity, as both reference and sample would be affected equally. The measured absorbance is the difference between the two transmitted beams of light recorded. Depending on the instrument, a second detector measures the intensity of the incoming beam, although some instruments use an arrangement where one detector measures the incoming and the transmitted intensity alternately. The latter design is better from an analytical point of view as it eliminates potential variations between the two detectors. At about 350 nm most instruments require a change of the light source from visible to UV light. This is achieved by mechanically moving mirrors that direct the appropriate beam along the optical axis and divert the other. When scanning the interval of 500–210 nm, this frequently gives rise to an offset of the spectrum at the switchover point.

Since borosilicate glass and normal plastics absorb UV light, such cuvettes can only be used for applications in the visible range of the spectrum (up to 350 nm). For UV measurements, quartz cuvettes need to be used. However, disposable plastic cuvettes have been developed that allow for measurements over the entire range of the UV/Vis spectrum.

12.2.4 Applications

The usual procedure for (colorimetric) assays is to prepare a set of standards and produce a plot of concentration versus absorbance called calibration curve. This should be linear as long as the Beer–Lambert law applies. Absorbances of unknowns are then measured and their concentration interpolated from the linear region of the plot. It is important that one never extrapolates beyond the region for which an instrument has been calibrated as this potentially introduces enormous errors.

To obtain good spectra, the maximum absorbance should be approximately 0.5 which corresponds to concentrations of about 50 μM (assuming $\varepsilon = 10\,000\,\mathrm{dm}^3\,\mathrm{mol}^{-1}\,\mathrm{cm}^{-1}$).

Qualitative and quantitative analysis

Qualitative analysis may be performed in the UV/Vis regions to identify certain classes of compounds both in the pure state and in biological mixtures (e.g. protein-bound). The application of UV/Vis spectroscopy to further analytical purposes is rather limited, but possible for systems where appropriate features and parameters are known.

Most commonly, this type of spectroscopy is used for quantification of biological samples either directly or via colorimetric assays. In many cases, proteins can be quantified directly using their intrinsic chromophores, tyrosine and tryptophan. Protein spectra are acquired by scanning from 500 to 210 nm. The characteristic features in a protein spectrum are a band at 278/280 nm and another at 190 nm (Fig. 12.6). The region from 500 to 300 nm provides valuable information about the presence of any prosthetic groups or coenzymes. Protein quantification by single wavelength measurements at 280 and 260 nm only should be avoided, as the presence of larger aggregates (contaminations or protein aggregates) gives rise to considerable Rayleigh scatter that needs to be corrected for (Fig. 12.6).

Difference spectra

The main advantage of difference spectroscopy is its capacity to detect small absorbance changes in systems with high background absorbance. A difference spectrum is obtained by subtracting one absorption spectrum from another. Difference spectra can be obtained in two ways: either by subtraction of one absolute absorption spectrum from another, or by placing one sample in the reference cuvette and another in the test cuvette. The latter method requires usage of a dual-beam instrument, the former method has become very popular due to most instruments being controlled

Example 2 **DETERMINATION OF CONCENTRATIONS**

Question (1) The concentration of an aqueous solution of a protein is to be determined assuming:
   (i) knowledge of the molar extinction coefficient ε
   (ii) molar extinction coefficent ε is not known.
(2) What is the concentration of an aqueous solution of a DNA sample?

Answer (1) (i) The protein concentration of a pure sample can be determined by using the Beer–Lambert law. The absorbance at 280 nm is determined from a protein spectrum, and the molar extinction coefficient at this wavelength needs to be experimentally determined or estimated:

$$\rho^* = \frac{A \times M}{\varepsilon \times d}$$

where ρ^* is the mass concentration in mg cm^{-3} and M the molecular mass of the assayed species in g mol^{-1}.

(ii) Alternatively, an empirical formula known as the Warburg–Christian formula can be used without knowledge of the value of the molar extinction coefficient:

$$\rho^* = (1.52 \times A_{280} - 0.75 \times A_{260})\,\text{mg cm}^{-3}$$

Other commonly used applications to determine the concentration of protein in a sample make use of colorimetric assays that are based on chemicals (folin, biuret, bicinchoninic acid or Coomassie Brilliant Blue) binding to protein groups. Concentration determination in these cases requires a calibration curve measured with a protein standard, usually bovine serum albumin.

(2) As we have seen above, the genetic bases have absorption bands in the UV/Vis region. Thus, the concentration of a DNA sample can be determined spectroscopically. Assuming that a pair of nucleotides has a molecular mass of $M = 660$ g mol^{-1}, the absorbance A of a solution with double-stranded DNA at 260 nm can be converted to mass concentration ρ^* by:

$$\rho^* = 50\,\mu\text{g cm}^{-3} \times A_{260}$$

The ratio A_{260}/A_{280} is an indicator for the purity of the DNA solution and should be in the range 1.8–2.0.

by computers which allows easy processing and handling of data. From a purist's point of view, the direct measurement of the difference spectrum in a dual-beam instrument is the preferred method, since it reduces the introduction of inconsistencies between samples and thus the error of the measurement. Figure 12.7 shows the two absolute spectra of ubiquinone and ubiquinol, the oxidised and reduced species of the same molecular skeleton, as well as the difference spectrum.

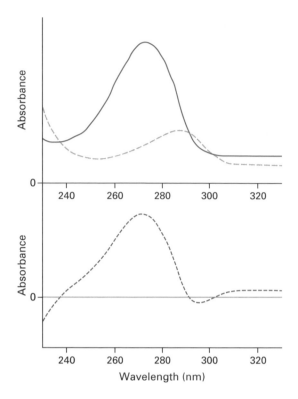

Fig. 12.7 Top: Absolute spectra of ubiquinone (solid curve) and ubiquinol (dotted curve). Bottom: Difference spectrum.

Difference spectra have three distinct features as compared to absolute spectra:

- difference spectra may contain negative absorbance values;
- absorption maxima and minima may be displaced and the extinction coefficients are different from those in peaks of absolute spectra;
- there are points of zero absorbance, usually accompanied by a change of sign of the absorbance values. These points are observed at wavelengths where both species of related molecules exhibit identical absorbances (isosbestic points), and which may be used for checking for the presence of interfering substances.

Common applications for difference UV spectroscopy include the determination of the number of aromatic amino acids exposed to solvent, detection of conformational changes occurring in proteins, detection of aromatic amino acids in active sites of enzymes, and monitoring of reactions involving 'catalytic' chromophores (prosthetic groups, coenzymes).

Derivative spectroscopy

Another way to resolve small changes in absorption spectra that otherwise would remain invisible is the usage of derivative spectroscopy. Here, the absolute absorption spectrum of a sample is differentiated and the differential $\delta^x A/\delta\lambda^x$ plotted against the wavelength. Since the algebraic relationship between A and λ is unknown, differentiation is carried out by numerical methods using computer software. The usefulness of this approach depends on the individual problem. Examples of successful applications include the binding of a monoclonal antibody to its antigen with second-order derivatives and the quantification of tryptophan and tyrosine residues in proteins using fourth-order derivatives.

Solvent perturbation

As we have mentioned above, aromatic amino acids are the main chromophores of proteins in the UV region of the electromagnetic spectrum. Furthermore, the UV absorption of chromophores depends largely on the polarity in its immediate environment. A change in the polarity of the solvent changes the UV spectrum of a protein by bathochromic or hypsochromic effects without changing its conformation. This phenomenon is called solvent perturbation and can be used to probe the surface of a protein molecule. In order to be accessible to the solvent, the chromophore has to be accessible on the protein surface. Practically, solvents like dimethyl-sulfoxide, dioxane, glycerol, mannitol, sucrose and polyethylene glycol are used for solvent perturbation experiments, because they are miscible with water. The method of solvent perturbation is most commonly used for determination of the number of aromatic residues that are exposed to solvent.

Spectrophotometric and colorimetric assays

For biochemical assays testing for time- or concentration-dependent responses of systems, an appropriate read-out is required that is coupled to the progress of the reaction (reaction coordinate). Therefore, the biophysical parameter being monitored (read-out) needs to be coupled to the biochemical parameter under investigation. Frequently, the monitored parameter is the absorbance of a system at a given wavelength which is monitored throughout the course of the experiment. Preferably, one should try to monitor the changing species directly (e.g. protein absorption, starting product or generated product of a reaction), but in many cases this is not possible and a secondary reaction has to be used to generate an appropriate signal for monitoring. A common application of the latter approach is the determination of protein concentration by Lowry or Bradford assays, where a secondary reaction is used to colour the protein. The more intense the colour, the more protein is present. These assays are called colorimetric assays and a number of commonly used ones are listed in Table 12.1.

12.3 FLUORESCENCE SPECTROSCOPY

12.3.1 Principles

Fluorescence is an emission phenomenon where an energy transition from a higher to a lower state is accompanied by radiation. Only molecules in their excited forms are able to emit fluorescence; thus, they have to be brought into a state of higher energy prior to the emission phenomenon.

We have already seen in Section 12.1.2 that molecules possess discrete states of energy. Potential energy levels of molecules have been depicted by different Lennard–Jones potential curves with overlaid vibrational (and rotational) states (Fig. 12.3). Such diagrams can be abstracted further to yield Jablonski diagrams (Fig. 12.8).

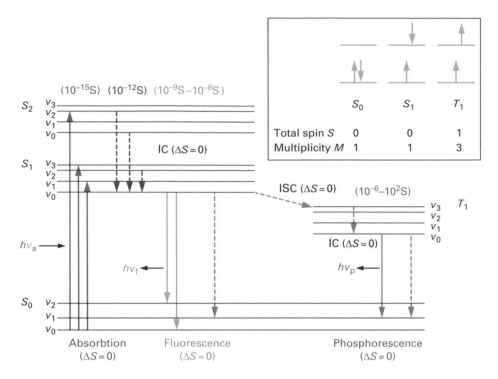

$(10^{-15}S)$ $(10^{-12}S)$ $(10^{-9}S - 10^{-8}S)$

S_2 v_3 v_2 v_1 v_0

IC $(\Delta S = 0)$

	S_0	S_1	T_1
Total spin S	0	0	1
Multiplicity M	1	1	3

S_1 v_3 v_2 v_1 v_0

ISC $(\Delta S = 0)$ $(10^{-6} - 10^2 S)$

T_1 v_3 v_2 v_1 v_0

$hv_a \longrightarrow$

IC $(\Delta S = 0)$

$hv_f \longleftarrow$ $hv_p \longleftarrow$

S_0 v_2 v_1 v_0

Absorbtion $(\Delta S = 0)$ Fluorescence $(\Delta S = 0)$ Phosphorescence $(\Delta S \neq 0)$

Fig. 12.8 Jablonski diagram. Shown are the electronic ground state (S_0), two excited singlet states (S_1, S_2) and a triplet state (T_1). Vibrational levels (v) are only illustrated exemplarily. Solid vertical lines indicate radiative transitions, dotted lines show non-radiative transitions. The inset shows the relationship between electron configurations, total spin number S and multiplicity M.

In these diagrams, energy transitions are indicated by vertical lines. Not all transitions are possible; allowed transitions are defined by the selection rules of quantum mechanics. A molecule in its electronic and vibrational ground state ($S_0 v_0$) can absorb photons matching the energy difference of its various discrete states. The required photon energy has to be higher than that required to reach the vibrational ground state of the first electronic excited state ($S_1 v_0$). The excess energy is absorbed as vibrational energy ($v > 0$), and quickly dissipated as heat by collision with solvent molecules. The molecule thus returns to the vibrational ground state ($S_1 v_0$). These relaxation processes are non-radiating transitions from one energetic state to another with lower energy, and are called internal conversion (IC). From the lowest level of the first electronic excited state, the molecule returns to the ground state (S_0) either by emitting light (fluorescence) or by a non-radiative transition. Upon radiative transition, the molecule can end up in any of the vibrational states of the electronic ground state (as per quantum mechanical rules).

If the vibrational levels of the ground state overlap with those of the electronic excited state, the molecule will not emit fluorescence, but rather revert to the ground state by non-radiative internal conversion. This is the most common way for excitation energy to be dissipated and is why fluorescent molecules are rather rare. Most molecules are flexible and thus have very high vibrational levels in the ground state. Indeed, most fluorescent molecules possess fairly rigid aromatic rings or ring systems. The fluorescent group in a molecule is called a fluorophore.

Since radiative energy is lost in fluorescence as compared to the absorption, the fluorescent light is always at a longer wavelength than the exciting light (Stokes shift). The emitted radiation appears as band spectrum, because there are many closely related wavelength values dependent on the vibrational and rotational energy levels attained. The fluorescence spectrum of a molecule is independent of the wavelength of the exciting radiation and has a mirror image relationship with the absorption spectrum. The probability of the transition from the electronic excited to the ground state is proportional to the intensity of the emitted light.

An associated phenomenon in this context is phosphorescence which arises from a transition from a triplet state (T_1) to the electronic (singlet) ground state (S_0). The molecule gets into the triplet state from an electronic excited singlet state by a process called intersystem crossing (ISC). The transition from singlet to triplet is quantum-mechanically not allowed and thus only happens with low probability in certain molecules where the electronic structure is favourable. Such molecules usually contain heavy atoms. The rate constants for phosphorescence are much longer and phosphorescence thus happens with a long delay and persists even when the exciting energy is no longer applied.

The fluorescence properties of a molecule are determined by properties of the molecule itself (internal factors), as well as the environment of the protein (external factors). The fluorescence intensity emitted by a molecule is dependent on the lifetime of the excited state. The transition from the excited to the ground state can be treated like a decay process of first order, i.e. the number of molecules in the excited state decreases exponentially with time. In analogy to kinetics, the exponential coefficient k_r is called rate constant and is the reciprocal of the lifetime: $\tau_r = k_r^{-1}$. The lifetime is the time it takes to reduce the number of fluorescence emitting molecules to N_0/e, and is proportional to λ^3.

The effective lifetime τ of excited molecules, however, differs from the fluorescence lifetime τ_r since other, non-radiative processes also affect the number of molecules in the excited state. τ is dependent on all processes that cause relaxation: fluorescence emission, internal conversion, quenching, fluorescence resonance energy transfer, reactions of the excited state and intersystem crossing.

The ratio of photons emitted and photons absorbed by a fluorophore is called quantum yield Φ (equation 12.3). It equals the ratio of the rate constant for fluorescence emission k_r and the sum of the rate constants for all six processes mentioned above.

$$\Phi = \frac{N(\text{em})}{N(\text{abs})} = \frac{k_r}{k} = \frac{k_r}{k_r + k_{IC} + k_{ISC} + k_{\text{reaction}} + k_Q c(Q) + k_{\text{FRET}}} = \frac{\tau}{\tau_r} \qquad (12.3)$$

The quantum yield is a dimensionless quantity, and, most importantly, the only absolute measure of fluorescence of a molecule. Measuring the quantum yield is a difficult process and requires comparison with a fluorophore of known quantum yield. In biochemical applications, this measurement is rarely done. Most commonly, the fluorescence emissions of two or more related samples are compared and their relative differences analysed.

12.3.2 Instrumentation

Fluorescence spectroscopy works most accurately at very low concentrations of emitting fluorophores. UV/Vis spectroscopy, in contrast, is least accurate at such

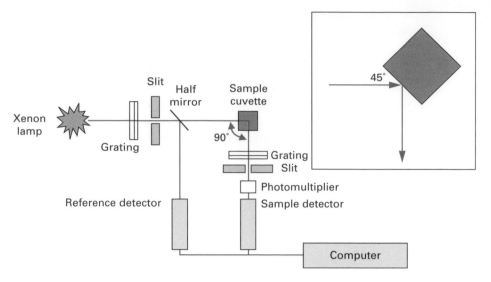

Fig. 12.9 Schematics of a spectrofluorimeter with 'T' geometry (90°). Optical paths are shown as green lines. Inset: Geometry of front-face illumination.

low concentrations. One major factor adding to the high sensitivity of fluorescence applications is the spectral selectivity. Due to the Stokes shift, the wavelength of the emitted light is different from that of the exciting light. Another feature makes use of the fact that fluorescence is emitted in all directions. By placing the detector perpendicular to the excitation pathway, the background of the incident beam is reduced.

The schematics of a typical spectrofluorimeter are shown in Fig. 12.9. Two monochromators are used, one for tuning the wavelength of the exciting beam and a second one for analysis of the fluorescence emission. Due to the emitted light always having a lower energy than the exciting light, the wavelength of the excitation monochromator is set at a lower wavelength than the emission monochromator. The better fluorescence spectrometers in laboratories have a photon-counting detector yielding very high sensitivity. Temperature control is required for accurate work as the emission intensity of a fluorophore is dependent on the temperature of the solution.

Two geometries are possible for the measurement, with the 90° arrangement most commonly used. Pre- and post-filter effects can arise owing to absorption of light prior to reaching the fluorophore and the reduction of emitted radiation. These phenomena are also called inner filter effects and are more evident in solutions with high concentrations. As a rough guide, the absorption of a solution to be used for fluorescence experiments should be less than 0.05. The use of microcuvettes containing less material can also be useful. Alternatively, the front-face illumination geometry (Fig. 12.9 inset) can be used which obviates the inner filter effect. Also, while the 90° geometry requires cuvettes with two neighbouring faces being clear (usually, fluorescence cuvettes have four clear faces), the front-face illumination technique requires only one clear face, as excitation and emission occur at the same face. However, front-face illumination is less sensitive than the 90° illumination.

12.3.3 Applications

There are many and highly varied applications for fluorescence despite the fact that relatively few compounds exhibit the phenomenon. The effects of pH, solvent composition and the polarisation of fluorescence may all contribute to structural elucidation. Measurement of fluorescence lifetimes can be used to assess rotation correlation coefficients and thus particle sizes. Non-fluorescent compounds are often labelled with fluorescent probes to enable monitoring of molecular events. This is termed extrinsic fluorescence as distinct from intrinsic fluorescence where the native compound exhibits the property. Some fluorescent dyes are sensitive to the presence of metal ions and can thus be used to track changes of these ions in *in vitro* samples, as well as whole cells.

Since fluorescence spectrometers have two monochromators, one for tuning the excitation wavelength and one for analysing the emission wavelength of the fluorophore, one can measure two types of spectra: excitation and emission spectra. For fluorescence excitation spectrum measurement, one sets the emission monochromator at a fixed wavelength (λ_{em}) and scans a range of excitation wavelengths which are then recorded as ordinate (x-coordinate) of the excitation spectrum; the fluorescence emission at λ_{em} is plotted as abscissa. Measurement of emission spectra is achieved by setting a fixed excitation wavelength (λ_{exc}) and scanning a wavelength range with the emission monochromator. To yield a spectrum, the emission wavelength λ_{em} is recorded as ordinate and the emission intensity at λ_{em} is plotted as abscissa.

Intrinsic protein fluorescence

Proteins possess three intrinsic fluorophores: tryptophan, tyrosine and phenylalanine, although the latter has a very low quantum yield and its contribution to protein fluorescence emission is thus negligible. Of the remaining two residues, tyrosine has the lower quantum yield and its fluorescence emission is almost entirely quenched when it becomes ionised, or is located near an amino or carboxyl group, or a tryptophan residue. Intrinsic protein fluorescence is thus usually determined by tryptophan fluorescence which can be selectively excited at 295–305 nm. Excitation at 280 nm excites tyrosine and tryptophan fluorescence and the resulting spectra might therefore contain contributions from both types of residues.

The main application for intrinsic protein fluorescence aims at conformational monitoring. We have already mentioned that the fluorescence properties of a fluorophore depend significantly on environmental factors, including solvent, pH, possible quenchers, neighbouring groups, etc.

A number of empirical rules can be applied to interpret protein fluorescence spectra:

- As a fluorophore moves into an environment with less polarity, its emission spectrum exhibits a hypsochromic shift (λ_{max} moves to shorter wavelengths) and the intensity at λ_{max} increases.
- Fluorophores in a polar environment show a decrease in quantum yield with increasing temperature. In a non-polar environment, there is little change.
- Tryptophan fluorescence is quenched by neighbouring protonated acidic groups.

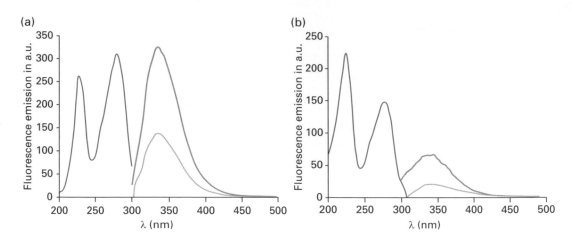

Fig. 12.10 Comparison of fluorescence excitation and emission spectra can yield insights into internal quenching. Excitation spectra with emission wavelength 340 nm are shown in dark green. Emission spectra with excitation wavelength 295 nm are shown in light green; emission spectra with excitation wavelength 280 nm are grey. (a) PDase homologue (*Escherichia coli*). (b) CPDase (*Arabidopsis thaliana*); in this protein, the fluorophores are located in close proximity to each other which leads to the effect of intrinsic quenching, as obvious from the lower intensity of the emission spectrum as compared to the excitation spectrum.

When interpreting effects observed in fluorescence experiments, one has to consider carefully all possible molecular events. For example, a compound added to a protein solution can cause quenching of tryptophan fluorescence. This could come about by binding of the compound at a site close to the tryptophan (i.e. the residue is surface-exposed to a certain degree), or due to a conformational change induced by the compound.

The comparison of protein fluorescence excitation and emission spectra can yield insights into the location of fluorophores. The close spatial arrangement of fluorophores within a protein can lead to quenching of fluorescence emission; this might be seen by the lower intensity of the emission spectrum when compared to the excitation spectrum (Fig. 12.10).

Extrinsic fluorescence

Frequently, molecules of interest for biochemical studies are non-fluorescent. In many of these cases, an external fluorophore can be introduced into the system by chemical coupling or non-covalent binding. Some examples of commonly used external fluorophores are shown in Fig. 12.11. Three criteria must be met by fluorophores in this context. Firstly, it must not affect the mechanistic properties of the system under investigation. Secondly, its fluorescence emission needs to be sensitive to environmental conditions in order to enable monitoring of the molecular events. And lastly, the fluorophore must be tightly bound at a unique location.

A common non-conjugating extrinsic chromophore for proteins is 1-anilino-8-naphthalene sulphonate (ANS) which emits only weak fluorescence in polar environment, i.e. in aqueous solution. However, in non-polar environment, e.g. when bound to hydrophobic patches on proteins, its fluorescence emission is significantly increased and the spectrum shows a hypsochromic shift; λ_{max} shifts from 475 nm to 450 nm. ANS

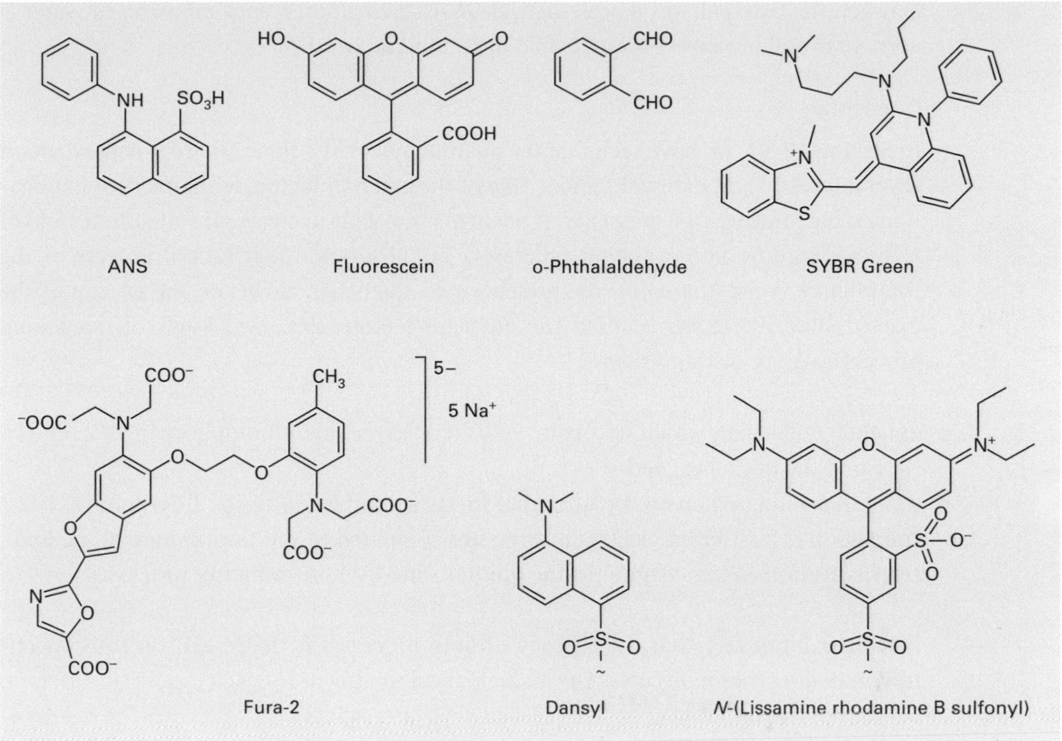

Fig. 12.11 Structures of some extrinsic fluorophores. Fura-2 is a fluorescent chelator for divalent and higher valent metal ions (Ca^{2+}, Ba^{2+}, Sr^{2+}, Pb^{2+}, La^{3+}, Mn^{2+}, Ni^{2+}, Cd^{2+}).

is thus a valuable tool for assessment of the degree of non-polarity. It can also be used in competition assays to monitor binding of ligands and prosthetic groups.

Reagents such as fluorescamine, o-phthalaldehyde or 6-aminoquinolyl N hydroxysuccinimidyl carbamate have been very popular conjugating agents used to derivatise amino acids for analysis (see Section 8.4.2). o-Phthalaldehyde, for example, is a non-fluorescent compound that reacts with primary amines and β-mercaptoethanol to yield a highly sensitive fluorophore.

Metal-chelating compounds with fluorescent properties are useful tools for a variety of assays, including monitoring of metal homeostasis in cells. Widely used probes for calcium are the chelators Fura-2, Indo-1 and Quin-1. Since the chemistry of such compounds is based on metal chelation, cross-reactivity of the probes with other metal ions is possible.

The intrinsic fluorescence of nucleic acids is very weak and the required excitation wavelengths are too far in the UV region to be useful for practical applications. Numerous extrinsic fluorescent probes spontaneously bind to DNA and display enhanced emission. While in earlier days ethidium bromide was one of the most widely used dyes for this application, it has nowadays been replaced by SYBR Green, as the latter probe poses fewer hazards for health and environment and has no teratogenic properties like ethidium bromide. These probes bind DNA by intercalation of the planar aromatic ring systems between the base pairs of double-helical DNA. Their fluorescence emission in water is very weak and increases about 30-fold upon binding to DNA.

Quenching

In Section 12.3.1, we have seen that the quantum yield of a fluorophore is dependent on several internal and external factors. One of the external factors with practical implications is the presence of a quencher. A quencher molecule decreases the quantum yield of a fluorophore by non-radiating processes. The absorption (excitation) process of the fluorophore is not altered by the presence of a quencher. However, the energy of the excited state is transferred onto the quenching molecules. Two kinds of quenching processes can be distinguished:

- dynamic quenching which occurs by collision between the fluorophore in its excited state and the quencher; and
- static quenching whereby the quencher forms a complex with the fluorophore. The complex has a different electronic structure compared to the fluorophore alone and returns from the excited state to the ground state by non-radiating processes.

It follows intuitively that the efficacy of both processes is dependent on the concentration of quencher molecules. The mathematical treatment for each process is different, because of two different chemical mechanisms. Interestingly, in both cases the degree of quenching, expressed as $I_0\,I^{-1}$, is directly proportional to the quencher concentration. For collisional (dynamic) quenching, the resulting equation has been named the Stern–Volmer equation (equation 12.4).

$$\frac{I_0}{I} - 1 = k_Q c_Q \tau_0 \tag{12.4}$$

$$\frac{I_0}{I} - 1 = K_a c_Q \tag{12.5}$$

The Stern–Volmer equation relates the degree of quenching (expressed as $I_0\,I^{-1}$) to the molar concentration of the quencher c_Q, the lifetime of the fluorophore τ_0, and the rate constant of the quenching process k_Q. In case of static quenching (equation 12.5), $I_0\,I^{-1}$ is related to the equilibrium constant K_a that describes the formation of the complex between the excited fluorophore and the quencher, and the concentration of the quencher. Importantly, a plot of $I_0\,I^{-1}$ versus c_Q yields for both quenching processes a linear graph with a y-intercept of 1.

Thus, fluorescence data obtained by intensity measurements alone cannot distinguish between static or collisional quenching. The measurement of fluorescence lifetimes or the temperature/viscosity dependence of quenching can be used to determine the kind of quenching process. It should be added, that both processes can also occur simultaneously in the same system.

The fact that static quenching is due to complex formation between the fluorophore and the quencher makes this phenomenon an attractive assay for binding of a ligand to a protein. In the simplest case, the fluorescence emission being monitored is the intrinsic fluorescence of the protein. While this is a very convenient titration assay when validated for an individual protein–ligand system, one has to be careful when testing unknown pairs, because the same decrease in intensity can occur by collisional quenching.

Highly effective quenchers for fluorescence emission are oxygen, as well as the iodide ion. Usage of these quenchers allows surface mapping of biological macromolecules. For instance, iodide can be used to determine whether tryptophan residues are exposed to solvent.

Fluorescence resonance energy transfer (FRET)

Fluorescence resonance energy transfer (FRET) was first described by Förster in 1948. The process can be explained in terms of quantum mechanics by a non-radiative energy transfer from a donor to an acceptor chromophore. The requirements for this process are a reasonable overlap of emission and excitation spectra of donor and acceptor chromophores, close spatial vicinity of both chromophores (10–100 Å), and an almost parallel arrangement of their transition dipoles. Of great practical importance is the correlation

$$\text{FRET} \propto \frac{1}{R_0^6} \qquad (12.6)$$

showing that the FRET effect is inversely proportional to the distance between donor and acceptor chromophores, R_0.

The FRET effect is particularly suitable for biological applications, since distances of 10–100 Å are in the order of the dimensions of biological macromolecules. Furthermore, the relation between FRET and the distance allows for measurement of molecular distances and makes this application a kind of 'spectroscopic ruler'. If a process exhibits changes in molecular distances, FRET can also be used to monitor the molecular mechanisms.

The high specificity of the FRET signal allows for monitoring of molecular interactions and conformational changes with high spatial (1–10 nm) and temporal resolution (< 1 ns). Especially the possibility of localising and monitoring cellular structures and proteins in physiological environments makes this method very attractive. The effects can be observed even at low concentrations (as low as single molecules), in different environments (different solvents, including living cells), and observations may be done in real time.

In most cases, different chromophores are used as donor and acceptor, presenting two possibilities to record FRET: either as donor-stimulated fluorescence emission of the acceptor or as fluorescence quenching of the donor by the acceptor. However, the same chromophore may be used as donor and acceptor simultaneously; in this case, the depolarisation of fluorescence is the observed parameter. Since non-FRET stimulated fluorescence emission by the acceptor can result in undesirable background fluorescence, a common approach is usage of non-fluorescent acceptor chromophores.

FRET-based assays may be used to elucidate the effects of new substrates for different enzymes or putative agonists in a quick and quantitative manner. Furthermore, FRET detection might be used in high-throughput screenings (see Sections 17.3.2 and 18.2.3), which makes it very attractive for drug development.

Example 3 **FRET APPLICATIONS IN DNA SEQUENCING AND INVESTIGATION OF MOLECULAR MECHANISMS**

BigDyes™ are a widely used application of FRET fluorophores (Fig. 12.12). Since 1997, these fluorophores are generally used as chain termination markers in automated DNA sequencing. As such, BigDyes™ are in major parts responsible for the great success of genome projects.

In many instances, FRET allows monitoring of conformational changes, protein folding, as well as protein–protein, protein–membrane and protein–DNA interactions. For instance, the three subunits of T4 DNA polymerase holoenzyme arrange around DNA in torus-like geometry. Using the tryptophan residue in one of the subunits as FRET donor and a coumarine label conjugated to a cysteine residue in the adjacent subunit (FRET acceptor), the distance change between both subunits could be monitored and seven steps involved in opening and closing of the polymerase could be identified. Other examples of this approach include studies of the architecture of *Escherichia coli* RNA polymerase, the calcium-dependent change of troponin and structural studies of neuropeptide Y dimers.

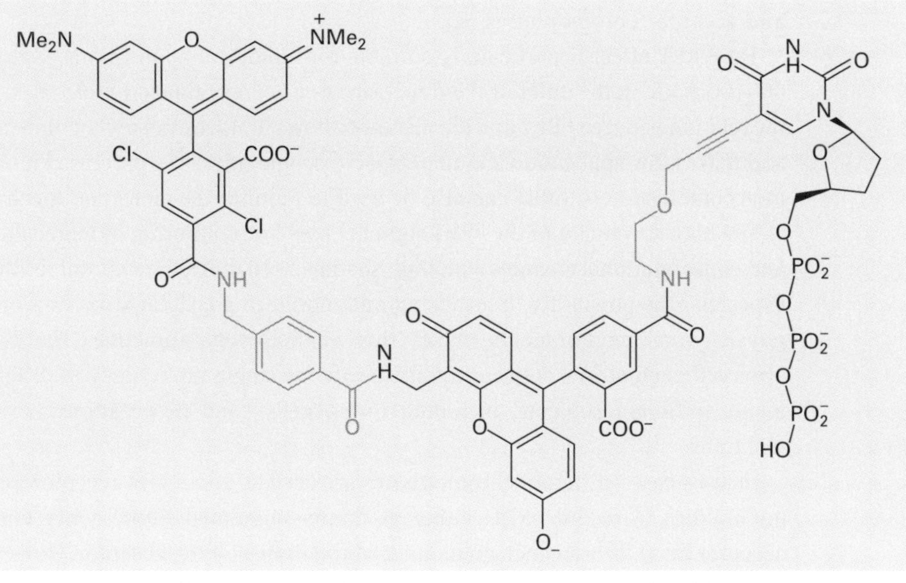

Fig. 12.12 Structure of one of the four BigDye™ terminators, ddT-EO-6CFB-dTMR. The moieties from left to right are: 5-carboxy-dichloro-rhodamine (FRET acceptor), 4-aminomethyl benzoate linker, 6-carboxy-4′-aminomethylfluorescein (FRET donor), propargyl ethoxyamino linker, dUTP.

Bioluminescence Resonance Energy Transfer (BRET)

Bioluminescence resonance energy transfer (BRET) uses the FRET effect with native fluorescent or luminescent proteins as chromophores. The phenomenon is observed naturally for example with the sea pansy *Renilla reniformis*. It contains the enzyme luciferase, which oxidises luciferin (coelenterazin) by simultaneously emitting light at λ_{exc}= 480 nm. This light directly excites green fluorescent protein (GFP), which, in turn, emits fluorescence at λ_{em}= 509 nm.

Fluorescence labelling of proteins by other proteins presents a useful approach to study various processes *in vivo*. Labelling can be done at the genetic level by generating fusion proteins. Monitoring of protein expression by GFP is an established technique and further development of 'living colours' will lead to promising new tools.

While nucleic acids have been the main players in the genomic era, the postgenomic/ proteomic era focusses on gene products, the proteins. New proteins are being discovered and characterised, others are already used within biotechnological processes. In particular for classification and evaluation of enzymes and receptors, reaction systems can be designed such that the reaction of interest is detectable quantitatively using FRET donor and acceptor pairs.

For instance, detection methods for protease activity can be developed based on BRET applications. A protease substrate is fused to a GFP variant on the N-terminal side and dsRED on the C-terminal side. The latter protein is a red fluorescing FRET acceptor and the GFP variant acts as a FRET donor. Once the substrate is cleaved by a protease, the FRET effect is abolished. This is used to directly monitor protease activity. With a combination of FRET analysis and two-photon excitation spectroscopy it is also possible to carry out a kinetic analysis.

A similar idea is used to label human insulin receptor (see Section 17.4.4) in order to quantitatively assess its activity. Insulin receptor is a glycoprotein with two α and two β subunits, which are linked by dithioether bridges. The binding of insulin induces a conformational change and causes a close spatial arrangement of both β subunits. This, in turn, activates tyrosine kinase activity of the receptor.

In pathological conditions such as diabetes, the tyrosine kinase activity is different than in healthy conditions. Evidently, it is of great interest to find compounds that stimulate the same activity as insulin. By fusing the β subunit of human insulin receptor to *Renilla reniformis* luciferase and yellow fluorescent protein (YFP) a FRET donor–acceptor pair is obtained, which reports the ligand-induced conformational change and precedes the signal transduction step. This reporter system is able to detect the effects of insulin and insulin-mimicking ligands in order to assess dose-dependent behaviour.

Fluorescence recovery after photo bleaching (FRAP)

If a fluorophore is exposed to high intensity radiation it may be irreversibly damaged and lose its ability to emit fluorescence. Intentional bleaching of a fraction of fluorescently labelled molecules in a membrane can be used to monitor the motion of labeled molecules in certain (two-dimensional) compartments. Moreover, the time-dependent monitoring allows determination of the diffusion coefficient. A well-established application is the usage of phospholipids labelled with NBD (e.g. NBD-phosphatidylethanolamine, Fig. 12.13b) which are incorporated into a biological or artificial membrane. The specimen is subjected to a pulse of high-intensity light (photo bleaching), which causes a sharp drop of fluorescence in the observation area (Fig. 12.13). Re-emergence of fluorescence emission in this area is monitored as unbleached molecules diffused into the observation area. From the time-dependent increase of fluorescence emission, the rate of diffusion of the

(a)

(b)

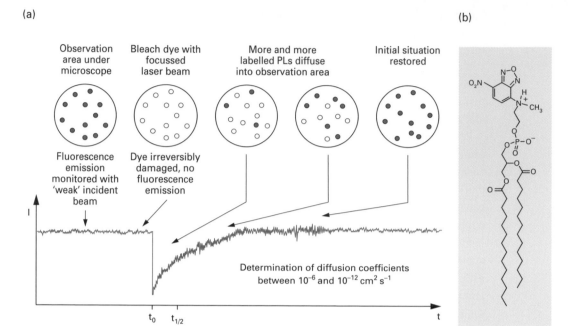

Fig. 12.13 (a) Schematic of a FRAP experiment. Time-based monitoring of fluorescence emission intensity enables determination of diffusion coefficients in membranes. (b) A commonly used fluorescence label in membrane FRAP experiments: chemical structure of phosphatidylethanolamine conjugated to the fluorophore NBD.

phospholipid molecules can be calculated. Similarly, membrane proteins such as receptors or even proteins in a cell can be conjugated to fluorescence labels and their diffusion coefficients can be determined.

Fluorescence polarisation

A light source usually consists of a collection of randomly oriented emitters, and the emitted light is a collection of waves with all possible orientations of the E vectors (non-polarised light). Linearly polarised light is obtained by passing light through a polariser that transmits light with only a single plane of polarisation; i.e. it passes only those components of the E vector that are parallel to the axis of the polariser (Fig. 12.14). The intensity of transmitted light depends on the orientation of the polariser. Maximum transmission is achieved when the plane of polarisation is parallel to the axis of the polariser; the transmission is zero when the orientation is perpendicular. The polarisation P is defined as

$$P = \frac{I_{\updownarrow} - I_{\leftrightarrow}}{I_{\updownarrow} + I_{\leftrightarrow}}$$

(12.7)

$I_{\updownarrow}$ and $I_{\leftrightarrow}$ are the intensities observed parallel and perpendicular to an arbitrary axis. The polarisation can vary between -1 and $+1$; it is zero when the light is unpolarised. Light with $0 < |P| < 0.5$ is called partially polarised.

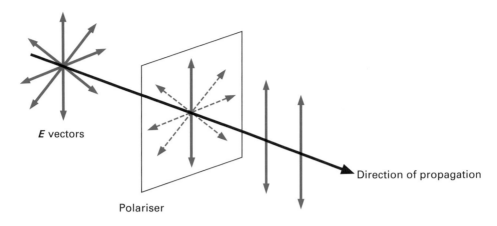

Fig. 12.14 Generation of linearly polarised light.

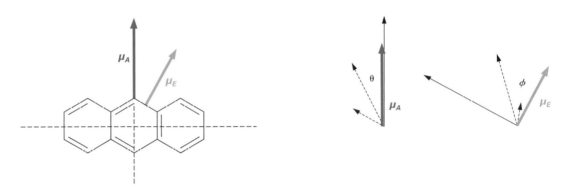

Fig. 12.15 Absorption dipole moment μ_A (describing the probability of photon absorption) and transition dipole moment μ_E (describing the probability for photon emission) for any chromophore are usually not parallel. Absorption of linearly polarised light varies with $\cos^2\theta$ and is at its maximum parallel to μ_A. Emission of linearly polarised light varies with $\sin^2\phi$ and is highest at a perpendicular orientation to μ_E.

Experimentally, this can be achieved in a fluorescence spectrometer by placing a polariser in the excitation path in order to excite the sample with polarised light. A second polariser is placed between the sample and the detector with its axis either parallel or perpendicular to the axis of the excitation polariser. The emitted light is either partially polarised or entirely unpolarised. This loss of polarisation is called fluorescence depolarisation.

Absorption of polarised light by a chromophore is highest when the plane of polarisation is parallel to the absorption dipole moment μ_A (Fig. 12.15). More generally, the probability of absorption of exciting polarised light by a chromophore is proportional to $\cos^2\theta$, with θ being the angle between the direction of polarisation and the absorption dipole moment. Fluorescence emission, in contrast, does not depend on the absorption dipole moment, but on the transition dipole moment μ_E. Usually, μ_A and μ_E are tilted against each other by about 10° to 40°. The probability of emission of polarised light at an angle ϕ with respect to the transition dipole moment is proportional to $\sin^2 \phi$, and thus at its maximum in a perpendicular orientation.

As a result if the chromophores are randomly oriented in solution, the polarisation P is less than 0.5. It is thus evident that any process that leads to a deviation from random orientation will give rise to a change of polarisation. This is certainly the case when a chromophore becomes more static. Furthermore, one needs to consider Brownian motion. If the chromophore is a small molecule in solution, it will be rotating very rapidly. Any change in this motion due to temperature changes, changes in viscosity of the solvent, or binding to a larger molecule, will therefore result in a change of polarisation.

Fluorescence cross-correlation spectroscopy

With fluorescence cross-correlation spectroscopy the temporal fluorescence fluctuations between two differently labelled molecules can be measured as they diffuse through a small sample volume. Cross-correlation analysis of the fluorescence signals from separate detection channels extracts information of the dynamics of the dual-labelled molecules. Fluorescence cross-correlation spectroscopy has thus become an essential tool for the characterisation of diffusion coefficients, binding constants, kinetic rates of binding and determining molecular interactions in solutions and cells (see also Section 17.3.2).

Fluorescence microscopy, high-throughput assays

Fluorescence emission as a means of monitoring is a valuable tool for many biological and biochemical applications. We have already seen the usage of fluorescence monitoring in DNA sequencing; the technique is inseparably tied in with the success of projects such as genome deciphering.

Fluorescence techniques are also indispensable methods for cell biological applications with fluorescence microscopy (see Sections 4.6 and 17.3.2). Proteins (or biological macromolecules) of interest can be tagged with a fluorescent label such as e.g. the green fluorescent protein (GFP) from the jelly fish *Aequorea victoria* or the red fluorescent protein from *Discosoma striata,* if spatial and temporal tracking of the tagged protein is desired. Alternatively, the use of GFP spectral variants such as cyan fluorescent protein (CFP) as a fluorescence donor and yellow fluorescent protein (YFP) as an acceptor allows investigation of mechanistic questions by using the FRET phenomenon. Specimens with cells expressing the labelled proteins are illuminated with light of the excitation wavelength, and then observed through a filter that excludes the exciting light and only transmits the fluorescence emission. The recorded fluorescence emission can be overlaid with a visual image computationally, and the composite image then allows for localisation of the labelled species. If different fluorescence labels with distinct emission wavelengths are used simultaneously, even co-localisation studies can be performed.

Time-resolved fluorescence spectroscopy

The emission of a single photon from a fluorophore follows a probability distribution. With time-correlated single photon counting, the number of emitted photons can be recorded in a time-dependent manner following a pulsed excitation of the sample.

By sampling the photon emission for a large number of excitations, the probability distribution can be constructed. The time-dependent decay of an individual fluorophore species follows an exponential distribution, and the time constant is thus termed the lifetime of this fluorophore. Curve fitting of fluorescence decays enables the identification of the number of species of fluorophores (within certain limits), and the calculation of the lifetimes for these species. In this context, different species can be different fluorophores or distinct conformations of the same fluorophore.

12.4 LUMINOMETRY

In the preceding section, we mentioned the method of bioluminescence resonance energy transfer (BRET) and its main workhorse, luciferase. Generally, fluorescence phenomena depend on the input of energy in the form of electromagnetic radiation. However, emission of electromagnetic radiation from a system can also be achieved by prior excitation in the course of a chemical or enzymatic reaction. Such processes are summarised as luminescence. Luminometry is not strictly speaking a spectrophotometric technique, but is included here due to its importance in the life sciences.

12.4.1 Principles

Luminometry is the technique used to measure luminescence, which is the emission of electromagnetic radiation in the energy range of visible light as a result of a reaction. Chemiluminescence arises from the relaxation of excited electrons transitioning back to the ground state. The prior excitation occurs through a chemical reaction that yields a fluorescent product. For instance, the reaction of luminol with oxygen produces 3-aminophthalate which possesses a fluorescence spectrum that is then observed as a chemiluminescence. In other words, the chemiluminescence spectrum is the same as the fluorescence spectrum of the product of the chemical reaction.

Bioluminescence describes the same phenomenon, only the reaction leading to a fluorescent product is an enzymatic reaction. The most commonly used enzyme in this context is certainly luciferase (see Section 15.3.2). The light is emitted by an intermediate complex of luciferase with the substrate ('photoprotein'). The colour of the light emitted depends on the source of the enzyme and varies between 560 nm (greenish yellow) and 620 nm (red) wavelengths. Bioluminescence is a highly sensitive method, due to the high quantum yield of the underlying reaction. Some luciferase systems work with almost 100% efficiency. For comparison, the incandescent light bulb loses about 90% of the input energy to heat.

Because luminescence does not depend on any optical excitation, problems with autofluorescence in assays are eliminated.

12.4.2 Instrumentation

Since no electromagnetic radiation is required as a source of energy for excitation, no light source and monochromator are required. Luminometry can be performed with a rather simple set-up, where a reaction is started in a cuvette or mixing chamber, and the resulting light is detected by a photometer. In most cases, a photomultiplier tube is needed to amplify the output signal prior to recording. Also, it is fairly important to maintain a strict temperature control, as all chemical, and especially enzymatic, reactions are sensitive to temperature.

12.4.3 Applications

Chemiluminescence

Luminol and its derivatives can undergo chemiluminescent reactions with high efficiency. For instance, enzymatically generated H_2O_2 may be detected by the emission of light at 430 nm wavelength in the presence of luminol and microperoxidase (see Section 15.3.2).

Competitive binding assays (see Section 15.2) may be used to determine low concentrations of hormones, drugs and metabolites in biological fluids. These assays depend on the ability of proteins such as antibodies and cell receptors to bind specific ligands with high affinity. Competition between labelled and unlabelled ligand for appropriate sites on the protein occurs. If the concentration of the protein, i.e. the number of available binding sites, is known, and a limited but known concentration of labelled ligand is introduced, the concentration of unlabelled ligand can be determined under saturation conditions when all sites are occupied. Exclusive use of labelled ligand allows the determination of the concentration of the protein and thus the number of available binding sites.

During the process of phagocytosis by leukocytes, molecular oxygen is produced in its singlet state (see Section 12.1.2) which exhibits chemiluminescence. The effects of pharmacological and toxicological agents on leukocytes and other phagocytic cells can be studied by monitoring this luminescence.

Bioluminescence

Firefly luciferase is mainly used to measure ATP concentrations. The bioluminescence assay is rapidly carried out with accuracies comparable to spectrophotometric and fluorimetric assays. However, with a detection limit of 10^{-15} M, and a linear range of 10^{-12} to 10^{-6} M ATP, the luciferase assay is vastly superior in terms of sensitivity. Generally, all enzymes and metabolites involved in ATP interconversion reactions may be assayed in this method, including ADP, AMP, cyclic AMP and the enzymes pyruvate kinase, adenylate kinase, phosphodiesterase, creatine kinase, hexokinase and ATP sulphurase (see Section 15.3.2). Other substrates include creatine phosphate, glucose, GTP, phosphoenolpyruvate and 1,3-diphosphoglycerate.

The main application of bacterial luciferase is the determination of electron transfer co-factors, such as nicotine adenine dinucleotides (and phosphates) and flavin mononucleotides in their reduced states, for example NADH, NADPH and $FMNH_2$. Similar to the firefly luciferase assays, this method can be applied to a whole range of coupled RedOx enzyme reaction systems. The enzymatic assays are again much more sensitive

Example 4 **ENZYMATIC CALCIUM MONITORING**

Question Calcium signalling is a common mechanism, since the ion, once it enters the cytoplasm, exerts allosteric regulatory affects on many enzymes and proteins. How can intracellular calcium be monitored?

Answer The EF-hand protein aequorin from *Aequorea* species (jellyfish) has been used for determination of intracellular calcium concentrations. Despite the availability of calcium-specific electrodes, this bioluminescence assay presents advantages due to its high sensitivity to and specificity for calcium. Since the protein is non-toxic, has a low leakage rate from cells and is not intracellularly compartmentalised, it is ideally suited for usage in living cells. Its disadvantages are the scarcity, large molecular size, consumption during the reaction and the non-linearity of the light emission relative to calcium concentration. The reaction is further sensitive to the chemical environment and the limited speed in which it can respond to rapid changes in calcium concentration, for example influx and efflux in certain cell types. The protein possesses a reflective yellow colour and is non-fluorescent in its apo- (non-calcium-bound) state. In the calcium-bound form, the prosthetic group coelenterazine, a molecule belonging to the luciferin family, is oxidised to coelenteramide and CO_2. Upon relaxation to the ground state, blue light of 469 nm wavelength is emitted.

than the corresponding spectrophotometric and fluorimetric assays, and a concentration range of 10^{-9} to 10^{-12} M can be achieved. The NADPH assay is by a factor of 20 less sensitive than the NADH assay.

12.5 CIRCULAR DICHROISM SPECTROSCOPY

12.5.1 Principles

In Section 12.3.3 we have already seen that electromagnetic radiation oscillates in all possible directions and that it is possible to preferentially select waves oscillating in a single plane, as applied for fluorescence polarisation. The phenomenon first known as mutarotation (described by Lowry in 1898) became manifest in due course as a special property of optically active isomers allowing the rotation of plane-polarised light. Optically active isomers are compounds of identical chemical composition and topology, but whose mirror images cannot be superimposed; such compounds are called chiral.

Linearly and circularly polarised light

Light is electromagnetic radiation where the electric vector (E) and the magnetic vector (M) are perpendicular to each other. Each vector undergoes an oscillation as the light

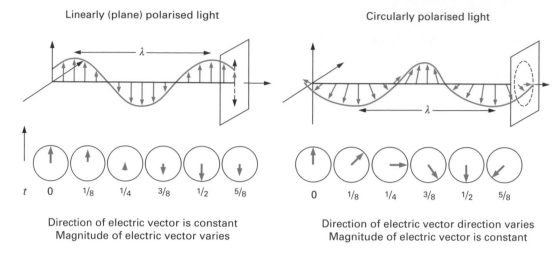

Fig. 12.16 Linearly (plane) and circularly polarised light.

travels along the direction of propagation, resulting in a sine-like waveform of each, the E and the M vectors. A light source usually consists of a collection of randomly oriented emitters. Therefore, the emitted light is a collection of waves with all possible orientations of the E vectors. This light is non-polarised. Linearly or plane-polarised light is obtained by passing light through a polariser that transmits light with only a single plane of polarisation, i.e. it passes only those components of the E vector that are parallel to the axis of the polariser (Fig. 12.14). If the E vectors of two electromagnetic waves are ¼ wavelength out of phase and perpendicular to each other, the vector that is the sum of the E vectors of the two components rotates around the direction of propagation so that its tip follows a helical path. Such light is called circularly polarised (Fig. 12.16).

While the E vector of circularly polarised light always has the same magnitude but a varying direction, the direction of the E vector of linearly polarised light is constant; it is its magnitude that varies. With the help of vector algebra, one can now reversely think of linearly polarised light as a composite of two circularly polarised beams with opposite handedness (Fig. 12.17a).

Polarimetry and optical rotation dispersion

Polarimetry essentially measures the angle through which the plane of polarisation is changed after linearly polarised light is passed through a solution containing a chiral substance. Optical rotation dispersion (ORD) spectroscopy is a technique that measures this ability of a chiral substance to change the plane-polarisation as a function of the wavelength. The angle α_λ between the plane of the resulting linearly polarised light against that of the incident light is dependent on the refractive index for left (n_{left}) and right (n_{right}) circularly polarised light. The refractive index can be calculated as the ratio of the speed of light *in vacuo* and the speed of light in matter. After normalisation against the amount of substance present in the sample (thickness of sample/cuvette length d, and mass concentration ρ^*), a substance-specific constant $[\alpha]_\lambda$ is obtained that can be used to characterise chiral compounds.

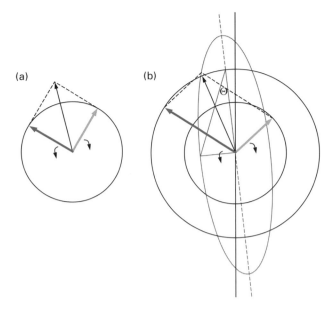

Fig. 12.17 (a) Linearly polarised light can be thought of consisting of two circularly polarised components with opposite 'handedness'. The vector sum of the left- and right-handed circularly polarised light yields linearly polarised light. (b) If the amplitudes of left- and right-handed polarised components differ, the resulting light is elliptically polarised. The composite vector will trace the ellipse shown in grey. The ellipse is characterised by a major and a minor axis. The ratio of minor and major axis yields tan Θ. Θ is the ellipticity.

Circular dichroism

In addition to changing the plane of polarisation, an optically active sample also shows unusual absorption behaviour. Left- and right-handed polarised components of the incident light are absorbed differently by the sample, which yields a difference in the absorption coefficients $\Delta\varepsilon = \varepsilon_{\text{left}} - \varepsilon_{\text{right}}$. This latter difference is called circular dichroism (CD). The difference in absorption coefficients $\Delta\varepsilon$ (i.e. CD) is measured in units of $\text{cm}^2\,\text{g}^{-1}$, and is the observed quantity in CD experiments. Historically, results from CD experiments are reported as ellipticity Θ_λ. Normalisation of Θ_λ similar to the ORD yields the molar ellipticity:

$$\theta_\lambda = \frac{M \times \Theta_\lambda}{10 \times \rho^* \times d} = \frac{\ln 10}{10} \times \frac{180°}{2\pi} \times \Delta\varepsilon \qquad (12.8)$$

It is common practice to display graphs of CD spectra with the molar ellipticity in units of $1°\,\text{cm}^2\,\text{dmol}^{-1} = 10°\,\text{cm}^2\,\text{mol}^{-1}$ on the ordinate axis (Fig. 12.18).

Three important conclusions can be drawn:

- ORD and CD are the manifestation of the same underlying phenomenon;
- if an optically active molecule has a positive CD, then its enantiomer will have a negative CD of exactly the same magnitude; and
- the phenomenon of CD can only be observed at wavelengths where the optically active molecule has an absorption band.

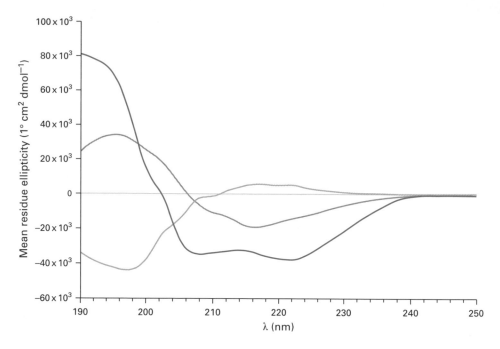

Fig. 12.18 Circular dichroism spectra for three standard secondary structures according to Fasman. An α-helical peptide is shown in dark green, a peptide adopting β-strand structure in grey, and a random coil peptide in light green.

The chromophores of protein secondary structure

In Section 12.2, we saw that the peptide bond in proteins possesses UV absorption bands in the area of 220–190 nm. The carbon atom vicinal to the peptide bond (the C_α atom) is asymmetric and a chiral centre in all amino acids except glycine. This chirality induces asymmetry into the peptide bond chromophore. Because of the serial arrangement of the peptide bonds making up the backbone of a protein, the individual chromophores couple with each other. The (secondary) structure of a polypeptide thus induces an 'overall chirality' which gives rise to the CD phenomenon of a protein in the wavelength interval 260–190 nm.

With protein circular dichroism, the molar ellipticity θ also appears as mean residue ellipticity θ_{res}, owing to the fact that the chromophores responsible for the chiral absorption phenomenon are the peptide bonds. Therefore, the number of chromophores of a polypeptide in this context is equal to the number of residues. Because of the law of Beer–Lambert (Equation 12.2), the number of chromophores is proportional to the magnitude of absorption, i.e. in order to normalise the spectrum of an individual polypeptide for reasons of comparison, the CD has to be scaled by the number of peptide bonds.

12.5.2 Instrumentation

The basic layout of a CD spectrometer follows that of a single-beam UV absorption spectrometer. Owing to the nature of the measured effects, an electro-optic modulator, as well as a more sophisticated detector are needed, though.

Generally, left and right circularly polarised light passes through the sample in an alternating fashion. This is achieved by an electro-optic modulator which is a crystal that transmits either the left- or right-handed polarised component of linearly polarised light, depending on the polarity of the electric field that is applied by alternating currents. The photomultiplier detector produces a voltage proportional to the ellipticity of the resultant beam emerging from the sample. The light source of the spectrometer is continuously flushed with nitrogen to avoid the formation of ozone and help to maintain the lamp.

CD spectrometry involves measuring a very small difference between two absorption values which are large signals. The technique is thus very susceptible to noise and measurements must be carried out carefully. Some practical considerations involve having a clean quartz cuvette, and using buffers with low concentrations of additives. While this is sometimes tricky with protein samples, reducing the salt concentrations to values as low as 5 mM helps to obtain good spectra. Also, filtered solutions should be used to avoid any turbidity of the sample that could produce scatter. Saturation of the detector must be avoided, this becoming more critical with lower wavelengths. Therefore, good spectra are obtained in a certain range of protein concentrations only where enough sample is present to produce a good signal and does not saturate the detector. Typical protein concentrations are 0.03–0.3 mg cm^{-3}.

In order to calculate specific ellipticities (mean residue ellipticities) and be able to compare the CD spectra of different samples with each other, the concentration of the sample must be known. Provided the protein possesses sufficient amounts of UV/Vis-absorbing chromophores, it is thus advisable to subject the CD sample to a protein concentration determination by UV/Vis as described in Section 12.2.3.

12.5.3 Applications

The main application for protein CD spectroscopy is the verification of the adopted secondary structure. The application of CD to determine the tertiary structure is limited, owing to the inadequate theoretical understanding of the effects of different parts of the molecules at this level of structure.

Rather than analysing the secondary structure of a 'static sample', different conditions can be tested. For instance, some peptides adopt different secondary structures when in solution or membrane-bound. The comparison of CD spectra of such peptides in the absence and presence of small unilamellar phospholipid vesicles shows a clear difference in the type of secondary structure. Measurements with lipid vesicles are tricky, because due to their physical extensions they give rise to scatter. Other options in this context include CD experiments at lipid monolayers which can be realised at synchrotron beam lines, or by usage of optically clear vesicles (reverse micelles).

CD spectroscopy can also be used to monitor changes of secondary structure within a sample over time. Frequently, CD instruments are equipped with temperature control units and the sample can be heated in a controlled fashion. As the protein undergoes its transition from the folded to the unfolded state, the CD at a certain wavelength (usually 222 nm) is monitored and plotted against the temperature, thus yielding a thermal denaturation curve which can be used for stability analysis.

Example 5 **DETERMINATION OF THE SECONDARY STRUCTURE CONTENT OF A PROTEIN SAMPLE**

Question You have purified a recombinant protein and wonder whether it adopts a folded structure. How might you address this problem?

Answer CD spectra of poly-L-amino acids as well as proteins with known three-dimensional structure have been obtained and are used as standards for deducing the secondary structure composition of unknown proteins. The simplest approach is a visual comparison of the shape of the CD spectrum with the three 'Fasman standard spectra' (Fig. 12.18), allowing conclusions as to α helix, β strand and random coil structure.

CD deconvolution is a curve-fitting process where the experimental CD spectrum is fitted with a given set of basis spectra using a weighting scheme. The estimated weighting coefficients determined in the fitting process reveal the percentage of each form of secondary structure in proteins. Different algorithms for deconvolution have been generated, ranging from a simple linear combination of three or five basis sets, to fitting procedures using 10–30 basis spectra and different mathematical algorithms (CONTIN fit, neural networks, etc.).

Further applications include the use of circular dichroism for an observable for kinetic measurements using the stopped flow technique (see Section 15.3).

12.6 LIGHT SCATTERING

The scattering of light can yield a number of valuable insights into the properties of macromolecules, including the molecular mass, dimensions and diffusion coefficients, as well as association/dissociation properties and internal dynamics. The incident light hitting a macromolecule is scattered into all directions with the intensity of the scatter being only about 10^{-5} of the original intensity. The scattered light is measured at angles higher than 0° and less than 180°. Most of the scattered light possesses the same wavelength as the incident light; this phenomenon is called elastic light scattering. When the scattered light has a wavelength higher or lower than the incident light, the phenomenon is called inelastic light scattering. The special properties of lasers (see Section 12.1.3) with high monochromaticity, narrow focus and strong intensity, make them ideally suited for light scattering applications.

12.6.1 Elastic (static) light scattering

Elastic light scattering is also known as Rayleigh scattering and involves measuring the intensity of light scattered by a solution at an angle relative to the incident laser beam. The scattering intensity of macromolecules is proportional to the squared molecular mass, and thus ideal for determination of M, since the contribution of small solvent molecules can be neglected. In an ideal solution, the macromolecules are entirely independent from each other, and the light scattering can be described as:

$$\frac{I_\theta}{I_0} \sim R_\theta = P_\theta \times K \times c \times M \qquad (12.9)$$

where I_θ is the intensity of the scattered light at angle θ, I_0 is the intensity of the incident light, K is a constant proportional to the squared refractive index increment, c is the concentration and R_θ the Rayleigh ratio. P_θ describes the angular dependence of the scattered light.

For non-ideal solutions, interactions between molecules need to be considered. The scattering intensity of real solutions has been calculated by Debye and takes into account concentration fluctuations. This results in an additional correction term comprising the second virial coefficient B which is a measure for the strength of interactions between molecules:

$$\frac{Kc}{R_\theta} = \frac{1}{P_\theta}\left(\frac{1}{M} + 2Bc\right) \qquad (12.10)$$

Determination of molecular mass with multi-angle light scattering

In solution, there are only three methods for absolute determination of molecular mass: membrane osmometry, sedimentation equilibrium centrifugation and light scattering. These methods are absolute, because they do not require any reference to molecular mass standards. In order to determine the molecular mass from light scattering, three parameters must be measured: the intensity of scattered light at different angles, the concentration of the macromolecule and the specific refractive index increment of the solvent. As minimum instrumentation, this requires a light source, a multi-angle light scattering (MALS) detector, as well as a refractive index detector. These instruments can be used in batch mode, but can also be connected to an HPLC to enable online determination of the molecular mass of eluting macromolecules. The chromatography of choice is size-exclusion chromatography (SEC), also called gel filtration (see Section 11.7), and the combination of these methods is known as SEC–MALS. Unlike conventional size-exclusion chromatography, the molecular mass determination from MALS is independent of the elution volume of the macromolecule. This is a valuable advantage, since the retention time of a macromolecule on the size-exclusion column can depend on its shape and conformation.

12.6.2 Quasi-elastic (dynamic) light scattering – photon correlation spectroscopy

While intensity and angular distribution of scattered light yields information about molecular mass and dimension of macromolecules, the wavelength analysis of scattered light allows conclusions as to the transport properties of macromolecules. Due to rotation and translation, macromolecules move into and out of a very small region in the solution. This Brownian motion happens at a timescale of microseconds to milliseconds, and the translation component of this motion is a direct result of diffusion, which leads to a broader wavelength distribution of the scattered light compared to the incident light. This analysis is the subject of dynamic light scattering, and yields the distribution of diffusion coefficients of macromolecules in solution.

The diffusion coefficient is related to the particle size by an equation known as the Stokes–Einstein relation. The parameter derived is the hydrodynamic radius, or Stokes radius, which is the size of a spherical particle that would have the same diffusion coefficient in a solution with the same viscosity. Most commonly, data from dynamic light scattering are presented as a distribution of hydrodynamic radius rather than wavelength of scattered light.

Notably, the hydrodynamic radius describes an idealised particle and can differ significantly from the true physical size of a macromolecule. This is certainly true for most proteins which are not strictly spherical and their hydrodynamic radius thus depends on their shape and conformation.

In contrast to size exclusion chromatography, dynamic light scattering measures the hydrodynamic radius directly and accurately, as the former method relies on comparison with standard molecules and several assumptions.

Applications of dynamic light scattering include determination of diffusion coefficients and assessment of protein aggregation, and can aid many areas *in praxi*. For instance, the development of 'stealth' drugs that can hide from the immune system or certain receptors relies on the PEGylation of molecules. Since conjugation with PEG (polyethylene glycol) increases the hydrodynamic size of the drug molecules dramatically, dynamic light scattering can be used for product control and as a measure of efficiency of the drug.

12.6.3 Inelastic light scattering – Raman spectroscopy

When the incident light beam hits a molecule in its ground state, there is a low probability that the molecule is excited and occupies the next higher vibrational state (Figs. 12.3, 12.8). The energy needed for the excitation is a defined increment which will be missing from the energy of the scattered light. The wavelength of the scattered light is thus increased by an amount associated with the difference between two vibrational states of the molecule (Stokes shift). Similarly, if the molecule is hit by the incident light in its excited state and transitions to the next lower vibrational state, the scattered light has higher energy than the incident light which results in a shift to lower wavelengths (anti-Stokes shift). These lines constitute the Raman spectrum. If the wavelength of the incident light is chosen such that it coincides with an absorption band of an electronic transition in the molecule, there is a significant increase in the intensity of bands in the Raman spectrum. This technique is called resonance Raman spectroscopy (see Section 13.2).

12.7 ATOMIC SPECTROSCOPY

So far, all methods have dealt with probing molecular properties. In Section 12.1.2, we discussed the general theory of electronic transitions and said that molecules give rise to band spectra, but atoms yield clearly defined line spectra. In atomic emission spectroscopy (AES), these lines can be observed as light of a particular wavelength (colour). Conversely, black lines can be observed against a bright background in atomic absorption spectroscopy (AAS). The wavelengths emitted from excited atoms may be identified using a spectroscope with the human eye as the 'detector' or a spectrophotometer.

12.7.1 Principles

In a spectrum of an element, the absorption or emission wavelengths are associated with transitions that require a minimum of energy change. In order for energy changes to be minimal, transitions tend to occur between orbitals close together in energy terms. For example, excitation of a sodium atom and its subsequent relaxation gives rise to emission of orange light ('D-line') due to the transition of an electron from the $3s$ to the $3p$ orbital and return (Fig. 12.19).

Electron transitions in an atom are limited by the availability of empty orbitals. Filling orbitals with electrons is subject to two major rules:

- one orbital can be occupied with a maximum of two electrons; and
- the spins of electrons in one orbital need to be paired in an antiparallel fashion (Pauli principle).

Together, these limitations mean that emission and absorption lines are characteristic for an individual element.

12.7.2 Instrumentation

In general, atomic spectroscopy is not carried out in solution. In order for atoms to emit or absorb monochromatic radiation, they need to be volatilised by exposing them to high thermal energy. Usually, nebulisers are used to spray the sample solution into a flame or an oven. Alternatively, the gaseous form can be generated by using inductively coupled plasma (ICP). The variations in temperature and composition of a flame make standard conditions difficult to achieve. Most modern instruments thus use an ICP.

Atomic emission spectroscopy (AES) and atomic absorption spectroscopy (AAS) are generally used to identify specific elements present in the sample and to determine their concentrations. The energy absorbed or emitted is proportional to the number of atoms in the optical path. Strictly speaking, in the case of emission, it is the number of excited atoms that is proportional to the emitted energy. Concentration determination with AES or AAS is carried out by comparison with calibration standards.

Sodium gives high backgrounds and is usually measured first. Then, a similar amount of sodium is added to all other standards. Excess hydrochloric acid is commonly added, because chloride compounds are often the most volatile salts. Calcium and magnesium emission can be enhanced by the addition of alkali metals and suppressed by addition of phosphate, silicate and aluminate, as these form non-dissociable salts. The suppression effect can be relieved by the addition of lanthanum and strontium salts. Lithium is frequently used as an internal standard. For storage of

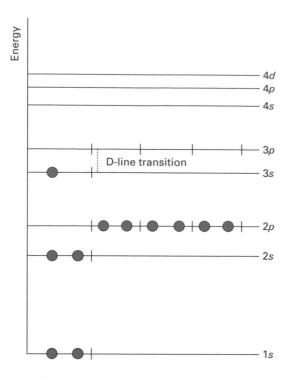

Fig. 12.19 Energy levels of atomic orbitals in the sodium atom. Each atomic orbital can be occupied by electrons following the rules of quantum chemistry until the total number of electrons for that element is reached (in case of sodium: 11 electrons). The energy gap between the $3s$ and the $3p$ orbitals in the sodium atom is such that it can be overcome by absorption of orange light.

samples and standards, polyethylene bottles are used, since glass can absorb and release metal ions, and thus impact the accuracy of this sensitive technique.

Cyclic analysis may be performed that involves the estimation of each interfering substance in a mixture. Subsequently, the standards for each component in the mixture are doped with each interfering substance. This process is repeated two or three times with refined estimates of interfering substance, until self-consistent values are obtained for each component.

Flame instability requires experimental protocols where determination of an unknown sample is bracketed by measurements of the appropriate standard, in order to achieve the highest possible accuracy.

Biological samples are usually converted to ash prior to determination of metals. Wet ashing in solution is often used, employing an oxidative digestion similar to the Kjeldahl method (see Section 8.3.2).

12.7.3 Applications

Atomic emission and atomic absorption spectrophotometry

Sodium and potassium are assayed at concentrations of a few p.p.m. using simple filter photometers. The modern emission spectrophotometers allow determination of about

20 elements in biological samples, the most common being calcium, magnesium and manganese. Absorption spectrophotometers are usually more sensitive than emission instruments and can detect less than 1 p.p.m. of each of the common elements with the exception of alkali metals. The relative precision is about 1% in a working range of 20–200 times the detection limit of an element.

AES and AAS have been widely used in analytical chemistry, such as environmental and clinical laboratories. Nowadays, the technique has been superseded largely by the use of ion-selective electrodes (see Section 16.2.2).

Atomic fluorescence spectrophotometry

Despite being limited to only a few metals, the main importance of atomic fluorescence spectrophotometry (AFS) lies in the extreme sensitivity. For example, zinc and cadmium can be detected at levels as low as 1–2 parts per 10^{10}.

AFS uses the same basic setup as AES and AAS. The atoms are required to be vaporised by one of three methods (flame, electric, ICP). The atoms are excited using electromagnetic radiation by directing a light beam into the vaporised sample. This beam must be intense, but not spectrally pure, since only the resonant wavelengths will be absorbed, leading to fluorescence (see Section 12.3.1).

12.8   SUGGESTIONS FOR FURTHER READING

General biophysics
Hoppe, W., Lohmann, W., Markl, H. and Ziegler, H. (1982). *Biophysik*, 2nd edn. Berlin: Springer-Verlag. (A rich and authorative compendium of the physical basics of the life sciences.)

WEBSITES
http://lectureonline.cl.msu.edu/%7Emmp/applist/Spectrum/s.htm
http://www.colorado.edu/physics/2000/lasers/index.html

Ultraviolet and visible light spectroscopy
Simonian, M. H. and Smith, J. A. (2006). Spectrophotometric and colorimetric determination of protein concentration. *Current Protocols in Molecular Biology*, Chapter 10, Unit 10.1A. New York: Wiley Interscience

WEBSITES
http://teaching.shu.ac.uk/hwb/chemistry/tutorials/molspec/uvvisab1.htm
http://www.cem.msu.edu/~reusch/VirtualText/Spectrpy/UV-Vis/spectrum.htm#uv1
http://www.sis.dl.ac.uk/VUV/
http://phys.educ.ksu.edu/vqm/html/absorption.html

Fluorescence spectroscopy
Brown, M. P. and Royer, C. (1997). Fluorescence spectroscopy as a tool to investigate protein interactions. *Current Opinion in Biotechnology*, 8, 45–49.
Groemping, Y. and Hellmann, N. (2005). Spectroscopic methods for the determination of protein interactions. *Current Protocols in Protein Science*, Chapter 20, Unit 20.8. New York: Wiley Interscience.
Hwang, L. C. and Wohland, T. (2007). Recent advances in fluorescence cross-correlation spectroscopy. *Cell Biochemistry and Biophysics*, 49, 1–13.

Lakowicz, J. R. (1999). *Principles of Fluorescence Spectroscopy* 2nd edn. New York: Kluwer/Plenum. (An authorative textbook on fluorescence spectroscopy.)

Langowski, J. (2008). Protein–protein interactions determined by fluorescence correlation spectroscopy. *Methods in Cell Biology*, **85**, 471–484.

Prinz, A., Reither, G., Diskar, M. and Schultz, C. (2008). Fluorescence and bioluminescence procedures for functional proteomics. *Proteomics*, **8**, 1179–1196.

Roy, R., Hohng, S. and Ha, T. (2008). A practical guide to single-molecule FRET. *Nature Methods*, **5**, 507–516.

VanEngelenburg, S. B. and Palmer, A. E. (2008). Fluorescent biosensors of protein function. *Current Opinion in Chemical Biology*, **12**, 60–65.

WEBSITES

http://www.invitrogen.com/site/us/en/home/support/Tutorials.html

http://www.microscopyu.com/tutorials/java/fluorescence/excitationbalancer/index.html

Luminometry

Bacart, J., Corbel, C., Jockers, R., Bach, S. and Couturier, C. (2008). The BRET technology and its application to screening assays. *Biotechnology Journal*, **3**, 311–324.

Deshpande, S. S. (2001). Principles and applications of luminescence spectroscopy. *Critical Reviews in Food Science Nutrition*, **41**, 155–224.

Jia, Y., Quinn, C. M., Kwak, S. and Talanian, R. V. (2008). Current in vitro kinase assay technologies: the quest for a universal format. *Current Drug Discovery Technology*, **5**, 59–69.

Meaney, M. S. and McGuffin, V. L. (2008). Luminescence-based methods for sensing and detection of explosives. *Analytical and Bioanalytical Chemistry*, **391**, 2557–2576.

WEBSITES

http://www.turnerbiosystems.com/doc/appnotes/998_2620.html

Circular dichroism spectroscopy

Fasman, G. D. (1996). *Circular Dichroism and the Conformational Analysis of Biomolecules.* New York: Plenum Press. (An authorative textbook on circular dichroism in biochemistry.)

Gottarelli, G., Lena, S., Masiero, S., Pieraccini, S. and Spada, G. P. (2008). The use of circular dichroism spectroscopy for studying the chiral molecular self-assembly: an overview. *Chirality*, **20**, 471–485.

Kelly, S. M. and Price, N. C. (2006). Circular dichroism to study protein interactions. *Current Protocols in Protein Science*, Chapter 20, Unit 20.10. New York: Wiley Interscience.

Martin, S. R. and Schilstra, M. J. (2008). Circular dichroism and its application to the study of biomolecules. *Methods in Cell Biology*, **84**, 263–293.

WEBSITES

http://www.cryst.bbk.ac.uk/cdweb/html/

http://www.ap-lab.com/circular_dichroism.htm

Light scattering

Lindner, P. and Zemb, T. (2002). *Neutron, X-rays and Light. Scattering Methods Applied to Soft Condensed Matter*, rev. edn. Amsterdam: North-Holland. (In-depth coverage of theory and applications of light scattering at expert level.)

Villari, V. and Micali, N. (2008). Light scattering as spectroscopic tool for the study of disperse systems useful in pharmaceutical sciences. *Journal of Pharmaceutical Science*, **97**, 1703–1730.

WEBSITES

http://www.ap-lab.com/light_scattering.htm

http://www.people.vcu.edu/~ecarpenter2/Tutorial.html

Atomic spectroscopy

L'vov, B. V. (2005). Fifty years of atomic absorption spectrometry. *Journal of Analytical Chemistry*, **60**, 382–392.

Zybin, A., Koch, J., Wizemann, H. D., Franzke, J. and Niemax, K. (2005). Diode laser atomic absorption spectrometry. *Spectrochimica Acta*, **B60**, 1–11.

WEBSITES

http://www.colorado.edu/physics/2000/quantumzone/index.html
http://zebu.uoregon.edu/nsf/emit.html

13 Spectroscopic techniques: II Structure and interactions

A. HOFMANN

13.1 INTRODUCTION

The overarching theme of techniques such as mass spectrometry (Chapter 9), electron microscopy and imaging (Chapter 4), analytical centrifugation (Chapter 3) and molecular exclusion chromatography (Chapter 11) is the aim to obtain clues about the structure of biomolecules and larger assemblies thereof. The spectroscopic techniques discussed in Chapters 12 and 13 are further complementary methods, and by assembling the jigsaw of pieces of information, one can gain a comprehensive picture of the structure of the biological object under study. In addition, the spectroscopic principles established in Chapter 12 are often employed as read-out in a huge variety of biochemical assays, and several more sophisticated technologies employ these basic principles in a 'hidden' way.

In the previous chapter, we established that the electromagnetic spectrum is a continuum of frequencies from the long wavelength region of the radio frequencies to the high-energy γ-rays of nuclear origin. While the methods and techniques discussed in Chapter 12 concentrated on the use of visible and UV light, there are other spectroscopic techniques that employ electromagnetic radiation of higher as well as lower energy. Another shared property of the techniques in this chapter is the higher level of complexity in undertaking. These applications are usually employed at a later stage of biochemical characterisation and aimed more at investigation of the three-dimensional structure, and in the case of proteins and peptides, address the tertiary and quaternary structure.

13.2   INFRARED AND RAMAN SPECTROSCOPY

13.2.1   Principles

Within the electromagnetic spectrum (Fig. 12.1), the energy range below the UV/Vis is the infrared region, encompassing the wavelength range of about 700 nm to 25 μm, and thus reaching from the red end of the visible to the microwave region. The absorption of infrared light by a molecule results in transition to higher levels of vibration (Fig. 12.3).

For the purpose of this discussion, the bonds between atoms can be considered as flexible springs, illustrating the constant vibrational motion within a molecule (Fig. 13.1). Bond vibrations can thus be either **stretching** or **bending** (deformation) actions. Theory predicts that a molecule with n atoms will have a total of $3n - 6$ fundamental vibrations ($3n - 5$, if the molecule is linear): $2n - 5$ bending, and $n - 1$ stretching modes (Fig. 13.2).

Infrared and Raman spectroscopy give similar information about a molecule, but the criteria for the phenomena to occur are different for each type. For asymmetric molecules, incident infrared light will give rise to an absorption band in the infrared spectrum, as well as a peak in the Raman spectrum. However, as shown in Fig. 13.2, symmetric molecules, such as for example CO_2, that possess a centre of symmetry show a selective behaviour: bands that appear in the infrared spectrum do not appear in the Raman spectrum, and vice versa.

An **infrared spectrum** arises from the fact that a molecule absorbs incident light of a certain wavelength which will then be 'missing' from the transmitted light. The recorded spectrum will show an absorption band.

A **Raman spectrum** arises from the analysis of scattered light, and we have already introduced the basics of inelastic light scattering in Section 12.6.3. The largest part of an incident light beam passes through the sample (transmission). A small part is scattered isotropically, i.e. uniformly in all directions (**Rayleigh scatter**), and possesses the same wavelength as the incident beam. The Raman spectrum arises from the fact

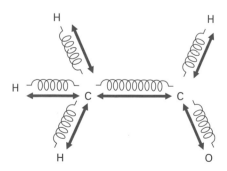

Fig. 13.1 Possible stretching vibrations in acetaldehyde.

Mode		Wavenumber	IR	Raman
Stretching, symmetric	$O{=}{=}C{=}{=}O$	1340 cm^{-1}	–	+
Stretching, asymmetric	$O{=}{=}C{=}{=}O$	2349 cm^{-1}	+	–
Deformation	$O{=}{=}C{=}{=}O$	667 cm^{-1}	+	–
Deformation	$O{=}{=}C{=}{=}O$	667 cm^{-1}	+	–

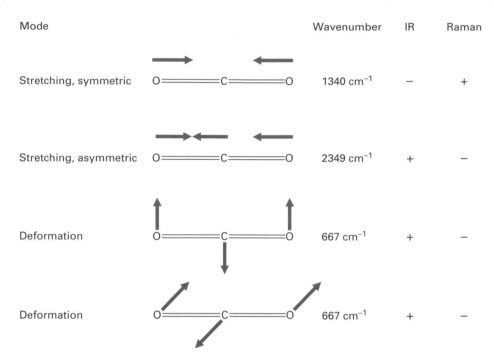

Fig. 13.2 Normal vibrational modes for CO_2. For symmetric molecules that possess a centre of symmetry, bands that appear in the IR do not appear in the Raman spectrum.

that a very small proportion of light scattered by the sample will have a different frequency than the incident light. As different vibrational states are excited, energy portions will be missing, thus giving rise to peaks at lower frequencies than the incident light (Stokes lines). Notably, higher frequencies are also observed (anti-Stokes lines); these arise from excited molecules returning to ground state. The emitted energy is dumped onto the incident light which results in scattered light of higher energy than the incident light.

The criterion for a band to appear in the infrared spectrum is that the transition to the excited state is accompanied by a change in dipole moment, i.e. a change in charge displacement. Conversely, the criterion for a peak to appear in the Raman spectrum is a change in polarisability of the molecule during the transition.

Infrared spectroscopy

The fundamental frequencies observed are characteristic of the functional groups concerned, hence the term fingerprint. Figure 13.3 shows the major bands of an FT–IR spectrum of the drug phenacetin. As the number of functional groups increases in more complex molecules, the absorption bands become more difficult to assign. However, groups of certain bands regularly appear near the same wavelength and may be assigned to specific functional groups. Such group frequencies are thus extremely helpful in structural diagnosis. A more detailed analysis of the structure of a molecule is possible, because the wavenumber associated with a particular functional group varies slightly, owing to the influence of the molecular environment.

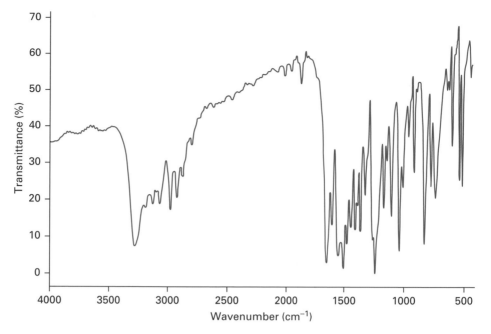

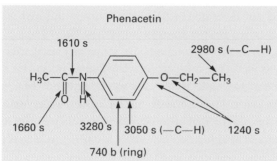

Fig. 13.3 FT–IR spectrum of phenacetin, the historically first synthetic fever reducer to go on the market. Bands at the appropriate wavenumbers (cm^{-1}) are shown, indicating the bonds with which they are associated, and the type (s, stretching; b, bending).

For example, it is possible to distinguish between C–H vibrations in methylene (–CH$_2$–) and methyl groups (–CH$_3$).

Raman spectroscopy

The assignment of peaks in Raman spectra usually requires consideration of peak position, intensity and form, as well as depolarisation. This allows identification of the type of symmetry of individual vibrations, but not the determination of structural elements of a molecule. The depolarisation is calculated as the ratio of two intensities with perpendicular and parallel polarisation with respect to the incident beam. The use of lasers as light source for Raman spectroscopy easily facilitates the use of linearly polarised light. Practically, the Raman spectrum is measured twice. In the second measurement, the polarisation plane of the incident beam is rotated by 90°.

13.2.2 Instrumentation

The most common source for infrared light is white-glowing zircon oxide or the so-called globar made of silicium carbide with a glowing temperature of 1500 K. The beam of infrared light passes a monochromator and splits into two separate beams: one runs through the sample, the other through a reference made of the substance the sample is prepared in. After passing through a splitter alternating between both beams, they are reflected into the detector. The reference is used to compensate for fluctuations in the source, as well as to cancel possible effects of the solvent. Samples of solids are either prepared in thick suspensions (mulls) such as nujol, and held as layers between NaCl planes or pressed into KBr disks. Non-covalent materials must be used for sample containment and in the optics, as these materials are transparent to infrared. All materials need to be free of water, because of the strong absorption of the O–H vibration.

Analysis using a Michelson interferometer enables Fourier transform infrared spectroscopy (FT–IR). The entire light emitted from the source is passed through the sample at once, and then split into two beams that are reflected back onto the point of split (interferometer plate). Using a movable mirror, path length differences are generated between both beams yielding an interferogram that is recorded by the detector. The interferogram is related with a conventional infrared spectrum by a mathematical operation called Fourier transform (see also Fig. 13.9).

For Raman spectroscopy, aqueous solutions are frequently used, since water possesses a rather featureless weak Raman spectrum. The Raman effect can principally be observed with bright, monochromatic light of any wavelength; however, light between the visible region of the spectrum is normally used due to few unwanted absorption effects. The ideal light source for Raman spectrometers is therefore a laser. Because the Raman effect is observed in light scattered off the sample, typical spectrometers use a 90° configuration.

13.2.3 Applications

The use of infrared and Raman spectroscopy is mainly in chemical and biochemical research of small compounds such as drugs, metabolic intermediates and substrates. Examples are the identification of synthesised compounds, or identification of sample constituents (e.g. in food) when coupled to a separating method such as gas chromatography (GC–IR).

FT–IR is increasingly used for analysis of peptides and proteins. The peptide bond gives rise to nine characteristic bands, named amide A, B, I, II, III, . . ., VII. The amide I ($1600-1700 \text{ cm}^{-1}$) and amide II ($1500-1600 \text{ cm}^{-1}$) bands are the major contributors to the protein infrared spectrum. Both bands are directly related to the backbone conformation and have thus been used for assessment of the secondary structure of peptides and proteins. The interpretation of spectra of molecules with a large number of atoms usually involves deconvolution of individual bands and second derivative spectra.

Time-resolved FT–IR (trFT–IR) enables the observation of protein reactions at the sub-millisecond timescale. This technique has been established by investigation of the light-driven proton pump bacteriorhodopsin. For instance, the catalytic steps in the proton pumping mechanism have been validated with trFT–IR, and involve transfer of a proton from the Schiff base ($R_1R_2C{=}N{-}R_3$) to a catalytic aspartate residue, followed by re-protonation of a second catalytic aspartate residue.

13.3 SURFACE PLASMON RESONANCE

13.3.1 Principles

Surface plasmon resonance (SPR) is a surface-sensitive method for monitoring smallest changes of the refractive index or the thickness of thin films. It is mainly used for monitoring the interaction of two components (e.g. ligand and receptor, Section 17.3.2) one of which is immobilised on a sensor chip surface (Fig. 13.4a), such as a hydrogel layer on a glass slide, via either biotin–avidin interactions or covalent coupling using amine or thiol reagents similar to those used for cross-linking to affinity chromatography resins (see Section 11.8). Typical surface concentrations of the bound protein component are in the range of 1–$5\,ng\,mm^{-2}$. The sensor chip forms one wall of a microflow cell so that an aqueous solution of the ligand can be pumped at a continuous, pulse-free rate across the surface. This ensures that the concentration of ligand at the surface is maintained at a constant value. Environmental parameters such as temperature, pH and ionic strength are carefully controlled, as is the duration of exposure of the immobilised component to the ligand. Replacing the ligand solution by a buffer solution enables investigation of the dissociation of bound ligand.

Binding of ligand to the immobilised component causes an increase in mass at the surface of the chip. Vice versa, dissociation of ligand causes a reduction of mass. These mass changes, in turn, affect the refractive index of the medium at the surface of the chip, the value of which determines the propagation velocity of electromagnetic radiation in that medium.

Plasmon is a term for a collection of conduction electrons in a metal or semiconductor. Excitation of a plasmon wave requires an optical prism with a metal film of about 50 nm thickness. Total internal reflection (TIR) occurs when a light beam travelling through a medium of higher refractive index (e.g. glass prism with gold-coated surface) meets at an interface with a medium of lower refractive index (e.g. aqueous sample) at an angle larger than the critical angle. TIR of an incident light beam at the prism–metal interface elicits a propagating plasmon wave by leaking an electrical field intensity, called an evanescent field wave, into the medium of lower refractive index where it decays at an exponential rate and effectively only travels one wavelength.

Since the interface between the prism and the medium is coated with a thin layer of gold, incident photons excite a vibrational state of the electrons of the conducting band of the metal. In thin metal films this propagates as a longitudinal vibration. The electrons vibrate with a resonance frequency (hence the term 'resonance') that is dependent on metal and prism properties, as well as the wavelength and the angle of

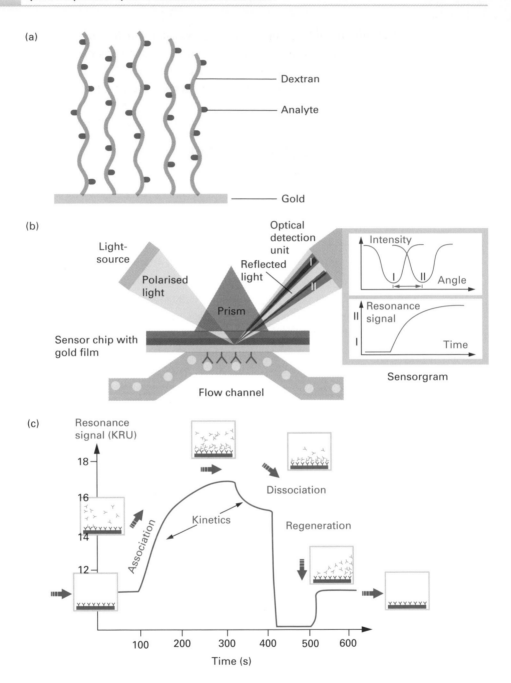

Fig. 13.4 The principles of surface plasmon resonance technology. (a) The sensor 'chip' surface. (b) The flow channel. Insert: change in intensity of reflected light as a function of angle of incidence of the light beam, and change in resonance signal as a function of time. (c) The sensorgram. (Reproduced by permission of GE Healthcare.)

the incident beam. Excitation of the plasmon wave leads to a decreased intensity of the reflected light. Thus, SPR produces a dip in the reflected light intensity at a specific angle of reflection. The propagating surface plasmon wave enhances the amplitude of the evanescent field wave, which extends into the sample region (Fig. 13.4b).

When binding to the chip occurs, the refractive index on the sample side of the interface increases. This alters the angle of incidence required to produce the SPR effect and hence also alters the angle of reflected light. The change in angle brings about a change in detector position, which can be plotted against time to give a sensorgram reading (Fig. 13.4c). The angle is expressed in resonance units (RU), such that 1000 RU corresponds to a change in mass at the surface of the chip of about 1 ng mm^{-2}.

Since in SPR instruments the angle and the wavelength of the incident beam are constant, a shift in the plasmon resonance leads to a change in the intensity of the reflected beam. The shift is restricted locally and happens only in areas where the optical properties have changed. The usage of an array detector as compared to a single detector cell therefore allows for measurement of an SPR image. SPR can detect changes in the refractive index of less than 10^{-4} or changes in layer heights of about 1 nm. This enables not only the detection of binding events between biomolecules but also binding at protein domains or changes in molecular monolayers with a lateral resolution of a few μm. For SPR, light of wavelengths between infrared (IR) and near-infrared (NIR) may be used. In general, the higher the wavelength of the light used the better the sensitivity but the less the lateral resolution. Vice versa, if high lateral resolution is required, red light is to be used because the propagation length of the plasmon wave is approximately proportional to the wavelength of the exciting light.

13.3.2 Applications

The SPR technique enjoys frequent use in modern life science laboratories, due to its general applicability and the fact that there are no special requirements for the molecules to be studied (label-free), such as fluorescent properties, spectral labels or radio labels. It can even be used with coloured or opaque solutions.

Generally, all two-component binding reactions can be investigated, which opens a variety of applications in the areas of drug design (protein–ligand interactions), as well as mechanisms of membrane-associated proteins (protein–membrane binding) and DNA-binding proteins. SPR has thus successfully been used to study the kinetics of receptor–ligand interactions, antibody–antigen and protein–protein interactions. The method is extensively used in proteomic research and drug development.

SPR imaging

The focus of SPR imaging experiments has shifted in recent years from characterisation of ultrathin films to analysis of biosensor chips, especially affinity sensor arrays. SPR imaging can detect DNA–DNA, DNA–protein and protein–protein interactions in a two-dimensional manner. The detection limit for such biosensor chips is in the order of nM to fM. Apart from the detection of binding events as such, the quality of binding (low affinity, high affinity) can also be assessed by SPR imaging. Promising future applications for SPR imaging include peptide arrays that can be prepared on modified gold surfaces. This can prove useful for assessing peptide–antibody interactions. The current time resolution of less than 1 s for an entire image also allows for high-throughput screenings and *in situ* measurements.

13.4 ELECTRON PARAMAGNETIC RESONANCE

Prior to any detailed discussion of electron paramagnetic resonance (EPR) and nuclear magnetic resonance (NMR) methods, it is worthwhile considering the more general phenomena applicable to both.

13.4.1 Magnetic phenomena

Magnetism arises from the motion of charged particles. This motion is controlled by internal forces in a system. For the purpose of this discussion, the major contribution to magnetism in molecules is due to the spin of the charged particle.

In chemical bonds of a molecule, the negatively charged electrons have a spin controlled by strict quantum rules. A bond is constituted by two electrons with opposite spins occupying the appropriate molecular orbital. According to the Pauli principle, the two electrons must have opposite spins, leading to the term paired electrons. Each of the spinning electronic charges generates a magnetic effect, but in electron pairs the effect is almost self-cancelling. In atoms, a value for magnetic susceptibility may be calculated and is of the order of -10^{-6} g^{-1}. This diamagnetism is a property of all substances, because they all contain the minuscule magnets, i.e. electrons. Diamagnetism is temperature independent.

If an electron is unpaired, there is no counterbalancing opposing spin and the magnetic susceptibility is of the order of $+10^{-3}$ to $+10^{-4}$ g^{-1}. The effect of an unpaired electron exceeds the 'background' diamagnetism, and gives rise to paramagnetism. Free electrons can arise in numerous cases. The most notable example is certainly the paramagnetism of metals such as iron, cobalt and nickel, which are the materials that permanent magnets are made of. The paramagnetism of these metals is called ferromagnetism. In biochemical investigations, systems with free electrons (radicals) are frequently used as probes.

Similar arguments can be made regarding atomic nuclei. The nucleus of an atom is constituted by protons and neutrons, and has a net charge that is normally compensated by the extra-nuclear electrons. The number of all nucleons (Z) is the sum of the number of protons (P) and the number of neutrons (N). P and Z determine whether a nucleus will exhibit paramagnetism. Carbon-12 (^{12}C), for example, consists of six protons (P = 6) and six neutrons (N = 6) and thus has Z = 12. P and Z are even, and therefore the ^{12}C nucleus possesses no nuclear magnetism. Another example of a nucleus with no residual magnetism is oxygen-16 (^{16}O). All other nuclei with P and Z being uneven possess residual nuclear magnetism.

The way in which a substance behaves in an externally applied magnetic field allows us to distinguish between dia- and paramagnetism. A paramagnetic material is attracted by an external magnetic field, while a diamagnetic substance is rejected. This principle is employed by the Guoy balance, which allows quantification of magnetic effects. A balance pan is suspended between the poles of a suitable electromagnet supplying the external field. The substance under test is weighed in air with the current switched off. The same sample is then weighed again with the current (i.e. external magnetic field) on. A paramagnetic substance appears to weigh more, and a diamagnetic substance appears to weigh less.

13.4.2 The resonance condition

In both EPR and NMR techniques, two possible energy states exist for either electronic or nuclear magnetism in the presence of an external magnetic field. In the low-energy state, the field generated by the spinning charged particle is parallel to the external field. Conversely, in the high-energy state, the field generated by the spinning charged particle is antiparallel to the external field. When enough energy is input into the system to cause a transition from the low- to the high-energy state, the condition of resonance is satisfied. Energy must be absorbed as a discrete dose (quantum) $h\nu$, where h is the Planck constant and ν is the frequency (see equation 12.1). The quantum energy required to fulfil the resonance condition and thus enable transition between the low- and high-energy states may be quantified as:

$$h\nu = g\beta B \qquad (13.1)$$

where g is a constant called spectroscopic splitting factor, β is the magnetic moment of the electron (termed the Bohr magneton), and B is the strength of the applied external magnetic field. The frequency ν of the absorbed radiation is a function of the paramagnetic species β and the applied magnetic field B. Thus, either ν or B may be varied to the same effect.

With appropriate external magnetic fields, the frequency of applied radiation for EPR is in the microwave region, and for NMR in the region of radio frequencies. In both techniques, two possibilities exist for determining the absorption of electromagnetic energy (i.e. enabling the resonance phenomenon):

- constant frequency ν is applied and the external magnetic field B is swept; or
- constant external magnetic field B is applied and the appropriate frequency ν is selected by sweeping through the spectrum.

For technical reasons, the more commonly used option is a sweep of the external magnetic field.

13.4.3 Principles

The absorption of energy is recorded in the EPR spectrum as a function of the magnetic induction measured in Tesla (T) which is proportional to the magnetic field strength applied. The area under the absorption peak is proportional to the number of unpaired electron spins. Most commonly, the first derivative of the absorption peak is the signal that is actually recorded.

For a delocalised electron, as observed e.g. in free radicals, the g value is 2.0023; but for localised electrons such as in transition metal atoms, g varies, and its precise value contains information about the nature of bonding in the environment of the unpaired electron within the molecule. When resonance occurs, the absorption peak is broadened owing to interactions of the unpaired electron with the rest of the molecule (spin–lattice interactions). This allows further conclusions as to the molecular structure.

High-resolution EPR may be performed by examining the hyperfine splitting of the absorption peak which is caused by interaction of the unpaired electron with adjacent nuclei, thus yielding information about the spatial location of atoms in the molecule. The proton hyperfine splitting for free radicals occurs in the range of $0–3*10^{-3}$ T, and yields data analogous to those obtained in high-resolution NMR (see Section 13.5).

The effective resolution of an EPR spectrum can be considerably improved by combining the method with NMR, a technique called electron nuclear double resonance (ENDOR). Here, the sample is irradiated simultaneously with microwaves for EPR and radio frequencies (RF) for NMR. The RF signal is swept for fixed points in the EPR spectrum, yielding the EPR signal height versus nuclear RF. This approach is particularly useful when there are a large number of nuclear levels that broaden the normal electron resonance lines.

The technique of electron double resonance (ELDOR) finds an application in the separation of overlapping multiradical spectra and the study of relaxation phenomena, for example chemical spin exchange. In ELDOR, the sample is irradiated with two microwave frequencies simultaneously. One is used for observation of the EPR signal at a fixed point in the spectrum, the other is used to sweep other parts of the spectrum. The recorded spectrum is plotted as a function of the EPR signal as a function of the difference of the two microwave frequencies.

13.4.4 Instrumentation

Figure 13.5 shows the main components of an EPR instrument. The magnetic fields generated by the electromagnets are of the order of 50 to 500 mT, and variations of less than 10^{-6} are required for highest accuracy. The monochromatic microwave radiation is produced in a klystron oscillator with wavelengths around 3 cm (9 GHz).

The samples are required to be in the solid state; hence biological samples are usually frozen in liquid nitrogen. The technique is also ideal for investigation of membranes and membrane proteins. Instead of plotting the absorption A versus B, it is the first-order differential (dA/dB) that is usually plotted against B (Fig. 13.6). Such a shape is called a 'line' in EPR spectroscopy. Generally, there are relatively few unpaired electrons in a molecule, resulting in fewer than 10 lines, which are not closely spaced.

13.4.5 Applications

Metalloproteins

EPR spectroscopy is one of the main methods to study metalloproteins, particularly those containing molybdenum (xanthine oxidase), copper (cytochrome oxidase, copper blue enzymes) and iron (cytochrome, ferredoxin). Both copper and non-haem iron, which do not absorb in the UV/Vis region, possess EPR absorption peaks in one

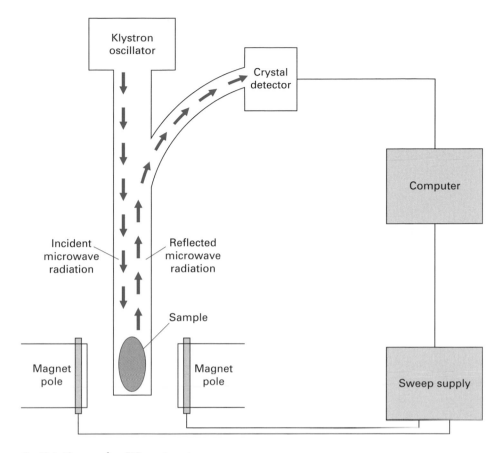

Fig. 13.5 Diagram of an EPR spectrometer.

of their oxidised states. The appearance and disappearance of their EPR signals are used to monitor the activity of these proteins in the multi enzyme systems of intact mitochondria and chloroplasts, as well as in isolated enzymes. In many metallo-proteins, the ligands coordinating the metal ion are the amino acid residues of the protein. Coordination chemistry requires a specific stereochemical structure of the ligands, and EPR studies show that the geometry is frequently distorted in proteins when compared to model systems. Such distortions may be related to biological function.

Spin labels

Spin labels are stable and non-reactive unpaired electrons used as reporter groups or probes for EPR. The procedure of spin labelling is the attachment of these probes to biological molecules that lack unpaired electrons. The label can be attached to either a substrate or a ligand. Often, a spin label contains the nitric oxide moiety. These labels enable the study of events that occur with a frequency of 10^7 to 10^{11} s^{-1}. If the motion is restricted in some directions, only anisotropic motion (movement in one particular direction) may be studied, for example in membrane-rigid spin labels in bilayers. Here, the label is attached so that the NO group lies parallel to the long axis of the lipid.

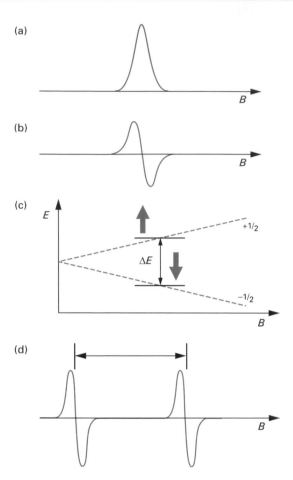

Fig. 13.6 Instead of the absorption signal (a), EPR spectrometry records its first derivative (b). (c) The energy of the two spin states of a free electron is shown as a function of the external magnetic field B. Resonance happens when the energy of the applied microwave radiation is the same as the energy difference ΔE. (d) Hyperfine splitting due to coupling of an unpaired electron with a nuclear spin of 1/2. For the hydrogen atom, the distance between the two signals is 50.7 mT.

Intramolecular motions and lateral diffusion of lipid through the membrane, as well as the effect of proteins and other factors on these parameters may be observed. Quantification of effects often involves calculation of the order parameter Z. Spin-labelled lipids are either concentrated into one region of the bilayer or randomly incorporated into model membranes. The diffusion of spin labels allows them to come into contact with each other, which causes line-broadening in the spectrum. Labelling of phospholipids with 2,2,6,6-tetramethylpiperidine-1-oxyl (TEMPOL) is used for measurement of the flip rate of phospholipids between inner and outer surfaces as well as lateral diffusion.

Free radicals

Molecules in their triplet states (Fig. 12.8) have unpaired electrons and thus are amenable to EPR spectroscopy. Such molecules possess the property of phosphorescence and EPR may deliver data complementary to the UV/Vis region of the spectrum. For instance, free radicals due to the triplet state of tryptophan have been observed in cataractuous lenses.

Spin trapping is a process whereby an unstable free radical is being stabilised by reaction with a compound such as 5,5-dimethylpyrroline-1-oxide (DMPO). Hyperfine splittings (Fig. 13.6) are observed that depend upon the nature of the radical.

Carcinogenesis is an area where free radicals have been implicated. While free radicals promote the generation of tumours through damage due to their high reactivity, there is, in general, a lower concentration of radicals in tumours than in normal tissue. Also, a gradient has been observed with higher concentrations of radicals in the peripheral non-necrotic surface layers than in the inner regions of the tumour. EPR has been used to study implanted tumours in mice, but also in evaluation of potential chemical carcinogens. Polycyclic hydrocarbons, such as naphthalene, anthracene and phenantrene, consist of multiple aromatic ring systems. These extended aromatic systems allow for single free electrons to be accommodated and thus yield long-lived free radicals, extending the periods of time in which damage can be done. Many of the precursors of these radicals exist in natural sources such as coal tar, tobacco smoke and other products of combustion, hence the environmental risk. Another source of free radicals is irradiation with UV light or γ-rays. Ozone is an oxygen radical that is

Example 1 FREE RADICALS IN THE HEALTHY ORGANISM

An important concern for humans today is environmental pollution and its effects on our bodies. Environmental pollutants – from auto exhaust, second-hand cigarette smoke, pesticides, or even ultraviolet radiation from the Sun – create what are known as free radicals in our bodies.

A lot of metabolic studies have made use of EPR spectroscopy. Free radicals are found in many metabolic pathways and as degradation products of drugs and toxins. Electron transfer mechanisms in mitochondria and chloroplasts involve paramagnetic species, such as the Fe-S centres. Other RedOx processes involving the flavin derivatives FAD, FMN and semiquinons lend themselves readily to exploration by EPR spectroscopy. The signal of $g = 2.003$ mainly stems from mitochondria. However, different cell lines show different intensities, because this phenomenon also depends on the metabolic state. Factors increasing the metabolic activity also lead to an increase in organic radical signal. Many studies in this context have focussed on the free radical polymer melanin (the skin pigment) and the ascorbyl radical (vitamin C metabolite).

Nitric oxide (NO) operates as a physiological messenger regulating the nervous, immune and cardiovascular systems. It has been implicated in septic (toxic) shock, hypertension, stroke and neurodegenerative diseases. Although NO is involved in normal synaptic transmission, excess levels are neurotoxic. Enzymes such as superoxide dismutase attenuate the neurotoxicity by removal of radical oxygen species, hence limiting their availability for reaction with NO to produce peroxynitrite.

present as a protective shield around the Earth, filtering the dangers of cosmic UV irradiation by complex radical chemistry. The pollution of the Earth's atmosphere with radical-forming chemicals has destroyed large parts of the ozone layer, increasing the risk of skin cancer from sun exposure. EPR can be used to study biological materials, including bone or teeth, and detect radicals formed due to exposure to high energy radiation.

Another major application for EPR is the examination of irradiated foodstuffs for residual free radicals, and it is mostly used to establish whether packed food has been irradiated.

13.5 NUCLEAR MAGNETIC RESONANCE

The essential background theory of the phenomena that allow NMR to occur have been introduced in Sections 13.4.1 and 13.4.2. However, the miniature magnets involved here are not electrons, but the nuclei. The specific principles, instrumentation and applications are discussed below.

13.5.1 Principles

Most studies in organic chemistry involve the use of ^1H, but NMR spectroscopy with ^{13}C, ^{15}N and ^{31}P isotopes is frequently used in biochemical studies. The resonance condition in NMR is satisfied in an external magnetic field of several hundred mT, with absorptions occurring in the region of radio waves (frequency 40 MHz) for resonance of the ^1H nucleus. The actual field scanned is small compared with the field strengths applied, and the radio frequencies absorbed are specifically stated on such spectra.

Similar to other spectroscopic techniques discussed earlier, the energy input in the form of electromagnetic radiation promotes the transition of 'entities' from lower to higher energy states (Fig. 13.7). In case of NMR, these entities are the nuclear magnetic spins which populate energy levels according to quantum chemical rules. After a certain time-span, the spins will return from the higher to the lower energy level, a process that is known as relaxation.

The energy released during the transition of a nuclear spin from the higher to the lower energy state can be emitted as heat into the environment and is called spin–lattice relaxation. This process happens with a rate of T_1^{-1}, and T_1 is termed the longitudinal relaxation time, because of the change in magnetisation of the nuclei parallel to the field. The transverse magnetisation of the nuclei is also subject to change over time, due to interactions between different nuclei. The latter process is thus called spin–spin relaxation and is characterised by a transverse relaxation time T_2.

The molecular environment of a proton governs the value of the applied external field at which the nucleus resonates. This is recorded as the chemical shift (δ) and is measured relative to an internal standard, which in most cases is tetramethylsilane

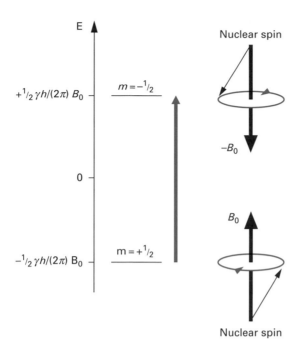

Fig. 13.7 Energy levels of a proton in the magnetic field B_0. The nuclear spin of a nucleus is characterised by its magnetic quantum number m. For protons, m can only adopt $+\frac{1}{2}$ and $-\frac{1}{2}$. The corresponding energies are calculated by $-m\gamma h/(2\pi)H_0$, where γ is a constant characteristic for a particular nucleus, h is the Planck constant, and H_0 is the strength of the magnetic field B_0.

(TMS; $(H_3C)_4Si$) because it contains 12 identical protons. The chemical shift arises from the applied field inducing secondary fields of about 0.15–0.2 mT at the proton by interacting with the adjacent bonding electrons.

- If the induced field opposes the applied field, the latter will have to be at a slightly higher value for resonance to occur. The nucleus is said to be shielded, the magnitude of the shielding being proportional to the electron-withdrawing power of proximal substituents.
- Alternatively, if the induced and applied fields are aligned, the latter is required to be at a lower value for resonance to occur. The nucleus is then said to be deshielded.

Usually, deuterated solvents such as $CDCl_3$ are used for sample preparation of organic compounds. For peptides and proteins D_2O is the solvent of choice. Because the stability of the magnetic field is critical for NMR spectroscopy, the magnetic flux needs to be tuned, e.g. by locking with deuterium resonance frequencies. The use of deuterated solvents thus eliminates the need for further experiments.

The chemical shift is plotted along the x-axis, and measured in p.p.m. instead of the actual magnetic field strengths. This conversion makes the recorded spectrum independent of the magnetic field used. The signal of the internal standard TMS appears at $\delta = 0$ p.p.m. The type of proton giving rise to a particular band may thus be identified by the resonance peak position, i.e. its chemical shift, and the area under each peak

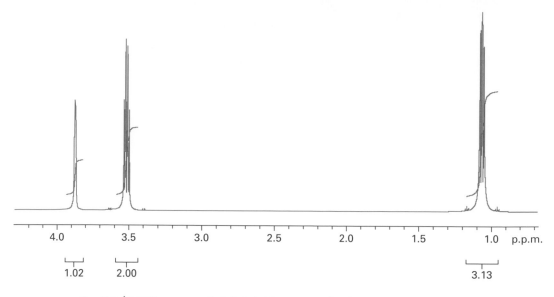

Fig. 13.8 ^{1}H NMR spectrum of ethyl alcohol (H_3C-CH_2-OH) with integrated peaks.

is proportional to the number of protons of that particular type. Figure 13.8 shows an ^{1}H NMR spectrum of ethyl alcohol, in which there are three methyl, two methylene and one alcohol group protons. The peak areas are integrated, and show the proportions $3:2:1$. Owing to the interaction of bonding electrons with like or different spins, a phenomenon called **spin–spin coupling** (also termed scalar or J-coupling) arises that can extend to nuclei four or five bonds apart. This results in the splitting of the three bands in Fig. 13.8 into several finer bands (hyperfine splitting). The hyperfine splitting yields valuable information about the near-neighbour environment of a nucleus.

NMR spectra are of great value in elucidating chemical structures. Both qualitative and quantitative information may be obtained. The advances in computing power have made possible many more advanced NMR techniques. Weak signals can be enhanced by running many scans and accumulating the data. Baseline noise, which is random, tends to cancel out whereas the signal increases. This approach is known as computer averaging of transients or CAT scanning, and significantly improves the signal-to-noise ratio.

Despite the value and continued use of such 'conventional' ^{1}H NMR, much more structural information can be obtained by resorting to pulsed input of radio frequency energy, and subjecting the output to Fourier transform. This approach has given rise to a wide variety of procedures using multidimensional spectra, ^{13}C and other odd-isotope NMR spectra and the determination of multiplicities and scan images.

Pulse-acquire and Fourier transform methods

In 'conventional' NMR spectroscopy, the electromagnetic radiation (energy) is supplied from the source as a continuously changing frequency over a preselected spectral range (continuous wave method). The change is smooth and regular between fixed limits. Figure 13.9a illustrates this approach. During the scan, radiation of

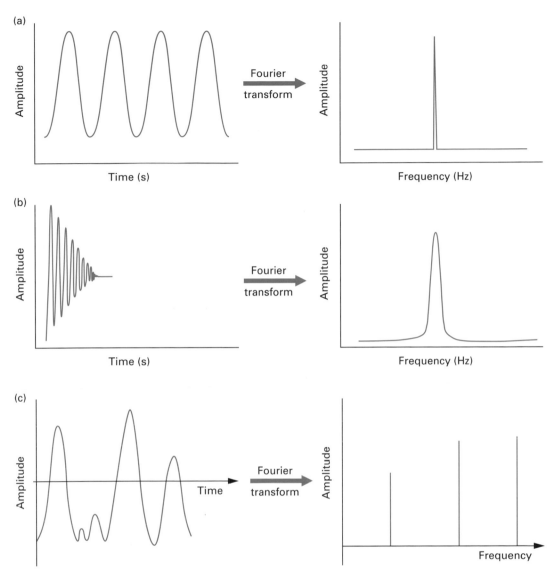

Fig. 13.9 Diagrammatic representation of the Fourier transformation of (a) a single frequency sine wave, (b) a single frequency FID, and (c) a three-sine-wave combination.

certain energy in the form of a sinc wave is recorded. By using the mathematical procedure of Fourier transform, the 'time domain' can be resolved into a 'frequency domain'. For a single-frequency sine wave, this procedure yields a single peak of fixed amplitude. However, because the measured signal in NMR is the re-emission of energy as the nuclei return from their high-energy into their low-energy states, the recorded radiation will decay with time, as fewer and fewer nuclei will return to the ground state. The signal measured is thus called the free induction decay (FID). Figure 13.9b shows the effect of the FID on the corresponding Fourier transform. The frequency band broadens, but the peak position and the amplitude remain the same. The resolved frequency peak represents the chemical shift of a nucleus resonating at this energy.

Alternatively, the total energy comprising all frequencies between the fixed limits can be put in all at the same time. This is achieved by irradiating the sample with a broadband pulse of all frequencies at one go. The output will measure all resonance energies simultaneously and will result in a very complicated interference pattern. However, Fourier transform is able to resolve this pattern into the constituting frequencies (Fig. 13.9c).

In the presence of an external magnetic field, nuclear spins precess around the axis of that field with the so-called Larmor frequency. The vector sum of all nuclear magnetic moments yields a magnetisation parallel to the external field, i.e. a longitudinal magnetisation. When a high-frequency pulse is applied, the overall magnetisation is forced further off the precession by a pulse angle. This introduces a new vector component to the overall magnetisation which is perpendicular to the external field; this component is called transverse magnetisation. The FID measured in pulse-acquired spectra is, in fact, the decay of that transverse magnetisation component.

Nuclear Overhauser effect

It has already been mentioned above that nuclear spins generate magnetic fields which can exert effects through space, for example as observed in spin–spin coupling. This coupling is mediated through chemical bonds connecting the two coupling spins. However, magnetic nuclear spins can also exert effects in their proximal neighbourhood via dipolar interactions. The effects encountered in the dipolar interaction are transmitted through space over a limited distance on the order of 0.5 nm or less. These interactions can lead to the nuclear Overhauser effects (NOEs), as observed in a changing signal intensity of a resonance when the state of a near neighbour is perturbed from the equilibrium. Because of the spatial constraint, this information enables conclusions to be drawn about the three-dimensional geometry of the molecule being examined.

^{13}C NMR

Due to the low abundance of the ^{13}C isotope, the chance of finding two such species next to each other in a molecule is very small (see Chapter 9). As a consequence, ^{13}C–^{13}C couplings (homonuclear couplings) do not arise. While ^{1}H–^{13}C interactions (heteronuclear coupling) are possible, one usually records decoupled ^{13}C spectra where all bands represent carbon only. ^{13}C spectra are thus much simpler and cleaner when compared to ^{1}H spectra. The main disadvantage though is the fact that multiplicities in these spectra cannot be observed, i.e. it cannot be decided whether a particular ^{13}C is associated with a methyl (H_3C), a methylene (H_2C) or a methyne (HC) group. Some of this information can be regained by irradiating with an off-resonance frequency during a decoupling experiment. Another routinely used method is called distortionless enhancement by polarisation transfer (DEPT), where sequences of multiple pulses are used to excite nuclear spins at different angles, usually 45°, 90° or 135°. Although interactions have been decoupled, in this situation the resonances exhibit positive or negative signal intensities dependent on the number of protons bonded to the carbon. In DEPT-135, for example, a methylene group yields a negative intensity, while methyl and methyne groups yield positive signals.

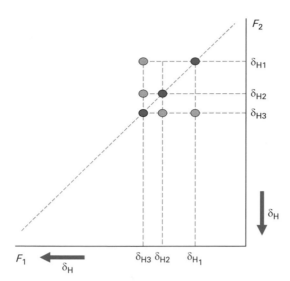

Fig. 13.10 Schematics of a correlated 2D-^{1}H NMR spectrum. H3 couples with H2 and H1. H1 and H2 show no coupling.

Multidimensional NMR

As we learned above, the observable in pulse-acquired Fourier transform NMR is the decay of the transverse magnetisation, called free induction decay (FID). The detected signal thus is a function of the detection time t_2. Within the pulse sequence, the time t_1 (evolution time) describes the time between the first pulse and signal detection. If t_1 is systematically varied, the detected signal becomes a function of both t_1 and t_2, and its Fourier transform comprises two frequency components. The two components form the basis of a two-dimensional spectrum.

Correlated 2D-NMR spectra show chemical shifts on both axes. Utilising different pulse sequences leads to different methods, such as correlated spectroscopy (COSY), nuclear overhauser effect spectroscopy (NOESY), etc. Such methods yield the homonuclear ^{1}H couplings. The 1D-NMR spectrum now appears along the diagonal and long-range couplings between particular nuclei appear as off-diagonal signals (Fig. 13.10).

Summary of NMR parameters

The parameters obtained from NMR spectra used to derive structural determinants of a small molecule or protein are summarised in Table 13.1.

13.5.2 Instrumentation

Schematically, an analytical NMR instrument is very similar to an EPR instrument, except that instead of a klystron generating microwaves two sets of coils are used to generate and detect radio frequencies (Fig. 13.11). Samples in solution are contained in sealed tubes which are rotated rapidly in the cavity to eliminate irregularities and imperfections in sample distribution. In this way, an average and uniform signal

Table 13.1 **NMR-derived structural parameters of molecules**

Parameter	Information	Example/Comment
Chemical shift	Chemical group	^1H, ^{13}C, ^{15}N, ^{31}P
	Secondary structure	
J-couplings (through bond)	Dihedral angles	^3J(amide-H, Hα), ^3J(Hα, Hβ), . . .
NOE (through space)	Interatomic distances	<0.5 nm
Solvent exchange	Hydrogen bonds	Hydrogen-bonded amide protons are protected from H/D exchange, while the signals of other amides disappear quickly
Relaxation/line widths	Mobility, dynamics, conformational/chemical exchange	The exchange between two conformations, but also chemical exchange, gives rise to two distinct signals for a particular spin
	Torsion angles	
Residual dipolar coupling	Torsion angles	^1H–^{15}N, ^1H–^{13}C, ^{13}C–^{13}C, . . .

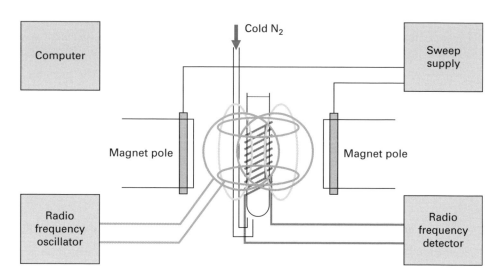

Fig. 13.11 Schematic diagram of an NMR spectrometer with cryoprobe.

is reflected to the receiver to be processed and recorded. In solid samples, the number of spin–spin interactions is greatly enhanced due to intermolecular interactions that are absent in dissolved samples due to translation and rotation movements. As a result, the resonance signals broaden significantly. However, high-resolution spectra can be obtained by spinning the solid sample at an angle of 54.7° (magic angle spinning). The sophisticated pulse sequences necessary for multidimensional NMR require a certain geometric layout of the radio frequency coils and sophisticated electronics. Advanced computer facilities are needed for operation of NMR instruments, as well as analysis of the acquired spectra.

13.5.3 Applications

Molecular structure determination

Traditionally, NMR spectroscopy is the main method of structure determination for organic compounds. The chemical shift provides a clue about the environment of a particular proton or carbon, and thus allows conclusions as to the nature of functional groups. Spin–spin interactions allow conclusions as to how protons are linked with the carbon skeleton. For structure determination, the fine structure usually is the most useful information because it provides a unique criterion while chemical shifts of some groups can vary over an extended range. Additionally, the signal intensity provides information as to how many protons contribute to a particular signal.

Solution structure of proteins and peptides

The structures of proteins up to a mass of about 50 kDa can be determined with biomolecular NMR spectroscopy. The development of magnets with very high field strengths (currently 900 MHz) continues to push the size limit. The preparation of proteins or selected domains for NMR requires recombinant expression and isotopic labelling to enrich the samples with ^{13}C and ^{15}N; ^{2}H labelling might be required as well. Sample amounts in the order of 10 mg used to be required for NMR experiments; however, the introduction of cryoprobe technology has reduced the sample amount significantly. Heteronuclear multidimensional NMR spectra need to be recorded for the assignment of all chemical shifts (^{1}H, ^{13}C, ^{15}N). For interproton NOEs, ^{13}C- and ^{15}N-edited 3D NOESY spectra are required. The data acquisition can take several weeks, after which spectra are processed (Fourier transformation) and improved with respect to digital resolution and signal-to-noise. Assignment of chemical shifts and interatomic distances is carried out with the help of software programs. All experimentally derived parameters are then used as restraints in a molecular dynamics or simulated annealing structure calculation. The result of a protein NMR structure is an ensemble of structures, all of which are consistent with the experimentally determined restraints, but converge to the same fold.

Magnetic resonance imaging

The basic principles of NMR can be applied to imaging of live samples. Because the proton is one of the more sensitive nuclides and is present in all biological systems abundantly, ^{1}H resonance is used almost exclusively in the clinical environment. The most important compound in biological samples in this context is water. It is distributed differently in different tissues, but constitutes about 55% of body mass in the average human. In soft tissues, the water distribution varies between 60% and 90%. In NMR, the resonance frequency of a particular nuclide is proportional to the

Example 2 ASSESSING PROTEIN CONFORMATIONAL EXCHANGE BY NMR

Question Identification of protein–protein interaction sites is crucial for understanding the basis of molecular recognition. How can such sites be identified?

Answer Apart from providing the absolute three-dimensional structure of molecules, NMR methods can also yield insights into protein interactions by mapping. In a technique called saturation transfer difference NMR, protein resonances can be selectively saturated. One then calculates the ^{1}H NMR difference spectrum of the ligand from the saturation experiment subtracted from the ligand spectrum without saturation of the protein. Intensities of protons in close contact with the ligand appear enhanced in the difference spectrum, allowing the identification of chemical groups of the ligand interacting with the protein. Using titration experiments, this technique also allows determination of binding constants.

Beyond mapping the flexibility of residues in known protein binding sites, NMR techniques can also be used to identify novel binding sites in proteins. Protein motions on the timescale of microseconds to milliseconds are accessible to NMR spectroscopy, and the diffusion constants for rotation around the three principal axes x, y and z (called rotational diffusion tensor) can be determined. The principal axes are fixed in the protein, and the principal components as well as the orientation can be derived from analysis of the ratio of the spin–spin and spin–lattice relaxation times T_2/T_1. Analysing these values for the protons of the rigid amide (CO–NH) groups allows a characterisation of the conformational exchange of proteins.

Residues constituting the ligand-binding interface often experience a different environment in the bound state as compared to the free state. The amide signals of these residues are thus broadened due to exchange between these two environments when the free and bound states are in equilibrium.

This approach has been successfully applied to identify the amino acids at the binding site of a 16 kDa protein that binds to and regulates the 251 kDa hydroxylase of the methane monooxygenase protein system. The free and bound forms of the regulatory protein exchange on the timescale of milliseconds.

Other examples include the identification of specific sites involved in the weak self-association of the N-terminal domain of the rat T-cell adhesion protein CD2 (CD2d1) using the concentration dependence of the T_2 values.

strength of the applied external magnetic field. If an external magnetic field gradient is applied then a range of resonant frequencies are observed, reflecting the spatial distribution of the spinning nuclei. Magnetic resonance imaging (MRI) can be applied to large volumes in whole living organisms and has a central role in routine clinical imaging of large-volume soft tissues.

The number of spins in a particular defined spatial region gives rise to the spin density as an observable parameter. This measure can be combined with analysis of

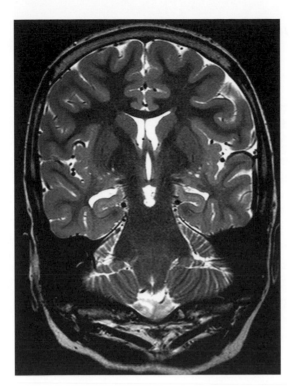

Fig. 13.12 Magnetic resonance imaging: 2-mm thick coronal T_2 weighted fast spin echo image at the level of the foramina monroi connecting the anterior horns of the lateral ventricles with the third ventricle. The sequence consisting of 40 images was acquired at a field strength of 3 tesla and generates $0.47 \times 0.64 \times 2$ mm voxels. (Image courtesy of Professor H. Urbach, University of Bonn.)

the principal relaxation times (T_1 and T_2). The imaging of flux, as either bulk flow or localised diffusion, adds considerably to the options available. In terms of whole-body scanners, the entire picture is reconstructed from images generated in contiguous slices. MRI can be applied to the whole body or specific organ investigations on head, thorax, abdomen, liver, pancreas, kidney and musculoskeletal regions (Fig. 13.12). The use of contrast agents with paramagnetic properties has enabled investigation of organ function, as well as blood flow, tissue perfusion, transport across the blood–brain barrier and vascular anatomy. Resolution and image contrast are major considerations for the technique and subject to continuing development. The resolution depends on the strength of the magnetic field and the availability of labels that yield high signal strengths. MRI instruments used for clinical imaging typically operate with field strengths of up to 3 T, but experimental instruments can operate at more than 20 T, allowing the imaging of whole live organisms with almost enough spatial and temporal resolution to follow regenerative processes continuously at the single-cell level. Equipment cost and data acquisition time remain important issues. On the other hand, according to current knowledge, MRI has no adverse effects on human health, and thus provides a valuable diagnostic tool, especially due to the absence of the hazards of ionising radiation.

13.6 X-RAY DIFFRACTION

13.6.1 Principles

The interaction of electromagnetic radiation with matter causes the electrons in the exposed sample to oscillate. The accelerated electrons, in turn, will emit radiation of the same frequency as the incident radiation, called the secondary waves. The superposition of waves gives rise to the phenomenon of interference. Depending on the displacement (phase difference) between two waves, their amplitudes either reinforce or cancel each other out. The maximum reinforcement is called constructive interference, the cancelling is called destructive interference. The interference gives rise to dark and bright rings, lines or spots, depending on the geometry of the object causing the diffraction. Diffraction effects increase as the physical dimension of the diffracting object (aperture) approaches the wavelength of the radiation. When the aperture has a periodic structure, for example in a diffraction grating, repetitive layers or crystal lattices, the features generally become sharper. Bragg's law (Fig. 13.13) describes the condition that waves of a certain wavelength will constructively interfere upon partial reflection between surfaces that produce a path difference only when that path difference is equal to an integral number of wavelengths. From the constructive interferences, i.e. diffraction spots or rings, one can determine dimensions in solid materials.

Since the distances between atoms or ions are on the order of 10^{-10} m (1 Å), diffraction methods used to determine structures at the atomic level require radiation in the X-ray region of the electromagnetic spectrum, or beams of electrons or neutrons with a similar wavelength. While electrons and neutrons are particles, they also possess wave properties with the wavelength depending on their energy (de Broglie hypothesis). Accordingly, diffraction can also be observed using electron and neutron beams. However, each method also has distinct features, including the penetration depth which increases in the series electrons – X-rays – neutrons.

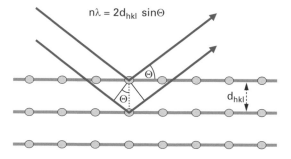

Fig. 13.13 Bragg's law. Interference effects are observable only when radiation interacts with physical dimensions that are approximately the same size as the wavelength of the radiation. Only diffracted beams that satisfy the Bragg condition are observable (constructive interference). Diffraction can thus be treated as selective reflection. *n* is an integer ('order'), λ is the wavelength of the radiation, *d* is the spacing between the lattice planes and Θ is the angle between the incident/reflected beam and the lattice plane.

13.6.2 **X-ray diffraction**

X-rays for chemical analysis are commonly obtained by rotating anode generators (in-house) or synchrotron facilities (Fig. 13.14). In rotating anode generators, a rotating metal target is bombarded with high-energy (10–100 keV) electrons that knock out core electrons. An electron in an outer shell fills the hole in the inner shell and emits the energy difference between the two states as an X-ray photon. Common targets are copper, molybdenum and chromium, which have strong distinct X-ray emission at 1.54 Å, 0.71 Å and 2.29 Å, respectively, that is superimposed on a continuous spectrum known as Bremsstrahlung. In synchrotrons, electrons are accelerated in a ring, thus producing a continuous spectrum of X-rays. Monochromators are required to select a single wavelength.

As X-rays are diffracted by electrons, the analysis of X-ray diffraction data sets produces an electron density map of the crystal. Since hydrogen atoms have very little electron density, they are not usually determined experimentally by this technique.

Unfortunately, the detection of light beams is restricted to recording the intensity of the beam only. Other properties, such as polarisation, can only be determined with rather complex measurements. The phase of the light waves is even systematically lost in the measurement. This phenomenon has thus been termed the phase problem

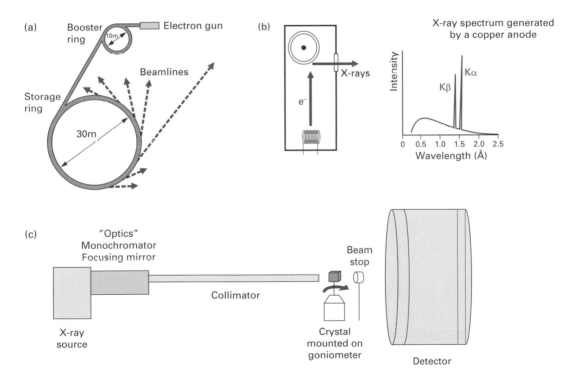

Fig. 13.14 Instrumentation for X-ray diffraction. The most common X-ray sources are (a) particle storage rings which produce synchrotron radiation, and (b) rotating anode tubes. The schematics of an X-ray diffractometer are shown in (c).

owing to the essential information contained in the phase in diffraction and microscopy experiments. The X-ray diffraction data can be used to calculate the amplitudes of the three-dimensional Fourier transform of the electron density. Only together with the phases can the electron density be calculated, in a process called Fourier synthesis.

Different methods to overcome the phase problem in X-ray crystallography have been developed, including:

- molecular replacement, where phases from a structurally similar molecule are used;
- experimental methods that require incorporation of heavy element salts (multiple isomorphous replacement);
- experimental methods where methionine has been replaced by seleno-methionine in proteins (multi-wavelength anomalous diffraction);
- experimental methods using the anomalous diffraction of the intrinsic sulphur in proteins (single wavelength anomalous diffraction);
- direct methods, where a statistical approach is used to determine phases. This approach is limited to very high resolution data sets and is the main method for small molecule crystals as these provide high-quality diffraction with relatively few numbers of reflections.

13.6.3 Applications

Single-crystal diffraction

A crystal is a solid in which atoms or molecules are packed in a particular arrangement within the unit cell which is repeated indefinitely along three principal directions in space. Crystals can be formed by a wide variety of materials, such as salts, metals, minerals and semiconductors, as well as various inorganic, organic and biological molecules.

A crystal grown in the laboratory is mounted on a goniometer and exposed to X-rays produced by rotating anode generators (in-house) or a synchrotron facility. A diffraction pattern of regularly spaced spots known as reflections is recorded on a detector, most frequently image plates or CCD cameras for proteins, and moveable proportional counters for small molecules.

An incident X-ray beam is diffracted by a crystal such that beams at specific angles are produced, depending on the X-ray wavelength, the crystal orientation and the structure of the crystal (i.e. unit cell).

To record a data set, the crystal is gradually rotated and a diffraction pattern is acquired for each distinct orientation. These two-dimensional images are then analysed by identifying the appropriate reflection for each lattice plane and measuring its intensity, measuring the cell parameters of the unit cell and determining the appropriate space group. If information about the phases is available, this data can then be used to calculate a three-dimensional model of the electron density within the unit cell using the mathematical method of Fourier synthesis. The positions of the atomic nuclei are then deduced from the electron density by computational refinement and manual intervention using molecular graphics.

Fibre diffraction

Certain biological macromolecules, such as DNA and cytoskeletal components, cannot be crystallised, but form fibres. In fibres, the axes of the long polymeric structures are parallel to each other. While this can be an intrinsic property, for example in muscle fibres, in some cases the parallel alignment needs to be induced. As fibres show helical symmetry, by analysing the diffraction from oriented fibres one can deduce the helical symmetry of the molecule, and in favourable cases the molecular structure. Generally, a model of the fibre is constructed and the expected diffraction pattern is compared with the observed diffraction.

Historically, fibre diffraction was of central significance in enabling the determination of the three-dimensional structure of DNA by Crick, Franklin, Watson and Wilkins.

Two classes of fibre diffraction patterns can be distinguished. In crystalline fibres (e.g. A form of DNA), the long fibrous molecules pack to form thin micro-crystals randomly arranged around a shared common axis. The resulting diffraction pattern is equivalent to taking a long crystal and spinning it about its axis during the X-ray exposure. All Bragg reflections are recorded at once. In non-crystalline fibres (e.g. B form of DNA), the molecules are arranged parallel to each other but in a random orientation around the common axis. The reflections in the diffraction pattern are now a result of the periodic repeat of the fibrous molecule. The diffraction intensity can be calculated via Fourier–Bessel transformation replacing the Fourier transformation used in single-crystal diffraction.

Powder diffraction

Powder diffraction is a rapid method to analyse multicomponent mixtures without the need for extensive sample preparation. Instead of using single crystals, the solid material is analysed in the form of a powder where, ideally, all possible crystalline orientations are equally represented.

From powder diffraction patterns, the interplanar spacings d of the lattice planes (Fig. 13.13) are determined and then compared to a known standard or to a database (Powder Diffraction File by the International Centre for Diffraction Data or the Cambridge Structural Database) for identification of the individual components.

13.7 SMALL-ANGLE SCATTERING

The characteristics of molecules at larger size scales are fundamentally different than at atomic scales. While atomic scale structures are characterised by high degrees of order (e.g. crystals), on the nano scale, the building blocks of matter are rarely well organised and are composed of rather complex building blocks (i.e. shapes). Consequently, sharp diffraction peaks are observed in X-ray diffraction from single crystals, but diffuse patterns are obtained from X-ray scattering from biological molecules or nano-structures.

In Section 12.6, we learned that incident light scattered by a particle in the form of Rayleigh scattering has the same frequency as the incident light. It is thus called elastic light scattering. The light scattering techniques discussed in Section 12.6 have used a combination of visible light and molecules, so that the dimension of the particle is smaller than the wavelength of the light. When using light of smaller wavelengths such as X-rays, the overall dimension of a molecule is large as compared to the incident light. Electrons in the different parts of the molecule are now excited by the incident beam with different phases. The coherent waves of the scattered light therefore show an interference that is dependent on the geometrical shape of the molecule. As a result

- in the forward direction (at 0°), there is no phase difference between the waves of the scattered light, and one observes maximum positive interference, i.e. highest scattering intensity;
- at small angles, there is a small but significant phase difference between the scattered waves which results in diminished scattering intensity due to destructive interference.

Small-angle X-ray or neutron scattering (SAXS or SANS) are experimental techniques used to derive size and shape parameters of large molecules. Both X-ray and neutron scattering are based on the same physical phenomenon, i.e. scattering due to differences in scattering mass density between the solute and the solvent or indeed between different molecular constituents. An advantage for protein structure determination is the fact that samples in aqueous solution can be assessed.

Experimentally, a monodisperse solution of macromolecules is exposed to either X-rays (wavelength λ = ca. 0.15 nm) or thermal neutrons (λ = ca. 0.5 nm). The intensity of the scattered light is recorded as a function of momentum transfer q ($q = 4\pi \sin\theta \, \lambda^{-1}$, where 2θ is the angle between the incident and scattered radiation). Due to the random positions and orientations of particles, an isotropic intensity distribution is observed that is proportional to the scattering from a single particle averaged over all orientations. In neutron scattering, the **contrast** (squared difference in scattering length density between particle and solvent) can be varied using H_2O/D_2O mixtures or selective deuteration to yield additional information. At small angles the scattering curve is a rapidly decaying function of q, and essentially determined by the **particle shape**. Fourier transformation of the scattering function yields the so-called **size distribution function** which is a histogram of interatomic distances. Comparison of the size distribution function with the particle form factor of regular geometrical bodies allows conclusions as to the shape of the scattering particle. Analysis of the scattering function further allows determination of the **radius of gyration** R_g (average distance of the atoms from the centre of gravity of the molecule), and the mass of the scattering particle from the scattering in the forward direction.

Shape restoration

Software programs have been developed that enable the calculation of three-dimensional structures from the one-dimensional scattering data obtained by SAXS. Due to the low resolution of SAXS data, the structural information is restricted to the shape of the scattering molecules. Furthermore, the scattering data do not imply a single, unique solution. The reconstruction of three-dimensional structures might thus result in a number of different models. One approach is to align and average a set of independently reconstructed models thus obtaining a model that retains the most persistent features.

13.8 SUGGESTIONS FOR FURTHER READING

General

Ciulli, A. and Abell, C. (2007). Fragment-based approaches to enzyme inhibition. *Current Opinion in Biotechnology*, 18, 489–496.

Infrared spectroscopy

Beekes, M., Lasch, P. and Naumann, D. (2007). Analytical applications of Fourier transform-infrared (FT–IR) spectroscopy in microbiology and prion research. *Veterinary Microbiology*, 123, 305–319.

Ganim, Z., Chung, H. S., Smith, A. W., Deflores, L. P., Jones, K. C. and Tokmakoff, A. (2008). Amide I two-dimensional infrared spectroscopy of proteins. *Accounts of Chemical Research*, 41, 432–441.

Tonouchi, M. (2007). Cutting-edge terahertz technology. *Nature Photonics*, 1, 97–105.

WEBSITES

http://www.cem.msu.edu/~reusch/VirtualText/Spectrpy/InfraRed/infrared.htm
http://www.chem.uic.edu/web1/ocol/spec/IR.htm
http://orgchem.colorado.edu/hndbksupport/irtutor/tutorial.html
http://www.umd.umich.edu/casl/natsci/slc/slconline/IR/
http://www.biophysik.uni-freiburg.de/Spectroscopy/Time-Resolved/spectroscopy.html

Raman spectroscopy

Benevides, J. M., Overman, S. A. and Thomas, G. J. Jr. (2004). Raman spectroscopy of proteins. *Current Protocols in Protein Science*, Chapter 17, Unit 17.8. New York: Wiley Interscience.

Wen, Z. Q. (2007). Raman spectroscopy of protein pharmaceuticals. *Journal of Pharmaceutical Sciences*, 96, 2861–2878.

WEBSITES

http://www.jobinyvon.com/Raman%20Tutorial%20Intro
http://pcople.bath.ac.uk/pysdw/newpage11.htm

Surface plasmon resonance

Anker, J. N., Hall, W. P., Lyandres, O., Shah, N. C., Zhao, J. and Van Duyne, R. P. (2008). Biosensing with plasmonic nanosensors. *Nature Materials*, 7, 442–453.

Campbell, C. T. and Kim, G. (2007). SPR microscopy and its applications to high-throughput analyses of biomolecular binding events and their kinetics. *Biomaterials*, 28, 2380–2392.

Majka, J. and Speck, C. (2007). Analysis of protein–DNA interactions using surface plasmon resonance. *Advances in Biochemical Engineering and Biotechnology*, 104, 13–36.

Neumann, T., Junker, H. D., Schmidt, K. and Sekul, R. (2007). SPR-based fragment screening: advantages and applications. *Current Topics in Medicinal Chemistry*, 7, 1630–1642.

Phillips, K. S. and Cheng, Q. (2007). Recent advances in surface plasmon resonance based techniques for bioanalysis. *Analytical and Bioanalytical Chemistry*, 387, 1831–1840.

WEBSITES

http://www.blacore.com/
http://www.uksaf.org/tech/spr.html
http://people.clarkson.edu/~ekatz/spr.htm

Electron paramagnetic resonance spectroscopy

Matsumoto, K., Subramanian, S., Murugesan, R., Mitchell, J. B. and Krishna, M. C. (2007). Spatially resolved biologic information from in vivo EPRI, OMRI, and MRI. *Antioxidants and Redox Signaling*, 9, 1125–1141.

Schiemann, O. and Prisner, T. F. (2007). Long-range distance determinations in biomacromolecules by EPR spectroscopy. *Quarterly Reviews in Biophysics*, 40, 1–53.

WEBSITES

http://hyperphysics.phy-astr.gsu.edu/hbase/molecule/esr.html
http://www.chemistry.nmsu.edu/studntres/chem435/Lab7/intro.html

Nuclear magnetic resonance spectroscopy

Blamire, A. M. (2008). The technology of MRI: the next 10 years? *British Journal of Radiology*, **81**, 601–617.

Ishima, R. and Torchia, D. A. (2000). Protein dynamics from NMR. *Nature Structural Biology*, **7**, 740–743.

McDermott, A. and Polenova, T. (2007). Solid state NMR: new tools for insight into enzyme function. *Current Opinion in Structural Biology*, **17**, 617–622.

Skinner, A. L. and Laurence, J. S. (2008). High-field solution NMR spectroscopy as a tool for assessing protein interactions with small molecule ligands. *Journal of Pharmaceutical Science*, **97**, 4670–4695.

Spiess, H. W. (2008). NMR spectroscopy: pushing the limits of sensitivity. *Angewandte Chemie International Edition (English)*, **47**, 639–642.

WEBSITES

http://www.cem.msu.edu/~reusch/VirtualText/Spectrpy/nmr/nmr1.htm#nmr1
http://arrhenius.rider.edu/nmr/NMR_tutor/pages/nmr_tutor_home.html
http://www.cis.rit.edu/htbooks/nmr/
http://teaching.shu.ac.uk/hwb/chemistry/tutorials/molspec/nmr1.htm
http://www.chem.queensu.ca/FACILITIES/NMR/nmr/webcourse/

X-ray diffraction

Hickman, A. B. and Davies, D. R. (2001). Principles of macromolecular X-ray crystallography. *Current Protocols in Protein Science*, Chapter 17, Unit 17.3. New York: Wiley Interscience.

Miao, J., Ishikawa, T., Shen, Q. and Earnest, T. (2008). Extending X-ray crystallography to allow the imaging of noncrystalline materials, cells, and single protein complexes. *Annual Reviews in Physical Chemistry*, **59**, 387–410.

Mueller, M., Jenni, S. and Ban, N. (2007). Strategies for crystallization and structure determination of very large macromolecular assemblies. *Current Opinion in Structural Biology*, **17**, 572–579.

Wlodawer, A., Minor, W., Dauter, Z. and Jaskolski, M. (2008). Protein crystallography for non-crystallographers, or how to get the best (but not more) from published macromolecular structures. *FEBS Journal*, **275**, 1–21.

WEBSITES

http://www.colorado.edu/physics/2000/xray/index.html
http://www.physics.upenn.edu/~heiney/talks/hires/hires.html
http://www.matter.org.uk/diffraction/x-ray/default.htm

Small-angle scattering

Lipfert, J. and Doniach, S. (2007). Small-angle X-ray scattering from RNA, proteins, and protein complexes. *Annual Reviews of Biophysical and Biomolecular Structure*, **36**, 307–327.

Neylon, C. (2008). Small angle neutron and X-ray scattering in structural biology: recent examples from the literature. *European Biophysics Journal*, **37**, 531–541.

Putnam, C. D., Hammel, M., Hura, G. L. and Tainer, J. A. (2007). X-ray solution scattering (SAXS) combined with crystallography and computation: defining accurate macromolecular structures, conformations and assemblies in solution. *Quarterly Reviews in Biophysics*, **40**, 191–285.

WEBSITES

http://www.ncnr.nist.gov/programs/sans/tutorials/index.html
http://www.isis.rl.ac.uk/largescale/loq/documents/sans.htm
http://www.embl-hamburg.de/workshops/2001/EMBO/

14 Radioisotope techniques

R. J. SLATER

14.1 WHY USE A RADIOISOTOPE?

When researchers contemplate using a radioactive compound there are several things they have to consider. First and foremost, they must ask the questions: is a radioisotope necessary, is there another way to achieve our objectives? The reason for this is that radioisotope use is governed by very strict legislation. The rules are based on the premise that radioactivity is potentially unsafe (if handled incorrectly) and should therefore only be used if there are no alternatives. Then, once it is decided that there is no alternative, the safest way of carrying out the work needs to be planned. Essentially this means using the safest isotope and the smallest amount possible.

But why do we use radioisotopes in the first place? There are very good reasons; here are some of them. Firstly, it is possible to detect radioactivity with exquisite sensitivity. This means that, for example, the progress of a chemical through a metabolic pathway or in the body of a plant or animal can be followed relatively easily. In short, much less of the chemical is needed, and the detection methods are simple. Secondly, it is possible to follow what happens in time. Imagine a metabolic pathway such as carbon dioxide fixation (the Calvin cycle). All the metabolites in the cycle are present simultaneously so a good way to establish the order of the metabolism is to add a radioactive molecule (e.g. ^{14}C-labelled sodium bicarbonate) and see what happens to it. Thirdly, it is possible to trace what happens to individual atoms in a pathway. This is done for example by creating compounds with ^{14}C in specific locations on the molecule. Fourthly, we can identify a part or end of a molecule, and follow reactions very precisely. This has been very useful in molecular biology, where it is often necessary to label one end of a DNA molecule (e.g. for techniques such as DNA footprinting, a method for investigating sequence-specific DNA-binding proteins).

Finally, there is a use that seems obvious after you have heard it; but you may never have thought of it yourself until now. In chemistry and biochemistry we are used to chemical reactions where one compound is turned into another. We can identify and measure ('assay') the reactants and products and learn something about the reaction. But what if the product of the reaction is identical to what we start with? You may have guessed the example already: DNA replication. To study this we need some method for detecting the product of the reaction, and this is often done with isotopes.

Having understood why we need a radioisotope we now need to understand what radioactivity is and how to use it. Read on.

14.2 THE NATURE OF RADIOACTIVITY

14.2.1 Atomic structure

An atom is composed of a positively charged central nucleus inside a much larger cloud of negatively charged electrons. The mass of an atom is concentrated in the nucleus, even though it accounts for only a small fraction of the total size of the atom. Atomic nuclei are composed of two major particles, protons and neutrons. Protons are positively charged with a mass approximately 1850 times greater than that of an electron. The number of protons present in the nucleus is known as the atomic number (Z), and it determines what the element is, for example six protons is carbon. Neutrons are uncharged particles with a mass approximately equal to that of a proton. The sum of protons and neutrons in a given nucleus is the mass number (A). Thus

$$A = Z + N$$

where N is the number of neutrons present.

Since the number of neutrons in a nucleus is not related to the atomic number, it does not affect the chemical properties of the atom. Atoms of a given element may not necessarily contain the same number of neutrons. Atoms of a given element with different mass numbers (i.e. different numbers of neutrons) are called isotopes. Symbolically, a specific nuclear species is represented by a subscript number for the atomic number, and a superscript number for the mass number, followed by the symbol of the element. For example:

$$^{12}_{6}\text{C} \quad ^{14}_{6}\text{C} \quad ^{16}_{8}\text{O} \quad ^{18}_{8}\text{O}$$

However, in practice it is more conventional just to cite the mass number (e.g. ^{14}C). The number of isotopes of a given element varies: there are three isotopes of hydrogen (^{1}H, ^{2}H and ^{3}H), seven of carbon (^{10}C to ^{16}C inclusive) and 20 or more of some of the elements of high atomic number.

14.2.2 Atomic stability and radiation

In general, the ratio of neutrons to protons will determine whether an isotope of an element is stable enough to exist in nature. Stable isotopes for elements with low

Table 14.1 **Properties of different types of radiation**

Alpha	Beta	Gamma, X-rays and Bremsstrahlung
Heavy charged particle	Light charged particle	Electromagnetic radiation (em)
More toxic than other forms of radiation	Toxicity same as em radiation per unit of energy	Toxicity same as beta radiation per unit of energy
Not penetrating	Penetration varies with source	Highly penetrating

atomic numbers tend to have an equal number of neutrons and protons, whereas stability for elements of higher atomic numbers requires more neutrons. Unstable isotopes are called radioisotopes. They become stable isotopes by the process of radioactive decay: changes occur in the atomic nucleus, and particles and/or electromagnetic radiation are emitted.

14.2.3 Types of radioactive decay

There are several types of radioactive decay; only those most relevant to biochemists are considered below. A summary of properties is given in Table 14.1.

Decay by negatron emission

In this case a neutron is converted to a proton by the ejection of a negatively charged beta (β) particle called a negatron (β^-):

Neutron $\rightarrow$ proton + negatron

To all intents and purposes a negatron is an electron, but the term negatron is preferred, although not always used, since it serves to emphasise the nuclear origin of the particle. As a result of negatron emission, the nucleus loses a neutron but gains a proton. The mass number, A, remains constant. An isotope frequently used in biological work that decays by negatron emission is ^{14}C.

$$^{14}_{6}C \rightarrow {}^{14}_{7}N + \beta^-$$

Negatron emission is very important to biochemists because many of the commonly used radionuclides decay by this mechanism. Examples are: ^3H and ^{14}C, which can be used to label any organic compound; ^{35}S used to label methionine, for example to study protein synthesis; and ^{33}P or ^{32}P, powerful tools in molecular biology when used as nucleic acid labels.

Decay by positron emission

Some isotopes decay by emitting positively charged β-particles referred to as positrons (β^+). This occurs when a proton is converted to a neutron:

Proton $\rightarrow$ neutron + positron

Positrons are extremely unstable and have only a transient existence. Once they have dissipated their energy they interact with electrons and are annihilated. The mass and energy of the two particles are converted to two γ-rays emitted at 180° to each other. This phenomenon is frequently described as back–to–back emission.

As a result of positron emission the nucleus loses a proton and gains a neutron, the mass number stays the same. An example of an isotope decaying by positron emission is ^{22}Na:

$$^{22}_{11}\text{Na} \rightarrow \,^{22}_{10}\text{Ne} + \beta^+$$

Positron emitters are detected by the same instruments used to detect γ-radiation. They are used in biological sciences to spectacular effect in brain scanning with the technique positron emission tomography (PET scanning) used to identify active and inactive areas of the brain.

Decay by alpha particle emission

Isotopes of elements with high atomic numbers frequently decay by emitting alpha (α) particles. An α-particle is a helium nucleus; it consists of two protons and two neutrons ($^4\text{He}^{2+}$). Emission of α-particles results in a considerable lightening of the nucleus, a decrease in atomic number of 2 and a decrease in the mass number of 4. Isotopes that decay by α-emission are not frequently encountered in biological work although they can be found in instruments such as scintillation counters and smoke alarms. Radium-226 (^{226}Ra) decays by α-emission to radon-222 (^{222}Rn), which is itself radioactive. Thus begins a complex decay series, which culminates in the formation of ^{206}Pb:

$$^{226}_{88}\text{Ra} \rightarrow \,^4_2\text{He}^{2+} + \,^{222}_{86}\text{Rn} \rightarrow\rightarrow\rightarrow \,^{206}_{82}\text{Pb}$$

Alpha emitters are extremely toxic if ingested, due to the large mass and the ionising power of the α-particle.

Electron capture

In this form of decay a proton captures an electron orbiting in the innermost K shell:

proton + electron $\rightarrow$ neutron + X-ray

The proton becomes a neutron and electromagnetic radiation (X-rays) is given out.
Example:

$$^{125}_{53}\text{I} \rightarrow \,^{125}_{52}\text{Te} + \text{X-ray}$$

Decay by emission of γ-rays

In some cases α- and β-particle emission also give rise to γ-rays (electromagnetic radiation similar to, but with a shorter wavelength than, X-rays). The γ-radiation has low ionising power but high penetration. For example, the radiation from ^{60}Co will penetrate 15 cm of steel. The toxicity of γ-radiation is similar to that of X-rays.

Example:

$$^{131}_{53}\text{I} \rightarrow \,^{131}_{54}\text{Xe} + \beta^- + \gamma$$

14.2.4 Radioactive decay energy

The usual unit used in expressing energy levels associated with radioactive decay is the electron volt. One electron volt (eV) is the energy acquired by one electron in accelerating through a potential difference of 1 V and is equivalent to 1.6×10^{-19} J. For the majority of isotopes, the term million or mega electron volts (MeV) is more applicable. Isotopes emitting α-particles are normally the most energetic, falling in the range 4.0 to 8.0 MeV, whereas β- and γ-emitters generally have decay energies of less than 3.0 MeV. The higher the energy of radiation the more it can penetrate matter and the more hazardous it becomes.

14.2.5 Rate of radioactive decay

Radioactive decay (measured as disintegrations per minute, d.p.m.) is a spontaneous process and it occurs at a rate characteristic of the source, defined by the rate constant (λ, the fraction of an isotope decaying in unit time, t^{-1}). Decay is a nuclear event so λ is not affected by temperature or pressure. The number of atoms disintegrating at any time is proportional to the number of atoms of the isotope (N) present at that time (t). Clearly, the number of atoms N, is always falling (as atoms decay) and so the rate of decay (d.p.m.) falls with time. Also, the slope of the graph of number of unstable atoms present, or rate of decay (d.p.m.) against time, similarly falls. This means that a graph of radioactivity against time shows a curve, called an exponential decay curve (Fig. 14.1). The mathematical equation that underpins the graph shown is as follows:

$$\ln N_t/N_0 = -\lambda t \tag{14.1}$$

where λ is the decay constant for an isotope, N_t is the number of radioactive atoms present at time t, and N_0 is the number of radioactive atoms orginally present. You will notice the natural logarithm (ln) in the equation; this means if we were to plot log d.p.m. against time we would get a graph with a straight line and a negative slope (gradient determined by the value of λ).

In practice it is more convenient to express the decay constant in terms of half-life ($t_{1/2}$). This is defined as the time taken for the activity to fall from any value to half that value (see Fig. 14.1). When N_t in equation 14.1 is equal to one-half of N_0 then t will equal the half-life of the isotope. Thus

$$\ln 1/2 = -\lambda t_{1/2} \tag{14.2}$$
$$\text{or } t_{1/2} = 0.693\lambda \tag{14.3}$$

The values of $t_{1/2}$ vary widely from over 10^{19} years for lead-204 (^{204}Pb) to 3×10^{-7} seconds for polonium-212 (^{212}Po). The half-lives of some isotopes frequently used in biological work are given in Table 14.2. The advantages and disadvantages of working with isotopes of differing half-lives are given in Table 14.3.

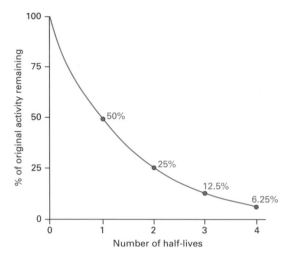

Fig. 14.1 Demonstration of the exponential nature of radioactive decay.

Example 1 **THE EFFECT OF HALF-LIFE**

Question The half-life of ^{32}P is 14.2 days. How long would it take a solution containing 42 000 d.p.m. to decay to 500 d.p.m.?

Answer Use equation 14.3 to calculate the value of λ. This gives a value of 0.0488 days^{-1}. Then use equation 14.1 to calculate the time taken for the counts to decrease. In this equation $N_0 = 42\,000$ and $N_t = 500$. This gives a value for t of 90.8 days. (You can check that this is right by doing an estimate calculation in your head; it is roughly 6 half-lives to get from 42 000 to 500.)

14.2.6 Units of radioactivity

The Système International d'Unités (SI system) uses the becquerel (Bq) as the unit of radioactivity. This is defined as one disintegration per second (1 d.p.s.). However, an older unit, not in the SI system and still frequently used, is the curie (Ci). This is defined as the quantity of radioactive material in which the number of nuclear disintegrations per second is the same as that in 1 g of radium, namely 3.7×10^{10} (or 37 GBq). For biological purposes this unit is too large and the micro-curie (mCi) and millicurie (mCi) are used. It is important to realise that the units Bq and Ci refer to the number of disintegrations actually occurring in a sample not to the disintegrations detected, which generally will be only a proportion of the disintegrations occurring. Detected decays are referred to as counts (i.e. counts per second or c.p.s.).

Table 14.2 **Properties of radioisotopes commonly used in the biological sciences**

Property	^{3}H	^{14}C	^{35}S	^{32}P	^{33}P	^{125}I	^{131}I
$t_{1/2}$	12.3 years	5730 years	87.4 days	14.3 days	25.4 days	59.6 days	8.04 days
Mode of decay	β	β	β	β	β	X (EC) and Auger electrons	γ and β
Max β energy (MeV)	0.019	0.156	0.167	1.709	0.249	Auger electrons 0.035	0.806
ALI[a]	480 (Mbq)[b]	34 (Mbq)	15 (Mbq)	6.3 (Mbq)	14 (Mbq)	1.3 (Mbq)[c]	0.9 (Mbq)[c]
Maximum range in air	6 mm	24 cm	26 cm	790 cm	49 cm	>10 m	>10 cm
Shielding required	None	1 cm acrylic	1 cm acrylic	1 cm acrylic	1 cm acrylic	Lead 0.25 m or lead-impregnated acrylic	Lead 13 mm
γ dose rate (μSv h^{-1} from 1 GBq at 1 m)	–	–	–	(β dose rate 760 μSv, 10 cm from 1 MBq)	–	41	51
Čerenkov counting	–	–	–	Yes	–	–	–

Notes: [a]Annual limit on intake, based on a dose limit of 20 mSv using the most restrictive dose coefficients for inhalation or ingestion.
[b]Bound ^{3}H.
[c]Based on dose equivalent limit of 500 mSv to thyroid.

Table 14.3 **The advantages and disadvantages of working with a short-half-life isotope**

Advantages	Disadvantages
High specific activity (see Section 14.3.3) makes the experiment more sensitive	Experimental design; isotope decays during time of experiment
Easier and cheaper to dispose of	Cost of replacement for further experiments
Lower doses likely (e.g. in diagnostic testing of human subjects)	Frequently need to calculate amount of activity remaining

For quick reference, a list of units and definitions frequently used in radioisotope work is provided in Table 14.6 at the end of the chapter.

14.2.7 Interaction of radioactivity with matter

α-Particles

These particles have a very considerable energy (3–8 MeV) and all the particles from a given isotope have the same amount of energy. They react with matter in two ways: they cause excitation (energy is transferred from the α-particle to orbital electrons of neighbouring atoms, these electrons being elevated to higher orbitals, but eventually fall back, emitting energy as photons of light) and they ionise atoms in their path (the target orbital electron is removed, thus the atom becomes ionised and forms an ion-pair, consisting of a positively charged ion and an electron). Because of their size, α-particles have slow movement and double positive charge. They cause intense ionisation and excitation and their energy is rapidly dissipated. Despite their initial high energy, α-particles frequently collide with atoms in their path and so the radiation is not very penetrating (a few centimetres through air).

Negatrons

Negatrons are very small and rapidly moving particles that carry a single negative charge. They interact with matter to cause ionisation and excitation exactly as with α-particles. However, due to their speed and size, they are less likely than α-particles to interact with matter and therefore are less ionising and more penetrating. Another difference between α-particles and negatrons is that negatrons are emitted over a range of energies. Negatron emitters have a characteristic energy spectrum (see Fig. 14.5b below). The maximum energy level (E_{max}) varies from one isotope to another, ranging from 0.018 MeV for ^{3}H to 4.81 MeV for ^{38}Cl. The difference in E_{max} affects the penetration of the radiation and therefore the safety measures that are required: β-particles from ^{3}H can travel only a few millimetres in air, whereas those from ^{32}P can penetrate over 1 m of air. Therefore radiation shields are needed when working with ^{32}P.

γ-Rays and X-rays

These rays (henceforth collectively referred to as γ-rays for simplicity) are electro-magnetic radiation and therefore have no charge or mass. They cause excitation and ionisation. They interact with matter to create secondary electrons that behave as per negatron emission.

Bremsstrahlung radiation

When high atomic number materials absorb high energy β-particles, the absorber gives out a secondary radiation, an X-ray, called bremsstrahlung radiation. For this reason, shields for ^{32}P use low-atomic-number materials such as acrylic.

14.3 DETECTION AND MEASUREMENT OF RADIOACTIVITY

There are three commonly used methods of detecting and quantifying radioactivity. These are based on the ionisation of gases, on the excitation of solids or solutions, and the ability of radioactivity to expose photographic emulsions (i.e. autoradiography).

14.3.1 Methods based upon gas ionisation

If a charged particle passes through a gas, its electrostatic field dislodges orbital electrons from atoms sufficiently close to its path and causes ionisation (Fig. 14.2). The ability to induce ionisation decreases in the order

$A > \beta > \gamma \ (10\,000 : 100 : 1)$

If ionisation occurs between a pair of electrodes enclosed in a suitable chamber (Fig. 14.2) a pulse (current) flows. Ionisation counters like those shown in Fig. 14.2 are sometimes called proportional counters ('proportional' because small voltage changes can affect the count rate). The Geiger–Müller counter (Figs. 14.3, 14.4a) has a cylindrical-shaped gas chamber and it operates at a high voltage. This makes the instrument less dependent on a stable voltage, so the counter is cheaper and lighter.

Example 2 COUNTING WITH IONISATION COUNTERS

Question Using the information described in Sections 14.2.3 and 14.2.7, what types of radiation can be detected by an ionisation counter?

Answer The counters will detect α-particles and medium- to high-energy β-particles; they will detect γ-radiation, but with lower efficiency. They will not detect tritium because the β-particles have very low energy and will not pass into the gas chamber.

In ionisation counters, the ions have to travel to their respective electrodes; other ionising particles entering the tube during this time (the so-called 'dead time') are not detected and this reduces the counting efficiency.

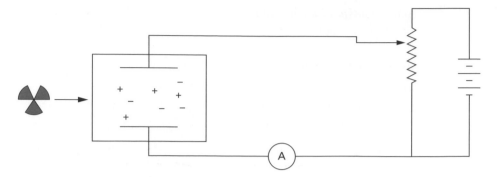

Fig. 14.2 Detection based on ionisation.

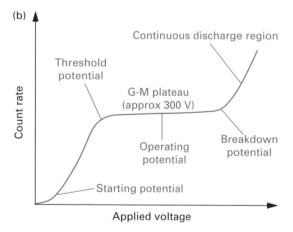

Fig. 14.3 (a) The Geiger–Müller (G–M) tube and (b) the effect of applied voltage on count rate.

Ionisation counters are used for routine monitoring of the laboratory to check for contamination. They are also useful in experimental situations where the presence or absence of radioactivity needs to be known rather than the absolute quantity, for example quick screening of radioactive gels prior to autoradiography, checking that

(a)

(b)

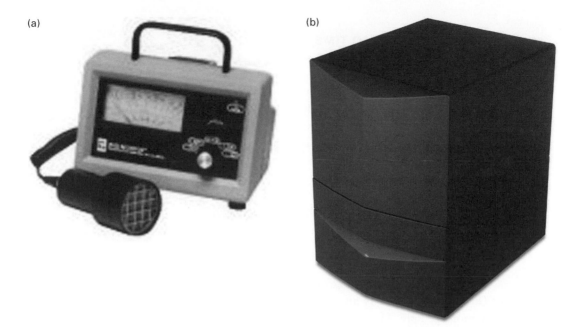

(c)

Fig. 14.4 (a) Geiger–Müller bench monitor; (b) liquid scintillation counter; (c) sample rack for liquid scintillation counter. (Reproduced with permission from LabLogic; instrument shown is the Hidex 300SI.)

Example 3 **THE EFFECT OF DEAD TIME**

Question   What do you think will happen to the counting efficiency of a Geiger–Müller counter as the count rate rises?

Answer   The efficiency will fall since there will be an increased likelihood that two or more β-particles will enter the tube during the dead time.

a labelled DNA probe is where you think it is (and not down the sink!) or checking chromatographic fractions for labelled components.

14.3.2   Methods based upon excitation

Radioactive isotopes interact with matter in two ways, ionisation and excitation. The latter effect leads an excited atom or compound (known as a fluor) to emit photons of light. The process is known as scintillation. When the light is detected by a photomultiplier, it forms the basis of scintillation counting. Essentially, a photomultiplier converts the energy of radiation into an electrical signal, and the strength of the electric pulse that results is directly proportional to the energy of the original radioactive event. This means that two, or even more, isotopes can be separately detected and measured in the same sample, provided they have sufficiently different emission energy spectra. The mode of action of a photomultiplier is shown in Fig. 14.5a, and the energy spectrum of a β-particle emitter in Fig. 14.5b.

Types of scintillation counting

There are two types of scintillation counting, which are illustrated diagrammatically in Fig. 14.6. In solid scintillation counting the sample is placed adjacent to a solid fluor (e.g. sodium iodide). Solid scintillation counting is particularly useful for γ-emitting isotopes. This is because they can penetrate the fluor. The counters can be small handheld devices with the fluor attached to the photomultiplier tube (Fig. 14.5.a), or larger bench-top machines with a well-shaped fluor designed to automatically count many samples (Fig. 14.6.a).

In liquid scintillation counting (Fig. 14.6b; see also Figs. 14.4b, c), the sample is mixed with a scintillation fluid containing a solvent and one or more dissolved fluors. This method is particularly useful in quantifying weak β-emitters such as ^{3}H, ^{14}C and ^{35}S, which are frequently used in biological work. Scintillation fluids are called 'cocktails' because there are different formulations, made of a solvent (such as toluene or diisopropylnaphthalene) plus fluors such as 2,5-diphenyloxazole (PPO), 1,4-bis(5-phenyloxazol-2-yl)benzene (nicknamed POPOP, pronounced as it reads: 'pop op') or 2-(4'-t-butylphenyl)-5-(4''-bi-phenyl)-1,3,4-oxydiazole (butyl-PBD). Cocktails can be designed for counting organic samples, or may contain detergent to facilitate counting of aqueous samples.

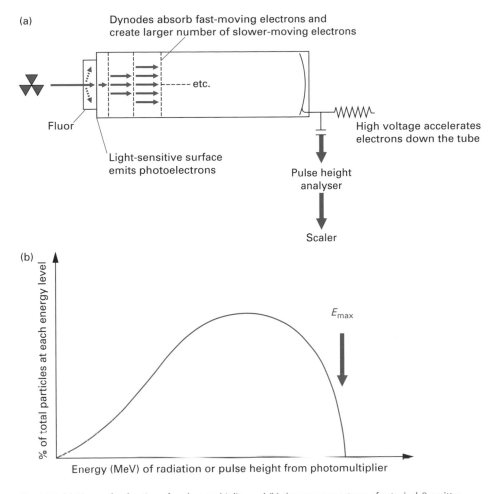

(a)

Dynodes absorb fast-moving electrons and create larger number of slower-moving electrons

etc.

Fluor

Light-sensitive surface emits photoelectrons

High voltage accelerates electrons down the tube

Pulse height analyser

Scaler

(b)

% of total particles at each energy level

E_{max}

Energy (MeV) of radiation or pulse height from photomultiplier

Fig. 14.5 (a) The mode of action of a photomultiplier and (b) the energy spectrum of a typical β-emitter.

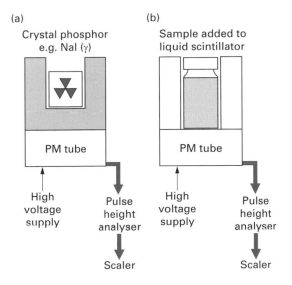

(a)

Crystal phosphor e.g. NaI (γ)

PM tube

High voltage supply

Pulse height analyser

Scaler

(b)

Sample added to liquid scintillator

PM tube

High voltage supply

Pulse height analyser

Scaler

Fig. 14.6 Diagrammatic illustration of (a) solid and (b) liquid scintillation counting methods.

Advantages of scintillation counting

Scintillation counting is widely used in biological work and it has several advantages over gas ionisation counting:

- fluorescence is very fast so there is effectively no dead time
- counting efficiencies are high (from about 50% for low-energy β-emitters to 90% for high-energy emitters)
- the ability to count samples of many types, including liquids, solids, suspensions and gels
- the general ease of sample preparation
- the ability to count separately different isotopes in the same sample (used in dual-labelling experiments)
- highly automated (hundreds of samples can be counted automatically and built-in computer facilities carry out many forms of data analysis, such as efficiency correction, graph plotting, radioimmunoassay calculations, etc.).

Disadvantages of scintillation counting

No technique is without disadvantages, so the following have to be considered or overcome in the design of the instruments:

- cost of the instrument and cost per sample (for scintillation fluid, the counting vials and disposal of the organic waste)
- potentially high background counts; this is due to photomultiplier noise but can be compensated for by using more than one tube (noise is random, but counts from a radioactive decay are simultaneous, the coincident counts only are recorded)
- 'quenching': this is the name for reduction in counting efficiency caused by coloured compounds that absorb the scintillated light, or chemicals that interfere with the transfer of energy from the radiation to the photomultiplier (correcting for quenching contributes significantly to the cost of scintillation counting)
- chemiluminescence: this is when chemical reactions between components of the samples to be counted and the scintillation cocktail produce scintillations that are unrelated to the radioactivity; modern instruments can detect chemiluminescence and subtract it from the results automatically
- phospholuminescence: this results from pigments in the sample absorbing light and re-emitting it; the solution is to keep the samples in the dark prior to counting.

Using scintillation counting for dual-labelled samples

Different β-particle emitters have different energy spectra, so it is possible to quantify two isotopes separately in a single sample, provided their energy spectra can be distinguished from each other. Examples of pairs of isotopes that can be counted together are: ^{3}H and ^{14}C, ^{3}H and ^{35}S, ^{3}H and ^{32}P, ^{14}C and ^{32}P, ^{35}S and ^{32}P. The principle of the method is illustrated in Fig. 14.7, where it can be seen that the spectra of two isotopes (referred to as S and T) overlap only slightly. By setting a pulse height analyser to reject all pulses of an energy below X (threshold X) and to reject all pulses of an energy above Y (window Y) and also to reject below a threshold of A and a

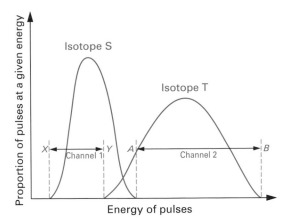

Fig. 14.7 Diagram to illustrate the principle of counting dual-labelled samples.

window of B, it is possible to separately count the two isotopes. A pulse height analyser set with a threshold and window for a particular isotope is known as a channel. Modern counters operate with a so-called multichannel analyser that records the entire energy spectrum simultaneously. This greatly facilitates multi-isotope counting and in particular allows the effect of quenching on dual-label counting to be assessed adequately.

Dual-label counting has proved to be useful in many aspects of molecular biology (e.g. nucleic acid hybridisation and transcription), metabolism (e.g. steroid synthesis) and drug development.

Determination of counting efficiency

When we detect radioactivity we usually need to know the actual rate of decay (the d.p.m.) (see Section 14.2.6). This is because we want to record and publish data that are independent of the types of equipment used to do the detection. To calculate d.p.m. we need to know the efficiency of counting. What's more, in liquid scintillation counting we have to contend with quenching. Samples may vary in nature so the levels of quenching may vary from one to the next. Therefore the efficiency of counting needs to be determined for every sample.

One way to do this is to use an internal standard (called a spike). The sample is counted (and gives a reading of, say, A c.p.m.), removed from the counter and a small amount of standard material of known disintegrations per minute (B d.p.m.) is added. The sample is then recounted (C c.p.m.) and the counting efficiency of the sample calculated:

$$\text{counting efficiency} = [100(C - A)/B]\%$$
(14.4)

Carefully carried out, it is the most accurate way of correcting for quenching. On the other hand, it is tedious since the process has to be done for every sample. Therefore automated methods have been devised; however, these all use the internal standard as the basis for establishing the parameters.

As a sample is quenched, the efficiency of light production falls, and it therefore creates an illusion that the radiation has a lower energy. The principle is straightforward: observe how the energy spectrum shifts as the efficiency falls, store this relationship in a computer, and then analyse the energy spectrum of the sample to determine the efficiency. However, there is a catch: different isotopes have different energies so we need to be sure that we are measuring the effect of quench and not just the fundamental differences in isotopes. This is resolved by using an 'external standard' source of radioactivity built into the counter. It's called an external standard because it is placed just outside the sample vial by a mechanical device in the instrument. The quenching in the sample is observed by counting this standard, the external source is then moved away, and the experimental sample counted. This is done for every sample in turn and the counter prints out corrected d.p.m. automatically. To set all this up a standard curve using a set of quenched standards is counted: the absolute amount of radioactivity in the standard is known and therefore the efficiency of counting can easily be determined. These are the data that are stored in the computer as a standard curve.

Some instruments (e.g. the Hidex 300 SL, shown in Figs. 14.4, and 14.8) on the market calculate counting efficiency in a totally different way. The counters use three photomultiplier tubes. The mathematics of the process are beyond the scope of this

Example 4 EXTERNAL STANDARD EFFICIENCY CALCULATIONS

A scintillation counter analyses the energy spectrum of an external standard, it records a point on the spectrum (the quench parameter, QP) and assesses the shift in spectrum as the efficiency falls. The efficiency of detecting ^{14}C in a scintillation counter is determined by counting a standard sample containing 105 071 d.p.m. at different degrees of quench (by adding increasing amounts of a quenching chemical such as chloroform). The results look like this:

c.p.m	QP
87 451	0.90
62 361	0.64
45 220	0.46
21 014	0.21

Then an experimental sample gives 2026 c.p.m. at a QP of 0.52. What is the true count rate?

Firstly, the counting efficiency of the quenched standards needs to calculated; the efficiency is the ratio of the c.p.m. to the d.p.m. ($\times 100$ if expressed as a %). Then the efficiency is plotted against the QP. The QP for the experimental sample (0.52) is put into the curve and the efficiency read (in this case 48%). This is then used to calculate the true d.p.m.: $2026 \times 100/48$. The answer is 4221 d.p.m.

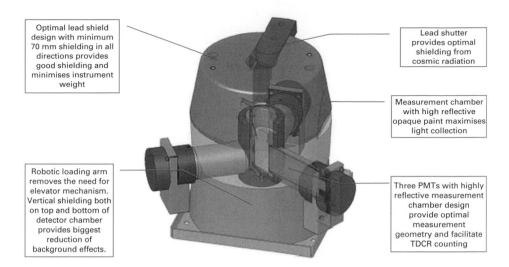

Optimal lead shield design with minimum 70 mm shielding in all directions provides good shielding and minimises instrument weight

Lead shutter provides optimal shielding from cosmic radiation

Measurement chamber with high reflective opaque paint maximises light collection

Robotic loading arm removes the need for elevator mechanism. Vertical shielding both on top and bottom of detector chamber provides biggest reduction of background effects.

Three PMTs with highly reflective measurement chamber design provide optimal measurement geometry and facilitate TDCR counting

Fig. 14.8 The arrangement of photomultiplier tubes in a liquid scintillation counter. (Reproduced with permission from LabLogic; instrument shown is the Hidex 300SL.)

chapter but the chances of either three or two tubes detecting a signal is affected by the extent of quench. Therefore the ratio of triple to double coincidence is related to the counting efficiency. Such counters are smaller, and do not require a built-in radioactive source.

Sample preparation

For solid scintillation counting, sample preparation is easy and only involves transferring the sample to a glass or plastic vial (or tube) compatible with the counter. In liquid scintillation counting, sample preparation is more complex and starts with a decision on the type of sample vial to be used (glass; low-potassium glass, with low levels of ^{40}K that reduce background count; or polyethylene, cheaper but not re-usable). Vials need to be chemically resistant, have good light transmission and give low background counts. The trend has been towards mini-vials, which use smaller volumes of scintillation fluid. Some counters are designed to accept very small samples in special polythene bags split into an array of many compartments; these are particularly useful to, for example, the pharmaceutical industry where there are laboratories that do large numbers of receptor binding assays. Accurate counting depends on the sample being in the same phase as the counting cocktail. As described above the scintillation fluid should be chosen as appropriate to aqueous or organic samples.

If colour quenching is a problem it is possible to bleach samples before counting. Solid samples such as plant and animal tissues may be counted after solubilisation by quaternary amines such as NCS solubiliser or Soluene. Not surprisingly these solutions are highly toxic and great care is required.

Radioactive compounds are often separated by HPLC. The output of an HPLC instrument can be connected to a flow cell system where scintillation fluid is added

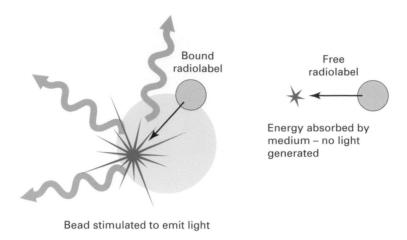

Bound
radiolabel

Free
radiolabel

Energy absorbed by
medium – no light
generated

Bead stimulated to emit light

Fig. 14.9 The concept behind SPA. (Reproduced by courtesy of Amersham Biosciences.)

to the effluent prior to entering a detector. This significantly increases automation and reduces waste.

Čerenkov counting

The Čerenkov effect occurs when a particle passes through a substance with a speed higher than that of light passing through the same substance. If a β-emitter has a decay energy in excess of 0.5 MeV, then this causes water to emit a bluish white light usually referred to as Čerenkov light. It is possible to detect this light using a typical liquid scintillation counter. Since there is no requirement for organic solvents and fluors, this technique is relatively cheap, sample preparation is very easy, and there is no problem of chemical quenching. Table 14.2 indicates which isotopes can be counted this way.

Scintillation proximity assay

Scintillation proximity assay (SPA) is an application of scintillation counting that facilitates automation and rapid throughput of experiments. It is therefore highly suited to work such as screening for biological activity in new drugs. The principle of SPA is illustrated in Fig. 14.9. The beads for SPA are constructed from polystyrene (or sometimes other materials) that combine a binding site for a molecule of interest with a scintillant. You need to remember that some types of radiation do not travel far, in particular β-particles from weak energy emitters such as ^{3}H and ^{14}C. If molecules containing such radioisotopes are in solution with a suspension of SPA beads, the radiation does not stimulate the scintillant in the beads and cannot be detected efficiently by a scintillation counter. This is because the radiation is absorbed by the solution; it does not reach the scintillant. If, on the other hand, the radioisotope becomes bound to the bead, it is close enough to stimulate the scintillant in the bead, so light is given out and the isotope is detected.

Table 14.4 **Advantages of scintillation proximity assay**

Versatile: use with enzyme assays, receptors, any molecular interactions
Works with a range of appropriate isotopes such as ^{3}H, ^{14}C, ^{35}S and ^{33}P
No need for separation step (e.g. free from bound ligand)
Less manipulation therefore reduced toxicity
Amenable to automation

There are many applications of this technology such as enzyme assays and receptor binding, indeed any situation where we want to investigate the interaction between two molecules. Take receptor binding as an example. In this case a receptor for a particular ligand (such as a drug or hormone) is attached to the SPA beads. The ligand is radiolabelled and mixed with the beads. Any ligand that binds will stimulate the scintillant and be counted. If the researcher wishes to investigate chemicals that might interface with this binding (which is the mode of action of many medicines), they can be added at increasing concentration to study the effect and, for example, determine optimum dosage (see also Section 16.3.2).

A summary of the advantages of SPA technology is shown in Table 14.4.

14.3.3 Methods based upon exposure of photographic emulsions

Ionising radiation acts upon a photographic emulsion or film to produce a latent image much as does visible light. This is called autoradiography. The emulsion or film contains silver halide crystals. As energy from the radioactive material is dissipated the silver halide becomes negatively charged and is reduced to metallic silver, thus forming a particulate latent image. Photographic developers show these silver grains as a blackening of the film, then fixers are used to remove any remaining silver halide and a permanent image results.

It is a very sensitive technique and has been used in a wide variety of biological experiments. A good example is autoradiography of nucleic acids separated by gel electrophoresis (see Fig. 14.10).

Suitable isotopes

In general, weak β-emitting isotopes (e.g. ^{3}H, ^{14}C and ^{35}S) are most suitable for autoradiography, particularly for cell and tissue localisation experiments. This is because the energy of the radiation is low. The sample must be close to the film, the radiation does not spread out very far and so a clear image results. Radiation with higher energy (e.g. ^{32}P) give faster results but poorer resolution because the higher-energy negatrons produce much longer track lengths, exposing a greater surface area of the film, and result in less discrete images. This is illustrated in Fig. 14.10, showing autoradiography with three different isotopes.

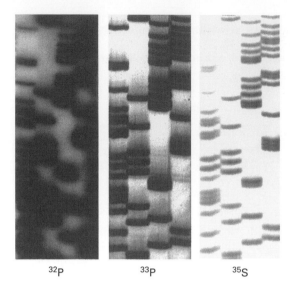

^{32}P ^{33}P ^{35}S

Fig. 14.10 Three autoradiographs showing the use of different radioisotopes in DNA sequencing. The isotope with the highest energy (^{32}P) leads to the poorest resolution because the radiation spreads out further, making the DNA bands appear thicker. The lowest energy radiation (from ^{35}S) gives the best resolution. (Reproduced with permission from M. W. Cunningham, A. Patel, A. C. Simmonds and D. Williams (2002), *In vitro* labelling of nucleic acids and proteins, in *Radioisotopes in Biology,* (2nd edn), R. J. Slater (ed.), Oxford University Press, Oxford.)

Choice of emulsion and film

Autoradiography emulsions are solutions of silver halide that can be made to set solid by the inclusion of materials such as gelatine. This can be used for example for autoradiography of microscope slides. X-ray film is the alternative and is used for gels (as shown in Fig. 14.10). Films differ in sensitivity; advice on what to use is provided by the manufacturers.

Direct autoradiography

In direct autoradiography, the X-ray film or emulsion is placed as close as possible to the sample and exposed at any convenient temperature. Quantitative images are produced until saturation is reached. The shades of grey in the image are related to a combination of levels of radiation and length of exposure until a black or nearly black image results. Isotopes with an energy of radiation equal to, or higher than, ^{14}C ($E_{\max}= 0.156$ MeV) are required. The higher the energy the quicker the results.

Fluorography

If low-energy β-emitters are used it is possible to enhance the sensitivity several orders of magnitude by using fluorography. A fluor (e.g. PPO or sodium silicate) can be used to enhance the image. The β-particles emitted from the isotope will cause the fluor to become excited and emit light, which will react with the film. This has been used for example for detecting radioactive nucleic acids in gels. The fluor is infiltrated into the gel following electrophoresis; the gel is dried and then placed in contact with a preflashed film (see below).

Intensifying screens

Intensifying screens are used when obtaining a fast result is more important than high resolution. It is useful for example in gel electrophoresis or analysis of membrane filters where high-energy β-emitters (e.g.^{32}P-labelled DNA) or γ-emitting isotopes (e.g. ^{125}I-labelled protein) are used. The intensifying screen consists of a solid phosphor, and it is placed on the other side of the film from the sample. High-energy radiation passes through the film, causes the phosphor to fluoresce and emit light, which in turn superimposes its image on the film. The reduction in resolution is due to the spread of light emanating from the screen.

Low-temperature exposure

When intensifying screens or fluorography are used the exposure should be done at low temperature. This is because the kinetics of the film's response are affected. The light is of low intensity and a back reaction occurs that cancels the latent image. Exposure at low temperature ($-70\,^\circ$C) slows this back reaction and will therefore provide higher sensitivity. There is no point in doing direct autoradiography at low temperature as the kinetic basis of the film's response is different.

Preflashing

The response of a photographic emulsion to radiation is not linear and usually involves a slow initial phase (lag) followed by a linear phase. Sensitivity of films may be increased by preflashing. This involves a millisecond light flash prior to the sample being brought into juxtaposition with the film and is often used where high sensitivity is required or if results are to be quantified.

Quantification

Autoradiography is usually used to locate rather than to quantify radioactivity. However, it is possible to obtain quantitative data directly from autoradiographs by using digital image analysis. Quantification is not reliable at low or high levels of exposure because of the lag phase (see preflashing above) or saturation, respectively. Preflashing combined with fluorography or intensifying screens create the best conditions for quantitative working.

14.4 OTHER PRACTICAL ASPECTS OF COUNTING RADIOACTIVITY AND ANALYSIS OF DATA

14.4.1 Self-absorption

Self-absorption is primarily a problem with low energy β-emitters: radiation is absorbed by the sample itself. Self-absorption can be a serious problem in the counting of low-energy radioactivity by scintillation counting if the sample is particulate or is, for instance, stuck to a membrane filter. Automated methods for calculating counting efficiency in a scintillation counter will not correct for self-absorption effects. Particulate samples should be digested or otherwise solubilised prior to counting if quench correction is required.

14.4.2 Specific activity

The specific activity of a radioisotope defines its radioactivity in relation to the amount of material, expressed by units such as $Bq\ mol^{-1}$, $Ci\ mmol^{-1}$ or $d.p.m.\ mmol^{-1}$. It is a very important aspect of the use of radioisotopes in biological work because the higher the specific activity the more sensitive the experiment. This is because the higher the specific activity the smaller the quantities of labelled substance that can be detected. The highest specific activities are associated with isotopes with short half-lives, since the rate of decay per unit mass (or mol) is higher.

Sometimes, it is not necessary to purchase the highest specific activity available. For example, enzyme assays *in vitro* often require a relatively high substrate concentration and so specific activity may need to be lowered. Consider the example below (for definitions of units, see Table 14.6 at the end of the chapter):

$[^3H]$Leucine is purchased with a specific activity of $5.55\ TBq\ mmol^{-1}$ ($150\ Ci\ mmol^{-1}$) and a concentration of $9.25\ MBq\ 250\ mm^{-3}$ ($250\ mCi\ 250\ mm^{-3}$). A $10\ cm^3$ solution of $250\ mM$ and $3.7\ kBq\ cm^{-3}$ ($0.1\ mCi\ cm^{-3}$) is required. It is made up as follows:

- $10\ cm^3$ at $3.7\ kBq\ cm^{-3}$ is $37\ kBq$ ($1\ mCi$), therefore pipette $1\ mm^3$ of stock radioisotope into a vessel (or, to be more accurate, pipette $100\ mm^3$ of a $\times 100$ dilution of stock in water).
- Add $2.5\ cm^3$ of a $1\ M$ stock solution of cold leucine, and make up to $10\ cm^3$ with distilled water.

Note that there is no need in this case to take into account the amount of leucine in the $[^3H]$leucine preparation; it is a negligible quantity due to the high specific activity.

If necessary (e.g. to manipulate solutions of relatively low specific activity), however, the following formula can be applied:

$$W = Ma[(1/A') - (1/A)] \tag{14.5}$$

where W is the mass of cold carrier required (mg), M is the amount of radioactivity present (MBq), a is the molecular weight of the compound, A is the original specific activity (MBq mmol^{-1}), and A' is the required specific activity (MBq mmol^{-1}).

14.4.3 Statistics

The emission of radioactivity is a random process. The spread of results forms a normal distribution. The standard deviation can be calculated very simply (using mathematics devised by Poisson) by taking a square root of the counts:

$$\sigma = \sqrt{\text{total counts taken}}$$

Essentially this means that the more counts we take the smaller the standard deviation is as a proportion of the mean count rate. Put simply, the more counts measured the more accurate the data.

Example 5 **MAKING UP A SOLUTION OF KNOWN ACTIVITY**

Question One litre of [^{3}H]uridine with a concentration of 100 mmol cm^{-3} and 50 000 c.p.m. cm^{-3} is required. If all measurements are made on a scintillation counter with an efficiency of 40%, how would you make up this solution if the purchased supply of [^{3}H]uridine has a radioactive concentration of 1 mCi cm^{-3} and a specific activity of 20 Ci mol^{-1}, 0.75 TBqmol^{-1}?
[NB: M_r uridine = 244; 1 Ci=22.2 × 10^{11} d.p.m.]

Answer This problem is similar to the leucine example given above. Correcting for the 48% counting efficiency: 50 000 c.p.m. is 125 000 d.p.m. Multiplying this by 10^3 for a litre gives a d.p.m. equivalent to 56.3 μCi (125 × 10^6/22.2 × 10^5 = 56.3 μCi). Given 20 Ci mol^{-1}, work out how many moles there are in 56.3 μCi (56.3/20 × 10^6 = 2.815 mmoles). 100 000 mmoles of uridine are required in a litre; from the molecular mass this is 24.4 g. The 2.815 mmoles from the radioactive input is only 0.685 mg and so can effectively be ignored. The answer is, therefore, 56.3 mm^3 (56.3 μCi, 2.08 MBq) of [^{3}H]uridine plus 24.4 g of uridine.

Consider these simple examples for a series of 1 min counts:

counts = 100	$\sigma = \sqrt{\text{total counts}} = 10$	σ is 10% of the mean
counts = 1000	$\sigma = \sqrt{\text{total counts}} = 33$	σ is 3% of the mean
counts = 10 000	$\sigma = \sqrt{\text{total counts}} = 100$	σ is 1% of the mean

It is common practice to count to 10 000 counts or for 10 min, whichever is the quicker, although for very low count rates longer counting times are required. Another common practice is to quote mean results plus or minus 2 standard deviations, since 95.5% of results lie within this range.

Example 6 **ACCURACY OF COUNTING**

Question A sample recording 564 c.p.m. was counted over 10 min. What is the accuracy of the measurement for 95.5% confidence?

Answer 5640 counts were recorded (564 × 10); the square root is 75. Therefore the range is 5640 ± 150 for 95.5% confidence, or 564 ± 15 c.p.m., an acceptable level of accuracy.

14.4.4 The choice of radionuclide

This is a complex question depending on the precise requirements of the experiment. A summary of some of the key features of radioisotopes commonly used in biological

Table 14.5 **The relative merits of commonly used radioisotopes**

Isotope	Advantages	Disadvantages
^{3}H	Relative safety	Low efficiency of detection
	High specific activity possible	Isotope exchange with environment
	Wide choice of positions in organic compounds	Isotope effect
	Very high resolution in autoradiography	
^{14}C	Relative safety	Low specific activity
	Wide choice of labelling position in organic compounds	
	Good resolution in autoradiography	
^{35}S	High specific activity	Short half-life
	Good resolution in autoradiography	Relatively long biological half-life
^{33}P	High specific activity	Lower specific activity than ^{32}P
	Good resolution in autoradiography	Less sensitive than ^{32}P
	Less hazardous than ^{32}P	Cost
^{32}P	Ease of detection	Short half-life affects costs and experimental design
	High specific activity	
	Short half-life simplifies disposal	High β energy so external radiation hazard
	Čerenkov counting	Poor resolution in autoradiography
^{125}I	Ease of detection	High penetration of radiation
	High specific activity	
	Good for labelling proteins	
^{131}I	Ease of detection	High penetration of radiation
	High specific activity	Short half-life

work is shown in Table 14.5. The key factors in the decision are often based on safety, the type of detection to be used, the sensitivity required (see section 14.4.2) and the cost. For example ^{33}P may be chosen for work with DNA because it has high enough energy to be detected easily, it is safer than ^{32}P and its half-life is short enough to give high specific activity but long enough to be convenient to use.

Although they undergo the same reactions, different isotopes may do so at different rates. This is known as the isotope effect. The different rates are approximately proportional to the differences in mass between the isotopes. This can be a problem in the case of ^1H and ^3H, but the effect is small for ^{12}C and ^{14}C and almost insignificant for ^{33}P and ^{32}P. The isotope effect may be taken into account when choosing which part of a molecule to label with ^3H.

14.5 SAFETY ASPECTS

The greatest practical disadvantage of using radioisotopes is the toxicity: they produce ionising radiations. When absorbed, radiation causes ionisation and free radicals form that interact with the cell's macromolecules, causing mutation of DNA and hydrolysis of proteins. The toxicity of radiation is dependent not simply on the amount present but on the amount absorbed by the body, the energy of the absorbed radiation and its biological effect. There are, therefore, a series of additional units used to describe these parameters.

A key aspect of determining toxicity is to know how much energy might be absorbed, just as too much sun gives you sunburn and potentially skin cancer. The higher the energy of the radiation the greater the potential hazard. The gray (Gy), an SI unit, is the unit used to describe this; 1 Gy is an absorption of $1 \, J \, kg^{-1}$ of absorber. The gray (Gy) is a useful unit, but it still does not adequately describe the hazard to living organisms. This is because different types of radiation are associated with differing degrees of biological hazard. It is, therefore, necessary to introduce a correction factor, which is calculated by comparing the biological effects of any type of radiation with that of X-rays. The unit of absorbed dose, which takes into account this weighting factor is the sievert (Sv) and is known as the equivalent dose. For β-radiation Gy and Sv are the same, but for α-radiation 1 Gy is 20 Sv. In other words α-radiation is 20 times as toxic to humans as X-rays for the same energy absorbed. Clearly, from the point of view of safety, it is advisable to use radioisotopes with low energy wherever possible.

Absorbed dose from known sources can be calculated from knowledge of the rate of decay of the source, the energy of radiation, the penetrating power of the radiation and the distance between the source and the laboratory worker. As the radiation is emitted from a source in all directions, the level of irradiation is related to the area of a sphere. Thus the absorbed dose is inversely related to the square of the distance (the radius of the sphere) from the source; or, put another way, if the distance is doubled the dose is quartered. A useful formula is:

$$dose_1 \times distance_1^2 = dose_2 \times distance_2^2 \tag{14.6}$$

The relationship between radioactive source and absorbed dose is illustrated in Fig. 14.11. The rate at which dose is delivered is referred to as the dose rate, expressed in $Sv \, h^{-1}$. It can be used to calculate your total dose. For example, a source may be delivering $10 \, m \, Sv \, h^{-1}$. If you worked with the source for 6 h, your total dose would be 60 mSv. Dose rates for isotopes are provided in Table 14.2.

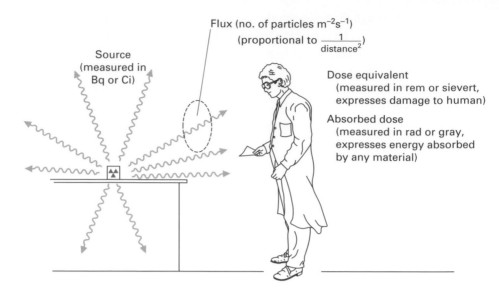

Fig. 14.11 The relationship between radioactivity of source and absorbed dose.

Example 7 **CALCULATION OF DOSE**

Question   A 1 mCi source of ^{125}I gives a dose of $10\,\mathrm{mSv\,h^{-1}}$ at 1 cm. What will be the dose rate at 5 cm?

Answer   Using the formula above $10 \times 1^2 = $ new dose $\times\, 5^2$.

Therefore the new dose is $10/25 = 0.4\,\mathrm{mSv\,h^{-1}}$.

Currently the dose limit for workers exposed to radiation is 20 mSv in a year to the whole body, but this is rarely ever approached by biologists because the levels of radiation used are so low. Limits are set for individual organs. The most important of these to know are for hands (500 mSv per year) and for lens of the eye (150 mSv per year).

Dose limits are constantly under review and, although limits are set, it is against internationally agreed guidelines to work up to such a limit, that is we are not allowed to assume that all is satisfactory if the limit is not exceeded. Instead, the ALARA principle is applied, to work always to a dose limit that is As Low As Reasonably Achievable. Work that may cause a worker to exceed three-tenths or one-tenth of the dose limit must be carried out in a controlled area or a supervised area, respectively. In practice, work in the biosciences rarely involves a worker receiving a measurable dose. Supervised areas are common but not always required (e.g. for ^{3}H or ^{14}C experiments). Controlled areas are required only in certain circumstances, for example for isotope stores or radioiodination work. A potential problem, however, in biosciences is the internal radiation hazard. This is caused by radiation entering the body, for example by inhalation, ingestion, absorption or puncture. This is a likely source of hazard where work involves open sources (i.e. liquids

Table 14.6 **Units commoly used to describe radioactivity**

Unit	Abbreviation	Definition
Counts per minute or second	c.p.m.	The *recorded* rate of decay
	c.p.s.	
Disintegrations per minute or second	d.p.m.	The *actual* rate of decay
	d.p.s.	
Curie	Ci	The number of d.p.s. equivalent to 1 g of radium $(3.7 \times 10^{10}$ d.p.s.)
Millicurie	mCi	$Ci \times 10^{-3}$ or 2.22×10^9 d.p.m.
Microcurie	μCi	$Ci \times 10^{-6}$ or 2.22×10^6 d.p.m.
Becquerel (SI unit)	Bq	1 d.p.s.
Terabecquerel (SI unit)	TBq	10^{12} Bq or 27.027 Ci
Gigabecquerel (SI unit)	GBq	10^9 Bq or 27.027 m Ci
Megabecquerel (SI unit)	MBq	10^6 Bq or 27.027 μCi
Electron volt	eV	The energy attained by an electron accelerated through a potential difference of 1 volt. Equivalent to 1.6×10^{-19} J
Roentgen	R	The amount of radiation that produces 1.61×10^{15} ion-pairs kg^{-1}
Rad	rad	The dose that gives an energy absorption of 0.01 J kg^{-1}
Gray	Gy	The dose that gives an energy absorption of 1 J kg^{-1}. Thus 1 Gy=100 rad
Rem	rem	The amount of radiation that gives a dose in humans equivalent to 1 rad of X-rays
Sievert	Sv	The amount of radiation that gives a dose in humans equivalent to 1 Gy of X-rays. Thus 1 Sv=100 rem

and gases); most work in biology involves manipulations of radioactive liquids. Control of contamination is assisted by:

- complying with local rules, written by an employer
- conscientious personal conduct in the laboratory
- regular monitoring
- carrying out work in some kind of containment.

A useful guide for internal risks is the annual limit on intake (ALI). The ingestion of one ALI results in a person receiving a dose limit to the whole body or to a particular organ. Some ALIs are shown in Table 14.2. Management of radiation protection is similar in most countries. In the USA, there is a Code of Federal Regulations. In the UK there is the Radiaoactive Substances Act (1993) and the Ionising Radiations Regulations (1999). Every institution requires certification (monitored by the Environmental Protection Agency in the USA or the Environment Agency in the UK) and employs a Radiation Protection Advisor.

When planning to use a radioisotope consider the following:

(1) Is a radioisotope necessary?

If the answer is no, then a non-radioactive method should be used.

(2) Which isotope to use?

Ideally the one with the lowest energy that can deliver your needs.

When handling radioisotopes the rules are to:

- Wear protective clothing, gloves and glasses
- Use the smallest amount possible
- Keep radioactive materials safe, secure and well labelled
- Work in defined areas in a spill tray
- Monitor your working area frequently
- Have no foods or drinks in the laboratory
- Wash and monitor hands after the work is done
- Follow all local rules such as for the dispensing of stock and the disposal of waste
- Do not create radioactive aerosols or dust

and for penetrating radiations (e.g. ^{32}P and λ-emitters):

- Maximise the distance between yourself and the source
- Minimise the time of exposure
- Maintain shielding at all times.

14.6 SUGGESTIONS FOR FURTHER READING

Billington, D., Jayson, G. G. and Maltby, P. J. (1992). *Radioisotopes*. Oxford: Bios Scientific. (A description of principles and applications in the biosciences, for undergraduates and research workers.)

Connor, K. J. and McLintock, I. S. (1994). *Radiation Protection Handbook for Laboratory Workers*. Leeds: HHSC. (A safety manual for laboratory work.)

Slater, R. J. (1996). Radioisotopes in molecular biology. In *Molecular Biology and Molecular Medicine*, ed. R. A. Myers, pp. 209–219. New York: VCH. (A summary of the application of radioisotopes to molecular biology.)

Slater, R. J. (2002). *Radioisotopes in Biology: A Practical Approach*, 2nd edn. Oxford: Oxford University Press. (A detailed account of the handling and use of radioactivity in biological research.)

Wolfe, R. R. and Chinkes D. L. (2004). *Isotope Tracers in Metabolic Research: Principles and Practice of Kinetic Analysis*, 2nd edn. New York: John Wiley. (A detailed description of the use of radioactivity for the study of metabolism.)

15 Enzymes

K. WILSON

15.1 CHARACTERISTICS AND NOMENCLATURE

15.1.1 Specificity and nomenclature

Enzymes are nature's biological catalysts possessing the ability to promote specific chemical reactions under the mild conditions that prevail in most living organisms. They are all proteins but range widely in their size from as few as 60–70 amino acid residues as in RNase to as many as several thousand. Generally they are much larger than their substrates and bind with them by means of active sites created by the specific three-dimensional folding of the protein. Interaction of specific functional groups in a small number of amino acid residues lining the active site with the substrate results in the formation of a transition state for which the activation energy barrier is significantly reduced relative to the non-enzyme-catalysed reaction. As a result, the reaction rate is increased by a factor of many millions relative to the uncatalysed reaction. Enzymes do not alter the position of equilibrium of reversible reactions that they catalyse but they do accelerate the establishment of the position of equilibrium for the reaction.

Many enzymes are members of coordinated metabolic or signalling pathways that collectively are responsible for maintaining a cell's metabolic needs under varying physiological conditions (Sections 15.5 and 17.4.5). The over- or under-expression of an enzyme can lead to cell dysfunction which we may recognise as a particular disease state. Enzyme inhibitors are widely used as therapeutic agents for the treatment of such conditions (Sections 15.2 and 18.1). Organ damage, for example heart muscle as a result of deprivation of oxygen following a heart attack, or the liver as a result of chemical damage as in alcoholic cirrhosis, results in the release of cellular enzymes into extracellular fluids and eventually into the blood. Such release can be clinically monitored to aid diagnosis of the organ damage and to make a prognosis for the patient's future recovery (Section 16.3).

Enzymes are believed to catalyse over 4 000 different reactions but individual enzymes are characterised by their specificity for a particular type of chemical reaction. As a generalisation, enzymes involved in biosynthetic or signalling reactions show a higher specificity than ones involved in degradation reactions. Bond specificity is characteristic of enzymes such as peptidases and esterases that hydrolyse specific bond types. The specificity of these enzymes is determined by the presence of specific functional groups within the substrate adjacent to the bond to be cleaved. Group specificity is characteristic of enzymes that promote a particular reaction on a structurally related group of substrates. As an example, the kinases catalyse the phosphorylation of substrates that have a common structural feature such as a particular amino acid (e.g. the tyrosine kinases, see Section 17.4.4) or sugar (e.g. hexokinase). DNA polymerase has a high specificity not only copying the base sequence of the DNA but also checking the product for accuracy afterwards. Enzymes may also display stereospecificity and be able to distinguish between optical and geometrical isomers of substrates. Enzymes have a high capacity for regulation in that the activity of enzymes that control the rate of a particular metabolic or signalling pathway can be enhanced or reduced in response to changing intracellular and extracellular demands. A range of regulatory mechanisms operates to allow short, medium and long-term changes in activity (Section 15.5.2).

Nomenclature and classification

By international convention, each enzyme is classified into one of six groups on the basis of the type of chemical reaction that it catalyses. Each group is divided into subgroups according to the nature of the chemical group and coenzymes involved in the reaction. In accordance with the Enzyme Commission (EC) rules, each enzyme can be assigned a unique four-figure code and an unambiguous systematic name based upon the reaction catalysed. The six groups are:

- *Group 1*: Oxidoreductases, which transfer hydrogen or oxygen atoms or electrons from one substrate to another. This group includes the dehydrogenases, reductases, oxidases, dioxidases, hydroxylases, peroxidases and catalase.
- *Group 2*: Transferases, which transfer chemical groups between substrates. The group includes the kinases, aminotransferases, acetyltransferases and carbamyltransferases.
- *Group 3*: Hydrolases, which catalyse the hydrolytic cleavage of bonds. The group includes the peptidases, esterases, phosphatases and sulphatases.
- *Group 4*: Lyases, which catalyse elimination reactions resulting in the formation of double bonds. The group includes adenylyl cyclase (also known an adenylate cyclase), enolase and aldolase.
- *Group 5*: Isomerases, which interconvert isomers of various types by intramolecular rearrangements. The group includes phosphoglucomutase and glucose-6-phosphate isomerase.

- *Group 6*: Ligases (also called *synthases*), which catalyse covalent bond formation with the concomitant breakdown of a nucleoside triphosphate, commonly ATP. The group includes carbamoyl phosphate synthase and DNA ligase.

As an example of the operation of these rules, consider the enzyme alcohol dehydrogenase which catalyses the reaction:

$$\text{alcohol} + \text{NAD}^+ \rightleftharpoons \text{aldehyde or ketone} + \text{NADH} + \text{H}^+$$

It has the systematic name alcohol : NAD oxidoreductase and the classification number 1:1:1:1. The first 1 indicates that it is an oxidoreductase, the second 1 that it acts on a CH–OH donor, the third 1 that NAD^+ or NADP^+ is the acceptor and the fourth 1 that it is the first enzyme named in the 1:1:1 subgroup. Systematic names tend to be user-unfriendly and for day-to-day purposes recommended trivial names are preferred. When correctly used they give a reasonable indication of the reaction promoted by the enzyme in question but they fail to identify fully all the reactants involved. For example, glyceraldehyde-3-phosphate dehydrogenase fails to identify the involvement of orthophosphate and NAD^+ and phosphorylase kinase fails to convey the information that it is the *b* form of phosphorylase that is subject to phosphorylation involving ATP.

Cofactors

The catalytic properties of an enzyme are often dependent upon the presence of non-peptide molecules called cofactors or coenzymes. These may be either weakly or tightly bound to the enzyme; in the latter case they are referred to as a prosthetic group. Examples of coenzymes include NAD^+, NADP^+, FMN and FAD, whilst examples of prosthetic groups include haem and oligosaccharides, and simple metal ions such as Mg^{2+}, Fe^{2+} and Zn^{2+}. DNA and RNA polymerases and many nucleases, for example, require two divalent cations for their active site. The cations correctly orientate the substrate and promote acid–base catalysis.

15.1.2 Isoenzymes and multienzyme complexes

Isoenzymes

Some enzymes exist in multiple forms called isoenzymes or isoforms that differ in amino acid sequence. An example is lactate dehydrogenase (LD) (EC 1:1:1:27) which exists in five isoforms. LD is a tetramer which can be assembled from two subunits, H (for heart) and M (for muscle). The five forms are therefore H4, H3M, H2M2, HM3 and M4 which can be separated by electrophoresis and shown to have different affinities for their substrates, lactate and pyruvate, and for analogues of these two compounds. They also have different maximum catalytic activities and tissue distributions, and as a consequence are important in diagnostic enzymology (Section 16.3).

Multienzyme complexes

Some enzymes that promote consecutive reactions in a metabolic pathway associate to form a multienzyme complex. Examples include the fatty-acid synthase (EC 2:3:1:86) (seven catalytic centres), pyruvate dehydrogenase (EC 2:7:1:99) (three catalytic centres) and DNA polymerase (EC 2:7:7:7) (three catalytic centres). Multienzyme complexes have a number of advantages over individual enzymes including a reduction in the transit time for the diffusion of the product of one enzyme to the catalytic site of the next, a reduction in the possibility of the product of one enzyme being acted upon by another enzyme not involved in the pathway, and the possibility of one enzyme activating an adjacent enzyme (Section 15.5.4).

Units of enzyme activity

Units of enzyme activity are expressed either in the SI units of katals (defined as the number of moles of substrate consumed or product formed per second) or international units (number of μ moles of substrate consumed or product formed per minute). Allied to activity units is specific activity which expresses the number of international units per mg protein or katals per kg protein (note: 60 international units per mg protein is equivalent to 1 katal (kg protein)$^{-1}$).

15.2 ENZYME STEADY-STATE KINETICS

15.2.1 Monomeric enzymes

Initial rates

When an enzyme is mixed with an excess of substrate there is an initial short period of time (a few hundred microseconds) during which intermediates leading to the formation of the product gradually build up (Fig. 15.1). This so-called pre-steady state requires special techniques for study and these are discussed in Section 15.3.3. After this pre-steady state, the reaction rate and the concentration of intermediates change relatively slowly with time and so-called steady-state kinetics exist. Measurement of the progress of the reaction during this phase gives the relationships shown in Fig. 15.2. Tangents drawn through the origin to the curves of substrate concentration and product concentration versus time allow the initial rate, ν_0, to be calculated. This is the maximum rate for a given concentration of enzyme and substrate under the defined experimental conditions. Measurement of the initial rate of an enzyme-catalysed reaction is a prerequisite to a complete understanding of the mechanism by which the enzyme works, as well as to the estimation of the activity of an enzyme in a biological sample. Its numerical value is influenced by many factors, including substrate and enzyme concentration, pH, temperature and the presence of activators or inhibitors.

For many enzymes, the initial rate, ν_0, varies hyperbolically with substrate concentration for a fixed concentration of enzyme (Fig. 15.3). The mathematical equation expressing this hyperbolic relationship between initial rate and substrate concentration is known as the Michaelis–Menten equation:

$$\nu_0 = \frac{V_{\max}[S]}{K_m + [S]}$$

(15.1)

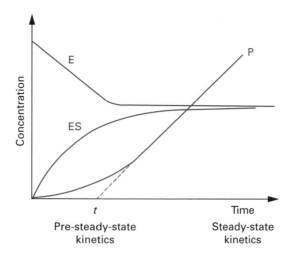

Fig. 15.1 Pre-steady-state progress curve for the interaction of an enzyme (E) with its substrate (S). P, product; t, induction time.

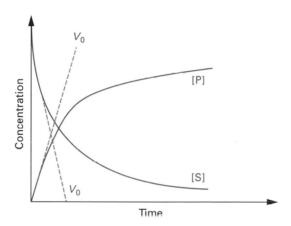

Fig. 15.2 Calculation of initial rate (ν_0) from the time-dependent change in the concentration of substrate (S) and product (P) of an enzyme-catalysed reaction.

where V_{max} is the limiting value of the initial rate when all the active sites are occupied, K_m is the Michaelis constant and [S] is the substrate concentration. At low substrate concentrations the occupancy of the active sites on the enzyme molecules is low and the reaction rate is directly related to the number of sites occupied. This approximates to first-order kinetics in that the rate is proportional to substrate concentration. At high substrate concentrations effectively all of the active sites are occupied and the reaction becomes independent of the substrate concentration since no more enzyme–substrate complex can be formed and zero-order or saturation kinetics are observed. Under these conditions the reaction rate is only dependent upon the conversion of the enzyme–substrate complex, ES, to products and the diffusion of the products from the enzyme.

It can be seen from equation 15.1 that when $\nu_0 = 0.5\,V_{max}$, $K_m = $ [S]. Thus K_m is numerically equal to the substrate concentration at which the initial rate is one-half of

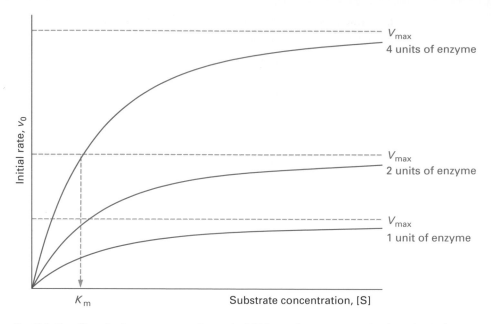

Fig. 15.3 The effect of substrate concentration on the initial rate of an enzyme-catalysed reaction in the presence of three different concentrations of enzyme. Doubling the enzyme concentration doubles the maximum initial rate, V_{max}, but has no effect on K_m.

the maximum rate (Fig. 15.3) and has units of molarity. Values of K_m are usually in the range 10^{-2} to 10^{-5} M and are important because they enable the concentration of substrate required to saturate all of the active sites of the enzyme in an enzyme assay to be calculated. When $[S] \gg K_m$, equation 15.1 reduces to $\nu_0 \approx V_{max}$, but a simple calculation reveals that when $[S] = 10\,V_{max}$, ν_0 is only 90% V_{max} and that when $[S] = 100\,K_m$, $\nu_0 = 99\%\,V_{max}$. Appreciation of this relationship is vital in enzyme assays.

As previously stated, enzyme-catalysed reactions proceed via the formation of an enzyme substrate complex in which the substrate (S) is non-covalently bound to the active site of the enzyme (E). The formation of this complex for the majority of enzymes is rapid and reversible and is characterised by the dissociation constant, K_s, of the complex:

$$E + S \underset{k_{-1}}{\overset{k_{+1}}{\rightleftharpoons}} ES$$

where k_{+1} and k_{-1} are the rate constants for the forward and reverse reactions. At equilibrium, the rates of the forward and reverse reactions are equal and the Law of Mass Action can be applied to the reversible process:

$$k_{+1}[E][S] = k_{-1}[ES] \qquad (15.2)$$

hence:

$$K_s = \frac{[E][S]}{[ES]} = \frac{k_{-1}}{k_{+1}} = \frac{1}{K_a}$$

where K_a is the **association** (or **affinity**) constant.

It can be seen that when K_s is numerically large, the equilibrium is in favour of unbound E and S, i.e. of non-binding, whilst when K_s is numerically small, the equilibrium is in favour of the formation of ES, i.e. of binding. Thus K_s is inversely proportional to the affinity of the enzyme for its substrate.

The conversion of ES to product (P) can be most simply represented by the irreversible equation:

$$ES \xrightarrow{k_{+2}} E + P$$

where k_{+2} is the first-order rate constant for the reaction.

In some cases the conversion of ES to E and P may involve several stages and may not necessarily be essentially irreversible. The rate constant k_{+2} is generally smaller than both k_{+1} and k_{-1} and in some cases very much smaller. In general, therefore, the conversion of ES to products is the rate-limiting step such that the concentration of ES is essentially constant but not necessarily the equilibrium concentration. Under these conditions the Michaelis constant, K_m, is given by:

$$K_m = \frac{k_{+2} + k_{-1}}{k_{+1}} = K_s + \frac{k_{+2}}{k_{+1}} \tag{15.3}$$

It is evident that under these circumstances, K_m must be numerically larger than K_s and only when k_{+2} is very small do K_m and K_s approximately equal each other. The relationship between these two constants is further complicated by the fact that for some enzyme reactions two products are formed sequentially, each controlled by different rate constants:

$$E + S \underset{}{\overset{k_{+2}}{\rightleftharpoons}} ES \rightarrow P_1 + EA \xrightarrow{k_{+3}} E + P_2$$

where P_1 and P_2 are products, and A is a metabolic product of S that is further metabolised to P_2. In such circumstances it can be shown that:

$$K_m = K_s \frac{k_{+3}}{k_{+2} + k_{+3}} \tag{15.4}$$

so that K_m is numerically smaller than K_s. It is obvious therefore that care must be taken in the interpretation of the significance of K_m relative to K_s. Only when the complete reaction mechanism is known can the mathematical relationship between K_m and K_s be fully appreciated and any statement made about the relationship between K_m and the affinity of the enzyme for its substrate.

Although the Michaelis–Menten equation can be used to calculate K_m and V_{max}, its use is subject to the difficulty of experimentally measuring initial rates at high substrate concentrations and hence of extrapolating the hyperbolic curve to give an accurate value of V_{max}. Linear transformations of the Michaelis–Menten equation are therefore commonly used alternatives. The most popular of these is the Lineweaver–Burk equation obtained by taking the reciprocal of the Michaelis–Menten equation:

$$\frac{1}{v_0} = \frac{K_m}{V_{max}} \times \frac{1}{[S]} + \frac{1}{V_{max}} \tag{15.5}$$

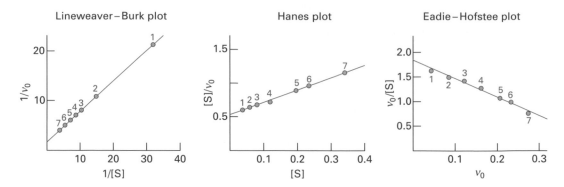

Fig. 15.4 Lineweaver–Burk, Hanes and Eadie–Hofstee plots for the same set of experimental data of the effect of substrate concentration on the initial rate of an enzyme-catalysed reaction.

A plot of $1/\nu_0$ against $1/[S]$ gives a straight line of slope K_m/V_{max}, with an intercept on the $1/\nu_0$ axis of $1/V_{max}$ and an intercept on the $1/[S]$ axis of $-1/K_m$. Alternative plots are based on the Hanes equation:

$$\frac{[S]}{\nu_0} = \frac{K_m}{V_{max}} + \frac{[S]}{V_{max}} \qquad (15.6)$$

so that $[S]/\nu_0$ is plotted against $[S]$, and on the Eadie–Hofstee equation:

$$\frac{\nu_0}{[S]} = \frac{V_{max}}{K_m} - \frac{\nu_0}{K_m} \qquad (15.7)$$

so that $\nu_0/[S]$ is plotted against ν_0. The relative merits of the Lineweaver–Burk, Hanes and Eadie–Hofstee equations for the determination of K_m and V_{max} are illustrated in Fig. 15.4 using the same set of experimental values of ν_0 for a series of substrate concentrations (for further details, see Example 1).

It can be seen that the Lineweaver–Burk equation gives an unequal distribution of points and greater emphasis to the points at low substrate concentration that are subject to the greatest experimental error whilst the Eadie–Hofstee equation and the Hanes equation give a better distribution of points. In the case of the Hanes plot, greater emphasis is placed on the experimental data at higher substrate concentrations and on balance it is the statistically preferred plot. In spite of their widespread use, these linear transformations of enzyme kinetic data are subject to error. Specifically, they assume that the scatter of points around the line follows a Gaussian distribution and that the standard deviation of each point is the same. In practice this is rarely true. With the advent of widely available non-linear regression software packages such as DynaFit (www.biokin.com) and BRENDA (www.brenda-enzymes.info), there are now strong arguments for their preferential use in cases where accurate kinetic data are required.

It is important to appreciate that whilst K_m is a characteristic of an enzyme for its substrate and is independent of the amount of enzyme used for its experimental determination, this is not true of V_{max}. It has no absolute value but varies with the amount of enzyme used. This is illustrated in Fig. 15.3 and is discussed further in Example 1. A valuable catalytic constant in addition to K_m and V_{max} is the turnover number, k_{cat}, defined as:

Example 1 **PRACTICAL ENZYME KINETICS**

Question The enzyme α-D-glucosidase isolated from *Saccharomyces cerevisiae* was studied using the synthetic substrate *p*-nitrophenyl-α-D-glucopyranoside (PNPG), which is hydrolysed to release *p*-nitrophenol which is yellow in alkaline solution (see Section 15.2.1 for further details). A 3 mM solution of PNPG was prepared and portions used to study the effect of substrate concentration on initial rate using a fixed volume of enzyme preparation. The total volume of each assay mixture was $10 \, cm^3$. A $1 \, cm^3$ sample of the reaction mixture was withdrawn after 2 min, and placed in $4 \, cm^3$ borate buffer pH 9.0 to stop the reaction and develop the yellow colour. The change in absorbance at 400 nm was determined and used as a measure of the initial rate. The following results were obtained:

PNPG (cm^3)	0.1	0.2	0.3	0.4	0.6	0.8	1.2
Initial rate	0.055	0.094	0.130	0.157	0.196	0.230	0.270

What kinetic constants can be obtained from these data?

Answer Subject to the calculation of the molar concentration of PNPG in each reaction mixture, it is possible to construct Lineweaver–Burk, Hanes and Eadie–Hofstee plots to obtain the values of K_m and V_{max}. The fact that a $1 \, cm^3$ sample of the reaction mixture was used to measure the initial rate is not relevant to the calculation of [S]. Lineweaver Burk, Hanes and Eadie–Hofstee plots derived from these data are shown in Fig. 15.6 in which v_0 measurements are expressed simply as the increase in absorption at 400 nm. The plots give K_m values of approximately 0.2 mM and V_{max} values of approximately 0.4.

As pointed out in Section 15.2.1, V_{max} values can be expressed in a variety of units and their experimental value is dependent on a number of variables particularly the concentration of enzyme. For comparative reasons, V_{max} is best expressed in terms of the number of moles of product formed in unit time. To do this, it is necessary to convert absorbance units to amount of product by means of a Beer–Lambert law plot. Data for such a plot in this experiment are given in Table A.

Table A

[PNP] (μM)			2.0	4.0	6.0	8.0	12.0	16.0	24.0
Absorbance (400 nm)			0.065	0.118	0.17	0.23	0.34	0.45	0.65

A plot of these data confirms that the Beer–Lambert law is held and enables the amount of product to be calculated. From this, v_0 values in units of $\mu mol \, min^{-1}$ can be calculated. The data for the three linear plots are presented in Table B.

Example 1 (*cont.*)

Table B

[S] (mM)	0.03	0.06	0.09	0.12	0.18	0.24	0.36
v_0 (µmol min^{-1})	0.054	0.096	0.138	0.168	0.210	0.251	0.294
$1/[S]$ (mM)$^{-1}$	33.33	16.67	11.11	8.33	5.55	4.17	2.78
$1/v_0$ (µmol min^{-1})$^{-1}$	18.52	10.42	7.25	5.95	4.76	3.98	3.40
$v_0/[S] \times 10^3$ (dm^3 min^{-1})	1.8	1.6	1.53	1.40	1.17	1.05	0.82
$[S]/v_0 \times 10^{-3}$ (min dm^{-3})	0.56	0.63	0.65	0.71	0.85	0.95	1.22

Data derived from the three linear plots are presented in Table C.

Table C

Plot	Regression coefficient	Slope	Intercept	K_m (mM)	V_{max} (µmol min^{-1})
Lineweaver–Burk	0.9997	0.499	1.91	0.26	0.52
Hanes	0.9970	1.990	0.489	0.25	0.50
Eadie–Hofstee	0.9930	−3.99	2.030	0.25	0.51

The agreement between the three plots for the values of K_m and V_{max} was good but the quality of the fitted regression line for the Lineweaver–Burk plot was noticeably better. However, the distribution of the experimental points along the line is the poorest for this plot (Fig. 15.6). The value for V_{max} indicates the amount of product released per minute, but of course this is for the chosen amount of enzyme and is for 10 cm^3 of reaction mixture. For the value of V_{max} to have any absolute value, the amount of enzyme and the volume of reaction mixture have to be taken into account. The volume can be adjusted to 1 dm^3 giving V_{max} of 51 µmol min^{-1}dm^{-3}, but it is only possible to correct for enzyme amount if it was pure and of known amount in molar terms. The enzyme is known to have a molecular mass of 68 kDa so if there was 3 µg of pure enzyme in each 10 cm^3 reaction mixture, its molar concentration would be 4.4×10^{-3} µM. This allows the value of the turnover number k_{cat} to be calculated (see equation 15.8):

$$k_{cat} = V_{max}/[E_t] = 51\ \mu M\ min^{-1}/4.4 \times 10^{-3}\ \mu M$$
$$= 11 \times 10^3\ min^{-1}\ or\ 1.8 \times 10^2\ s^{-1}$$

k_{cat} is a measure of the number of molecules of substrate (PNPG in this case) converted to product per second by the enzyme under the defined experimental conditions. The value of 180 is in the mid-range for the majority of enzymes. It is also possible to calculate the specificity constant that is a measure of the efficiency with which the enzyme converts substrate to product at low (K_m) substrate concentrations:

$$k_{cat}/K_m = 1.8 \times 10^2\ s^{-1}/0.25\ mM = 7.2 \times 10^2\ mM^{-1}s^{-1}\ or\ 7.2 \times 10^5\ M^{-1}s^{-1}$$

Example 1 (*cont.*)

Note that the units of the specificity constant are that of a second-order rate constant, effectively for the conversion of E + S to E + P. Its value in this case is typical of many enzymes and is lower than the limiting value.

$$k_{cat} = \frac{V_{max}}{[E_t]} \tag{15.8}$$

where $[E_t]$ is the total concentration of enzyme. The turnover number is the maximum number of moles of substrate that can be converted to product per mole of enzyme in unit time. It has units of reciprocal time in seconds. Its values range from 1 to $10^7 \, s^{-1}$. Catalase has a turnover number of $4 \times 10^7 \, s^{-1}$ and is one of the most efficient enzymes known. The catalytic potential of high turnover numbers can only be realised at high (saturating) substrate concentrations and this is seldom achieved under normal cellular conditions. An alternative constant, termed the specificity constant, defined as k_{cat}/K_m, is a measure of how efficiently an enzyme converts substrate to product at low substrate concentrations. It has units of $M^{-1}s^{-1}$.

For a substrate to be converted to product, molecules of the substrate and of the enzyme must first collide by random diffusion and then combine in the correct orientation. Diffusion and collision have a theoretical limiting rate constant value of about $10^9 \, M^{-1}s^{-1}$ and yet many enzymes, including acetylcholine esterase, carbonic anhydrase, catalase, β-lactamase and triosephosphate isomerase, have specificity constants approaching this value indicating that they have evolved to almost maximum kinetic efficiency. Since specificity constants are a ratio of two other constants, enzymes with similar specificity constants can have widely different K_m values. As an example, catalase has a specificity constant of $4 \times 10^7 \, M^{-1}s^{-1}$ with a K_m of 1.1 M (very high), whilst fumerase has a specificity constant of 3.6×10^7 $M^{-1} \, s^{-1}$ with a K_m of 2.5×10^{-5} M (very low). Multienzyme complexes overcome some of the diffusion and collision limitations to specificity constants. The product of one reaction is passed directly by a process called channelling to the active site of the next enzyme in the pathway as a consequence of its juxtaposition in the complex, thereby eliminating diffusion limitations (Section 15.4.2).

Effect of enzyme concentration

It can be shown that for monosubstrate enzymatic reactions that obey simple Michaelis–Menten kinetics:

$$v_0 = \frac{k_{+2}[E][S]}{K_m + [S]}$$

and hence that

$$v_0 = \frac{k_{+2}[E]}{(K_m/[S]) + 1} \tag{15.9}$$

Thus, when the substrate concentration is very large, equation 15.9 reduces to $v_0 = k_{+2} [E]$, i.e. the initial rate is directly proportional to the enzyme concentration. This is the basis of the experimental determination of enzyme activity in a particular biological sample (Section 15.3). Figure 15.3 illustrates the importance of the correct measurement of initial rate.

15.2.2 Inhibition of monomeric enzyme reactions

Competitive reversible inhibition

Reversible inhibitors combine non-covalently with the enzyme and can therefore be readily removed by dialysis. Competitive reversible inhibitors combine at the same site as the substrate and must therefore be structurally related to the substrate. An example is the inhibition of succinate dehydrogenase by malonate:

```
CH2COOH                          CH2COOH
|                                |
CH2COOH                          COOH
succinic acid (substrate)        malonic acid (inhibitor)
  ↓↑ succinate dehydrogenase            ↓↑
CHCOOH                            no reaction
||
CHCOOH
fumaric acid (product)
```

All types of reversible inhibitors are characterised by their dissociation constant K_i, called the inhibitor constant, which may relate to the dissociation of EI (K_{EI}) or of ESI (K_{ESI}). For competitive inhibition the following two equations can be written:

$$E + S \rightleftharpoons ES \longrightarrow E + P$$
$$E + I \rightleftharpoons EI \longrightarrow \text{no reaction}$$

Since the binding of both substrate and inhibitor involves the same site, the effect of a competitive reversible inhibitor can be overcome by increasing the substrate concentration. The result is that V_{max} is unaltered but the concentration of substrate required to achieve it is increased so that when $v_0 = 0.5 V_{max}$ then:

$$[S] = K_m \left(1 + \frac{[I]}{K_i}\right) \tag{15.10}$$

where [I] is the concentration of inhibitor.

It can be seen from equation 15.10 that K_i is equal to the concentration of inhibitor that apparently doubles the value of K_m. With this type of inhibition, K_i is equal to K_{EI} whilst K_{ESI} is infinite because no ESI is formed. In the presence of a competitive inhibitor, the Lineweaver–Burk equation (15.5) becomes:

$$\frac{1}{v_0} = \frac{K_m}{V_{max}} \times \frac{1}{[S]} \left(1 + \frac{[I]}{K_i}\right) + \frac{1}{V_{max}} \tag{15.11}$$

Application of this equation allows the diagnosis of competitive inhibition (Fig. 15.5a). The numerical value of K_i can be calculated from Lineweaver–Burk plots for the uninhibited and inhibited reactions. In practice, however, a more accurate

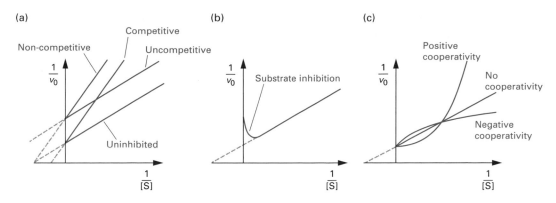

Fig. 15.5 Lineweaver–Burk plots showing (a) the effects of three types of reversible inhibitor, (b) substrate inhibition and (c) homotropic cooperativity.

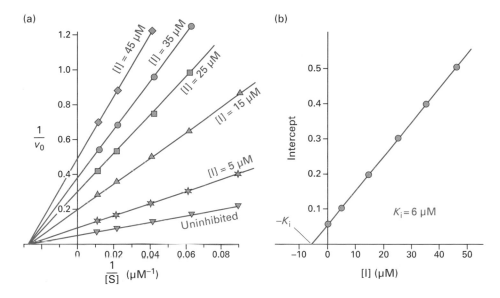

Fig. 15.6 (a) Primary Lineweaver–Burk plots showing the effect of a simple linear non-competitive inhibitor at a series of concentrations and (b) the corresponding secondary plot that enables the inhibitor constant K_i to be calculated.

value is obtained from a secondary plot (Fig. 15.6). The reaction is carried out for a range of substrate concentrations in the presence of a series of fixed inhibitor concentrations and a Lineweaver–Burk plot for each inhibitor concentration constructed. Secondary plots of the slope of the primary plot against the inhibitor concentration or of the apparent K_m, (K'_m) (which is equal to $K_m (1 + [I]/K_i)$ and which can be calculated from the reciprocal of the negative intercept on the I/[S] axis) against inhibitor concentration, will both have intercepts on the inhibitor concentration axis of $-K_i$. Sometimes it is possible for two molecules of inhibitor to bind at the active site. In these cases, although all the primary double reciprocal plots are linear, the secondary plot is parabolic. This is referred to as **parabolic competitive inhibition** to distinguish it from normal **linear competitive inhibition**.

Non-competitive reversible inhibition

A non-competitive reversible inhibitor combines at a site distinct from that for the substrate. Whilst the substrate can still bind to its catalytic site, resulting in the formation of a ternary complex ESI, the complex is unable to convert the substrate to product and is referred to as a dead-end complex. Since this inhibition involves a site distinct from the catalytic site, the inhibition cannot be overcome by increasing the substrate concentration. The consequence is that V_{max}, but not K_m, is reduced because the inhibitor does not affect the binding of substrate but it does reduce the amount of free ES that can proceed to the formation of product. With this type of inhibition K_{EI} and K_{ESI} are identical and K_i is numerically equal to both of them. In this case the Lineweaver–Burk equation (15.5) becomes:

$$\frac{1}{v_0} = \frac{K_m}{V_{max}} \times \frac{1}{[S]} + \frac{1}{V_{max}}\left(1 + \frac{[I]}{K_i}\right) \tag{15.12}$$

Once non-competitive inhibition has been diagnosed (Fig. 15.5a), the K_i value is best obtained from a secondary plot of either the slope of the primary plot or of $1/V'_{max}$ (which is equal to the intercept on the $1/v_0$ axis) against inhibitor concentration. Both secondary plots will have an intercept of $-K_i$ on the inhibitor concentration axis (Fig. 15.6).

Uncompetitive reversible inhibition

An uncompetitive reversible inhibitor can bind only to the ES complex and not to the free enzyme, so that inhibitor binding must be either at a site created by a conformational change induced by the binding of the substrate to the catalytic site or directly to the substrate molecule. The resulting ternary complex, ESI, is also a dead-end complex.

$$\text{E} + \text{S} \Longleftrightarrow \quad \text{ES} \quad \longrightarrow \text{E} + \text{P}$$
$$+\text{I} \downarrow\uparrow -\text{I}$$
$$\text{ESI} \quad \longrightarrow \quad \text{no reaction}$$

As with non-competitive inhibition, the effect cannot be overcome by increasing the substrate concentration, but in this case both K_m and V_{max} are reduced by a factor of $(1 + [I]/K_i)$. An inhibitor concentration equal to K_i will therefore halve the values of both K_m and V_{max}. With this type of inhibitor, K_{EI} is infinite because the inhibitor cannot bind to the free enzyme so K_i is equal to K_{ESI}. The Lineweaver–Burk equation (15.5) therefore becomes:

$$\frac{1}{v_0} = \left(\frac{K_m}{V_{max}} \times \frac{1}{[S]} + \frac{1}{V_{max}}\right)\left(1 + \frac{[I]}{K_i}\right) \tag{15.13}$$

The value of K_i is best obtained from a secondary plot of either $1/V'_{max}$ or I/K'_m (which is equal to the intercept on the $I/[S]$ axis) against inhibitor concentration. Both secondary plots will have an intercept of $-K_i$ on the inhibitor concentration axis.

Mixed reversible inhibition

For some inhibitors either the ESI complex has some catalytic activity or the K_{EI} and K_{ESI} values are neither equal nor infinite. In such case so-called mixed inhibition kinetics are obtained. Mixed inhibition is characterised by a linear Lineweaver–Burk plot that does not fit any of the patterns shown in Fig. 15.5a. The plots for the uninhibited and inhibited reactions may intersect either above or below the 1/[S] axis. The associated K_i can be obtained from a secondary plot of the slope either of the primary plot or of $1/V_{max}$ for the primary plots against inhibitor concentration. In both cases the intercept on the inhibitor concentration axis is $-K_i$. Non-competitive inhibition may be regarded as a special case of mixed inhibition.

Substrate inhibition

A number of enzymes at high substrate concentration display substrate inhibition characterised by a decrease in initial rate with increased substrate concentration. The graphical diagnosis of this situation is shown in Fig. 15.5b. It is explicable in terms of the substrate acting as an uncompetitive inhibitor and forming a dead-end complex.

End-product inhibition

The first enzymes in an unbranched metabolic pathway are commonly regulated by end-product inhibition. Here the final product of the pathway acts as an inhibitor of the first enzyme in the pathway thus switching off the whole pathway when the final product begins to accumulate. The inhibition of aspartate carbamyltransferase by cytosine triphosphate (CTP) in the CTP biosynthetic pathway is an example of this form of regulation. In branched pathways, product inhibition usually operates on the first enzyme after the branch point.

Irreversible inhibition

Irreversible inhibitors, such as the organophosphorus and organomercury compounds, cyanide, carbon monoxide and hydrogen sulphide, combine with the enzyme to form a covalent bond. The extent of their inhibition of the enzyme is dependent upon the reaction rate constant (and hence time) for covalent bond formation and upon the amount of inhibitor present. The effect of irreversible inhibitors, which cannot be removed by simple physical techniques such as dialysis, is to reduce the amount of enzyme available for reaction. The inhibition involves reactions with a functional group, such as hydroxyl or sulphydryl, or with a metal atom in the active site or a distinct allosteric site. Thus the organophosphorus compound, diisopropyl-phosphofluoridate, reacts with a serine group in the active site of esterases such as acetylcholinesterase, whilst the organomercury compound p-hydroxymercuribenzoate reacts with a cysteine group, in both cases resulting in covalent bond formation and enzyme inhibition. Such inhibitors are valuable in the study of enzyme active sites (Section 15.4.1).

Applications of enzyme inhibition

The study of the classification and mechanism of enzyme inhibition is of importance in a number of respects:

- it gives an insight into the mechanisms by which enzymes promote their catalytic activity (Section 15.4.1);

- it gives an understanding of the possible ways by which metabolic activity may be controlled *in vivo*;
- it allows specific inhibitors to be synthesised and used as therapeutic agents to block key metabolic pathways underlying clinical conditions (Section 18.1.2).

15.2.3 **Effect of temperature and pH on enzyme reactions**

Effect of temperature

The initial rate of an enzyme reaction varies with temperature according to the Arrhenius equation:

$$\text{rate} = Ae^{-E/RT} \tag{15.14}$$

where A is a constant known as the pre-exponential factor, which is related to the frequency with which molecules of the enzyme and substrate collide in the correct orientation to produce the enzyme–substrate complex, E is the activation energy (J mol^{-1}), R is the gas constant ($8.2 \text{ J mol}^{-1}\text{K}^{-1}$), and T is the absolute temperature (K).

Thus a plot of the natural logarithm of the initial rate (or better k_{cat}) against the reciprocal of the absolute temperature allows the value of E to be determined.

Equation 15.14 explains the sensitivity of enzyme reactions to temperature as the relationship between reaction rate and absolute temperature is exponential. The rate of most enzyme reactions approximately doubles for every 10 °C rise in temperature (Q_{10} value). At a temperature characteristic of the enzyme, and generally in the region 40 to 70 °C, the enzyme is denatured and enzyme activity is lost. The activity displayed in this 40 to 70 °C temperature range depends partly upon the equilibration time before the reaction is commenced. The so-called optimum temperature, at which the enzyme appears to have maximum activity, therefore arises from a combination of thermal stability, temperature coefficient and incubation time and for this reason is not normally chosen for the study of enzyme activity. Enzyme assays are routinely carried out at 30 or 37 °C (Section 15.3). Interestingly, recent work with enzymes from mesophiles and thermophiles have indicated that some have a genuine temperature optimum in that above a certain temperature the enzyme becomes reversibly less active but not as a consequence of denaturation. The nature of the structural changes responsible for such observations has yet to be determined.

Enzymes work by facilitating the formation of a transition state, which is a transient intermediate in the formation of the product(s) from the substrate(s), that has a lower energy barrier than that for the non-catalysed reaction. This results in a decrease in the activation energy (E_{act}) for the reaction relative to that for the non-enzyme-catalysed reaction (Fig. 15.7). A decrease in the energy barrier of as little as 5.7 kJ mol^{-1}, equivalent in energy terms to the strength of a hydrogen bond, will result in a 10-fold increase in reaction rate. The energy barrier is, of course, lowered equally for both the forward and reverse reactions, so that the position of equilibrium is unchanged. As an extreme example of the efficiency of enzyme catalysis, the enzyme catalase decomposes hydrogen peroxide 10^{14} times faster than occurs in the uncatalysed reaction! Figure 15.7 shows a simple energy profile for the conversion of

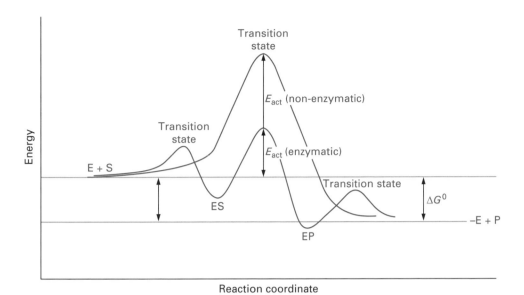

Fig. 15.7 Energy profile of a simple enzyme-catalysed reaction. The formation of ES and EP and the subsequent release of E + P proceeds via several transition states. The activation energy for the overall reaction is dictated by the initial free energy of E and S and the highest energy transition state. The non-enzyme-catalysed reaction proceeds via a higher energy transition state and hence the reaction has a higher activation energy than the enzyme-catalysed reaction.

a substrate to product as a function of the reaction coordinate that measures the time-related progress of the reaction. The number of energy barriers in the profile will depend upon the number of kinetically important stages in the reactions. For the majority of enzyme-catalysed reactions the major energy barrier, which dictates the activation energy for the overall reaction and hence its rate, is the formation of one or more intermediates in which covalent bonds are being made and broken and which cannot be isolated. However, for a few enzymes, notably ATP synthase, the energy-requiring step is the initial binding of the substrate(s) and the subsequent release of the product(s).

The thermodynamic constants ΔG^0, ΔH^0 and ΔS^0 for the binding of substrate to the enzyme can be calculated from a knowledge of the binding constant, K_a $(=1/K_s)$. ΔG^0 can be obtained from the equation:

$$\Delta G^0 = -RT \ln K_a \qquad (15.15)$$

If K_a is measured at two or more temperatures, a plot of $\ln K_a$ versus $1/T$, known as the van't Hoff plot, will give a straight line slope $-\Delta H^0/R$ with intercept on the y axis of $\Delta S^0/R$, the relevant equation being:

$$\ln K_a = \frac{\Delta S^0}{R} - \frac{\Delta H^0}{RT} \qquad (15.16)$$

A small number of enzymes appear to operate by a mechanism that does not rely on the formation of a transition state. Studies with the enzyme methylamine dehydrogenase, which promotes the cleavage of a C–H bond, have shown that the reaction is independent of temperature and hence is inconsistent with transition state theory.

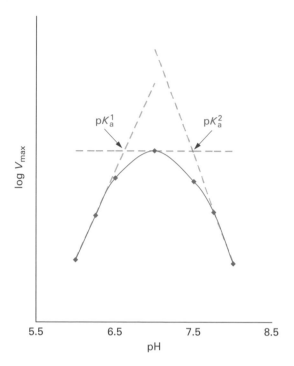

Fig. 15.8 The effect of pH on V_{max} of an enzyme-catalysed reaction involving two ionisable groups in the active site of the enzyme. The construction of tangents to the experimental line allows the pK_a values of the ionisable groups to be estimated.

The observation is explained in terms of enzyme–catalysed quantum tunnelling. Under this mechanism, rather than overcoming the potential energy barrier, the reaction proceeds through the barrier (hence 'tunnelling') at an energy level near that of the ground state of the reactants. Concerted enzyme and substrate vibrations are coupled in such a way as to reduce the width and height of the potential energy barrier and facilitate the cleavage of the C–H bond by the process of quantum mechanical tunnelling. This phenomenon is known to occur with some chemical reactions but only at low temperatures. The fine detail of precisely how enzymes promote this process remains to be elucidated.

Effect of pH

The state of ionisation of amino acid residues in the catalytic site of an enzyme is pH dependent. Since catalytic activity relies on a specific state of ionisation of these residues, enzyme activity is also pH dependent. As a consequence, plots of log K_m and log V_{max} (or better, k_{cat}) against pH are either bell-shaped (indicating two important ionisable amino acid residues in the active site), giving a narrow pH optimum, or they have a plateau (one important ionisable amino acid residue in the active site). In either case, the enzyme is generally studied at a pH at which its activity is maximal. By studying the variation of log K_m and log V_{max} with pH, it is possible to identify the pK_a values of key amino acid residues involved in the binding and catalytic processes (Fig. 15.8).

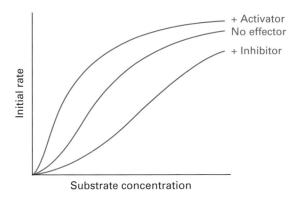

Fig. 15.9 Effect of activators and inhibitors on the sigmoidal kinetics of an enzyme subject to allosteric control.

15.2.4 **Allosterism and cooperativity**

The discussion of the mechanism of enzyme action so far has been based on the assumption that successive substrate molecules bind to the enzyme with the same ease (affinity). This is not true with some enzymes. With some enzymes successive binding occurs with either progressive greater ease or with reduced ease and the enzymes are said to be subject to allosteric control. Historically, all such enzymes were believed to possess quaternary structure and consist of several protein subunits (oligomers), which could be identical or different, and to possess multiple catalytic sites. Proteins subject to allosteric control are not confined to enzymes. Other examples are haemoglobin and many cell membrane receptors especially those of the G-protein-coupled receptors (GPCRs) type (Section 17.4.3). Oligomeric enzymes may either display simple Michaelis–Menten kinetics in which case all substrate molecules bind with equal ease, as is the case with the tetramer lactate dehydrogenase (Section 15.1.2), or they display a characteristic sigmoidal relationship between initial rate and substrate concentration (Fig. 15.9) indicative of an allosteric enzyme. Examples of such enzymes are aspartate carbamoylase and phosphofructokinase. Progressive binding of the substrate molecules to the subunits of an allosteric enzyme may result in either increased (positive cooperativity) or decreased (negative cooperativity) activity towards the binding of further substrate molecules. In such cases the substrate molecules are said to display a homotropic effect. Changes in catalytic activity towards the substrate may also be brought about by the binding of molecules other than the substrate at distinct allosteric binding sites on one or more subunit. Compounds that induce such changes are referred to as heterotropic effectors. They are commonly key metabolic intermediates such as ATP, ADP, AMP and P_i. Heterotropic activators increase the catalytic activity of the enzyme, making the curve less sigmoidal and moving it to the left, whilst heterotropic inhibitors cause a decrease in activity, making the curve more sigmoidal and moving it to the right (Fig. 15.9). The diagnosis of cooperativity by use of the Lineweaver–Burk plot is shown in Fig. 15.5c. The operation of cooperative effects may be confirmed by a Hill plot, which is based on the equation:

$$\log \frac{v_0}{V_{\max} - v_0} = h \log[S] + \log K \qquad (15.17)$$

where h is the Hill constant or coefficient, and K is an overall binding constant related to the individual binding constants for n sites. The Hill constant, which is equal to the slope of the plot, is a measure of the cooperativity between the sites such that: if $h = 1$, binding is non-cooperative and normal Michaelis–Menten kinetics exist; if $h > 1$, binding is positively cooperative; and if $h < 1$, binding is negatively cooperative. At very low substrate concentrations that are insufficient to fill more than one site and at high concentrations at which most of the binding sites are occupied, the slopes of Hill plots tend to a value of 1. The Hill coefficient is therefore taken from the linear central portion of the plot. One of the problems with Hill plots is the difficulty of estimating $V_{\max}$ accurately.

The Michaelis constant K_m is not used with allosteric enzymes. Instead, the term $S_{0.5}$, which is the substrate concentration required to produce 50% saturation of the enzyme, is used. It is important to appreciate that sigmoidal kinetics do not confirm the operation of allosteric effects because sigmoidicity may be the consequence of the enzyme preparation containing more than one enzyme capable of acting on the substrate. It is easy to establish the presence of more than one enzyme, as there will be a discrepancy between the amount of substrate consumed and the expected amount of product produced.

Two classical models have been proposed to interpret allosteric regulation. They are both based on the assumption that the allosteric enzyme consists of a number of subunits (protomers) each of which can bind substrate and exist in two conformations referred to as the R (relaxed) and T (tense) states. It is assumed that the substrate binds more tightly to the R form. The first such model was due to Jacques Monod, Jeffries Wyman and Jean-Pierre Changeux, and is referred to as the symmetry model. It assumes that conformational change between the R and T states is highly coupled so that all subunits must exist in the same conformation. Thus binding of substrate to a T state protomer, causing it to change conformation to the R state, will automatically switch the other protomers to the R form, thereby enhancing reactivity (Fig. 15.10). The second model of Daniel Koshland, known as the induced-fit or sequential model, does not assume the tightly coupled concept and hence allows protomers to exist in different conformations but in such a way that binding to one protomer modifies the reactivity of others.

Recent research has shown that allosterism is not confined to oligomeric proteins since some monomeric proteins may also display the behaviour (Fig. 15.10). The emerging opinion is that allostery is a consequence of the flexibility of proteins such that the continuous folding and unfolding in localised regions of the protein gives rise to a population of conformations that interconvert on various timescales, which differ in their affinity for certain ligands and, in the case of enzymes, in their catalytic activity. The interconverting conformations have similar energies and their mixture constitutes the 'native state' of the protein. The binding of an allosteric effector at its distinct site results in the redistribution of the conformational ensembles as a result of the alteration of their rates of interconversion, and as a consequence, the conformation and hence activity of the active site is modified. Integral to this redistribution

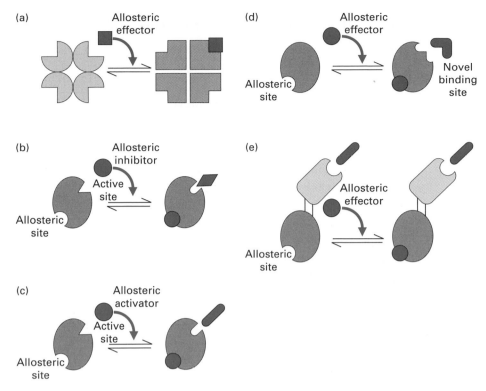

Fig. 15.10 Different modes of allosteric behaviour. (a) A representation of the Monod–Wyman–Changeux model of allosteric transitions. A symmetric, multimeric protein can exist in one of two distinct conformational states – the active and inactive conformations. Each subunit has a binding site for an allosteric effector as well as an active site or binding site. (b) A monomeric, allosterically inhibited protein. The binding of an allosteric inhibitor alters the active site or binding site geometry in an unfavourable way, thereby decreasing affinity or catalytic efficiency. (c) A monomeric, allosterically activated protein. The binding of an allosteric activator results in an increased affinity or activity in the second site. (d) The binding of an allosteric effector might introduce a new binding site to a protein. Binding of a ligand to this new binding site could lead to changes in active site geometry, providing an indirect mechanism of allosteric control. This type of effect is of great interest in the design of allosteric drugs and can be considered as a subset of the example in (c). (e) The fusion of an enzyme to a protein under allosteric control. This type of construct can act as an allosteric switch because the activity of the enzyme is indirectly under allosteric control via the bound protein with an allosteric site. Such constructs are both present in nature and the target of protein engineering studies. (Reproduced from Nina M. Goodey and Stephen J. Benkovic (2008). Allosteric regulation and catalysis emerge via a common route. *Nature Chemical Biology*, **4**, 474–482, by permission of the Nature Publishing Group.)

of protein conformations is the existence of amino acid networks that facilitate communication between the different sites and that are linked to the mechanism of catalysis. NMR relaxation dispersion (Section 13.5.2) and isothermal titration calorimetry (Section 15.3.3) studies have shown that the timescale of linked amino acid networks is milliseconds to microseconds and the timescale of binding site change is microseconds to nanoseconds. These values compare with the very fast rate of atomic fluctuations (nanoseconds to picoseconds).

15.3 ANALYTICAL METHODS FOR THE STUDY OF ENZYME REACTIONS

15.3.1 General considerations

Enzyme assays are undertaken for a variety of reasons, but the most common are:

- to determine the amount (or concentration) of enzyme present in a particular preparation (this is particularly important in diagnostic enzymology, Section 16.3);
- to gain an insight into the kinetic characteristics of the reaction and hence to determine a range of kinetic constants such as K_m, V_{max} and k_{cat};
- to study the effect of pH, temperature, inhibitors, etc. on the enzyme and to make comparative studies of other enzymes that may be involved in a metabolic or signalling pathway of which the enzymes are members. The study of enzyme inhibition is fundamental to the development of new drugs.

Analytical methods for enzyme assays may be classified as either continuous (kinetic) or discontinuous (fixed-time). Continuous methods monitor some property change (e.g. absorbance or fluorescence) in the reaction mixture, whereas discontinuous methods require samples to be withdrawn from the reaction mixture and analysed by some convenient technique. The inherent greater accuracy of continuous methods commends them whenever they are available.

For simplicity, initial rates are sometimes determined experimentally on the basis of a single measurement of the amount of substrate consumed or product produced in a given time rather than by the tangent method. This approach is valid only over the short period of time when the reaction is proceeding effectively at a constant rate. This linear rate section comprises at the most the first 10% of the total possible change and clearly the error is smaller the earlier the rate is measured. In such cases, the initial rate is proportional either to the reciprocal of the time to produce a fixed change (fixed change assays) or to the amount of substrate reacted in a given time (fixed time assays). The potential problem with fixed-time assays is illustrated in Fig. 15.11, which represents the effect of enzyme concentration on the progress of the reaction in the presence of a constant initial substrate concentration (Fig. 15.11a). Measurement of the rate of the reaction at time t_0 (by the tangent method) to give the true initial rate or at two fixed times, t_1 and t_2, gives the relationship between initial rate and enzyme concentration shown in Fig. 15.11b. It can be seen that only the tangent method gives the correct linear relationship. Since the correct determination of initial rate means that the observed changes in the concentration of substrate or product are relatively small, it is inherently more accurate to measure the increase in product concentration because the relative increase in its concentration is significantly larger than the corresponding decrease in substrate concentration.

15.3.2 Analytical methods for steady-state studies

Visible and ultraviolet spectrophotometric methods

Many substrates and products absorb light in the visible or ultraviolet region and the change in the absorbance during the reaction can be used as the basis for the enzyme

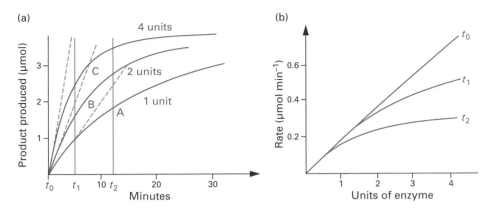

Fig. 15.11 The importance of measuring the initial rate in the assay of an enzyme. (a) Time-dependent variation in the concentration of products in the presence of 1, 2 and 4 units of enzyme; (b) variation of reaction rate with enzyme concentration using true initial rate (v_0) and two fixed time assays (t_1 and t_2).

assay. It is essential that the substrate and product do not absorb at the same wavelength and that the Beer–Lambert law (Section 12.2.2) is obeyed for the chosen analyte. A large number of common enzyme assays are based on the interconversion of NAD(P)$^+$ and NAD(P)H. Both of these nucleotides absorb at 260 nm but only the reduced form absorbs at 340 nm. Enzymes that do not involve this interconversion can be assayed by means of a coupled reaction that involves two enzyme reactions linked by means of common intermediates. The assay of 6-phosphofructokinase (PFK) (EC 2:7:1:11) coupled to fructose-bisphosphatase aldolase (FBPA) (EC 4.1.2.13) and glyceraldehyde-3-phosphate dehydrogenase (G3PDH) (EC 1:2:1:12) illustrates the principle:

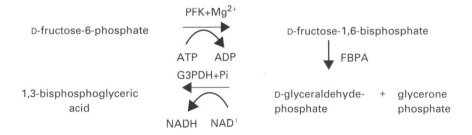

The assay mixture would contain D-fructose-6-phosphate, ATP, Mg^{2+}, FBPA, G3PDH, NAD$^+$ and P$_i$ all in excess so that the reaction would go to completion and the rate of reduction of NAD$^+$ and the production of NADH and hence the increase in absorbance at 340 nm, would be determined solely by the activity of PFK added to the reaction mixture in a known volume of the test enzyme preparation. In principle there is no limit to the number of reactions that can be coupled in this way provided that the enzyme under investigation is always present in limiting amounts.

The number of units of enzyme in the test preparation can be calculated by applying the Beer–Lambert law to calculate the amount of product formed per second:

$$\text{enzyme units (katals per cm}^3 \text{ test solution)} = \frac{\Delta E_{340}}{\varepsilon_\lambda} \times \frac{a}{1000} \times \frac{1000}{x}$$

where ΔE_{340} is the control-corrected change in the absorbance at 340 nm per second, a is the total volume (cm^3) of reaction mixture (generally about 3 cm^3) in a cuvette of 1 cm light path, x is the volume (mm^3) of test solution added to the reaction mixture and ε_λ is the molar extinction coefficient for NADH at 340 nm ($6.3 \times 10^3 \, \text{M}^{-1} \, \text{cm}^{-1}$). By dividing the above equation by the total concentration of protein in the test enzyme preparation, the specific activity (katals kg^{-1}) of the preparation can be calculated.

The scope of visible spectrophotometric enzyme assays can be extended by the use of synthetic substrates that release a coloured product. Many such artificial substrates are available commercially particularly for the assay of hydrolytic enzymes. The favoured coloured products are phenolphthalein and p-nitrophenol both of which are coloured in alkaline solution. An extension of this approach is the use of synthetic dyes for the study of oxidoreductases. The oxidised and reduced forms of these dyes have different colours. Examples are the tetrazolium dyes, methylene blue, 2,6-dichlorophenol indophenol and methyl and benzyl viologen.

Spectrofluorimetric methods

Fluorimetric enzyme assays have the significant practical advantage that they are highly sensitive and can therefore detect and measure enzymes at low concentrations. NAD(P)H is fluorescent and so enzymes utilising it can be assayed either by their absorption at 340 nm or by their fluorescence (primary wavelength 340 nm, reference wavelength 378 nm). Synthetic substrates that release a fluorescent product are also available for the assay of some enzymes. An example is the assay of β-D-glucuronidase (EC 3:2:1:31) using 4-methylumbelliferyl-β-D-glucuronide as substrate and assaying 4-methyl-umbelliferone as the fluorescent product. The large commercial interest in the development of inhibitors of kinases, phosphatases and proteases for their therapeutic potential has stimulated the development of assays for these enzymes using fluorogenic substrates to allow kinetic measurements of enzyme activity to be undertaken in vivo using the principle of fluorescence resonance energy transfer (FRET). These substrates contain two fluorochromes situated less than 100 Å apart joined by a 'linker' that is cleaved by the test enzyme and such that the emission wavelength of the donor fluorochrome overlaps with the excitation wavelength of the acceptor fluorochrome allowing the former to transfer energy to the latter. The most commonly used fluors are cyan fluorescent protein, red fluorescent protein and yellow fluorescent protein.

Luminescence methods

Bioluminescence reactions are commonly used as the basis for an enzyme assay due to their high sensitivity. The assay of luciferase is an example:

$$\text{luciferin} + \text{ATP} + O_2 \xrightleftharpoons{\text{luciferase}} \text{oxyluciferin} + \text{AMP} + \text{PP}_i + CO_2 + \textit{light}$$

The assay can be used to assay ATP and enzymes that utilise ATP by means of coupled reactions. The use of excess reagents would ensure that each reaction went to completion.

Immunochemical methods

Monoclonal antibodies raised to a particular enzyme can be used as a basis for a highly specific ELISA-based assay for the enzyme. Such assays can distinguish between isoenzyme forms, which make the assay attractive for diagnostic purposes. An important clinical example is creatine kinase. It is a dimer based on two different subunits, M and B. The MB isoenzyme is important in the diagnosis of myocardial infarction (heart attack) and an immunological assay is important in its assay.

15.3.3 Analytical methods for pre-steady-state studies

The experimental techniques discussed in the previous section are not suitable for the study of the progress of enzyme reactions in the short period of time (commonly milliseconds) before steady-state conditions, with respect to the formation of enzyme-substrate complex, are established. Figure 15.1 shows the progress curves for this pre-steady-state initial stage of an enzymic reaction. The induction time t is related to the rate constants for the formation and dissociation of the ES complex. Two main types of method are available for the study of this pre-steady-state.

Rapid mixing methods

In the continuous flow method, separate solutions of the enzyme and substrate are introduced from syringes, each of $10 \, cm^3$ maximum volume, into a mixing chamber typically of $100 \, mm^3$ capacity. The mixture is then pumped at a preselected speed through a narrow tube that is illuminated by a light source and monitored by a photomultiplier detector. Flow through the tube is fast, typically $10 \, m \, s^{-1}$, so that it is turbulent thus ensuring that the solution is homogeneous. The precise flow time from the time of mixing to the observation point can be calculated from the known flow rate. By varying the flow rate the reaction time at the observation point can be varied, allowing the extent of reaction to be studied as a function of time. From these data the various rate constants can be calculated. The technique uses relatively small amounts of reactants and is limited only by the time required to mix the two reactants.

The stopped-flow method is a variant of the continuous flow method in that shortly after the reactants emerge from the mixing chamber the flow is stopped and the detector triggered to continuously monitor the change in the experimental parameter such as absorbance or fluorescence (Fig. 15.12). Special flow cells are used together with a detector that allows readings to be taken 180° to the light source for absorbance, transmittance or circular dichroism measurements, or at 90° to the source for fluorescence, fluorescence anisotropy or light scattering measurements.

A variant of the stopped-flow method is the *quenching method*. In this technique the reactants from the mixing chamber are treated with a quenching agent from a third syringe. The quenching agent, such as trichloroacetic acid, stops the reaction that is then monitored by an appropriate analytical method for the build-up of intermediates.

SFM-400 Stopped-flow mode

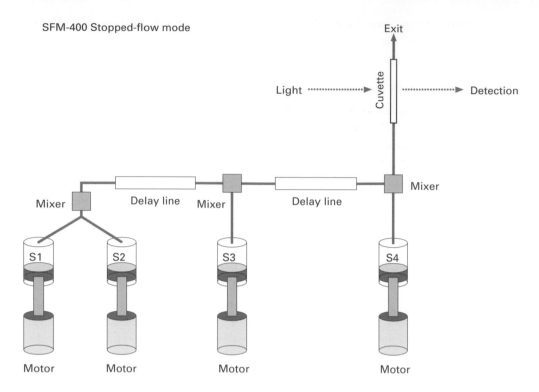

SFM-400 Quenched-flow mode

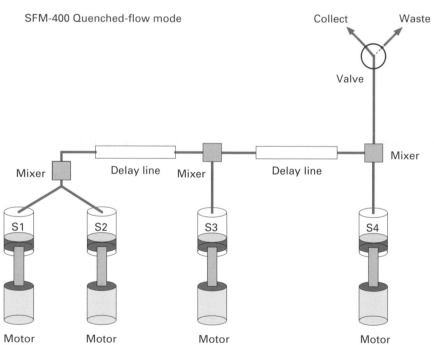

Fig. 15.12 The BioLogic stopped-flow and quenched-flow apparatus. The reactants are placed in separate syringes each driven by a microprocessor-controlled stepping motor capable of delivering 0.01 to 10.00 cm^3 min^{-1} with a minimum injection volume of 10–30 mm^3. The reactants are pre-mixed before they

By varying the time between mixing the reactants and adding the quenching reagent, the kinetics of this build-up can be studied. A disadvantage of this approach is that it uses more reactants than the stopped-flow method since the kinetic data are acquired from a series of studies rather than by following one reaction for a period of time. Both methods have difficulty in monitoring the first millisecond of reaction due to the need to allow mixing to take place, but this problem can be partly solved by changing the pH or temperature in order to slow down the reaction. Both methods commonly use synthetic substrates that release a coloured product or give rise to a coloured acyl or phosphoryl intermediate.

Relaxation methods

The limitation of the stopped-flow method is the dead time during which the enzyme and substrate are mixed. In the relaxation methods an equilibrium mixture of the reactants is preformed and the position of equilibrium altered by a change in reaction conditions. The most common procedure for achieving this is the temperature jump technique in which the reaction temperature is raised rapidly by 5–10 °C by the discharge of a capacitor or infrared laser. The rate at which the reaction mixture adjusts to its new equilibrium (relaxation time τ, generally a few microseconds) is inversely related to the rate constants involved in the reaction. This return to equilibrium is monitored by one or more suitable spectrophotometric methods. The recorded data enable the number of intermediates to be deduced and the various rate constants calculated from the relaxation times.

These pre-steady-state techniques have shown that the enzyme and its substrate(s) associate very rapidly, with second-order rate constants for the formation of ES in the range 10^6 to 10^8 M^{-1} s^{-1} and first-order rate constants for the dissociation of ES in the range 10 to 10^4 s^{-1}. The upper limit of these values is such that for some enzymes virtually every interaction between an enzyme and its substrate leads to the formation of a complex. The stopped-flow and quenching methods have also been used to study other biochemical processes that are kinetically fast and may involve transient intermediates. For example, the stopped-flow method has been applied to the study of protein folding, protein conformational changes and receptor–ligand binding, and the quenching method to the study of second messenger pathways (Section 17.4.1).

Isothermal titration calorimetry

This is a general method for studying the thermodynamics of any binding (association) process. It detects and quantifies small heat changes associated with the binding and has the advantages of speed, accuracy and not requiring either of the reacting species to be

Caption for fig. 15.12 (*cont.*)

enter the delay line (variable volume between 25 and 1000 mm³) and then the flow cell cuvette with a minimum dead time of 0.6 ms. The flow can be stopped at any predetermined time either by stopping the stepping motor or by closing the outlet from the reaction cuvette. The reaction can be studied by visible, ultraviolet, fluorescence or circular dichroism spectroscopy. The optical path length can be varied between 0.8 and 10 mm. In quench-flow mode the minimum ageing time is <2 ms. The quenching agent is added from the third or fourth syringe. (Reproduced by permission of BioLogic Science Instruments, France: website www.bio-logic.info.)

(a)

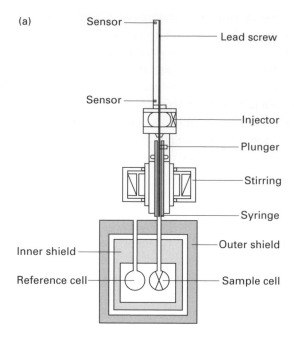

(b)

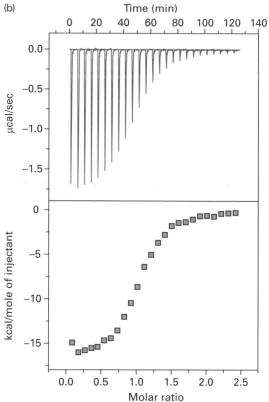

Fig 15.13 Isothermal titration calorimetry. (a) Diagram of MicroCal ITC cell and syringe. A thermoelectric device measures the temperature difference between the two cells and a second device measures the temperature difference between the two cells and the jacket. The plunger is computer-controlled and injects precise volumes

chemically modified or immobilised. The apparatus consists of a pair of matched cells (sample and reference) of approximately $2\,cm^3$ volume contained in a microcalorimeter (Fig. 15.13a). One of the reactants (say the enzyme preparation) is added to the sample cell and the ligand (substrate, inhibitor or effector) added via a stepper-motor-driven syringe. The mixture is stirred to ensure homogeneity. The reference cell contains an equal volume of reference liquid. A constant power of less than 1 mW is applied to the reference cell. This directs a feedback circuit activating a heater attached to the sample cell. The addition of the ligand solution causes a heat change due to the binding process and the dilution of both the enzyme and ligand preparations. If the reaction is exothermic less energy is required to maintain the cell at constant temperature. If the reaction is endothermic more energy is required. The power required to maintain a constant temperature is recorded as a series of spikes as a function of time (Fig. 15.13b). Each spike is integrated to give $\mu cal\ s^{-1}$ and summed to give the total heat exchange per injection. The study is repeated with a series of increasing ligand concentrations and control experiments carried out replacing the ligand with buffer solution to allow the heat exchange (ΔH) associated solely with the addition of ligand to be calculated. A plot is then made of enthalpy change against the molar ratio of the ligand to enzyme. The plot is hyperbolic from which it is possible to calculate enthalpy, free energy, and entropy changes associated with the ligand binding and hence the dissociation constant, K_d, and stoichiometry of binding n. Isothermal titration calorimetry has been successfully used in the study of the thermodynamics of the interconversion of protein conformations and the elucidation of the mechanism of allosterism.

15.3.4   Analytical methods for *in vivo* studies

The increasing importance of genome sequencing studies, particularly in the context of drug development, has stimulated the development of techniques for the study of enzymes in intact cells and whole organisms. *In vitro* methods have the disadvantage that they lead to the disruption of organelles and micro-departments, commonly result in the release of activators or inhibitors and invariably use assay conditions that are not representative of the *in vivo* situation. One of the most successful analytical techniques for studying enzymology in individual cells and in whole organisms is nuclear magnetic resonance spectroscopy (NMR). This non-invasive technique allows the measurement of steady-state metabolite concentrations and of metabolic flux using simple proton NMR, or the redistribution of a ^{13}C label among glycolytic intermediates or the use of ^{31}P NMR to measure ATP turnover and flux. Evidence for enzyme–enzyme interaction has been obtained by studying conformational changes in the enzyme protein. This approach requires the protein to be labelled in some appropriate way. One of the most attractive

Caption for fig. 15.13 (*cont.*)

of the ligand whilst the syringe rotates to provide continuous mixing. Heat is added or removed from the sample cell, as appropriate, and the associated power required to maintain constant temperature recorded in units of $\mu cal\ s^{-1}$. (b) Data for the binding of 2′ CMP to RNase. Top panel, energy exchange; bottom panel, the binding isotherm from which the value of n (1), K_d (0.85 μM) and ΔH (16.7 kcal mol^{-1}) can be calculated. (Reproduced by permission of MicroCal Europe, Milton Keynes, UK: website www.microcal.com.)

methods is to insert a fluorine atom into the molecule. From an NMR point of view this is an excellent label, since it is a spin-half nuclide that is readily studied by NMR. The chemical shift change of the fluorine nucleus is large, making it very sensitive to its local environment in the protein. Moreover, its size is very similar to that of a proton, so that it is unlikely to modify the enzyme's structure. Since fluorine is very rare in biological systems, the NMR signal from the label can be interpreted unambiguously. By studying the relaxation times associated with the fluorine nucleus it is possible to detect restricted motion of the enzyme in a cell due to protein–protein aggregation.

15.3.5 Analytical methods for substrate assays

Enzyme-based assays are very convenient methods for the estimation of the amount of substrate present in a biological sample. The principle of using excess enzyme (i.e. the substrate concentration should be less than the K_m) and relating the substrate concentration in the test solution to the observed initial rate can be used. It is essential that the reaction goes to completion in a relatively short time. If the reaction is freely reversible, then it is necessary to change the experimental conditions, such as pH or by chemically trapping the product, so that the reaction does approach completion. Coupled reactions are commonly used in substrate assays and they have the attraction that they help in the displacement of reversible reactions. The sensitivity of this initial rate method to substrate assay depends upon the value of the molar extinction coefficient for the analyte being assayed and also on the K_m for the substrate. In practice these two factors place a constraint on the level of substrate that can be assayed. Several approaches are available to overcome this problem. The end–point technique avoids the measurement of initial rate by converting all the substrate to product and then computing the amount present by correlating it with the total change in parameter such as absorbance or fluorescence. The sensitivity of an assay can also be significantly increased by the technique of enzymic cycling. In this method the substrate is regenerated by means of a coupled reaction and the total change in absorbance, etc. in a given time measured. Precalibration using a range of substrate concentrations with all the other reactants in excess allows the substrate concentration in a test solution to be computed. This method has a 10^4- to 10^5-fold increase in sensitivity relative to the end-point technique.

Enzyme-based assays are commonly used in clinical biochemistry to measure substrates in biological samples. For example, the three most common assays for serum glucose are those based on the use of hexokinase, glucose oxidase and glucose dehydrogenase. The first two are based on the coupled reaction technique:

- *Hexokinase method*: This couples the reaction to that of glucose-6-phosphate dehydrogenase and measures the absorbance at 340 nm due to NADH:

D-glucose + ATP $\rightleftharpoons$ D-glucose-6-phosphate + ADP

D-glucose-6-phosphate + NAD$^+$ $\rightleftharpoons$ D-glucono-1,5-lactone-6-phosphate + NADH + H$^+$

- *Glucose oxidase method*: This couples the reaction to peroxidase and measures the absorption of the oxidised dye in the visible region or uses an oxygen electrode to measure the oxygen consumption directly:

$$\text{D-glucose} + H_2O + O_2 \rightleftharpoons \text{D-gluconic acid} + H_2O_2$$
$$H_2O_2 + \text{dye}_{reduced} \rightleftharpoons H_2O + \text{dye}_{oxidised}$$

The glucose oxidase method uses β-D-glucose as substrate but blood glucose contains an equilibrium mixture of it and the α-isomer. Fortunately, preparations of glucose oxidase contain an isomerase that interconverts the two isomers thus allowing the assay of total D-glucose. Examples of the commonly used dye are 4-aminophenazine and *o*-dianisidine.

- *Glucose dehydrogenase method*: This requires no coupled reaction, but simply measures the increase in absorption at 340 nm:

$$\beta\text{-D-glucose} + NAD^+ \rightleftharpoons \text{D-glucono-1,5-lactone} + NADH + H^+$$

Another blood substrate commonly assayed by enzyme-based techniques in clinical biochemistry is cholesterol:

- *Cholesterol*: This is used as an indicator of atherosclerosis and susceptibility to coronary heart disease (Section 18.2.2). It uses cholesterol oxidase and peroxidase and measures the absorption in the visible region due to the oxidised dye, for example 4-aminophenazine:

$$\text{cholesterol} + O_2 \rightleftharpoons \text{4-cholesten-3-one} + H_2O_2$$
$$H_2O_2 + \text{dye}_{reduced} \rightleftharpoons H_2O + \text{dye}_{oxidised}$$

Any cholesterol ester in the sample is hydrolysed to free cholesterol by the inclusion of cholesterol esterase in the reaction mixture.

Biosensors (Sections 1.3.5 and 16.2.2) such as those for glucose and cholesterol are based on the above reactions and afford a simple means for the fast measurement of these substrates.

15.4 ENZYME ACTIVE SITES AND CATALYTIC MECHANISMS

15.4.1 Enzyme active sites

As previously pointed out, enzymes are characterised by their high specificity, catalytic activity and capacity for regulation. These properties must reflect the specific three-dimensional interaction between the enzyme and its substrate. A complete understanding of the way enzymes work must therefore include the elucidation of the mechanism underlying the binding of a substrate(s) to the enzyme catalytic site and the subsequent conversion of the substrate(s) to product(s). The mechanism must include details of the nature of the binding and catalytic sites, the nature of the intermediate enzyme–substrate complex(es), and the associated electronic and stereo-chemical events that result in the formation of the product. A wide range of strategies and analytical techniques has been adopted to gain such an understanding:

- *X-ray crystallographic studies:* These are capable of giving, either directly or indirectly, decisive information about the mechanism of enzyme action. X-ray diffraction patterns enable the position of each amino acid in the protein to be located and the details deduced of how the substrate binds and undergoes reaction. Such deductions are facilitated by the study of crystals grown in the presence and absence of the substrate, competitive inhibitor or of effector molecules. The Catalytic Site Atlas (www.ebl.ac.uk/thornton-srv/databases) and Protein Data Bank (www.wwpdp.org) list the active sites and catalytic residues of enzymes whose three-dimensional structure has been determined. As knowledge of protein structures and catalytic mechanisms has increased, computer programs have become available that enable the chemical and stereochemical conformations of the substrate(s) to be modelled and a prediction made of the three-dimensional structure of the enzyme that promotes the formation of product(s). This approach is now widely used in the pharmaceutical industry to identify 'lead' compounds for the development of new drugs (Section 18.2).

- *Irreversible inhibitor and affinity label studies:* Irreversible inhibitors act by forming a covalent bond with the enzyme. By locating the site of the binding of the inhibitor, information can often be obtained about the identity of specific amino acids in the binding site. A development of this approach is the use of photoaffinity labels that structurally resemble the substrate but which contain a functional group, such as azo ($-N=N-$), which on exposure to light is converted to a reactive functional group, such as a carbene or nitrene, which forms a covalent bond with a neighbouring functional group in the active site. It is common practice to tag the inhibitor or photoaffinity label with a radioisotope so that its location in the enzyme protein can easily be established experimentally.

- *Kinetic studies:* This approach is based on the use of a range of substrates and/or competitive inhibitors and the determination of the associated K_m, k_{cat} and K_i values. These allow correlations to be drawn between molecular structure and kinetic constants and hence deductions to be made about the structure of the active site. Further information about the structure of the active site can be gained by studying the influence of pH on the kinetic constants. Specifically, the effect of pH on K_m (i.e. on binding of E to S) and on V_{max} or k_{cat} (i.e. conversion of ES to products) is studied. Plots are then made of the variation of log K_m with pH and of log V_{max} or log k_{cat} with pH. The intersection of tangents drawn to the curves gives an indication of the pK_a values of ionisable groups involved in the active site (Fig 15.8). These are then compared with the pK_a values of the ionisable groups known to be in proteins. For example, pH sensitivity around the range 6–8 could reflect the importance of one or more imidazole side chains of a histidine residue in the active site.

- *Isotope exchange studies:* The replacement of the natural isotope of an atom in the substrate by a different isotope of the same element and the study of the impact of the isotope replacement on the observed rate of enzymatic reaction and its associated stereoselectivity, often enables deductions to be made about the mechanism of the reaction. Two examples illustrate the principle. Firstly, alcohol dehydrogenase (AD) that oxidises ethanol to ethanal using NAD^+ : NADH:

$$CH_3CH_2OH + NAD^+ \xrightleftharpoons{\text{AD}} CH_3CHO + NADH + H$$
$$\text{ethanol} \qquad\qquad\qquad\qquad \text{ethanal}$$

The two hydrogen atoms on the methylene (CH_2) group of ethanol are chemically indistinguishable, but if one is replaced by a deuterium or tritium atom the carbon atom becomes a chiral centre and the resulting molecule can be identified as either *R* or *S* configuration according to the Cahn–Ingold–Prelog rule for defining the stereochemistry of asymmetric centres. Studies have shown that alcohol dehydrogenase exclusively removes the hydrogen atom in the pro*R* configuration, i.e. (*R*) CH_3CHDOH always loses the D isotope in its conversion to ethanal but (*S*) CH_3CHDOH retains it. Such a finding can only be interpreted in terms of the specific orientation of the ethanol molecule at the binding site such that the two hydrogen atoms are effectively not equivalent. All dehydrogenases have been shown to display this type of stereospecificity and can be classified as either A-side dehydrogenases (e.g. alcohol dehydrogenase, lactate dehydrogenase, malate dehydrogenase) or B-side dehydrogenases (e.g. glycerol 3-phosphate dehydrogenase, glucose dehydrogenase, glyceraldehyde-3-phosphate dehydrogenase). Interestingly, the class type is independent of the hydrogen acceptor being NAD^+ or $NADP^+$. Secondly, the hydrolysis of esters by esterases that convert the ester to a mixture of acid and alcohol simultaneously incorporating a molecule of water into the products:

$$RCOO^*R' + H_2O \xrightarrow{\text{esterase}} RCOOH + R'O^*H$$
$$\text{ester} \qquad\qquad\qquad \text{acid} \qquad \text{alcohol}$$

In this reaction the oxygen atom identified as O* can either be retained in the acid or in the alcohol depending upon which side of the labelled oxygen atom the bond is broken, with water providing the second oxygen atom in the products. Labelling the oxygen atom in question as ^{18}O and studying, by mass spectrometry, its location after hydrolysis enables details to be drawn about the mechanism of the hydrolysis of the ester by the esterase. In practice, the labelled oxygen is found in the alcohol which supports the view that the reaction mechanism involves initial attack by water, acting as a nucleophile, on the carbonyl carbon atom and the subsequent elimination of the R'O* group.

- *Site-directed mutagenesis studies:* Advances in molecular biology and particularly in the ability to clone genes and express them in a particular vector have opened up the possibility of producing variants of the enzyme in which a particular amino acid residue, thought to be involved in substrate binding and catalysis, is replaced by another amino acid. By studying the impact of the replacement of an ionisable or nucleophilic amino acid with an unreactive one on the catalytic properties of the enzyme, conclusions can be drawn about the role of the amino acid residue that has been replaced. In principle it is also possible to produce variants that are more active than the native enzyme. Such studies are based on knowledge of the protein structure, function and mechanism of course and assume that the impact of the single amino acid replacement is confined to the active site and has not affected other aspects of the enzyme's structure. This needs to be confirmed by complementary structural studies, for example, spectroscopic techniques. This rational redesign approach has resulted in the generation of a superoxide dismutase with an enhanced activity relative to the native enzyme and an isocitrate dehydrogenase with specificity different to that of the native form.

15.4.2 Catalytic mechanisms

The application of the various strategies outlined above to a wide range of enzymes has enabled mechanisms to be deduced for many of them. Crystallographic and site-directed mutagenesis studies have been particularly successful in providing detailed information about the stereochemical and electronic events involved in substrate binding and product formation. The most commonly occurring amino acid residues in enzyme catalytic sites are the imidazole group of histidine, the guanidinium group of arginine, the carboxylate groups of glutamate and aspartate, the amino group of lysine, the hydroxyl groups of serine, threonine and tyrosine and the thiol group of cysteine. Of these nine groups, the imidazole group of histidine appears to be quantitatively the most important. These specific amino acid side chains commonly occur in pairs or triplets within catalytic sites. In the lipases and peptidases the catalytic triad serine, aspartate and histidine occurs very commonly. In chymotrypsin for example, the three amino acids are located in positions 195, 102 and 95 respectively and are brought into juxtaposition by the three-dimensional folding of the protein chain. Their roles include:

- to activate the substrate by forming a hydrogen bond thereby lowering the reaction activation energy barrier. Such a hydrogen bond has been referred to as a low–barrier hydrogen bond;
- to provide a nucleophile to attack the substrate;
- to provide the components for the acid–base catalysis of the substrate of the type well known in conventional organic chemistry;
- to stabilise the transition state of the reaction.

The quantitative importance of histidine can be explained by the facts that it is the only residue that has a pK_a near neutral so that it can easily function as an acid–base catalyst, and that it can also easily function as a nucleophile and use its charged form to stabilise transition states. Whilst these amino acid side chains provide the components for catalysis, the specificity of the reaction is determined by the three-dimensional structure of the enzyme and the microenvironment it creates within the active site.

The experimental techniques discussed above have successfully identified key amino acids involved in the catalytic process. However, the question remains as to the nature of the factors limiting the rate of catalysis. Recent studies, particularly those based on NMR relaxation dispersion experiments, have provided clear evidence that a key to the rate of catalysis is the interconversion of a population of protein conformations (Section 15.2.4). For example, studies with the enzyme dihydrofolate reductase have identified five conformations involved in the catalytic process that interconvert at rates that are identical to those previously determined by conventional kinetic experiments. Similarly, studies with the enzyme triosephosphate isomerase have shown that the opening and closing of a loop in the catalytic site has a vital role in the catalytic cycle. The loop has been shown to open and close at a rate of $10\,000\,s^{-1}$, a value that is the same as the catalytic rate constant. This correlation between rate-limiting steps in catalysis and rates of conformational change suggests that as the substrates bind and are converted to products, the population of conformations adjust and act to drive catalysis along the product pathway.

Multienzyme complexes

Studies on multienzyme complexes, including tryptophan synthase and carbamyl phosphate synthase, have demonstrated that the active site of one enzyme is coupled to that of the next enzyme in the metabolic sequence by means of allosteric conformational changes. The reaction products are channelled from one active site to the next by means of an intermolecular tunnel. In the case of tryptophan synthase, which is a $(\alpha\beta)_2$ complex in which the α and β subunits catalyse separate reactions, the tunnel is approximately 25 Å in length whereas that in carbamoyl phosphate synthase is approximately 100 Å long. The tunnels protect reactive intermediates from coming in contact with the external environment and reduce their transit time to the next active site. In the case of both enzymes the tunnels are formed prior to the binding of the initial substrates but with some other multienzyme complexes the tunnels are formed after the substrates bind to the active site.

Closely related to multienzyme complexes are the megasynthases responsible for the synthesis of antibiotics such as penicillin and vancomycin. They are large multifunctional enzymes consisting of clusters of active sites known as modules. They promote a series of reactions starting with simple organic intermediates that are progressively converted to the antibiotic. During the synthetic process, the intermediates are tethered to carrier proteins that shuttle them in sequence to the designated active site (referred to as client enzymes) in the module. NMR evidence indicates that the enzymes and the carrier proteins exist in an ensemble of dynamic conformations and that successful docking between the two to promote the next synthetic stage relies on the selection of the correct conformation in which the enzyme's carrier protein binding site is exposed.

15.5 CONTROL OF ENZYME ACTIVITY

15.5.1 Control of the activity of individual enzymes

The activity of an enzyme can be regulated in two basic ways:

- by alteration of the kinetic conditions under which the enzyme is operating;
- by alteration of the amount of the active form of the enzyme present by promoting enzyme synthesis, enzyme degradation or the chemical modification of the enzyme.

The latter option is inherently a long-term one and will be discussed later. In contrast, there are several mechanisms by which the activity of an enzyme can be altered almost instantaneously:

- *Product inhibition:* Here the product produced by the enzyme acts as an inhibitor of the reaction so that unless the product is removed by further metabolism the reaction will cease. An example is the inhibition of hexokinase by glucose-6-phosphate. Hexokinase exists in four isoenzyme forms I, II, III and IV. The first three isoforms all have a low K_m for glucose (about $10–100\,\mu M$) and are inhibited by glucose-6-phosphate, whereas isoform IV has a higher K_m ($10\,mM$) and is not inhibited by glucose-6-phosphate. Isoform IV is confined to the liver where its higher K_m allows it to deal with high glucose concentrations following a carbohydrate-rich meal. The other three isoforms are distributed widely and do not encounter such high glucose concentrations as those found in the liver. Thus their lower K_m values allow them to work optimally under their prevailing physiological conditions.
- *Allosteric regulation:* Here a small molecule that may be a substrate, product or key metabolic intermediate such as ATP or AMP alters the conformation of the catalytic site as a result of its binding to an allosteric site. A good example is the regulation of 6-phosphofructokinase discussed earlier.
- *Reversible covalent modification:* This may involve adenylation of a Tyr residue by ATP (e.g. glutamine synthase), the ADP-ribosylation of an arginine residue by NAD^+ (e.g. nitrogenase) but most frequently involves the phosphorylation of specific Tyr, Ser or Thr residues by a protein kinase. Most significantly, phosphorylation is reversible by the action of a phosphatase. Phosphorylation introduces the highly polar γ-phosphate group of ATP that is capable of inducing conformational changes in the enzyme structure such as to either activate or deactivate the enzyme. Reversible covalent modification is quantitatively the most important of the three mechanisms. A characteristic feature of many of these kinases is that they are involved in a cascade of enzyme reactions such as glycogenolysis and glycogenesis that will be discussed in the following section. Such cascades offer the opportunity for fine metabolic control and a large amplification of the original signal received by the membrane receptor.

15.5.2 Control of metabolic pathways

A large proportion of the thousands of enzymes in a cell are involved in the promotion of coordinated chemical pathways such as glycolysis, the citric acid cycle and the biosynthesis of fatty acids and steroids. Enzymes linked in a coordinated pathway are frequently clustered in one of three ways namely:

- by being located in the same compartment of the cell;
- by being physically associated as a multienzyme complex such as that of the fatty acid synthase of *E. coli*;
- by being membrane-bound such as the enzymes of electron transport.

This clustering facilitates the transport of the product of one enzyme to the next enzyme in the pathway.

Identification of rate-controlling enzymes

The individual enzymes in a metabolic pathway combine to produce a given flow of substrates and of products through the pathway. This flow is referred to as the metabolic flux. Its value is determined by factors such as the availability of starting substrate and cofactors but above all by the activity of the individual enzymes. Studies have revealed that the enzymes in a given pathway do not all possess the same activity. As a consequence, one or at most a small number with the lowest activity determine the overall flux through the pathway. In order to identify these rate-controlling enzymes three types of study need to be carried out:

- *in vitro* kinetic studies of each individual enzyme conducted under experimental conditions as near as possible to those found *in vivo* and such that the enzyme is saturated with substrate (i.e. such that $[S] > 10\,K_m$);
- studies to determine whether or not each individual enzyme stage operates at or near equilibrium *in vivo*;
- studies to determine the flux control coefficient, C, for each enzyme. This is a property of the enzyme that expresses how the flux of reactants through a pathway is influenced by a change in the activity (note: *not* concentration) of the enzyme under the prevailing physiological conditions. Such a change may be induced by allosteric activators or inhibitors or by feedback inhibition. Values for C can vary between 0 and 1. A flux control coefficient of 1 means that the flux through the pathway varies in proportion to the increase in the activity of the enzyme whereas a flux control coefficient of 0 means that the flux is not influenced by changes in the activity of that enzyme. The sum of the C values for all the enzymes in a given pathway is 1 so that the higher a given C value the greater the impact of that enzyme on the flux through the pathway. These C values are therefore highest for the rate-determining enzymes.

A reaction that is not at or near equilibrium and which is therefore associated with a large free energy change, is potentially a rate-limiting enzyme since the most probable reason for the non-establishment of equilibrium is the lack of adequate enzyme activity. To test for a non-equilibrium reaction it is necessary to analyse the concentration of each substrate and product *in vivo*. This is normally done by stopping all further reactions by denaturing the enzymes by the addition of a suitable denaturant to the *in vivo* test system and then analysing the analytes by a technique such as chromatography or NMR.

The application of these three tests to the enzymes in the glycolytic pathway shows that three of the ten enzymes, hexokinase, 6-phosphofructokinase and pyruvate kinase, have a potential rate-limiting role and do not achieve equilibrium but are associated with a large negative free energy change and are therefore effectively irreversible. The same three enzymes have the largest C values. Studies have revealed that all three enzymes are subject to various control mechanisms of their activity and all contribute to the control of flux through the glycolytic pathway. The actual quantitative values for the three test parameters vary between cell types in a given organism and between cells of a given type in different organisms.

Hexokinase exists in four isoenzyme forms, the first three of which are subject to inhibition by glucose-6-phosphate, the product of the reaction. Isoenzyme IV (also known as glucokinase) is not subject to this type of inhibition and has a higher K_m for glucose than have the other three forms. It is confined mainly to the liver where it is able to metabolise high concentrations of glucose, the resulting glucose-6-phosphate being diverted to glycogen biosynthesis via glucose-1-phosphate. In some tissues the limiting activity of hexokinase is bypassed by the provision of glucose-6-phosphate from glycogen via glucose-1-phosphate.

The activity of pyruvate kinase is regulated allosterically, being inhibited by ATP and activated by AMP and fructose-1,6-bisphosphate. In muscle, pyruvate kinase is present in large amounts, hence minimising its rate-limiting constraint. The fact that pyruvate kinase is located at the end of the pathway makes it unlikely that it will have a major role in the regulatory control of glycolysis.

6-Phosphofructokinase (PFK) is subject to allosteric control by a number of allosteric effectors that are related to the energy status of the cell. The principal activators are AMP and fructose-2,6-bisphosphate whilst ATP is an activator at low concentrations but an inhibitor at higher concentrations (1 mM). AMP activates the enzyme by releasing it from the inhibitory control of ATP disturbing the equilibrium away from the state that contains the ATP inhibitor site. The balance of control exerted by ATP and AMP is thus determined by their relative concentrations. This in turn is influenced by the enzyme adenylate kinase which catalyses the reaction:

$$2ADP \rightleftharpoons ATP + AMP$$

ATP is normally present in a cell at much higher concentrations that the other two nucleotides and as a consequence a small decrease in the concentration of ATP that is too small to relieve the inhibitory effect of ATP on PFK, results in a proportionally much larger change in AMP concentration that is normally only about 2% of that of ATP. This large percentage increase in the concentration of AMP allows it to exert a powerful activator effect on PFK hence facilitating increased glycolytic flux.

Additional control of glycolytic flux by PFK is exerted by its involvement in a substrate cycle with the enzyme fructose bisphosphatase (FBP) which is part of the gluconeogenesis pathway from pyruvate to glucose (Fig. 15.14). Both reactions are strongly exergonic and essentially irreversible. Whereas AMP acts as a powerful activator of PFK, it acts as a potent inhibitor of FBP and hence plays a reciprocal role in the control of these two opposing pathways. PFK converts D-fructose-6-phosphate to D-fructose-1,6-bisphosphate and simultaneously converts ATP to ADP, while FBP converts D-fructose-1,6-bisphosphate to D-fructose-6-phosphate and inorganic phosphate. The net result is apparently only the hydrolysis of ATP but in fact it results in a proportionally large increase in AMP concentration via adenylate kinase. As discussed above, this produces a large increase in flux through the glycolytic pathway by the activation of PFK and inhibition of FBP. A two-fold increase in AMP concentration can increase the glycolytic flux by 200-fold. However, the regulatory importance of changes in AMP concentration is not confined to its stimulation of glycolytic flux. Equally important is the fact that decreases in AMP concentration result in the ATP-inhibition of PFK activity becoming dominant, resulting in the virtual switching off of the glycolytic pathway and a concomitant increase in glycogen biosynthesis.

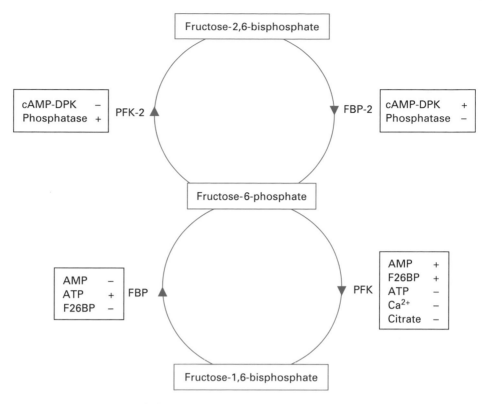

Fig. 15.14 Regulation of phosphofructokinase (PFK). Two substrate cycles each centred on fructose-6-phosphate are involved. Different enzymes promote the forward and reverse reactions of each cycle so that all reactions are exergonic (negative free energy changes). Each enzyme is subject to activation (+) or inhibition (−) either by allosteric effectors or by phosphorylation/dephosphorylation (cAMP-DPK, cyclic AMP-dependent protein kinase). The importance of each regulatory mechanism varies between organisms and between different tissues in a given organism.

Isoforms of PFK-2 and FBP-2 have been identified in different tissues. They differ in their affinity (K_s) for their substrates and in their sensitivity to regulation by phosphorylation/dephosphorylation. This rationalises the observation that the fine mechanistic detail for the control of PFK activity and hence of the regulation of glycolysis varies between different mammalian tissues.

In general, substrate cycles are an important means by which the activity of metabolic pathways is controlled. They operate at the expense of energy (ATP) and may simultaneously determine the relative importance of branch points in bidirectional pathways.

15.5.3 Signal amplification

The substrate cycles discussed above enabled opposing pathways to be controlled and small changes in the concentration of ATP to be amplified in terms of concomitant changes in AMP that is a key allosteric regulator of rate-limiting enzymes. This concept of amplification is important in the fine control of metabolic pathways and in the response of cells to hormone and neurotransmitter signals. Amplification is commonly achieved by a series of stages in which linked enzymes are themselves the substrate of a reaction, commonly based on phosphorylation or dephosphorylation, as a result of which the enzymes are either activated or deactivated. Such a series of reactions is referred to as a metabolic cascade and its merit is that it affords the opportunity for a large amplification of an original biochemical signal. The mobilisation of glycogen as glucose-1-phosphate by phosphorylase provides a good illustration of this principle. The components of this phosphorylase cascade are a membrane receptor that receives the original signal in the form of a hormone, neurotransmitter or similar, a G_s protein, adenylyl cyclase, cAMP-dependent protein kinase, phosphorylase kinase, phosphorylase and glycogen. cAMP released from adenylate cyclase as a result of its activation by a G_s protein (see Section 17.4.3) activates cAMP-dependent protein kinase, which in its inactive form is a tetramer consisting of two regulatory (R) and two catalytic (C) subunits (R_2C_2). Two cAMP molecules bind to each of the R subunits in a positively cooperative manner causing them to dissociate:

$$R_2C_2 + 4cAMP \longrightarrow 2R\text{-}2cAMP + 2C$$
$$\text{inactive form} \qquad\qquad\quad \text{active form}$$

The intracellular concentration of cAMP determines the proportion of cAMP-dependent protein kinase that is present in the active form and it is this that acts as a kinase that in the presence of ATP phosphorylates and thereby activates phosphorylase kinase:

$$\text{phosphorylase kinase} + \text{ATP} \xrightarrow{\text{C unit}} \text{phosphorylated phosphorylase kinase} + \text{ADP}$$
$$\text{inactive form} \qquad\qquad\qquad\qquad\qquad\qquad \text{active form}$$

Phosphorylase kinase is a tetrameric protein with four different subunits, α, β, γ and δ. The γ subunit contains the catalytic kinase site, the other three subunits having a regulatory role. The δ subunit is calmodulin, a Ca^{2+}-binding protein that contains two Ca^{2+} binding sites. Phosphorylase kinase is activated by the phosphorylation of the α and β subunits and by the binding of two Ca^{2+} ions to the δ subunit. The binding of Ca^{2+} ions to the δ subunit promotes the autophosphorylation of the enzyme at a different site to that phosphorylated by cAMP-dependent kinase. The activated phosphorylase kinase activates phosphorylase b (a dimer) by the phosphorylation of Ser14 on each subunit causing conformational changes and dimerisation to a tetramer to give phosphorylase a which degrades glycogen to glucose-1-phosphate. Most interestingly, phosphorylase b can also be activated allosterically by AMP, two molecules of which are capable of inducing conformation changes to give phosphorylase a but by a different induction mechanism to that brought about by phosphorylation. ATP and glucose-6-phosphate can induce the reverse allosteric change, deactivating the enzyme.

At each step in the phosphorylase cascade there is amplification of at least 100-fold. Thus occupation of only a very small percentage of the membrane receptors is needed to produce a final metabolic response approaching the maximum. It is evident that the larger the number of components in the cascade, the greater the potential for amplification. The mobilisation of glycogen is reversed by glycogen synthase which is inactivated by phosphorylation by phosphorylase kinase and activated by phospho-protein phosphatase-1 which simultaneously inactivates phosphorylase kinase and glycogen phosphorylase *a*. Phosphoprotein phosphatase-1 is itself subject to control by phosphorylation/dephosphorylation. As is discussed in more detail in Section 17.4.4, receptor-linked cascades seldom operate in isolation but rather form intricate networks that better allow the fine control necessary for the maintenance of homeostasis.

15.5.4 Long-term control of enzyme activity

The forms of control of enzyme activity discussed so far are essentially short- to medium-term control in that they are exerted in a matter of seconds or a few minutes at the most. However, control can also be exerted on a longer timescale. Long-term control, exerted in hours, operates at the level of enzyme synthesis and degradation. Whereas many enzymes are synthesised at a virtually constant rate and are said to be constitutive enzymes, the synthesis of others is variable and is subject to the operation of control mechanisms at the level of gene transcription and translation. One of the best-studied examples is the induction of β-galactosidase and galactoside permease by lactose in *E. coli*. The expression of the *lac* operon is subject to control by a repressor protein produced by the repressor gene (the normal state) and an inducer, the presence of which causes the repressor to dissociate from the operator allowing the transcription and subsequent translation of the *lac* genes. The lac repressor protein binds to the lac operator with a $K_i = 10^{-13}$ M and a binding rate constant of 10^7 M^{-1} s^{-1}. This rate constant is greater than that theoretically possible for a diffusion-controlled process and indicates that the process is facilitated in some way, possibly by DNA.

The metabolic degradation of enzymes is the same as that of other cellular proteins including membrane receptors. It is a first order process characterised by a half-life. The half-life of enzymes varies from a few hours to many days. Interestingly, enzymes that exert control over pathways have relatively short half-lives. The precise amino acid sequence of a protein is thought to influence its susceptibility to proteolytic degradation. N-terminal Leu, Phe, Asp, Lys and Arg, for example, appear to predispose the protein to rapid degradation. Proteins for proteolytic degradation are initially 'tagged' by a small protein (76 amino acids), called ubiquitin (Ub), which requires ATP and is able to form an enzyme-catalysed peptide-like bond with the C-terminal end of the protein to be degraded. Ubiquitin may either monoubiquitinate or polyubiquitinate a protein and the functional consequences vary. Monoubiquitination leads to the 'trafficking' of the protein, a process that is fundamental to the cycling of receptors (Section 17.5.2), whereas polyubiquitination leads to degradation. More than 12 ubiquitin-binding domains have been identified on proteins but they all bind to the same hydrophobic patch of ubiquitin which contains Ile44 as a central residue. There is increasing evidence that proteins to be degraded contain specific degradation signals, referred to as degrons, and that in some cases these signals are controlled by the protein folding or assembly so that biosynthetic errors and misfolding can be recognised and the protein removed by degradation. Misfolded proteins are ubiquitinated and directed to a juxtanuclear intracellular compartment where proteasomes (see below) are also concentrated.

The interaction between ubiquitin and a protein involves a series of enzymes and stages (Fig 15.15):

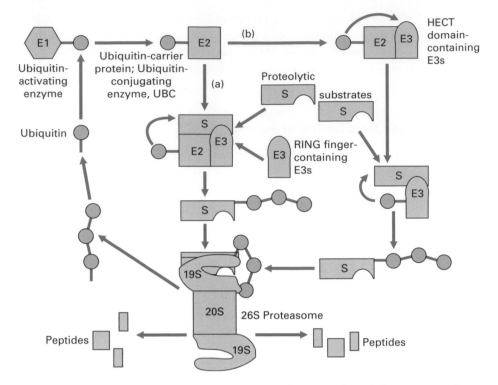

Fig.15.15 The ubiquitin–proteasome system (UPS). Ubiquitin is first activated to a high-energy intermediate by E1. It is then transferred to a member of the E2 family of enzymes. From E2 it can be transferred directly to the substrate (S) that is bound specifically to a member of the ubiquitin ligase family of proteins, E3 (a). This occurs when the E3 belongs to the RING finger family of ligases. In the case of a HECT-domain-containing ligase (b), the activated ubiquitin is transferred first to the E3 before it is conjugated to the E3-bound substrate. Additional ubiquitin moieties are added successively to the previously conjugated moiety to generate a polyubiquitin chain. The polyubiquitinated substrate binds to the 26S proteasome complex, the substrate is degraded to short peptides, and free and reusable ubiquitin is released through the activity of deubiquitinating enzymes (DUBs). (Reproduced from A. Ciechanover and R. Ben-Saadon (2004). N-terminal ubiquitination: more protein substrates join in. *Trends in Cell Biology*, **14**, 103–106, by permission of Elsevier Science.)

- *Ubiquitin-activating enzyme* (E1s): This requires ATP and involves the initial formation of a ubiquitin–adenylate intermediate. Ubiquitin is then transferred to a cysteine residue in the active site of the E1 with the concomitant release of AMP.
- *Ubiquitin-conjugating enzymes* (E2s): These receive the ubiquitin from the E1 by a transthioesterification reaction. 20–30 different E2s are known.
- *Ubiquitin protein ligases* (E3s): These interacts with both the E2 and the substrate by means of either a HECT or RING binding domain on the E3 resulting in the transfer of the ubiquitin to a lysine residue in the substrate. Results from the Human Genome Project provide evidence for the existence of several hundred of these E3 enzymes.
- The process either stops after monoubiquitination or is repeated resulting in the attachment of four or more ubiquitin residues to the substrate, a process called polyubiquitination.

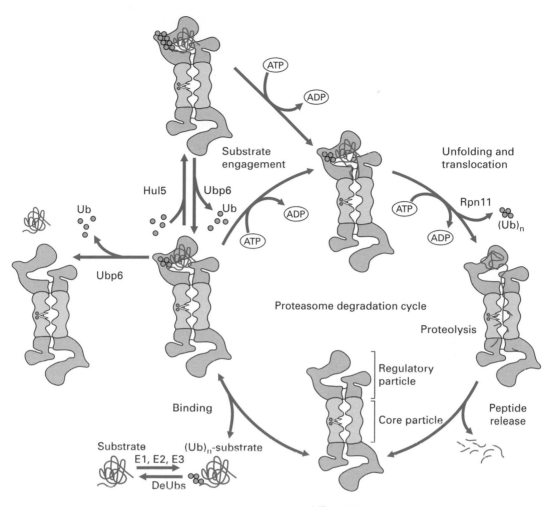

Fig. 15.16 Proteasome-degrading cycle, incorporating Hul5 and Ubp6 activities. Substrates are polyubiquitinated through the action of E1, E2 and E3 enzymes. Following binding to the proteasome, substrates are either engaged, unfolded and degraded in an ATP-dependent manner, with concomitant removal of the polyubiquitin chain by Rpn11, or are released from the proteasome. Release can be accelerated through the deubiquitinating action of Ubp6, which decreases substrate affinity. Further elongation of the ubiquitin chain by Hul5 increases affinity of the substrate for the proteasome, decreasing the release rate and thereby increasing degradation. The balance of Ubp6 and Hul5 activities can therefore affect the partitioning between binding and release by modifying the polyubiquitin chain length and thereby affecting the release rate. (Reproduced from D. A. Kraut, S. Prakash and A. Matouschek (2007). To degrade or release: ubiquitin-chain remodelling. *Trends in Cell Biology*, **17**, 419–421, by permission of Elsevier Science.)

The polyubiquitinated substrate is degraded by a multicatalytic complex based on a 20S proteasome (the S stands for Svedberg, see Section 3.5.3). Proteasomes are multisubunit proteases with a cylindrical core that has a 'lid' at both ends. The catalytic sites are within the core cylinder. The 20S proteasome consists of 14 α-subunits and 14 β-subunits arranged in four rings each of seven units. The proteolytic activity is located in the β–subunits at five sites that lie in the core. Entry of the substrate protein into the cylindrical core is controlled by a number of activators and may either proceed sequentially, starting from one end of the protein, or may involve a 'hairpin' conformation of the protein entering the proteasome allowing limited proteolysis of an internal segment. Proteolysis is ATP dependent and involves an additional 19S regulatory complex unit that consists of approximately 20 subunits, six of which have ATPase sites. These ATPases have similar structures but distinct functions that include the capture of the protein to be degraded, unfolding its structure and injection into the proteasome. This 19S complex unit combines with both ends of the 20S proteasome cylinder to form a 26S proteasome that promotes the cleavage of the peptide bonds with the concomitant hydrolysis of ATP. The recruitment and degradation of a protein relies on the presence of two subunits, Rpn1 and Rpn2, in the 19S unit. Two other proteins, Hul5 and Ubp6, when bound to the proteasome also regulate the degradation process (Fig. 15.16).

The balance between enzyme *de novo* synthesis and proteolytic degradation coupled with the regulation of enzyme activity enables the amount and activity of enzymes present in a cell to be regulated to meet fluctuating cell and whole organism needs. There is growing evidence to indicate that ubiquitination/deubiquitination is as important as phosphorylation/dephosphorylation for cellular homeostasis and cell cycle control. Dysfunction of the ubiquitin–proteasome pathway has been implicated in a number of disease states. For example, there is evidence that the accumulation of abnormal or damaged proteins due to impairment of the pathway contributes to a number of neurodegenerative diseases including Alzheimer's. In contrast, deliberately blocking the pathway in cancer cells could lead to a disruption of protein regulation that in turn could cause the apoptosis of the malignant cells. Accordingly, proteasome inhibitors have been developed for evaluation as anti-tumour agents against selected cancers. Bortezomib is one such inhibitor that targets the 26S proteasome, and in combination with other chemotherapeutic agents was shown to have therapeutic potential.

15.6 SUGGESTIONS FOR FURTHER READING

General texts
Frey, P. and Hegeman, A. (2007). *Enzymatic Reaction Mechanisms*. Oxford: Oxford University Press. (Discusses over 100 case studies of enzyme mechanisms.)

Review articles
Barglow, K. T. and Cravatt, B. F. (2007). Activity-based protein profiling for the functional annotation of enzymes. *Nature Methods*, **4**, 822–827.
English, B. P., Min, W., van Oijen, A. M., Lee, K. T., Luo, G., Sun, H., Cherayil, B. J., Kou, S. C. and Xie, S. (2006). Ever-fluctuating single enzyme molecules: Michaelis–Menten equation revisited. *Nature Reviews Chemical Biology*, **2**, 87–94.
Furnham, N., Garavelli, J. S., Apweiler, R. and Thornton, J. M. (2009). Missing in action: enzyme functional annotations in biological databases. *Nature Chemical Biology*, **5**, 521–525.
Komander, D., Clague, M. J. and Urbe, S. (2009). Breaking the chains: structure and function of the deubiquitinases. *Nature Reviews Molecular Cell Biology*, **10**, 550–563.
Kapure, S. and Khosia, C. (2008). Fit for an enzyme. *Nature*, **454**, 832–833.
Ravid, T. and Hochstrasser, M. (2008). Diversity of signals in the ubiquitin–proteasome system. *Nature Reviews Molecular Cell Biology*, **9**, 679–689.
Ye, Y. and Rape, M. (2009). Building ubiquitin chains: Ezenzymes at work. *Nature Reviews Molecular Call Biology*, **10**, 755–764.
Zalatan, J. G. and Herschlag, D. (2009). The far reaches of enzymology. *Nature Chemical Biology*, **5**, 516–520.

16 Principles of clinical biochemistry

J. FYFFE AND K. WILSON

16.1 Principles of clinical biochemical analysis

16.2 Clinical measurements and quality control

16.3 Examples of biochemical aids to clinical diagnosis

16.4 Suggestions for further reading

16.5 Acknowledgements

16.1 PRINCIPLES OF CLINICAL BIOCHEMICAL ANALYSIS

16.1.1 Basis of analysis of body fluids for diagnostic, prognostic and monitoring purposes

Underlying most human diseases is a change in the amount or function of one or more proteins that in turn triggers changes in cellular, tissue or organ function. The dysfunction is commonly characterised by a significant change in the biochemical profile of body fluids. The application of quantitative analytical biochemical tests to a large range of biological analytes in body fluids and tissues is a valuable aid to the diagnosis and management of the prevailing disease state. In this section the general biological and analytical principles underlying these tests will be discussed and related to the general principles of quantitative chemical analysis discussed in Section 1.3.

Body fluids such as blood, cerebrospinal fluid and urine in both healthy and diseased states contain a large number of inorganic ions and organic molecules. Whilst the normal biological function of some of these chemical species lies within that fluid, for the majority it does not. The presence of this latter group of chemical species within the fluid is due to the fact that normal cellular secretory mechanisms and the temporal synthesis and turnover of individual cells and their organelles within the major organs of the body, both result in the release of cell components, particularly those located in the cytoplasm, into the surrounding extracellular fluid and eventually into the blood circulatory system. This in turn transports them to the main excretory organs, namely the liver, kidneys and lungs, so that these cell components and/or their degradation products are eventually excreted in faeces, urine, sweat and expired air. Examples of cell components in this category include enzymes, hormones, intermediary metabolites and small organic and inorganic ions.

The concentration, amount or activity of a given cell component that can be detected in these fluids of a healthy individual at any point in time depends on many factors that can be classified into one of three categories, namely chemical characteristics of the component, endogenous factors characteristic of the individual and exogenous factors that are imposed on the individual.

- *Chemical characteristics*: Some molecules are inherently unstable outside their normal cellular environment. For example, some enzymes are reliant on the presence of their substrate and/or coenzyme for their stability and these may be absent or in too low concentrations in the extracellular fluid. Molecules that can act as substrates of catabolic enzymes found in extracellular fluids, in particular blood, will also be quickly metabolised. Cell components that fall into these two categories therefore have a short half-life outside the cell and are normally present in low concentrations in fluids such as blood.
- *Endogenous factors*: These include age, gender, body mass and pregnancy. For example:
  (a) serum cholesterol concentrations are higher in men than premenopausal women but the differences decreases post-menopause;
  (b) serum alkaline phosphatase activity is higher in children than in adults and is raised in women during pregnancy;
  (c) serum insulin and triglyceride concentrations are higher in obese individuals than in the lean;
  (d) serum creatinine, a metabolic product of creatine important in muscle metabolism, is higher in individuals with a large muscle mass;
  (e) serum sex hormone concentrations differ between males and females and change with age.
- *Exogenous factors*: These include time, exercise, food intake and stress. Several hormones are secreted in a time-related fashion. Thus cortisol and to a lesser extent thyroid stimulating hormone (TSH) and prolactin all show a diurnal rhythm in their secretion. In the case of cortisol, its secretion peaks around 9.00 am and declines during the day reaching a trough between 11.00 pm and 5.00 am. The secretion of female sex hormones varies during a menstrual cycle and that of 25-hydroxycholecalciferol (vitamin D_3) varies with the seasons peaking during the late summer months. The concentrations of glucose, triglycerides and insulin in blood rise shortly after the intake of a meal. Stress, including that imposed by the process of taking a blood sample by puncturing a vein (venipuncture), can stimulate the secretion of a number of hormones and neurotransmitters including prolactin, cortisol, adrenocorticotropic hormone (ACTH) and adrenaline.

The influence of these various factors on the extent of release of cell components into extracellular fluids inevitably means that even in healthy individuals there is a considerable intraindividual variation (i.e. variation from one occasion to another) in the value of any chosen test analyte of diagnostic importance and an even larger interindividual variation (i.e. variation between individuals). More importantly, the superimposition of a disease state onto these causes of intra- and interindividual variation will result in an even greater variability between test occasions.

Many clinical conditions compromise the integrity of cells located in the organs affected by the condition. This may result in the cells becoming more 'leaky' or, in more severe cases, actually dying (necrosis) and releasing their contents into the surrounding extracellular fluid. In the vast majority of cases the extent of release of specific cell components into the extracellular fluid, relative to the healthy reference range, will reflect the extent of organ damage and this relationship forms the basis of diagnostic clinical biochemistry. If the cause of the organ damage continues for a prolonged time and is essentially irreversible (i.e. the organ does not undergo self-repair), as is the case in cirrhosis of the liver for example, then the mass of cells remaining to undergo necrosis will progressively decline so that eventually the release of cell components into the surrounding extracellular fluid will decrease even though organ cells are continuing to be damaged. In such case the measured amounts will not reflect the extent of organ damage.

Clinical biochemical tests have been developed to complement in four main ways a provisional clinical diagnosis based on the patient's medical history and clinical examination:

- *To support or reject a provisional diagnosis* by detecting and quantifying abnormal amounts of test analytes consistent with the diagnosis. For example, serum myoglobin, troponin-I (part of the cardiac contractile muscle), creatine phosphokinase (specifically the CK-MB isoform) and aspartate transaminase all rise following a myocardial infarction (heart attack) that results in cell death in some heart tissue. The released cellular components also cause cell inflammation (leakiness) in surrounding cells causing an amplification of total cellular component release. Tests can also help a differential diagnosis, for example in distinguishing the various forms of jaundice (yellowing of the skin due to the presence of the yellow pigment bilirubin, a metabolite of haem) by the measurement of alanine transaminase (ALT) and aspartate transaminase (AST) activities and by determining whether or not the bilirubin is conjugated with β-glucuronic acid.
- *To monitor recovery following treatment* by repeating the tests on a regular basis and monitoring the return of the test values to those within the reference range. Following a myocardial infarction, for example, the raised serum enzyme activities referred to above usually return to reference range values within 10 days (Section 16.3.2, Fig. 16.3). Similarly, the measurement of serum tumour markers such as CA125 can be used to follow recovery or recurrence after treatment for ovarian cancer.
- *To screen for latent disease* in apparently healthy individuals by testing for raised levels of key analytes. For example, measuring serum glucose for diabetes mellitus and immunoreactive trypsin for cystic fibrosis. It is now common for serum cholesterol levels to be used as a measure of the risk of the individual developing heart disease. This is particularly important for individuals with a family history of the disease. An action limit of serum cholesterol >5.2 mM has been set by the British Hyperlipidaemia Association for an individual to be counselled on the importance of a healthy (low fat) diet and regular exercise and a higher action limit of serum cholesterol >6.6 mM for cholesterol-lowering 'statin' drugs to be prescribed (Section 18.2.2) and clinical advice given.
- *To detect toxic side effects of treatment*, for example in patients receiving hepatotoxic drugs, by undertaking regular liver function tests. An extension of this is therapeutic drug monitoring in which patients receiving drugs such as phenytoin and

carbamezepine (both of which are used in the treatment of epilepsy) that have a low therapeutic index (ratio of the dose required to produce a toxic effect relative to the dose required to produce a therapeutic effect) are regularly monitored for drug levels and liver function to ensure that they are receiving effective and safe therapy.

16.1.2 Reference ranges

For a biochemical test for a specific analyte to be routinely used as an aid to clinical diagnosis, it is essential that the test has the required performance indicators (Section 16.2), especially specificity and sensitivity. Sensitivity expresses the proportion of patients with the disease who are correctly identified by the test. Specificity expresses the proportion of patients without the disease who are correctly identified by the test. These two parameters may be expressed mathematically as follows:

$$\text{sensitivity} = \frac{\text{true positive tests} \times 100\%}{\text{total patients with the disease}}$$

$$\text{specificity} = \frac{\text{true negative tests} \times 100\%}{\text{total patients without the disease}}$$

Ideally both of these indicators for a particular test should be 100% but this is not always the case. This problem is most likely to occur in cases where the change in the amount of the test analyte in the clinical sample is small compared with the reference range values found in healthy individuals. Both of these indicators express the performance of the test but it is equally important to be able to quantify the probability that the patient with a positive test has the disease in question. This is best achieved by the predictive power of the test. This expresses the proportion of patients with a positive test who are correctly diagnosed as disease positive:

$$\text{positive predictive value} = \frac{\text{true positive patients}}{\text{total positive tests}}$$

$$\text{negative predictive value} = \frac{\text{true negative patients}}{\text{total negative tests}}$$

The concept of predictive power can be illustrated by reference to foetal screening for Down's syndrome and neural tube defects. Preliminary tests for these conditions in unborn children are based on the measurement of α-fetoprotein (AFP), human chorionic gonadotropin (hCG) and unconjugated oestriol (uE$_3$) in the mother's blood. The presence of these conditions results in an increased hCG and decreased AFP and uE$_3$ relative to the average in healthy pregnancies. The results of the tests are used in conjunction with the gestational and maternal ages to calculate the risk of the baby suffering from these conditions. If the risk is high, further tests are undertaken including the recovery of some foetal cells for genetic screening from the amniotic fluid surrounding the foetus in the womb by inserting a hollow needle into the womb (amniocentesis). The three tests detect two out of three cases (67%) of Down's syndrome and four out of five cases (80%) of neural tube defects. Thus the performance indicators of the tests are not 100% but they are sufficiently high to justify their routine use.

The correct interpretation of all biochemical test data is heavily dependent the use of the correct reference range against which the test data are to be judged. As previously pointed out, the majority of biological analytes of diagnostic importance are subject to considerable inter- and intraindividual variation in healthy adults, and the analytical method chosen for a particular analyte assay will have its own precision, accuracy and selectivity that will influence the analytical results. In view of these biological and analytical factors, individual laboratories must establish their own reference range for each test analyte using their chosen methodology and a large number (hundreds) of 'healthy' individuals. The recruitment of individuals to be included in reference range studies presents a considerable practical and ethical problem due to the difficulty of defining 'normal' and of using invasive procedures, such as venipuncture, to obtain the necessary biological samples. The establishment of reference ranges for children, especially neonates, is a particular problem.

Reference ranges are most commonly expressed as the range that covers the mean ± 1.96 standard deviations of the mean of the experimental population. This range covers 95% of the population. The majority of reference ranges are based on a normal distribution of individual values but in some cases the experimental data are asymmetric often being skewed to the upper limits. In such cases it is normal to use logarithmic data to establish the reference range but even so, the range may overlap with values found in patients with the test disease state. Typical reference ranges are shown in Table 16.1.

16.2 CLINICAL MEASUREMENTS AND QUALITY CONTROL

16.2.1 The operation of clinical biochemistry laboratories

The clinical biochemistry laboratory in a typical general hospital in the UK serves a population of about 400 000 containing approximately 60 General Practitioner (GP) groups depending upon the location in the UK. This population will generate approximately 1200 requests from GPs and hospital doctors each weekday for clinical biochemical tests on their patients. Each patient request will require the laboratory to undertake an average of seven specific analyte tests. The result is that a typical general hospital laboratory will carry out between 2.5 and 3 million tests each year. The majority of clinical biochemistry laboratories offers the local medical community as many as 200 different clinical biochemical tests that can be divided into eight categories as shown in Table 16.2.

Most of the requests for biochemical tests will arise on a routine daily basis but some will arise from emergency medical situations at any time of the day. The large number of daily test samples coupled with the need for a 24-hour 7-day week service dictates that the laboratory must rely heavily on automated analysis to carry out the tests and on information technology to process the data.

Table 16.1 **Typical reference ranges for biochemical analytes**

Analyte	Reference range	Comment
Sodium	133–145 mM	
Potassium	3.5–5.0 mM	Values increased by haemolysis or prolonged contact with cells.
Urea	3.5–6.5 mM	Range varies with sex and age e.g. values up to 12.1 will be found in males over the age of 70.
Creatinine	75–115 mM (males) 58–93 mM (females)	Creatinine (a metabolite of creatine) production relates to muscle mass and is also a reflection of renal function. Values for both sexes increase by 5–20% in the elderly.
Aspartate transaminase (AST)	$<40\,IU\,dm^{-3}$	Perinatal levels are $<80\,IU\,dm^{-3}$ and fall to adult values by the age of 18. Some slightly increased values up to $60\,IU\,dm^{-3}$ may be found in females over the age of 50. Results are increased by haemolysis.
Alanine transaminase (ALT)	$<40\,IU\,dm^{-3}$	Higher values are found in males up to the age of 60.
Alkaline phosphatase (AP)	$<122\,IU\,dm^{-3}$ (adults) $<455\,IU\,dm^{-3}$ (children $<12\,y$)	Significantly raised results of up to two- or three-fold would be experienced during growth spurts through teenage years. Slightly raised levels also seen in the elderly and in women during pregnancy.
Cholesterol	No reference range but recommended value of <5.2 mM	The measurement of cholesterol in an adult 'well' population does not show a Gaussian distribution but a very tailed distribution with relatively few low results. The majority are <10 mM but there is a long tail up to over 20 mM. There is a tendency for males to have higher cholesterol than females of the same age but after the menopause female values revert to those of males. Generally values increase with age.

To achieve an effective service, a clinical biochemistry laboratory has three main functions:

- *to advise* the requesting GP or hospital doctor on the appropriate tests for a particular medical condition and on the collection, storage and transport of the patient samples for analysis;
- *to provide* a quality analytical service for the measurement of biological analytes in an appropriate and timely way;
- *to provide* the requesting doctor with a data interpretation and advice service on the outcome of the biochemical tests and possible further tests.

Table 16.2 **Examples of biochemical analytes used to support clinical diagnosis**

Type of analyte	Examples
Foodstuffs entering the body	Cholesterol, glucose, fatty acids, triglycerides
Waste products	Bilirubin, creatinine, urea
Tissue-specific messengers	Adrenocorticotropic hormone (ACTH), follicle-stimulating hormone (FSH), luteinising hormone (LH), thyroid-stimulating hormone (TSH)
General messengers	Cortisol, insulin, thyroxine
Response to messengers	Glucose tolerance test assessing the appropriate secretion of insulin; tests of pituitary function
Organ function	Adrenal function – cortisol, ACTH; renal function – K^+, Na^+; urea, creatinine; thyroid function – free thyroxine (FT_4), free tri-iodothyronine (FT_3), TSH
Organ disease markers	Heart – troponin-I, creatine kinase (CKMB), AST, lactate dehydrogenase (LD); liver – ALT, AP, γ-glutamyl transferase (GGT), bilirubin, albumin
Disease-specific markers	Specific proteins ('tumour markers') secreted from specific organs – prostate-specific antigen (PSA), CA-125 (ovary), calcitonin (thyroid), $\alpha_1$fetoprotein (liver)

The advice given to the clinician is generally supported by a User Handbook, prepared by senior laboratory personnel, that includes a description of each test offered, instructions on sample collection and storage, normal laboratory working hours and the approximate time it will take the laboratory to undertake each test. This turnaround time will vary from less than one hour to several weeks depending upon the speciality of the test. The vast majority of biochemical tests are carried out on serum or plasma derived from a blood sample. Serum is the preferred matrix for biochemical tests but the concentrations of most test analytes are almost the same in the two fluids. Serum is obtained by allowing the blood to clot and removing the clot by centrifugation. To obtain plasma it is necessary to add an anticoagulant to the blood sample and remove red cells by centrifugation. The two most common anticoagulants are heparin and EDTA, the choice depending on the particular biochemical test required. For example, EDTA complexes calcium ions so that calcium in EDTA plasma would be undetectable. For the measurement of glucose, fluoride-oxalate is added to the sample not as an anticoagulant but to inhibit glycolysis during the transport and storage of the sample. Special vacuum collection tubes containing specific anticoagulants or other additives are available for the storage of blood samples. Collection tubes are also available containing clot enhancers to speed the clotting process for serum preparation. Many containers incorporate a gel with a specific gravity designed to float the gel between cells and serum providing a barrier between the two for up to 4 days. During these 4 days, the cellular component will experience lysis, so any subsequent contamination of the serum will include intracellular components. Biochemical tests may also be carried out on whole blood, urine, cerebrospinal fluid (the fluid surrounding the spinal cord and brain), faeces, sweat, saliva and amniotic fluid. It is essential that the samples are collected in the appropriate container at the correct time (particularly important if the test is for the measurement of hormones such as cortisol subject to diurnal release) and labelled with appropriate patient and biohazard details. Samples submitted to the laboratory for biochemical tests are accompanied by a request form, signed by the requesting clinician, which gives details of the tests required and brief details of the reasons for the request to aid data interpretation and to help identify other appropriate tests.

Laboratory reception

On receipt in the laboratory both the sample and the request form will be assigned an acquisition number usually in an optically readable form but with a bar code. A check is made of the validity of the sample details on both the request form and sample container to ensure that the correct container for the tests required has been used. Samples may be rejected at this stage if details are not in accordance with the set protocol. Correct samples are then split from the request form and prepared for analysis typically by centrifugation to prepare serum or plasma. The request form is processed into the computer system that identifies the patient against the sample acquisition number, and the tests requested by the clinician typed into the database. It is vital at this reception phase that the sample and patient data match and that the correct details are placed in the database. These details must be adequate to uniquely identify the patient bearing in mind the number of potential patients in the catchment area, and will include name, address, date of birth, CHI number (a unique identifier for each individual in the UK for health purposes), hospital or Accident and Emergency number and acquisition number.

16.2.2 Analytical organisation

The analytical organisation of the majority of clinical biochemical laboratories is based on three work areas:

- auto-analyser section,
- immunoassay section,
- manual section.

Auto-analyser section

Auto-analysers dedicated to clinical biochemical analysis are available from many commercial manufacturers. The majority of analysers are fully automated and have carousels for holding the test samples in racks each carrying up to 15 samples, one or two carousels each for up to 60 different reagents that are identified by a unique bar code, carousels for sample washing/preparation and a reaction carousel containing up to 200 cuvettes for initiating and monitoring individual test reactions. The Abbott ARCHITECT c1600 (Fig. 16.1) utilises three methodologies: spectrophotometry, immunoturbidometry and potentiometry. The spectrophotometry system is capable of measuring at 16 wavelengths simultaneously whilst the potentiometric system, based on the use of ion-selective electrodes (ISEs) (Section 1.3.5) that are

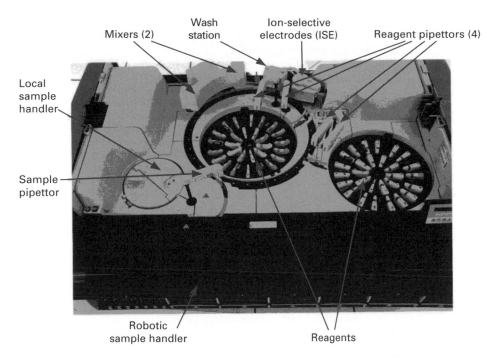

Fig. 16.1 The Abbott ARCHITECT c1600 clinical chemical analyser. (Reproduced by permission of Abbott Diagnostics UK & Northern Ireland, Maidenhead, UK.)

combined into a single unit, is used to measure electrolyte concentrations of Na^+, K^+ and Cl^- simultaneously with an analytical time of less than 4 min on a sample size of only 25 mm^3.

The reaction carousel contains 330 cuvettes and before each assay each cuvette is automatically cleaned and dried using an eight-stage wash mechanism. Supplementary washes are available to reduce carryover when necessary. Reagents for up to 62 assays based on different analytical techniques (Table 16.3) may be loaded at the same time and to maximise throughput, testing is organised in two parallel lines each one of which is resource and analysis-controlled separately. The analyser uses a fixed cycle time. Every 4.5 seconds the reaction carousel moves approximately one-quarter turn and readings are taken each full turn. The total analysis time is dependent upon the assay being performed and is determined by the assay parameters. All tests are complete in less than 10 minutes and results are reported immediately whilst complete results for an individual patient sample are recorded as soon as the final assay is complete. The theoretical throughput for the analyser is 1800 tests per hour. As with the majority of analysers, the c1600 is an 'open system' in that in addition to the programmed tests, 'in house' tests may also be incorporated into the routine by laboratory personnel. These may include therapeutic drug monitoring, drugs of abuse tests and specific protein assays. Each laboratory will have at least two analysers each offering a similar analytical repertoire so that one can back up the other. The analyser reads the bar code acquisition number for each sample and on the basis of the reading interrogates the host computer database to identify the tests to be carried out on the sample. The identified

Table 16.3 **Examples of analytical techniques used to quantify analytes by auto-analysers**

Analytical technique	Examples of analytes
Ion-selective electrodes	K^+, Na^+, Li^+, Cl^-
Visible and UV spectrophotometry	Urea, creatinine, calcium, urate
Turbidimetry	IgG, IgA, IgM, D-dimer (a metabolic product of fibrinogen)
Reaction rate	Enzymes – AST, ALT, GGT, AP, CK, LD,
Enzyme multiplier immunoassay test (EMIT)	Therapeutic drug monitoring – phenytoin, carbamezapine

tests are then automatically prioritorised into the most efficient order and the analyser programmed to take the appropriate volume of sample by means of a sampler that may also be capable of detecting microclots in the sample, add the appropriate volume of reagents in a specified order and to monitor the progress of the reaction. Internal quality control samples are also analysed on an identical basis at regular intervals. The analyser automatically monitors the use of all reagents so that it can identify when each will need replenishing. When the test results are calculated, the operator can validate them either on the analyser or on the main computer database. When appropriate, the results can also be checked against previous results on the same patient.

Micro sensor analysers

Recent advances in micro sensor technology have stimulated the development of miniaturised multi-analyte sensor blocks that are widely used in the clinical biochemistry laboratory for the routine measurement of pH, pCO_2, Na^+, K^+, Ca^{2+}, Cl^-, glucose and lactate (Section 1.3.5). The basic component of the block is a 'cartridge' that contains the analyte sensors, a reference electrode, a flow system, wash solution, waste receptacle and a process controller. The block is thermostatically controlled at $37\,^\circ C$ and its surface provides an interface to the analyte sensors (Fig. 16.2a). The sensors are embedded in three layers of plastic, the size and shape of a credit card. Each card may contain up to 24 sensors. A metallic contact under each sensor forms the electrical interface with the cartridge. As the test sample passes over the sensors, a current is generated by mechanisms specific to the individual analyte and recorded. The size of the current is proportional to the concentration of analyte in the fluid in the sample path. Calibration of the sensors with standard solutions of the analyte allows the concentration of the test sample to be evaluated. The sensor card and the sample path are automatically washed after each test sample and can be used for the analysis of up to 750 whole blood samples before being discarded.

The glucose and lactate sensors have a platinum amperometric electrode with a positive potential relative to the reference electrode. In the case of the measurement of glucose, the glucose oxidase reacts with the glucose and oxygen to generate hydrogen

(a)

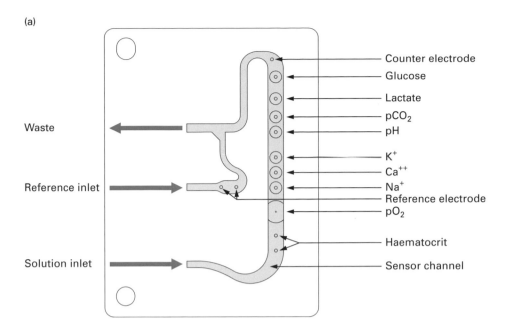

Counter electrode
Glucose
Lactate
pCO_2
pH
K^+
Ca^{++}
Na^+
Reference electrode
pO_2
Haematocrit
Sensor channel

Waste

Reference inlet

Solution inlet

(b)

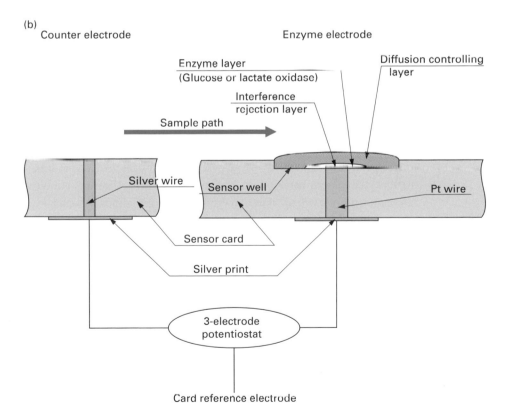

Fig. 16.2 Micro sensor analyte detectors. (a) The GEM 3000 Sensor Card; (b) the Amperometric Glucose/Lactate Analyte Sensors. (Reproduced by permission of IL Critical Care, Lexington, USA.)

peroxide which then diffuses through a controlling layer (Fig. 16.2b) and is oxidised by the platinum electrode to release electrons and create a current flow, the size of which is proportional to the rate of hydrogen peroxide diffusion.

The GEM Premier 4000 system contains an active quality control process controller that monitors the operation of the system, validates the integrity of the cartridge and monitors the electrode response to detect microclots in the test sample that may invalidate the analytical results.

Immunoassay section

Immunoassay procedures undertaken by modern auto-analysers are mostly based on fluorescence or polarised fluorescence techniques. The range of analytes varies from manufacturer to manufacturer but usually involves basic endocrinology (e.g. thyroid function tests), therapeutic drugs (theophylline, digoxin) and drugs of abuse (opiates, cannabis). The operation of auto-analysers in immunoassay mode is similar to that described above and the results are generally reported on the same day, and are generally compared with the previous set of results for the patient.

Manual assays section

This approach to biochemical tests is generally more labour intensive than the other two sections and covers a range of analytical techniques such as acetate or gel electrophoresis, immunoelectrophoresis and some more difficult basic spectrophotometric assays. Examples include the assays for catecholamines (for the diagnosis of phaeochromocytoma), 5-hydroxyindole acetic acid (for the diagnosis of carcinoid syndrome) or HbA_{1c} (for the monitoring of diabetes).

Result reporting

The instrument operator or the section leader initially validates analytical results. This validation process will, in part, be based on the use of internal quality control procedures for individual analytes. Quality control samples are analysed at least twice daily or are included in each batch of test analytes. The analytical results are then subject to an automatic process which identifies results that are either significantly abnormal or require clinical comment or interpretation against rules set by senior laboratory staff.

Neonatal screening

Neonatal or newborn screening is the process of testing newborn babies for certain potentially dangerous disorders. If these conditions are detected early, preventative measures can be adopted that help to protect the child from the disorders. However, such testing is not easy due to the difficulty of obtaining adequate samples of biological fluids for the tests. The development of tandem MS techniques has significantly alleviated this problem. It is possible to screen for a range of metabolic diseases using a single dried bloodspot 3 mm in diameter. There is no need for pre HPLC separation of the sample but the technique is used simply to deliver the sample to the MS. A large number of inherited metabolic diseases can be screened by the technique including aminoacidopathies such as phenylketonuria (PKU) caused by a deficiency of the enzyme phenylalanine hydroxylase, fatty acid oxidation defects such as medium chain acyl-CoA dehydrogenase deficiency (MCADD) and organic acidaemias such as propionic acidaemia caused by a deficiency of the enzyme propionyl-CoA carboxylase.

16.2.3 Quality assessment procedures

In order to validate the analytical precision and accuracy of the biochemical tests conducted by a clinical biochemistry department, the department will participate in external quality assessment schemes in addition to routinely carrying out internal quality control procedures that involve the repeated analysis of reference samples covering the full analytical range for the test analyte. In the UK there are two main national clinical biochemistry external quality assessment schemes: the UK National External Quality Assessment Scheme (UK NEQAS: www.ukneqas.org.uk) coordinated at the Queen Elizabeth Medical Centre, Birmingham and the Wales External Quality Assessment Scheme (WEQAS: www.weqas.com) coordinated at the University Hospital of Wales, Cardiff. The majority of UK hospital clinical biochemistry departments subscribe to both schemes. UK NEQAS and WEQAS distribute test samples on a fortnightly basis, the samples being human serum based. In the case of UK NEQAS the samples contain multiple analytes each at an undeclared concentration within the analytical range. The concentration of each analyte is varied from one distribution to the next. In contrast, WEQAS distributes four or five test samples containing the test analytes at a range of concentrations within the analytical range. Both UK NEQAS and WEQAS offer a number of quality assessment schemes in which the distributed test samples contain groups of related analytes such as general chemistry analytes, peptide hormones, steroid hormones and therapeutic drug monitoring analytes. Participating laboratories elect to subscribe to schemes relevant to their analytical services.

The participating laboratories are required to analyse the external quality assessment samples alongside routine clinical samples and to report the results to the organising centre. Each centre undertakes a full statistical analysis of all the submitted results and reports them back to the individual laboratories on a confidential basis. The statistical data record the individual laboratory's data in comparison with all the submitted data and with the compiled data broken down into individual methods (e.g. the glucose oxidase and hexokinase methods for glucose) and for specific manufacturers' systems. Results are presented in tabular, histogram and graphical form and are compared with the results from recent previously submitted samples. This comparison with previous performance data allows longer-term trends in analytical performance for each analyte to be monitored. Laboratory data that are regarded as unsatisfactory are identified and followed up. Selected data from typical UK NEQAS reports are presented in Table 16.4 and Fig. 16.3 and from a WEQAS report in Table 16.5.

16.2.4 Clinical audit and accreditation

In addition to participating in external quality assessment schemes, clinical laboratories are also subject to clinical audit. This is a systematic and critical assessment of the general performance of the laboratory against its own declared standards and procedures

Table 16.4 **Selected UK NEQAS quality assessment data for serum glucose (mM)** © **The data are reproduced by permission of UK NEQAS, Wolfson EQA Laboratory, Birmingham**

Analytical method	n	Mean	SD	CV (%)
All methods	521	16.04	0.48	3.0
Dry slide	78	15.45	0.51	3.3
OCD (J & J) slides [1JJ]	78	15.45	0.51	3.3
Glucose oxidase electrode	59	15.74	0.32	2.0
Beckman reagents [11BK]	56	15.77	0.28	1.8
Hexokinase + G6PDH	347	16.03	0.44	2.8
Abbott reagents [15AB]	89	16.14	0.43	2.7
Olympus reagents [15OL]	113	16.06	0.40	2.5
Roche Modular reagents [15BO]	70	15.93	0.35	2.2
Glucose oxidase/dehydrogenase	112	16.20	0.47	2.9
Roche Modular reagents [4BO]	60	16.12	0.37	2.3

Notes: The Beckman Glucose Analyser uses the glucose oxidase method and measures oxygen consumption using an oxygen electrode. The Vitros method is a so-called 'dry slide' method that involves placing the sample on a slide, similar to a photographic slide, that has the reagents of the glucose oxidase method impregnated in the emulsion. A blue colour is produced and its intensity measured by reflected light.

These data are for a laboratory that used the hexokinase method and reported a result of 16.0 mM. UK NEQAS calculates a Method Laboratory Trimmed Mean (MLTM) as a target value. It is the mean value of all the results returned by all laboratories using the same method principle with results ±2 SD outside the mean omitted. Its value was 16.03. On the basis of the difference between the MLTM and the laboratory's result UK NEQAS also calculates a score of the specimen accuracy and bias together with a measurement of the laboratory's consistency of bias. This involves aggregating the bias data from all specimens of that analyte submitted by the laboratory within the previous 6 months representing the 12 most recent distributions. This score is an assessment of the tendency of the laboratory to give an over-positive or under-negative estimate of the target MLTM values. The score indicated that the laboratory was performing consistently in agreement with the MLTM.

The results embodied in this table are shown in histogram form in Fig. 16.3.

and against nationally agreed standards. In the context of analytical procedures, the audit evaluates the laboratory performance in terms of the appropriateness of the use of the tests offered by the laboratory, the clinical interpretation of the results and the procedures that operate for the receipt, analysis and reporting of the test samples. Thus whilst it includes the evaluation of analytical data, the audit is primarily concerned with processes leading to the test data with a view to implementing change and improvement. The ultimate objective of the audit is to ensure that the patient receives the best possible care and support in a cost-effective way. The audit is normally undertaken by junior doctors, laboratory staff or CPA (see below) assessors from the hospital, lasts for several days, and involves interaction with all laboratory personnel.

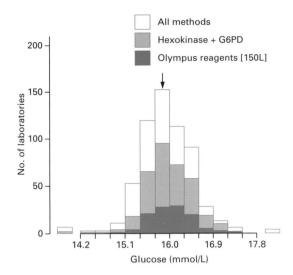

Fig.16.3 Histogram of UK NEQAS quality assessment data for serum glucose based on data in Table 16.2. The arrow indicates the location of the value submitted by the participating laboratory. (Reproduced by permission of UK NEQAS, Wolfson EQA Laboratory, Birmingham, UK.)

Closely allied to the process of clinical audit is that of accreditation. However, whereas clinical audit is carried out primarily for the local benefit of the laboratory and its staff and ultimately for the patient, accreditation is a public and national recognition of the professional quality and status of the laboratory and its personnel. The accreditation process and assessment is the responsibility of either a recognised public professional body or a government department or agency. Different models operate in different countries. In the UK, accreditation of clinical biochemistry laboratories is required by government bodies and is carried out by either Clinical Pathology Accreditation (UK) Ltd (CPA) or less commonly the United Kingdom Accreditation Service (UKAS). In the USA accreditation is mandatory and may be carried out by one of a number of 'deemed authorities' such as the College of American Pathologists. Accreditation organisations also exist for non-clinical analytical laboratories. Examples in the UK include the National Measurement Accreditation Service (NAMAS) and the British Standards Institution (BSI). The International Accreditation Co-operation (ILAC), the European Co-operation for Accreditation (EA) and the Asia-Pacific Laboratory Accreditation Co-operation (APLAC) are three of many international fora for the harmonisation of national standards of accreditation for analytical laboratories.

Assessors appointed by the accreditation body assess the compliance by the laboratory with standards set by the accreditation body. The standards cover a wide range of issues such as those of accuracy and precision, timeliness of results, clinical relevance of the tests performed, competence to carry out the tests as judged by the training and qualifications of the laboratory staff, health and safety, the quality of administrative and technical support systems and the quality of the laboratory management systems and document control. The successful outcome of an assessment is the national recognition

Table 16.5 **WEQAS quality assessment data for serum glucose (mM)** © **The data are reproduced by permission of WEQAS, Directorate of Laboratory Medicine, University Hospital of Wales, Cardiff**

Analytical method		Sample number			
		1	2	3	4
Reported result		7.2	3.7	17.2	8.7
Hexokinase	Mean	6.9	3.6	17.0	8.4
	SD	0.2	0.1	0.5	0.3
	Number	220	221	219	219
Aeroset	Mean	7.2	3.8	17.3	8.5
	SD	0.15	0.08	0.045	0.2
	Number	8	8	9	8
Overall	Mean	7.0	3.7	17.2	8.6
	SD	0.28	0.20	0.54	0.33
	Number	388	392	388	389
WEQAS SD		0.26	0.16	0.6	0.3
SDI		1.15	0.63	0.33	1.0

Notes: SD, standard deviation; SDI, standard deviation index.

These data are for a laboratory that used the hexokinase method for glucose using an Aeroset instrument. Accordingly, the WEQAS report includes the results submitted by all laboratories using the hexokinase method and all results for the method using an Aeroset. The overall results refer to all methods irrespective of instrument. All the data are 'trimmed' in that results outside ± 2 SD of the mean are rejected which explains why the total number for each test sample varies slightly. WEQAS SD is calculated from the precision profiles for each analyte and the SDI (Standard Deviation Index) is equal to (laboratory result – method mean result)/WEQAS SD at that level. SDI is a measure of total error and includes components of inaccuracy and imprecision. The four SDIs for the laboratory are used to calculate an overall analyte SDI, in this case 0.78. A value of less than 1 indicates that all estimates were within ± 1 SD and is regarded as a good performance. An unacceptable performance would be indicated by a value greater than 2.

that the laboratory is in compliance with the standards and hence provides quality healthcare. The accreditation normally lasts for 4 years.

16.3 EXAMPLES OF BIOCHEMICAL AIDS TO CLINICAL DIAGNOSIS

16.3.1 Principles of diagnostic enzymology

The measurement of the activities or masses of selected enzymes in serum is a long-established aid to clinical diagnosis and prognosis. The enzymes found in serum can be divided into three categories based on the location of their normal physiological function:

- *Serum-specific enzymes*: The normal physiological function of these enzymes is based in serum. Examples include the enzymes associated with lipoprotein metabolism and with the coagulation of blood.
- *Secreted enzymes*: These are closely related to the serum-specific enzymes. Examples include pancreatic lipase, prostatic acid phosphatase and salivary amylase.
- *Non-serum-specific enzymes*: These enzymes have no physiological role in serum. They are released into the extracellular fluid and consequently appear in serum as a result of normal cell turnover or more abundantly as a result of cell membrane damage, cell death or morphological changes to cells such as those in cases of malignancy. Their normal substrates and/or cofactors may be absent or in low concentrations in serum.

Serum enzymes in this third category are of the greatest diagnostic value. When a cell is damaged the contents of the cell are released over a period of several hours with enzymes of the cytoplasm appearing first since their release is dependent only on the impairment of the integrity of the plasma membrane. The release of these enzymes following cell membrane damage is facilitated by their large concentration gradient, in excess of a thousand-fold, across the membrane. The integrity of the cell membrane is particularly sensitive to events that impair energy production, for example by the restriction of supply of oxygen. It is also sensitive to toxic chemicals including some drugs, microorganisms, certain immunological conditions and genetic defects. Enzymes released from cells by such events may not necessarily be found in serum in the same relative amounts as were originally present in the cell. Such variations reflect differences in the rate of their metabolism and excretion from the body and hence of differences in their serum half-lives. This may be as short as a few hours (intestinal alkaline phosphatase, glutathione S-transferase, creatine kinase) or as long as several days (liver alkaline phosphatase, alanine aminotransferase, lactate dehydrogenase).

The clinical exploitation of non-serum-specific enzyme activities is influenced by several factors:

- *Organ specificity*: Few enzymes are unique to one particular organ but fortunately some enzymes are present in much larger amounts in some tissues than in others. As a consequence, the relative proportions (pattern) of a number of enzymes found in serum are often characteristic of the organ of origin.
- *Isoenzymes*: Some clinically important enzymes exist in isoenzyme forms and in many cases the relative proportion of the isoenzymes varies considerably between tissues so that measurement of the serum isoenzymes allows their organ of origin to be deduced.
- *Reference ranges*: The activities of enzymes present in the serum of healthy individuals are invariably smaller than those in the serum of individuals with a diagnosed clinical condition such as liver disease. In many cases, the extent to which the activity of a particular enzyme is raised by the disease state is a direct indicator of the extent of cellular damage to the organ of origin.
- *Variable rate of increase in serum activity*: The rate of increase in the activity of released enzymes in serum following cell damage in a particular organ is a characteristic of each enzyme. Moreover, the rate at which the activity of each enzyme decreases towards the reference range following the event that caused cell damage and the subsequent

treatment of the patient is a valuable indicator of the patient's recovery from the condition.

The practical implication of these various points to the applications of diagnostic enzymology is illustrated by its use in the management of heart disease and liver disease.

16.3.2 Ischaemic heart disease and myocardial infarction

The healthy functioning of the heart is dependent upon the availability of oxygen. This oxygen availability may be compromised by the slow deposition of cholesterol-rich atheromatous plaques in the coronary arteries (Section 18.2.2). As these deposits increase, a point is reached at which the oxygen supply cannot be met at times of peak demand, for example at times of strenuous exercise. As a consequence, the heart becomes temporarily ischaemic ('lacking in oxygen') and the individual experiences severe chest pain, a condition known as angina pectoris ('angina of effort'). Although the pain may be severe during such events, the cardiac cells temporarily deprived of oxygen are not damaged and do not release their cellular contents. However, if the arteries become completely blocked either by the plaque or by a small thrombus (clot) that is prevented from flowing through the artery by the plaque, the patient experiences a myocardial infarction ('heart attack') characterised by the same severe chest pain but in this case the pain is accompanied by the irreversible damage to the cardiac cells and the release of their cellular contents. This release is not immediate, but occurs over a period of many hours. From the point of view of the clinical management of the patient, it is important for the clinician to establish whether or not the chest pain was accompanied by a myocardial infarction. In about one-fifth of the cases of a myocardial event the patient does not experience the characteristic chest pain ('silent myocardial infarction') but again it is important for the clinician to be aware that the event has occurred. Electrocardiogram (ECG) patterns are a primary indicator of these events but in atypical presentations ECG changes may be ambiguous and additional evidence is sought in the form of changes in serum enzyme activities. The activities of three enzymes are commonly measured:

- *Creatine kinase* (CK): This enzyme converts phosphocreatine (important in muscle metabolism) to creatine. CK is a dimeric protein composed of two monomers, one denoted as M (muscle), the other as B (brain), so that three isoforms exist: CK-MM, CK-MB and CK-BB. The tissue distribution of these isoenzymes is significantly different such that heart muscle consists of 80–85% MM and 15–20% MB, skeletal muscle 99% MM and 1% MB and brain, stomach, intestine and bladder predominantly BB. CK activity is raised in a number of clinical conditions but since the CK-MB form is almost unique to the heart, its raised activity in serum gives unambiguous support for a myocardial infarction even in cases in which the total CK activity remains within the reference range. A rise in total serum CK activity is detectable within 6 hours of the myocardial infarction and the serum activity reaches a peak after 24–36 hours. However, a rise in CK-MB is detectable within 3–4 hours, has 100% sensitivity within 8–12 hours and reaches a peak within 10–24 hours. It remains raised for 2–4 days.

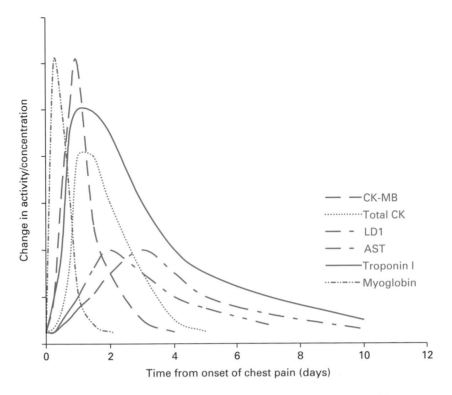

Fig. 16.4 Serum enzyme activity and myoglobin and troponin-I concentration changes following a myocardial infarction. Changes are expressed as a multiple of the upper limit of the reference range. Values vary according to the severity of the event, but the time course of each profile is characteristic of all events.

- *Aspartate aminotransferase* (AST): This is one of a number of transaminases involved in intermediary metabolism. It is found in most tissues but is abundant in heart and liver. Its activity in serum is raised following a myocardial infarction and reaches a peak between 48 and 60 hours. It has little clinical value in the early diagnosis of heart muscle damage but is of use in the case of delayed presentation with chest pain.
- *Lactate dehydrogenase* (LD): This is a tetrameric protein made of two monomers denoted as H (heart) and M (muscle) so that five isoforms exist: LD-1 (H4), LD-2 (H3M), LD-3 (H2M2), LD-4 (HM3) and LD-5 (M4). LD-1 predominates in heart, brain and kidney and LD-5 in skeletal muscle and liver. Total LD activity and LD-1 activity in serum increases following a myocardial infarction and reaches a peak after 48–72 hours. The subsequent decline in activity is much slower than that of CK or AST. The diagnostic value of LD activity measurement is mainly confined to monitoring the patient's recovery from the myocardial infarction event.

Typical changes in the activities of these three enzymes following a myocardial infarction are shown in Fig. 16.4. All three enzymes are assayed by an automated method based on the following reactions.

- *Total CK activity*: This is assessed by coupled reactions (Section 15.3.2) with hexokinase and glucose-6-phosphate dehydrogenase in the presence of *N*-acetylcysteine as activator, and the measurement of increase in absorbance at 340 nm or by fluorescence polarisation (primary wavelength 340 nm, reference wavelength 378 nm):

$$\text{creatine phosphate} + \text{ADP} \rightleftharpoons \text{creatine} + \text{ATP}$$

$$\text{ATP} + \text{D-glucose} \rightleftharpoons \text{ADP} + \text{D-glucose-6-phosphate}$$

$$\text{D-glucose-6-phosphate} + \text{NAD(P)}^+ \rightleftharpoons \text{D-6-phosphogluconate}$$
$$+ \text{NAD(P)H} + \text{H}^+$$

- *CK-MB activity*: This is assessed by the inhibition of the activity of the M monomer by the addition to the serum sample of an antibody to the M monomer. This inhibits CK-MM and the M unit of CK-MB. The activity of CK-BB is unaffected but is normally undetectable in serum hence the remaining activity in serum is due to the B unit of CK-MB. It is assayed by the above coupled assay procedure and the activity doubled to give an estimate of the CK-MB activity. An alternative assay uses a double antibody technique: CK-MB is bound to anti-CK-MB coated on microparticles, the resulting complex washed to remove non-bound forms of CK and anti-CK-MM conjugated to alkaline phosphatase added. It binds to the antibody–antigen complex, is washed to remove unbound materials and assayed using 4-methylumbelliferone phosphate as substrate, the released 4-methylumbelliferone being measured by its fluorescence and expressed as a concentration (ng cm^{-3}) rather than as activity.

- *Aspartate aminotransferase activity*: This is assessed by a coupled assay with malate dehydrogenase and the measurement of the decrease in absorbance at 340 nm:

$$\text{L-aspartate} + \text{2-oxoglutarate} \rightleftharpoons \text{oxaloacetate} + \text{L-glutamate}$$

$$\text{oxaloacetate} + \text{NADH} + \text{H}^+ \rightleftharpoons \text{malate} + \text{NAD}^+$$

- *Lactate dehydrogenase*: The measurement of total activity is based on the measurement of the increase in absorbance at 340 nm using lactate as substrate. The measurement of LD-1 is based on the use of 2-hydroxybutyrate as substrate since only LD-1 and LD-2 can use it:

$$\text{Total LD:} \quad \text{lactate} + \text{NAD}^+ \rightleftharpoons \text{pyruvate} + \text{NADH} + \text{H}^+$$

$$\text{LD-1 : 2-hydroxybutyrate} + \text{NAD}^+ \rightleftharpoons \text{2-ketobutyrate} + \text{NADH} + \text{H}^+$$

The clinical importance of obtaining early unambiguous evidence of a myocardial infarction has encouraged the development of markers other than enzyme activities and currently two tests are commonly run alongside enzyme activities. These are based on myoglobin and troponin-I:

- *Myoglobin*: Concentrations in serum, assayed by HPLC or immunoassay, increase more rapidly than CK-MB after a myocardial infarction. An increase is detectable within 1–2 hours, has 100% sensitivity and reaches a peak within 4–8 hours and returns to normal within 12–24 hours. However, myoglobin changes are not specific for myocardial infarction since similar changes also occur in other syndromes such as muscle damage or crush injury such as that following a road accident.

- *Troponin-I*: This is one of three proteins (the others being troponin-T and troponin-C) of a complex which regulates the contractility of the myocardial cells. Its activity in serum increases at the same rate as CK-MB after a myocardial infarction, has a similar time for 100% sensitivity and for peak time, but it remains raised for up to 4 days after the onset of symptoms. Its reference range is less than 1 ng cm^{-3} but its concentration in serum is raised to up to 30–50 ng cm^{-3} within 24 hours of a myocardial infarction event. It is assayed by a 'sandwich' immunological assay in which the antibody is labelled with alkaline phosphatase. Using 4-methylumbelliferone phosphate as substrate, the release of 4-methylumbelliferone is measured by fluorescence. The measurement of serum troponin-I is widely used to exclude cardiac damage in patients with chest pain since it remains raised for several days following a myocardial infarction, but the timing of the test sample is important as a sample taken too early may give a false negative result. A limitation of its use is that its release into serum is not specific to myocardial infarction; an increase in serum mass may occur following a crush injury.

The measurement of enzyme activities and myoglobin and troponin-I concentrations, together with plasma potassium, glucose and arterial blood gases, is routinely used to monitor the recovery of patients following a myocardial infarction. A patient may experience a second myocardial infarction within a few days of the first. In such cases the pattern of serum enzymes shown in Fig. 16.4 is repeated, the pattern being superimposed on the remnants of the first profile. CK-MB is the best initial indicator of a second infarction since the levels of troponin-I may not reflect a secondary event.

The sensitivity and specificity (Section 16.1.2) of ECG and diagnostic enzymology in the management of heart disease are complementary. Thus the specificity of ECG is 100% whilst that of enzyme measurements is 90%, and the sensitivity of ECG is 70% whilst that of enzyme measurements is 95%.

16.3.3 Liver disease

Diagnostic enzymology is routinely used to discriminate between several forms of liver disease including:

- *Hepatitis*: General inflammation of the liver most commonly caused by viral infection but which may also be a consequence of blood poisoning (septicaemia) or glandular fever. It results in only mild necrosis of the hepatic cells and hence of a modest release of cellular enzymes.
- *Cirrhosis*: A general destruction of the liver cells and their replacement by fibrous tissue. It is most commonly caused by excess alcohol intake but is also a result of prolonged hepatitis, various autoimmune diseases and genetic conditions. They all result in extensive cell damage and release of hepatic cell enzymes.
- *Malignancy*: Primary and secondary tumours.
- *Cholestasis*: The prevention of bile from reaching the gut due either to blockage of the bile duct by gallstones or tumours or to liver cell destruction as a result of cirrhosis or prolonged hepatitis. This gives rise to obstructive jaundice (presence of bilirubin, a yellow metabolite of haem, in the skin).

Patients with these various liver diseases often present to their doctor with similar symptoms and a differential diagnosis needs to be made on the basis of a range of investigations including imaging techniques especially ultrasonography (ultrasound), magnetic resonance imaging (MRI), computerised tomography (CT) scanning, microscopic examination of biopsy samples and liver function tests. Four enzymes are routinely assayed to aid differential diagnosis:

- *Aspartate aminotransferase* (*AST*) and *alanine aminotransferase* (*ALT*): As previously stated, these enzymes are widely distributed but their ratios in serum are characteristic of the specific cause of liver cell damage. For example, an AST/ALT ratio of less than 1 is found in acute viral hepatitis and fresh obstructive jaundice, a ratio of about 1 in obstructive jaundice caused by viral hepatitis and a ratio of greater than 1 in cases of cirrhosis.
- *γ-Glutamyl transferase* (*GGT*): This enzyme transfers a γ-glutamyl group between substrates and may be assayed by the use of γ-glutamyl-4-nitroaniline as substrate and monitoring the release of 4-nitroaniline at 400 nm. GGT is widely distributed and is abundant in liver, especially bile canaliculi, kidney, pancreas and prostate but these do not present themselves by contributing to serum levels. Raised activities are found in cirrhosis, secondary hepatic tumours and cholestasis and tend to parallel increases in the activity of alkaline phosphatase especially in cholestasis. Its synthesis is induced by alcohol and some drugs also cause its serum activity to rise.
- *Alkaline phosphatase* (*AP*): This enzyme is found in most tissues but is especially abundant in the bile canaliculi, kidney, bone and placenta. It may be assayed by using 4-nitrophenylphosphate as substrate and monitoring the release of 4-nitrophenol at 400 nm. Its activity is raised in obstructive jaundice and when measured in conjunction with ALT can be used to distinguish between obstructive jaundice and hepatitis since its activity is raised more than that of ALT in obstructive jaundice. Decreasing serum activity of AP is valuable in confirming an end of cholestasis. Raised serum AP levels can also be present in various bone diseases and during growth and pregnancy.

16.3.4 Kidney disease

The kidneys, together with the liver, are the major organs responsible for the removal of waste material from the body. The kidneys also have other specific functions including the control of electrolyte and water homeostasis, and the synthesis of erythropoietin. Each of the two kidneys contains approximately 1 million nephrons that receive the blood flowing to the kidneys. Blood flowing to the kidneys is first presented to the glomerulus of each nephron which filters the plasma water to produce the ultrafiltrate or primary urine removing all the contents of the plasma except proteins. Each nephron produces approximately $100 \, mm^3$ of primary urine per day giving a total production of primary urine by the two kidneys of approximately $100–140 \, cm^3$ per minute or $200 \, dm^3$ per day in a healthy adult person. This is referred to as the glomerular filtration rate (GFR). The primary urine then encounters the tubule of the nephron that is the site of the reabsorption of water and the active and passive reabsorption of lipophilic compounds and cellular nutrients such as sugars and amino acids and the active secretion of others. These two processes in combination result in the production of approximately $2 \, dm^3$ of urine per day that is collected in the urinary bladder.

Glomerular filtration rate is the accepted best indicator of kidney function. Any pathology of the kidneys is reflected in a decreased GFR and this in turn has serious physiological consequences including anaemia and severe cardiovascular disease. Kidney disease is a progressive one, proceeding through subacute or intrinsic renal disease such as glomerular nephritis into chronic kidney disease (CKD). Complete kidney failure leads to the need for kidney dialysis and kidney transplantation. There is evidence that the incidence of CKD is increasing in developed countries and is associated with increasing risk of diabetes and an increasingly elderly population. There is thus a great clinical demand for accurate measurements of GFR in order to detect the onset of kidney disease, to assess its severity and to monitor its subsequent progression.

Measurement of glomerular filtration rate

The measurement of GFR is based on the concept of renal clearance which is defined as the volume of serum cleared of a given substance by glomerular filtration in unit time. It therefore has units of $cm^3 min^{-1}$. In principle any endogenous or exogenous substance that is subject to glomerular filtration and is not reabsorbed could form the basis of the measurement. The polysaccharide inulin meets these criteria and is subject to few variables or interferences but because it is not naturally occurring in the body is inconvenient for routine clinical use but is commonly used as a standard for alternative methods. In practice serum creatinine is the most commonly used marker. It is the end product of creatine metabolism in skeletal muscle and meets the excretion criteria so that its serum concentration is inversely related to GFR. However, it is subject to a number of non-renal variables including:

- *Muscle mass*: Serum values are influenced by extremes of muscle mass as in athletes and in individuals with muscle-wasting disease or malnourished patients.
- *Gender*: Serum creatinine is higher in males than females for a given GFR.
- *Age*: Children under 18 years have a reduced serum creatinine and the elderly have an increased value.
- *Ethnicity*: African–Caribbeans have a higher serum creatinine for a given GFR than have Caucasians.
- *Drugs*: Some commonly used drugs such as cimetidine, trimethoprim and cephalosporins interfere with creatinine excretion and hence give elevated GFR values.
- *Diet:* Recent intake of red meats and oily fish can raise serum creatinine levels.

Routine laboratory estimations of GFR (referred to as *e*GFR) are based on the measurement of serum creatinine concentration and the calculation of *e*GFR from it using an equation that makes corrections for four of the above variables. Serum creatinine is routinely measured by one of two ways:

- *Spectrophotometric method based on the Jaffe reaction*: This involves the use of alkaline picric acid reagent which produces a red-coloured product that is measured at 510 nm. A limitation is that the reagent also reacts with some non-creatinine chromogens such as ketones, ascorbic acid and cephalosporins and as a result gives high values.

- *Coupled enzyme assay*: The method uses either creatinine kinase, pyruvate kinase or lactate dehydrogenase and measuring the change in absorption at 340 nm:

$$\text{creatinine} + H_2O \longleftrightarrow \text{creatine}$$

$$\text{creatine} + ATP \longleftrightarrow \text{phosphocreatine} + ADP$$

$$ADP + \text{phosphoenolpyruvate} \longleftrightarrow ATP + \text{pyruvate}$$

$$\text{pyruvate} + NADH + H^+ \longleftrightarrow \text{L-lactate} + NAD^+$$

or an alternative coupled assay using creatinine iminohydrolase and glutamate dehydrogenase.

For research, two other methods are available:

- *HPLC or GC/MS:* HPLC uses a C18 column and water/acetonitrile (95:5 v/v) eluent containing 1-octanesulphonic acid as a cation-pairing agent. GC-MS is based on the formation of the *t*-butyldimethylsilyl derivative of creatinine.
- *Isotopic dilution*: This method is coupled with mass spectrometry (ID-MS). This involves the addition of ^{13}C- or ^{15}N-labelled creatinine to the serum sample, isolation of creatinine by ion-exchange chromatography and quantification by mass spectrometry using selective ion monitoring. The lower limit of detection is about 0.5 ng.

The lack of an internationally or even nationally agreed standard assay for creatinine leads to significant inter-laboratory differences in both bias and imprecision so that national external quality assurance schemes, such as UK NEQAS and WEQAS (Section 16.2.3) have important roles in alerting laboratories to assays that stray outside national control values. UK NEQAS provides clinical laboratories that participate in the *e*GFR scheme with an assay-specific adjustment factor (F) to correct for methodological variations in estimations of serum creatinine. The factor is obtained using calibration against a GC-MS creatinine assay. It is updated at 6-monthly intervals. A number of equations have been derived to calculate *e*GFR from serum creatinine values but the one currently used throughout the UK is the four-variable Modification of Diet in Renal Disease (MDRD) Study equation:

*e*GFR

$$= F \times 175 \times (\text{serum creatinine}/88.4)^{-1.154} \times \text{age}^{-0.203} \times 0.742(\text{if female}) \times 1.212(\text{if black})$$

Serum creatinine concentrations are expressed in μM to the nearest whole number and are adjusted for variations in body size by normalising using a factor for body surface area (BSA) correcting to a BSA value of 1.73 m^2. The units of *e*GFR are therefore cm^3 min^{-1} 1.73 m^{-2} and values are reported to one decimal place. The equation has been validated in a large-scale study against the most accurate method based on the use of ^{125}I-iothalamate GFR. Alternative equations exist for use with children.

Reference values for *e*GFR are 130 cm^3 min^{-1} 1.73 m^{-2} for males in the age range 20–30 and 125 cm^3 min^{-1} 1.73 m^{-2} for females of the same age. Values decline with increasing age becoming 95 and 85 cm^3 min^{-1} 1.73 m^{-2} for males and females respectively in the age range 50–60 and 70 and 65 cm^3 min^{-1} 1.73 m^{-2} for males and females in the age range 70–80.

Table 16.6 **Stages of chronic kidney disease (CKD)**

CKD stage	eGFR (cm^3 min^{-1} 1.73 m^{-2})	Clinical relevance
1	>90	Regard as normal unless other symptoms present[a]
2	60–89	Regard as normal unless other symptoms present[a]
3	30–59	Moderate renal impairment
4	15–29	Severe renal impairment
5	<15	Advanced renal failure

Note: [a]Symptoms include persistent proteinuria, haematuria, weight loss, hypertension.

Clinical assessment of renal disease

Acute renal failure (ARF)

Acute renal failure is the failure of renal function over a period of hours or days and is defined by increasing serum creatinine and urea. It is a life-threatening disorder caused by the retention of nitrogenous waste products and salts such as sodium and potassium. The rise in potassium may cause ECG changes and a risk of cardiac arrest. Acute renal failure may be classified into pre-renal, renal and post-renal. Prompt identification of pre- or post-renal factors and appropriate treatment action may allow correction before damage to the kidneys occurs. Pre-renal failure occurs due to a lack of renal perfusion. This can occur in volume loss in haemorrhage, gastrointestinal fluid loss and burns or because of a decrease in cardiac output caused by cardiogenic shock, massive pulmonary embolus or cardiac tamponade (application of pressure) or other causes of hypertension such as sepsis. Post-renal causes include bilateral uretic obstruction because of calculi or tumours or by decreased bladder outflow/urethral obstruction e.g. urethral stricture or prostate enlargement through hypertrophy of carcinoma. Correction of the underlying problem can avoid any kidney damage. Renal causes of acute renal failure include glomerular nephritis, vascular disease, severe hypertension, hypercalcaemia, invasive disorders such as sarcoidosis or lymphoma and nephrotoxins including animal and plant toxins, heavy metals, aminoglycosides, antibiotics and non-steroidal anti-inflammatory drugs.

Chronic kidney disease (CKD)

CKD is a progressive condition characterised by a declining eGFR (Table 16.6). All CKD patients are subject to regular clinical and laboratory assessment and once Stage 3 has been reached to additional clinical management. This is aimed at attempting to reverse or arrest the disease by drug therapy saving the patient the inconvenience and the paying authority the cost of dialysis or transplantation.

16.3.5 **Endocrine disorders**

Endocrine hormones are synthesised in the brain, adrenal, pancreas, testes and ovary, and most importantly in the hypothalamus and pituitary, but they act elsewhere in

Table 16.7 **Examples of hormones of the hypothalamus–pituitary axis**

Secreted hormone	Pituitary effect	Gland effect
Thyrotropin-releasing hormone (TRH)	Release of thyroid-stimulating hormone (TSH)	Release of thyroxine (T4) and triiodothyronine (T3) by thyroid gland
Growth-hormone-releasing hormone (GHRH)	Release of growth hormone (GH)	Stimulates cell and bone growth
Corticotropin-releasing hormone (CRH)	Release of adrenocorticotropic hormone (ACTH)	Stimulates production and secretion of cortisol in adrenal cortex
Gonadotropin-releasing hormone (GnRH)	Release of follicle-stimulating hormone (FSH) and luteinising hormone (LH)	FSH – maturation of follicles/spermatogenesis; LH – ovulation/production of testosterone

the body as a result of their release into the circulatory system. The result is a **hormonal cascade** that incorporates an amplification of the amount of successive hormone released into the circulatory system, increasing from micrograms to milligrams, as well as a negative feedback that operates to control the cascade when the level of the 'action' hormone has reached its optimum value. Most signals originate in the central nervous system as a result of an environmental (external) signal, such as trauma or temperature, or an internal signal. The response is a signal to the hypothalamus and the release of a hormone such as corticotropin-releasing hormone (CRH). This travels in the bloodstream to the anterior pituitary gland where it acts on its receptor and results in the release of a second hormone, adrenocorticotropic hormone (ACTH). It circulates in the blood to reach its target gland, the adrenal cortex, where it acts to release the 'action' hormone, cortisol, known as the stress hormone. The released cortisol raises blood pressure and blood glucose and is subject to a natural diurnal variation, peaking in early morning and being lowest around midnight. It has a negative feedback effect on the pituitary and adrenal cortex. Glands linked by the action of successive hormones are referred to as an axis, e.g. the hypothalamus–pituitary–adrenal axis. These coordinated cascades regulate the growth and function of many types of cell (Table 16.7). The hormones released act at specific receptors, commonly of the GPCR type discussed in Section 17.4.3, which trigger the release of second messengers such as cAMP, cGMP, inositol triphosphate, Ca^{2+} and protein kinases. Diseases of the endocrine system result in dysregulated hormone release, inappropriate signalling response or, in extreme cases, the destruction of the gland. Examples include diabetes mellitus, Addison's disease, Cushing's syndrome, hyper- and hypothyroidism and obesity. Such medical conditions are characterised by their long-term nature. Laboratory tests are commonly employed to measure hormone levels in order to assist in the diagnosis of the condition and the subsequent care of the patient.

Thyroid function tests

Approximately 1% of the population suffer from some form of thyroid disease although in many cases the symptoms may be non-specific. Even so, over 1 million thyroid function tests are conducted annually in the UK. As shown in Table 16.7, the hypothalamus releases thyrotropin-releasing hormone (TRH) which acts directly on the pituitary to produce thyroid-stimulating hormone (TSH) which in turn stimulates the thyroid gland to produce two thyroid hormones, thyroxine (T4) and triiodothyronine (T3). The gland produces approximately 10% of the circulating T3, the remainder being produced by the metabolism of T4 mainly in the liver and kidney. The majority of T4 and T3 are bound to thyroxine-binding globulin (TBG) but only the free unbound forms (fT4, fT3) are biologically active. Although the concentration of T3 is approximately one-tenth of that of T4, T3 is ten times more active. Both hormones act on nuclear receptors to increase cell metabolism and both have a negative feedback effect on the hypothalamus to switch off the secretion of TRH and on the pituitary to switch off TSH secretion. Hyperthyroidism is a consequence of the overproduction of the two hormones and common causes are thyroiditis, Grave's disease and TSH-producing pituitary tumours. Hypothyroidism, characterised by weakness, fatigue, weight gain and joint or muscle pain, may be primary due to the undersecretion of T4 and T3, possibly due to irradiation or drugs such as lithium, or secondary due to damage to the hypothalamus or pituitary. Normal laboratory tests for these conditions are based on the measurement of TSH and either total (bound and unbound) T4 and total T3 or fT4 and fT3 all by immunoassay.

16.3.6 Hypothalamus–pituitary–gonad axis

In both sexes, the hypothalamus produces gonadotropin-releasing hormone (GRH) that stimulates the pituitary to release luteinising hormone (LH) and follicle-stimulating hormone (FSH). In males, the release of LH and FSH is fairly constant, whereas in females the release is cyclical. In males LH stimulates Leydig cells in the testes to produce testosterone which together with FSH causes the production of sperm. The testosterone has a negative feedback effect on both the hypothalamus and the pituitary thereby controlling the release of GRH. The testosterone acts on various body tissues to give male characteristics. In females, FSH acts on the ovaries to produce both oestradiol and the development of the follicle. The oestradiol and LH then act to stimulate ovulation. Oestradiol has a negative feedback effect on the hypothalamus and the pituitary and acts on body tissues to produce female characteristics.

16.3.7 Diabetes mellitus

Diabetes is the most common metabolic disorder of carbohydrate, fat and protein metabolism, and is primarily due to either a deficiency or complete lack of the secretion of insulin by the β-cells of the islets of Langerhans in the pancreas. It affects 1–2% of Western populations and 5–10% of the population over the age of 40. It is characterised

Case study HYPOTHYROID CASE

A 59-year-old woman presented with a history of lethargy, cold intolerance and weight gain. On examination, the doctor noticed that the patient's hair appeared thin and her skin dry. Several tests were requested including thyroid function tests the results of which were:

$$TSH = 46.9 \, mU \, dm^{-3} \text{ (normal range } 0.4 - 4.5 \, mU \, dm^{-3})$$
$$fT4 = 5.6 \, pM \text{ (normal range } 9.0 - 25 \, pM)$$

These results indicate overt primary hypothyroidism. As the patient suffered from cardiovascular disease the doctor commenced thyroxine replacement therapy at an initial dose of 25 µg daily. After 2 weeks, the tests were repeated:

$$TSH = 37.6 \, mU \, dm^{-3}$$
$$fT4 = 8.2 \, pM$$

These results remain abnormal so it was agreed that the tests should be repeated in 6 weeks' time. At this stage the results were:

$$TSH = 19.1 \, mU \, dm^{-3}$$
$$fT4 = 11.8 \, pM$$

These results confirm that either the thyroxine replacement dosage was inadequate or that compliance was poor. As the patient confirmed that she had been taking the therapy as prescribed, the doctor increased the dose to 50 µg per day. After a further 8 weeks the tests were repeated:

$$TSH = 1.5 \, mU \, dm^{-3}$$
$$fT4 = 14.8 \, pM$$

The patient reported feeling much better and had improved clinically.

Comment   The above results are typical of patients with hypothroidism, also referred to as **myxoedema**. Primary hypothyroidism due to thyroid gland dysfunction is by far the most common cause of the condition but secondary (pituitary) and tertiary (hypothalamic) causes also exist. In these latter two cases the main biochemical abnormality is a low fT4. TSH may be low or within the reference range in secondary and tertiary hypothyroidism, i.e. it does not respond to low fT4. Patients with primary hypothroidism require lifelong therapy. Patients with cardiovascular disease, as in this case, must be initiated at a lower dose than normal as over-treatment can lead to angina, cardiac arrhythmia and myocardial infarction. Elderly patients are also started at a lower dose for the same reason. Once therapy has been commenced, thyroid function tests should be carried out after 2–3 months to check for steady-state conditions and thereafter repeated on an annual basis.

by hyperglycaemia (elevated blood glucose level) leading to long-term complications. Diabetes can be classified into a number of types:

● *Insulin-dependent diabetes* (Type 1) (also called *juvenile diabetes* and *brittle diabetes*) is due to the autoimmune destruction of β-cells in the pancreas. Generally it has a rapid onset with a strong genetic link.

Case study **PREGNANCY**

A 28-year-old female PE teacher presented to her GP with non-specific symptoms of increased tiredness, nausea, stomach cramps and amenorrhoea with a last menstrual period 3 months previously. She was a previously fit, healthy lady who had a normal menstrual history. Having recently moved house, she thought that stress might be the cause of her symptoms. Her GP requested routine biochemistry tests including thyroid function tests, all of which were within normal reference range. A urine pregnancy test was also performed and to the patient's surprise was positive and confirmed by laboratory serum β-human chorionic gonadotropin (β-hCG) of 150 640 IU dm^{-3}.

Comment These results confirm that this lady was approximately 10 weeks pregnant. The serum β-hCG levels during pregnancy are shown in Fig. 16.5. A level >25 IU dm^{-3} is indicative of pregnancy. Implantation of the developing embryo into the endometrial lining of the uterus results in the secretion of β-hCG and as pregnancy continues its synthesis increases at an exponential rate, doubling every 2 days, and reaching a peak of 100 000–200 000 IU dm^{-3} at 60–90 days (1st trimester). Levels then decline to approximately 1000 IU dm^{-3} at around 20 weeks pregnancy (during 2nd trimester) to a stable plateau for the remainder of the pregnancy. Oestradiol, oestrone, oestriol and progesterone all increase in the early stages of pregnancy as a result of the action of β-hCG on the corpus luteum of the ovaries. Unlike β-hCG levels, the levels of the three oestrogens and progesterone continue to rise during pregnancy playing a vital role in the sustenance and maintenance of the foetus. At the end of pregnancy, the placental production of progesterone falls, stimulating contractions leading to birth.

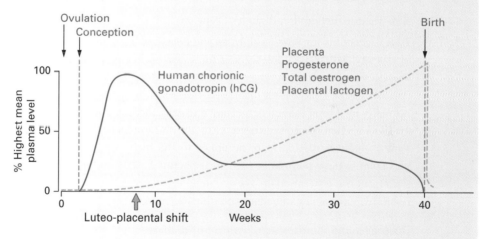

Fig. 16.5 Hormonal profile during pregnancy. (Adapted with permission from Professor Alan S. McNeilly, MRC Human Reproductive Science Unit, Edinburgh, UK.)

- *Non-insulin-dependent diabetes* (Type 2) (also called *adult-onset diabetes* and *maturity-onset diabetes*), is a complex progressive metabolic disorder characterised by β-cell failure and variable insulin resistance. A subtype is *maturity-onset diabetes of the young* (MODY) which usually occurs before the age of 25 years. It is the first form of diabetes for which a genetic cause and molecular consequence have been established. Mutations of the genes for hepatocyte nuclear factor 4α (MODY1), glucokinase (MODY2), HNF1α (MODY3), insulin promotor factor 1 (MODY4), HNF1β (MODY5) and neurogenic differentiation factor 1 (MODY6) have all been characterised.

- *Impaired glucose tolerance* where there is an inability to metabolise glucose in the 'normal' way but not so impaired as to be defined as diabetes.
- *Gestational diabetes* that is any degree of glucose intolerance developed during pregnancy. It is characterised by a decrease in insulin sensitivity and an inability to compensate by increased insulin secretion. The condition is generally reversible after the termination of pregnancy, but up to 50% of women who develop it are prone to develop Type 2 diabetes later in life.
- *Other types* which include certain genetic syndromes, pancreatic disease, endocrine disease and drug or chemical induced diabetes.

Insulin-dependent diabetes (Type 1)

Between 5% and 10% of all diabetics have the insulin-dependent form of diabetes requiring regular treatment with insulin. Type 1 develops in young people with a peak incidence of around 12 years of age. In this type of diabetes the degree of insulin deficiency is so severe that only insulin replacement can avoid the complications of diabetes that are discussed later. Dietary control or oral drugs are not sufficient. The disease is caused by the autoimmune destruction of β-cells in the pancreas thus reducing the ability of the body to produce insulin. Islet cell antibodies (ICA), antibodies IA-2 and IA-2β to transmembrane protein tyrosine phosphatases in islet cells, autoantibodies to glutamic acid decarboxylase (GAD) found in β-cells and insulin autoantibodies (IAA) are all used as diagnostic markers of the disease.

Non-insulin-dependent diabetes (Type 2)

Type 2 accounts for 90% of all cases and develops later in life and can be exacerbated by obesity. MODY versions account for 1–5% of all cases and are not associated with obesity. From population screening studies it is thought that only half of those individuals with Type 2 have been diagnosed. Control of blood glucose levels in this group is normally by a combination of diet and oral drug therapy but occasionally it may require insulin injection. There is growing evidence that the increasing worldwide incidence of Type 2 diabetes may, in part, be linked to the increasing concentration of so-called persistent organic pollutants, such as bisphenol A, DDT and polychlorinated biphenyls (PCBs), in the environment. These compounds suppress adiponectin, a hormone that regulates fatty acid catabolism and glucose metabolism.

Diagnosis and monitoring of control of diabetes

Diabetes is frequently recognised by the symptoms it causes but can be confirmed by clinical biochemical measurements based on World Health Organisation (WHO) recommendations:

- a fasting (12 hours) plasma glucose level greater than 7.0 mM;
- a random plasma glucose level greater than 11.1 mM;
- application of an oral glucose tolerance test in which a 75 g dose of glucose is administered and the plasma level measured after 2 hours. Diabetes is characterised by a value greater than 11.1 mM.

The diagnostic cut-off values of 7.0 and 11.1 mM are based on the level at which retinopathy begins to appear in a population. The clinical aim in the treatment of Type 1 diabetes is to maintain plasma glucose levels in the healthy range of 4–6 mM. This is typically monitored by patients themselves by measuring their blood glucose at predetermined times that are interrelated to their meal times during the day. For example, the lowest blood glucose of the day is likely to be after the longest fast before breakfast and the highest blood glucose of the day is likely to be 1 hour after the main meal. By manipulating treatment around these highs and lows, good glycaemic control is generally maintained. The patients measure their blood glucose using hand-held, portable blood glucose meters based on glucose oxidase using dry stick technology to measure finger-prick blood samples.

Another measure of glycaemic control is by using haemoglobin A1c (Hb_{a1c}) measurements. This testing strategy works on the basis that most proteins (in this case haemoglobin A) will bind glucose dependent on the length of time they are in contact with glucose, the temperature and the concentration of glucose. Hence haemoglobin, having a typical half-life of 120 days and a standardised body temperature of $37\,^{\circ}C$, will bind the appropriate amount of glucose depending on the concentration of glucose. Hb_{a1c} is typically measured in the clinic using HPLC to separate the different haemoglobin pigments and is expressed as a percentage of total haemoglobin. In 2009 the International Federation of Clinical Chemistry and Laboratory Medicine (IFCC) published recommendations for the standardisation of Hb_{a1c} using IFCC standards that allow traceability of the method back to the IFCC reference method. This caused a change in units from % to mM. This was introduced in June 2009 in the UK although most laboratories report results in both units for education purposes. The lower the result the better the control. This test is extremely useful in measuring long-term control of diabetes but is not without its pitfalls. For example, if the patient has very brittle diabetes having equal numbers of hypoglycaemic and hyperglycaemic periods (see below), then the hypos will cancel out the hyper periods and the Hb_{a1c} will appear to show that the patient is in good glycaemic control.

Complications of diabetes

The diabetic patient needs to have regular intake of carbohydrate to maintain their blood glucose level and appropriate levels of insulin treatment. If these are not in balance then hypoglycaemia or hyperglycaemia may take place. In hypoglycaemia the patient will become cold, clammy and sweaty and may become confused or even unconscious. Giving a sweet drink easily treats this complication. The major complication of hyperglycaemia is diabetic ketoacidosis. Almost one-third of insulin-dependent diabetic patients present for the first time with ketoacidosis that is often precipitated by infection. The biochemical features of presentation are a high or very high blood glucose level, glycosuria (glucose in the urine) and ketonuria (ketones in the urine). The patient's breath will often smell of acetone. Treatment consists of administration of fluids and an insulin infusion but this can often lead to precipitate falls in serum potassium and this must be monitored at all times.

All types of diabetes are also associated with several types of long-term complication. These can largely be split into macro-vascular disease, micro-vascular disease and others. Macro-vascular disease involves accelerated atherosclerosis in the large and medium-sized vessels. Macro-vascular disease accounts for most of the excessive mortality seen in diabetes. Micro-vascular disease leads to diabetic retinopathy and diabetic nephropathy and diabetic foot that may turn to gangrenous ulcers of the feet. The other major complications of diabetes are conditions such as gout, fatty changes in the liver, hypertension and diabetic dyslipidaemias (raised blood lipid levels).

16.3.8 Plasma proteins

Plasma contains a very large number of proteins many of which are present only in trace amounts. The ones that have their main physiological role in plasma have three main functions:

- osmotic regulation;
- transport of ligands such as hormones, metal ions, bilirubin, fatty acids, vitamins and drugs;
- response to infection or foreign bodies entering the body.

All plasma proteins are synthesised in the liver with the exception of the immuno-globulins which are synthesised in the bone marrow. Plasma proteins are readily separated by electrophoresis and this technique forms the basis of several clinical diagnostic tests. The tests are normally recorded subjectively, but a densitometer may be used to get a semi-quantitative result.

Albumin

Albumin is the commonest plasma protein making up some 50% of all plasma protein. Its half-life in plasma is about 20 days and in a good nutritional state the liver produces about 15 g a day to replace this loss. Albumin is the main regulator of the osmotic pressure of plasma but also acts as a transporter of haem, bilirubin (a metabolite of haem), biliverdin (a metabolite of bilirubin), free fatty acids, steroids and metal ions (e.g. Cu^{2+}, Fe^{3+}). It also binds some drugs. Other specialist proteins found in plasma are also involved in transport, for example of steroids e.g. cortisol-binding globulin, sex-hormone-binding globulin (androgens and oestrogens) and metal ions, e.g. ceruloplasmin (Cu^{2+}) and transferrin (Fe^{3+}). Other transport plasma proteins include thyroid-binding globulin (thyroxine T4 and triiodothyronine T3) and haptoglobin (haemoglobin dimers).

Immunoglobulins

Immunoglobulins are synthesised in bone marrow in response to the exposure to a specific foreign body (Chapter 7). Immunoglobulins share a common Y-shaped struc-ture of two heavy and two light chains, the light chains forming the upper arms of the Y. There are two types of light chains, these are either kappa (κ) or lambda (λ), and each are found in all classes of the immunoglobulins. The class of immunoglobulin is determined by the heavy chain that gives rise to five types – IgG, IgA, IgM, IgD and IgE.

IgG accounts for approximately 75% of the immunoglobulins present in the plasma of adults and has a half-life of approximately 22 days. It is present in extracellular fluids and appears to eliminate small proteins through aggregation and the reticuloendothelial system. IgA is the secretory immunoglobulin protecting the mucosal surfaces. IgA is synthesised by mucosal cells and represents approximately 10% of plasma immuno-globulins and has a half-life of 6 days. It is found in bronchial and intestinal secretions and is a major component of colostrum (the form of milk produced by the mammary gland immediately after giving birth). IgA is the primary immunological barrier against pathogenic invasion of the mucosal membranes. IgM is found in the intravascular space and its role is to eliminate circulating microorganisms and antigens. IgM accounts for about 8% of plasma immunoglobulins and has a half-life of 5 days. IgM is the first antibody to be synthesised after an antigenic challenge. IgD and IgE are minor immunoglobulins whose roles are not clear since a deficiency of either seems to be associated

Case study   MYELOMA

A 72-year-old woman presented to her GP with a 3 week history of painful hips, chest and shoulders, and with shortness of breath on exertion. She was constipated and had lost 12.5 kg over the last 6 months. She complained that she was very thirsty and had to get up to urinate during the night, something that she had previously never had to do.

Initial laboratory investigations found her to be hypercalcaemic, dehydrated and anaemic. Her biochemistry results showed marked renal impairment, with raised urea and creatinine. Her alkaline phosphatase (AP) was within the reference range. Her serum protein concentration was raised, despite having a low albumin. Serum protein electrophoresis showed a large monoclonal band in the gamma globulin region. By comparison of area under the peak with total protein, the band was quantified as 61 g dm^{-3}. The band was typed as IgAκ by immunofixation, using antisera specific against individual immunoglobulin subclasses to bind the monoclonal protein to the electrophoresis before staining. An early morning urine sample was requested for Bence Jones protein analysis by electrophoresis. This detected a large band of free κ light chains in the urine. Her immunoglobulins were quantified by turbidimetry, a method that measures the refraction of light by antibody complexes. The results indicated that her other immunoglobulins were suppressed, leaving her susceptible to infection. An isotope bone scan using diphosphonates labelled with ^{99}Te demonstrated osteolytic lesions (bone loss) that are characteristic of multiple myeloma.

Comment   The two most common causes of hypercalcaemia are primary hyperparathyroidism and malignancy. The signs and symptoms in a person of this age are typical of multiple myeloma, especially in the context of hypercalcaemia with a normal AP, which is raised in primary hyperparathyroidism. Hypercalcaemia results from stimulation of osteoclasts (a type of bone cell) released by the myeloma cells and can cause polyuria, polydipsia (need to drink excessive fluid) and dehydration. The impaired renal function in this lady may be a result of hypercalcaemia and Bence Jones protein as both are nephrotoxic. As the malignant plasma cells proliferate throughout the bone marrow, the bone marrow has a reduced capacity to produce normal cells, causing anaemia and immunosuppression. The difference between the concentration of serum total protein and albumin is attributed to the Bence Jones proteins. These proteins contribute to this fraction and consequently myeloma patients can have a high total protein concentration in the presence of normal or low albumin.

The patient was rehydrated with intravenous saline and started on a bisphosphonate to lower her calcium. Her renal failure resolved over time, although some patients with advanced myeloma will require haemodialysis. A bone marrow aspiration was performed, which showed >80% infiltration of plasma cells, confirming the diagnosis of multiple myeloma. A serum β2-microglobulin was requested as a prognostic indicator. Once she had stabilised, she was commenced on a course of dexamethasone, a synthetic glucocorticoid that binds immunoglobulins and hence relieves some of the symptoms of the malignancy. Her response to treatment was monitored by regular quantification of her monoclonal band.

with no obvious pathology. IgE plays a major part in allergy and may be significantly raised in situations of allergic response, for example in hay fever and atopic eczema.

Myeloma

Myeloma, also called multiple myeloma, is a malignant pathology of plasma cells in which there is a proliferation of a single β-cell clone in the bone marrow effectively behaving as a tumour. The replication of the cell is unregulated so it proliferates. The cells produce large quantities of a single identical antibody which runs as a single dense band in the gamma globulin region on electrophoresis of a serum sample. The protein is called the para protein and has been shown to be an immunoglobulin with two light chains and two heavy chains. Some myelomas produce an excess of light chains that appear in the serum and because they are small they also appear in the urine. They are detected by electrophoresis and are referred to as Bence Jones proteins. Their detection is a bad prognosis as they indicate that the cell line may be more aggressive and replicating faster. In rare cases of myeloma, the marrow cells only produce light chains.

Acute phase response

Following a stimulus of tissue injury or infection, the body will respond by producing an acute phase response characterised by the release from the liver of a number of acute phase proteins which cause a change in the pattern of plasma protein electrophoresis. There will be an increased synthesis of some proteins such as α-1-antitrypsin, a proteinase inhibitor that down regulates inflammation, fibrinogen and prothroylian (coagulation) complement and C-reactive protein (CRP). These are referred to as positive acute phase proteins. There will also be a decrease in the production of other proteins such as albumin, transferrin and transcortin. These are known as negative acute phase proteins. The clinical measurement of acute phase proteins, particularly CRP, by immunoassay is widely used as a marker of inflammation in a variety of clinical conditions.

16.4 SUGGESTIONS FOR FURTHER READING

Basic principles
Saunders, G. C. and Parkes, H. C. (1999). *Analytical Molecular Biology*. Teddington: LGC. (Contains an excellent chapter on quality in the molecular biology laboratory.)

Clinical biochemistry
Beckett, G. I., Walker, S. W., Rae, P. and Ashby, P. (2005). *Lecture Notes on Clinical Biochemistry*, 6th edn. Oxford: Blackwell Science. (An excellent reference text for all aspects of clinical biochemistry.)
Bruns, D. E. and Ashwood, E. R. (2007). *Tietz Fundementals of Clinical Chemistry*, 6th edn. Philadelphia: W. B. Saunders. (A comprehensive coverage of the principles and practice of clinical biochemistry.)

Data analysis
Jones, R. and Payne, B. (1997). *Clinical Investigation and Statistics in Laboratory Medicine*. London: ACB Ventures. (Written specifically for analytical studies in clinical biochemistry.)

Newborn screening
Blau, N., Duran, M., Blaskovics, M. E. and Gibson, K. M. (eds.) (2003). *Physician's Guide to the Laboratory Diagnosis of Metabolic Diseases*, 2nd edn. Berlin: Springer-Verlag.

Chace, D. H. and Kalas, T. A. (2005). A biochemical perspective on the use of tandem mass spectrometry for newborn screening and clinical testing. *Clinical Biochemistry*, **38**, 296–309.

16.5 ACKNOWLEDGEMENTS

We are grateful to the following colleagues at Yorkhill Hospital and the Southern General Hospital in Glasgow for their help in the preparation of the case studies presented in this chapter: Dr Jane McNeilly (pregnancy), Mr Graeme Chalmers (thyroid) and Mr Neil Squires (myeloma). We would also like to thank Dr Susan Bonham Carter for her advice on neonatal clinical measurements.

17 Cell membrane receptors and cell signalling

K. WILSON

17.1 RECEPTORS FOR CELL SIGNALLING

17.1.1 Intercellular signal transduction

Cells in multicellular organisms need to be able to communicate with each other in order to respond to external stimuli and to coordinate their activities to achieve homeostasis. Such communication is termed intercellular signalling and is achieved by:

- The release by the 'signalling' cells of signalling molecules, referred to as endogenous agonists or first messengers.
- The specific recognition and binding of these agonists by receptor molecules, simply referred to as receptors, located either in the cell membrane or in the cytoplasm of the 'target' cell. Each cell membrane contains between 10^3 and 10^6 molecules of a given receptor. Binding of the agonist to a specific binding domain on the receptor changes the receptor from its inactive, resting state, to an active state. Exceptionally, a receptor may possess activity in the absence of agonist. Such receptors are said to possess constitutive activity.
- The initiation of a sequence of molecular events commonly involving interaction between the active receptor and other, so-called effector molecules, the whole process being referred to as intracellular signal transduction which terminates in the final cellular response.

Agonist signalling molecules range from the gas nitric oxide, amines, amino acids, nucleosides, nucleotides and lipids to hormones, growth factors, interleukins, interferons and cytokines. These molecules are either lipophilic or hydrophilic. Lipophilic signalling molecules such as the steroid hormones (progesterone, oestrogen and testosterone) and non-steroid hormones (thyroxine and triiodothyronine) can readily cross the cell membrane, in which case they bind to receptors located in the cytoplasm. These receptors have two binding domains: an agonist-binding domain and a DNA-binding domain. Binding of the lipophilic agonist to its site results in a receptor–agonist complex that is able to pass through pores in the nuclear membrane into the nucleus where it interacts with a specific DNA sequence, termed a hormone response element, to regulate (activating or repressing) the transcription of down-stream genes. For this reason, the receptors are referred to as nuclear receptors. In the majority of cases of intercellular signalling, however, the signalling agonist is hydrophilic and therefore incapable of diffusing across the cell membrane. The receptors for such agonists are therefore embedded in, and span, the cell membrane with the agonist-binding domain exposed on the extracellular side. This chapter will consider the molecular nature and mode of action of such cell surface receptors.

Cell membrane receptor proteins possess three distinct domains:

- *Extracellular domain*: This protrudes from the external surface of the membrane and contains all or part of the agonist-binding domain known as the orthosteric agonist-binding site.
- *Transmembrane domain*: This is inserted into the phospholipid bilayer of the membrane and may consist of several regions that loop repeatedly back and forth across the membrane. In some cases these loops form a channel for the 'gating' (hence the channel may be open or closed) of ions across the membrane, whilst in other receptors the loops create part of the orthosteric site.
- *Intracellular domain*: This region of the protein has to respond to the extracellular binding of the agonist to initiate the transduction process. In some cases it is the site of the activation of enzyme activity within the receptor protein, commonly kinase activity, or is the site that interacts with effector proteins.

The existence of three domains within receptor proteins reflects their amphipathic nature in that they contain regions of 19 to 24 amino acid residues possessing polar groups that are hydrophilic, and similar sized regions that are rich in non-polar groups that are hydrophobic and hence lipophilic. The hydrophobic regions, generally in the form of α-helices, are the transmembrane regions that are inserted into the non-polar, long-chain fatty acid portion of the phospholipid bilayer of the membrane. Superfamilies of receptor proteins can be recognised from the precise number of transmembrane regions each possesses. In contrast, the hydrophilic regions of the receptor are exposed on the outside and inside of the membrane where they interact with the aqueous, hydrophilic environment.

17.1.2 Classification of cell membrane receptors

Studies of the structure of membrane receptors and of the mechanisms of their signal transduction have led to the identification of three main classes of cell membrane receptors:

- *Ligand-gated ion-channel receptors*: These are responsible for the selective movement of ions such as Na^+, K^+ and Cl^- across membranes. Binding of the agonist triggers the gating (opening) of a channel and the movement of ions across the membrane. This ion movement is a short-term, fast response that results in the propagation of a membrane potential wave. It may be excitatory and result in the depolarisation of the cell (e.g. the nicotinic acetylcholine and ionotropic glutamate receptors), or inhibitory (e.g. the γ-aminobutyric acid A ($GABA_A$) receptor). All receptors in this class consist of four or five homo- or heteromeric subunits. Responses produced by this class of receptors occur in fractions of a second.

- *G-protein-coupled receptors (GPCRs)*: Receptors in this class are linked to a G-protein that is trimeric (i.e. it has three subunits). Receptor activation by agonist binding triggers its interaction with a G-protein located within the cell membrane and protruding into the cytoplasm resulting in the exchange of GTP for GDP (hence the name G-protein) on one subunit that dissociates from the trimer causing the activation of an effector molecule such as adenylyl cyclase (also known as adenylate cyclase) that is part of an intricate network of intracellular transduction pathways. Responses produced by GPCRs occur in the timescale of minutes.

- *Protein kinase receptors*: These receptors all undergo agonist-stimulated autophosphorylation in their intracellular domain. This activates a kinase activity within this domain. The majority of activated receptor kinases catalyse the transfer of the γ-phosphate of ATP to the hydroxyl group of a tyrosine in the target effector protein, hence the term receptor tyrosine kinases (RTKs). This phosphorylation process controls the activity of many vital cell processes. Members of a minor subgroup of protein kinase receptors transfer the phosphate group of ATP to a serine or threonine group rather than tyrosine, hence the term receptor serine (or threonine) kinases. Responses produced by protein kinase receptors occur over a timescale of minutes to hours. Closely related to the protein kinase receptors is a group of receptors that lack intrinsic protein kinase activity but which recruit a non-receptor tyrosine kinase after agonist binding. The recruited kinase phosphorylates tyrosine residues in the receptors' intracellular domains then act as recognition sites for other effector proteins which, when activated, translocate to the cell nucleus where, in association with other regulatory proteins, they modify gene expression. Responses to these receptors occur within a timescale of minutes to hours.

Further details of the mechanism by which each of these three classes of receptor induce the transduction process are discussed in Section 17.5. There is evidence that many receptors exist in multiple isoforms that have subtly different physiological roles and that many receptors form homo- and/or heterodimers or oligomers that are sensitive to allosteric regulation and which function as partners to initiate interactions between the downstream signalling molecules triggered by each receptor. This cross-talk between receptors allows cells to integrate signalling information originating from various external sources and to respond to it with maximum regulatory efficiency.

Signalling co-receptors

In addition to the three groups of signalling receptors discussed above, there is a group of membrane receptors that bind agonists but which do not directly transduce a cellular signal but to do so they form a complex with a receptor from one of the above three classes. These receptors are termed co-receptors or accessory receptors. These co-receptors are:

- used by a large number of ligands including interleukin, epidermal growth factor and fibroblast growth factor;
- expressed ubiquitously within a given organism such that they are often the most abundant receptor for the agonists they bind;
- expressed with conserved structural features on the cell surface of a diverse range of organisms.

Eight families of co-receptors have been identified, each containing up to eight members. The agonists that they bind are promiscuous in that a given co-receptor may bind up to nine different agonists and a given agonist can bind to more than one co-receptor. Their central role in the regulation of cellular processes is evident from the observation that their mutation and/or altered expression is associated with such human diseases as certain cancers, inflammation and ischaemic heart disease. Mutations commonly cause a loss of co-receptor function leading to an autosomal dominant or recessive inherited disorder (see cytokine receptors, Section 17.4.4).

17.2 QUANTITATIVE ASPECTS OF RECEPTOR–LIGAND BINDING

17.2.1 Dose–response curves

The response of membrane receptors in their resting inactive state to exposure to an increasing concentration (dose) of agonist is a curve that has three distinct regions:

- an initial threshold below which little or no response is observed;
- a slope in which the response increases rapidly with increasing dose;
- a declining response with further increases in dose and a final maximum response.

Since such plots commonly span several hundred-fold variations in agonist concentration, they are best expressed in semi-logarithmic form (Fig. 17.1).

A dose–response curve for an inverse agonist (Section 17.2.2) acting on a receptor with constitutive activity would be a mirror image of that shown in Fig. 17.1 resulting in a progressive decrease in receptor activity.

Dose–response studies coupled with the observed transduction pathway have enabled molecules (ligands) that bind to a receptor to be placed into one or more of the following classes:

- *Full agonists*: These ligands increase the activity of the receptors and produce the same maximal response but they differ in the dose required to achieve it (Fig. 17.1).

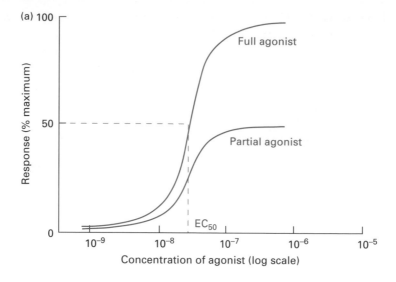

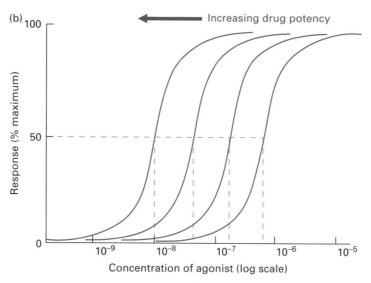

Fig. 17.1 Dose–response curves for receptor agonists. (a) The biological effect (% maximum response) and the concentration of a full agonist are plotted on a logarithmic scale. An equipotent partial agonist has a lower efficacy than a full agonist – it cannot achieve the maximum response even when all the receptors are occupied. EC_{50} is the concentration of agonist that produces 50% maximum effect. (b) Dose–response curves for four full agonist drugs of different potencies but equal efficacy. (Reproduced from Maxwell, S. R. J. and Webb, D. J. (2008). Receptor functions. *Medicine*, **36**, 344–349, by permission of Elsevier Science.)

- *Partial agonists*: These ligands also increase the activity of the receptors but do not produce the maximal response shown by full agonists even when present in large excess such that all the receptors are occupied (Fig. 17.1).
- *Inverse agonists*: These agonists decrease the activity of constitutively active receptors to their inactive state.

- *Partial inverse agonists*: These agonists decrease the activity of constitutively active receptors but not to their inactive state.
- *Protean agonists*: These agonists acting on receptors possessing constitutive activity display any response ranging from full agonism to full inverse agonism depending on the level of constitutive activity in the system and the relative efficacies of the constitutive activity and that induced by the agonist.
- *Biased agonists*: This form of agonist behaviour is found with receptors that can couple to two or more different G-proteins and as a consequence the agonist preferentially selects one of them thus favouring one specific transduction pathway.
- *Antagonists*: In the absence of agonists these ligands produce no change in the activity of the receptors. Three subclasses have been identified using the antagonist in the presence of an agonist and using receptors not possessing constitutive activity:
  - (i) *Competitive reversible antagonists*: The antagonist competes with the agonist for the orthosteric sites so that the effect of the antagonist can be overcome by increasing the concentration of agonist (Fig. 17.2).
  - (ii) *Non-competitive reversible antagonists*: The antagonist binds at a different site on the receptors to that of the orthosteric site so that the effect of the antagonist cannot be overcome by increasing the concentration of agonist.
  - (iii) *Irreversible competitive antagonists*: The antagonist competes with the agonist for the orthosteric site but the antagonist forms a covalent bond with the site so that its effect cannot be overcome by increasing the concentration of agonist.
- *Allosteric modulators*: These bind to a site distinct from that of the orthosteric site and can only be detected using functional, as opposed to ligand-binding, assays. They are discussed more fully in Section 17.2.3.

As a result of this classification, ligand action on receptors can be characterised by a number of parameters:

- *Intrinsic activity*: This is a measure of the ability of an agonist to induce a response by the receptors. It is defined as the maximum response to the test agonist relative to the maximum response to a full agonist acting on the same receptors. All full agonists, by definition, have an intrinsic activity of 1 whereas partial agonists have an intrinsic activity of less than 1.
- *Efficacy (e)*: This is a measure of the inherent ability of an agonist to initiate a physiological response following binding to the orthosteric site. The initiation of a response is linked to the ability of the agonist to promote the formation of the active conformation of the receptors whereas for inverse agonists it is linked to their ability to promote the formation of the inactive conformation. While all full agonists must have a high efficacy their efficacy values will not necessarily be equal, in fact values of e have no theoretical maximum value. Partial agonists have a low efficacy, antagonists have zero efficacy and inverse agonists have negative efficacy.
- *Collateral efficacy*: This relates to the ability of the agonist to preferentially select one of the two or more possible transduction pathways displayed by the binding

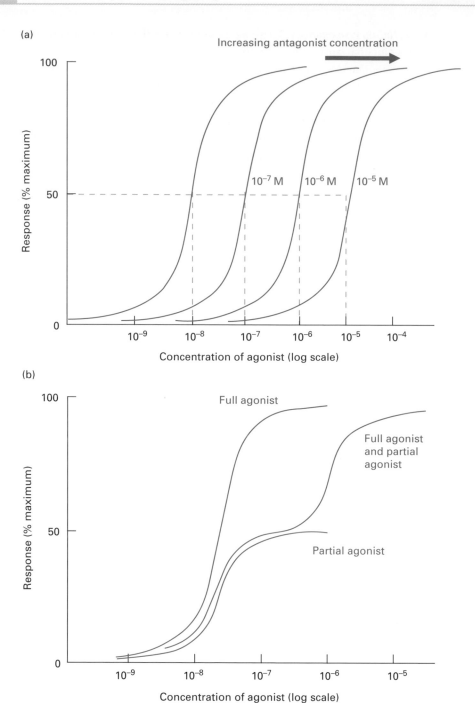

Fig. 17.2 Dose–response curves for receptor agonist. (a) In the presence of a competitive reversible antagonist. In the presence of a reversible antagonist, the dose–response curve for a full agonist is shifted to the right because the full agonist competes for receptor binding. The higher the concentration of antagonist the greater the shift, but the maximum response remains unchanged. below (b) In the presence of a partial agonist, the dose-response curve of the full agonist is shifted to the right. (Reproduced from Maxwell, S. R. J. and Webb, D. J. (2008). Receptor functions. *Medicine*, **36**, 344–349, by permission of Elsevier Science.)

receptors. It primarily relates to G-protein-coupled receptors that can bind to two or more G-proteins.

- *Potency*: This is a measure of the concentration of agonist required to produce the maximum effect; the more potent the agonist the smaller the concentration required. The potency of an agonist is related to the position of the sigmoidal curve on the log dose axis. It is expressed in a variety of forms including the effective dose or concentration for 50% maximal response, ED_{50} or EC_{50}. On a semi-logarithm plot, the value emerges as pED_{50} or pEC_{50} value (i.e. $-\log_{10} ED_{50}$). Thus an agonist with EC_{50} of 3×10^{-5} M would have a pEC_{50} of 4.8. The potency of a reversible antagonist is expressed by its pA_2 value, defined as negative logarithm of the concentration of antagonist that will produce a two-fold shift in the concentration–response curve for an agonist.

- *Affinity*: This is a measure of the concentration of agonist required to produce 50% binding. As will be shown in the following section, affinity is a reflection of both the rate of association of the ligand with the receptors and the rate of dissociation of the resulting complexes. The rate of association is a reflection of the three-dimensional interaction between the two and the rate of dissociation a reflection of the strength of binding within the complexes. Affinity of an agonist can be expressed by an affinity or binding constant, K_a, but is more commonly expressed as a dissociation constant, K_d, of the receptor–ligand complex where K_d is equal to the reciprocal of K_a. The affinity of receptors for an antagonist is expressed by the corresponding dissociation constant K_b.

- *Selectivity*: This is a measure of the ability of an agonist to discriminate between receptor subtypes. This is particularly important from a therapeutic perspective.

- *Functional selectivity*: This is a measure of the ability of the agonist to induce selective response from receptors capable of promoting more than one transduction activity.

17.2.2 Constitutive receptor activity, inverse agonists and receptor activation

The long-held view of receptor–agonist interaction was based on a two-state model that visualised that the binding of the agonist (A) by the receptor to form a receptor-agonist complex triggers a conformational change in the receptor that converts it from a dormant or resting inactive state (R) to an active state (R*):

$$R + A \longrightarrow AR \longrightarrow AR^* \longrightarrow \text{transduction via}$$

| inactive | inactive | active | effector |

The formation of the active state of the receptor initiates a transduction (linking) process in which the receptor activates an effector protein. The effector protein may be the receptor itself or a distinct protein that is either attached to the inside of the membrane or free in the cytoplasm. This activated effector either allows the passage of selected ions across the membrane thereby changing the membrane potential or it produces a second messenger which initiates a cascade of molecular events, involving molecules located on and/or at the internal surface of the cell membrane or within the cell cytoplasm, that terminate in the final target cell response. Examples of second messengers are Ca^{2+}, cAMP, cGMP, 1,2-diacylglycerol and inositol-1,4,5-trisphosphate.

However, in 1989 it was discovered that opioid receptors in NG108 cells possessed activity in the absence of agonist. This receptor activity in the absence of agonist was termed constitutive activity, and synthetic ligands were identified that could bind to the receptor and decrease its constitutive activity in the absence of physiological agonist. Such ligands were termed inverse agonists. Subsequent *in vivo* and *in vitro* investigations with a wide range of receptor types, many of which were G-protein-coupled receptors, identified other examples of receptors with constitutive activity and showed that this activity may have a physiological role thereby confirming that constitutive activity is not solely a consequence of a mutation or overexpression of a receptor gene. Of particular interest was the observation that certain receptor mutations (known as constitutively active mutants, CAMs) were associated with such clinical disorders as retinitis pigmentosa, hyperthyroidism and some autoimmune diseases.

The conformational selection model of receptor action was formulated to rationalise the concomitant existence of active and inactive conformations. The model envisages that in the absence of agonist, receptors exist as an equilibrium mixture of inactive (R) and active (R*) forms and that the relative proportion of the two forms is determined by the associated equilibrium constant. An introduced ligand will preferentially bind to one conformation, thereby stabilising it, and causing a displacement of the equilibrium between the two forms. Agonists will preferentially bind to the R* state displacing the equilibrium to increase the proportion of the R* form. Partial agonists are deemed to have the ability to bind to both forms with a preference for the R* form again resulting in an increase in the R* form but by a smaller amount than that produced by full agonists. Inverse agonists preferentially bind to the R conformation, displacing the equilibrium and increasing the proportion of R. Partial inverse agonists can bind to both states with a preference for the R resulting in decrease in the proportion of R* form. Unlike the two-state model, the conformational selection model does not require the binding ligand to cause a conformational change in the receptor in order to alter the activity of the receptor.

Prior to the discovery of constitutive activity in some receptors, ligands were classified either as agonists or antagonists. This classification formed the basis of the understanding of the pharmacological action of many therapeutic agents and the development of new ones by the pharmaceutical industry. However, retrospective evaluation of ligands previously classified as antagonists but using receptors produced by cloning techniques and shown to possess constitutive activity revealed that many were actually inverse agonists and possessed negative efficacy whilst others were neutral antagonists in that they neither increased nor decreased the receptor activity. To date approximately 85% of all antagonists that have been re-evaluated have been shown to be inverse agonists. These observations can be rationalised in that:

- *agonism* is a behaviour characteristic of a particular ligand and can be demonstrated in the absence of any other ligand for the receptor,
- *antagonism* can only be demonstrated in the presence of an agonist,
- *inverse agonism* can only be demonstrated when the receptor possesses constitutive activity that can be reduced by the agent. In the absence of constitutive activity inverse agonists can only demonstrate simple competitive antagonism of a full or partial agonist.

Further understanding of the nature of constitutive activity and the mode of action of different agonists on a given receptor has come from studies on the histamine H_3 receptor (H_3R). This is a G-protein-coupled receptor, specifically coupling to G_i/G_o proteins (see Section 17.4.3). Constitutive activity has been found in both rat and human brain in which the activity inhibits histamine release from synaptosomes. Studies using the ligand proxyfan, previously classified as an antagonist, have assigned to it a spectrum of activities ranging from full agonist through partial agonist to partial inverse agonist and full inverse agonist. Such behaviour by a ligand has been classified as protean agonism and proxyfan as a protean agonist. The precise behaviour of proxyfan in a given study correlated with the level of constitutive activity of the system and the relative efficacy of the constitutively active state R^* and that induced by the ligand, AR^*. Thus in the absence of any R^* or in the presence of AR^* with a lower efficacy than that of R^*, the ligand will display agonist activity. When both R^* and AR^* states are present with equal efficacy the ligand will display neutral antagonism and when the AR^* state has a higher efficacy than R^* the ligand will display inverse agonism. Such behaviour can only be explained by a multistate model in which multiple R^* and AR^* states of the receptor can be formed. Furthermore, studies using a range of agonists on the histamine H_3 receptor indicated that different ligands could promote the creation of distinct active states that can display differential signalling. A wider understanding of the mechanism by which receptors are activated by agonists is linked to the discovery, initially made with the α_{1B} adrenergic receptor, that certain mutations in the sequence of G-protein-coupled receptors caused a large increase in the constitutive activity of the receptor. This observation had the implication that there may be domains in the receptor that are crucial to the conservation of a receptor not displaying constitutive activity, and that the action of agonists was to release these constraints creating the active receptor. In mutant receptors possessing constitutive activity these constraints have been released as a result of the mutation. Studies have shown that the activation of inactive receptors by agonists proceeds by a series of conformational changes. The question as to whether agonists and inverse agonists switch the receptor in a linear 'on–off' scale or whether they operate by different mechanisms has been studied using the α_{2A} adrenergic receptor and a fluorescence resonance energy transfer (FRET) based approach. Differences in the kinetics and character of the conformational changes induced by these two classes of agonists provided clear evidence for distinct types of molecular switch. Moreover, full agonists and partial agonists also showed distinct differences indicating that receptors do not operate by a simple 'on–off' switch but rather that they have several distinct conformational states and that these states can be switched with distinct kinetics by the various classes of agonist.

As previously pointed out, the pharmaceutical industry seeks to identify receptor agonists or antagonists that can be used for the treatment of specific clinical conditions. The potential role of inverse agonists in this respect remains to be fully evaluated. However, it is apparent that clinical conditions caused by a mutant receptor that has constitutive activity in contrast to the normal receptor that is only active in the presence of the physiological agonist could be treated with an inverse agonist that would eliminate the constitutive activity. Equally, use of an inverse agonist may be advantageous in conditions resulting from the overstimulation of the receptor due to the overproduction of the signalling agonist. At the present time many inverse agonists are used clinically although at the time of their development they were believed to be competitive antagonists.

17.2.3 **Allosteric modulators**

Studies using functional screening assays have demonstrated that many receptors possess allosteric sites distinct from the orthosteric agonist site. Such sites have been identified in monomeric receptors as well as those that form homo- and heterodimers. These allosteric sites are capable of binding allosteric modulators that exert one of three distinct effects (Fig. 17.3):

- alteration of the affinity of the agonist for its orthosteric site; or
- alteration of the efficacy (i.e. ability to produce the response of the receptor via its various effectors) of the agonist; or
- display an efficacy independent of the presence of an agonist. Such efficacy could be of the agonist or inverse agonist variety.

The binding to the allosteric site is characterised by its dissociation constant and by a cooperativity factor α, which is a thermodynamic measure of the strength of interaction between this site and the orthosteric site. Since these two sites are distinct, allosteric modulators induce unique conformational changes in the receptor and at least in principle these may alter the signalling, desensitisation and internalisation states induced by agonists binding at the orthosteric site. Recently, evidence has been obtained to indicate that allosteric interaction between distinct sites may operate by ligand-dependent changes in the dynamic properties of the receptor rather than simple conformational changes. This idea is based on the recognition that receptors, like enzymes, exist as assemblies of conformations the balance between which can be altered by ligand binding (Section 15.2.4). Ligands acting as allosteric modulators can be classified into one of four types on the basis of the effects they produce on the activity of the orthosteric site:

- *Allosteric agonist*: This is a ligand that binds to an allosteric site and mediates the activation of the receptor in the absence of the physiological agonist.
- *Allosteric enhancer*: This is a ligand that enhances the affinity or efficacy of an agonist acting on the orthosteric site without having any activity of its own.
- *Allosteric modulator*: This is a ligand that alters (increases or decreases) the activity of an agonist or antagonist acting on the orthosteric site without having any activity of

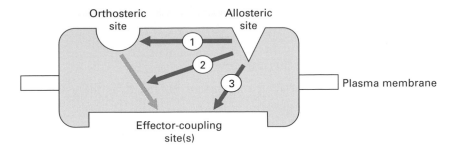

Fig. 17.3 Types of allosteric modulator. Allosteric ligands can affect receptor function in three general ways. (1) Allosteric modulation of orthosteric ligand-binding affinity; (2) allosteric modulation of orthosteric ligand efficacy; (3) direct allosteric agonism. (Reproduced from Langmead, C. J. and Christopoulos, A. (2006). Allosteric agonists of 7TM receptors: expanding the pharmacological toolbox. *Trends in Pharmacological Sciences*, **27**, 475–481, by permission of Elsevier Science.)

its own in the absence of the agonist or antagonist. Allosteric enhancers are a subgroup of these allosteric modulators;

- *Ago-allosteric modulator*: This is a ligand that can act as both an allosteric agonist and as an allosteric modulator altering the efficacy and/or the potency of agonists acting on the orthosteric site.

The presence of allosteric sites is important in drug discovery since in principle drugs acting at this site may modify the physiological response transduced by the receptor (Section 18.2.2).

17.2.4 Quantitative characterisation of receptor–ligand binding

The previous discussion has shown that ligands capable of binding to a receptor may be of a variety of types. To fully characterise any ligand it is essential that its binding be expressed in quantitative terms. This is achieved by binding studies. If under the conditions of the binding studies the total concentration of ligand is very much greater than that of receptor (so-called saturation conditions), changes in ligand concentration due to receptor binding can be ignored but changes in the free (unbound) receptor concentration cannot. Hence if:

[R_t] is the total concentration of receptor that determines the maximum binding capacity for the ligand

[L] is the free ligand concentration

[RL] is the concentration of receptor–ligand complex

then [R_t − RL] is the concentration of free receptor.

At equilibrium, the forward and reverse reactions for ligand binding and dissociation will be equal:

$$k_{+1}([R_t] - [RL])[L] = k_{-1}[RL]$$

where k_{+1} is the association rate constant and k_{-1} is the dissociation rate constant. Therefore:

$$\frac{k_{-1}}{k_{+1}} = K_d = \frac{1}{K_a} = \frac{([R_t] - [RL])[L]}{[RL]} \qquad (17.1)$$

where K_d is the dissociation constant for RL and K_a is the association or affinity constant. Rearranging gives:

$$[RL] = \frac{[L][R_t]}{K_d + [L]} \qquad (17.2)$$

Determination of K_d

Equation 17.2 is of the form of a rectangular hyperbola which predicts that ligand binding will reach a limiting value as the ligand concentration is increased and therefore that receptor binding is a saturable process. The equation is of precisely the same form as equations 15.1 and 15.2 that define the binding of the substrate to its enzyme in terms of K_m and V_{max}. For the experimental determination of K_d, equation 17.2 can be used directly by analysing the experimental data by non-linear regression curve-fitting programs (see Section 15.2.1). However, several linear transformations of equation 17.2 have been developed. One such is the Scatchard equation (17.3):

$$\frac{[RL]}{[L]} = \frac{[R_t]}{K_d} - \frac{[RL]}{K_d} \qquad (17.3)$$

This equation predicts that a plot of [RL]/[L] against [RL] will be a straight-line slope $-1/K_d$ allowing K_d to be calculated. However, in many studies the relative molecular mass of the receptor protein is unknown so that the concentration term [RL] cannot be calculated in molar terms. In such cases it is acceptable to express the extent of ligand binding in any convenient unit (B), e.g. pmoles 10^{-6} cells, pmoles mg^{-1} protein or more simply as an observed change, for example in fluorescence (ΔF), under the defined experimental conditions. Since maximum binding (B_{max}) will occur when all the receptor sites are occupied, i.e. when $[R_t] = B_{max}$, equation 17.3 can be written in the form:

$$\frac{B}{[L]} = \frac{B_{max}}{K_d} - \frac{B}{K_d} \qquad (17.4)$$

Hence a plot of $B/[L]$ against B will be a straight line, slope $-1/K_d$ and intercept on the y-axis of B_{max}/K_d (Fig. 17.4). In cases where the relative molecular mass of the receptor protein is known the Scatchard equation can be expressed in the form:

$$\frac{B}{[L]B_{max}} = \frac{n}{K_d} - \frac{B}{B_{max}K_d} \qquad (17.5)$$

where n is the number of independent ligand-binding sites on the receptor. The expression B/B_{max} is the number of moles of ligand bound to one mole of receptor. If this expression is defined as r, then:

$$\frac{r}{[L]} = \frac{n}{K_d} - \frac{r}{K_d} \qquad (17.6)$$

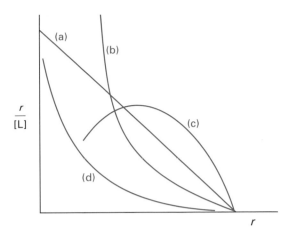

Fig. 17.4 Scatchard plot for (a) a single set of sites with no cooperativity, (b) two sets of sites with no cooperativity, (c) a single set of sites with positive cooperativity, and (d) a single set of sites with negative cooperativity.

In this case, a plot of $r/[L]$ against r will again be linear with a slope of $-1/K_d$ but in this case the intercept on the x-axis will be equal to the number of ligand-binding sites, n, on the receptor.

Alternative linear plots to the Scatchard plot are:

Lineweaver–Burk plot

$$\frac{1}{B} = \frac{1}{B_{max}} + \frac{K_d}{B_{max}[L]}$$ (17.7)

Hanes plot

$$\frac{[L]}{B} = \frac{K_d}{B_{max}} + \frac{[L]}{B_{max}}$$ (17.8)

In practice, Scatchard plots are most commonly carried out although statistically they are prone to error since the experimental variable B occurs in both the x and y terms so that linear regression of these plots overestimates both K_d and B_{max}. There is a view that linear transformations of the three types above are all inferior to the non-linear regression analysis of equation 17.2 since they all distort the experimental error. For example, linear regression assumes that the scatter of experimental points around the line obeys Gaussian distribution and that the standard deviation of the points is the same. In practice this is rarely true and as a consequence values of the slope and intercept are not the 'best' value. It can be seen from equation 17.2 that when the receptor sites are half saturated, i.e. $B = B_{max}/2$, then $[L] = K_d$. Hence K_d will have units of molarity.

The derivation of equation 17.2 is based on the assumption that there is a single set of homogeneous receptors and that there is no cooperativity between them in the binding of the ligand molecules. In practice, two other possibilities arise namely that there are two distinct populations of receptors each with different binding constants and secondly that there is cooperativity in binding within a single population. In both cases the Scatchard plot will be curvilinear (Fig. 17.4). If cooperativity is suspected, it should be confirmed by a Hill plot which, in its non-kinetic form, is:

$$\log\left(\frac{Y}{1-Y}\right) = h\log[\text{L}] - \log K_\text{d} \tag{17.9}$$

or

$$\log\left(\frac{B}{B_\text{max} - B}\right) = h\log[\text{L}] - \log K_\text{d}$$

where Y is the fractional saturation of the binding sites (from $0-1$) and h is the Hill constant. For a receptor with multiple binding sites that function independently $h = 1$, whereas for a receptor with multiple sites which are interdependent, h is either greater than 1 (positive cooperativity) or less than 1 (negative cooperativity). Scatchard plots that are biphasic due to ligand multivalence (i.e. multiple binding sites) rather than receptor cooperativity, are sometimes taken to indicate that the two extreme, and approximately linear, sections of the curvilinear plots represent high affinity (high bound/free ratio at low bound values) and low affinity (low bound/free ratio at high bound values) sites and that tangents drawn to these two sections of the curve can be used to calculate the associated K_d and B_max values. This is incorrect, and the correct values can only be obtained from the binding data by means of careful mathematical analysis generally undertaken by the use of special computer programs many of which are commercially available.

Determination of rate constants

To determine the dissociation rate constant, k_{-1} (units: time^{-1}) of any ligand for the receptor, some of the receptor–ligand complex (B_0) is allowed to form usually using the ligand labelled with a radioactive isotope. The availability of the remaining unoccupied receptors to the labelled ligand is then blocked by the addition of at least 100-fold excess of the unlabelled ligand or competitive antagonist and the rate of release (B_t) of the radiolabelled ligand from its binding site monitored as a function of time. This generally necessitates the separation of the bound and unbound fractions. The rate is given by the expression:

$$\frac{\text{d}B_0}{\text{d}t} = -k_{-1}B_0$$

and the equation governing the release by the expression:

$$B_t = B_0\text{e}^{-k_{-1}t}$$

hence:

$$\log B_t = \log B_0 - 2.303 k_{-1} t \tag{17.10}$$

Thus a plot of $\log B_t$ against time will give a straight line with a slope of $-2.303 k_{-1}$ allowing k_{-1} to be estimated.

The association rate constant, k_{+1} (units: $M^{-1}time^{-1}$), is best estimated by the approach to equilibrium method by which the extent of agonist binding is monitored continuously until equilibrium is reached under conditions that are such that $[L] \gg [R_t]$ (this gives pseudo first-order conditions rather than second order; under these conditions $[R_t]$ decreases with time but $[L]$ remains constant.) Ligand binding increases asymptotically such that:

$$\log\left(\frac{B_{eq}}{B_{eq} - B_t}\right) = 2.303(k_{+1}[L] + k_{-1})t \qquad (17.11)$$

Thus a plot of $\log B_{eq}/(B_{eq} - B_t)$ against time will be linear with a slope of $2.303(k_{+1}[L] + k_{-1})$ where B_{eq} and B_t are the ligand binding at equilibrium and time t respectively. From knowledge of k_{-1} (obtained by the method discussed above) and $[L]$, the value of k_{+1} can be calculated from the slope.

The K_d values observed for a range of receptors binding to their physiological agonist are in the range $10^{-6}-10^{-11}$ M, which is indicative of a higher affinity than is typical of enzymes for their substrates. The corresponding k_{+1} rate constants are in the range 10^5-10^8 $M^{-1}min^{-1}$ and k_{-1} in the range $0.001-0.5$ min^{-1}. Studies with G-protein-coupled receptors that form a tertiary complex (AR*G) have shown that the tertiary complex has a higher affinity for the agonist than has the binary complex (AR*). Receptor affinity for its agonist is also influenced by receptor interaction with various adaptor protein molecules present in the intracellular cell membrane. This is discussed more fully later.

It is relatively easy to calculate the number of receptors on cell membranes from binding data. The number is in the range 10^3-10^6 per cell. Although this may appear

Example 1   ANALYSIS OF LIGAND-BINDING DATA

Question The extent of the binding of an agonist to its membrane-bound receptor on intact cells was studied as a function of ligand concentration in the absence and presence of a large excess of unlabelled competitive antagonist. In all cases the extent of total ligand binding was such that there was no significant change in the total ligand concentration. What quantitative information about the binding of the ligand to the receptor can be deduced from this data?

	[Ligand] (nM)							
	40	60	80	120	200	500	1000	2000
Total ligand bound (pmoles 10^{-6} cells)	0.284	0.365	0.421	0.547	0.756	1.269	2.147	2.190
Ligand binding in presence of competitive antagonist (pmoles 10^{-6} cells)	0.054	0.068	0.084	0.142	0.243	0.621	1.447	1.460

Example 1 (*cont.*)

Answer To address this problem it is first necessary to calculate the specific binding of the ligand to the receptor (B_s). The use of a large excess of unlabelled competitive antagonist enables the non-specific binding to be measured. The difference between this and the total binding gives the specific binding. Once this is known, various graphical options are open to evaluate the data. The simplest is a plot of the specific ligand binding as a function of the total ligand binding. More accurate methods are those based on linear plots such as a Scatchard plot (equation 17.4) and a Lineweaver–Burk plot (equation 17.7). In addition, it is possible to carry out a Hill plot (equation 17.9) to obtain an estimate of the Hill constant, h. The derived data for each of these three plots are shown in the following table:

	[Ligand] (nM)							
	40	60	80	120	200	500	1000	2000
Total bound ligand (pmol 10^{-6} cells)	0.284	0.365	0.421	0.547	0.756	1.269	2.147	2.190
Non-specific binding (B_{ns}) (pmol 10^{-6} cells)	0.054	0.068	0.084	0.142	0.243	0.621	1.447	1.460
Specific binding (B_s) (pmol 10^{-6} cells)	0.230	0.297	0.337	0.405	0.513	0.648	0.700	0.730
$B_s/[L] \times 10^3$ (dm^3 10^{-6} cells)	5.75	4.95	4.21	3.37	2.56	1.30	0.70	0.43
$1/[B_s]$ (pmol 10^{-6} cells)$^{-1}$	4.35	3.37	2.97	2.47	1.95	1.54	1.43	1.37
$1/[L]$ (nM)$^{-1}$	0.0250	0.0170	0.0125	0.0083	0.0050	0.0020	0.0010	0.0005
$(B_{max} - B_s)$ (pmol 10^{-6} cells)	0.52	0.45	0.413	0.345	0.237	0.102	0.050	0.020
$B_s/(B_{max} - B_s)$	0.44	0.66	0.816	1.174	2.164	6.35	14.00	36.50
log $B_s/(B_{max} - B_s)$	−0.356	−0.180	−0.088	0.070	0.335	0.803	1.146	1.562
log [L]	1.60	1.78	1.90	2.08	2.30	2.70	3.00	3.30

The hyperbolic plot allows an estimate to be made of the maximum ligand binding, B_{max}. It is approximately 0.75 pmoles 10^{-6} cells. An estimate can then be made of K_d by reading the value of [L] that gives a ligand binding value of 0.5 B_{max} (0.375 pmoles 10^{-6} cells). It gives an approximate value for K_d of 100 nM.

A Scatchard plot obtained by regression analysis gives a correlation coefficient, r, of 0.996, B_{max} of 0.786 pmoles 10^{-6} cells and K_d of 97.3 nM. A Lineweaver–Burk plot gives a correlation coefficient, r, of 0.998, B_{max} of 0.746 pmoles 10^{-6} cells and K_d of 90.5 nM. Note that there is some variation between these three sets of calculated values and the ones given by the Lineweaver–Burk plot are more likely to be correct since, as previously pointed out, the Scatchard plot overestimates both values when the binding data are subjected to linear regression analysis.

The Hill plot based on a value of B_{max} of 0.75 pmoles 10^{-6} cells, gave a correlation coefficient of 0.998 and a value for the slope of 1.13. This is equal to the Hill constant, h.

Table 17.1 **G-protein coupling of 5-hydroxtryptamine (serotonin) receptors**

Family	G-protein coupling	Response
5-HT$_1$ (5 isoforms)	G$_i$/G$_o$ coupled	Decreasing levels of cAMP
5-HT$_2$ (3 isoforms)	G$_q$/G$_{11}$ coupled	Increasing cellular levels of IP$_3$ and DAG
5-HT$_3$ (not GPCR)	Ligand-gated Na$^+$ and K$^+$ cation channel	Depolarising plasma membrane
5-HT$_4$	G$_s$ coupled	Increasing levels of cAMP
5-HT$_{5A}$	G$_i$/G$_o$ coupled?	Inhibition of adenylyl cyclase
5-HT$_6$	G$_s$ coupled	Stimulates adenylyl cyclase
5-HT$_7$	G$_s$ coupled	Increasing levels of cAMP

large, it actually represents a small fraction of the total membrane protein. This partly explains why receptor proteins are sometimes difficult to purify. From knowledge of receptor numbers and the K_d values for the agonist, it is possible to calculate the occupancy of these receptors under normal physiological concentrations of the agonist. In turn it is possible to calculate how the occupancy and the associated cellular response will respond to changes in the circulating concentration of the agonist. The percentage response change will be greater the lower the normal occupancy of the receptors. This is seen from the shape of the dose–response curve within the physiological range of the agonist concentration. It is clear that if the normal occupancy is high, the response to change in agonist concentration is small. Under such conditions, the response is likely to be larger if the receptor–agonist binding is a positively cooperative process.

Receptor subclasses

Binding studies using both agonists and inverse agonists have identified receptor subclasses for a given endogenous agonist. As an example, over a dozen types of 5-hydroxytryptamine (5-HT, serotonin) receptor have been identified (Table 17.1). An interesting feature of some receptor subclasses is that not only do they have different binding characteristics, but they may also trigger opposing cellular responses. Thus there are three subclasses (β_1, β_2 and β_3) of β-adrenergic receptors with different amino acid composition, affinities for agonists and physiological responses but all of which are activated by adrenaline and noradrenaline. β_1-Adrenergic receptors mediate cardiac responses, β_2-adrenergic receptors are involved in skeletal and smooth muscle function, and β_3-adrenergic receptors are involved in metabolic responses. Selective synthetic agonists, such as salbutamol used in the treatment of asthma, readily discriminate between the three subclasses.

Quantitative characterisation of competitive antagonists

The ability of a competitive antagonist to reduce the response of a receptor to a given concentration of agonist can be quantified in two main ways:

- IC_{50} *value*: The antagonist concentration that reduces the response of the receptor, in the absence of the antagonist, to a given concentration of agonist by 50%.
- K_b *value*: The dissociation constant for the binding of the antagonist.

To determine an IC_{50} value, the standard procedure is to study the effect of increasing concentrations of antagonist on the response to a fixed concentration of agonist. In the absence of antagonist the response will be a maximum. As the antagonist concentration is increased the response will decrease in a manner that is a mirror image of a dose–response curve (Fig. 17.1). From the curve the antagonist concentration required to reduce the response by half (IC_{50}) can be determined. If this study is repeated for a series of increasing fixed agonist concentrations it will be evident that the IC_{50} value is critically dependent on the agonist concentration used, i.e. it is not an absolute value. In spite of this, it is commonly used, because of its simplicity, particularly in the screening of potential therapeutic agents. From knowledge of the IC_{50} value, in principle it is possible to calculate the K_b value using the *Cheng–Prusoff equation*:

$$K_b = \frac{IC_{50}}{(1 + [L]/K_d)} \tag{17.12}$$

where [L] is the concentration of agonist and K_d is its dissociation constant. It is evident from this equation that IC_{50} only approximates to K_b when [L] is very small and the denominator approaches 1. Although this equation is commonly used to calculate K_b values its application is subject to reservations primarily because inhibition curves do not confirm the nature of the antagonism but also because the application of the equation is subject to the concentration of agonist used relative to its EC_{50} value. Antagonist equilibrium constants, K_b, are best determined by application of the **Schildt equation**:

$$r = 1 + [B]/K_b \tag{17.13}$$

or in its logarithmic form:

$$\log(r - 1) = \log[B] - \log K_b \tag{17.14}$$

where [B] is the concentration of antagonist, and *r* is the *dose factor* that measures the amount by which the agonist concentration needs to be increased in the presence of the antagonist to produce the same response as that obtained in the absence of antagonist, the assumption being made that the same fraction of receptors needs to be activated in the presence and absence of the antagonist to produce a given response. Experimentally, the dose ratio is equal to the dose of agonist required to give 50% response in the presence of the given antagonist concentration divided by the EC_{50} value.

To carry out a Schildt plot, the receptor response to increasing concentrations of agonist for a series of fixed concentrations of antagonist is studied and the dose ratio for each concentration of antagonist calculated. The maximum response in all cases should be the same. Equation 17.14 then predicts that a plot of the $\log(r - 1)$ against $\log[B]$ should be a straight line of slope unity with an intercept on the abscissa equal to $\log K_b$. If the slope is not unity then either the antagonist is not acting competitively or more complex interactions are occurring between the antagonist and the receptor, possibly allosteric in nature. It is important to note from the Schild equation that the value of K_b, unlike that of IC_{50}, is independent of the precise agonist used to generate the data and is purely a characteristic of the antagonist for the specific receptor. A limitation of a Schildt plot is that any error in measuring EC_{50} (i.e. the response in the absence of antagonist) will automatically influence the value of all the derived dose ratios. The intercept of a Schildt plot on the y-axis also gives the pA_2 value for the antagonist (p for negative logarithm; A for antagonist; and 2 for the dose ratio when the concentration of antagonist equals pA_2). The pA_2 will be equal to the value of $-\log K_b$ since at an antagonist concentration that gives a dose ratio of 2, the Schildt equation reduces to $\log[B] = \log K_b$. pA_2 is a measure of the potency of the antagonist. Software programs are commercially available for the analysis of ligand-binding data. Examples include: *Prism* (www.graphpad.com/curvefit); *Sigmaplot* (www.systat.com/products/sigmaplot); *Origin8* and *OriginPro 8* (www.originlab.com) and *Calcusyn* (www.biosoft.com).

Example 2 **SCHILDT PLOT – CALCULATION OF A K_b AND pA_2 VALUE**

Question Use the data in Fig. 17.2 to construct a Schildt plot. The left-hand plot is the receptor response to agonist binding in the absence of antagonist. The next three responses are in the presence of 10^{-7} M, 10^{-6} M and 10^{-5} M antagonist. Read off from the graph the concentration of agonist required to produce 50% maximum response in the absence and presence of the antagonist and calculate the dose ratio (r) at each of the three antagonist concentrations. Then plot a graph of $\log(r - 1)$ against $\log$ [antagonist] and hence calculate both pA_2 and K_b.

[Antagonist] (M)	10^{-7}	10^{-6}	10^{-5}
EC_{50} (M)	10^{-7}	10^{-6}	10^{-5}
r	10	100	1000
$r - 1$	9	99	999
$\log (r - 1)$	0.954	1.9956	2.999
$\log [B]$	-7	-6	-5

Answer You will see that the Schildt plot is linear ($r = 0.9999$) and that it has a slope of 1.02 confirming the competitive nature of the antagonist. The extrapolation of the line to the y-axis gives a value of -8.12. This is equal to $-\log K_b$, hence $K_b = 7.58 \times 10^{-9}$ M and $pA_2 = 8.12$.

17.3 LIGAND-BINDING AND CELL-SIGNALLING STUDIES

17.3.1 Selection of ligand and receptor preparation

Receptor preparations

Preparations of receptors for ligand-binding studies may either leave the membrane intact or involve the disruption of the membrane and the release of the receptor with or without membrane fragments, some of which could form vesicles with variable receptor orientation and control mechanisms. Membrane receptor proteins show no or very little ligand-binding properties in the absence of phospholipid so that if a purified receptor protein is chosen, it must be introduced into a phospholipid vesicle for binding study purposes. The range of receptor preparations available for binding studies is shown in Table 17.2.

Kinetic studies aimed at the determination of individual rate constants are best carried out using isolated cells whilst studies of the number of receptors in intact tissue are best achieved by labelling the receptors with a radiolabel preferably using an irreversible competitive antagonist and applying the technique of quantitative autoradiography.

Ligands

A common technique for the study of ligand–receptor interaction is the use of a radiolabelled ligand with isotopes such as ^{3}H, ^{14}C, ^{32}P, ^{35}S and ^{135}I. Generally a high-specific-activity ligand is used as this minimises the problem of non-specific binding (see below). If a large number of ligands are being studied, such as in the screening of potential new therapeutic agents, the cost and time of producing the radiolabelled forms becomes virtually prohibitive and experimental techniques such as fluorescence spectroscopy and surface plasmon resonance spectroscopy have to be used. However, the use of radiolabelled ligands remains attractive as a means of distinguishing between orthosteric and allosteric ligands.

The technique of using radiolabelled ligand generally requires the separation of bound and unbound ligand once equilibrium has been achieved. This is most commonly achieved by techniques such as equilibrium dialysis and ultrafiltration exploiting the inability of receptor-bound ligand to cross a semi-permeable membrane, and by simple centrifugation exploiting the ability of the receptor-bound ligand to be pelleted by an applied small centrifugal field.

For the study of ligand–receptor interactions that occur on a sub-millisecond timescale, special approaches such as stopped-flow and quench-flow methods (Section 15.3.3) need to be adopted to deliver the ligand to the receptor. An alternative approach is the use of so-called caged compounds. These possess no inherent ligand properties but on laser flash photolysis with light of a specific wavelength, a protecting group masking a key functional group is instantaneously cleaved releasing the active ligand.

Table 17.2 **Receptor preparations for the study of receptor–ligand binding**

Receptor preparation	Comments
Tissue slices	5–50 μm thick, generally adhered to gelatine-coated glass slide. Good for study of receptor distribution.
Cell membrane preparation	Disrupt cells (from tissue or cultures) by sonication and isolate membrane fraction by centrifugation. Lack of cytoplasmic components may compromise receptor function. Used to study ligand binding and receptor distribution in lipid rafts and caveolae. Increasingly commonly used with cell lines transfected with human receptor genes.
Solubilised receptor preparation	Disrupt membrane with detergents and purify receptors by affinity chromatography using an immobilised competitive antagonist. Isolation from other membrane components may compromise studies.
Isolated cells	Release cells from tissue by mechanical or enzymatic (collagenase, trypsin) means. Cells may be in suspension or monolayers. May be complicated by presence of several cell types. Widely used for the study of a range of receptor functions. Allows ligand binding and cellular functional responses to be studied under the same experimental conditions.
Cultured cell lines	Very popular. Has advantage of cell homogeneity and ease of replication.
Recombinant receptors	Produced by cloning or mutagenesis techniques and inserted into specific cultured cell line including ones of human origin. Popular for the study of the effect of mutations on receptor function such as constitutive activity (CAMs) and cell signalling. Care needed to ensure that receptors have same functional characteristics (e.g. post-translational modification) as native cells.

17.3.2 Experimental procedures for ligand-binding studies

The general experimental approach for studying the kinetics of receptor–ligand binding and hence to determine the experimental values of the binding constants and the total number of binding sites is to incubate the receptor preparation with the ligand under defined conditions of temperature, pH and ionic concentration for a specific period of time that is sufficient to allow equilibrium to be attained. The importance of allowing the system to reach equilibrium cannot be overstated as equations 17.1 to 17.11 do not hold if equilibrium has not been attained. Using an appropriate analytical procedure, the bound and unbound forms of the ligand are then quantified or some associated change measured. This quantification may necessitate the separation of the bound and unbound fractions. The study is then repeated for a

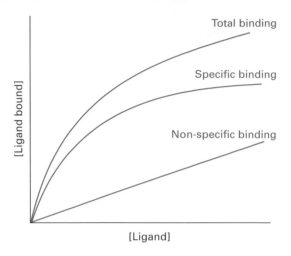

Fig. 17.5 Specific and non-specific binding of a ligand to a membrane receptor. Specific binding is normally hyperbolic and shows saturation. Non-specific binding is linear and is not readily saturated.

series of ligand concentrations to cover 10–90% of maximum binding at a fixed receptor concentration. The binding data are then analysed using equations 17.4 to 17.11 often in the form of a computer program many of which are available commercially. Most perform linear or non-linear least-squares regression analysis of the experimental data. Measurements may be made on a steady-state (single measurement) basis or a time-resolved (multiple measurements over a period of time) basis most commonly by stopped-flow or quench-flow procedures (Section 15.3.3).

Non-specific binding

A general problem in the study of receptor–ligand binding is the non-specific binding of the ligand to sites other than the orthosteric or allosteric binding sites. Such non-specific binding may involve the membrane lipids and other proteins either located in the membrane or released by the isolation procedure. The characteristic of non-specific binding is that it is non-saturable but is related approximately linearly to the total concentration of the ligand. Thus the observed ligand binding is the sum of the saturable (hyperbolic) specific binding to the receptor and the non-saturable (linear) binding to miscellaneous sites. The specific binding component is usually obtained indirectly either by carrying out the binding studies in the presence of an excess of non-labelled ligand (agonist or antagonist) if a labelled ligand is being used, or by using a large excess of an agonist or competitive antagonist in other studies. The presence of the excess unlabelled or competitive ligand will result in the specific binding sites not being available to the ligand under study and hence its binding would be confined to non-specific sites (Fig. 17.5). In practice, a concentration of the competitive ligand of at least 1000 times its K_d or K_b must be used and confirmation that under the conditions of the experiment, non-specific binding was being studied would be sought by repeating the study using a range of different and structurally dissimilar competitive ligands that should give consistent estimates of the non-specific binding.

Experimental techniques for the study of the binding of a ligand to a membrane receptor

Equations 17.1 to 17.11 allow the calculation of quantitative parameters that characterise the binding of a ligand to a receptor. Such parameters are fundamental to the understanding of the mechanism of the binding and its relationship to the subsequent cellular response. They also allow comparisons to be made of the comparative efficacy and affinity of a series of ligands for a common receptor, a process that is essential in the development of new drugs (Chapter 18). Numerous techniques are available for the study of ligand binding but those that are amenable to automation, do not require the bound and unbound fraction of ligand to be separated and do not require the use of radiolabelled ligand are generally the preferred methods. Examples are as follows.

Fluorescence spectroscopy

Fluorescence-based techniques are ideal for the study of ligand–receptor binding as they are ultra sensitive, being capable of studying binding involving a few or even individual ligand molecules and single receptors. The general principles of fluorescence spectroscopy are discussed in Chapter 12. The methods are based on either changes in the intrinsic fluorescence of the receptor protein tagged with a suitable fluorescent marker (fluor) or the induction of fluorescence in either the ligand or the receptor protein as a result of receptor–ligand binding. Commonly used fluors include fluorescein, rhodamine and the dye Fluo-3 (Table 4.3.) but a better alternative is the green fluorescent protein (GFP) of the jellyfish *Aequorea victoria* or the red fluorescent protein of *Discosoma striata* either of which can be attached to receptor proteins by gene cloning without altering the normal function of the protein. The main advantage of using either of these two autofluorescent proteins is that no cofactors are required for fluorescence to occur hence the study protocols are relatively simple.

The most common forms of fluorescence spectroscopy applied to the study of receptor-ligand binding are:

- *Fluorescence resonance energy transfer* (FRET): This relies on the presence of two fluors in distinct locations within the receptor protein such that the emission spectrum of one and the excitation spectrum of the other overlap. In such circumstances, the emission light of one fluor may be absorbed by the second (hence energy transfer) and be emitted as part of its emission. The extent to which this may occur is proportional to $1/R^6$, where R is the distance between the two fluors and which is changed as a result of ligand binding.
- *Fluorescence anisotropy*: Anisotropy is the directional variation in optical properties, in this case fluorescence, along perpendicular and parallel axes. In this technique fluorescence is induced by plane-polarised (blue) light. Molecules of the fluor orientated parallel to this plane of polarisation will be excited preferentially. However, if some of these molecules rotate after the absorption of the light but before the fluorescence has time to occur some of the resulting fluorescence will be depolarised (i.e. no longer in one plane). The extent to which this occurs can be used to deduce information about the size, shape and flexibility of the protein carrying the fluor. It can also be used to monitor the binding of a ligand to the protein. The fluorescence

intensity is measured parallel (i.e. in the same plane) to the absorbed plane-polarised light and at right angles to it. From the two measurements it is possible to calculate the degree of fluorescence depolarisation and hence the fluorescence anisotropy, both of which are expressed in terms of the difference between the fluorescence parallel to the absorbed plane-polarised light and that perpendicular to it, the difference being expressed as a function of the sum of the fluorescence in the two planes.

- *Fluorescence cross-correlation spectroscopy* (FCCS): This technique differs from other forms of fluorescence spectroscopy in that it is not primarily concerned with fluorescence intensity, but rather with small spontaneous fluorescence fluctuations induced by molecules diffusing into and out of a small focal volume in aqueous solution. Such fluctuations are related to changes in the diffusion coefficient of each probe and hence can be correlated with receptor–ligand binding. Both the receptor and the ligand are labelled with spectrally distinct fluors and their interaction studied by confocal microscopy (Section 4.3.1). Two lasers are aligned to the same confocal point and used to excite the two fluors. Following binding, the receptor–ligand complex emits in both fluorescent wavelengths that are monitored to give a cross-correlation signal that is directly related to the concentration of the receptor–ligand complex.

Surface plasmon resonance (SPR) spectroscopy

The principles and experimental details of this technique are discussed in Section 13.3. The advantages of the technique are that it does not require the molecules to be fluorescent or radiolabelled; it can be used to study molecules as small as 100 Da and can be used with coloured or opaque solutions. Such studies with G-protein-coupled receptors have shown that agonists, inverse agonists and antagonists can readily be distinguished by the conformational changes they induce in the membrane in the region of the receptor. Specifically, agonists and inverse agonists increase membrane thickness (agonists more so than inverse agonists) by causing an elongation of the receptor whereas antagonists cause no change. Plasmon-waveguide resonance (PWR) spectroscopy is closely related to SPR spectroscopy. It is more sensitive than SPR spectroscopy and can be used to study receptor conformational changes in lipid bilayers. Thus a single lipid bilayer is deposited on the resonator surface and the receptor protein inserted from a detergent-solubilised solution. A solution containing the ligand is then passed over the layer allowing the binding process to be studied.

Isothermal titration calorimetry

This method is based on the measurement of the heat change, positive or negative, associated with the binding and of the relationship between the enthalpy change ($\triangle H$), Gibbs' free energy change ($\triangle G$), the entropy change ($\triangle S$), the number of molecules of ligand bound to each receptor protein (stoichiometry) (n) and the binding constant (K_a). The experimental details are discussed in Section 15.3.3. Its practical advantages are that it can be applied to the study of any receptor–ligand pair without the need for radiolabelling or the attachment of fluors, or the need to separate bound and unbound fractions.

Protein microarray technology

This approach to the study of receptor–ligand binding exploits the principle that assay systems that use a small amount of capture molecules (the ligand) and a small amount of target molecules (the receptor) can be more sensitive than systems that use a hundred times more material (Section 8.5.5). In this miniaturisation approach, the ligand is immobilised onto a small area of a solid phase, commonly a derivatised glass slide. The resulting 'microspot' contains a high density (concentration) of ligand but a very small amount of it. It is then incubated with the receptor, commonly fluorescently tagged, resulting in the binding of some of the receptor molecules. Since the microspot covers a small area there is effectively no change in the concentration of the unbound ligand in the sample even if its concentration was low and the binding affinity was high. This is true provided that $<0.1\ K_d$ of the ligand molecules are bound in the complex, where K_d is the dissociation constant for the complex. The ligand–receptor complex is then quantified by fluorimetric methods and the procedure repeated for a series of increasing microspot sizes (increasing ligand concentration) but such that the density of the capture molecules is constant. After each incubation excess receptor protein is washed away and the remaining complex analysed by surface-enhanced laser desorption/ionisation (SELDI) mass spectrometry, a variant of MALDI (Section 9.3.8). In practice, microspots are immobilised in rows on the solid support allowing the simultaneous analysis of hundreds or even thousands of samples.

Other methods used to study ligand–receptor binding

Other experimental methods that can be used to quantify receptor–ligand binding and hence characterise both the receptor and the ligand include:

- analytical ultracentrifugation by the sedimentation velocity (Section 3.5.1),
- scintillation proximity assay (Section 14.3.2) commonly using ^{3}H- or ^{125}I-labelled ligand,
- NMR (Section 13.5) observing either changes in the signal of the protein or of the ligand induced by receptor–ligand binding,
- X-ray crystallography (Section 13.6) either by co-crystallising the receptor and ligand or by soaking crystals of the receptor in solutions of the ligand.

17.4 MECHANISMS OF SIGNAL TRANSDUCTION

17.4.1 Cell signalling assays

In order to determine the details of a specific transduction pathway it is necessary to identify the proteins and other small effectors whose activity or concentration change in response to the activation of the receptor. This is possible using functional screening assays that are of two main types: cell-free (biochemical) and cell-based (cellular) (see Section 18.2.3 for further details). In practice, whilst cell-free systems are rapid and relatively easy to interpret, cellular assays have several advantages:

- they do not require any pre-purification of the receptor protein which make them attractive for the screening of so-called orphan receptors whose cellular function and physiological ligand have yet to be determined;
- the conformation and hence the activity of the receptor protein is most likely to better reflect the physiological situation than is the case for biochemical cell-free assays;
- their use allows screening for potential therapeutic agents so as to quickly identify and reject compounds that possess any cytotoxic properties;
- they are readily adaptable to robotically controlled high-throughput screening or high-content screening protocols both of which are commonly based on fluorescence microscopy using cells in microtitre plates (Section 18.2.3).

Much of the work using these assays is aimed at the discovery of new drugs and so cells of human origin are the preferred targets. In practice such cells are expensive to culture and are not readily adapted to automated assays. For such reasons, micro-organisms are commonly used, with yeast being most commonly chosen because of the high degree of conservation of basic molecular and cellular mechanisms between yeast and human cells. Moreover, it is relatively easy to engineer yeast cells to incorporate human receptor proteins such as GPCRs, RTKs and ion-channels.

Numerous analytical techniques have been used to probe signalling pathways (Fig. 17.6). The majority are adaptations of standard techniques for studying protein structure, protein–ligand interaction and protein–protein interaction. As a result of their high sensitivity, fluorimetric techniques, particularly fluorescence resonance energy transfer and fluorescence correlation spectroscopy, feature prominently in these techniques. Techniques are now available for the site-specific post-translational labelling of proteins with small fluorescent tags. Commercial companies offer reagents, kits and services to facilitate the rapid identification of proteins associated with the activation of a particular receptor. The most common approaches include:

- The use of monoclonal antibodies for western blots, protein purification and immunocytochemistry. Western blots can identify changes in the expression of a specific protein in the pathway; immunocytochemistry can detect movement within the cell as a result of the activation of a pathway and the use of phosphospecific antibodies can detect the phosphorylation of a particular protein.
- The use of *knock-out* and *knock-in* strategies using mice or cell lines, commonly embryonic stem cells. In these techniques either the endogenous locus of the receptor gene is manipulated or a modified receptor gene with an appropriate promoter is expressed in the host. The effect of such action is then studied by techniques such as Southern and western blots.
- The use of microarray techniques – 'chips' are now commercially available on which all the proteins of the *Saccharomyces cerevisiae* proteome or a large proportion of those of the human proteome have been individually deposited (see Sections 8.5.4 and 8.5.5 for further details). The approach is particularly suitable for identifying

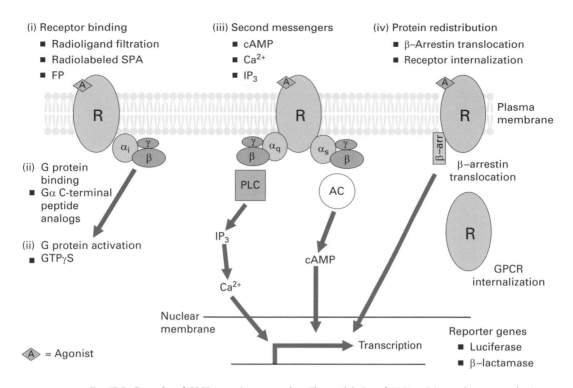

(i) Receiver binding
- Radioligand filtration
- Radiolabeled SPA
- FP

(ii) G protein binding
- Gα C-terminal peptide analogs

(ii) G protein activation
- GTPγS

(iii) Second messengers
- cAMP
- Ca²⁺
- IP₃

(iv) Protein redistribution
- β–Arrestin translocation
- Receptor internalization

Plasma membrane

β–arrestin translocation

GPCR internalization

Nuclear membrane

Transcription

Reporter genes
- Luciferase
- β–lactamase

= Agonist

Fig. 17.6. Examples of GPCR screening approaches. The modulation of GPCR activity can be measured using a variety of assays, including (i) receptor binding; (ii) G-protein binding and activation; (iii) second-messenger signalling, including reporter systems that measure the response downstream; and (iv) protein redistribution. (i) Receptor binding can be assayed using filtration binding of a radiolabelled ligand using a radioactive ligand and a scintillation proximity assay (SPA), or using a fluorescently labelled ligand and fluorescence polarisation (FP) to monitor specific binding between the ligand and the receptor. Receptor activity can also be monitored using (ii) G-protein binding and activation with high-affinity peptides that mimic the Gα C-terminus or using GTPγS assays. Subsequently, Gα and Gβγ go on to activate their effector molecules, and (iii) second-messenger signalling pathways can be monitored. For example, the Gα$_q$ subunit can interact with phospholipase C (PLC), causing enzyme activation and the catalysis of phosphotidylinositol-4,5-bisphosphate to inositol-1,4, 5-trisphosphate (IP₃) which can then mobilise Ca²⁺. Alternatively, the Gα$_s$ subunit can interact with and promote the activity of adenylyl cyclase (AC), resulting in the production of cAMP. When assaying for (iv) protein redistribution, either the translocation of β-arrestin (β-arr) or the internalisation of the GPCR can be monitored. (Reproduced from Gilchrist, A. (2007). Modulating G-protein-coupled receptors: from traditional pharmacology to allosterics. *Trends in Pharmacological Sciences*, **28**, 431–438, by permission of Elsevier Science.)

the likely substrates of the very large number of protein kinases involved in cell signalling and to study the role of ubiquitination in receptor recycling.

- The use of the enzyme fragment complementation (EFC) technique. This is commonly carried out using the enzyme β-galactosidase. The technique uses a small fragment of the enzyme and an inactive deletion mutant of the enzyme both of which are enzymically inactive. The two forms are attached either by chemical conjugation or by a recombination procedure to two proteins suspected of interacting in a particular signalling pathway. If the two proteins do interact they allow the two attached galactosidase forms to complement each other generating a heteromeric complex capable of acting on an added substrate of the enzyme to release a fluorescent or chemiluminescent signal that can readily be detected and quantified. The technique has been particularly successful in the study of pathways that result in a nuclear signal since the resulting fluorescent signal will be located in the nucleus. In an alternative format, EFC can be used to monitor second messengers such as cAMP and hence be used to study GPCRs. In this case, cAMP is chemically conjugated to the galactosidase fragment and an antibody that binds to the conjugated cAMP added to the assay. The antibody sterically prevents the two galactosidase forms combining to form the active enzyme and hence produce a signal. However, as free cAMP is generated by the GPCR–adenylyl cyclase system it displaces the conjugated cAMP from the antibody allowing the two galactosidase fragments to complement and generate a signal thus allowing the cAMP to be quantified. Similar assays have been developed for tyrosine kinases and serine threonine kinases and the technique can be used to monitor protein expression changes in response to receptor activation. The assays are easily miniaturised and automated to give high throughput precision data.

17.4.2 Signal transduction through ligand-gated ion channels

Ligand-gated ion channels constitute one of the mechanisms for the control of the transmembrane movement of ions down their concentration gradient resulting in a change in membrane potential. This control of ion movement is exerted on the basis of ion type (anion or cation), ion charge and ion size. Binding of the ligand to the resting state of the receptor induces a conformational change in the receptor protein that results in the opening of the channel and the movement (gating) of ions. The channel remains open until either the ligand is removed or when, in the continued presence of the ligand, the receptor protein changes to its desensitised state in which the channel is closed. Since this mechanism of transduction is independent of any other membrane component or intracellular molecule, the cellular response to ligand binding is almost instantaneous. This class of membrane receptors includes numerous receptors that are involved in signal transmission between neurons, between glia and neurons and between neurons and muscles.

Four superfamilies of ligand-gated ion channels, classified on the basis of the number of transmembrane (TM) segments within the subunits (2TM, 3TM, 4TM and 6TM) have been identified. The 4TM family has been the most thoroughly investigated and all members shown to consist of five subunits (pentameric). Members of the family include:

- nicotinic acetylcholine receptors (nAChR) that are the primary excitatory receptors in skeletal muscle and the peripheral nervous system;
- serotonin (5-hydroxytryptamine) (5HT) receptors located in neurons;
- γ-aminobutyric acid receptors (GABA$_A$ and GABA$_B$) found in the cortex and which are inhibitory transmitters;
- glycine receptors found in the spinal cord and brainstem and which like the GABA receptors are inhibitory transmitters.

The five subunits of the nAChR receptor are of four types, α, β, γ and δ with a stoichiometry of $\alpha_2\beta\gamma\delta$. All five span the membrane four times mainly with α-helical structure, but with some β-structure. Each α-helical region has been designated TM1–TM4 and the experimental evidence supports the view that the TM2 regions of each of the five subunits line the ion channel with the TM1 and TM3 regions forming a scaffold to support the channel. Binding studies have provided evidence for the allosteric binding of two molecules of agonist with a Hill constant of about 2. Affinity labelling studies have shown that the ligand-binding sites are in a cleft between two subunits. In both cases an α-subunit has the principal role in binding with the γ and δ subunits playing a minor role. The first ligand molecule binds to the α_1 subunit which is in contact with the δ subunit and the second ligand molecule to α_2 which is in contact with the γ subunit. Thus the two binding sites are not structurally identical. Genetically engineered variants of the five subunits have shown that the absence of the α subunit results in the lack of binding of acetylcholine thereby confirming the importance of this subunit.

Electron microscopic study of the nAChR on the *Torpedo* electric organ postsynaptic membrane has given an indication of the three-dimensional structure of the channel. It is funnel-shaped with a large proportion of the receptor outside the membrane protruding into the postsynaptic cleft. The channel is 25–30 Å wide at the entrance and only 6.4 Å wide at its narrowest point. Three rings of negatively charged amino acid residues, all on the four TM2 helices, line the narrow part of the channel and appear to determine its selectivity. The importance of these amino acid residues has been confirmed by mutagenicity studies. Detailed electrophysiological studies have revealed that channel opening involves the subtle rearrangement of three transmembrane α-helices, one in each of the TM1, TM2 and TM3 segments. The whole process is kinetically complex but the essential features may be represented as follows:

$$\text{channel}: \quad \begin{array}{ccccccccc} A + R & \rightleftharpoons & AR + A & \rightleftharpoons & A_2R & \rightleftharpoons & A_2R^* & \rightleftharpoons & A_2R^{**} \\ \text{closed} & & \text{closed} & & \text{closed} & & \text{open} & & \text{desensitised} \end{array}$$

where A is acetylcholine and R the receptor containing two binding sites for acetylcholine, one on each α subunit.

Measurement of the numerous rate constants for these reversible processes has revealed that the rate constant for the opening of the channel is greater than the rate constants for the corresponding reverse process (i.e. the reversion to the closed conformation), for the dissociation of a ligand molecule from the closed conformation and for the transition from the open active state to the desensitised state. The consequence of this is that many opening and closing events of the channel occur before either the transition to the desensitised state or a molecule of ligand dissociates from the binding site. The mean channel open time for torpedo nAChR is 3.0 ms and the mean closed time 94 μs within the bursts of opening and closing activity. The desensitised receptor (R^{**}) eventually reverts to the closed resting state (R).

Desensitisation may be linked to phosphorylation. All five subunits contain amino acid residues located between the TM3 and TM4 regions that are potential sites for phosphorylation, and phosphorylation of the receptor has been shown to occur at two serine residues on each of the γ and δ subunits. Each phosphate group introduces two negative oxygen atoms that could induce important conformational changes in the receptor structure and desensitisation. Mutagenesis studies of these serine residues have shown that their replacement by non-polar amino acids minimises the susceptibility of the receptor to acetylcholine-induced desensitisation. In contrast, replacement of the serines by glutamates, which contain negatively charged carboxyl groups, permanently desensitises the receptor. This phosphorylation-induced modulation of receptor function is found in many other types of ion-transport proteins indicating a common mechanism, but it is not clear whether or not phosphorylation is an essential prerequisite for receptor desensitisation.

17.4.3 Signal transduction through G-protein-coupled receptors (GPCRs)

The fact that over 800 human G-protein-coupled receptors have been identified and verified emphasises their cellular importance. They are the largest group of signalling transduction receptors and are responsible for a wide range of physiological processes ranging from the transmission of light and odorant signals to hormonal action and neurotransmission. Dysregulation of GPCRs is associated with several clinical conditions such that many currently used drugs target these receptors (Section 18.2).

Although these receptors share a common 7TM structure (Fig. 17.7), their agonist-binding (orthosteric) domains vary considerably. For small agonists (e.g. adrenaline, histamine, dopamine, serotonin) the domain is partially embedded within a transmembrane helical structure but for large agonists, including the neuropeptides and chemokines, the domain may span the extracellular loops or be located near the N-terminal region.

Based on phylogenetic criteria, GPCRs have been classified into five families:

- *Rhodopsin family*: This is the largest and most widely studied family. It currently contains 672 verified members which are highly heterogeneous and have therefore been divided into four subgroups, α, β, γ and δ. The α subgroup contains many receptors that are involved in basic physiological functions and hence are targets for drug therapy. Examples include the H_1 and H_2 histamine receptors, the serotonin receptors 5-HT_{1A}, 5-HT_{1D} and 5-HT_{2A}, the adrenergic receptors 1A, 2A, B1 and B2 and muscarinic receptors.
- *Secretin family*: This family is characterised by a long N-terminal tail containing six cysteine residues linked by disulphide bridges and involved in ligand binding. The family has 15 members which include the glucagon, calcitonin and secretin receptors. They all activate adenylyl cyclase and couple through the same G_s protein. Their overexpression is commonly linked to human tumours.
- *Adhesion family*: This family contains 33 members all of which bind cell adhesion molecules such as integrins, cadherins and selectins, which are involved in the control of mitogenesis, differentiation and the immune system.
- *Glutamate family*: This family has 22 members all of which bind either glutamate or GABA. They are involved in neurotransmission in the brain by controlling the movement of ions such as Ca^{2+}, Na^+ and Cl^- through ion channels. Their action may be either excitatory or inhibitory. Members include the so-called metabotropic glutamate receptors (mGluR) of which eight have been identified and each shown

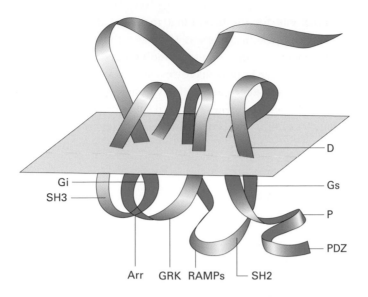

Fig. 17.7 Schematic diagram of a hypothetical G-protein-coupled receptor. Labels denote general regions of interaction of the receptor with other cellular proteins including different G-proteins (G_i and G_s), PDZ, SH2 and SH3-domain proteins, receptor-activity-modifying proteins (RAMPs), arrestin (Arr), G-protein-coupled receptor kinase (GRK), sites for dimerisation with other GPCRs (D), and phosphorylation sites that lead to uncoupling and internalisation (P). Any one of these active processes could be considered a form of expression of efficacy. The figure is a general description of various loci for protein interactions, but does not represent accurate locations, as, in most cases, these are not well characterised. (Reproduced from Kenakin, T. (2002). Efficacy at G-protein-coupled receptors. *Nature Reviews Drug Discovery*, **1**, 103–110, by permission of Nature Publishing Group.)

to be distributed in specific regions of the brain. In the resting state the channels are closed, but the binding of glutamate leads to the production of a second messenger that triggers the opening of the channel and the movement of ions. The channel then becomes desensitised and closes as the glutamate begins to dissociate returning the receptor to its resting state. The receptors are targets for a number of drugs such as the benzodiazepines. mGluRs contain an allosteric site that is a potential target for drugs involved in the treatment of Parkinson's disease (mGluR4), schizophrenia (mGluR5) and addiction (mGluR2).

- *Frizzled/Taste family*: These two subgroups of the same family have 11 and 25 members respectively. The Frizzled receptors are named after the *Drosophila* tissue polarity gene known as *frizzled*. The receptors bind Wnt glycoproteins involved in transduction pathways for the control of gene expression in embryonic development and regulation.

G-protein structure

G-proteins are heterotrimeric, consisting of one of each of three subunits, α (40–45 kDa), β (36–40 kDa) and γ (8 kDa), which are loosely attached to the inner surface of the cellular membrane through lipophilic tails on the α and γ subunits. The β and γ subunits are firmly attached to each other but the linkage to the α subunit is weaker. In addition to the binding site for the β subunit, the α subunit has a binding site near the N-terminal end for the C-terminal region of the receptor and a guanine nucleotide-binding site that also possess GTPase activity. Although the α subunit contains the binding site for the receptor, binding only occurs when the α subunit is bound to the $\beta\gamma$ dimer. Studies have revealed a very complex picture of the G-proteins. Sixteen human genes and splice variants encode α subunits giving rise to at least 28 distinct types; five genes encode β subunits and 14 genes the γ subunits. The potential number of different $G\alpha\beta\gamma$ functional trimers is therefore very large.

G-protein subgroups

The most important are:

- the G_s subgroup which stimulate adenylyl cyclase. It includes G_{olf} coupled to olfactory receptors;
- the G_i subgroup which inhibit adenylyl cyclase and activate some Ca^{2+} and K^+ channels;
- the G_q subgroup which couple receptors to calcium mobilisation through phospholipase C_β that in turn generates the two second messengers inositol trisphosphate (IP_3) and diacylglycerol (DAG);
- the G_o subgroup which reduce the probability of opening of some voltage-gated Ca^{2+} channels involved in neurotransmitter release;
- the G_t subgroup which stimulate phosphodiesterase following light stimulation of the retina involving transducin; and
- the $G_{12/13}$ subgroup which is involved in the regulation of the cytoskeleton and processes related to movement. The subgroup activates inducible nitric oxide synthetase and the Na^+/H^+ exchanger. The action involves the low-molecular-weight protein Rho.

The G-protein α subunit family is divided into four subgroups based on sequence homology: $G_S\alpha$, $G_{i/o}\alpha$, $G_q\alpha$ and $G_{12/13}\alpha$. Each subgroup has been further divided into specific isotypes. The $G\alpha$ subunit nomenclature is used to classify GPCRs, hence GPCRs are referred to as G_S-, G_i- or G_q- etc. reflecting their primary signalling pathway as discussed below. Since the number of human GPCRs is far greater than the number of human G-proteins it is obvious that each member of the $G\alpha$ subgroups must be able to interact with many GPCRs.

The G-protein cycle

Agonist binding to the GPCR triggers a G-protein cycle:

- In the normal resting state, the trimeric G-protein has a molecule of GDP bound to the α subunit. At this stage the G-protein is not coupled to the receptor but it is firmly attached to the inner face of the cell membrane.
- An agonist binds to its binding site on the GPCR and induces a rapid allosteric conformational change that activates the G-protein binding site located in intracellular loops, resulting in a GPCR–G-protein complex. The complex interacts by diffusion translocation with a G-protein–adaptor complex that binds to the α subunit.

- Binding of the GPCR–G-protein complex to the G-protein–adaptor complex induces an allosteric conformational change in the guanine nucleotide-binding site on the α subunit of the G-protein causing the site to be more accessible to the cytosol where [GTP] > [GDP] resulting in the dissociation of the GDP and the formation of a transient 'empty state'.
- GTP binding to the nucleotide-binding site on the α subunit triggers a second, rapid conformational change in the α subunit causing the dissociation of GTPα subunit and leaving the G$\beta\gamma$ subunits as a dimer. Both the GTPα subunit and the G$\beta\gamma$ dimer remain attached to the cell membrane.
- GTPα subunit and/or the G$\beta\gamma$ dimer bind to an inactive effector molecule causing its activation or inhibition. Examples of such effector molecules include adenylyl cyclase, phospholipase C, cyclic nucleotide phosphodiesterases and a number of ion channels.
- Hydrolysis of GTP to GDP by the GTPase site of the α subunit, with the involvement of RGS proteins (see below for further details), terminates the activation or inhibition by reversing the conformational change originally induced by the receptor–agonist complex. This facilitates the dissociation of the α subunit from the effector and its reassociation with the G$\beta\gamma$ dimer, thus completing the cycle. Concomitantly, the binding of the receptor–agonist complex to the G-protein reduces the affinity of the receptor for its agonist encouraging its dissociation from its binding site resulting in the reformation of the inactive conformation of the receptor and hence terminating its binding to the G-protein.

Two important examples of the role of Gα-GTP as a transducer are the activation of the key enzymes adenylyl cyclase, that converts ATP to the second messenger cAMP, and phosphodiesterase (phospholipase C) that cleaves phosphatidylinositol-4, 5-bisphosphate (PIP_2), a component of the cytoplasmic side of the cell membrane, to two second messengers – inositol-1,4,5-trisphosphate and diacylglycerol. Most examples of the transducer role of the G$\beta\gamma$ dimer are linked to the activation of G$_i$ and G$_o$. Examples include the activation of β-adrenergic receptor kinase (βARK), phospholipase A$_2$ and the K$^+$ channel GIRK (G-protein-activated inwardly rectifying potassium channel).

A given G-protein may be activated by a large number of different receptors (referred to as G-protein promiscuity), whilst a given receptor may interact with different G-proteins and/or produce more than one response (referred to as receptor promiscuity). A receptor capable of activating more than one type of G-protein and hence of initiating more than one response is referred to as a pleiotropic receptor (meaning it has multiple phenotypic expressions). An example is the human adenosine receptor that can couple to G$_i$, G$_s$ and G$_q$. Some GPCRs are capable of binding several agonists each of which can induce a specific conformational change that preferentially selects a specific G-protein that in turn leads it to activate a specific transduction pathway. Examples are the 5-HT receptors, 13 of which have been identified to date and 12 shown to be GPCRs (Table 17.1). Such multiple roles for a given agonist have lead to the concepts of functional selectivity (Fig. 17.8) and biased agonism (Fig. 17.9).

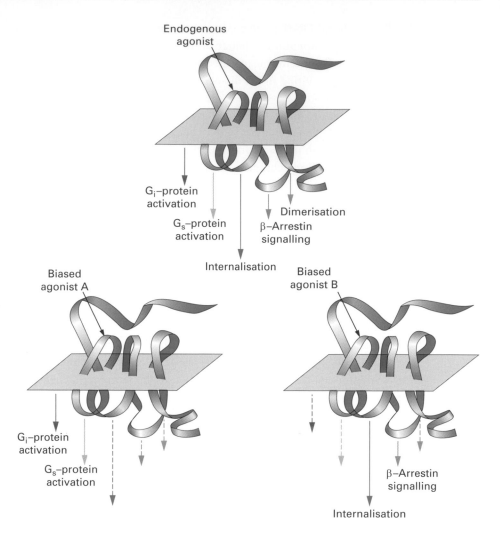

Fig. 17.8 Biased agonism. Whereas a natural agonist can activate the receptor to express all of its behaviour towards its cellular host, some agonists can stabilise active receptor conformations that trigger only some of these behaviours. Thus biased agonist A activates primary G-protein signalling pathways, whereas biased agonist B activates mainly G-protein-independent β-arrestin signalling and internalisation. (Reproduced from Kenakin, T. (2007). Collateral efficacy in drug discovery: taking advantage of the good (allosteric) nature of 7TM receptors. *Trends in Pharmacological Sciences*, **28**, 407–412, by permission of Elsevier Science.)

GPCR dimerisation

Evidence has been obtained to indicate that GPCRs interact with other GPCRs to form either homo- or heterodimers. In the case of the receptors in families B and C, the evidence is that the functional activity (phenotype) of the receptors is linked to these forms but in the case of family A receptors this link is less clear. The functional unit of both the mGluR1 and GABA$_B$ receptors in family C, for example, is a homodimer and X-ray crystallographic studies on the glutamate receptor have indicated that the dimer exists as a dynamic equilibrium between two conformations, one 'open' the

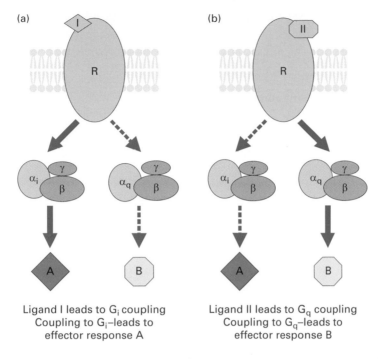

Fig. 17.9 Functional selectivity. GPCRs often couple to multiple G-proteins. Functional selectivity is seen when ligand binding influences which G-protein associates with the receptor by promoting distinct coupling efficiencies. For example, binding of a distinct ligand (I) leads to activation of G_i and effector responses initiated through this G protein (a), whereas binding of a different ligand (II) to the same GPCR leads to activation of G_q and to an alternative set of effector responses driven through this G-protein (b). (Reproduced from Gilchrist, A. (2007). Modulating G-protein-coupled receptors: from traditional pharmacology to allosterics. *Trends in Pharmacological Sciences*, **28**, 431–437, by permission of Elsevier Science.)

other 'closed', and that the role of glutamate is to stabilise the closed, active form. In contrast, the $GABA_B$ receptor is a heterodimer involving two receptors from family B. The explanation for the importance of this dimerisation of GPCRs is a topic of current research but a number of points have emerged so far:

- Firstly, it appears to occur early in the biosynthetic pathway of the receptors, specifically in the endoplasmic reticulum, and is essential for the trafficking of the receptors into the membrane.
- Secondly, molecular modelling studies with the rhodopsin receptor have demonstrated that it can only effectively interact with subunits of a G-protein if it is a homodimer emphasising the importance of molecular size and conformation in receptor activity.
- Thirdly, the formation of heterodimers facilitates the opportunity for crosstalk between the two protomers. Such crosstalk has been demonstrated for a number of receptors including histamine H_1 receptors and $GABA_B$ receptors and there is growing evidence of the general importance of such receptor crosstalk in maintaining specificity from signal to cellular response.

- Fourthly, the dimerisation affords the potential for further allosteric modification of ligand (agonist, G-protein, other proteins) binding on the individual protomers that may be linked to receptor crosstalk and hence cellular response (Section 17.2.3).

GPCR association with other proteins

In addition to forming dimers with other receptors, GPCRs also form associations with several groups of other proteins (Fig. 17.6) that are crucial to the signal activity and regulation of the receptor probably by allosteric and cooperative effects.

- *Receptor activity modifying proteins* (RAMPs): Three RAMPs (RAMPs1, 2 and 3) have been characterised and shown to be relatively small (RAMP1 is a 140 amino acid protein) with a single membrane-spanning domain, a large extracellular domain and a small intracellular domain. The RAMP–receptor heterodimer determines the specificity of the functional receptor. The dimers are formed in the endoplasmic reticulum and the RAMP remains associated with the receptor for the whole of the receptor's lifetime. RAMPs appear to be most important for the class B GPCRs.
- *G-protein-coupled receptor kinases (GRKs) and β-arrestins*: These two families of proteins are intimately involved in the control of GPCR activity. There are seven members of the GRK family (GRK 1–7) and four members of the β-arrestin family (β-arrestin 1–4). Their actions are coordinated in that the GRK phosphorylates the agonist-activated receptor at serine and threonine residues in the intracellular domain (Fig. 17.10) and this stimulates the binding of β-arrestin. This in turn uncouples the GPCR from its G-protein thereby desensitising the receptor in spite of the continuing presence of the agonist and simultaneously targets the receptor to clathrin-coated pits in the membrane and subsequent endocytosis (Section 17.5.2).
- *GPCR-interacting proteins* (GIPs): These proteins are involved in a number of key processes including (a) targeting GPCRs to specific cellular compartments, (b) the assembly of GPCRs into functional complexes called receptosomes and (c) the fine-tuning of the signalling of the GPCRs. Examples of these GIPs include the multi-PDZ proteins, the Shank family of proteins and the Homer proteins. As their name implies, multi-PDZ proteins possess a number of PDZ (PSD-95, Dig and ZO-1/2) domains each of which can bind to the C-terminal region of different receptor and effector proteins involved in the transduction of a given signal. The Shank proteins possess several protein–protein interaction motifs including the SH2 (Src-homology domain 2) motif which recognises and binds tyrosine-phosphorylated sequences (this includes some receptors with intrinsic protein kinase activity (Section 17.4.3) and the SH3 (Src-homology domain 3) motif which recognises and binds sequences that are rich in the amino acid proline). GPCRs therefore possess domains capable of recognising and binding to these various motifs (Fig. 17.6).

In addition to the three groups of proteins discussed above, a fourth group is also involved in the regulation of the GPCR-transduced signal, but rather than associating with the GPCR directly, proteins in this group interact with the associated G-protein and are therefore referred to as regulation of G-protein signalling proteins

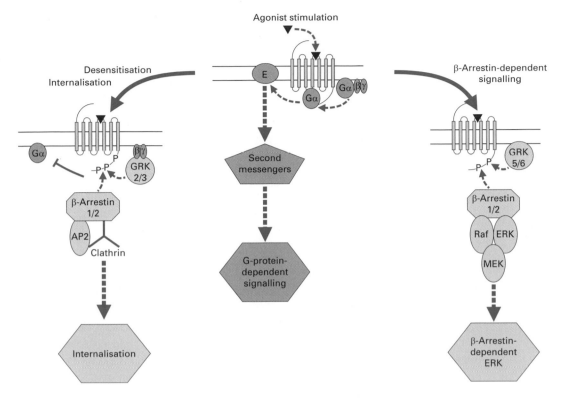

Fig. 17.10 Functional specialisation of the different GRKs. Upon agonist stimulation, GRK2 and GRK3 are recruited to the plasma membrane by interacting with Gβγ subunits. They have a predominant role in receptor phosphorylation, β-arrestin recruitment, desensitisation and internalisation. GRK5 and GRK6 are constitutively associated with the plasma membrane. They are both required for β-arrestin-dependent ERK activation, although the locus of their action might be at the receptor or downstream. In some systems, GRK5 and GRK6 can also mediate desensitisation and internalisation. E, second-messenger-generating enzyme. (Reproduced from Reiter, E. and Lefkowitz, R. J. (2006). GRKs and β-arrestins: roles in receptor silencing, trafficking and signalling. *Trends in Endocrinology and Metabolism*, 17, 159–165, by permission of Elsevier Science.)

(RGS). This group of over 20 proteins can be divided into five subfamilies based on sequence homology. They all have two actions namely they reduce the binding of the $G\alpha$ GTP to the effector and they act as GTPases accelerating the hydrolysis of GTP to GDP by a factor of over 2000-fold, hence they are also referred to as GTP-accelerating proteins (*GAPs*). As a consequence of this latter activity, RGSs effectively regulate the duration of the response resulting from the activation of a GPCR.

Kinetics of the activation of and signalling by GPCRs

Spectroscopic techniques, especially FRET, have been used to evaluate details of the kinetics and mechanisms of GPCR-mediated signals. Such studies have shown that agonist binding, the activation of the receptor and interaction with the G-protein all occur within less than 50 ms, that the activation of the G-protein and its interaction with the effector protein occur within 500 ms, that the formation of second messengers such as cAMP can take up to 20 s, that the interaction of the GRK-phosphorylated receptor with β-arrestin can take up to 50 s and that a conformational change in β-arrestin may take up to 5 min. These results indicate that G-protein activation is the rate-limiting step in GPCR signalling. Interestingly, whereas full agonists activate the G-protein within 50 ms, partial agonists take up to 1 s and inverse agonists take about 1 s to reduce the intrinsic activity of a receptor. Such studies have also shown that some G_i-like G-proteins can initiate a signal without the need to dissociate whereas other G-proteins do appear to need to dissociate first.

17.4.4　Signal transduction through receptors with intrinsic protein kinase activity

It has been appreciated for over 20 years that phosphorylation coupled with dephosphorylation represents an important mechanism for the regulation of protein activity. A large number of intracellular kinases and phosphatases have been characterised and their regulatory action linked to conformational changes induced in the target protein as a result of the introduction or removal of a phosphate group. Control of protein activity by the kinase/phosphatase principle is found in a broad range of organisms, indicating its early evolution. It operates with the net consumption of ATP, but with the considerable gain in sensitivity, amplification and flexibility that more than compensate for the ATP consumed.

There are two main classes of receptor kinases based on the specificity of the induced protein kinase, namely receptor tyrosine kinases and serine/threonine receptor kinases.

Receptor tyrosine kinases

Twenty subclasses of receptor tyrosine kinases (RTKs) are known, based on their extra- and intracellular structures. The human genome contains approximately 90 RTK genes. Three of the best characterised are the epidermal growth factor receptor (EGFR) (also known as ErbB and HER), platelet derived growth factor (PDGF) receptor and the insulin receptor (IR). The vast majority of RTKs, including EGFR and PDGF, are single-chain, monomeric proteins in the absence of their agonist but they dimerise on agonist binding. However, a few, including the insulin receptor and the closely related insulin-like growth factor-1 (IGF-1) receptor, are permanently dimeric. Each monomer of the insulin receptor consists of an α and a β chain that are linked via a disulphide bridge making the functional dimer an $\alpha_2\beta_2$ tetramer (Fig. 17.11). The insulin receptor is involved in the regulation of lipid and protein metabolism in addition to its role in the maintenance of glucose homeostasis. EGFR regulates aspects of development including that of the nervous system, but mutant forms are linked to cancer.

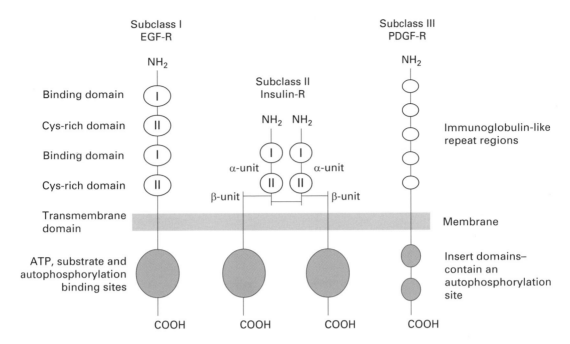

Fig. 17.11 Diagrammatic representation of three receptor tyrosine kinase subclasses. The EGF subclass contains two ligand-binding domains that are located in juxtaposition so that the ligands bind in a cleft between the two domains. The two cysteine-rich domains are both located near the membrane surface. On ligand binding, both the EGF and PDGF subclass receptors dimerise so that the intracellular tyrosine kinase domains possess elevated activity and enhanced binding affinity relative to the monomeric forms. The insulin subclass receptors are effectively dimeric, but, as with subclasses I and III, there is allosteric interaction between the two αβ halves of each receptor on ligand binding. The tyrosine kinase domains of the three subclasses show the greatest degree of homology between the three subclasses.

Example 3   RECEPTORS AND CANCER THERAPY

Cancers are caused by a mutation that stimulates a cell to grow in an uncontrolled fashion. Inherited mutations cause a relatively small proportion of cancers; approximately 10% of breast cancers, for example, are the result of inherited mutations such as those caused by the *BRCA1* and *BRAC2* genes. A diversity of spontaneous mutations underlie most cancers and a given type of cancer may be the consequence of a range of mutations including single-letter changes to a codon, gene deletions, insertions and duplications and chromosomal rearrangements. The implication of this diversity of cause is that a given type of cancer such as breast cancer may be the result of changes in different molecular pathways and hence will need a range of drugs to be available to the clinician for the treatment of patients. The implication of this is that for the effective treatment of a given patient it is vital that the underlying cause is identified. Most current drug treatments target either a membrane receptor or an enzyme. Examples of receptor targets are given below:

Example 3   (*cont.*)

Drug	Target	Mechanism
Gefitinib (Iressa) – a small molecule	Binds to the ATP binding site of the tyrosine kinase domain of epidermal growth factor receptor (EGFR) in non-small-cell lung cancer	Inhibits the activation of the Ras signal transduction cascade thereby inhibiting tumour growth
Trastuzumab (Herceptin) – an antibody	HER2 receptor, a member of the EGFR family	Blocks HER2 receptor preventing it from forming dimers with similar receptors that cause a signal to the nucleus to stimulate the tumour to grow and divide
Tositumomab (Bexxar) – an antibody carrying ^{131}I	CD20 receptor on β-cells	Binds to receptor and the radioactive iodine isotope that it carries kills cancerous lymphoma cells
Rituximab (MabThera) – a chimeric monoclonal antibody	CD20 receptor on β-cells	Triggers the immune system to destroy healthy and lymphoma cells
Tamoxifen (Nolvadex) – a small molecule	Oestrogen receptor	Decreases DNA synthesis and inhibits oestrogen effects that cause the tumour to grow
Sorafenib (Nexavar) – a small prodrug molecule	Vascular endothelial growth factor receptor (VEGFR) and platlet-derived growth factor receptor (PDGFR) on cells lining blood vessels	Inhibits protein kinase activity of receptors. Also binds to c-kit receptors on cancer cell involved in cancer growth

Like the treatment for HIV/AIDS (Section 18.2.2) cancer therapy most commonly involves the use of a cocktail of drugs acting by different mechanisms.

Binding of insulin to its binding site (one on each of the α chains) causes a conformational change that activates the kinase activity. In the case of EGFRs, agonist binding causes either a conformational change that allows the occupied receptors to recognise each other resulting in their association to form homo- or heterodimers or stabilises a pre-existing equilibrium between monomers and dimers and induces conformational changes within the dimer. At this stage the kinase action of the receptor is inactive, but the dimerisation is rapidly followed by the mutual cross-autophosphorylation of one to three Tyr residues in the tyrosine kinase domains on each monomer forming SH2 or phosphotyrosine binding (PTB) domains that serve as docking sites for effector proteins that possess SH2 sites causing the phosphorylation and activation of the effector. One such effector protein for the insulin receptor is insulin receptor substrate-1 (IRS). Effector proteins that bind either possess enzyme activity such as kinase, phosphatase, lipase and GTPase or act as adaptors that act to link the activated and phosphorylated receptor with other effectors. In addition, the effector may possess other binding domains to which further effectors can bind. This multiplicity of binding domains and bound effectors enable a web of signalling pathways and cascades to be established with the potential for transduction pathway branching to meet prevailing cellular demands.

Receptor serine and threonine kinases

Complementary to the tyrosine kinase group of receptors is a second group of protein kinase receptors characterised by their ability to autophosphorylate serine and threonine residues in the intracellular domain of the receptor. These protein serine/threonine kinase receptors are specific for members of the transforming growth factor (TGF)-β superfamily of receptors which regulate growth, differentiation, migration and cell adhesion. They are classified into a number of subgroups on the basis of their structure, particularly their serine/threonine kinase domain. They are all single transmembrane receptors that on ligand binding form hetero-oligomeric complexes between subgroup types. This stimulates autophosphorylation and activation of the serine/threonine kinase activity towards other cytosolic proteins that are components of the transduction pathway.

Like the G-protein-coupled receptors, receptor protein kinases stimulate numerous transduction pathways. The downstream members of these transduction pathways include the phospholipases and phosphoinositide kinases that are also involved in the G-protein transduction pathways. Receptor protein kinases in turn are subject to regulation by ubiquitin ligases, protein kinases and phosphatases and various adaptor proteins. Importantly, 30% of all RTKs are repeatedly found either mutated or overexpressed in a wide range of different human malignancies. EGFR, for example, is associated with breast cancers. One of a number of effectors unique to the receptor protein kinases is Ras, a membrane-bound guanosine binding protein with intrinsic GTPase activity, involved in cell growth and development in all eukaryotes. Activated Ras triggers the mitogen-activated protein kinase (MAP kinase cascade) which in turn phosphorylates multiple proteins in the nucleus including transcription factors that regulate gene expression in cell division, cell adhesion, apoptosis and cell migration. The Ras gene was the first oncogene to be discovered to be associated with a human cancer. It is the target for the treatment of various forms of cancer by the development of monoclonal antibodies.

Cytokine receptors

A group of receptors closely related to the protein kinase receptors are the cytokine receptors. Cytokines are a group of peptides and proteins intimately involved in cell signalling. They play key roles in the immune system and are involved in immunological, inflammatory and infectious diseases. They have also been termed lymphokines, interleukins and chemokines. They act by binding to specific cytokine receptors that structurally resemble protein kinase receptors in that they exist as functional homo- or hetero-oligomers but they lack intrinsic kinase activity. Binding of the cytokine activates the receptor enabling it to recruit a non-receptor tyrosine kinase, such as a member of the src family or the Janus kinases family, to the cytoplasmic side of the receptor. Activation of this kinase by the receptor enables the receptor–kinase complex to recruit other effector proteins that trigger a signalling cascade linked to the up- or downregulation of genes and their transcription factors.

Protein tyrosine phosphatases

Crucial to the control of cell signalling involving receptor kinases is the existence of a group of protein tyrosine and serine/threonine phosphatases that can either deactivate or activate pathways by dephosphorylation of receptors or effectors. The human genome includes approximately 100 genes for tyrosine phosphatases. This is a similar number of genes as for RTKs suggesting that the two families are partners in the regulation of the signalling response with the kinases controlling the amplitude of the responses and the phosphatases their rate and duration. Some of the phosphatases are purely cytoplasmic whilst others are receptor-like and are referred to as receptor tyrosine phosphatases (RTPs). Most protein tyrosine phosphatases have two phosphatase domains for reasons that are not clear but their specificity may be linked to interaction between the two sites. The activity of the phosphatases appears to be linked to their own phosphorylation and a significant number have a SH2 domain similar to that of the receptor tyrosine kinases. There is currently great interest in protein tyrosine phosphatases as many have been shown to act a tumour suppressors in contrast to protein tyrosine kinases some of which act as oncoproteins.

17.4.5 Dynamics of signalling pathways

The classical view of receptor signal transduction is that the process is a linear cascade:

$$\text{agonist} \longrightarrow \text{receptor} \longrightarrow \text{effector(s)} \longrightarrow \text{transduction} \longrightarrow \text{response}$$

However, the current evidence is that such linear cascades are an oversimplification of the actual organisation. Most receptors appear to operate in complex and highly integrated networks that control several linked processes. GPCRs and PTKs that control many physiologically important pathways commonly merge to form integrated networks. Receptors in such networks share key components or nodes that mediate and modify the receptors' signals and which are the sites of crosstalk between the receptors that can result in signal divergence. Crosstalk between GPCRs and RTKs is common and may be bidirectional in that the GPCR may be 'upstream' or 'downstream' of the RTK signal transduction. This has given rise to the concept of transactivation by which a given receptor is activated by a ligand of a heterologous receptor that may belong to a different class of receptor with respect to the signal transduction mechanism. Thus an activated RTK may initiate GPCR signalling by causing the dissociation of the Gα subunit from the Gβγ subunit allowing the Gα subunit to initiate signalling, for example by activating adenylyl cyclase, or the agonist-bound RTK may directly associate with a GPCR through scaffold molecules such as RGS, allowing the RTK to use components of GPCR signalling such as GRK/β-arrestin. A good example is the regulation of glucose metabolism involving the β-adrenergic receptor (a GPCR) and its regulation by the insulin receptor. The ability of the insulin receptor to counter the release of glucose by the action of catecholamines on the adrenergic receptor leads to the tight control of serum glucose levels. This control by the insulin receptor of the β-adrenergic receptor appears to operate by two (at least) mechanisms, one involving the internalisation of the β-adrenergic receptors via GRK/β-arrestin hence decreasing the sensitivity of the cell to circulating catecholamine, the other at the level of the activation of RGSs hence terminating the action of the activated β-adrenergic receptor. Crosstalk between the two GPCRs μ-opioid receptor (MOR) and $α_{2A}$-adrenergic receptor ($α_{2A}$AR) has been shown to proceed via a direct conformational change-induced inactivation of the noradrenaline-occupied $α_{2A}$AR by the binding of morphine to the MOR. The two receptors form a heterodimer and activate common transduction pathways mediated through the inhibition of G_i and G_o. The inhibition of the $α_{2A}$AR occurs within a subsecond of the morphine binding and terminates a downstream MAP kinase cascade induced by the $α_{2A}$AR.

Nodes characteristically consist of a group of related proteins that are essential for the receptor-mediated signal but such that two or more of these proteins have unique roles within the network and are therefore the source of divergence within the network (Fig. 17.12). Many networks contain cascades of cycles formed by two or more interconvertable forms of a signalling protein. This protein is modified by two opposing enzymes, commonly a kinase and a phosphatase for phosphoproteins or a guanosine nucleotide exchange factor (GEF) and a GTPase-activating protein (GAP) for G-proteins. Such cascades afford the property of ultrasensitivity to the input signal particularly under conditions in which the enzymes are operating near saturation.

17.5 RECEPTOR TRAFFICKING

17.5.1 Membrane structure

Detailed studies of cell membranes and of the area of the membrane occupied by signalling receptors have revealed that the membrane is very patchy with segregated regions or microdomains of different protein and lipid structure. One such type of these microdomains are lipid rafts that are dynamic structures rich in sphingolipids, glycosphingolipids, sphingomyelin and cholesterol and as a result are less fluid than the remainder of the membrane. Lipid rafts are involved in the organisation of receptors and their associated signal-transduction pathways. A subset of lipid rafts are caveolae characterised by their invaginated morphology produced as a result of cross-linking between a constituent protein called caveolin. There is evidence that a given receptor accumulates in either lipid rafts or caveolae but not both.

d Network with critical nodes

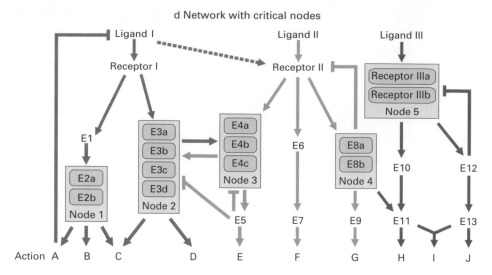

Fig. 17.12 Signalling pathways with critical nodes. Ligand binding to a receptor (I, II and III) activates the receptor and in turn activates (full arrows) the linked effectors (numbers beside an effector E indicate distinct proteins, the small letters indicate different isoforms of a given protein) leading to the transduction of one or more actions (A to J) involving different nodes. The actions may result in the inhibition of the action of the initiating ligand or of the receptor (blocked arrows). A ligand may also activate a second receptor but with less intensity (dotted arrows). Activated effectors may alternatively either network with other effectors (arrows) or inhibit other effectors (blocked arrows). (Adapted from Taniguchi, C. M., Emanuelli, B. and Kakn, C. R. (2006). Critical nodes on signalling pathways: insights into insulin action. *Nature Reviews Molecular Cell Biology*, **7**, 85–96, by permission of Nature Publishing Group.)

17.5.2 **Receptor endocytosis**

Endocytosis, also referred to as **endosomal trafficking**, is the process by which lipids and proteins, including receptors, are taken into the cell from the plasma membrane. It is a complex and dynamic process for which multiple pathways exist, each involving characteristic scaffold and accessory proteins. The most important pathways for membrane receptors are the clathrin-dependent and clathrin-independent pathways.

The internalisation and trafficking of receptors has been most thoroughly investigated using G-protein-coupled receptors, particularly the β_2-adrenergic receptor. Studies have revealed that the processes involve three stages:

- The recruitment of the receptors, normally with agonist bound to its orthosteric site (**agonist-dependent internalisation**) but in some cases in the absence of bound agonist (**agonist-independent internalisation**), to discrete endocytic sites such as rafts and caveolae in the membrane. In the case of the agonist-dependent route, the receptor is phosphorylated by a G-protein-coupled receptor kinase (GRK) and associated with β-arrestin prior to internalisation.
- The internalisation of the receptors to form an **early endosome**, also referred to as a **sorting endosome**.

- The sorting of the receptor proteins into specific domains within the endosome for either subsequent recycling to the membrane for reuse or for their degradation in lysosomes.

In the case of G-protein-coupled receptors, after binding the agonist and G-protein, the receptors undergo agonist-dependent phosphorylation by a GRK and this stimulates interaction with one of the β-arrestin family of protein adaptor complexes. These cytosolic adaptors facilitate the disruption of the interaction of the receptors with G-proteins and the recruitment of the receptors into coated pits in the membrane where they link the receptor to a protein called clathrin via a second adaptor protein. A large number of such second adaptors have been identified and shown to belong to one of two classes: multimeric adaptor proteins of which AP-2 is best known, and monomeric adaptor proteins (also known as CLASPs). This group is numerically the largest and includes Epsin1, AP180, Dab2 and CALM as well as β-arrestin. The adaptor protein interacts with phosphoinositides and promotes both the assembly of the coated pits and the recruitment of the activated receptors to them. There is evidence that scaffold proteins such as Eps15 are also involved in the assembly of the clathrin cage. Coated pits are regions in the membrane that are rich in clathrin that is located on the cytoplasmic side of the membrane. Clathrin consists of three heavy and three light chains which can polymerise to form a polymeric cage-like structure or lattice that links to the C-terminal end of the receptor. The polymeric clathrin network drives membrane deformation and the 'budding' of the coated pit to create an endocytic vesicle. This budding process involves the actin cytoskeleton that is a dynamic network of over 100 structural and regulatory proteins, and the GTPase dynamin. With the expenditure of GTP and in collaboration with a number of BAR proteins that are inserted into the membrane, the actin and dynamin promote the removal of the budding vesicle from the membrane. Once the free vesicle has formed, the clathrin coat is lost as a result of the action of one of the proteins of the heat shock protein 70 family. The vesicles then fuse to form an early endosome in which the bound agonist is removed by the prevailing mildly acidic (pH 6.2) conditions (Fig. 17.13).

Recent evidence indicates that RTKs also undergo endocytosis from special localised regions of the membrane termed dorsal waves that are large, circular protrusions in the cell membrane. This endocytic process does not involve clathrin but it does require specific adaptors such as Grb2, dynamin and components of the actin cytoskeleton that combine to release the early endosome by a mechanism similar to that for GPCRs.

There are two possible fates for the receptors within the early endosome:

- recycling to the membrane thereby restoring a functional receptor;
- proteolysis resulting in a net decrease in membrane receptor function, a process known as downregulation.

The pathways for endocytic sorting are determined by the operation of sorting signals. The main sorting signal appears to reside in the cytoplasmic region of the receptor itself. Thus for the β_2-adrenergic receptor, a PDZ-binding domain in the C-terminal region interacts with a protein called EBP50 with the result that the receptor undergoes recycling, the process also involving β-arrestin and a protein

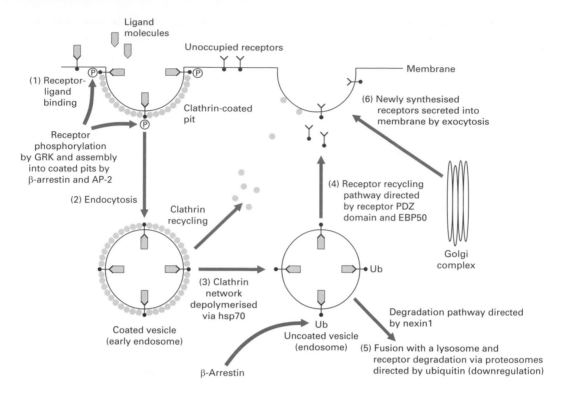

Fig. 17.13 Pathway of agonist-dependent G-protein-coupled receptor internalisation and endocytic sorting.
(1) Occupied receptors are phosphorylated (P) by a G-protein-coupled receptor kinase (GRK) leading to the
recruitment of β-arrestins. Arrestins serve as adaptor proteins by linking phosphorylated receptors to
components of the transport machinery such as clathrin and adaptor protein AP-2 and their recruitment to
clathrin-coated pits. (2) The coated pit 'buds' into the cytoplasm aided by the clathrin which forms a network
leading to the formation of an endosome. Note that the cytoplasmic domain of the receptor remains exposed to
the cytoplasm following endocytosis. (3) The clathrin network is depolymerised and the clathrin recycled to the
inner membrane. (4) The receptors are dephosphorylated and as a result of the interaction of EBP50 with a PDZ
domain on the receptor, traffic back to the cell surface resulting in functional resensitisation. Alternatively,
(5) dephosphorylated receptors are tagged with ubiquitin (Ub) and enter the degradation pathway. Here they
interact with nexin1 and ESCRTs which promote the fusion of the endosome with a lysosome and the
degradation of the receptor by a number of proteasomes – a process known as downregulation. (6) The Golgi
complex secretes newly synthesised receptor molecules to the outer membrane surface by exocytosis. The
balance between receptor cycling, receptor degradation and receptor synthesis and exocytosis determines the
number of functionally active receptors on the membrane surface at any time.

known as NSF. Chimeras of the receptor lacking the PDZ domain are directed to the
degradative pathway. GPCRs sorted for recycling are first dephosphorylated by an
endosome-associated phosphatase and recycled back to the membrane via the Golgi
complex. In the case of receptors directed to degradation, the sorting signal directs the
receptors to so-called late endosomes that have a multivesicular appearance referred
to as multivesicular bodies (MVBs) that contain a number of protein complexes
known as endosomal sorting complex required for transport (ESCRT-I-III) which
promote the fusion of the endosome with a lysosome. The resulting decrease in pH
within the vesicle to 5.3 facilitates the downregulation of the receptor by proteolysis.

There is evidence that this downregulation process is also dependent upon the ubiquitination of the receptor, a process that may include an active role for β-arrestin. Ubiquitin is known to 'tag' proteins for degradation, the process involving the action of a number of proteasomes (Section 15.5.4).

It has yet to be established just how universal the clathrin-linked and the non-clathrin-linked endocytotic pathways are to the large number of receptors of various cell types. What is very clear is that the expression, activation, regulation, desensitisation and endocytosis of receptors are all dependent on numerous protein–protein interactions, many of which occur at the plasma membrane interface, that each cause crucial conformational changes in the receptor and/or their regulators such as to couple receptor activity to current cellular and whole organism demands. It is equally evident that reversible multi-site phosphorylation plays a vital role in the regulation of the activity of receptors and their effectors.

The temporal variation in the number of cell surface receptors available for agonist binding is the net result of receptor trafficking and of new receptor synthesis, which takes place in the rough endoplasmic reticulum. A leader sequence in the protein results in its recognition and transport to the Golgi complex where it is glycosylated, packaged into coated vesicles and inserted into the membrane by exocytosis, in which clathrin plays a vital role. The balance between receptor synthesis, recycling and degradation is subject to various control mechanisms so that free receptor availability in the outer membrane meets current physiological needs. Temporal variation in cell membrane receptor numbers is also of significance in the clinical response to chronic drug administration that leads to the downregulation of receptor numbers, and in neurodegenerative conditions in which the release of the physiological agonist is deficient resulting in upregulation of receptor numbers.

17.6 SUGGESTIONS FOR FURTHER READING

Experimental protocols
Ali, H. and Haribabu, B. (eds.) (2006). *Transmembrane Signalling Protocols*, 2nd edn, *Methods in Molecular Biology*, vol. 332. New York: Humana Press.
Nienhaus, G. U. (ed.) (2005). *Protein-Ligand Interactions: Methods and Applications, Methods in Molecular Biology*, vol. 305. New York: Humana Press.
Willars, G. B. and Challiss, R. A. J. (eds.) (2004). *Receptor Signal Transduction Protocols*, 2nd edn, *Methods in Molecular Biology*, vol. 259. New York: Humana Press.

Review articles
Arrang, J.-M., Morisset, S. and Gbahou, F. (2007). Constitutive activity of the histamine H_3 receptor. *Trends in Pharmacological Sciences*, 28, 350–357.
Baker, J. G. and Hill, S. J. (2007). Multiple GPCR conformations and signalling pathways: implications for antagonist affinity estimates. *Trends in Pharmacological Sciences*, 28, 374–380.
Delcourt, N., Bockaert, J. and Marin, P. (2007). GPCR-jacking: from a new route in RTK signalling to a new concept in GPCR activation. *Trends in Pharmacological Sciences*, 28, 602–608.
De Meyts, P. (2008). The insulin receptor: a prototype for dimeric, allosteric membrane receptors? *Trends in Biochemical Sciences*, 33, 378–384.
Grant, B. D. and Donaldson, J. G. (2009). Pathways and mechanisms of endocytic recycling. *Nature Reviews Molecular Cell Biology*, 10, 597–608.
Gurevich, V. V. and Gurevich, E. V. (2008). How and why do GPCRs dimerize? *Trends in Pharmacological Sciences*, 29, 234–240.

Jaiswal, J. K. and Simon, S. M. (2007). Imaging single events at the cell membrane. *Nature Chemical Biology*, **3**, 92–98.

Kirkbride, K. C., Ray, B. N. and Blobe, G. C. (2005). Cell-surface co-receptors: emerging roles in signaling and human disease. *Trends in Biochemical Sciences*, **30**, 611–621.

Maldonado-Baez, L. and Wendland, B. (2006). Endocytic adaptors: recruiters, coordinators and regulators. *Trends in Cell Biology*, **16**, 505–512.

Parameswaren, N. and Spielman, W. S. (2006). RAMPs: the past, present and future. *Trends in Biochemical Sciences*, **31**, 632–638.

Schwartz, T. W. and Holst, B. (2007). Allosteric enhancers, allosteric agonists and ago-allosteric modulators: where do they bind and how do they act? *Trends in Pharmacological Sciences*, **28**, 366–372.

Sorkin, A. and von Zastrow, M. (2009). Endocytosis and signalling: intertwining molecular networks. *Nature Reviews Molecular Cell Biology*, **10**, 609–622.

Tonks, N. K. (2006). Protein tyrosine phosphatases: from genes, to function, to disease. *Nature Reviews Molecular Cell Biology*, **7**, 833–845.

Van der Goot, F. G. and Gruenberg, J. (2006). Intra-endosomal membrane traffic. *Trends in Cell Biology*, **16**, 514–521.

Wyllie, D. J. A. and Chen, P. E. (2007). Taking the time to study competitive antagonism. *British Journal of Pharmacology*, **150**, 541–551.

Useful websites

G-PROTEINS

www.gpcr.org/7tm

CELL SIGNALLING

www.cellsignal.com

www.signaling-gateway.org

RECEPTOR ENDOCYTOSIS

www.cytochemistry.net/cell-biology/recend.htm

18 Drug discovery and development

K. WILSON

18.1 HUMAN DISEASE AND DRUG THERAPY

18.1.1 Human disease

The wide range of diseases to which humans are exposed have in common the fact that each is the result of either some physiological dysfunction caused by a gene mutation or incorrect expression of the related protein, or of the exposure of the individual to an environmental factor, such as pesticides, diet, or bacterial, fungal or viral infection. The dysfunction gives rise to characteristic medical symptoms that enable the condition to be diagnosed, commonly by diagnostic tests of the type described in Chapter 16, and an evaluation made of the severity of the condition and the future prospects of the patient making a full recovery from it. Underlying many of the conditions at a molecular level is a change in the amount, function or activity of one or more proteins that in turn trigger changes in cellular, tissue or organ function. A large part of current worldwide medical research is aimed at the elucidation of the molecular mechanisms underlying diseases such as the various forms of cancer and neurological conditions such as Parkinson's disease, motor neurone disease and multiple sclerosis, in order to identify key proteins involved in the disease process with a view to selecting one of the proteins as a target for the development of a new drug and thereby to minimise or eliminate the symptoms.

18.1.2 The nature of drugs and their target proteins

At the present time there are just over 800 drugs in current use worldwide. The majority are organic molecules with a molecular weight of less than 500. However, other possibilities for the nature of the drug are receiving increasing attention. One such option is to develop a monoclonal antibody as the drug to target the protein. Thus an increasing number of monoclonal antibodies are being developed for the

709

treatment of specific forms of cancer. An example is transtuzumab (Herceptin®) used in the treatment of breast cancer. An alternative approach is to develop drugs to modify the expression of the gene producing the target protein rather than the protein itself. Our knowledge of gene replication and transcription has advanced to the stage where it has become possible to target the DNA or RNA responsible for the biosynthesis of a specific protein. Strategies based on the interference with the translation of mRNA, a process referred to as RNA interference (RNAi) have shown considerable potential in model studies (Section 6.8.5). Short interfering RNAs (siRNAs), for example, can be synthesised with a specific base sequence designed to complement and inactivate specific mRNAs or whole gene families. They work by activating a sequence-specific RNA-induced silencing complex (RISC) that cleaves the corresponding functional mRNA within the cell. Some siRNA 'libraries' are now available commercially. The main challenge with RNAi therapy is to devise an effective delivery system for the siRNAs as conventional oral administration would inevitably lead to their premature metabolism. A related potential therapy is the use of so-called DNAzymes. These are synthetic, single-stranded deoxyribonucleotides with the ability to bind and cleave RNA and thereby to suppress the expression of pathophysiologically active genes, for example in a number of cardiovascular states. They have the advantage over antisense molecules that they are less sensitive to nuclease activity.

Advances in recombinant DNA technology, particularly the discovery of restriction endonucleases, polynucleotide ligase and DNA polymerase (Section 5.5), created the technique of gene replacement therapy to correct inherited or acquired genetic defects affecting the availability of a specific protein underlying a disease state. DNA can be introduced into targeted cells, most commonly by incorporating it into a vector such as a modified virus, so there is strong reason to believe that this will become an increasingly important form of therapy in the future (see Table 6.9). Equally, cell-based therapies, particularly those based on the use of stem cells (Section 2.6), have proved to be successful in animal models and there is every reason to expect their future adaptation to the treatment of human disease.

The following discussion will concentrate on the discovery and development of small organic molecules as drugs but many of the principles and challenges discussed are equally applicable to these alternative forms of therapy. The vast majority of these small organic drugs target one of three specific types of proteins namely enzymes, membrane or nuclear receptors and transporters. It has been estimated that the current total number of different protein targets used by marketed drugs in humans is approximately 500. Nearly three-quarters of these targets are human proteins, the remainder are proteins in infecting organisms. Of those aimed at human targets nearly one-third are aimed at G-protein-coupled receptors (GPCRs) (Section 17.4.3) and one-third at enzymes.

18.1.3 Case studies

To illustrate how drug development is based on the targeting of a specific protein, three common human disorders – hypertension (high blood pressure), dyspepsia (heartburn) and bacterial infection – will be considered briefly. Two others, cardiovascular disease and HIV/AIDS are considered in greater detail in Section 18.2.2.

Case study 1 HYPERTENSION

Hypertension, also referred to as high blood pressure (bp), is defined as a systolic bp >140 mm Hg and a diastolic bp >90 mm Hg. Its cause may be primary or secondary to a range of conditions such as kidney disease. If unchecked, hypertension can lead to strokes and heart attacks. It can be reduced by a number of drugs acting by significantly different mechanisms:

- $\beta_1$adrenergic receptor antagonists such as propranolol and labetalol and the $\alpha_1$adrenergic receptor antagonist prazosin involve the blocking of the action of GPCRs.
- Inhibitors, such as captopril, of angiotensin converting enzyme (ACE), which converts angiotensin I to angiotensin II that in turn leads to an increase in blood pressure by its action on angiotensin II receptor, are the preferred first choice therapy to lower bp.
- Antagonists of the GPCR angiotensin II receptor, such as telmisartan, are a related drug therapy to that of ACE inhibitors.
- Antagonists of the dihydropyridine Ca^{2+} channel, such as nifedipine and verapamil, that block the movement of Ca^{2+} ions into smooth muscle cells lining coronary arteries and thereby lower bp, are also valuable therapeutic agents for the treatment of hypertension.
- Inhibitors of phosphodiesterases (PDE) found in vascular smooth muscle and involved in contractility, also reduce blood pressure. One such inhibitor is silendafil (Viagra®) but it specifically inhibits PDE5 which hydrolyses cGMP to 5′-GMP and thereby enhances the action of nitric oxide induced penile erection and is therefore widely prescribed for erection dysfunction and not for the treatment of hypertension!

Case study 2 DYSPEPSIA (INDIGESTION)

Dyspepsia presents as upper abdominal pain and is associated with excess production of acid in the stomach. If simple antacids are inadequate for its alleviation, drugs are available to reduce the acid secretion. In the 1970s it was shown that antagonists of the GPCR histamine H_2-receptor successfully inhibit stomach acid production. The first clinically used antagonist was cimetidine but it was soon replaced by ranitidine due to its better tolerability, longer action and greater activity. Cimetidine is also a significant inhibitor of several key cytochrome P450s involved in the metabolism of other drugs (see Table 18.1 below). However, the preferred choice of treatment for dyspepsia is now the use of a proton pump inhibitor (PPI) such as omeprazole and lansoprazole. These inhibit the action of the H^+/K^+ ATPase 'pump' that transports protons across membranes and which is the terminal stage in gastric acid secretion into the gastric lumen. Omeprazole (Losec® and Prilosec®) is one of the largest selling drugs ever produced. It is administered as a racemate of the R and S forms. The R form is inactive but is converted to the S form *in vivo* by the cytochrome enzyme CYP2C19 (Table 18.1).

It is evident from these case studies that a specific therapeutic outcome can be achieved by targeting one of a number of possible proteins. The challenge in the process of discovering and developing a new drug is firstly to identify the possible targets and then to take an informed decision on which one to select for the discovery process.

Case study 3 **BACTERIAL INFECTION**

Unlike the situation in the previous two cases, the drug target for the treatment of bacterial infection is not one of the patient's proteins, all of which are presumed to be functioning normally, but rather one of the proteins in the infecting organism. The aim of the therapy is either to prevent the replication of the infecting organism by administering a bacteriostatic drug, or to cause its death using a bacteriocidal drug. From knowledge of the mechanism of bacterial replication, five types of antibiotics have been developed:

- *Cell wall biosynthesis inhibitors*: Penicillin was the first such drug. It is one of a number of β-lactams that inhibit the enzyme DD-transpeptidase that is involved in the formation of peptidoglycan cross-links in bacterial cell walls. As a result of the absence of cross-links, the cell undergoes lysis and dies so the drug is bacteriocidal in its action. Penicillin G and penicillin V are commonly used to treat a wide range of streptococcal infections. The cephalosporins are also β-lactams and have a similar action to that of the penicillins.

- *Folic acid antagonists*: Trimethoprim, widely used in the treatment of urinary tract infections, acts by inhibiting the enzyme tetrahydrofolate reductase thereby inhibiting the synthesis of tetrahydrofolic acid, an essential precursor in the synthesis of the nucleotide thymidine. It is thus a bacteriostatic agent. The enzyme is an excellent target as this particular pathway is absent in humans for whom folic acid is an essential vitamin for the synthesis of thymidine. The sulphonamides act in a related way. They inhibit the enzyme dihydropteroate synthetase, an enzyme in the pathway to folic acid and hence thymidine.

- *Protein synthesis inhibitors*: The tetracyclines inhibit prokaryotic 30S ribosomes (not found in eukaryotes) by binding to aminoacyl-tRNA. Streptomycin acts in a related way. It binds to the bacterial ribosome 16S rRNA and thus inhibits the binding of formyl-methionyl-tRNA to the 30S subunit. It is very effective against tuberculosis. Erythromycin also inhibits aminoacyl translocation but in this case it binds to the 50S subunit of the 70S rRNA complex, thereby blocking protein synthesis. All these drugs are bacteriostatic.

- *RNA synthesis inhibitors*: Rifampicin acts against mRNA synthesis by inhibiting DNA-dependent RNA polymerase by blocking the β subunit, thus preventing transcription to mRNA and subsequent translation to proteins. Like the protein synthesis inhibitors, it is a bacteriostatic agent.

- *DNA synthesis inhibitors*: The quinolone bacteriocidal drugs inhibit the enzymes DNA gyrase and topoisomerase IV, neither of which are found in eukaryotic cells, and thereby inhibit bacterial DNA replication and transcription. Examples are ciprofloxacin and norfloxacin.

18.1.4 **Basic characteristics of drugs**

Pharmacological parameters

A number of parameters characterise the interaction of a drug with its target protein. Drugs acting as receptor agonists target the orthosteric sites, to which the physiological agonist binds, of the receptors that are present in very large numbers on each cell membrane. The three most important parameters of agonist action are efficacy, potency and selectivity (see Section 17.2 for full details). Efficacy is a measure of the ability of the drug to produce the maximum response from the receptors. Potency is a measure of the dose of the drug required to produce one-half (50%) of the maximum response from the receptors and is expressed as an ED_{50}. Its value influences the clinical dose of a drug. Selectivity is a measure of the ability of the drug to discriminate between the target receptors and other receptors including isoforms of the target receptor. It therefore influences the side effects exerted by the drug. Effectiveness describes the ability of the drug to alleviate the symptoms in a large group of heterogeneous patients with apparently similar symptoms. Drugs acting as competitive receptor antagonists also bind to the orthosteric site of the receptors thereby blocking the action of the physiological agonist and reducing the receptor response. They are quantified by an IC_{50} value, which like ED_{50} measures the concentration required to reduce the receptor response by 50%. The majority of enzyme inhibitors compete with the natural substrate for the active site of the enzyme (Section 15.2.2). Alternatively, a drug may bind at an allosteric site on the enzyme producing a conformational change that either increases or decreases the activity of the substrate-binding site (Section 15.2.4).

Pharmacokinetic parameters

Pharmacokinetics relates to the way a drug is absorbed, distributed, metabolised and eliminated (hence the term ADME studies) from the body. For a drug to exert its desired effect it must be delivered to the site of action and normally this has to be achieved by its distribution about the body by the blood circulatory system. Most drugs are administered by the oral route, but the inhalation (via the lungs), sublingual (beneath the tongue), transdermal (across the skin) and subcutaneous injection (beneath the skin) routes are more appropriate for the administration of some drugs due to their particular physical properties. Drugs administered orally must be absorbed from the gastrointestinal tract in order to enter the circulatory system. For the majority of drugs passive diffusion is the mechanism by which this transfer occurs. The optimum requirement for this passive diffusion is that the drug is lipophilic, i.e. it possesses adequate lipid solubility and therefore is unionised. Many drugs are weak bases and as such exist in an ionised and hence hydrophilic state at the pH of 1 prevailing in the stomach. Thus, generally speaking, the extent of absorption of drugs from the stomach is low. In contrast, in the small intestine, with a pH of 7, most drugs are unionised and hence suitable for passive diffusion across the gut wall. In these cases, gastric emptying rate is normally the limiting factor for absorption and this is influenced by the food content of the stomach. On entering the circulatory system from the gut the drug is taken by the hepatic portal vein to the liver which is the major site of drug metabolism. Some drugs are readily metabolised by hepatic enzymes and therefore subject to significant inactivation by metabolism before they have entered the general circulatory system. Loss of active drug at this stage is referred to as the first pass effect – hence the importance of assessing the susceptibility of a candidate drug to hepatic metabolism in the early stages of drug discovery. The proportion of an oral dose of drug reaching the systemic circulation from its site of administration is referred to as its bioavailability (β). It is possible to avoid the first pass effect by administering the drug by routes such as sublingual and transdermal.

For an orally administered drug, its concentration in the blood increases as absorption from the gut continues. It eventually reaches a plateau, at which point the rate of absorption and the rate of loss of the drug are equal, and then declines as the drug is

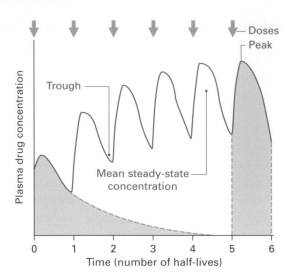

Fig. 18.1 Plasma drug concentration following repeated oral dosing. Following each dose the plasma concentration increases, reaches a peak and declines normally at an exponential rate. When the next dose is administered the plasma concentration profile is superimposed on the existing profile but if the dosing is given after one half-life the peak plasma concentration gradually reaches a maximum and thereafter remains at this level following further repeated dosing. This maximum is generally achieved after five doses. (Adapted from McLeod, H. L. (2008). Pharmacokinetics for the prescriber. *Medicine*, **36**, 350–354, by permission of Elsevier Science.)

eliminated from the body (Fig. 18.1). For a drug to exert its desired pharmacological effect its concentration in blood needs to exceed a threshold value referred to as the minimum effective concentration. At a higher concentration the drug may begin to display toxic side effects, referred to as the toxicity threshold. The ratio of these two thresholds is referred to as the therapeutic index or therapeutic ratio. The closer the index to 1, the more difficult the drug is in clinical use. Drugs such as the anti-epileptics phenytoin, carbamazepine and phenobarbitone, with an index in the region of 1, are often subject to therapeutic drug monitoring to ensure that the patient is not exposed to potential toxic effects. The aim of repeat dosing with any drug is to maintain the concentration of the drug in the therapeutic range or window in which toxic effects are not observed. If the dosing interval is adjusted correctly in relation to the plasma half-life (see below) of the drug, the drug plasma concentration will oscillate within the therapeutic range (Fig. 18.1). Once in the general circulation, the drug may bind to a plasma protein, especially albumin, and if the extent of binding is high (>90%) the ability of the drug to cross membranes to reach its site of action and exert a pharmacological effect may be impaired. Due to the high density of endothelial cells lining the brain, the so-called blood–brain barrier, drug entry into the brain from blood is slower than to other regions of the body and so drugs targeting the brain often exploit endogenous carrier-mediated transport systems rather than passive diffusion.

The pharmacokinetic parameters of a drug quantify the non-pharmacological behaviour of the drug from the time of its administration to the time of its removal from the body. There are three main parameters:

- *Intrinsic clearance, Cl_{int}:* This is defined as the volume of plasma apparently cleared of drug per unit time by all routes. It has units of $cm^3\,min^{-1}$. Drug is removed from the body by two main routes – metabolism, normally by hepatic enzymes, to one or more polar metabolites that generally lack pharmacological activity and which are readily excreted by the kidneys, and renal excretion of unchanged drug. Intrinsic clearance is therefore the sum of hepatic clearance Cl_{hep} (the volume of plasma apparently cleared of drug in unit time by hepatic metabolism) and renal clearance Cl_r (the volume of plasma apparently cleared of drug in unit time by renal excretion of unchanged drug). The value of Cl_{int} can be calculated by administering a dose (units: mg) of the drug by the oral route and dividing the dose by the area under the resulting plasma concentration/time curve (AUC) (units: $mg.min\,cm^{-3}$) making an allowance for the bioavailability of the drug:

$$Cl_{int} = \frac{\text{dose} \times \beta}{\text{AUC}} \qquad (18.1)$$

Bioavailability is calculated from the ratio of the AUC values for an oral dose and for the same dose administered intravenously and which is not subject to first pass loss. It is expressed as a percentage and can vary from 0 to 100%.

- *Apparent volume of distribution, V_d:* This is defined as the volume of body fluid in which the drug appears to be distributed. It has units of dm^3. In an adult body the total volume of water is about $42\,dm^3$, and is made up of $3\,dm^3$ plasma water, $14\,dm^3$ extracellular water and $25\,dm^3$ intracellular water. The value of V_d is measured by administering a dose by bolus (fast) intravenous injection (to avoid the first pass loss) and using the equation:

$$V_d = \frac{\text{dose}}{\text{peak plasma concentration}} \qquad (18.2)$$

Many drugs have V_d values in the range of $42\,dm^3$ indicating that they are fully distributed in body water. Abnormally high values are the result of a low plasma concentration caused by the deposition of the drug in some particular tissue, most commonly fat tissue for highly lipophilic drugs. Generally speaking, this is an undesirable property of a drug. Equally, an abnormally low V_d due to a high plasma concentration is indicative of a poor ability of the drug to penetrate lipid barriers.

- *Plasma* or *elimination half-life, $t_{1/2}$:* This is defined as the time required for the plasma concentration to decline by 50% following its intravenous administration. It is a so-called hybrid constant as its value is linked to both Cl_{int} and V_d by the equation:

$$t_{1/2} = \frac{0.693 \times V_d}{Cl_{int}} \qquad (18.3)$$

Values of $t_{1/2}$ normally range from 1 to 24 hours. Clinically it is important as its value determines the frequency with which the drug needs to be administered to maintain the plasma concentration in the therapeutic range. Thus drugs with a short half-life need to be administered frequently whereas drugs with a long half-life can be given on a daily basis.

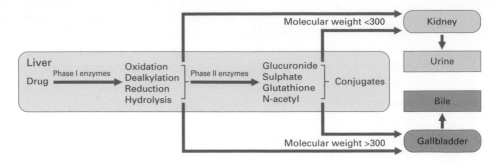

Fig. 18.2 Drug metabolism. Phase I enzymes catalyse the modification of existing functional groups in drug molecules (oxidation reactions). Conjugating enzymes (Phase II) facilitate the addition of endogenous molecules such as sulphate, glucuronic acid and glutathione to the original drug or its Phase I metabolites. (Adapted from McLeod, H. L. (2008). Pharmacokinetics for the prescriber. *Medicine*, **36**, 350–354, by permission of Elsevier Science.)

Knowledge of all of these pharmacokinetic properties of a drug is fundamental to its clinical use as they dictate the dose size and frequency. Hence pharmacokinetic studies form a vital part of the drug discovery and development processes.

Drug metabolism

Most drugs are sufficiently lipophilic to be poorly excreted by the kidneys and hence would be retained by the body for very long periods of time were it not for the intervention of metabolism, mainly in the liver. Metabolism occurs in two phases (Fig. 18.2). Phase I mainly involves oxidation reactions and Phase II conjugation of either the drug or its Phase 1 metabolites with glucuronic acid, sulphate or glutathione to increase the polarity of the drug or its metabolite(s) and hence ease of renal excretion. The oxidation reactions are carried out by a group of haem-containing enzymes collectively known as cytochrome P450 monooxygenases (CYP), so-called because of their absorption maximum at 450 nm when combined with CO. They are membrane-bound and associated with the endoplasmic reticulum. They operate in conjunction with a single NADPH-cytochrome P450 reductase and are capable of oxidising drugs at C, N and S atoms. More than 50 CYP human genes have been sequenced and divided into four families (CYP1–4) of which CYP2 is the largest. Five CYPs appear to be responsible for the metabolism of the majority of drugs in humans: CYP1A2, CYP2C9, CYP2C19, CYP2D6 and CYP3A4 (Table 18.1). The genes for these cytochromes have been cloned and expressed in cell lines suitable for drug metabolism studies. Several of these cytochromes are expressed polymorphically (i.e. they exist in several forms due to their expression by related genes) in humans resulting in considerable interindividual variation in the rate of metabolism of some drugs. There are also ethnic variations in the expression of some of these isoforms. This can be critically important in the use of drugs that have a narrow therapeutic index. In principle, it is possible to genotype individual patients for their CYP activity and hence to 'personalise' drug dosage but in practice this has yet to be put into widespread clinical practice. One other problem associated with the clinical use of

Table 18.1 **Action of some drugs on the major human cytochrome P450 isoforms**

P450 isoform	Substrate	Inhibitor	Inducer
CYP1A2	Imipramine, oestradiol, paracetamol, verapramil, propranolol	Fluvoxamine,[a] cimetidine, ciprofloxacin	Omeprazole, cigarette smoke
CYP2C9	Fluvastatin, ibuprofen, phenytoin, amitriptyline, tamoxifen	Fluconazole,[a] fluvastatin, lovastatin, sulphaphenazole, phenylbutazone	Rifampicin, secobarbital
CYP2C19	Diazepam, propranolol, amitriptyline, omeprazole, lansoprazole	Lansoprazole, omeprazole, cimetidine	Carbamazepine, rifampacin, prednisone
CYP2D6,	Amitriptyline, imipramine, propranolol	Quinidine,[a] bupropion,[a] cimetidine, ranitidine	Rifampicin, dexamethazone
CYP2E1	Paracetamol, theophylline, ethanol	Cimetidine, disulfiram	Ethanol
CYP3A4	Indinavir, diazepam, lansoprazole, saquinavir, lovastatin	Ketoconazole, indinavir,[a] nelfinavir,[a] ritonavir[a]	Carbamazepine, nevirapine, phenytoin

Note: [a]Strong inhibitors that cause at least 80% decrease in clearance.

some drugs is that they either inhibit or induce one or more of the cytochrome P450s (Table 18.1). Inhibition means that the intrinsic clearance of the drug and that of other concomitantly administered drugs is impaired and this may have toxicological consequences whilst enzyme induction results either in increased clearance of the drug that may render its therapy ineffective and/or in the production of toxic metabolites.

18.1.5   Desirable properties of a new drug

From the foregoing discussion it is evident that the drug discovery and development processes must lead to a drug that meets a number of criteria:

- *Chemical structure*: It must possess structural features that allow it to specifically interact with and bind to the target protein. To this end it must possess flexibility and hydrogen binding potential and be of an appropriate molecular size.
- *Physical properties*: It must possess some aqueous solubility and adequate lipophilicity to allow it to cross membranes to access the target site. Linked to these properties, its pK_a must be such that it exists *in vivo* predominantly in the unionised state.
- *Pharmacological properties*: It must have acceptable potency, efficacy, selectivity and effectiveness for receptor agonists or binding properties for enzyme inhibitors or receptor antagonists. It must also fill an unmet clinical niche. The chemical structure and pharmacological properties must be novel to allow patent protection.

- *Pharmacokinetic properties*: It must have an acceptable rate of absorption, bioavailability, clearance, volume of distribution and plasma half-life to ensure that its dosing size and frequency and onset and duration of action meet patient needs.
- *Toxicological properties*: Ideally it must possess a large therapeutic index, but in some forms of therapy, notably with cytotoxic drugs, this is not possible.

Most candidate drugs in clinical development fail to reach the clinical market for one of four reasons – inappropriate pharmacokinetics, lack of efficacy, unacceptable toxicology and adverse effects in humans.

18.2 DRUG DISCOVERY

18.2.1 Drug discovery and development processes

The launching of a new drug onto the clinical market is the culmination of three distinct processes – drug discovery, drug development and drug marketing. Schematically the processes can be represented as follows:

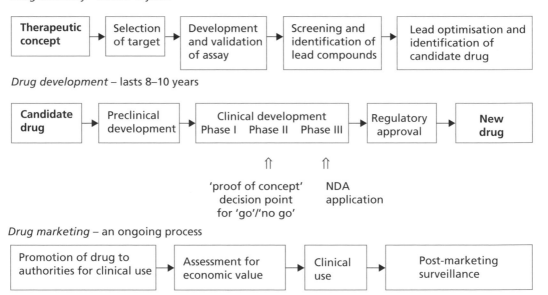

Drug discovery – lasts 2–5 years

Therapeutic concept → Selection of target → Development and validation of assay → Screening and identification of lead compounds → Lead optimisation and identification of candidate drug

Drug development – lasts 8–10 years

Candidate drug → Preclinical development → Clinical development Phase I Phase II Phase III → Regulatory approval → New drug

⇑ 'proof of concept' decision point for 'go'/'no go' ⇑ NDA application

Drug marketing – an ongoing process

Promotion of drug to authorities for clinical use → Assessment for economic value → Clinical use → Post-marketing surveillance

Although presented as three linear stages, in practice it is essentially a parallel activity supporting a 'learn and confirm' approach with the 'proof of concept' as an important decision point. The high cost of drug discovery and development, variously estimated at \$500 m to \$3 bn, and the long-term nature of drug discovery and development coupled with the competitive need to get the drug licensed and in clinical use as quickly as possible, drive the management strategy underlying the discovery and development processes. To minimise potential development losses there is a need to ensure that if the candidate drug is to fail to meet its pharmacokinetic, pharmacological and toxicological targets it should do so early in its development phase ('fail early fail cheaply'). A working deadline for the final 'go ahead' with drug development is typically the end of Phase IIa clinical studies in which the drug is administered to patients for the first time (Section 18.3.2). The strategic decision taken at this time is referred to as proof of concept. In the recent past there have been several examples of drugs being withdrawn from the market after they have been awarded a marketing licence by which time maximum development costs have been incurred with no prospect for future sales. For the above reasons, multidisciplinary management teams are created to define the critical path through the development process and to expedite the process in a time and cost effective manner.

18.2.2 Selection of the drug target

Four main approaches are employed in drug discovery to identify potential protein targets for a drug discovery programme:

- *Pathophysiological approach*: This is based on the elucidation of the biochemical mechanism underlying the selected disease state. The approach has been successful for such conditions as atherosclerosis and HIV/AIDS but is inherently slow in complex conditions such as cancer, Parkinson's and Alzheimer's diseases.
- *Gene expression profiling*: This approach screens the gene profile of patients with the selected disease in an attempt to identify genes that are either up- or down-regulated relative to control subjects (see Sections 6.8, 6.9 and 6.10).
- *Gene knock-out screening*: This commonly uses transgenic strains of mice in which specific genes have been deleted and the metabolic consequences studied (see Section 6.8.4). An increasingly common alternative to mice is the zebrafish (*Danio*). However, approaches using both species are slow relative to the use of the yeast *Saccharomyces cerevisiae* that has a strong genome similarity to that of the human genome and which is much easier to manipulate genetically.
- *Genetic approach*: This approach uses either whole animals or isolated cells and involves the use of antisense oligonucleotides or RNA interference (RNAi). The former bind to mRNA and prevent its translation whilst the latter destroy by cleavage the functional mRNA (see Sections 6.8.5, 6.8.6 and 6.9.2). The technique of looking for single nucleotide polymorphisms (SNPs) in patients with a common disease is growing in importance as a way of gaining quick insight into the genetic background for a disease.

The application of such approaches coupled with knowledge gained from the current scientific literature, commonly identify a number of potential protein targets and strategic decisions have to be made to select one. Two contrasting case studies illustrate this point:

Case study Target selection: Atherosclerosis

This multifactorial condition is associated with the thickening of the arterial walls due to the ingress of atherogenic lipoproteins into the intima where they are oxidised

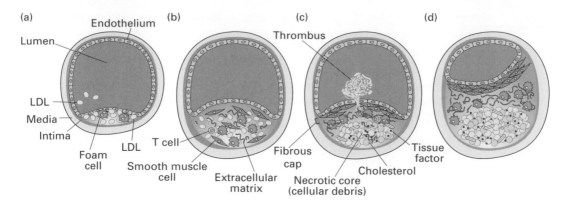

Fig. 18.3 Initiation and progression of atherosclerosis. (a) Low density lipoprotein (LDL) particles enter the intima where they are oxidised and aggregate within the extracellular intimal space. They are then phagocytosed by macrophages eventually leading to the formation of lipid-laden foam cells and fatty streaks, the initial lesion leading to the development of atherosclerotic lesions. (b) Smooth muscle cells that secrete matrix components such as collagen facilitate the formation of these lesions by increasing the retention of LDL. T cells are recruited to the lesion and perpetuate a state of chronic inflammation. The diameter of the lumen gradually increases. (c) Foam cells eventually die releasing cellular debris and crystalline cholesterol. The smooth muscle cells form a fibrous cap that walls off the plaque from the blood. This further promotes the recruitment of inflammatory cells. The plaque may rupture resulting in the formation of a thrombus in the lumen. If large enough, the thrombus may block the artery and cause a heart attack. (d) If the plaque does not rupture, it continues to grow and eventually blocks the lumen. (Adapted from Rader, D. J. and Daugherty, A. (2008). Translating molecular discoveries into new therapies for atherosclerosis. *Nature*, **451**, 904–912, by permission of the Nature Publishing Group.)

and subsequently phagocytosed by macrophages to form foam cells loaded with lipid, mainly cholesterol, and eventually fatty streaks and an atherosclerotic plaque (Fig. 18.3). The progressive growth of the plaque leads either to a restriction of blood flow and eventually a total blockage or if the plaque eventually ruptures it may cause a blood clot again causing a blockage and as a result a heart attack or stroke (Section 16.3.2).

Body cholesterol originates from dietary cholesterol and *de novo* synthesis in cells, most importantly in the liver. Cholesterol is distributed about the body in the form of five types of lipoprotein particles: chylomicrons, very low density lipoprotein (VLDL), intermediate density lipoprotein (IDL), low density lipoprotein (LDL) and high density lipoprotein (HDL) that differ in size and composition and which are linked metabolically via the enzyme lipoprotein lipase. The two most abundant types of particle are LDL and HDL. LDL is the main carrier of cholesterol from the liver to peripheral cells including those within the developing atherosclerotic plaque and has been referred to as 'bad' cholesterol. HDL is the carrier of cholesterol from peripheral cells to the liver, a process referred to as reverse cholesterol transport (RCT), hence HDL has been termed 'good' cholesterol (Fig. 18.4).

Patients with homozygous familial hypercholesterolaemia suffer from premature atherosclerosis due to a mutation in the gene coding for the LDL receptor located

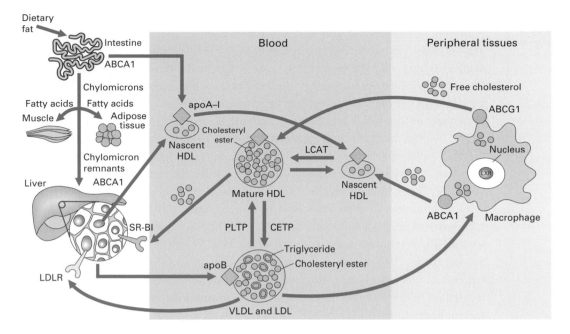

Fig. 18.4 Lipoprotein metabolism. The intestine absorbs dietary fat and packages it into chylomicrons that are transported to peripheral tissues through the blood. In muscle and adipose tissues the enzyme lipoprotein lipase breaks down the chylomicrons, and fatty acids enter these tissues. The liver takes up the chylomicron remnants, loads lipids onto apoB and secretes VLDL that undergoes lipolysis by lipoprotein lipase to form LDL. LDLs are taken up by LDL receptors (LDLR) on liver cells. The intestine and liver generate HDL particles through the secretion of lipid-free apoA1. This recruits cholesterol through the action of the transporter ABCA1 forming nascent HDLs. The free cholesterol in nascent HDLs is esterified by the enzyme lecithin cholesterol acyltransferase (LCAT) creating mature HDLs. The cholesterol in mature HDLs is returned to the liver directly through the receptor SR-B1 and indirectly by transfer to LDLs and VLDLs through the enzyme cholesteryl ester transfer protein (CETP). The lipid content of HDLs is also altered by the enzymes hepatic lipase and endothelial lipase and the phospholipids transfer protein (PLTP). (Adapted from Rader, D. J. and Daugherty, A. (2008). Translating molecular discoveries into new therapies for atherosclerosis. *Nature*, **451**, 904–912, by permission of the Nature Publishing Group.)

on the hepatocyte surface and which is the main mechanism for the removal of LDL from plasma. This observation stimulated the research for drugs that would increase LDL receptor expression and a reduction in plasma LDL. However, to date the main attention for the therapeutic reduction of plasma cholesterol has focussed on the metabolic pathway, starting with acetyl CoA, which leads to its biosynthesis. The controlling enzyme in this pathway is hydroxymethylglutaryl CoA reductase (HMG CoA reductase):

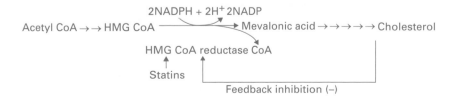

Twelve drugs, collectively called *statins*, have been developed as competitive inhibitors of this enzyme and approved for therapeutic use. They have proved to be very effective in producing a substantial reduction in plasma cholesterol so that statins are currently the most widely used drugs in the developed world. However, they are associated with a number of side effects such as muscle cramps and disturbance of liver enzymes and less commonly acute renal failure. Some statins are known to damage mitochondria and this may underlie some of these side effects.

One obvious target for a new candidate drug is HDL, the aim being to increase its plasma levels and hence facilitate increased reverse cholesterol transport. Some Japanese individuals have raised HDL due to a genetic deficiency of the enzyme cholesteryl ester transfer protein (CETP) (Table 6.4), and this led to the development of CETP inhibitors. One such candidate drug, torcetapib, was developed but had to be withdrawn from Phase III clinical trials (Section 18.3.2) because although it increased serum HDL it resulted in an increased death rate. Whether or not this is a general property of all CETP inhibitors or indeed of all agents that raise HDL, or one specific to torcetapib, remains to be investigated. Two transporter proteins, ABCG1 and ABCA1, located on cell membranes have been shown to promote the efflux of cholesterol from macrophages to form nascent HDL and hence to stimulate reverse cholesterol transport. The genes encoding these two transporters are stimulated by liver X receptors (LXR) that are nuclear transcription factor receptors, and agonists of this receptor have been shown to increase HDL cholesterol by 48% and to improve atherosclerosis in animals. Some of these agonists are now in human clinical trials. Other research has indicated that raising the amount of apoA-1 in HDL could be beneficial provided they do not modify its structure and some novel candidate drugs that may increase apoA-1 are currently in development. Recent research has also indicated that the protein proprotein convertase subtilisin/kexin type 9 (PCSK9) plays an important role in the regulation of the LDL receptor. Claims have been made that the use of iRNA to silence the gene for PCSK9 in mice reduces plasma cholesterol levels by half. Whether or not this approach can be applied to humans remains to be seen.

Case study Target selection: HIV/AIDS therapy

These conditions are induced by the human immunodeficiency viruses HIV-1 and HIV-2 that were discovered in 1983 and 1985 respectively. HIV-1 is the most virulent and most prevalent in the worldwide pandemic: 33 million people worldwide are believed to be living with HIV. They are retroviruses and have a RNA genome. They attack the immune system, especially T-helper cells, resulting in an increased rate of cell apoptosis and death. The virus attacks these cells via attachment of the viral envelope glycoprotein gp120 to a CD4 receptor on the host cell. Membrane penetration by the virus requires the additional involvement of either one of two cytokine co-receptors CCR5 and CXCR4 on the surface of the patient's cells that are linked to GPCRs. Once attached to the CD4 and the co-receptor, the envelope protein undergoes a conformational change that allows penetration into the host cell. The virus then injects various enzymes including a reverse transcriptase, protease, integrase and RNase into the cell. The reverse transcriptase promotes the synthesis of a DNA copy

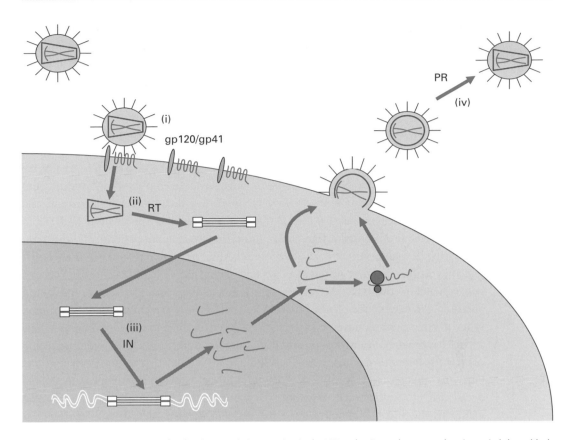

Fig. 18.5 Targets for the therapeutic intervention in the HIV cycle. Currently approved antiretroviral drugs block HIV infection at different steps of the viral life cycle: (i) virus entry through their interaction with gp120 or gp41; (ii) reverse transcription (i.e. reverse transcriptase inhibitors); (iii) integration (i.e. integrase (N) inhibitors); and (iv) maturation (i.e. inhibitors (PR) that block the conversion of immature virus into mature infectious virions). (Reproduced from Menendez-Arias, L. and Tozser, J. (2008). HIV-1 protease inhibitors: effects on HIV-2 replication and resistance. *Trends in Pharmacological Sciences*, 29, 47–49, by permission of Elsevier Science.)

of the viral RNA genome and integrase inserts it into the chromosome of the host cell; the protease is crucial to the replication process. The viral genome carries nine genes, three of which code for structural proteins and six for regulatory proteins that control the ability of the virus to infect cells. One of the features of the action of reverse transcriptase is that it frequently mismatches bases and this is exploited in the action of the nucleotide reverse transcriptase inhibitors (see below) which mimic natural bases and are incorporated into the DNA strand thereby terminating the chain. The enzyme makes an average of one mistake every time it copies the RNA so that thousands of variants, some viable some not, are produced daily.

From a detailed knowledge of the molecular mechanisms involved in virus infection and replication, numerous potential targets for drug discovery are evident (Fig. 18.5) but to date development has concentrated on the three viral-specific enzymes – reverse transcriptase, protease and integrase. Inhibitors of each of these enzymes have been developed and approved for therapeutic use against HIV-1 and their use has led to the slowing of the development of AIDS. The first anti-HIV drug to be licensed was the HIV nucleotide reverse transcriptase inhibitor (NRTI) azidothymidine (AZT) in 1987. Its discovery and development paralleled that of acyclovir for the inhibition of DNA replication of herpes simplex virus (HSV) in the late 1970s. In the 1990s the discovery that the protease was crucial for the replication of the HIV lead to the development of HIV protease inhibitors (PI). Whilst individually these two classes of drugs are effective in many patients, by the combination of a PI with two NRTIs (so-called highly active antiretroviral therapy, HAART) better reductions in viral burden have been achieved. A third class of drugs, the non-nucleotide reverse transcriptase inhibitors (NNRTI), has been added to these combinations, and modern treatment is based on combination therapy with drugs such as Atrivir®, Trizivir® and Combivir®. A one dose per day formulation, Atripla®, which contains one NNRTI and two NRTIs, is now a popular first choice drug. However, many patients on long-term antiretroviral therapy experience adverse effects and develop resistance to these drugs and new drugs aimed at different targets are being developed. Drug resistance is a consequence of the fact that reverse transcriptase does not carry out proofreading allowing the emergence of mutants that are not sensitive to the therapy. Two main types of new therapy have so far been developed in an attempt to circumvent this resistance problem. Maturation inhibitors interfere with the assembly and maturing of new virons prior to their release by the host cell. One such agent, Bevirimat®, is currently undergoing clinical trials. Fusion inhibitors block the binding of the virus to a receptor on the host cell. Selzentry® is a co-receptor antagonist for the receptor CCR5 involved in the attachment of the virus to the host cell. It was licensed for human use in 2007. Recent research has shown that the infection process is more complex than originally believed and that over 250 human proteins, many of which are as yet unidentified, are needed by the HIV to enable it to spread throughout the body, so potentially a large number of other drug targets are available for exploitation. Of particular interest is the observation that human cells possess antiviral activity that inhibits the release of retrovirus particles including HIV-1 and that this activity is antagonised by the HIV-1 accessory protein Vpu. The antiviral activity is due to protein-based 'tethers' that can be induced by interferon-α. The protein CD317 has been identified as one such tether. It is obvious that Vpu and CD317 are potential future therapeutic targets. It is hoped that the use of new classes of drugs, in combination with existing therapies, will lead to a reduction in the number of cases of drug resistance. Whether or not this will have a beneficial effect on patients on long-term therapy remains to be seen as will the consequence of changing a drug protein target from the virus to the human host.

In both of the above cases studies the state of knowledge of the underlying biochemistry is considerable and there are numerous existing drugs for therapeutic use. Any decision to select either a new target or a new combination of existing drugs for new drug development in either of these two therapeutic areas is far from straightforward and requires the combined expertise, experience and intuition of the strategic multidisciplinary development team to make an informed judgement. Marketing potential increasingly drives such strategic decisions. A new drug may be the 'first in class' in which case it will have no competitors and may become a 'blockbuster' drug with annual sales in excess of $1bn. However, the risk with such a drug programme is that the concept underlying its action will not previously have been confirmed. Alternatively, the drug may be in an existing class in which case it will have competitors and the aim will be to make it 'best in class'.

18.2.3 Selection of a screening target and assay procedure

In silico pharmacology

It has been estimated that in the process of discovering and developing the 800 drugs that are currently licensed for clinical use, pharmaceutical companies have synthesised and evaluated against their particular narrow target field over 10 million compounds. Individual pharmaceutical companies currently hold chemical 'libraries' of up to 2 or 3 million compounds. Such a large resource is available for screening against new therapeutic targets. Moreover, in the last decade huge advances have been made in compiling computer databases of both the structural, physical and biological properties of existing drugs and structural classification of all known protein targets. These databases are highly sophisticated. For example, the ligand database will contain such information as molecular weight, molecular structure, H-bond donor numbers, polarity, surface area, lipophilicity (expressed as a log P value, where P is a measure of the partition coefficient between water and an organic solvent usually 1-octanol), three-dimensional conformation, spectral properties and known biological targets (receptors, enzymes), quantitative structure–activity relationships (QSARs) and structure–absorption, distribution, metabolism and excretion relationships. Similarly, the protein database will contain all known information about primary, secondary, tertiary and quaternary structure, binding site details of the protein and structural details of known ligands. Knowledge of the amino acid sequence of a target receptor enables a computer model of the structure of the agonist-binding site to be made.

The availability of such data sets has given rise to the interdisciplinary field of chemogenomics which aims to predict gene/protein/ligand relationships and hence enable predictions to be made of potential ligands for new targets long before any screening has taken place. Such predictions will be based on the assumptions that chemically similar compounds should share common targets and targets sharing similar ligands should share similarities in their binding sites. This use of computer-based data to aid drug discovery is known under the names of *in silico* pharmacology, computational pharmacology and computational therapeutics. Such computer-based predictions are used to identify potential drug targets and to test the models. They are run in parallel with high-throughput screening techniques in drug discovery programmes.

High-throughput screening

The screening of large chemical libraries for a specific type of biological activity is generally based either on binding activity of the test compound against a selected target preparation or the ability of the compound to activate or inhibit some particular signalling pathway in a test preparation (Section 17.4.1). Assays need to be adaptable to automation to enable very large numbers of compounds to be screened. Such high-throughput screening is generally performed robotically and involves detection systems such as fluorescence-based techniques that do not require the separation of reaction product from substrate. In high-content screening automated microscopes are used to take multiple images of the fluorescent cells either in microtitre plates or in a flow cytometer. The images are analysed using image analysis algorithms that enable both biochemical and morphological parameters in the assay system to be evaluated simultaneously (see Sections 4.4 and 4.7).

In recent years high-throughput screening has been adapted to include fragment-based approaches to lead discovery. This involves screening the target against a library of small molecules rather than drug-sized molecules. The philosophy behind this approach is that it tests potential binding interactions in fewer assays than does the use of drug-sized molecules. The idea is then to combine the structural elements involved in the binding of these small molecules into a larger lead compound. The difficulties with the technique are firstly that the smaller molecules bind more weakly than do larger molecules so the assay system has to display the sensitivity to detect this weaker interaction with the target, and secondly that the approach will only work if the two or more fragments that are combined to give the larger molecule bind to the same or compatible conformational microstates of the target protein.

Assay types

The assay chosen for a particular target needs to be as specific and relevant to the desired final physiological response as possible. High-throughput screening can be carried out using cell-free assays or cell-based assays. In the former case, the target is isolated and may be a receptor preparation (Section 17.3.2) or enzyme preparation. They are commonly used for GPCR targets (Fig. 17.6). These *in vitro* systems are simple, unambiguous, cheap and easy to interpret. On the other hand, they give relatively little specific information and are insensitive to issues such as membrane permeability but are effective in the early phase of screening programmes. Cell-based assays, in contrast, give a better appreciation of a likely biological response, allow several responses, such as protein–protein interactions and signalling pathways, to be monitored simultaneously and can distinguish between receptor agonists and antagonists. On the negative side, they require more of the test compound and are generally more expensive. This is an important consideration in the early phase of screening when up to a million candidates may need to be screened.

Screening and developing 'hits'

In the early stages of screening, mixtures of compounds rather than individual compounds are interacted with the target. After the assay is complete, bound and unbound compounds are separated, commonly by size exclusion chromatography (Section 11.7), the bound compounds then released and their identity obtained by LC-MS. This gives rise to a number of 'hits'. To check the authenticity of these hits, the compounds in question will be re-screened individually in duplicate or triplicate. The compounds will then be re-screened using different concentrations and varying experimental conditions. It is also common for them to be tested in fully functional assays and at the same time checks will be made on their potential patentability. The molecular structure of the confirmed hits will then be compared to identify common structural features that can then be built into a newly synthesised series of candidate drugs. The technique of combinatorial chemistry is widely used by medicinal chemists to synthesise a large number of structurally related compounds on a small scale. These compounds are then screened for activity to identify further important structural features and the process repeated on an iterative basis. At the same time, attention will be paid to the physical properties of the emerging 'lead' compounds. For a compound to be an effective therapeutic agent it needs to possess a number of important physical and biological properties. These include a balanced water/lipid solubility expressed by its log P value, the extent to which it binds to human albumin, its interaction with the human cytochrome P450 system and its lack of toxicity. Simple *in vitro* screening methods are available for these properties without resorting to animal studies. The potential selectivity of the compounds will also be evaluated against other related physiological targets. The application of these various tests enables compounds with inappropriate properties to be rejected at an early stage.

Pharmacological and toxicological profiling

The purpose of these studies is to evaluate the full pharmacodynamic profile of the compounds in order to establish that they possess the required molecular and cellular properties in tissues and animals, that they possess adequate potency, efficacy and selectivity, appropriate pharmacokinetic properties and that repeated dosing with the compound or its withdrawal after repeated dosing do not lead to toxic effects. Such profiling will be undertaken using both *in vitro* and *in vivo* methods. Studies are likely to include evaluation in animal cell lines expressing cloned human genes and animal models to study both acute and chronic physiological, pharmacological and toxicological properties. The outcome of the drug discovery phase is the identification of a drug candidate, with a few 'back-up' candidates, for development as a licensed drug.

18.3 DRUG DEVELOPMENT

At the end of the drug discovery process the management team will know that the candidate drug has displayed encouraging properties in the screening programme and has appropriate physical and pharmacokinetic properties for clinical use. The team will also have reason to believe that the intellectual property represented by the drug can be protected by international patents. At this stage, however, the group has no evidence that it will display acceptable efficacy, potency and selectivity in human subjects. Thus the priority is to obtain 'proof of concept' that the drug will demonstrate the anticipated clinical properties. To achieve this, data from human subjects are needed and this is the initial priority of the development phase. Unlike the discovery process, the development phase is highly prescribed in that it has to meet national and international regulations and standards and is subject to review by regulatory and licensing bodies. It is a long and expensive process that accounts for up to two-thirds of the total cost of launching a new drug. It has two main phases:

- *Preclinical phase*: This involves the establishment of small-scale production of the test drug and the creation of a formulation for its clinical use; the establishment of the pharmacokinetic profile of the drug and the evaluation of its acute *in vitro* and *in vivo* toxicity in the rat and normally one other species; the evaluation of its genetic and reproductive toxicology, and the development of analytical methods for the drug and its metabolites.
- *Clinical phase*: This involves the first studies in healthy volunteers (Phase Ia and Ib) paralleled by chronic toxicology evaluation in animals and followed by first studies in patients (Phase IIa and IIb), further toxicology studies and large-scale studies in patients (Phase III and Phase IV).

The aim is to establish that the drug acts in the expected way, that it has therapeutic value and can be produced in a cost-effective way. 'Proof of concept' is therefore sought by the end of the Phase IIa clinical studies before large-scale human studies are commenced.

18.3.1   Preclinical phase

Pharmaceutical development

At the beginning of the development process the candidate drug is likely to only have been prepared in gram quantities. An initial task of the medicinal chemists is to develop a synthetic route capable of delivering the drug in kilogram quantities. The emphasis will be on purity and cost. A parallel task is to produce a formulation for pharmacological and safety evaluation for the chosen administrative route in human studies. In the case of drugs to be administered orally, the formulation will include excipients (pharmacologically inert materials) that will produce an appropriate bulk (bearing in mind that the active component in a tablet is likely to be a few milligrams at most) and stability that will ensure that when it has been administered it will release the active component at the required rate and physiological location (e.g. stomach or small intestine). Stability and dissolution tests will also need to be carried out on the formulation. As the development of the drug progresses there will be a need for manufacture scale-up coupled with a refined specification and finally there will be the need for mass production.

Pharmacokinetic studies

The preclinical phase of drug development includes both *in vitro* and whole-animal kinetic and metabolism studies. The ability of the drug to cross membranes by passive diffusion can be assessed by use of Caco-2 cells in monolayer culture on permeable supports. The drug is placed on one side of the layer and the rate at which it crosses to the other side measured. Comparisons are then made with standard reference compounds. Values are generally in good agreement with *in vivo* data. The possible role of carrier-mediated transport in drug distribution can be studied using an assay system based on the use of cells containing a cloned transporter.

As discussed in Section 18.1.4, the key pharmacokinetic parameters that dictate the clinical use of a drug are V_d, Cl_{int} and $t_{1/2}$. In addition, the nature and enzymology of the drug's metabolism are crucially important to the safe use of the drug. Isolated hepatocytes have proved to be a good model for the *in vitro* study of drug pharmacokinetics and metabolism. Using these cells, generally obtained from rat and /or dog and occasionally man, it is possible to study the metabolism of candidate drugs and to determine the constants K_m and V_{max} by established graphical means (Section 15.2.1). Cl_{int} is equal to V_{max}/K_m and from knowledge of it, Cl_{hep} can be estimated. From knowledge of the weight of the liver from which the cells were obtained, values can then be scaled to predict *in vivo* hepatic clearance. The values of these *in vitro* derived constants are generally in good agreement with the *in vivo* values obtained in whole animal and human studies. Automated multi-well plate-based assays coupled to fluorescence or LC-MS detection are widely available for the evaluation of the interaction of candidate drugs with the key five cytochrome P450s important to the drug discovery process. The technique of UPLC (Section 11.3.3) is now widely used in support of metabolic studies in drug discovery and development because of its increased speed and sensitivity relative to conventional HPLC.

The possibility of carrying out human pharmacokinetic studies as early as the drug discovery phase has arisen as a result of developments in accelerator mass spectrometry (AMS) (Section 9.3.6). Unlike other forms of mass spectrometry, the ions are accelerated to high kinetic energies before they are mass analysed. As a result, the technique is exceptionally sensitive and accurate in its analysis so that it can be applied to human dosing studies using the principle of microdosing, also known as Phase Zero studies. The European Medicines Evaluation Agency (EMEA) and the US Food and Drug Administration (FDA) have jointly defined a microdose as one-hundredth of the pharmacological dose and never greater than 100 µg. This is sufficient to enable AMS to study the absorption, distribution, metabolism and excretion of an investigational dose. Moreover, such studies have been defined as research rather than clinical studies and as a result are not subject to the regulatory requirements of normal clinical studies. Following microdosing by the chosen route in a volunteer, plasma or other biological fluid samples are taken at periodic intervals and analysed by HPLC or UPLC coupled to AMS. Quantities of drug and metabolites in the attomole range can be studied. A full pharmacokinetic profile can thus be obtained. Such data are invaluable in the selection of the best candidate drug. It is therefore easy in the early stages of the drug discovery process to eliminate candidate drugs that have unacceptable pharmacokinetic characteristics including either being rapidly metabolised by the cytochrome P450s or acting as activators or inhibitors of them.

Assessment of drug safety

Drug safety considerations are paramount throughout the whole drug discovery and development processes. *In silico* methods can give an advanced warning of potential safety problems associated with a specific chemical structure but many forms of safety problems are unpredictable and have to be identified by both *in vitro* and *in vivo* methods from a very early stage in a drug's evolution. A battery of exploratory *in vitro* screens has been developed to assess mutagenicity, cytotoxicity, immunotoxicity, hepatotoxicity, embryotoxicity and genotoxicity. Such assays are commenced early in the discovery phase and are used in the selection of the candidate drug in order to give confidence that the selected candidate drug will meet regulatory standards in the later stages of drug development. Once the strategic decision is taken to proceed to the preclinical phase of drug development, regulatory toxicology requirements come into play. These commence with 28-day repeat dose studies in two species, one of which is non-rodent. Rat and marmoset are most commonly selected. As the clinical phase is reached, the toxicology requirements progress to 3–12 month chronic studies in two species and reproductive toxicology studies, generally in rabbit, begin. These latter studies cover fertility and implantation, foetal development and pre- and postnatal effects. Two-year carcinogenicity studies in two species begin after the 'go' decision has been taken for the continued development of the drug. There are regulatory recommendations for the number of animals and species to be studied in all of these safety studies.

Safety pharmacology

In addition to a range of pharmacological studies being undertaken to fully establish the efficacy and selectivity of the candidate drug, studies are required by regulatory bodies to detect any undesirable effects of the drug. These include studies on the central nervous system, cardiovascular system, respiratory system and the autonomic nervous system.

18.3.2 Clinical phase

In order to obtain 'proof of concept' for the drug, clinical studies in human volunteers is essential. Human studies take place in four highly regulated stages:

- *Phase I*: This involves the evaluation of the drug in healthy volunteers, generally male in the age range 18–45. The emphasis of the initial studies (Phase Ia) is on its tolerability and basic pharmacokinetics rather than on the efficacy. The studies start with single sub-therapeutic doses (as predicted by animal studies) that are gradually increased as safety is demonstrated. A satisfactory outcome to this stage leads to Phase Ib in which volunteers receive repeat doses in a randomised, placebo-controlled and 'double-blind' trial (referred to as a RCT); the investigator is unaware of which subjects are receiving the drug and which placebo. Each stage involves about 50 volunteers and typically each takes 6 months to complete. Only if the drug is a cytotoxic agent, intended for treatment of specific types of cancer, are patients instead of healthy volunteers involved in this phase.
- *Phase II*: At this stage the volunteers are patients with the medical condition to be treated and the emphasis is on the assessment of the drug's efficacy but pharmacokinetic data will also be collected. Studies are again normally placebo and double-blind controlled, and would also involve comparative studies with an established drug (if one exists) to evaluate its therapeutic and economic

advantages. The dosage regimen is based on the outcome of the Phase I studies. In Phase IIa up to 200 patients are studied lasting up to 2 years. The results of this phase are then used to assess the efficacy and tolerability of the treatment and thus to either establish or reject the 'proof of concept' and hence to make a 'go'/'no-go' decision. A 'go' decision leads to Phase IIb studies in which a further 200–500 patients are studied for a further 2–3 years. The main aim of these studies is to confirm the results of Phase IIa and then establish the optimum dose and dosing regimen. The total number of subjects involved in Phase II trials must be such that the eventual statistical evaluation of the results has sufficient 'power calculation' to make the outcome, relative to the controls, unambiguous.

- *Phase III*: These studies are designed to extend the efficacy, tolerability and pharmacokinetic studies begun in Phase II and are carried out to a similar design but this time the emphasis is on safety and economic advantages relative to other 'in class' drugs. The selection of the volunteer patients for this stage will not be so restrictive and exclusive as in Phase II studies. The studies will therefore expose the drug to a wider patient population in terms of age and ethnicity. Up to 5000 patients may be involved with the studies being carried out in multiple centres often in different countries and taking 2–5 years to complete.

- *Phase IV*: By this stage the drug will have received a marketing license and its use will be at the discretion of the medical profession. The aims of Phase IV trials include the wish of the manufacturer to extend the therapeutic uses of the drug and to undertake a large-scale study of the mortality associated with the drug if it is a 'first in class' drug. Closely associated with Phase IV trials is the concept of pharmacovigilance – the monitoring of unexpected side effects that may only occur in a small minority of patients and hence to improve the safe use of the drug. It relies on general practitioners and pharmacists reporting the adverse drug reactions experienced by their patients but although there are recognised procedures for this reporting, it is estimated that only a small percentage, possibly as low as 10%, are reported. Licence holders are required to provide the licensing authority with regular updates on safety information relating to the use of the drug. This may lead to restrictions on the use of the drug. In 2004 the anti-inflammatory drug rofecoxib (Vioxx®) was withdrawn from the market after over 80 million patients had received it. Intended for the treatment of acute pain, concern was expressed over its association with increased risk of heart attack and stroke in patients on long-term therapy. Since 2004 a number of other high-profile new drugs have also been withdrawn from the market for similar concerns.

It is inevitable that with thousands of individuals involved in the clinical trials of a new drug that there will be reports of side effects. Symptoms such as headache, dizziness, feeling tired, constipation, muscle pain and confusion are quite commonly reported and in many cases it is difficult to establish that they are actually caused by the drug under evaluation but will be listed as 'possible side effects' for the marketed drug. Of greater concern are the occasional 'adverse reactions' of the heart attack and liver damage type. These cannot be ignored and, as in the case of Vioxx®, can lead to the withdrawal of a drug even though it affects an extremely small number of the total number of patients receiving it. In an attempt to address this problem, a group of seven major pharmaceutical companies has formed the International Severe Adverse Events Consortium (SAEC) to identify genetic DNA variants that may predispose an individual to drug adverse reactions. A parallel strategy that is also being considered is to establish a patient register to record all users of a particular drug so that should fear of an adverse reaction arise it may be possible to trace patients potentially at risk. Such an approach might prevent a drug such as Vioxx® from being withdrawn from the market to the disadvantage of the majority of its users.

Regulatory and ethical approval for clinical trials

Although the decision to enter a candidate drug into clinical trials is taken by the company, responsibility for authorising the studies lies with the national regulatory authority. In European Union countries, the European Medicines Evaluation Agency (EMEA) coordinates the control of clinical trials. The European Union Drug Regulating Authorities Clinical Trials (EudraCT) is the database for all clinical trials commencing in the Community. Each member state has a 'competent authority' with the power to award a Clinical Trial Authorisation (CTA). In the UK, that power rests with the Medicines and Health Regulatory Authority (MHRA). In the US the Food and Drug Administration (FDA) has the responsibility for issuing an Investigational New Drug Application (IND). Applications for a CTA/IND have to be supported by details of the proposed clinical trial's protocol which must define the formulation to be used, its route of administration, maximum dose and whether or not the drug is of biological origin and if it is whether or not it is of recombinant technology or a gene transfer product. In addition, details of the location of the clinical trial, including its emergency facilities, and the identity and expertise of the principal investigators have to be provided together with details of the insurance cover in the event of adverse effects on the trial subjects. Paralleling the award of a CTA/IND, a clinical trial has to be approved by an independent Research Ethics Committee (REC) or Institutional Review Board (IRB). These are nationally approved bodies that are subject to both national and international guidelines and regulations. In Europe there is the European Clinical Trials Directive (2001) and the EC Good Clinical Practice Directive (2005) both of which define the responsibilities, duties and functions of all those involved in clinical trials, all of which are enforced by law. All human studies are subject to the Declaration of Helsinki first signed in 1964 but subsequently modified on several occasions. The declaration identified the basic importance of 'risk assessment' and 'informed consent' in the review and approval of protocols for clinical trials. Essentially, benefits must outweigh risk and the interests of the study subjects must take precedence over potential advances in medicine. Subjects must be informed in lay language about the design, objectives and potential risks of the study and must give their written informed consent to participate. Moreover, they must be free to withdraw from a study at any point without giving a reason and all their personal details collected for the study must remain confidential and not put into the public domain. RECs and IRBs consist of 'expert' (medical and scientific) and 'lay' members and their task is to assess the study protocol, to evaluate its design and risks and to ensure that the interests of the recruited subjects are protected. Since Phase I studies mainly involve healthy subjects who by definition cannot personally gain from the study, the risk/benefit balance is difficult to judge. Healthy subjects are allowed to receive financial payment that should be proportionate to their time and inconvenience incurred, but such payments are sometimes judged to be an incentive to recruitment. Phase I studies are carried out either in a designated commercial or non-commercial unit or in a hospital. All Phase II to IV studies are carried out in a hospital. Once a CTA/IND and ethical approval have been granted investigators are not permitted to deviate from the approved design and must immediately report any adverse events to the approving bodies. Both the regulatory authority and the ethics committee have the power to stop a clinical trial if adverse events justify such action. Equally, the regulatory authority can agree to the premature termination of a trial if the therapeutic advantages of the trial drug over the control drug are statistically clear-cut.

The CTA system for approval of clinical trials means that the number of such trials worldwide can be monitored (www.ClinicalTrials.gov). At the end of 2007 more than 5000 trials were in progress at over 127 000 sites. The six most common therapeutic areas were oncology, central nervous system disorders, cardiology, infectious diseases, endocrinology and respiratory diseases. These six areas represented 68% of all protocols and 74% of all sites.

18.3.3 Patent protection

The large sums of money involved in drug discovery and development can only be recouped by the protection of the intellectual property inherent in the process thereby giving the company a monopoly over the sales of the drug subject to the approval of the appropriate licensing authority. For a drug to be patentable, it must be novel (i.e. new and not covered by previous patents), its discovery must involve an 'inventive step' represented by the pharmacological, biochemical and chemical research underlying the development, and the drug must have a 'utility' represented by its therapeutic use for a particular medical condition. One of the dilemmas facing the developing company is the timing of the filing of the patent application. Too early and the range of chemical structures possessing the particular pharmacological property may not have been fully defined thus allowing competitors the potential opportunity for developing a 'me-too' drug without infringing the patent. In addition, the earlier the application is filed the earlier it will subsequently expire. Too late and there is a risk that a competitor may 'prior date' with an identical application. In practice, most new drug applications are made towards the end of the discovery phase when the candidate drug is being identified. Most countries in the world now recognise the Patent Cooperation Treaty (PCT) under which filing a single application in one signatory country gives protection in the remainder. Once the application has been filed, officers of the World Intellectual Property Office in Geneva are charged with the task of producing a search report assessing the case for granting a patent. Once granted, a patent lasts for 20 years from the priority date. During this period the company must pay an annual fee that increases with the life of the patent to keep the patent in force. In the case of patents for new drugs this 20-year term may subsequently be extended by up to 5 years if the company can demonstrate that due to the long development time it has not had a reasonable opportunity to exploit the patent. By the time the patent protection of most drugs expires, more therapeutically effective drugs will have been licensed so lack of protection is not necessarily a major concern.

18.3.4 Marketing authorisation

When the Phase III clinical trials are complete, the sponsor must submit to the regulatory authority, such as the MHRA or FDA, a New Drug Application (NDA) for permission to market the drug. In the EU there is a mutual recognition approval procedure coordinated by the EMEA through its Committee on Human Medicinal Products (CHMP) whereby one member country assesses the application and seeks to gain recognition in others. Applications for marketing authorisation must be accompanied by large amounts of documentation supporting all the clinical trials data that underpin the application. Marketing authorisation is granted on a risk–benefit analysis basis. The authorities recognise that no drug is ever risk-free and their task is to assess whether the balance of the evidence submitted by the company is in favour of the benefits. Approval may take up to 2 years but the authorities do attempt to accelerate the process for drugs that represent totally new therapies. In the UK, the National Committee for Health and Clinical Excellence (NICE) has the additional power to decide whether or not the National Health Service (NHS) will agree to make a newly authorised drug available to UK patients. Decisions by NICE depend not only on the drug's novelty but also on the economic benefits ensuing from its use. This latter criterion is particularly important in approvals for expensive new therapies such as monoclonal antibodies and is an increasingly important factor in the clinical acceptability for all new medicines in many countries.

18.4 SUGGESTIONS FOR FURTHER READING

General texts

Anonymous (2007). *Guidelines on the Practice of Ethics Committees in Medical Research with Human Participants*, 4th edn. London: Royal College of Physicians.

Crooke, S. T. (ed.) (2008). *Antisense Drug Technology: Principles, Strategies and Applications*, 2nd edn. New York: CRC Press. (Gives a review of the mechanisms of action, pharmacokinetics and therapeutic potential of antisense drugs in the context of a wide range of clinical conditions.)

Kshirsagar, T. (ed.) (2008). *High-Throughput Lead Optimization in Drug Discovery*. New York: CRC Press. (Uses real examples to illustrate the application of the technique in modern drug development.)

Rang, H. P. (ed.) (2006). *Drug Discovery and Development: Technology in Transition*. London: Elsevier. (An in-depth coverage written in an authoritative style by experts from the pharmaceutical industry.)

Review articles

Cross, D. M. and Bayliss, M. K. (2000). A commentary on the use of hepatocytes in drug metabolism studies during drug discovery. *Drug Metabolism Reviews*, **32**, 219–240.

Eichler, H.-G., Pinatti, F., Flamion, B., Leufkens, H. and Breckenridge, A. (2008). Balancing early market access to new drugs with the need for benefit/risk data: a mounting dilemma. *Nature Reviews Drug Discovery*, **7**, 818–826.

Ekins, S., Mestres, J. and Tasta, B. (2007). In silico pharmacology for drug discovery: methods for virtual ligand screening and profiling. *British Journal of Pharmacology*, **152**, 9–20.

Gomez-Hens, A. and Aguillar-Caballos, M. P. (2007). Modern analytical approaches to high-throughput drug discovery. *Trends in Analytical Chemistry*, **26**, 171–180.

Imming, P., Sinning, C. and Meyer, A. (2006). Drugs, their targets and the nature and number of drug targets. *Nature Reviews Drug Discovery*, **5**, 821–832.

Joy, T. and Hegele, R. A. (2008). Is raising HDL a futile strategy for atheroprotection? *Nature Reviews Drug Discovery*, **7**, 143–154.

Karlberg, J. P. E. (2008). Trends in disease focus of drug development. *Nature Reviews Drug Discovery*, **7**, 639–640.

Lagerstrom, M. C. and Schioth, H. B. (2008). Structural diversity of G-protein coupled receptors and significance for drug discovery. *Nature Reviews Drug Discovery*, **7**, 339–357.

Lindsay, A. C. and Choudhury, R. P. (2008). Form to function: current and future roles for atherosclerosis imaging in drug discovery. *Nature Reviews Drug Discovery*, **7**, 517–530.

Neil, S. J., Zang, T. and Bieniasz, P. D. (2008). Tetherin inhibits retrovirus release and is antagonised by HIV-1 Vpu. *Nature*, **451**, 425–431.

Rader, D. J. and Daugherty, A. (2008). Translating molecular discoveries into new therapies for atherosclerosis. *Nature*, **451**, 904–912.

Rognan, D. (2007). Chemogenomic approaches to rational drug design. *British Journal of Pharmacology*, **152**, 38–52.

Rose, A. D. (2008). Pharmacogenetics in drug discovery and development: a translational perspective. *Nature Reviews Drug Discovery*, **7**, 807–817.

Starkuviene, V. and Pepperkok, R. (2007). The potential of high-content high-throughput microscopy in drug discovery. *British Journal of Pharmacology*, **152**, 62–71.

Wang, C.-Y., Liu, P. Y. and Liao, J. K. (2008). Pleiotropic effects of statin therapy: molecular mechanisms and clinical results. *Trends in Molecular Medicine*, **14**, 32–44.